Textbook of *in vivo* Imaging in Vertebrates

Textbook of *in vivo* Imaging in Vertebrates

Editors

Vasilis Ntziachristos

Department of Radiology, Harvard University HMS/MGH, Charlestown, USA

Anne Leroy-Willig

U2R2M, CNRS and Université Paris-Sud, Orsay, France

Bertrand Tavitian

Unité d'Imagerie de l'Expression des Genes, INSERM, Orsay, France

BICENTENNIAL 1807 WILEY 2007 BICENTENNIAL

John Wiley & Sons, Ltd

Telephone (+44) 1243 779777

Email (for orders and customer service enquiries): cs-books@wiley.co.uk
Visit our Home Page on www.wiley.com

Other Wiley Editorial Offices

John Wiley & Sons Inc., 111 River Street, Hoboken, NJ 07030, USA

Jossey-Bass, 989 Market Street, San Francisco, CA 94103-1741, USA

Wiley-VCH Verlag GmbH, Boschstr. 12, D-69469 Weinheim, Germany

John Wiley & Sons Australia Ltd, 42 McDougall Street, Milton, Queensland 4064, Australia

John Wiley & Sons (Asia) Pte Ltd, 2 Clementi Loop #02-01, Jin Xing Distripark, Singapore 129809

John Wiley & Sons Canada Ltd, 6045 Freemont Blvd, Mississauga, ONT, L5R 4J3, Canada

Wiley also publishes its books in a variety of electronic formats. Some content that appears in print may not be available in electronic books.

Anniversary Logo Design: Richard J. Pacifico

Library of Congress Cataloging-in-Publication Data

Ntziachristos, Vasilis.
Textbook of in vivo imaging in vertebrates / Vasilis Ntziachristos, Anne Leroy-Willig, Bertrand Tavitian.
p. ; cm.
Includes bibliographical references.
ISBN 978-0-470-01528-5 (alk. paper)
1. Diagnostic imaging. 2. Animal models in research. I. Leroy-Willig, Anne. II. Tavitian, Bertrand. III. Title.
[DNLM: 1. Diagnostic Imaging–methods. 2. Animals, Laboratory. 3. Models, Animal. WN 180 N961t 2007]
RC78.7.D53N82 2007
616.07'54–dc22 2007011297

British Library Cataloguing in Publication Data

A catalogue record for this book is available from the British Library

ISBN 978-0-470-01528-5

Typeset in 10/12 pt Sabon by Thomson Digital Noida, India
Printed and bound in Trento, Italy by Printer Trento SRL
This book is printed on acid-free paper responsibly manufactured from sustainable forestry in which at least two trees are planted for each one used for paper production.

Contents

Contributors xi

Introduction xv

1 Nuclear Magnetic Resonance Imaging and Spectroscopy 1
Anne Leroy-Willig and Danielle Geldwerth-Feniger

1.0 Introduction 1
1.1 Magnets and magnetic field 1
1.2 Nuclear magnetization 4
1.3 Excitation and return to equilibrium of nuclear magnetization 8
1.4 The NMR hardware: RF coils and gradient coils (more technology) 14
1.5 NMR spectroscopy: the chemical encoding 16
1.6 How to build NMR images: the spatial encoding 21
1.7 MRI and contrast 31
1.8 Sensitivity, spatial resolution and temporal resolution 38
1.9 Contrast agents for MRI 41
1.10 Imaging of 'other' nuclei 45
1.11 More parameters contributing to MRI contrast 46
1.12 More about applications 53

2 High Resolution X-ray Microtomography: Applications in Biomedical Research 57
Nora De Clerck and Andrei Postnov

2.0 Introduction 57
2.1 Principles of tomography 57
2.2 Implementation 62
2.3 Contribution of microtomography to biomedical imaging 65

3 Ultrasound Imaging 79
S. Lori Bridal, Jean-Michel Correas and Geneviève Berger

3.1 Principles of ultrasonic imaging and its adaptation to small laboratory animals 79
3.2 Pulse-echo transmission 81
3.3 Ultrasonic transducers 84
3.4 From echoes to images 88
3.5 Blood flow and tissue motion 91
3.6 Non-linear and contrast imaging 94
3.7 Discussion 99

4 *In Vivo* Radiotracer Imaging 103
Bertrand Tavitian, Régine Trébossen, Roberto Pasqualini and Frédéric Dollé

4.0 Introduction 103
4.1 Radioactivity 104
4.2 Interaction of gamma rays with matter 106

4.3 Radiotracer imaging with gamma emitters 109
4.4 Detection of positron emitters 114
4.5 Image properties and analysis 118
4.6 Radiochemistry of gamma-emitting radiotracers 121
4.7 Radiochemistry of positron-emitting radiotracers 134
4.8 Major radiotracers and imaging applications 139

5 Optical Imaging and Tomography **149**
Antoine Soubret and Vasilis Ntziachristos

5.0 Introduction 149
5.1 Light – tissue interactions 153
5.2 Light propagation in tissues 166
5.3 Reconstruction and inverse problem 173
5.4 Fluorescence molecular tomography (FMT) 176

6 Optical Microscopy in Small Animal Research **183**
Rakesh K. Jain, Dai Fukumura, Lance Munn and Edward Brown

6.0 Introduction 183
6.1 Confocal laser scanning microscopy 183
6.2 Multiphoton laser scanning microscopy 184
6.3 Variants for *In vivo* imaging 185
6.4 Surgical preparations 185
6.5 Applications 187

7 New Radiotracers, Reporter Probes and Contrast Agents **191**
Coordinated by Bertrand Tavitian

7.0 Introduction 191
Bertrand Tavitian

7.1 New radiotracers 192
Bertrand Tavitian, Roberto Pasqualini and Frédéric Dollé

7.2 Multimodal constructs for magnetic resonance imaging 199
Willem J.M. Mulder, Gustav J. Strijkers and Klaas Nicolay

7.3 Fluorescence reporters for biomedical imaging 203
Benedict Law and Ching-Hsuan Tung

7.4 New contrast agents for NMR 211
Silvio Aime

7.5 Imaging techniques – reporter gene imaging agents 215
Huongfeng Li and Andreas H. Jacobs

8 Multi-Modality Imaging **223**
Coordinated by Vasilis Ntziachristos

8.0 Introduction 223
Vasilis Ntziachristos

8.1 Concurrent imaging versus computer-assisted registration 223
Fred S. Azar

8.2 Combination of SPECT and CT 226
Jan Grimm

8.3 FMT registration with MRI 231
Vasilis Ntziachristos

9 Brain Imaging 233
Coordinated by Anne Leroy-Willig

9.0 Introduction 233
Anne Leroy-Willig

9.1 Bringing amyloid into focus with MRI microscopy 233
Greet Vanhoutte and Annemie Van der Linden

9.2 Cerebral blood volume and BOLD contrast MRI unravels brain responses to ambient temperature fluctuations in fish 236
Annemie Van der Linden

9.3 Assessment of functional and neuroanatomical re-organization after experimental stroke using MRI 239
Jet P. van der Zijden and Rick M. Dijkhuizen

9.4 Brain activation and blood flow studies with speckle imaging 242
Andrew K. Dunn

9.5 Manganese-enhanced MRI of the songbird brain: a dynamic window on rewiring brain circuits encoding a versatile behaviour 245
Vincent Van Meir and Annemie Van der Linden

9.6 Functional MRI in awake behaving monkeys 248
Wim Vanduffel, Koen Nelissen, Denis Fize and Guy A. Orban

9.7 Multimodal evaluation of mitochondrial impairment in a primate model of Huntington's disease 252
Vincent Lebon and Philippe Hantraye

10 Imaging of Heart, Muscle, Vessels 257
Coordinated by Yves Fromes

10.0 Introduction 257
Yves Fromes

10.1 Cardiac structure and function 258
Yves Fromes

10.2 Evaluation of therapeutic approaches in muscular dystrophy using MRI 260
Valérie Allamand

10.3 Canine muscle oxygen saturation: evaluation and treatment of M-type phosphofructokinase deficiency 264
Kevin McCully and Urs Giger

10.4 *In vivo* assessment of myocardial perfusion by NMR technology 267
Jörg. U.G. Streif, Matthias Nahrendorf and Wolfgang R. Bauer

10.5 Ultrasound microimaging of strain in the mouse heart 270
F. Stuart Foster

10.6 MR imaging of experimental atherosclerosis 272
Willem J.M. Mulder, Gustav J. Strijkers, Zahi A. Fayad and Klaas Nicolay

11 Tumor Imaging 277
Coordinated by Vasilis Ntziachristos

11.0 Introduction 277
Vasilis Ntziachristos

11.1 Dynamic contrast-enhanced MRI of tumour angiogenesis 277
Charles André Cuénod, Laure Fournier, Daniel Balvay, Clément Pradel, Nathalie Siauve and Olivier Clement

11.2 **Liver tumours: Evaluation by functional computed tomography** 281
Charles André Cuénod, Laure Fournier, Nathalie Siauve and Olivier Clément

11.3 **Early detection of grafted Wilms' tumours** 285
Erwan Jouannot

11.4 **Angiogenesis study using ultrasound imaging** 287
Olivier Lucidarme

11.5 **Nuclear imaging of apoptosis in animal tumour models** 291
Silvana Del Vecchio and Marco Salvatore

11.6 **Optical imaging of tumour-associated protease activity** 296
Benedict Law and Ching-Hsuan Tung

11.7 **Tumour angiogenesis and blood flow** 299
Rakesh K. Jain, Dai Fukumura, Lance L. Munn and Edward B. Brown

11.8 **Optical imaging of apoptosis in small animals** 301
Eyk Schellenberger

11.9 **Fluorescence molecular tomography (FMT) of angiogenesis** 305
Xavier Montet, Vasilis Ntziachristos, and Ralph Weissleder

11.10 **High resolution X-ray microtomography as a tool for imaging lung tumours in living mice** 307
Nora De Clerck and Andrei Postnov

12 **Other Organs** **311**
Coordinated by Anne Leroy-Willig

12.0 **Introduction** 311
Anne Leroy-Willig

12.1 **3D imaging of embryos and mouse organs by Optical Projection Tomography** 311
James Sharpe

12.2 **Visualizing early Xenopus development with time lapse microscopic MRI** 315
Cyrus Papan and Russell E. Jacobs

12.3 **Ultrasonic quantification of red blood cells development in mice** 318
Johann Le Floc'h

12.4 **Placental perfusion MR imaging with contrast agent in a mouse model** 320
Nathalie Siauve, Laurent Salomon and Charles André Cuénod

12.5 **Characterization of nephropathies and monitoring of renal stem cell therapies** 323
Nicolas Grenier, Olivier Hauger, Yahsou Delmas and Christian Combe

12.6 **Optical imaging of lung inflammation** 328
Jodi Haller

12.7 **Optical imaging in rheumatoid arthritis** 330
Andreas Wunder

13 **Gene Therapy** **333**
Markus Klein and Andreas H. Jacobs

13.0 Introduction 333
13.1 Expression systems for genes of interest (GOI) 334
13.2 Gene delivery systems (vectors) 334
13.3 Suicide gene therapy 335
13.4 Non-suicide gene therapy 336
13.5 Imaging of gene expression 337
13.6 Diseases targeted by gene therapy 340

14 Cellular Therapies and Cell Tracking 347
Coordinated by Yves Fromes

14.0 Introduction 347
Yves Fromes

14.1 Are stem cells attracted by pathology? The case for cellular tracking by serial *in vivo* MRI 348
Michel Modo

14.2 Cell tracking using MRI 352
Vít Herynek

14.3 Cell labelling strategies for *in vivo* molecular MR imaging 354
Mathias Hoehn

14.4 Animal imaging and medical challenges - cell labelling and molecular imaging 360
Yannic Waerzeggers, and Andreas H. Jacobs

Index 369

Contributors

Silvio Aime Department of Chemistry IFM and Molecular Imaging Center, University of Torino, Via. P. Giuria 7, I-10125 Turin, Italy
e-mail: silvio.aime@unito.it

Valérie Allamand INSERM U582, Institut de Myologie, Groupe Hospitalier Pitié-Salpétrière, 75651 Paris, France
e-mail: v.allamand@myologie.chups.jussieu.fr

Fred S. Azar Imaging and Visualization, Siemens Corporate Research, Princeton, NJ 08540, USA
e-mail: fred.azar@siemens.com

Daniel Balvay Laboratoire de Recherche en Imagerie, Necker Université Paris V Descartes, Department de Radiologie, Hospital Europeen Georges Pompidou, 75015 Paris, France
e-mail: Daniel.balvay@egp.aphp.fr

Wolfgang R. Bauer Medizinische Universitätsklinik 1, Josef Schneider Strasse 2, D-97080 Würzburg, Germany
e-mail: w.bauer@medizin.uni-wuerzburg.de

Geneviève Berger Laboratoire d'Imagerie Paramétrique, UMR 7623 C.N.R.S.- Université Paris 6, 15 rue de l'Ecole de Médecine, 75006 Paris, France
e-mail: genevieve.berger@lip.bhdc.jussieu.fr

S. Lori Bridal Laboratoire d'Imagerie Paramétrique, UMR 7623 C.N.R.S.-Université Paris 6, 15 rue de l'Ecole de Médecine, 75006 Paris, France
e-mail: bridal@lip.bhdc.jussieu.fr

Edward B. Brown Department of Biomedical Engineering, Box 639 University of Rochester Medical Center, 601 Elmwood Avenue, Rochester, NY, 14642, USA
e-mail: Edward_Brown@URMC.Rochester.edu

Olivier Clement Laboratoire de Recherche en Imagerie, Necker Université Paris V Descartes, Radiology Department Hospital Européen Georges Pompidou, 75015 Paris, France
e-mail: clement@necker.fr

Nora De Clerck Microtomography, University of Antwerp, Department of Biomedical Sciences, Universiteitsplein 1 B-2610 Antwerp, Belgium
e-mail: nora.declerck@ua.ac.be

Christian Combe INSERM E362, Université Victor Segalen-Bordeaux 2, 33076 Bordeaux, France
e-mail: christian.combe@chu-bordeaux.fr

Jean-Michel Correas Laboratoire d'Imagerie Paramétrique, UMR 7623 C.N.R.S.- Université Paris 6, 15 rue de l'Ecole de Médecine, 75006 Paris, France
e-mail: jean-michel.correas@nck.aphp.fr

Charles André Cuénod Laboratoire de Recherche en Imagerie, Necker Université Paris V Descartes, Department de Radiologie, Hospital Europeen Georges Pompidou, 75015 Paris, France
e-mail: charles-andre.cuenod@egp.aphp.fr

Yahsou Delmas ERT CNRS, Imagerie Moléculaire et Fonctionnelle, INSERM E362, Université Victor Segalen-Bordeaux 2,146 rue Léo-Saignat – 33076 Bordeaux, France
e-mail: yahsou.delmas@chu-bordeaux.fr

Rick M. Dijkhuizen Image Sciences Institute, University Medical Center Utrecht, Bolognalaan 50, 3584 CJ Utrecht, The Netherlands
e-mail: rick@invivonmr.uu.nl

Fréderic Dollé Groupe de Radiochimie, Laboratoire d'Imagerie Moléculaire Expérimentale, CEA, Direction des Sciences du Vivant, Institut d'Imagerie Biomédicale Médicale, Service Hospitalier Frédéric Joliot, 4 place du Général Leclerc, 91401 Orsay, France
e-mail: frederic.dolle@cea.fr

Andrew K. Dunn Biomedical Engineering Department, University of Texas at Austin, 1 University Station, C0800, Austin, TX 78712, USA
e-mail: adunn@mail.utexas.edu

Zahi A. Fayad Imaging Science Laboratories, Department of Radiology, Zena and Michael A. Wiener Cardiovascular Institute, Box 1234, One Gustave L. Levy Place, New York, NY 10029, USA
e-mail: Zahi.Fayad@mssm.edu

Danielle Geldwerth-Feniger INSERM U770, Hôpital Kremlin-Bicêtre, 78, rue du General Leclerc 94275, Le Kremlin-Bicêtre, France
e-mail: danielle.geldwerth@kb.inserm.fr

Denis Fize Centre de Recherche Cerveau et Cognition, CNRS-UPS UMR 5549, Université Paul Sabatier, 31062 Toulouse, France
e-mail: Denis.Fize@cerco. ups-tlse.fr

Johann Le Floc'h Department of Electronics, University of Roma Tre, Via della Vasca Navale, 84 00146 Rome, Italy
e-mail: lefloch@uniroma3.it

F. Stuart Foster Department of Medical Biophysics, Ontario Cancer Institute University of Toronto, 610 University Avenue, Toronto M5G 2M9, Canada
e-mail: stuart.foster@swri.ca

Laure Fournier Laboratoire de Recherche en Imagerie, Necker Université Paris V Descartes, Radiology Department, Hospital Européen Georges Pompidou, 75015 Paris, France
e-mail: laure.fournier@egp.aphp.fr

Yves Fromes Inserm U582, Institut de Myologie, Université Pierre et Marie Curie-Paris 6, IFR14, Groupe Hospitalier Paris Saint Joseph, Department of Cardiac Surgery, F-75014 Paris, France
e-mail: y.fromes@myologie.chups.jussieu.fr

Dai Fukumura E. L. Steele Laboratory for Tumor Biology, Department of Radiation Oncology Massachusetts General Hospital, Boston, MA 02114, USA
e-mail: dai@steele.mgh.edu

Urs Giger Section of Medical Genetics, University of Pennsylvania School of Veterinary Medicine, Philadelphia, PA 19104, USA
e-mail: giger@vet.upenn.edu

Nicolas Grenier ERT CNRS, Imagerie Moléculaire et Fonctionnelle, Université Victor Segalen-Bordeaux 2, 146 rue Léo-Saignat - 33076 Bordeaux, France
e-mail: nicolas.grenier@chu-bordeaux.fr

Jan Grimm Laboratory for Bio-optics and Biological Imaging, MGH-CMIR, Building 149 Room 5406, 13th Street, Charlestown, MA 02129-2060, USA
e-mail: jgrimm@helix.mgh.harvard.edu

Jodi Haller Laboratory for Bio-optics and Molecular Imaging, Center of Molecular Imaging Research, Massachusetts General Hospital, Harvard Medical School, Building 149, 13th Street, Room 5406, Charlestown, MA 02129-2060, USA
e-mail: jhaller@partners.org

Philippe Hantraye CEA-UIIBP and URA CEACNRS 2210, Institut d'Imagerie Biomédicale Médicale, Service Hospitalier Frédéric Joliot, 4 place du Général Leclerc, 91401 Orsay, France
e-mail: Philippe.hantraye@cea.fr

Olivier Hauger ERT CNRS, Imagerie Moléculaire et Fonctionnelle, Université Victor Segalen-Bordeaux 2 146 rue Léo-Saignat - 33076 Bordeaux, France
e-mail: olivier.hauger@chu-bordeaux.fr

Vít Herynek MR Unit, Department of Diagnostics and Interventional Radiology, Institute for Clinical and Experimental Medicine, Videnska 1958/9, 140 21 Prague 4, Czech Republic
e-mail: vit.herynek@medicon.cz

Mathias Hoehn In vivo NMR Laboratory, Max Planck Institute for Neurological Research, Center for Molecular Medicine and Department of Neurology, University of Cologne Gleuelerstrasse 50, 50931 Cologne, Germany
e-mail: mathias@mpinkoeln.mpg.de

Andreas H. Jacobs Laboratory for Gene Therapy and Molecular Imaging at the Max Planck Institute for Neurological Research, Center for Molecular Medicine and Department of Neurology, University of Cologne, Gleuelerstrasse 50, 50931 Cologne, Germany
e-mail: Andreas.Jacobs@nf.mpg.de

Russell E. Jacobs Beckman Institute, California Institute of Technology, Pasadena, CA 91125, USA
e-mail: rjacobs@caltech.edu

Rakesh K. Jain E. L. Steele Laboratory for Tumor Biology, Massachusetts General Hospital, Boston, MA 02114, USA
e-mail: jain@steele.mgh.edu

Erwan Jouannot Laboratoire d'Imagerie Parametrique, UMR 7623 C.N.R.S. - Universite Paris 6, 15 rue de l'école de médicine, 75006 Paris, France
e-mail: jouannot@sanofi-aventis.com

Markus Klein Laboratory for Gene Therapy and Molecular Imaging at the Max Planck Institute for Neurological Research, Center for Molecular Medicine and Department of Neurology, University of Cologne Gleuelerstrasse 50, 50931 Cologne, Germany
e-mail: info@nf.mpg.de

Benedict Law Center for Molecular Imaging Research, Massachusetts General Hospital, Harvard Medical School, Charlestown, MA 02129, USA
e-mail: blaw@partners.org

Vincent Lebon CEA-UIIBP and URA CEA-CNRS 2210, Institut d'Imagerie Biomédicale Médicale, Service Hospitalier Frédéric Joliot, 4 place du Général Leclerc, 91401 Orsay, France
e-mail: Vincent.lebon@cea.fr

Anne Leroy-Willig U2R2M (UMR8081 C.N.R.S.), Batiment 220, Université Paris-Sud, Faculté d'Orsay, 91405 Orsay, France
e-mail: anne.leroywillig@u2r2m.u-psud.fr

Huongfeng Li Laboratory for Gene Therapy and Molecular Imaging at the Max Planck Institute for Neurological Research, Center for Molecular Medicine and Department of Neurology, University of Cologne, Gleuelerstrasse 50, 50931 Cologne, Germany
e-mail: info@nf.mpg.de

Annemie Van der Linden Bio-Imaging Lab, Department of Biomedical Sciences Groenenborgerlaan 171, University of Antwerp, 2020 Antwerp, Belgium
e-mail: annemie.vanderlinden@ua.ac.be

Olivier Lucidarme Radiology Department, Pitié-Salpêtriére hospital, 47-83 boulevard de l'Hôpital, 75651 Paris, France
e-mail: Olivier.lucidarme@psl.aphp.fr

Kevin McCully Department of Kinesiology, University of Georgia, Athens, GA 30602, USA
e-mail: mccully@uga.edu

Vincent Van Meir Bio-Imaging Lab, University of Antwerp, Campus Middelheim Groenenborgerlaan 171 2020 Antwerp, Belgium
e-mail: Vincent.vanmeir@ua.ac.be

Michel Modo Centre for the Cellular Basis of Behaviour, The James Black Centre, Kings College London, Institute of Psychiatry, 125 Coldharbour Lane, London SE5 9NU, United Kingdom
e-mail: mike.modo@iop.kcl.ac.uk

Xavier Montet Geneva University Hospital, Radiology Department, Rue Micheli-du-Crest 24, 1205 Geneva, Switzerland
e-mail: xavier.montet@hcuge.ch

Willem J.M. Mulder Biomedical NMR, Eindhoven University of Technology, PO Box 513, Eindhoven 5600 MB, The Netherlands
e-mail: w.j.m.mulder@tue.nl

Lance L. Munn E. L. Steele Laboratory for Tumor Biology, Massachusetts General Hospital, Boston, MA 02114, USA
e-mail: lance@steele.mgh.harvard.edu

Matthias Nahrendorf Center for Molecular Imaging Research, Massachusetts General Hospital, Harvard Medical School, Room 5406, 149 13th Street, Charlestown, MA 02129, USA
e-mail: MNahrendorf@partners.org

Koen Nelissen Laboratorium voor Neuro- en Psychofysiologie, K.U.Leuven Medical School, Campus Gasthuisberg Herestraat 49, B-3000 Leuven, Belgium
e-mail: Koen.Nelissen@med.kuleuven.be

Klaas Nicolay Biomedical NMR, Eindhoven University of Technology, PO Box 513, Eindhoven 5600 MB, The Netherlands
e-mail: k.nicolay@tue.nl

Vasilis Ntziachristos Laboratory for Bio-optics and Molecular Imaging, Center of Molecular Imaging Research, Massachusetts General Hospital, Harvard Medical School, Building 149, 13th Street, Room 5406, Charlestown, MA 02129-2060, USA
e-mail: vasilis@helix.mgh.harvard.edu

Guy A. Orban Laboratorium voor Neuro- en Psychofysiologie, K.U.Leuven Medical School, Campus Gasthuisberg, Herestraat 49 B-3000 Leuven, Belgium.
e-mail: Guy.Orban@med.kuleuven.be

Cyrus Papan Institute of Bioengineering and Nanotechnology, The Nanos #04-01, 31 Biopolis Way, Singapore 138669
e-mail: cpapan@ibn.a-star.edu.sg

Roberto Pasqualini Research and Development, CIS bio international, Schering BP 32, 91192 Gif sur Yvette, France
e-mail: Rpasqualini@cisbiointernational.fr

Andrei Postnov Microtomography, Department of Biomedical Sciences, University of Antwerp, Universiteitsplein 1, B-2610 Antwerp, Belgium
e-mail: andrei.postnov@ua.ac.be

Clement Pradel Laboratoire de Recherche en Imagerie, Necker Université Paris V Descartes, Department de Radiologie, Hospital Europeen Georges Pompidou, 75015 Paris, France
e-mail: clement.pradel.@egp.aphp.fr

Laurent Salomon Laboratoire de Recherche en Imagerie, Université Paris V Descartes, Radiology Department, Hospital Européen Georges Pompidou, 75015 Paris, France
e-mail: Laurent.salomon@egp.aphp.fr

Marco Salvatore Dipartimento di Scienze Biomorfologiche e Funzionali, Universita' Federico II, Vias S Pansini 5, 80131 Naples, Italy
e-mail: marco.salvatore@unina.it

Eyk Schellenberger Department of Radiology, Institut für Radiologie, Charité-Universitätsmedizin Berlin, Campus Charité Mitte, Schumannstrasse 20/21, 10117 Berlin, Germany
e-mail: eyk.schellenberger@charite.de

James Sharpe ICREA Research Professor, EMBL/CRG Systems Biology Unit, Centre for Genomic Regulation (CRG), Dr. Aiguader 88, 08003 Barcelona, Spain
e-mail: James.Sharpe@crg.es

Nathalie Siauve Laboratoire de Recherche en Imagerie, Necker Université Paris V Descartes, Radiology Department, Hospital Europeen Georges Pompidou, 75015 Paris, France
e-mail: siauve@necker.fr

Antoine Soubret Novartis Pharma A.G., Modeling & Simulation Biology, WSJ-27.1.026, CH-4056 Basel, Switzerland
e-mail: soubreta@yahoo.fr

Jörg. U.G. Streif Physikalisches Institut, Lehrstuhl für Experimentelle Physik V (Biophysik), Universität Wüerzburg, Am Hubland, 97094 Wüerzburg, Germany
e-mail: streif@physik.uni-wuerzburg.de

Gustav J. Strijkers Biomedical NMR, Eindhoven University of Technology, PO Box 513, Eindhoven 5600 MB, The Netherlands
e-mail: g.j.strijkers@tue.nl

Bertrand Tavitian INSERM U803, Imagerie de l'expression des gènes, Laboratoire d'Imagerie Moléculaire Expérimentale, CEA, Direction des Sciences du Vivant, Institut d'Imagerie Biomédicale Médicale, Service Hospitalier Frédéric Joliot, 4 place du Général Leclerc, 91401 Orsay, France
e-mail: bertrand.tavitian@cea.fr

Régine Trébossen Groupe Instrumentation et Traitement d'Images, Laboratoire d'Imagerie Moléculaire Expérimentale, CEA, Direction des Sciences du Vivant, Institut d'Imagerie Biomédicale Médicale, Service Hospitalier Frédéric Joliot, 4 place du Général Leclerc, 91401 Orsay, France
e-mail: regine.trebossen@cea.fr

Ching-Hsuan Tung Center of Molecular Imaging Research, Massachusetts General Hospital, Harvard Medical School, 149 13th Street, Room 5406, Charlestown, MA 02129, USA
e-mail: tung@helix.mgh.harvard.edu

Wim Vanduffel Massachusetts General Hospital, Massachusetts Institute of Technology, Harvard Medical School, Athinoula A. Martino's Center for Biomedical Imaging, Charlestown, MA 02129, USA
e-mail: wim@nmr.mgh.harvard.edu

Greet Vanhoutte Bio-Imaging lab, Campus Middelheim Grenenborgerlaan 171 University of Antwerp, 2020 Antwerp, Belgium
e-mail: greetje.vanhoutte@ua.ac.be

Silvana Del Vecchio Instituto di Biostrutture e Bioimmagini, Consiglio Nazionale delle Ricerche, Universita' Federico II, Via S. Pansini 5, 80131 Naples, Italy
e-mail: delvecc@unina.it

Yannic Waerzeggers Laboratory for Gene Therapy and Molecular Imaging, Max Planck Institute for Neurological Research, Center for Molecular Medicine (CMMC) and Department of Neurology, University of Cologne, Gleuelerstrasse 50, 50931 Cologne, Germany
e-mail: info@nf.mpg.de

Ralph Weissleder Center for Molecular Imaging Research, Massachusetts General Hospital, Harvard Medical School, Room 5406, 149 13th Street Charlestown, MA 02129, USA
e-mail: weissleder@helix.mgh.harvard.edu

Andreas Wunder Molecular Imaging Group, Experimental Neurology, Charité Hospital, Schumannstrasse 20/21, 10098 Berlin, Germany
e-mail: andreas.wunder@charite.de

Jet P. van der Zijden Image Sciences Institute, University Medical Center Utrecht, Bolognalaan 50, 3584 CJ Utrecht, The Netherlands
e-mail: sjet@invivonmr.uu.nl

Introduction

Traditional biomedical methods study "life" on dead specimens. To address limitations of in-vitro assays, in-vivo imaging of vertebrates has emerged as a powerful tool used in virtually all forms of modern biomedical research and drug discovery. In-vivo imaging fulfils a basic necessity to dynamically and spatially resolve anatomical, functional and molecular events as they occur in live tissues. How would we quantify the cardiac function and adaptation to a stimulus other than in a live animal? How can the effects of a sensory stimulus on the cerebral cortex be accurately described if not by using in-vivo observations? Similar motivations arise in many other aspects of the life sciences such as examining complex molecular pathways in disease evolution or the longitudinal assessment of treatment.

Following this fundamental requirement for *in-vivo* assessment of tissue characteristics in the biomedical sciences, significant progress has been made towards non-invasive imaging of animals, from the embryonic stage to fully developed adult stage. This progress has seen three major traits. One approach has been the adaptation of clinical imaging methods to the animal dimensions for obtaining optimal imaging characteristics in the smaller volumes examined. A second trait has been the evolution or development of new methods, primarily based on photonic technologies, which are well suited for small animal research. The third trait has been the engineering of important new chemistry and biotechnology methods that impart significant ability to identify and report on a magnitude of cellular and sub-cellular functions, a capacity that was previously unavailable to traditional medical imaging. These newer imaging technologies have opened the possibility for visualizing proteins, genes and their function in entire animals in-vivo and non-invasively. Imaging of entire intact animals has therefore emerged as one of the important biomedical tools in the post genomic era. Similarly to the significant gains seen by the introduction of the microscope in biology, imaging of entire intact animals enables unprecedented insights at the system level and offers new found capabilities of accurate visualization of structure, physiology and molecular function.

The fundamental principles of interaction and image formation differ significantly between the imaging modalities used in vertebrate imaging. Combined with elaborate methods of inducing biological contrast, the plurality of technologies and the diverse performance characteristics may at times appear daunting not only for the biologist but even for the medical imaging specialist. This book intends to summarize the wealth of imaging technologies and applications that have emerged for in vivo imaging of animals and to serve as a reference to the biologist and biomedical investigator. It serves the dual role of 1) describing the basic underlying principles of image formation using different energies of the electromagnetic radiation spectrum and acoustic waves and of 2) exemplifying representative applications in studying living vertebrates, with the exception of humans. The ultimate goal is to explain the different types of information gained by modern *in vivo* imaging techniques and illustrate the potential to replace the accurate but destructive histological techniques with high-throughput imaging strategies. The utility of multimodality imaging is also scrutinized as it allows for optimal combination of complementary tissue parameters measured on the same animal/organ and position. Key characteristics and limitations of the different imaging approaches including the specificity and the sensitivity achieved in retrieving various biomarkers, the speed of acquisition for dynamic measurements, the easiness of "bench-top" or "cageside" application and the appropriateness by which to examine key biological problems is also presented. Minimally invasive imaging modalities increasingly used in biomedical research are also described. By combining expert descriptions of the most widespread imaging approaches for vertebrate imaging, we hope that this book will contribute in collectively describing the most important of imaging approaches in order to categorize them and describe in a concise manner.

Accelerating biomedical discovery

By enabling longitudinal studies, non-invasive imaging comes with increased observation accuracy; each animal can serve as its own control, thus reducing the sources of experimental variability. Moreover, since a single animal yields observations at multiple time points, smaller animal numbers are required in order to build meaningful statistics. This practice overall reduces research cost and the time required to reach meaningful conclusions. With this capacity, animal imaging can significantly accelerate biomedical discovery by enabling expeditious tests of agents, drugs and hypotheses. Imaging can be pivotal for example in accelerating drug discovery or the identification of potent diagnostic agents by utilising the animal as the test bed in the pre-clinical *in-vivo* assessment of treatment efficiency, targeting sensitivity and specificity, biodistribution and long term effects. Correspondingly, in-vivo imaging is in par with modern legislation that wisely incites researchers to spare animal life. The rule of the three R's – replacement, reduction and refinement of animal experimentation – enounced by Russel and Burch in 1959 is at best respected when atraumatic experimentation is exercised.

Similar benefits can be found when imaging the rising numbers of genetically modified animals, mostly mice, which come with the need for quick screen for phenotypes that correspond to human disease. Transgenic, knock out and knock in techniques can yield a significant number of animal model variants of unknown disease traits. Imaging plays an important role in identifying and comparing different phenotypes to human disease and can accelerate the traditional observations of biochemical testing, physiological inspection and molecular analyses.

In-vivo imaging of animals can further serve as a common framework for animal and human observations and yield a bridge between traditional biomedical research and improving human health. This can be achieved at many different levels. Technologies developed for the assessment of drugs in mice can be translated to imaging efficacy in humans as well, utilizing the imaging experience gained from animal imaging. Similarly, some of the most potent detection technologies, tested in mice, can be then employed diagnostically in humans using the same imaging modality. With modern imaging serving as the common denominator, quick pre-clinical screens and accurate clinical evaluations at the structural, physiological and molecular levels can be facilitated efficiently and at no significant additional technological expense.

Image formation and contrast mechanisms

All modalities used for *in-vivo imaging* utilize some wave form which non-destructively interacts with tissue. Information on the internal characteristics of tissues is obtained by recording the response to this interaction and is then utilized in forming images. Most imaging modalities use a part of the electromagnetic spectrum to form images, with the exception of ultrasound that uses acousto-mechanical waves. The most typical distinction of different imaging methods, is by means of the particular electromagnetic energy used. Shown in Fig. 1 is the correspondence of the most common imaging methods with the electromagnetic spectrum. The particular physical parameters of the wave used are ultimately responsible for the particular characteristics of each technology. There are three major types of information that can be assessed with modern imaging methods as summarized in Fig. 2:

Anatomical imaging is the traditional radiological approach, largely facilitated by X-ray imaging, X-ray CT, Magnetic Resonance Imaging, Ultrasound and

Figure 1 Wavelength, energy of photon and frequency of electromagnetic radiations used for *in vivo* imaging
Energy of electromagnetic radiation is indicated by the energy of one photon,

$$E = h.F = h.c/\lambda.$$

where h is the Planck's constant equal to 6.62 10^{-23} Joule s, F is the frequency of the wave and λ is its wavelength.
Here E is plotted in eV. Photons with energy higher than 1 eV can ionise molecules and then have biological effects

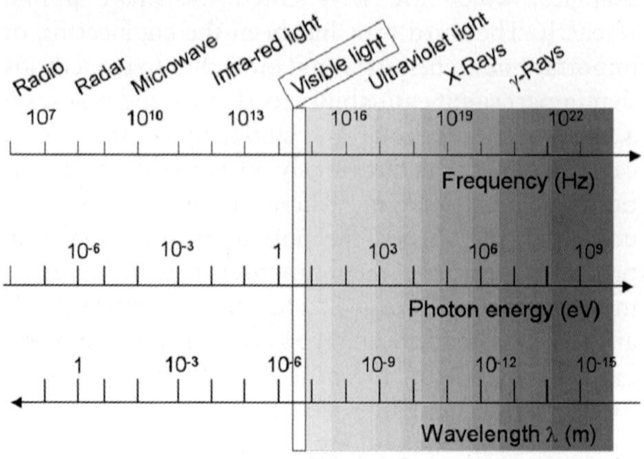

Figure 2 Which kind of information may be-given by the main imaging techniques

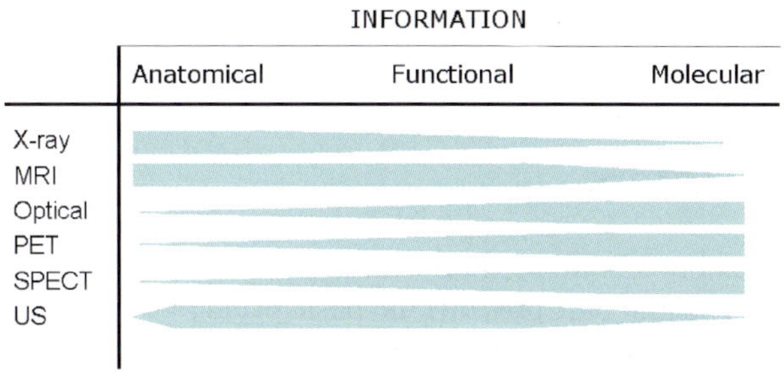

Optical Imaging, the latter when superficial structures are considered. Generally, the information and contrast visualized and the corresponding information conveyed by the image can be found in an anatomy textbook. This anatomical information, or the changes found from the expected known anatomy, relate to development and disease. Typically, these are high resolution images and the contrast imaged is endogenous, i.e. the attenuation of X-ray beams by bone or cancer-related calcifications or the differences between the concentration and motility of water molecules by MRI. However the use of contrast agents is occasionally used to improve the contrast in anatomical structures, for example in resolving the structure of the vascular system or better visualizing a suspicious lesion.

Functional imaging is used to study the function of organs, under physiological or pharmacological stimulations. It typically requires fast measurement techniques and resolves contrast parameters found in your physiology book. It can visualize for example organ movement, fluid flow, membrane permeability and the function of tissue bio-molecules associated with basic tissue function such as haemoglobin or oxygen. Imaging is based either on endogenous contrast or the administration of exogenous agents. All imaging modalities have been used for functional imaging with varying resolutions, often using a high-resolution anatomical image as reference. Examples of functional imaging include the visualization of deoxy haemoglobin changes during functional cortex studies by MRI or optical methods or blood flow measurement during the cardiac cycle by MRI or ultrasound.

Molecular imaging is the most recent of the imaging sciences and it refers to the visualization of biological processes at the cellular and molecular level. Molecular imaging is based on the combination of advanced chemistries, transgenic strategies and imaging technologies in order to resolve engineered contrast specific to particular cellular and sub-cellular processes. Generally it is used to visualize processes found in a molecular biology book and associated fields of science and offers the widest versatility over the two previous methods in terms of the contrast mechanisms that can be achieved and the technologies utilized. The images are typically low resolution and a high-resolution anatomical or functional image is used for reference. Typically all the standard radiological imaging modalities have been used for molecular imaging, except X-ray CT, that does not up to this point offer sufficient sensitivity.

The classification of anatomical, functional and molecular imaging is often associated with particular contrast mechanisms and strategies, but it does not impose strict boundaries. Anatomical imaging for example can be performed after administration of a contrast agent that can better outline architectural features. Similarly, molecular imaging can operate in the absence of exogenous contrast enhanced strategies; for example Magnetic Resonance spectroscopic imaging resolves the relative concentrations of various intrinsic molecules and correspondingly relays information on tissue and disease molecular status, at the absence of extrinsic contrast agents. However, while endogenous contrast can be used as a biomarker in many applications, it is the use of versatile exogenous contrast strategies that brings a new paradigm into animal imaging. There are several different classes of enhancing or generating contrast associated with particular tissue function and molecular activity. The classical approach follows the clinical radiological paradigm where a *contrast agent* is intravenously injected to enhance the capacity to detect disease. This agent preferentially distributes at the site of interest, or demarcates the vascular structure of an organ of interest. Examples include the injection of an iodinated agent or a super-paramagnetic contrast agent for imparting contrast on X-ray CT or MR images respectively. Another example is the injection of common molecules labeled with radioactive isotopes, or the use of labeled moieties such as antibodies or peptides. This latter approach is a change in paradigm as contrast is in this case engineered for specific biomolecules. This basic example of engineered contrast is significantly augmented in molecular imaging by sophisticated techniques that can mark virtually any protein, an increasingly large number of diverse cellular functions and cell traffic. Collectively, these engineered technologies are referred to as *reporter* technologies, since they report on specific targets and functions. There are two fundamental reporter approaches, i.e. direct and indirect imaging.

Direct imaging uses exogenously administered probes that are engineered to report on specific molecular process (e.g., a receptor target imaged with a ligand molecular imaging probe). This approach is similar to the nuclear imaging example discussed in the previous paragraph, but is significantly enhanced for use with different modalities (i.e optical, MRI etc) and using different design principles. Importantly, engineered probes used for direct imaging can be categorized to *active probes,* i.e. probes that carry an active reporting component and *activatable* probes, which carry an inactive reporting component which is activated through interaction with a molecular target, or more generally changes some of its own physical parameters after interaction with a specific target. Activatable probes are also known as molecular beacons, switches or smart probes and they are so far available for fluorescence, bioluminescence and MRI. An important distinction of probes vs. contrast agents is that the former have specificity against a gene or gene-expression product.

Indirect imaging refers to methods that utilize a *reporter trans-gene* which is inserted in the animal's DNA. Contrast is generated after transcription of the reporter gene. The product of the transcription and translation can be a reporter probe directly (for example a fluorescent protein) or otherwise a functional cellular change that facilitates preferential uptake of an exogenously administered probe, for example upregulation of an enzyme or receptor that is in turn responsible for accumulating or trapping a radionuclei-based agent into a cell or the cellular surface. Reporter gene imaging is a generalizable platform that in contrast to the direct imaging method, only one or few well validated reporter-gene & reporter probe pairs can be used to image many different molecular and genetic processes. On the downside is the introduction of foreign proteins and genes which limits applicability to animals.

Chapters

This book is divided into three parts.

The first part presents the basic principles of operation of the most common imaging techniques used in small animal imaging. Chapter 1 is devoted to Nuclear Magnetic Resonance Imaging (and Spectroscopy); Chapter 2 to X-Ray Tomography, Chapter 3 to Ultrasound Imaging. Chapter 4 is devoted to Nuclear Imaging (PET and SPECT) and to the production of radioactive tracers. Chapter 5 is devoted to Optical Imaging, and Chapter 6 to in vivo Optical Microscopy. Chapter 7 shows the newest radioactive tracers, reporters and contrast agents that are proposed in each imaging domain, and Chapter 8 presents the potentialities offered by the combination of several imaging techniques.

The second part is made from reports that each show how a given technique optimally adresses a specific biological question, with four chapters showing illustrations related respectively to brain (Chapter 9), heart vessels and muscle (Chapter 10), tumours (Chapter 11), other organs (Chapter 12).

The third part is devoted to the review of two domains where in vivo imaging has brought new insights: Gene therapies (Chapter 13) and cellular therapies (Chapter 14).

Vasilis Ntziachristos
Anne Leroy-Willig
Bertrand Tavitian

1 Nuclear Magnetic Resonance Imaging and Spectroscopy

Anne Leroy-Willig and
Danielle Geldwerth-Feniger

1.0 Introduction

Nuclear magnetic resonance (NMR) detects the magnetic moments of nuclei using their orientation in a strong magnetic field and their response at a specific resonance frequency. Discovered in 1946 by Bloch and Purcell, NMR spectroscopy (MRS), at first used for chemical and physical studies, quickly became a major tool for spectroscopic analysis of complex molecules and further of biochemical systems. Then in the 1980s, NMR gave rise to magnetic resonance imaging (MRI), a medical imaging technique very attractive despite its cost, from the profusion of anatomical and physiological information available.

In biomedical research, the two modalities, imaging (MRI) and spectroscopy (MRS) are increasingly used for *in vivo* animal studies, with benefit from the technical developments carried out for human studies. These two modalities give access to various data ranging from three dimentional (3D) anatomy to physiological and biochemical information, and many applications are available via specific measurement techniques that we will shortly explain here.

NMR is fully based upon quantum physics. Here we give a simplified and then by some ways approximate description, mixing classic and quantum physics, in paragraphs one to six; the later paragraphs are oriented towards *in vivo* explorations.

In this chapter, several levels of information are given, which are as follows: readers can jump the paragraphs labelled as 'more physics' or 'more technology'; also they may read only key points before coming to the following paragraph. For those who wish to know more about MRI and MRS, more complete descriptions are given in a free access Web book (Hornack, 2005), in books by Webb (2003) and Bushberg *et al.* (2001), and concerning the toolbox of MRI sequences, in Ness Aiver (1997) with a fully graphic presentation. Gadian (1995) wrote an excellent introduction to *in vivo* MRS.

1.1 Magnets and magnetic field

In everyday life, a magnet is a piece of a material which attracts or repels another magnet and creates a magnetic field. For example, the magnetic bar shown in Figure 1.1.1(a) has two poles; the magnetic field it creates goes from the North Pole to the South Pole. The magnetic field all around this magnet can be probed by its action, which is the force exerted on another magnet. For example, the weak earth's magnetic field acts upon a needle compass: The compass rotates and lines along the magnetic field pointing towards the magnetic North.

Magnetism is the fundamental property of matter. The magnetism of nuclei is weak, hidden behind the stronger contribution of electrons, and one may easily ignore its existence. Magnetism is the result of moving electrical charges (mostly the electrons). The magnetic field, the mediator of magnetic force, is created either by electric current flowing in a wire or by the microscopic electric circuits, which exist inside materials like iron, at the atomic scale.

Figure 1.1.1 How to create a magnetic field.
(a) The magnet bar, a rod made with iron, is a permanent source of magnetic field throughout space. The field lines (black curves), that indicate the direction of the magnetic field, go from North Pole to South Pole. A small needle compass lines along the direction of the magnetic field.
(b) A loop of copper wire, fed with electrical current, creates a magnetic field with similar spatial distribution at long distance as indicated by field lines.
(c) A solenoidal winding is made by many conductive wire loops winded upon a cylinder. In the central part of the cylinder, the magnetic field is lined along the axis of the cylinder (white lines) and is homogeneous

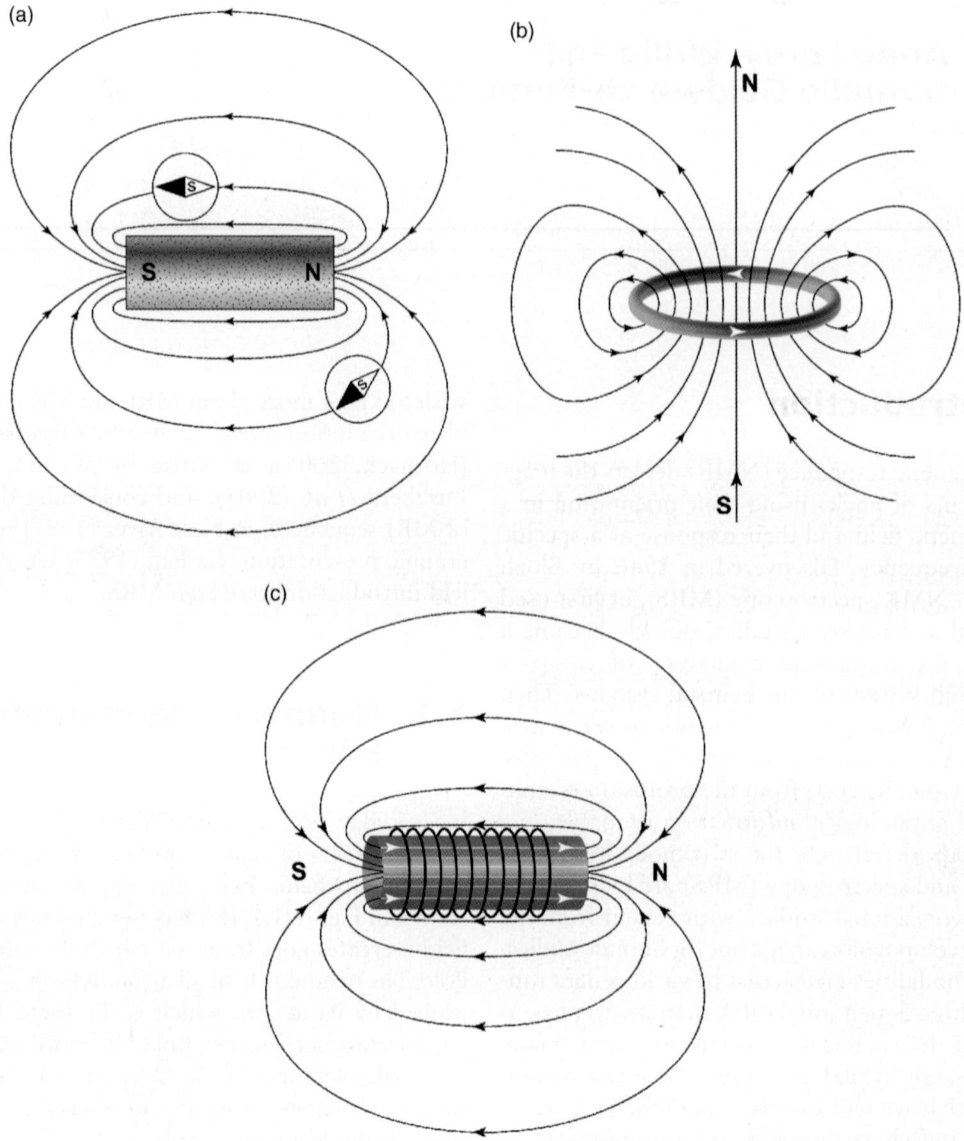

The magnetic field is measured in Tesla or in Gauss, with $1\,T = 10^4\,G$. (*Note that we make a rather loose use of magnetic units, forgetting the difference between magnetic field and magnetic induction, only needed when studying ferromagnetic materials.*)

Another simple magnet is made by a circular loop of copper wire fed with electric current, shown in Figure 1.1.1(b). A basic physical law tells us that the magnetic field created by a current rotates around the wire where the electric current is flowing. Then the magnetic field is perpendicular to the circle at its centre; elsewhere its intensity and its direction vary through space. The solenoid is made with multiple loops of wire coiled upon a cylinder (Figure 1.1.1(c)). The magnetic field inside the cylinder is very homogeneous.

1.1.1 More technology: The 'perpetual' magnet

Nearly all magnets for NMR are solenoids, made of supraconductive wire winded upon a hollow cylindrical support. Electric current circulates in the circuit that is immersed in a cryostat filled with liquid helium at temperature −269°C. Since the supraconductive wire has zero electrical resistance at low temperature, no electrical power is dissipated. This system creates a very stable magnetic field that may be disconnected from a power supply, as long as the temperature is kept low enough. The low temperature is maintained by high vacuum insulation that reduces liquid helium boil off.

Besides the high intensity, high homogeneity and stability of the magnetic field are also needed. The magnet is the more heavy and expensive piece of NMR hardware. There is a growing demand for high field magnets dedicated to biomedical research, but few centres can buy very high field magnets for large animals.

Big magnets delivering magnetic fields between 0.3 and 3 T are currently used for NMR human studies. For smaller animals, smaller magnets delivering higher fields (1.5 to 11 T) are currently used. *In vitro* experiments are done at still higher fields. For comparison, the earth magnetic field is 5×10^{-4} T (or 0.5 G).

Magnets for small animals are either vertical (as those commonly used for *in vitro* studies) or horizontal, yielding wider access and allowing more physiological housing of animals during NMR examination (as shown by Figure 1.1.2).

Figure 1.1.2 The supraconductive magnet used for NMR experiments. (a) High field supraconductive magnet for rats and mice NMR examination. This horizontal magnet, weighting 2 tons, delivers a magnetic field of 7 T inside a cylindrical access 30 cm wide. After installation of the shim and gradient coils, the access available for small animals is 15 cm wide. The chimney above the magnet is used for liquid helium refill. (b) Examination bed. The small animal is lain inside an anaesthesia chamber. The bed is positioned at the centre of the magnet bore during the examination (Courtesy of Bruker, SA, Ettlingen, Germany)

(a)

(b)

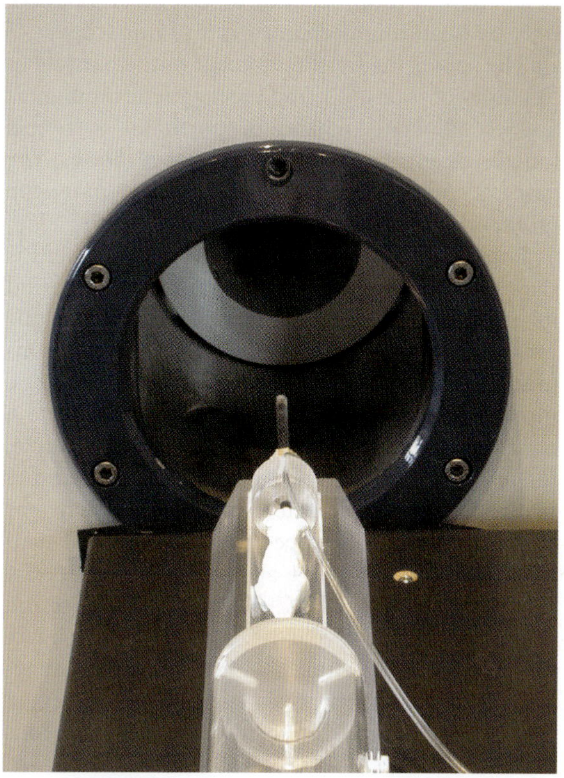

1.2 Nuclear magnetization

1.2.1 The magnetic moment of the nucleus

Key points
The nuclei that bear a net magnetic moment (such as ^{1}H, ^{31}P, ^{13}C) can be detected by NMR. Hydrogen nuclei, that bear the largest magnetic moment amongst stable nuclei, are detected to build *in vivo* NMR images. Hydrogen, phosphorus, sodium, fluorine nuclei are currently detected to build *in vivo* NMR spectra.

All elementary particles (electron, proton, neutron and others) bear a spin. The spin is purely quantic without strict correspondence in classical physics, but it can be described as a quantity of rotation of the particle spinning about one axis, where each spin $\vec{s}$ is associated with an elementary magnetic moment $\vec{\mu}$, related to the spin by a number, the gyromagnetic factor γ.

$$\vec{\mu} = \gamma . \vec{s} \qquad (1.1)$$

The elementary magnetic moment may be described as a tiny magnet that we will represent as an arrow; a more accurate description is possible only by quantum mechanics, out of our scope.

The spin is the kinetic moment of the particle (a 'quantity of rotation'), and the magnetic moment is always associated – and proportional – to this kinetic moment. (*Note that in many books, the word "spin" is written instead of 'magnetic moment'.)*

For one given nucleus, the magnetic moment is the sum of the magnetic moments of its protons and its neutrons. Hydrogen nucleus is made of one proton (Figure 1.2.1). When protons or neutrons are associated as pairs with their magnetic moments in opposed direction, these pairs have a net magnetic moment equal to zero. For example, the carbon nucleus ^{12}C (with 6 protons and 6 neutrons) cannot be detected by NMR, whereas the less abundant isotope ^{13}C (6 protons, 7 neutrons) has a detectable magnetic moment. *In vivo* NMR spectroscopy of ^{13}C allows the quantification of molecules such as glucose, acetate and glycogen.

Electrons bear a much larger elementary magnetic moment, nearly two thousand times bigger than that of protons. In most molecules, electrons are associated as pairs with their magnetic moments in opposed direction, and these pairs have nearly net zero magnetic moment. The iron atom has several non-paired electrons and then bears a large magnetic moment from its electrons, so that iron is a good material to experience what is magnetism, or to make magnets, and also NMR contrast agents (see paragraph 1.9).

1.2.2 The motion of a magnetic moment around the magnetic field and the resonance frequency

Key points
A magnetic moment rotates around the direction of the magnetic field $\vec{B}_o$ as does a spinning top. Its longitudinal component, along $\vec{B}_o$, is constant, whereas its transverse component, perpendicular to $\vec{B}_o$, rotates at the frequency F_o. F_o is proportional to the magnetic field intensity $\vec{B}_o$ and to the gyromagnetic factor characteristic of the nucleus, γ. The gyromagnetic factor γ has a characteristic value for each nucleus, so that at a given field value each kind of nucleus rotates at a specific frequency.

A magnetic moment $\vec{\mu}$ in presence of a magnetic field $\vec{B}_o$ is submitted to a torque: It rotates along a cone around the direction of the magnetic field, as does a spinning top. This special rotation is named precession (it is the name for the motion of a gyroscope when a torque is applied upon it). Then the longitudinal component of $\vec{\mu}$, μz, along $\vec{B}_o$, keeps a constant value, and the transverse component, μt, perpendicular to $\vec{B}_o$, rotates (Figure 1.2.2). The precession takes place at a well-defined frequency, F_o, proportional to the magnetic field intensity B_o and to the gyromagnetic factor, γ, characteristic of the nucleus.

F_o is the resonance frequency of this nucleus:

$$F_o = \gamma / 2\pi . B_o \qquad (1.2)$$

The gyromagnetic factor γ is determined by the internal quantum structure of the nucleus. It has a characteristic value for each nucleus, so that at a given field value each kind of nucleus rotates at a specific frequency as shown in Table 1.2.1.

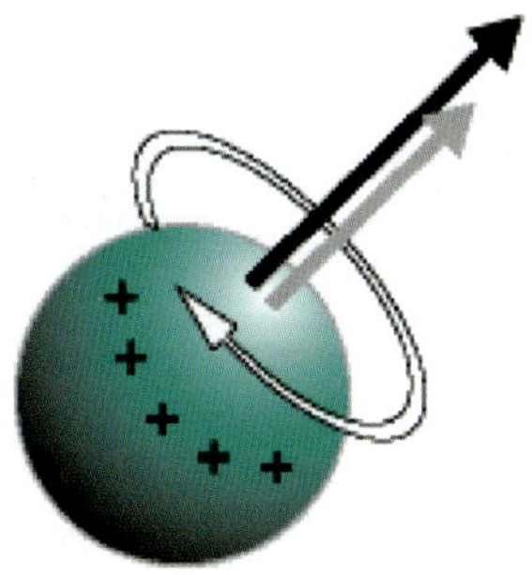

Figure 1.2.1 The hydrogen nuclear magnetic moment. The proton, with mass rotating upon itself, has some analogy with a spinning top. The positive charge rotating can also be described as some current flowing in a circuit and then behaves as a small magnet. Rotation (spin) is symbolized by the black arrow, magnetic moment by the grey arrow. The magnetic moment $\vec{\mu}$ and the spin $\vec{s}$ of a proton are collinear, and they are related by: $\vec{\mu} = \gamma.\vec{s}$, where γ is the gyromagnetic factor

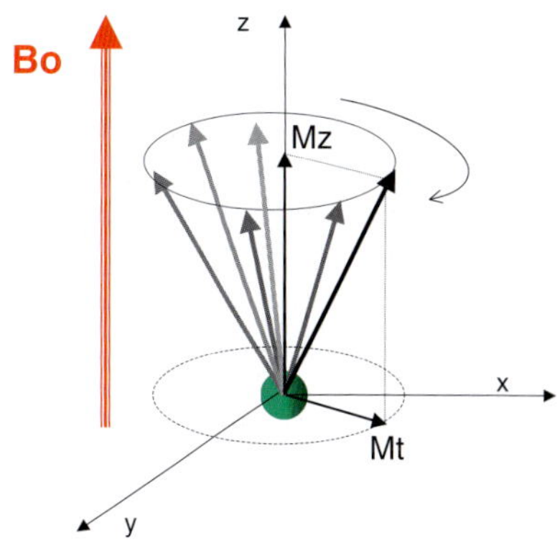

Figure 1.2.2 Precession of a magnetic moment around the magnetic field. The magnetic moment of a proton rotates around the field $\vec{B_0}$ tangentially to a cone. The angle between $\vec{\mu}$ and $\vec{B_0}$ is constant; the projection of $\vec{\mu}$ on the direction of $\vec{B_0}$, μz – named the longitudinal component, has a fixed value. The projection upon the plane perpendicular to $\vec{B_0}$, μt – named the transverse component, rotates at the frequency F_0

1.2.3 Resonance frequencies of nuclei of biological research

Amongst the stable nuclei, the hydrogen nucleus has the highest gyromagnetic factor and then the highest resonance frequency at a given magnetic field. NMR signals of hydrogen are currently detected at frequencies between 64 and 900 MHz (corresponding to magnetic field intensity between 1.5 T and 21.13 T). Other nuclei resonate at lower frequencies, because they have lower magnetic moments. These resonance frequencies are in the range used for radio, telephones and radars. In Table 1.2.1, the gyromagnetic factor γ of nuclei is expressed by their resonance frequency at $B_0 = 4.7$ T (the field of many NMR spectrometers used for small animal examinations).

Table 1.2.1 Nuclear magnetic resonance frequencies at Bo = 4.7 Tesla for nuclei of biological interest

Nucleus	Frequency at 4.7 Tesla (MHz)	Natural abundance (%)	Sensitivity*
^{1}H	200	99.98	1
^{3}He	152.4	$1.3 \cdot 10^{-4}$	6.10^{-5}**
^{13}C	50.2	1.1	$0.18 \cdot 10^{-3}$
^{19}F	188.2	100	0.85
^{23}Na	52.9	100	0.136
^{31}P	80.9	100	0.063

*The sensitivity for a given nucleus is the ratio of its signal to the signal of hydrogen, at same number of atoms (taking into account the natural abundance of the isotope detected), at the same magnetic field. The sensitivity for ^{13}C is low because ^{13}C nuclei are only 1.1% of all carbon nuclei. The sensitivity varies as the square of the gyromagnetic ratio of the nucleus.

**This nucleus is detected at abundance higher than its weak natural abundance, after separation from ^{4}He, and after hyperpolarization (cf paragraph 1.10.1).

1.2.4 The nuclear magnetization

> **Key points**
> Nuclear magnetization is the sum of the individual magnetic moments per unit volume.
>
> In presence of the external magnetic field B_o, individual magnetic moments are lined either parallel or anti-parallel to B_o, corresponding to two energy levels. The weak difference between the populations in these two energy levels determines the nuclear magnetization. At equilibrium, the nuclear magnetization is parallel to B_o, and its value M_o is proportional to the number of nuclei N and to B_o.

The magnetization is the sum of the individual magnetic moments in one unity of volume. These magnetic moments are borne by nuclei and electrons.

Here we consider only the magnetization from the nuclei, that we call 'nuclear magnetization', and only the contribution from the nuclei to be detected (very often hydrogen nuclei).

Let us consider a water sample of volume V that contains N hydrogen nuclei. In the absence of external magnetic field, the individual nuclear magnetic moments are oriented randomly with zero sum, and then the total magnetization is equal to zero (Figure 1.2.3(a)).

In the magnetic field $\vec{B_o}$, they do not behave as a classic magnet: A compass needle would always align with the field. Here they orientate either along or opposite the magnetic field (Figure 1.2.3(b)). Their z-component μz is quantified, taking values $+\mu$ or $-\mu$.

The magnetic energy of a magnetic moment $\vec{\mu}$ in the field $\vec{B_o}$ is given by

$$E = -\vec{\mu}.\vec{B_o} \tag{1.3}$$

The two orientations relative to B_o determine two energy levels. The energy of the lower level is $E^+ = -\mu.B_o$, for $\vec{\mu}$ parallel to $\vec{B_o}$ (assuming μ is positive).

The energy of the upper level is $E^- = +\mu.B_o$, for $\vec{\mu}$ in the direction opposite to B_o.

The two levels are separated by

$$\Delta E = 2.\mu.B_o \tag{1.4}$$

If the N hydrogen nuclei were reparted equally between these two levels, magnetization would still be zero. From thermal agitation, the hydrogen nuclei are continually jumping from one energy level to the other. At equilibrium, N^+ nuclei are in the lower level (which is slightly more populated) and N^- nuclei are in the upper level as drawn in Figure 1.2.4.

The magnetization, the sum of individual moments, is parallel to the magnetic field $\vec{B_o}$, and has the value M_o:

$$M_o = (N^+ - N^-).\mu/V \tag{1.5}$$

This magnetization is much lower than $N.\mu/V$, which would be its value if all magnetic moments were in the lowest energy level. The magnetization at equilibrium

Figure 1.2.3 Nuclear magnetic moments and magnetization. The sum of the elementary magnetic moments in a unit volume is the magnetization. (a) At zero magnetic field, the magnetic moments are randomly oriented. (b) At the magnetic field intensity B_o, the magnetic moments are oriented either parallel or anti-parallel to $\vec{B_o}$ and their sum is parallel to $\vec{B_o}$

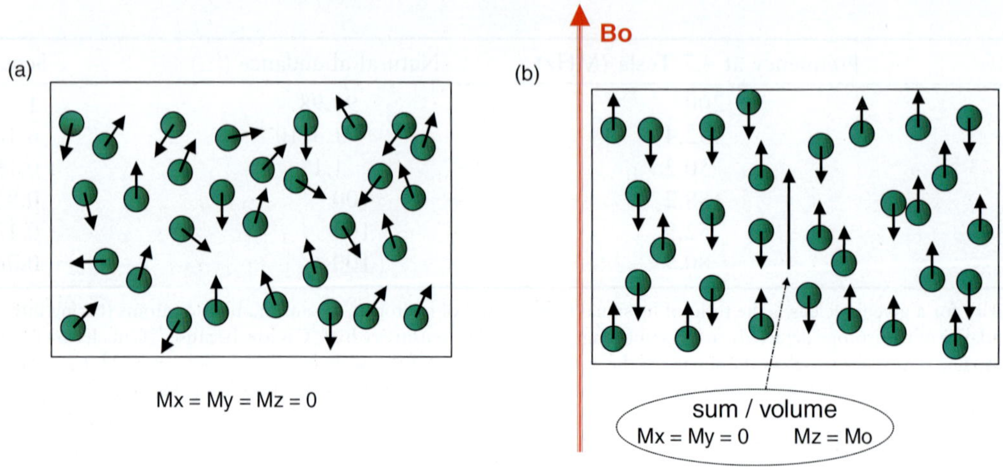

Figure 1.2.4 Population of the nuclear energy levels.
(a) Magnetic moments at equilibrium in the magnetic field B_o. The energy levels corresponding to the two orientations relative to B_o are separated by $\Delta E = 2\,\mu\,B_o$. The lower level contains N^+ hydrogen nuclei; the upper level contains N^- hydrogen nuclei. The net magnetization of the sample is $M_o = (N^+ - N^-)\,\mu/V$.
(b) Excitation of nuclear magnetic resonance. Photons from the electromagnetic field B_1, that have the energy $h\,F_o = \Delta E$, are absorbed and allow magnetic moments in the lower level to reach the upper level: The populations N^+ and N^- are modified by the absorption of photons. When $N^+ = N^-$ the longitudinal magnetization is equal to zero, while the photons have brought their polarization to the transverse magnetization that is no longer equal to zero.
(c) Energy levels and longitudinal relaxation. The recovery of M_z to equilibrium, or longitudinal relaxation, derives from rebuilding the difference between the populations N^+ and N^- of the two energy levels of hydrogen nuclei magnetic moments. The hydrogen nuclei that have been previously excited to the upper level have to emit the excess of energy in order to return to the lower level

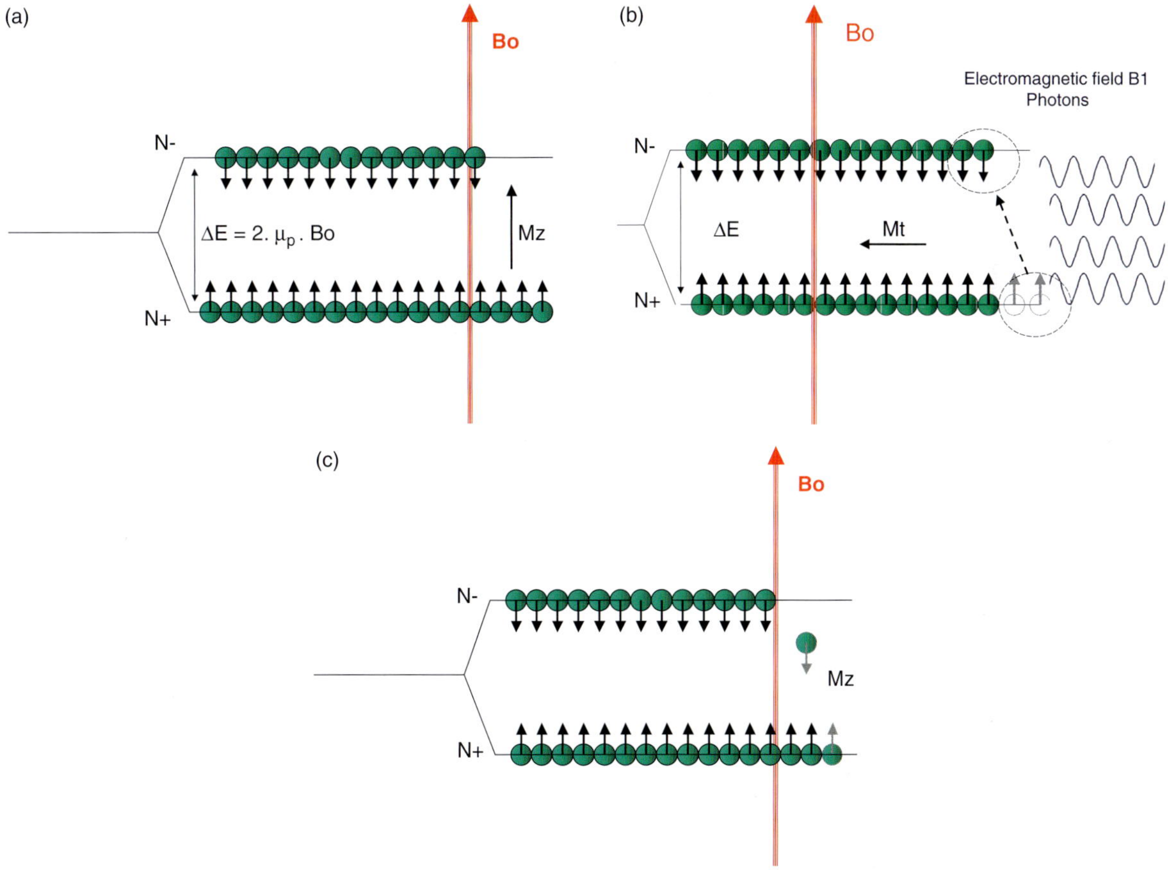

is calculated from the polarization P of nuclei that quantifies how much the magnetic moments are oriented by the magnetic field.

The polarization of nuclei by the magnetic field B_o, P is the ratio between the difference of populations of the two energy levels, $\Delta N = N^+ - N$, and the total population of nuclei, $N = N^+ + N^-$

$$P = \Delta N/N, \quad \text{so that } M_o = N\,P\mu/V. \tag{1.6}$$

At equilibrium, P depends on the ratio of the magnetic energy μBo (the source of magnetic order) to the thermal energy $k\,T$ (the source of disorder), where k is the Boltzman constant and T the temperature.

This ratio is very low in usual *in vivo* conditions. The polarization of hydrogen nuclei is equal to 3×10^{-6} at 1 T, at 300°K. We shall later see that the signal from nuclei is related to the polarization. A benefit of stronger magnetic field is the higher polarization of nuclear magnetic moments. Special

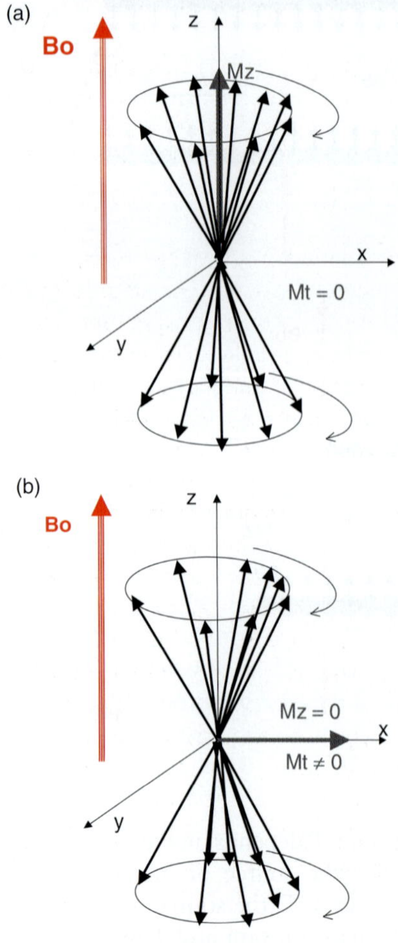

Figure 1.2.5 Nuclear magnetization precesses around the magnetic field.
(a) At equilibrium, all individual magnetic moments experience precession around B_o at the frequency F_o. The individual magnetic moments are distributed randomly across each cone. The upper cone (corresponding to the lower energy level) is more populated than the lower cone: there is a net longitudinal magnetization, $M_z = M_o$. The transverse components of the magnetic moments rotate; their rotations are not coherent, so that there is no net sum along the other directions. Then for the nuclear magnetization of the sample $M_x = M_y = 0$.
(b) When the absorption of photons from the RF field B_1 has equalized the populations of the two cones and has modified the transverse orientations of the magnetic moments, $M_z = 0$ and M_t gets a net value (M_x and $M_y \neq 0$)

techniques allow to increase very strongly the polarization of nuclei such as Xenon, Helium and Hydrogen (see paragraph 1.10.1).

Let us come back to our small water sample and complete the description of magnetic moments.

At equilibrium, all the individual magnetic moments experience precession around B_o at the frequency F_o as displayed by Figure 1.2.2, but their transverse components are reparted randomly in the plane perpendicular to B_o and the sum of transverse components, M_t, is equal to zero (Figure 1.2.5(a)). After excitation of nuclear magnetic resonance, their transverse components are oriented in the plane perpendicular to B_o and the sum of transverse components, M_t, can be detected (Figure 1.2.5(b)).

1.3 Excitation and return to equilibrium of nuclear magnetization

Key points
Excitation of NMR is done by irradiation of the sample with a magnetic field oscillating at the resonance frequency F_o. This magnetic field tips the nuclear magnetization away from its initial orientation along B_o. While the transverse nuclear magnetization M_t rotates, it can be easily detected. The receiver probe picks the weak magnetic signal created by the rotation of M_t and generates a voltage oscillating at the frequency F_o. Detection can be done during a time limited by the decay of M_t, measured by the transverse relaxation time T2. One has to wait for the return to equilibrium of the longitudinal nuclear magnetization, during a time related to the longitudinal relaxation time $T1$, before repeating excitation and detection.

The magnetization at equilibrium, parallel to B_o, cannot be measured directly: Magnetic forces are small and difficult to measure. Conversely, when the global magnetization rotates around B_o at the resonance frequency, measurement of an electrical signal is possible.

Here we describe how to excite resonance, how to detect NMR signal, and the way nuclear magnetization returns to its initial equilibrium.

An oscillating or rotating physical phenomenon can be described by its amplitude, its frequency and its phase.

Both the RF magnetic field B_1 and the transverse magnetization M_t are vectors perpendicular to B_o that rotate or oscillate at the resonance frequency F_o (for definition of phase, see Figure 1.3.1).

At best, the field B_1 used for NMR excitation is a rotating field; however, the experiment is often driven

Figure 1.3.1 Oscillating/rotating vectors. The rotating vector M rotates in the plane XOY, as the needle of a clock; Ma, the amplitude of M is like the length of the needle. The angle of this vector with the reference axis OX is the phase ϕ. The rotation takes place at the frequency F (measured in turns per second or hertz). The phase at time zero is ϕ_o; later at time t the phase is written as

$$\phi = 2\,\pi\,F\,t + \phi_o. \tag{1.7}$$

The components of the vector M are

$$M_x = M_a \cos(\phi) = M_a \cos(2\,\pi\,Ft + \phi_o), \tag{1.8}$$
$$M_y = M_a \sin(\phi) = M_a \sin(2\,\pi\,Ft + \phi_o), \tag{1.9}$$

M_x is an oscillating quantity, also characterized by its amplitude M_a, its frequency of oscillation F and its phase at $t = 0$, ϕ_o

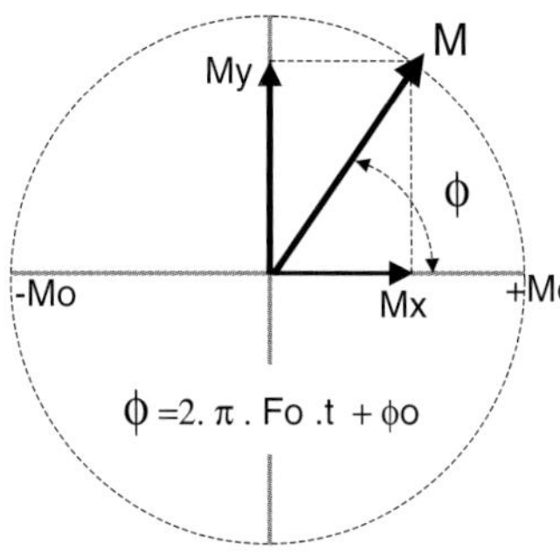

by a linear oscillating magnetic field that can be discomposed into two rotating fields: One of them rotates clockwise and the other counterclockwise. One of them is efficient to excite nuclear magnetic resonance and the other is not efficient.

The three characteristics of the transverse magnetization M_t are its amplitude, its precession frequency and its phase. They intervene in the generation of the NMR signal: The intensity of signal is proportional to the amplitude of the local magnetization, whereas the frequency and phase of the signal inform upon the spatial localization.

1.3.1 The excitation of nuclear magnetic resonance

To excite nuclear resonance means to set nuclear magnetization out of equilibrium by using *a second magnetic field*, the RF magnetic field $\vec{B_1}$ (RF means 'radio frequency'). This is done by irradiating the sample with an electromagnetic field rotating at the resonance frequency F_o. This field is created by sending current oscillating at the frequency F_o in a coil around the sample. (Note that this irradiation by an electromagnetic field is usually fully devoid of biological effects, except the thermal effects due to heating, because the energy of photons is more than 1×10^6 times smaller than any energy of ionisation: At the highest field used for MRI, 17.6 T, the photons of frequency 748 MHz have an energy equal to 3×10^{-6} eV. These photons can only heat tissues.)

The RF magnetic field $\vec{B_1}$ is perpendicular to $\vec{B_o}$ and rotates around the direction of $\vec{B_o}$. From the equivalence between electromagnetic field and photons, here again, there are two complementary descriptions of the excitation of resonance.

1.3.1.1 *In terms of energy levels and populations*

An electromagnetic wave of frequency F can also be described as made by photons of elementary energy

$E = h\ F$ (where h is the Planck's constant equal to 6.634×10^{-27} J.s).

The RF magnetic field $\vec{B_1}$ at the resonance frequency F_o is the magnetic component of an electromagnetic wave. The energy of the corresponding photons, equal to $h\ F_o$, is exactly equal to the difference between the magnetic energy of magnetic moments in the two energy levels, $\Delta E = 2\ \mu\ B_o$. Such photons convey exactly the energy needed to raise one magnetic moment from the lower level up to the higher level, whence the name of resonance frequency (Figure 1.2.4(b)).

When the two populations get equal, M_z is equal to zero and $M_t = M_o$. This is described geometrically as a 90° flip of the vector $\vec{M}$ that rotates around the direction of the RF field $\vec{B_1}$ (Figure 1.3.2).

The photons of the electromagnetic field at frequency F_o are fully polarized: This means that for every photon, the direction and phase of $\vec{B_1}$ is well defined and identical. When the photons are absorbed by the magnetic moments they give a well-defined value to the transverse component of the elementary magnetic moments, so that the transverse magnetization is no more equal to zero.

1.3.1.2 *In terms of vectors and forces*

The magnetic field $\vec{B_1}$ is perpendicular to $\vec{M}$ and exerts a force upon it. Under this force, the orientation of $\vec{M}$ is modified: $\vec{M}$ is tipped away the z axis.

Figure 1.3.2 Flip of the magnetization induced by the RF magnetic field B_1. The magnetic field B_1, perpendicular to the magnetization M and to B_o, exerts a torque upon M and modifies its direction. Since B_1 rotates around B_o at the same frequency F_o than does M, it keeps an adequate angle with M during the precession, and its action goes on during the rotation of M. Note that this drawing does not shows the fast rotation of M and B_1 at the frequency F_o: The observer is 'in the rotating frame' that rotates at the frequency F_o. Many of the following graphs are drawn with the same convention

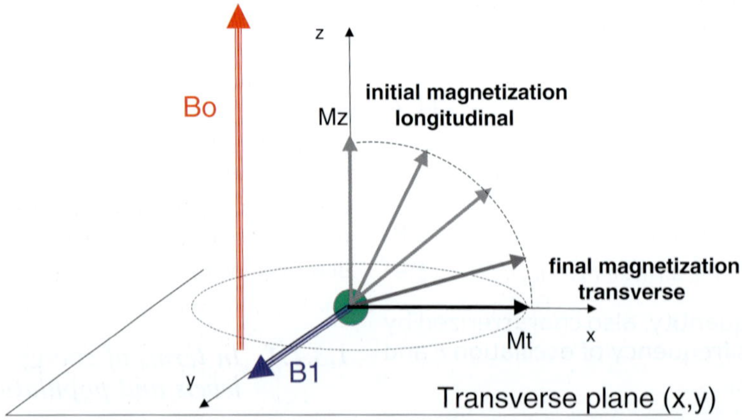

Both $\vec{M}$ and $\vec{B_1}$ rotate at the frequency F_o: The force exerted by $\vec{B_1}$ upon $\vec{M}$ also rotates, so that $\vec{M}$ goes on tipping away from z axis, and, while continuing its precession around $\vec{B_o}$, rotates around $\vec{B_1}$ (Figure 1.3.2).

The angulation of $\vec{M}$ relative to $\vec{B_o}$, induced by application of the RF field $\vec{B_1}$, is measured by the angle between the two vectors and is named as *the flip angle*. The magnitude of the flip angle depends on the amplitude of B_1 and the duration of its application τ, and also on the gyromagnetic factor γ, according to the relation

$$\alpha = \gamma B_1 \tau. \tag{1.10}$$

When the flip angle reaches 90°, so that $\vec{M}$ is perpendicular to $\vec{B_o}$, the RF field $\vec{B_1}$ is shut down. Now the magnetization is fully 'transverse': The vector $\vec{M}$ lies in the plane x-y and rotates around $\vec{B_o}$ at the frequency F_o. The transverse component M_t is the largest possible at the end of a 90° flip: Then M_t is equal to the value of M_z before the application of $\vec{B_1}$ and $M_z = 0$.

The sample is ready for detection of the rotating transverse nuclear magnetization.

Usually B_1 is applied during a very short time, at high intensity: This is called a pulse of RF magnetic field. The RF pulse, which makes a 90° flip angle, is called a '*90° RF pulse*'.

Other trajectories are possible with a flip angle smaller than 90° (M_z is smaller but positive at the end of the RF pulse) or larger than 90° (M_z is negative at the end of the RF pulse).

More physics: the 180° RF pulse.
Initially, the nuclear magnetization is at equilibrium ($M_z = M_o$) corresponding to the difference ΔN between the populations N^+ and N^-. After irradiation by the RF field B_1 at the resonance frequency, when the number of photons absorbed by the nuclear magnetic moments is twice of that corresponding to a 90° pulse, the difference between populations N^+ and N^- is inverted: The upper level is more populated and the longitudinal magnetization has the value $-M_o$. Magnetization has been inverted; geometrically, this corresponds to a flip angle of 180° around the direction of B_1.

A 180° RF pulse is also applied in order to refocus the transverse magnetization and hence to generate a spin echo (see Section 1.3.4). Then its effect is to invert the component of M_t perpendicular to the RF field B_1 as shown in Figure 1.3.3(b).

1.3.2 How to detect the nuclear magnetization?

Nuclear magnetization can be detected while it rotates at a well-defined frequency after excitation. Voltage at the same frequency is induced in a receiver coil.

When a magnet bar rotates next to a loop of conducting wire, a voltage is induced and current flows in the loop. The simplest receiver coil is a loop of conductive wire designed to deliver a large voltage when it 'sees' a small magnetic field oscillating at the frequency F_o.

Let us consider a small sample containing water, in proximity to the receiver coil. After excitation of NMR

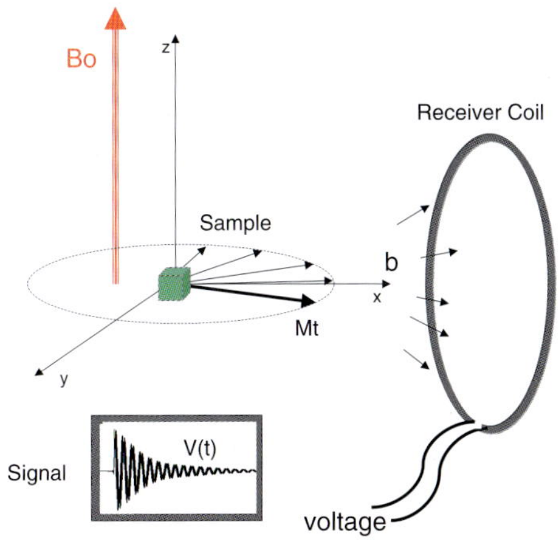

Figure 1.3.3 Signal induced in the receiver coil by the rotation of the transverse magnetization. The magnetization of the sample creates a small magnetic field b at vicinity. The flux of this field $\vec{b}$ through the receiver coil is modulated by the rotation of the transverse magnetization $\vec{M}_t$ of the sample at resonance. Thus, a voltage v is induced in the receiver coil. This voltage is the NMR signal. It is modulated at the frequency of rotation of magnetic moments, F_o. The decay of the transverse magnetization causes the NMR signal decay

for hydrogen nuclei, the transverse magnetic moment M_t, the sum of hydrogen magnetic moments in the sample, rotates around B_o and creates a variable magnetic field $\vec{b}$ across the receiver coil as shown in Figure 1.3.4.

ψ is the flux of this magnetic field through the coil, written as the integral of the magnetic field $\vec{b}$ over the receiver coil surface. Faraday's law states that when the magnetic flux varies, a voltage v is induced in the coil, given by

$$v = -\mathrm{d}\psi/\mathrm{d}t. \tag{1.11}$$

As M_t rotates with frequency F_o, the voltage v also oscillates at the frequency F_o. It is proportional to the magnetization M_t, M_t itself being proportional to M_o. From the derivation of the flux versus time, the voltage is proportional to the frequency F_o. (Remember that M_o and F_o are both proportional to B_o.)

The hydrogen magnetic moments in other adjacent samples also contribute to the total magnetic flux, and then to the total voltage V induced in the coil.

This voltage is named the free induction decay signal (FID) signal. The measurement of the FID signal is the simplest way to detect the nuclear magnetic resonance from nuclei in a sample.

1.3.3 The return to equilibrium of the nuclear magnetization

Key points

After excitation, the nuclear magnetization, set perpendicular to the external magnetic field, is out of equilibrium. Processes that bring back the nuclear magnetization to its equilibrium state take place. The recovery to equilibrium, named relaxation, is described by different evolutions for the longitudinal component M_z and the transverse component M_t of the vector magnetization M.

The longitudinal component M_z returns to its equilibrium M_o with the time constant $T1$, named the longitudinal relaxation time.

The transverse component M_t returns to its equilibrium value zero with the time constant $T2$, named the transverse relaxation time. Its decay is caused by the dephasing of magnetic moments due to the occurrence of different precession frequencies.

The local static inhomogeneity of magnetic field also contributes to the decay of M_t, often measured by the apparent relaxation time $T2^*$.

1.3.3.1 The longitudinal relaxation time T1

After emission of a 90° RF pulse at $t = 0$, the longitudinal component of magnetization, M_z, has been set to zero by the excitation, and M_z returns, or 'relaxes' towards its value M_o, by exchanging energy with its surrounding. Its evolution is described by

$$M_z = M_o[1 - \exp(-t/T1)], \tag{1.12}$$

where $T1$, the longitudinal relaxation time, is the time constant characteristic of this exponential process of return to equilibrium. It depends on the probability of energy transfers between the nuclear magnetic moments and their environment, needed to rebuild the difference of population $N^+ - N^-$ at equilibrium as shown in Figure 1.2.4(c).

1.3.3.2 The transverse relaxation time T2

In a perfectly homogeneous magnetic field B_o, the transverse magnetization $\vec{M}_t$, while rotating in the plane perpendicular to $\vec{B}_o$, decays towards its equilibrium value which is zero. This decay is exponential:

Figure 1.3.4 The return to equilibrium of the longitudinal magnetization M_z. Curves are drawn for $T1$=2000 ms.
Full curve, M_z recovery starting from zero after excitation by a 90° RF pulse
M_z tends to its equilibrium value M_o according to Eq. (1.12). At $t=T1$, M_z/M_o=63%; at $t=3T1$, M_z/M_o=95%; at t=5, $T1$ M_z/M_o=99%.
Dotted curve, M_z recovery after inversion by a 180° RF pulse at t=0. The recovery of M_z, starting from the value $-M_o$, is written as

$$M_z = M_o[1 - 2\exp(-t/T1)]. \qquad (1.13)$$

At $t = 0.693\ T1$, $M_z = 0$. The span of variation of M_z is doubled during recovery. The inversion of M_z is used to generate images with higher $T1$ weighting or to suppress the signal of a given tissue as illustrated by Figure 1.7.2(b).

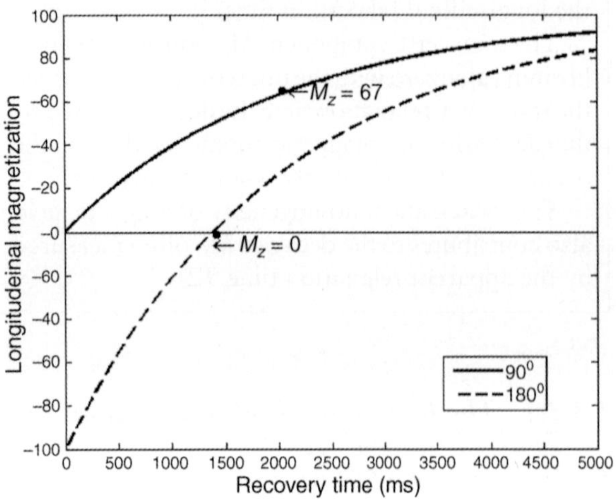

At time t after the creation of the transverse magnetization with amplitude M_{t0}, M_t is given by

$$M_t = M_{t0}\exp(-t/T2). \qquad (1.14)$$

The *transverse relaxation time* $T2$ is the time constant characteristic of this exponential decay.

The decay of the transverse magnetization occurs because each microscopic magnetic moment feels a local microscopic magnetic field created by the other nuclei at proximity, which is added to the external magnetic field $\vec{B}_o$. In a water sample, the thermal agitation of water molecules causes random fluctuations of this small magnetic field created by neighbouring nuclei and then causes small modifications of its resonance frequency. This mechanism spreads the microscopic vectors rotating around $\vec{B}_o$ as shown in Figure 1.3.5. The rotating transverse magnetizations of the nuclei are dephased. This causes the decay of transverse magnetization. The stronger these magnetic interactions between the neighbouring nuclei, the shorter is the relaxation time $T2$.

*1.3.3.3 The apparent relaxation time T2**

When the external magnetic field is not perfectly homogeneous in the volume of interest (the sample or the fraction of sample that constitutes a voxel), then the magnetic moments in this volume have slightly different resonance frequencies according to their location. If the spatial variation of the external magnetic field is larger than the fluctuating microscopic magnetic fields that cause the transverse relaxation, then in the volume of interest the transverse magnetizations are spread (or dephased) more efficiently and the resultant signal decreases more rapidly. This mechanism that accelerates the signal decrease is static: It can be reversed by the realization of a spin echo as seen in the following section.

The resulting apparent relaxation time is called $T2^*$. As seen further (as explained in paragraph 1.5.1), it is shorter inside tissues with heterogeneous structure and in proximity to magnetic agents enclosed in cells or blood vessels.

1.3.4 Dephasing and rephasing of the transverse magnetization: the spin echo

Key points

An echo is made by refocusing the transverse magnetization during its precession at a given time, the echo time. Some components of the transverse magnetization that underwent previous dephasing are rephrased; they add coherently to generate a larger signal. Dispersion of transverse magnetization is reduced at the time of the echo, where signal is maximal.

The spin echo can be used to measure the transverse relaxation time $T2$.

The spin echo can be compared to a race where some runners are faster and some slower (they have different velocities, i.e. precession frequencies). At time τ, they are ordered to start in the opposite direction. At time 2τ, runners arrive altogether at the starting point. Here the race is slightly different: The refocusing pulse does not invert the rotation of magnetization components, but puts them at modified positions on the track, with the same final result; all component have the same phase at the time 2τ (Figure 1.3.6).

The measurement of the transverse relaxation time $T2$ is done by acquisition of several echoes at different echo times, either by spectroscopic or by imaging experiments. The adjustment of these signals to an exponential model allows to determine $T2$.

Figure 1.3.5 The return to equilibrium of transverse magnetization. (a) After excitation, in the transverse plane, the small magnetic moments, the sum of which determines M_t, rotate at different frequencies and their phases become increasingly different: They are more and more dephased. In a very homogeneous magnetic field, the weak microscopic magnetic fields created by neighbouring magnetic moments at the level of one nucleus induce weak variation of the resonance frequency. The individual magnetizations are spread progressively while they rotate. The corresponding time constant is *T2*. (b) When the magnetic field is not homogeneous, the dephasing between individual magnetizations is faster. The corresponding time constant is *T2**, shorter than *T2*. (c) The decay of M_t during its precession is drawn for $T2 = 80$ ms in homogeneous field (full line) and for $T2^* = 20$ ms (dotted line). At $t = T2$ or $t = T2^*$, the ratio M_t/M_{t0} is equal to $e^{-1} = 0.367$

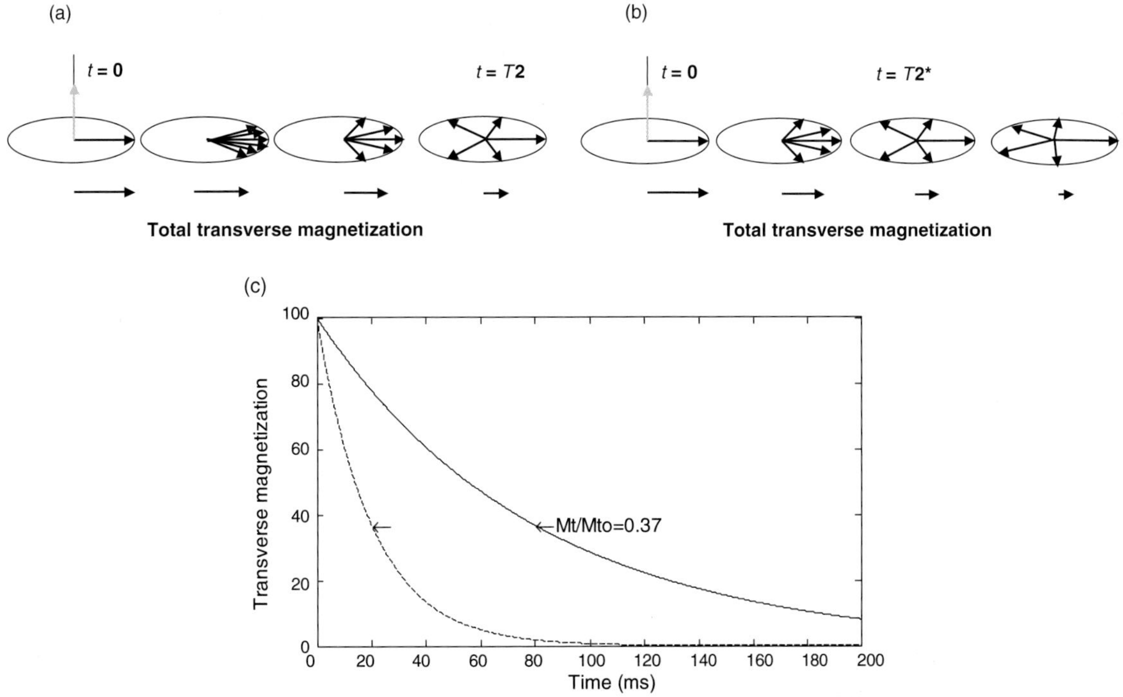

Figure 1.3.6 The spin echo. (a) The spin echo is done by applying a 180° pulse of the RF field B_1 at a given time τ between excitation and signal acquisition of the signal. At time 0, after the 90° RF pulse, the transverse magnetization is lined along OX and begins its rotation in the transverse plane. While it rotates, the spread of resonance frequencies causes quick dephasing of its components. At time τ, the RF field emitted along the *y* axis flips the magnetization components over the *x*-*y* plane, symmetrically to the axis OX. At time $TE=2\tau$, fast and slow rotating magnetizations are gathered and the signal goes through a maximum: This is the echo. (b) Variation of the NMR signal after excitation of resonance and through the echo. The FID takes place at beginning and decays with the time constant *T2**. The refocusing RF field is applied at 100 ms and the echo takes place at 200 ms. Before and after the centre of the echo at TE, the decay of signal on both sides of the echo depends on the shorter time *T2**

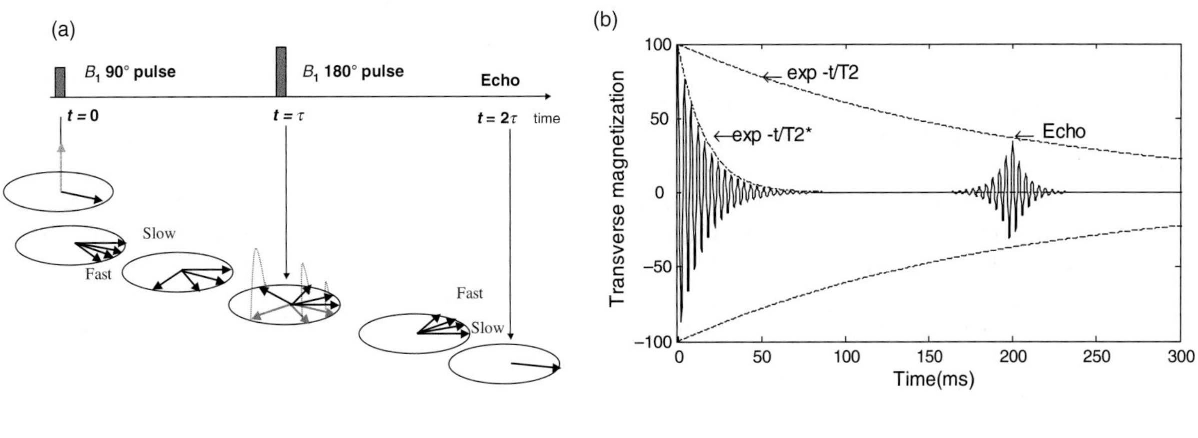

1.3.5 The global evolution of the vector nuclear magnetization

The two components of the nuclear magnetization have different kinetics of return to equilibrium (there is some quantum mechanics hidden behind). After excitation of resonance, the evolution of the magnetization M associates the fast precession of M at the frequency F_o around B_o, the slow increase of M_z along B_o and the faster decrease of the transverse component with the time constant $T2$ or $T2^*$. In most biological tissues $T2$ is much shorter than $T1$: The decay of M_t is much shorter that the recovery of M_z. Considering the handling and measurement of nuclear magnetization, we will go on to describe the evolutions of M_z and M_t as independent phenomena.

1.4 The NMR hardware: RF coils and gradient coils (more technology)

1.4.1 The probe (or RF coil)

A probe (also named as 'coil' because it is often built with coiled copper wire) is an electrical circuit built to emit a magnetic field oscillating at the resonance frequency F_o. The probe emits the 'radio-frequency field' B_1 used to excite nuclear magnetic resonance in the sample. It also receives the magnetic field created by the transverse nuclear magnetization during precession.

Each probe is built to excite or detect NMR at one given frequency (sometimes at two frequencies in order to detect two different nuclei).

The resonance frequency, proportional to B_o and depending on the nucleus observed, lies in the range 10–800 MHz; the frequency range for FM radio broadcasting and GSM telephony.

Probes are built with copper or silver wire or copper sheet to get high electrical conductivity, and with fixed or adjustable amagnetic capacitors.

They are resonating circuits adjusted for a sharp response with high current flowing at the resonance frequency. The current oscillating at the frequency F_o is injected in the circuit of adequate geometry, in order to create an intense magnetic field B_1 in the zone of interest. Homogeneity of B_1 over the zone of interest is useful, but not always obtained.

At the beginning of experiments, the probe response is optimised, usually by trimming adjustable capacitors, to emit maximal B_1 at a given power and at the required frequency. Power of the amplifier that delivers current at the frequency F_o is in the range 100 W–10 kW depending on coils and magnet size.

A probe also is the device used to receive the resonance signal from the sample, at the same resonance frequency. Then homogeneity is a less stringent need, and very small coils at immediate vicinity of the region of interest may offer better sensitivity for detection.

Often probes are used for emission and reception (then named transceivers probes). Some probes are used for B_1 emission only: They are named transmitter probes. Other probes, usually smaller, are used for signal reception: They are receiver probes, positioned in proximity to the organ under examination, around it or at its surface (then named surface coils).

A way to increase sensitivity and spatial extent of measurements is to combine multiple surface coils by building an array of coils around the object under study. Then it is possible to increase the speed of acquisition by parallel imaging (paragraph 6.8.3)

1.4.2 The two simplest RF coil geometries

The surface coil is often built as a circular loop of copper wire. When it is fed with current oscillating at frequency F_o, it generates a magnetic field B_1 also oscillating. The field B_1 has a complex topography across space, except at proximity of the coil centre, as shown in Figure 1.1.2. The quick variation of B_1 amplitude can be used as a tool to select a restricted volume where NMR excitation is done. The surface coil is often built as a circular loop, yielding optimal sensitivity in proximity to the coil centre, within a distance comparable to the coil radius.

The cylindrical coil offer good homogeneity of the RF field B_1 created inside; it is very convenient to house a rat or a mouse inside the cylindrical tunnel of the magnet. The field B_1 is created by several parallel wires (at least four wires are needed); it is homogeneous in the central zone of the structure. The saddle coil, the discrete cosine coil and the birdcage coil are cylindrical coils (Mispelter, Lupu and Briguet, 2006). They can be used for B_1 emission and signal reception, or for B_1 emission only while using a smaller receiver coil.

The efficiency of the coil, as an emitter or a receiver, is higher when the coil is small and is near the region of interest. Increasing the receiver coil efficiency directly increases the signal to the noise of

Figure 1.4.1 RF probes for small animal imaging. (a) Surface coil built to resonate at phosphorus frequency for rat leg muscle spectroscopy. (b) Cylindrical coil for rat half body examination, built according to the discrete cosine geometry (Bolinger, Prammer and Leigh, 1988) (*courtesy of C. Wary, Laboratoire de RMN, Institut de Myologie (AFM-CEA), Paris*)

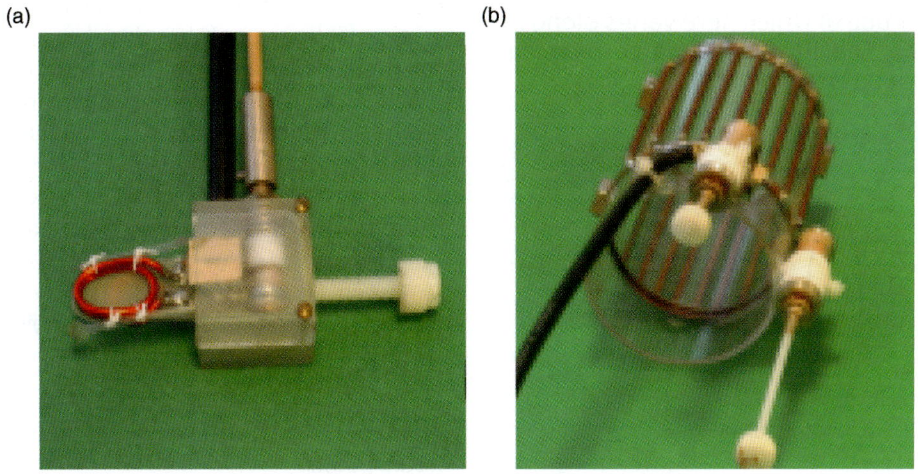

measurements. That is the reason why most labs involved in small animal imaging build their own coils, in order to optimise each experiment as illustrated by Figure 1.4.1. The way to build one's own coils is explained with much practical detail in (Mispelter, Lupu and Briguet, 2006).

1.4.3 The gradient coils

The NMR imaging system includes three distinct gradient coil sets. Each gradient coil set is built in order to create an additional magnetic field parallel to B_0 and varying along one axis, either the x-axis or the y-axis or the z-axis as illustrated in Figure 1.4.2. The geometry of the other coils is schematized in Ness Aiver (1997) or Webb (2003).

In presence of this additional field, the proton resonance frequencies vary as a function of location. This makes possible either to select in the sample a slice where NMR excitation takes place or to read a NMR signal with frequencies reflecting the object structure: Spatial encoding of signal is performed.

Each gradient coil set is fed with current during the time interval when signal labelling along the corresponding axis is performed. Strong gradient intensity, obtained by high current flowing in the gradient coil, is needed to obtain high spatial resolution of the sample image.

While current flows in a gradient coil, the force exerted by the magnetic field B_0 on the wire where current flows induces vibrations of the winding and its support: This is the cause of the strong acoustic noise during MRI acquisitions, though the gradient coils are firmly fixed along the inner wall of the magnet bore.

The magnetic field from one gradient coil set is risen at the needed value in a fraction of millisecond, the rise time. The rise time of magnetic field gradient is an important characteristic of the hardware: If this time is long, milliseconds are lost while current goes

Figure 1.4.2 The z-gradient coil. The z-gradient coil is made of two (or four) symmetric coils, fed with opposite currents. It creates an additional magnetic field ΔB parallel to B_0, which is proportional to z. At the centre of the magnet ($z = 0$), ΔB is equal to zero. For z positive (resp. negative), ΔB is positive (resp. negative). In the central zone, ΔB does not vary with x and y coordinates and varies linearly with z: $\Delta B = Gz \cdot z$

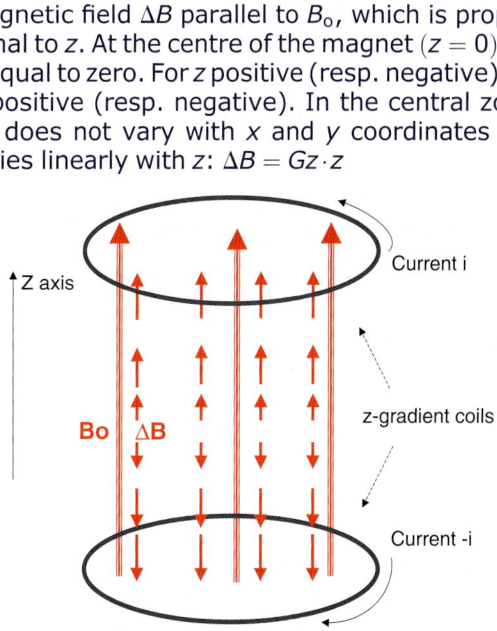

Figure 1.4.3 The three magnetic fields needed for MRI measurements. A mouse is laid upon a surface coil at the centre of the magnet bore. The main magnetic field B_0 created by the magnet is directed along z and the additional magnetic field ΔB (the amplitude of which here varies along z) is parallel to B_0. The RF field B_1 is perpendicular to B_0. The mouse is scaled up.

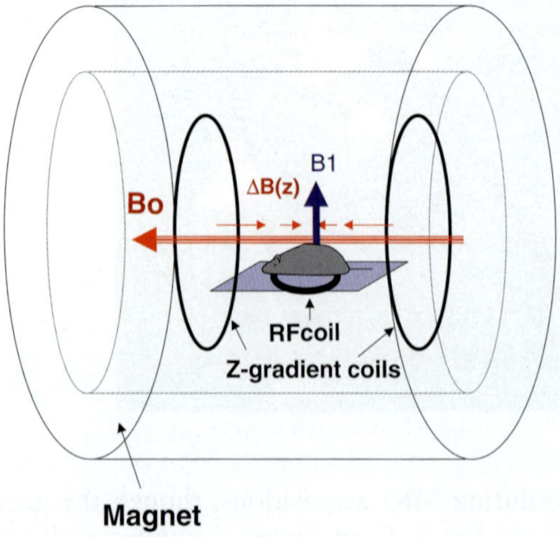

up or down in the gradient coils, and this limits the ability to perform fast imaging.

Gradient strength and gradient rise time are limiting factors for acquisitions with high spatial or temporal resolution. It is easier to obtain high gradient in small gradient coils built for small animals imaging than in larger gradient coils built for medical imaging; typical figures are gradient intensity equal to 200 mT/m and rise time equal to 100 µs, for current intensity peaking at 100 A and voltage at 150 V.

Figure 1.4.3 shows the set of different coils that create B_0, the additional magnetic field varying along $z$axis and the RF magnetic field.

1.5 NMR spectroscopy: the chemical encoding

In vitro NMR spectroscopy, a powerful analytical technique, allows identification and quantification of molecules in a test tube and is widely used to study the structure and dynamics of complex molecules.

In vivo NMR spectroscopy is not so efficient, due to the complexity and the heterogeneity of living systems and less optimal instrumental conditions (lower magnetic field, larger volumes and shorter acquisition times). Its sensitivity is low: Roughly, only metabolites at concentrations above 1 mM can be detected *in vivo*. However, NMR spectroscopy may distinguish several metabolites of a same nucleus that would not be separated by nuclear imaging techniques (PET or SPECT) since their scheme of disintegration is the same. It has then been applied to many physiological studies. An important characteristic of NMR techniques is the possibility to obtain successively or sequentially spectra and images encoding different parameters.

The nuclei ^{31}P, ^{1}H, ^{13}C provide most *in vivo* applications of NMR spectroscopy (Gadian, 1995).

1.5.1 NMR signal and NMR spectrum

Key points
The NMR signal registered during the acquisition time is transformed into a spectrum, which displays the different resonance frequencies of the nuclei present in the sample. As the frequency of resonance of a nucleus depends on its chemical environment, the nuclei in different molecules give rise to different peaks, more or less separated. When no spatial encoding is done, the NMR spectrum reflects the chemical composition of the sample under study. The surface of each peak is proportional to the number of molecules contributing to the peak. High magnetic field homogeneity allows obtaining narrower and higher resonance peaks.

What is a spectrum? The notion of spectrum is familiar: A drop of water, or a prism made with glass, spreads the different coloured components of white light, so that the spectrum of light is made visible. A spectrum is the display of components, at different frequencies, which contribute to a physical phenomenon such as light or sound, or, here, the NMR signal.

When the magnetic moments in the sample resonate at different frequencies (this means that they feel different values of magnetic field, a point explained later), the spectrum of its NMR signal displays these different components as peaks at each frequency. The mathematical operation that gives the spectrum of a NMR signal is the Fourier transform. This operation done by a computer can be compared to the capacity of our ears and brain to identify distinct musical notes played simultaneously.

Molecules can be detected and identified when the corresponding spectral peak is sharp, this being

Figure 1.5.1 Spectroscopy sequence and NMR signal. Sequence of measurement showing the excitation by the RF B_1 pulse, the signal registration during the read-out time and the waiting time for recovery of M_z, named the repetition time T_R. The signal results from the addition of several components with different amplitudes, frequencies and decay times. It is digitised at low frequency after demodulation at a frequency $\approx F_o$

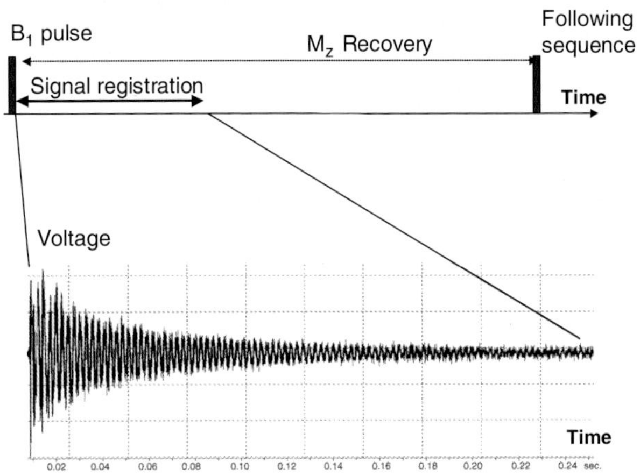

related to a long enough relaxation time $T2$; macromolecules that have fast transverse relaxation yield broad peaks that cannot be observed easily.

When the sample contains several kinds of molecules bearing the nucleus under study (most frequent *in vivo*!), the NMR spectrum of this nucleus contains several peaks. If the peaks have sufficient intensity and are well separated, these molecules can be identified and their respective concentrations in the tissue can be measured from the peak areas in the spectrum. Also if a molecule contains several times the observed nucleus, as the three phosphorus nuclei of the ATP molecule, this molecule has several resonance frequencies more or less separated.

When large samples, such as part of living organisms, are examined, the selection of a volume contributing to the spectrum can be done by using a small surface coil that registers signal from its proximity. Also it is possible to select a volume by using the imaging technique named selective irradiation (paragraph 6.1). Figure 1.5.1 shows how to register the NMR signal.

How to obtain a NMR spectrum?

The aim of the experiment is the excitation of resonance for one nucleus contained in several molecular species that have different resonance frequencies around the value $F_o = \gamma/2\pi\, B_o$. Typically, the acquisition is done after emission of the RF field B_1 at the resonance frequency F_o as an intense pulse of short duration (less than 1 ms). This pulse of duration τ excites the resonance frequencies in the interval $\delta F = 1/\tau$ around the frequency F_o, and then nuclei of several molecules with slightly different frequencies can be excited and detected.

The voltage induced in the receiver probe by the precession of magnetization is named the free induction decay signal (FID). It is collected immediately after the emission of the RF field B_1, during the precession of the transverse magnetization, without the additional manipulations needed to build an image. It decays exponentially with a time constant $T2$ or $T2^*$ and is registered as long as possible up to having decreased at the level of the electronic noise.

When the elementary signal is too weak, as is often the case for nuclei with low gyromagnetic factor γ, or molecules at weak concentration, several signals are accumulated, while one has to wait for the repetition time T_R between successive measurements in order to recover longitudinal magnetization M_z.

After amplification, filtering and numerisation, the signal is transformed into a spectrum by a mathematical calculation, the Fourier Transform. The spectrum displays the frequential content of the signal around the frequency F_o.

1.5.2 More physics: the width and height of a resonance peak

The NMR signal is proportional to the number of hydrogen nuclei excited by the RF B_1 pulse and detected by the receiver coil. After Fourier transform, the surface of each peak is proportional to the corresponding number of nuclei. One can estimate simply this surface by the product of height and width of the peak.

When the magnetic field is perfectly homogeneous, the spectrum of the peak corresponding to a given molecule (here for example water) is related to the corresponding transverse relaxation time $T2$. The longer the $T2$, that is the weaker the magnetic interactions between the magnetic moment of the nucleus detected and the neighbouring magnetic moments, the narrower is this peak. The line-width (defined as the full width of the peak at half height) is written as

$$\delta F_o = 1/\pi T2. \tag{1.15}$$

When the local value of the magnetic field varies inside the sample, the decay of transverse magnetization is faster. During the rotation of the transverse magnetization, the magnetic moments that feel slightly different values of the external magnetic field B_o rotate at different frequencies, and then their sum quickly decreases.

In first approximation the signal decay can be described as exponential, with a time constant analogous to $T2$. The apparent relaxation time that describes the decay of transverse magnetization $T2^*$ is written as

$$1/T2^* = 1/T2 + \pi\gamma\delta B_o, \quad (1.16)$$

where δB_o is the range of variation of B_o through the sample.

The spectral width of the peak is then

$$\delta F_o = 1/\pi T2^*. \quad (1.17)$$

It depends on the intrinsic value of $T2$ and on the extrinsic parameter δB_o. It is then important to obtain high magnetic field homogeneity in the zone under study in order to get minimal spectral width and maximal peak height as illustrated in Figure 1.5.2. By sending current in dedicated windings, it is possible to improve the magnetic field homogeneity through the sample. This operation named 'shimming' is needed to obtain high quality spectra. It is performed either manually or automatically with dedicated software.

1.5.3 More physics: Why do nuclei in the different molecules give rise to different peaks in the NMR spectrum?

In addition to the external magnetic field, each nucleus inside a molecule feels a magnetic field created by neighbouring electrons, which depends on chemical bonds between the atom and its neighbours. This additional magnetic field $\vec{B}_{el}$, which originates from

Figure 1.5.2 NMR FID signal and spectrum. (a) FID signal plotted at the relative precession frequency $F - F_o = 50$ Hz, with decay time $T2 = 100$ ms. (b) Spectrum, the peak line width measured at half-height is $\delta f = 3.3$ Hz. (c) FID signal at same relative precession frequency $F - F_o = 50$ Hz, with shorter decay time $T2^* = 50$ ms. (d) Spectrum, the peak area is unchanged, its width is doubled and its height is halved

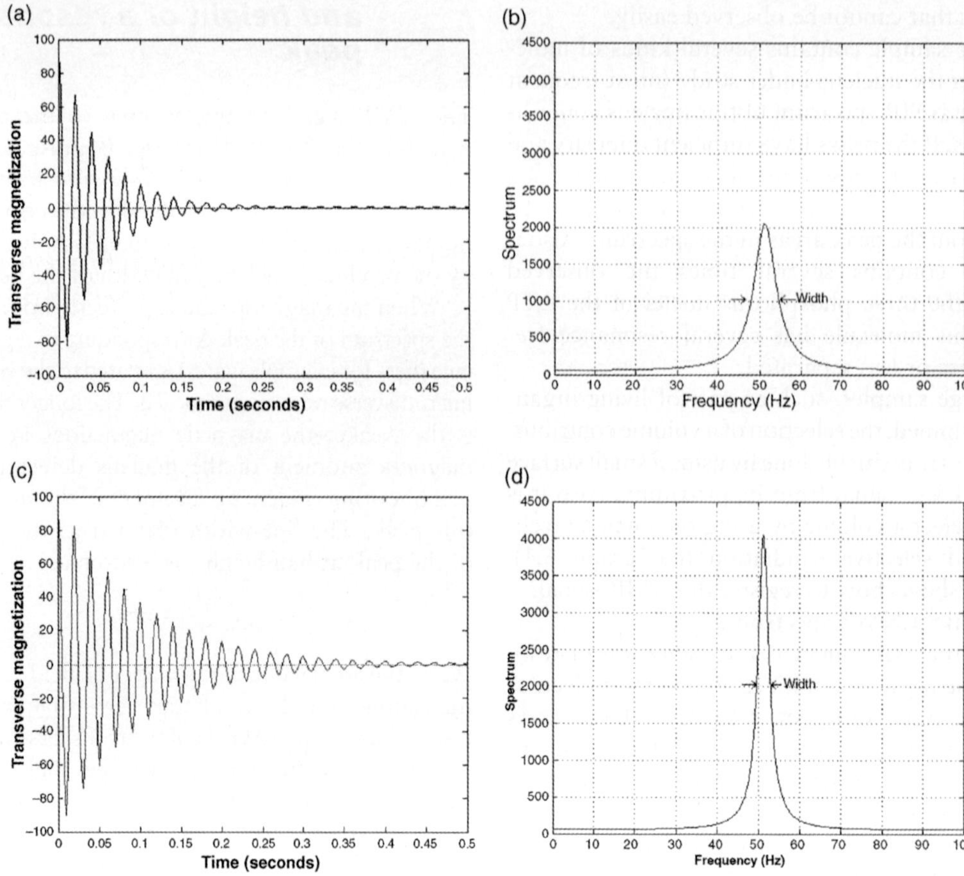

the magnetic polarization of the electrons, is proportional to the external magnetic field B_o.

The frequency of resonance is calculated as a function of this additional magnetic field of neighbouring electrons, written as

$$\vec{B_{el}} = -\sigma.\vec{B_o}. \tag{1.18}$$

Then the resonance frequency is written as

$$F = \gamma/2\pi.(B_o + B_{el}) = \gamma/2\pi.(1 - \sigma).B_o, \tag{1.19}$$

$$F = F_o.(1\ \sigma), \tag{1.20}$$

where σ is the chemical shift of the resonance frequency, independent of B_o value, and usually given in parts per million (ppm).

Each type of chemical bond corresponds to a value of σ: For example in the ATP molecules the three phosphorus nuclei have different resonance frequencies.

Spectra are displayed along a relative frequency scale calibrated in ppm (1 ppm = 10^{-6}).

$$\delta = (F - F_o)/F_o, \tag{1.21}$$

where δ is the displacement of the resonance frequency F relative to the reference frequency F_o.

Using this relative scale, the position of the peaks is independent of the magnetic field intensity, even though they are better separated at higher field, the reason why spectroscopic measurements are done at high magnetic field.

1.5.4 Phosphorus spectroscopy

The phosphorus nucleus ^{31}P is almost 100% naturally abundant and resonates at a frequency around 40% of that of hydrogen. From its lower gyromagnetic factor, its signal is then less intense (see Table 1.2.1); also different RF coils are used for phosphorus spectroscopy and hydrogen imaging.

Phosphorus spectroscopy gives access to the detection of the high-energy phosphorylated metabolites ATP and phosphocreatine (PCr), of the phosphate ion (Pi), and of phosphomonoesters and phosphodiesters. These compounds have intracellular concentrations in the range 1–30 mM. Their relative amounts at rest give an insight into the metabolic status of an organ. The dynamic follow-up of their *in vivo* concentrations can be performed non-invasively with a temporal resolution of a few seconds. The metabolism of organs such as skeletal muscle, myocardium, kidney, brain, and liver can be studied under normal conditions of temperature and pH, at rest and under stimulation. Phosphorus spectroscopy is widely used to perform *in vivo* fully atraumatic biochemical studies (Gadian, 1995).

The surface of each peak in the spectrum of a sample is proportional to the number of molecules of the corresponding metabolites in the sample, but it also depends on their relaxation times, $T1$ and $T2$, and on instrumental factors. This makes the absolute quantification of metabolite concentration difficult. It is easier to follow-up the variation of concentration of a molecule, by measuring the ratio of its peak area to that of a reference compound.

The ratio of phosphocreatine (PCr) peak to the sum of all phosphorylated metabolite peaks is widely used as an index to monitor variations of phosphocreatine concentration in skeletal muscle during exercise and recovery. As oxidative phosphorylations take place in mitochondriae, NMR phosphorus spectroscopy is an important tool for *in vivo* quantification of mitochondrial energy production (Balaban, 1984).

Figure 1.5.3 Phosphorus spectrum of rat leg muscle at rest. Acquisition is done at 4T by using a small 15 mm diameter surface coil. The spectrum exhibits five well-separated peaks: The peak of phosphorus in phosphocreatine (PCr) is the highest. The three phosphorus nuclei in the molecule of ATP are detected at different resonance frequencies located at −2.7 ppm, −7.5 ppm and −16 ppm from the PCr peak. The peak of the phosphorus nuclei of the phosphate ion is located at 4.8 ppm from the PCr peak, corresponding to muscle pH = 7. Another smaller and broader peak at 6.5 ppm is that of phosphorus from phosphomonoesters, mostly sugar phosphates (*courtesy of P. Carlier, Laboratoire de RMN, Institut de Myologie (AFM-CEA), Paris*)

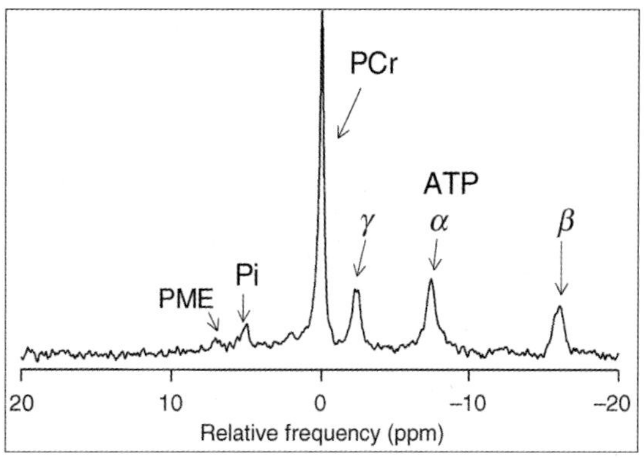

Figure 1.5.4 Rat brain hydrogen spectrum. Acquisition is done at 9.4 T with a spin-echo sequence (TR = 3500 ms, TE = 8 ms). The selection of a small cubic volume located in basal ganglia, with dimensions 3 × 3 × 3 mm is done, is based upon anatomical images by selective irradiation (PRESS sequence). The large water signal is suppressed prior to the acquisition of smaller signals from other less concentrated molecules. From the much weaker concentration of these molecules, the signal of corresponding peaks is low, so that 512 signals are averaged to improve the quality of the spectrum, and the acquisition time is 30 min. The largest peak, at 3 ppm, is that of the methyl protons of NAA. The second highest peak is that of the methyl protons of creatine and phosphocreatine, labelled Cr + PCr. The peak of myo-inositol (Ins) at 4.05 ppm corresponds to a concentration about 5 mM (*courtesy of Bruker SA, Ettlingen Germany*)

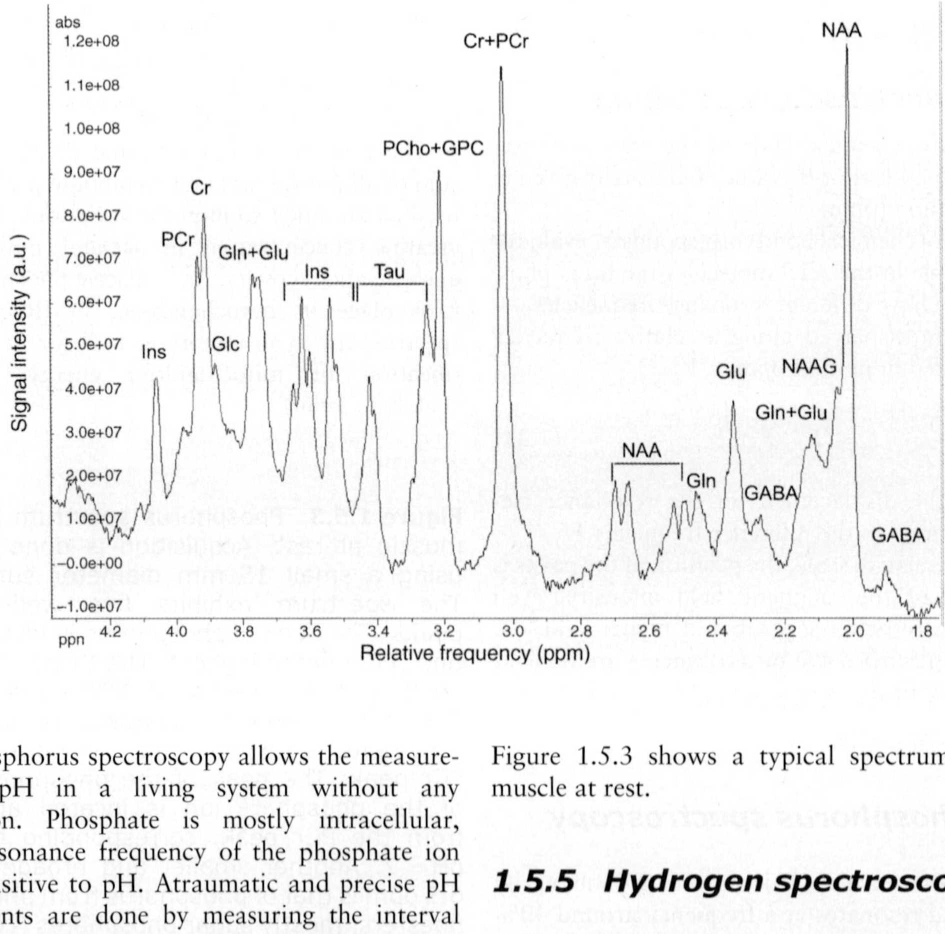

NMR phosphorus spectroscopy allows the measurement of pH in a living system without any perturbation. Phosphate is mostly intracellular, and the resonance frequency of the phosphate ion peak is sensitive to pH. Atraumatic and precise pH measurements are done by measuring the interval between Pi and PCr peaks. The intracellular pH is related to the concentrations of the $H_2PO_4^-$ and HPO_4^- ions and to the dissociation constant of phosphoric acid ($pK_a = 6.75$). The resonance frequencies of the two phosphoric ions differ by 2.42 ppm. In water, the inter-conversion between the two phosphoric ions is so fast that their peaks are collapsed into one single peak, the frequency of which reflects the proportion of the two forms and thus the pH. The relation that links the pH to the interval between Pi and PCr peaks, δ (in ppm), established by Moon and Richards (1973), is then

$$\mathrm{pH} = 6.75 + \log((\delta - 3.27)/(5.69 - \delta)). \quad (1.22)$$

Figure 1.5.3 shows a typical spectrum of rat leg muscle at rest.

1.5.5 Hydrogen spectroscopy

NMR hydrogen spectroscopy benefits from the highest sensitivity and from the value of the hydrogen gyromagnetic factor. However, the range of hydrogen chemical shifts for biological compounds is narrower, and the large signals of water and fat hide the signals of the less abundant metabolites. Also the number of molecules that contain hydrogen is huge, and spectra are often obscured by the multiplicity of superimposed peaks. Only a limited number of peaks are identified easily for biological studies. Most applications of hydrogen spectroscopy concern brain studies (Gadian, 1995). As brain is a complex and heterogeneous organ, brain studies require to combine precise localization and spectroscopic analysis. A volume is selected by using the selective irradiation, and then its NMR signal is read without magnetic field gradient. As water is at concentration circa 80 M in brain, the huge water signal that would obscure smaller signals has to be suppressed. The large peak of *N*-acetylaspartate (NAA) is a marker of neurons in mature brain. Creatine and choline peaks are also easily detected. The peak of lactate is a marker of brain metabolic disorders. Figure 1.5.4 shows a typical spectrum of rat brain.

1.5.6 Spectroscopic imaging

Magnetic resonance spectroscopy and imaging are combined into spectroscopic imaging, also named chemical shift imaging (CSI), where spectra are obtained for each voxel in the plane or the volume of interest (Gadian, 1995). Several operations of phase encoding (see paragraph 6.3) are performed along two or three spatial directions, similarly to 3D imaging (paragraph 6.7), before reading a signal in the absence of magnetic field gradient. Acquisition times can be very long unless sophisticated fast acquisition techniques are applied. Spectroscopic imaging is mostly applied to brain studies, where a precise localization of biochemical abnormalities is needed.

1.6 How to build NMR images: the spatial encoding

Key points

In order to build images of hydrogen magnetization, the magnetic field in the magnet is made to vary linearly along one axis, so that the resonance frequencies in the sample are labelled according to the location of nuclei along that axis. Three operations allow to build images:

The selective irradiation around a precise resonance frequency, done in presence of the first magnetic field gradient, excites only the magnetic moments located inside a slice perpendicular to this axis, at a selected position.

After excitation of resonance inside this slice, the second, phase-encoding, magnetic field gradient is installed and the precession of magnetization begins; its frequency is a function of the position of nuclei along the second gradient axis. At the end of this step of preparation, the magnetic moments in the selected slice have an initial phase labelled along the second coordinate axis.

Then the signal of nuclei in the slice is acquired in presence of the third, frequency-encoding, magnetic field gradient and the resonance frequency is labelled along the other coordinate axis.

The contributions of all voxels in the slice are registered altogether.The Fourier transform separates the different frequencies contained in the signal, spread by the frequency-encoding gradient. It also separates the phases of the different voxels prepared by the application of the phase-encoding gradient.

To localize the origin of NMR hydrogen signals, the basic operation consists into spreading the resonance frequencies by application of a non-homogeneous magnetic field B that is the sum of B_o and an additional field varying linearly along one direction through the object, for example

$$B = B_o + \Delta B = B_o + G_x\, x.$$

Then a linear relation between the resonance frequency and the position along one direction is created, because the resonance frequency of nuclei at x is

$$F = F_o + \gamma/2\pi . G_x .\, x.$$

What is a gradient?

A gradient is the variation through space, along one direction, of a physical parameter: A gradient of temperature or pressure, or magnetic field. Here the amplitude of the external magnetic field (not its direction, the magnetic field is always lined along z) is modified, so that the resonance frequency of hydrogen nuclei is modified correspondingly.

The magnetic field gradient along the x axis is $G_x = dB/dx$.

This additional magnetic field is made by sending current in a dedicated winding; here it is the x-gradient coil (see paragraph 4.3). However, if the magnetic field delivered by the main magnet is not homogeneous, some magnetic field gradients are present permanently, a source for image distortions.

Practically, three gradients along the x-, y- and z-axis are applied successively in order to label each point of an object and to build 2D or 3D images of this object. For 2D imaging, several slices are selected and measured successively, and a stack of images, each mapping a given slice, is reconstructed throughout a volume inside the object. For 3D imaging, a volume is selected and measured, and then slices of arbitrary orientation are reconstructed and displayed.

The three operations are described below with more details, in the case of axial slices (perpendicular to the axis z which is the direction of B_o), with the phase-encoding gradient along OY and the frequency-encoding gradient along OX. Then Figure 1.6.3 will display the three successive operations. Any other slice orientation (perpendicular to another axis, or oblique) is obtained easily by permutation of axis or by combination of several magnetic field gradients: A sagittal slice is obtained if selective irradiation is done using the gradient G_x.

1.6.1 The slice selection by selective irradiation

The slice selection relies upon excitation by the RF field $\vec{B}_1$ and simultaneous application of a magnetic field gradient.

The additional field $\Delta\vec{B} = \vec{G}_z.\,z$, created by current flowing in the z-gradient coil, is lighted on immediately before and during the application of the RF field at the frequency $F1$, as shown in Figure 1.6.1.

Figure 1.6.1 Slice selection with a gradient of magnetic field B along the z-axis. The two symmetrical z gradient coils create an additional magnetic field $\Delta B = G_z.\ z$, lined along B_o and proportional to z in a zone at centre of the magnet. During application of the gradient G_z, the resonance frequency at z is

$$F(z) = \gamma/2\pi.(B_o + \Delta B) = F_o + \gamma/2\pi.G_z.z.$$

The RF field $B1$ is applied at the precise frequency $F1$, so that resonance is excited only in the slice around $z1$: Only the magnetic moments in this slice are set in transverse orientation, while those out of the slice are not modified

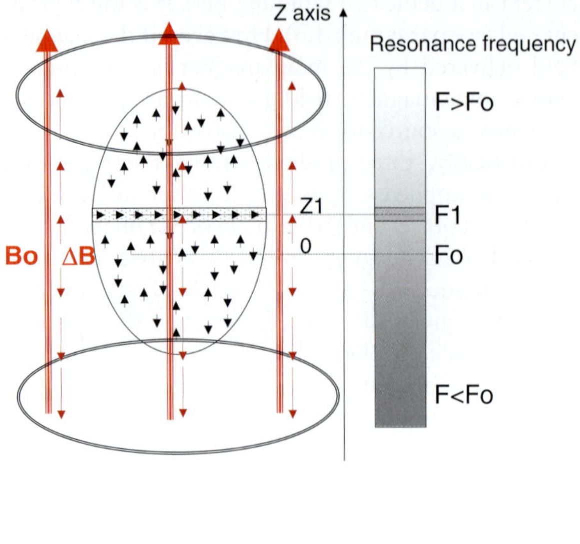

The magnetic field is

$$\vec{B} = \vec{B}_o + \Delta\vec{B} = \vec{B}_o + \vec{G}_z.z. \tag{1.23}$$

Then the nuclei located at the level z resonate at the frequency

$$F(z) = \gamma/2\pi.(B_o + \Delta B) = F_o + \gamma/2\pi.G_z. \tag{1.24}$$

At $z1$, the resonance frequency is

$$F1 = F_o + \gamma/2\pi.G_z.z1. \tag{1.25}$$

The magnetization in the slice at $z1$, which resonates around the frequency $F1$, is excited by the RF field $B1$ and is driven to the transverse plane. Other slices of the object also see the field $B1$ but their magnetization is not modified by this RF field that oscillates at a 'wrong' frequency.

This operation is named the selective irradiation.

Afterwards, the magnetic moments of nuclei inside the selected slice begin precession, and their signal can be registered.

More technology: selectivity of a RF pulse

The RF magnetic field is efficient through a narrow and well-defined frequency interval if it is applied with a carefully modulated shape during a rather long duration; this is a 'selective pulse'. Conversely, to excite a broad spectrum a short and intense pulse of RF magnetic field (a hard pulse) is efficient.

Note that in the spin-echo imaging sequence described further, the 180° refocusing pulse used to refocus the transverse magnetization is also selective and then it is applied in presence of a magnetic field gradient, so that only the moments in the excited slice are refocused, and those out of the slice do not feel the 180° RF field pulse.

It is also possible, before excitation, when the magnetization is longitudinal, to invert it inside a slice by a selective RF pulse giving a flip angle equal to 180°, as is done for the inversion-recovery sequence (as illustrated by Figure 1.7.2(b)).

The selection of a volume can be performed by selective irradiation of a slice along the three orthogonal directions, successively (for example one 90° pulse for excitation, and two successive refocusing 180° pulses). The signal read after this sequence comes from the cube at the intersection of the three orthogonal planes. This is the basis of single voxel brain spectroscopy where the signal of a small volume selected inside brain is measured, after localization from scout images.

Figure 1.6.2 (a) Signal acquisition and frequency encoding along the *x* axis. The frequency-encoding gradient is applied during the detection of the NMR signal of the whole slice selected. The magnetization of one elementary volume around the abscissae x resonates at the frequency

$$F = F_o + \gamma/2\pi.Gx.x.$$

The signals of all elementary volumes add altogether; Fourier transform allows to assign their specific contributions along the *x* axis in the spectrum.
(b) Signal preparation and phase encoding along the *x* axis. The phase-encoding gradient is applied during the preparation time T_P. The magnetization of one elementary volume around the coordinate *y* rotates in the transverse plane at the frequency

$$F = F_o + \gamma/2\pi.Gy.y$$

and its phase after the time TP depends on *y*.
The second step of 2D Fourier transform allows to assign each phase to the position along *y* axis

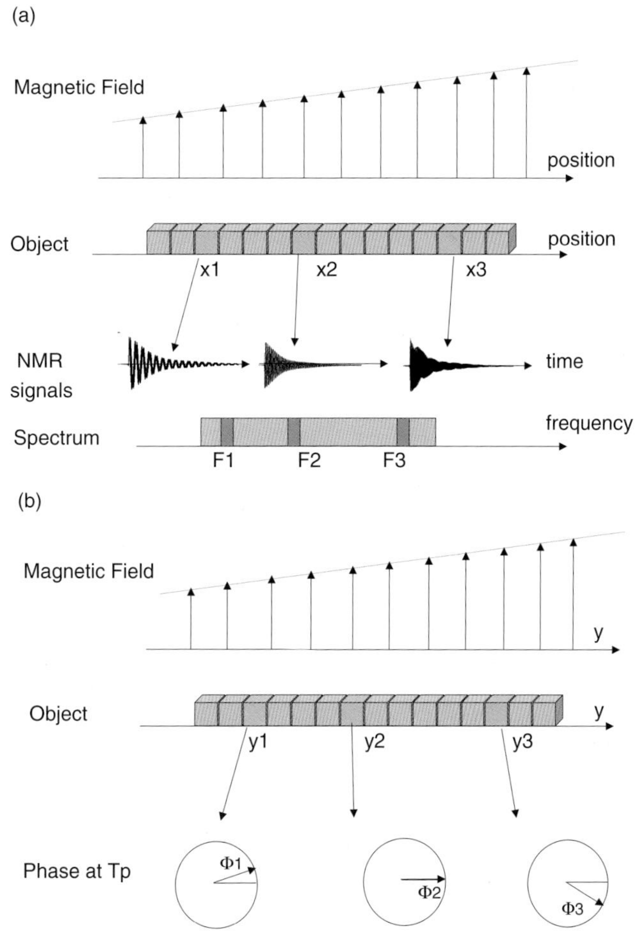

1.6.2 The signal acquisition with frequency encoding

During this operation, the resonance frequency is linearly related to the position of nuclei along the x axis. The additional field parallel to B_o and varying with x is as follows: $\vec{\Delta B} = \vec{G_x}.\ x$.

It is applied by setting current in the x-gradient coils, during the precession of transverse magnetization, as shown by Figure 1.6.2(a).

The resonance frequency is then

$$F = \gamma/2\pi.(B_o + \Delta B) = F_o + \gamma/2\pi.G_x.x. \qquad (1.26)$$

Around the position $x1$, inside a volume of width δx delimited inside the selected slice, the hydrogen nuclei resonate around the frequency

$$F1 = F_o + \gamma/2\pi.G_x.x1 \qquad (1.27)$$

in the interval

$$\delta F = \gamma/2\pi . G_x . \delta x. \quad (1.28)$$

The NMR signal registered comes from all magnetic moments in the slice; it is the sum of signals of magnetic moments at different abscissae and then at different frequencies.

From this signal the Fourier transform yields a spectrum. In this spectrum, the selection of the interval $\delta F = \gamma/2\pi.\ G_x.\ \delta x$ around $F1$ allows to estimate the signal which originates from this rod around the abscissa $x1$.

The spectrum of the total signal is a one-dimensional image of the object.

1.6.3 The signal preparation by phase encoding

The position in the plane along the other axis, here y, is encoded by another operation, the phase encoding, done after slice selection and immediately before reading the signal. Between these two operations, immediately after the extinction of the RF field $B1$, the gradient G_y along y is applied during the preparation time Tp (a few milliseconds) in order to modify the frequency of precession. At the end of Tp the gradient G_y is shut down, the gradient G_x is lighted on and the signal acquisition begins. The magnetic moments keep the memory of their precession during T_p, which took place at a frequency depending on y and on the value of the gradient G_y. They have rotated in the transverse plane, during T_p, by an angle ϕ depending on their coordinate y (Figure 1.6.2(b)).

F is related to y according to

$$F = F_o + \gamma/2\pi . G_y . y. \quad (1.29)$$

The phase of a magnetic moment with frequency F starting from $\phi = 0$ at $t = 0$, at time T_p is

$$\phi = 2\pi . F . T_p. \quad (1.30)$$

It is related to the position of nuclei along the y axis by:

$$\phi = 2\pi . F_o . T_p + \gamma . T_p . G_y . y. \quad (1.31)$$

Figure 1.6.3 Succession of the three operations that allow spatial labeling of the NMR signal. Slice selection is done by selective excitation of resonance at the frequency $F1$, that is the resonance frequency in the slice at $z1$ in presence of the magnetic field gradient G_z. Phase encoding is done by application of the magnetic field gradient G_y during the preparation time T_p. Then the phase of M_t in the transverse plane at the time T_p depends on the coordinate y. Frequency encoding is done by application of the magnetic field gradient G_x during the signal acquisition. Then the frequency of precession of M_t in the transverse plane during the signal acquisition depends on the coordinate x

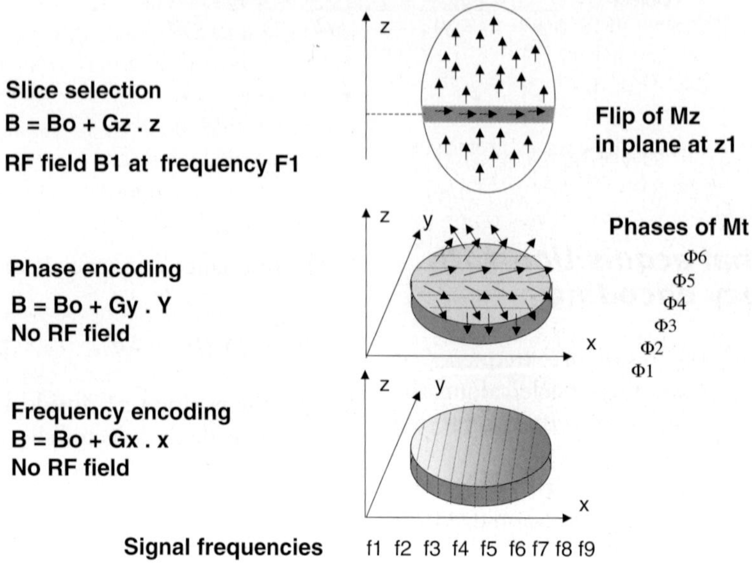

The first term of the phase is the same for all magnetic moments in the slice; the second term varies linearly according their position y.

The phase encoding operation is repeated N_y times, before each signal acquisition, in order to obtain N_y signals, encoded with increasing values of the gradient G_y at each successive step. Each signal acquisition is separated from the previous one by the time interval T_R.

This waiting time T_R is needed for the recovery of M_z before the next excitation and acquisition of the following signal.

1.6.3.1 *The phase artefacts: a specific feature of MRI*

The motions of organs such as heart, lungs and digestive structures often are not synchronized with the phase encoding. This is the source of some image degradation, because phase encoding gives information upon the localization along the y axis at times separated by the interval T_R (the repetition time), not synchronized with physiologic motion. If the localization of a given structure is not stable at the time of encoding, artefacts (often ghost-like images of the moving structures) degrade the image quality as illustrated in Figure 1.6.4. It is possible to reduce these artefacts by synchronizing the NMR acquisition with the motion (for example by triggering acquisition with an electrocardiographic signal to perform heart imaging).

Figure 1.6.4 Phase artefacts. Coronal image through the thorax and abdomen of an anaesthetised mouse, done at 1.5 T with a spin-echo sequence, T_R 600 ms, T_E 18 ms, acquisition time 4 min 15 s. The read gradient is applied along the head-tail direction and the phase encoding gradient along the left-right direction. The respiratory motions induce ghost-like images of organs in the direction of phase encoding, more visible for high intensity structures such as stomach and kidneys

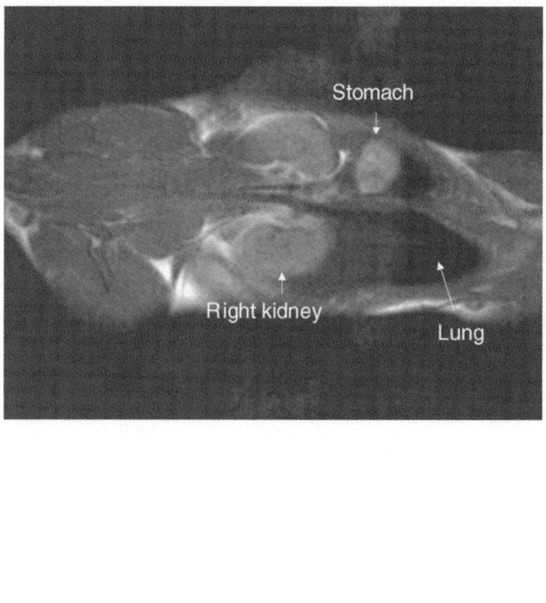

1.6.4 *How to increase the amplitude of the NMR signal: the echo*

Key points

An echo is made by refocusing the transverse magnetization during its precession at a given time, the echo time. The different components of transverse magnetization contained in one voxel, which were dephased, add more coherently so that their sum generates a larger signal. At the time of the echo the signal goes through a maximum. The value of this maximum is limited by the intrinsic decay of transverse magnetization, expressed by the transverse relaxation time, $T2$ or $T2^*$.

The spin echo efficiently decreases the dephasing caused by the local static magnetic field inhomogeneity and by the read gradient; the spin-echo signal at echo time T_E decreases as $\exp(-T_E/T2)$.

The gradient echo only removes the dephasing by the read gradient; the gradient-echo signal at echo time T_E decreases as $\exp(-T_E/T2^*)$.

The formation of an echo increases signal intensity and then image quality.

The acquisition of each NMR signal is delayed by the time needed for phase encoding. Magnetic field gradients, either resulting from the local inhomogeneity of the static magnetic field or applied for frequency or phase encoding operations, contribute to faster signal decay during the precession. This loss of signal can be partly recovered by the realisation of an echo. The echo is a signal that evolves through a maximum when the dephasing of its components is counterbalanced at a precise time, the echo time (as an opened fan that is closed).

The spin echo is done, as explained in paragraph 1.3.4, and Figure 1.3.6, by applying a 180° pulse of the RF field $B1$ at mid-time between the excitation and the centre of the signal acquisition. At time 0, the transverse magnetization begins its rotation in the transverse plane. While it rotates, the spread of resonance frequencies causes quick dephasing of its components. At time τ, the RF field flips the magnetization over the x-y plane, symmetrically to the axis OX. At time $T_E = 2\tau$, fast and slow rotating magnetizations are gathered: the signal goes through a maximum at the echo time T_E.

The signal acquisition is done on both sides of the echo.

The frequency-encoding gradient, which is an important source of magnetic inhomogeneity, is applied symmetrically before and after the refocusing pulse. The 'spontaneous' static field gradients from the magnet or from the object structure are applied permanently, and then they are also symmetrical with regard to the refocusing pulse. *Note that the dephasing induced by the phase-encoding gradient must not be refocused: The corresponding information would be lost.*

At T_E the decay of the signal is governed by the transverse relaxation time, and the signal is proportional to

$$M_t(T_E) = M_t(0).e^{-T_E/T2}. \quad (1.33)$$

When the homogeneity of local magnetic field is low, the acquisition of spin echoes greatly enhances the quality of images; however, the refocusing RF pulse takes some time and delays the signal acquisition.

With spectroscopy or imaging techniques, several successive echoes can be generated at times T_{E1}, T_{E2}, T_{E3}, ... This is used to measure $T2$ by sampling the decay of M_t along time and adjusting data to an exponential model.

The gradient echo is done by applying successively two intensities of the frequency-encoding magnetic field gradient, with opposite signs. The first gradient causes quick dephasing of the magnetic moments that have precession frequencies depending on their positions according to the corresponding direction; in presence of the second gradient the magnetic moment that had the higher precession frequency now have the lower one and vice versa. This allows a refocusing of magnetic moments located at different positions along the gradient axis, and then allows to measure a larger signal at time TE. This refocusing compensates 'only' for the dispersion of phases induced by this gradient, not for the other mechanisms of dephasing.

According to the local inhomogeneity of the magnetic field from other sources than this frequency encoding gradient, the transverse magnetization M_t decays with the time constant $T2^* < T2$, and the signal at the echo time T_E is proportional to

$$M_t(T_E) = M_t(0).e^{-T_E/T2^*}. \quad (1.34)$$

The echo time can be made very short because no refocusing RF pulse is applied. Short echo times are needed to perform fast acquisition with short repetition time T_R.

When a small structure inside the object is a local source of magnetic inhomogeneity, $T2^*$ is shortened locally, and at a given echo time T_E the signal around this structure is weakened. This is used to detect local magnetic structures, such as capillaries filled by deoxyhemoglobin in red blood cells, or cells magnetically labelled with iron. Conversely the magnetic inhomogeneity caused by large structures such as the interface between air-filled lungs and heart also shorten $T2^*$ through large zones and decrease image quality (Figure. 1.6.5(c)).

1.6.5 Duration and chronogram of image acquisition

Since N_y phase encoding operations are needed, the basic acquisition lasts for a time equal to N_yT_R. Adding several identical measurements in order to improve image quality can be needed. Then the acquisition time is multiplicated by N_a.

In conventional 'spin-echo' sequences, the repetition time T_R is in the range 0.1–3 s and acquisition lasts a few minutes. Gradient-echo sequences with shorter T_R values are widely used to reduce acquisition time to seconds. Still faster modalities are available (see Sections 1.6.8. and 1.7.2.).

Usually several slices are measured during one acquisition as shown in Figure 1.6.6(b), sparing a lot of time: Nuclei in all slices are detected sequentially by repeating successively selective irradiation and signal registration for each slice.

1.6.6 Image reconstruction and image matrix

Image reconstruction is done via a 2D Fourier transform (FT) that transforms the information contained in the frequency and the phase of signals into an information related to the location in the plane (x, y) of proton magnetization. As a comparison, when listening to the orchestra playing, the conductor can identify the sounds from different musical instruments played at the same time and can also detect that one given instrument played a note with a small temporal lag (its phase).

For each slice, N_y signals have been measured. Each signal, after numerisation, is made of N_x values. All the signals that have been measured are recombined

Figure 1.6.5 (a) **The spin-echo**: The frequency encoding gradient is applied symmetrically before and after the 180° RF pulse (applied at time τ). Faster and slower rotating magnetizations have their phases inverted by the 180° RF pulse, and then are rephased at the echo time $TE = 2.\tau$, where the signal is maximal.
(b) **The gradient-echo**: The frequency encoding gradient is applied with successive opposite polarities to obtain the gradient echo at TE. The precession is faster for magnetic moments at $x2 > x1$ when the magnetic field gradient is positive, slower when it is negative. Inverting the gradient creates a rephasing of magnetic moments and then an echo.
(c) **Images of a lemon done with a spin-echo sequences**. $TR = 1000$ ms, $TE = 14$ ms. Flip angle value is 90°, the acquisition time 512 s.
(d) **Images of a lemon done with a gradient-echo sequences**. $TR = 100$ ms, $TE = 9$ ms. Flip angle value is 25°, the acquisition time is 51 s. The two images are done with the same geometrical data and slice location but differ by their acquisition time and by the signal-to-noise ratio. The gradient-echo is done ten times faster, at the optimum flip angle as explained in paragraph 1.7.2.1. The two images also differ by the magnetic artifacts, clearly visible on the gradient-echo image. Magnetic inhomogeneities, induced by the small difference in local magnetic field between the water of lemon pulp and the denser tissue of fibrous borders, or between lemon structures and air, have high visibility in the gradient-echo image. The central region is filled with air that yields no MNR signal (no protons!) and also has a magnetization different from that of water. Then, from local shortening of T2*, the signal of water is destroyed at the border of the zone filled with air. This artifact is strongly reduced in the spin-echo image.

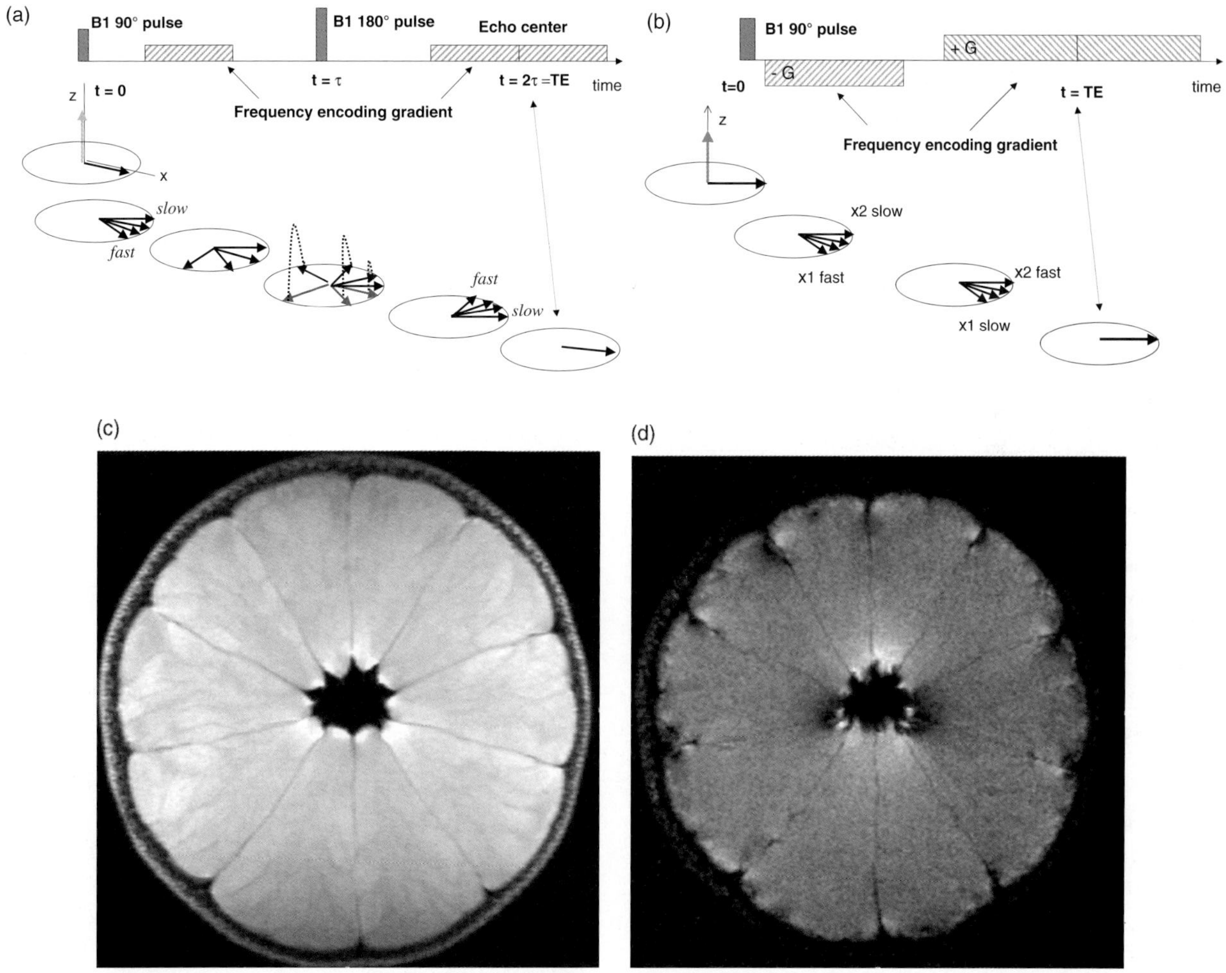

Figure 1.6.6 More technology Chronograms of MRI acquisition sequences. Each line of a chronogram shows a distinct physical operation. (a) Gradient-echo sequence. At $t = 0$, the RF $B1$ pulse is emitted and the slice selection gradient is applied to select the first slice. Then the phase encoding gradient is applied: here, 16 successive values of the gradient are drawn, the 10th value is applied. Simultaneously, the predephasing read gradient is applied. Then the signal of the slice is read before and after the centre of the gradient-echo at the time T_E. N_x signal values are registered during this read time. The longitudinal magnetization is let to recover during the repetition time T_R and then the following sequence is performed.
(b) Multislice gradient-echo sequence. The operations described for slice 1 are done consecutively for the other slices during the time T_R, while the magnetization in each slice returns to equilibrium after its own excitation. After excitation of the first slice, phase encoding and signal acquisition, the z gradient is lighted on again, RF field $B1$ is applied at the second frequency $F2$, the second slice at $z2$ is selected and so on. During the following T_R interval, the gradient G_y is increased and the same operations are performed, here again beginning with the first slice. Each of the N slices is then excited at intervals equal to T_R. At end of acquisition the sequence has been repeated N_y times. For each slice, N_y signals are registered successively, with N_y stepwise increasing values of the signal phase, obtained by stepwise increasing the value of the encoding phase gradient G_y.
(c) Spin-echo sequence. Slice selection and phase encoding are done in a similar way. The refocusing 180° pulse is applied at $T_E/2$ in order to rephase selectively the magnetic moments in the slice (this needs a selective RF impulsion applied in presence of the gradient G_z). The frequency-encoding gradient is applied symmetrically on each side of the refocusiong 180° pulse, and the echo is obtained at T_E.
(d) 3D gradient-echo sequence. The RF $B1$ pulse is applied without a selection gradient to flip magnetization of all the volume. Two phase-encoding gradients are applied. Here, each of them varies across 16 steps; the 9th step along the first axis, and the 4th step along the second axis are being applied. 16 × 16 phase-encoding steps will be done

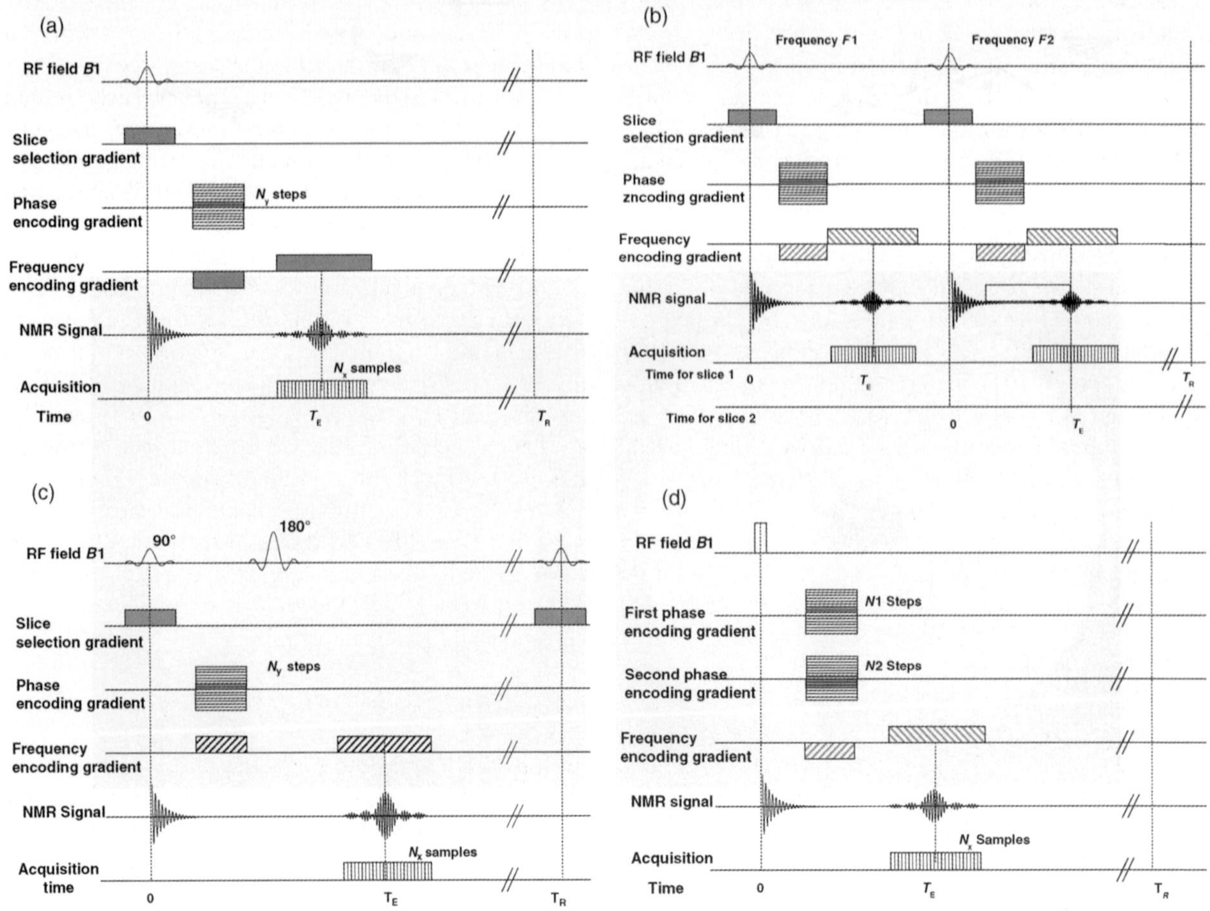

by the Fourier transform to calculate the magnetization of a given pixel of the image.

The matrix of the image is N_x and N_y: This means that $N_x.N_y$ voxels are identified, each giving rise to a signal, and that the corresponding image will be made of $N_x.N_y$ corresponding pixels.

The first Fourier transform gives the spectra of each of the N_y signals registered with frequency encoding along OX, after phase encoding along OY. Each of these N_y spectra is an image of the slice along the axis, distorted by phase variations. The second FT makes the same calculation for the N_y spectra, along the direction of phase encoding: Each step of phase encoding gradient is equivalent to a point along a pseudo-time scale.

Often an image acquired with matrix N_x, N_y is reconstructed with a larger matrix N_x, $2N_y$ by interpolation of data, an improvement of the display without additional information (Figure 1.6.7(b)).

1.6.7 Three-dimensional imaging

Three-dimensional (3D) imaging associates to each voxel of a selected volume, a number proportional to the local value of magnetisation.

From this collection of data, the images of slices with any orientation can be further displayed (since 2D visualisation is easier), without the need of another acquisition.

The first step of 3D acquisition is the excitation of resonance in the volume under study. The phase encoding is performed along two directions, before signal acquisition in presence of the frequency-encoding gradient. The number of phase encoding steps is defined by the matrix in the two corresponding directions: Typically, one may envisage that N_z (e.g. 64) phase-encoding steps are done along the first axis, N_y (e.g. 128) phase encoding steps are done along the second axis and Nx (e.g. 256) points are sampled according to the third axis during each signal acquisition (Figure 1.6.6(d)). Then the volume under study is divided into N_x. N_y. N_z (e.g. $256 \times 128 \times 64$) voxels.

All the possible combinations of the two phase-encoding gradient (their number is N_y. Nz) are applied successively and for each a signal is read.

The acquisition time is at least N_y. N_z. T_R multiplied by the number of averages of identical signals N_a if needed. The large number of signals measured successively (N_a. N_y. N_z instead of N_a. N_y in 2D imaging) contributes to efficient noise averaging (as explained further at paragraph 8.3).

From the large number of signals needed, 3D acquisition is usually performed at short values of T_R and T_E, except in MRI microscopy where acquisition times of several hours are often needed. 3D imaging is mostly used to display complex anatomy (brain, embryonic structure, articulations), when physiologic motion does not interfere with image quality. Figure 1.6.8 shows one slice of a 3D volumic acquisition.

1.6.8 From slow imaging to snapshot imaging

1.6.8.1 *Snaphot imaging*

In the two fastest acquisition sequences, echo planar imaging (EPI) or fast spin-echo imaging (FSE), N_y successive spin-echo or gradient-echo are read within a very short time, each one with a distinct phase encoding, in order to obtain all the information needed to build the image in one single signal acquisition (Stark and Bradley, 1992; Ness Aiver, 1997). This acquisition has to be performed within a time comparable to the time for signal decay, $T2^*$ or $T2$, typically 50–100 ms. Each elementary echo is read more quickly than in usual conditions, and gradients for encoding are set very quickly to strong intensities, needing powerful hardware. Another full image can be built after the recovery time T_R.

EPI is very sensitive to magnetic inhomogeneity, a problem when images are done at high magnetic field. The images obtained by snapshot acquisition are not degraded by motion, but their spatial resolution is low: N_y and N_x have low values, typically 64–128. The time between successive echoes is very short. The effective echo time TE_{eff} (often equal to the time at middle of the echo train) determines the influence of $T2$ on the signal, and the images are strongly $T2$-weighted (for FSE) or $T2^*$-weighted (for EPI). Then magnetic field inhomogeneities have strong influence on the quality of EPI images.

1.6.8.2 *Segmented imaging*

In less fast 'segmented' acquisitions with EPI or FSE sequences, a train of N echoes is read within one acquisition, each one at a distinct phase encoding. At the following T_R, another echo train is read, until the N_y signals needed for completion

Figure 1.6.7 Display of the signals registered to build an image. (a) Data before reconstruction. N_y signals (here 64), one for each phase encoding step, are displayed. Each signal contains N_x values read during 4 ms around the echo time T_E. Each signal is maximal at mid acquisition, at the echo time T_E. Larger signals are registered around the central line ($G_y = 0$). The whole of the slice contributes to each of the signals displayed. All the signals registered contribute to each pixel of the reconstructed image. (b) Image obtained by 2D Fourier transform of these data. The acquisition matrix is 256×64 (64 signals each with 256 samples). After Fourier transform, the image is interpolated: The display matrix is 256×256. This gradient-echo image shows the magnetic distortion induced by an air bubble at the middle of a sphere filled with water. The magnetic field distortion caused by the difference between air and water susceptibilities has a spatial extent much larger than the diameter of the air bubble, with two lateral lobes. Also the bubble is distorted in the phase encoding direction, because the resolution is lower and then the encoding gradient lower

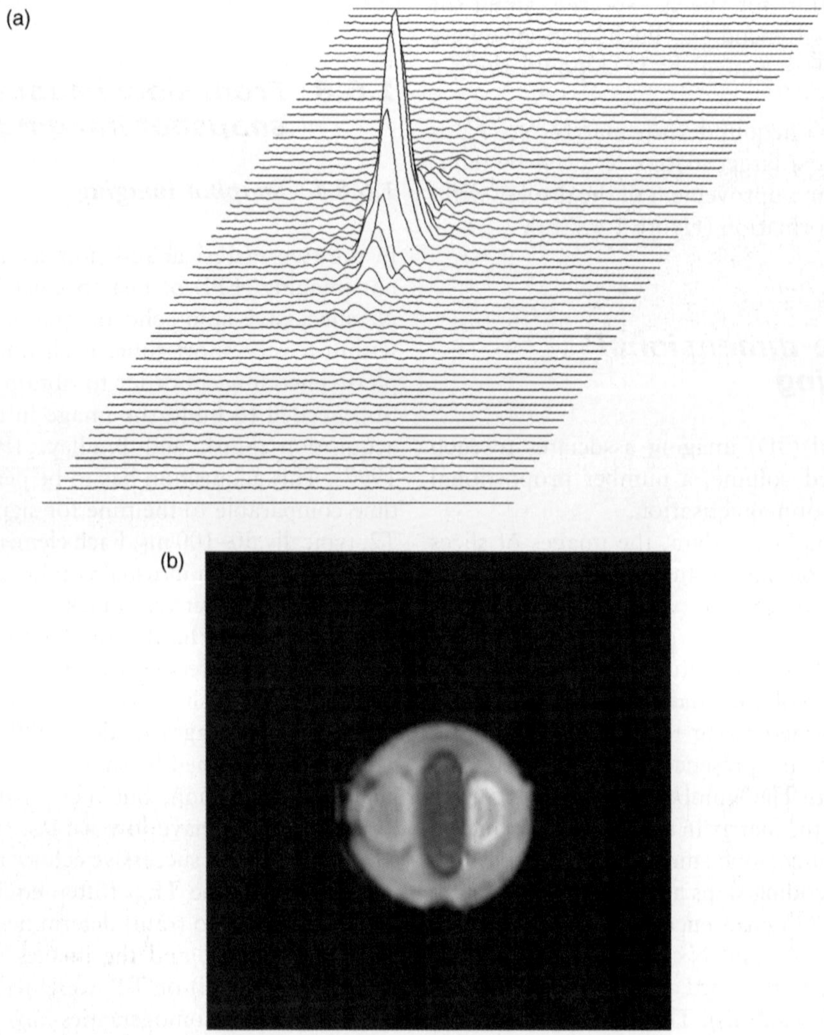

of image acquisition are obtained. Then the acquisition time is shortened N times and lasts for $(N_y/N).T_R$. This trade off between speed of acquisition and spatial resolution is less technically demanding and less subject to artifacts, and also images are less heavily $T2$ or $T2^*$ weighted. Figure 1.6.9 illustrates these snapshot and segmented spin-echo imaging techniques.

1.6.8.3 Parallel imaging

Parallel imaging, now available on all medical MRI systems, is another way to speed the acquisition of images (Hornack, 2005). Several (typically 4–12) receiver coils, positioned around the object, collect signals from distinct regions weakly overlapping. This makes possible to perform a lower number of phase encoding

Figure 1.6.8 *Ex vivo* 3D imaging of articular structures in rabbit knee at 4.7 T. 3D imaging is done at 4.7 T with a spin-echo sequence ($T_R = 500$ ms, $T_E = 10$ ms). Voxel size is $100\mu \times 200\mu \times 400\mu$. Acquisition time is 3 h. The knee structures are displayed on 48 thin contiguous slices. Fat is visualised with a high signal; cartilages have higher signal than bone and muscle. Cortical bone is dark; spongious bone appears heterogeneous, due to the magnetic heterogeneity induced by bone/fat interfaces (*courtesy of G. Guillot, U2R2M, Orsay France*)

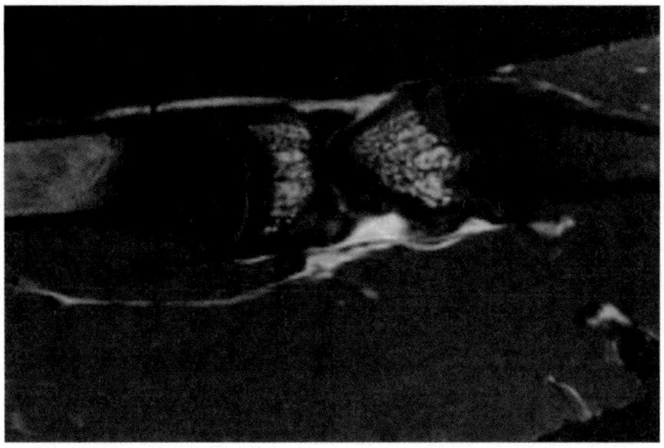

steps (e.g. $N_y/4$ instead of N_y). Then complex algorithms allow to reconstruct the full image with matrix $N_x.N_y$.

1.7 MRI and contrast

MRI is an imaging modality that offers huge versatility as so many different parameters may contribute to the contrast of images.

MRI is a black and white technique: The signal intensity is proportional to the local value of the proton transverse magnetization, modified by local values of parameters such as relaxation times, water diffusion coefficient, blood motion, blood oxygenation, iron load and so on. Sometimes the information derived from a special sequence is overlaid in 'false colours' upon an anatomical black and white image displaying anatomy.

Increasing the contrast means weakening the signal of some constituents: An image where all voxels

Figure 1.6.9 Fast spin-echo imaging. Transverse section of a fruit obtained at 1.5 T.
(a) Segmented acquisition: 16 groups of 8 echoes, each for one step of the phase encoding gradient, are obtained with T_R. 3s: The acquisition lasts 48 s, eight times shorter than the conventional acquisition done with same T_R). Matrix is 128×128. Spatial resolution = 0.63 mm × 0.63 mm × 3 mm. The effective echo time, corresponding to the middle of the echo train, is $TE_{eff} = 15$ ms. This image done with long T_R and short T_E has weak $T1$ and $T2$ weighting.
(b) Single-shot acquisition: 64 echoes, each for one step of the phase encoding gradient, are registered during 170 ms, that is the total acquisition time. Matrix is 64×64. Spatial resolution 1.25 mm × 1.25 mm × 3 mm. The effective echo time, corresponding to the middle of the echo train, is $TE_{eff} = 80$ ms. This low-resolution snapshot image is strongly $T2$ weighted: The variation of hydration between the different fruit layers is highlighted

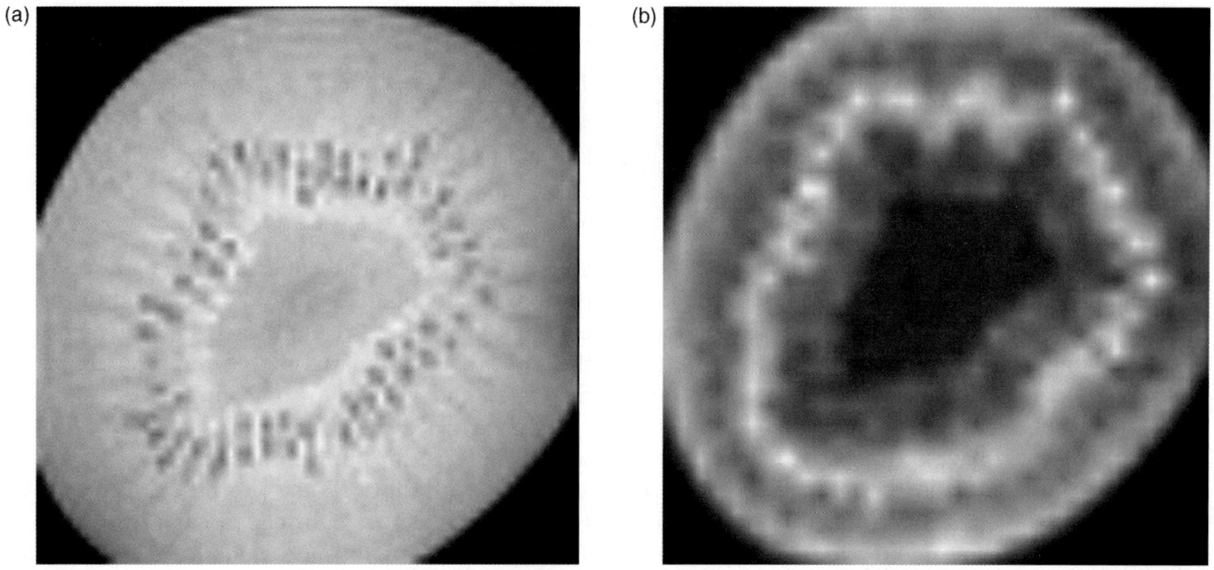

would have maximal intensity would not be very informative.

Let us consider now the basic parameters currently used in anatomical imaging.

The proton density is the first parameter that determines magnetization. The maximal signal intensity is that of pure water with maximal proton concentration, at least if the interval between measurements is much longer than water $T1$ and if the decay of transverse magnetization at the time of echo is negligible. The proton density does not vary so much between soft tissues (the water content is in the range 65–85% in non-adipose tissues; in adipose tissue, the fat content is around 85%, corresponding to a high proton density from -CH2- and -CH3- groups).

The other determinants of contrast are more effective. The amplitude of variation of the relaxation times in biological objects is much larger than that of the proton density, so that the contrast is modulated mostly by $T1$ and $T2$ values in each voxel.

The way to adjust contrast is to play with the times T_R and T_E of the acquisition sequence, the influence of which is related to relaxation times $T1$ and $T2$.

The repetition time T_R is the time waited between successive measurements of a slice to rebuild M_z.

The echo time T_E is the time at which some refocusing of dispersed magnetizations is done during signal acquisition.

Key points

The three main determinants of contrast are the proton density, which determines M_o, and the relaxation times $T1$, $T2$ of hydrogen nuclei. The repetition time T_R between measurements conveys the sensitivity to $T1$, and the echo time T_E conveys the sensitivity to $T2$.

When a spin-echo acquisition is done with short T_R and short T_E, the tissues with short $T1$ values (such as fat) appear as bright and tissues with long $T1$ (such as water) appear as dark. Images are mostly dependent on $T1$ values and are qualified as '$T1$ weighted'.

When a spin-echo acquisition is done with long T_R and long T_E, the tissues with long $T2$ values (water, other fluids, oedema) appear as bright, and those with short $T2$ values (muscle, tendon, bone) appear as dark. Images are mostly dependent on $T2$ values and are qualified as '$T2$ weighted'.

When fast imaging is performed at short value of T_R, other determinants of contrast are the flip angle and the management of the residual transverse relaxation.

Often several images of a same organ with different sequences are registered in order to obtain better characterization.

1.7.1 Contrast with spin-echo sequences

Let us write the expression of signal in the case of a classic spin-echo sequence (the flip angle for NMR excitation is 90°; the refocusing is done by a 180° RF pulse).

In a given voxel with coordinates x, y, z, the transverse magnetization $M(x, y, z)$ is proportional to the local density of protons that determines M_o.

It also depends on the time interval T_R between two measurements, during which M_z is rebuilt:

In the simple case where M is flipped with 90° angle at the beginning of the sequence, then $M_z = 0$ at time 0, and after waiting T_R, it is rebuilt to the value.

$$M_z(T_R) = M_o[1 - \exp(-T_R/T1)]. \quad (1.35)$$

At the following excitation, this longitudinal magnetization becomes transverse

$$M_t \text{ initial} = M_z(T_R). \quad (1.36)$$

M_t decreases during its precession, and the signal is read around the echo time T_E

$$M_t(T_E) = M_t \text{ initial}.\exp(-T_E/T2). \quad (1.37)$$

The spin-echo signal is proportional to

$$M_t(T_E) = M_o.[1 - \exp(-T_R/T1)].\exp(-T_E/T2), \quad (1.38)$$

where M_o depends on the local proton tissue concentration; local values of $T1$ and $T2$ depend on many complex parameters related to tissue structure, and the contrast is modulated by the choice of the parameters T_R and T_E, within some limits.

Figure 1.7.1 displays the variations of M_z and M_t with T_R and T_E for brain components.

Images with short T_R and short T_E values (such as Figure 1.9.3) display as bright the tissues with short $T1$ values (such as fat) and as dark the tissues with

Figure 1.7.1 Contrast: Influence of M_o, $T1$, $T2$. Values of M_z and M_t are calculated for $T1$, $T2$ and water percentages representative of brain white matter grey matter and cerebrospinal fluid (CSF) at 3 T.

Tissue	Water percentage (%)	$T1$ (ms)	$T2$ (ms)
White matter	80	600	60
Grey matter	90	900	90
CSF	100	2500	500

(a) Playing with T_R and $T1$. M_z recovery curves are plotted as a function of the repetition time T_R, according equation 1.35M_z values at long T_R reflect the tissue water percentages. At $T_R = 1000$ ms the values of M_z mostly reflect $T1$ differences between the three brain components. At time T_R much larger than $T1$ values (e.g. 10 000 ms), M_z is lower in white matter from its lower water percentage.
(b) Playing with T_E and $T2$. Signals for $T_R = 1000$ ms are plotted as a function of the echo time T_E, according equation 1.38. At short T_E, the values of signals mostly reflect $T1$ differences. At T_E above 20 ms WM and GM are differentiated from their $T2$ values. CSF appear as brighter only above $T_E = 80$ ms, due to its longer $T1$

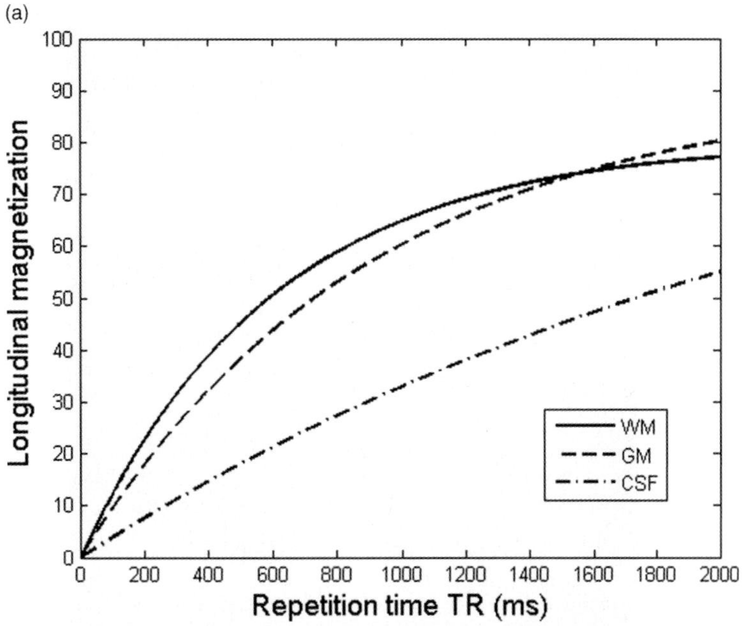

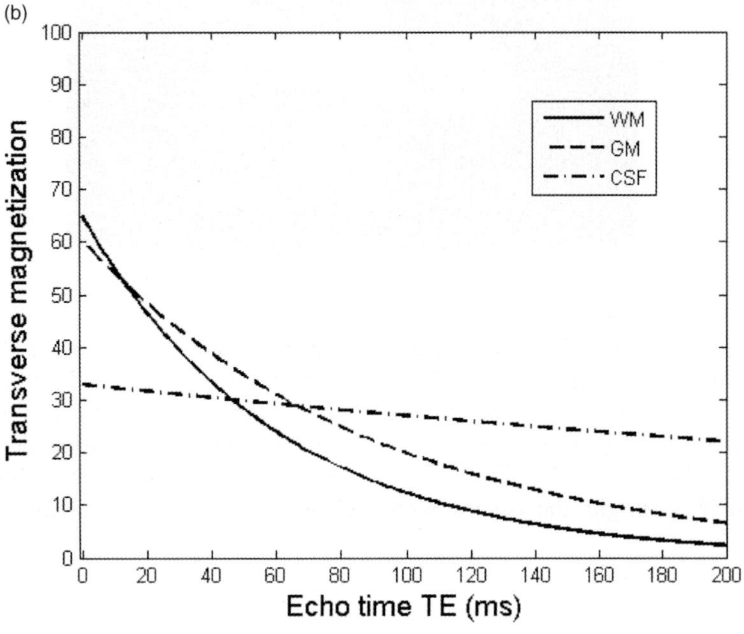

Figure 1.7.2 Contrast: Influence of M_0, $T1$, $T2$. Coronal images of normal Macaque brain, done at 3 T. A 8 cm × 5 cm surface coil laid at top of the skull is used for excitation and reception: the RF field $B1$ varies with the distance from the coil; this variation is the cause of image quality degradation near the coil, stronger for the spin-echo sequence. The same anatomical plane through striatal structures is visualised with different parameters.
(a) M_0-weighted gradient-echo image ($T_R = 2000$ ms, $T_E = 4$ ms, flip angle 25°). The contrast between brain structures is low; white matter tracts are delineated, from the lower proton density: A large fraction of myelin protons, strongly bound to membrane phospholipids, have very short $T2$ values and yield no signal even at this short echo time.
(b) $T1$-weighted inversion-recovery gradient-echo image ($T_R = 2800$ ms, $T_I = 880$ ms, $T_E = 4$ ms). The inversion time is chosen to suppress CSF signal as shown by Figure 1.3.3(b). From their markedly different $T1$ values, there is a high contrast between the white matter tracts with high signal, grey matter in cerebral cortex, caudate (arrow) and globus pallidus (arrowhead) with medium signal and CSF which appears dark.
(c) $T2$-weighted spin-echo image ($T_R = 2000$ ms, $T_E = 70$ ms). From their markedly different $T2$ values, there is a high contrast between CSF ($T2 = 500$ ms) in ventricles and subarachnoidal spaces (white), and brain ($T2$ about 60–80 ms). Globus pallidus appears slightly darker from the higher iron concentration. The temporal muscles with $T2$ about 30 ms are still darker (*courtesy of F. Boumezbeur and V. Lebon, Service Hospitalier Frédéric Joliot, CEA, Orsay France*)

(a)
(b)
(c)

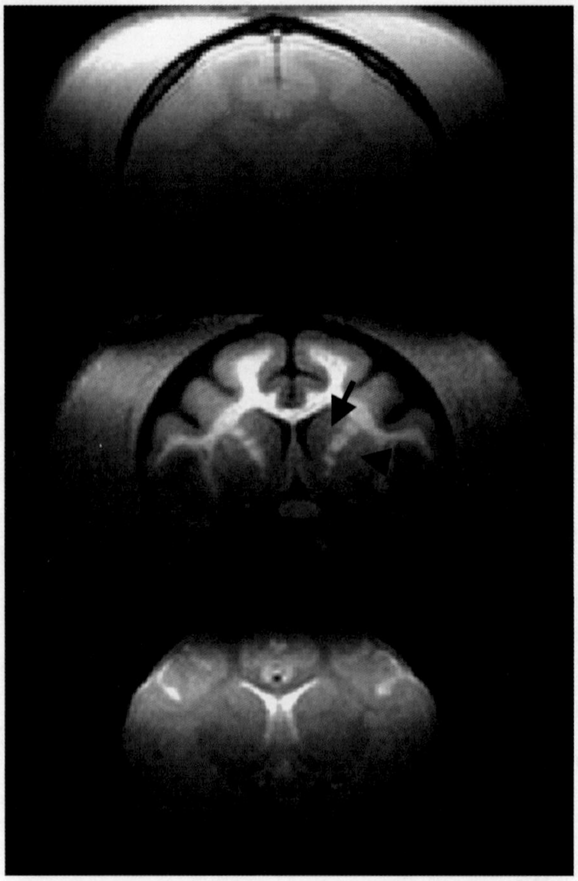

long $T1$ values (such as pure water). They are mostly weighted by $T1$ values and are qualified as '***T*1weighted images**'. The inversion-recovery image of Figure 1.7.2(b) also is $T1$-weighted.

Images with long T_R and long T_E values (Figure 1.7.2(c)) display as bright the tissues with long $T2$ values (water, other fluids, oedema). They are mostly weighted by $T2$ values and are qualified as '$T2$-weighted images'. However, we should keep in mind that image weighting depending on only one parameter is very difficult to obtain; some degree of $T1$ weighting is often present.

Long values of T_R directly cost long acquisition time; small values of T_R mean weak signals.

Short T_E values are limited by the time needed for phase encoding and by the 180° RF pulse duration. Long T_E values mean weak signals from exponential decay of M_t.

1.7.2 Contrast with gradient-echo sequences

Acquisition using a gradient-echo sequence can be done at much shorter echo time than when using a spin-echo sequence: T_E range is 0.5–5 ms depending on gradient hardware. The calculation of contrast is more complex and reflects the diversity of gradient-echo sequences.

The decay of M_t is determined by $T2^*$ instead of $T2$, because the gradient echo does not remove the influence of static magnetic field inhomogeneity. This influence can be utilised (see blood oxygen level dependent (BOLD) contrast at paragraph 11.2) or minimised.

The contrast of gradient-echo images done at very short T_E values and moderate T_R values is similar to that of $T1$-weighted spin-echo images done at the same T_R, but shorter T_E values are needed to limit signal decay and magnetic distortions (see Figure 1.7.2(a,b)).

When the repetition time T_R is shorter than some of the $T2$ values in the sample, some transverse magnetization did not decay to zero at the end of the time T_R. The way to destroy or to recycle this residual transverse magnetization modifies the NMR signal.

This complex way to play with T_R, T_E, the flip angle α and the residual transverse magnetization M_t is explained in Ness Aiver(1997) and Stark and Bradley(1992). The influence of the two last parameters is very briefly described below.

1.7.2.1 *Fast imaging at small flip angle*

Rapid gradient-echo imaging is often performed with a flip angle α smaller than 90°. Then M_z is less strongly decreased by the excitation at smaller flip angle, and a shorter value of the repetition time T_R is sufficient to rebuild M_z up to a given steady state value. The transverse magnetization available for measurement is lower, being equal to $M_z \sin\alpha$. The trade off is to collect fewer signals while waiting less between measurements.

At a very short T_R value, imaging with a small but optimal flip angle allows to obtain the highest possible

Figure 1.7.3 Signal variation with T_R and flip angle in gradient-echo imaging. Evolution of the signal calculated for $T1 = 1000$ ms, at T_R values from 10 to 1000 ms, and flip angles between 10° and 90°. The curves show that at low flip angles the signal reaches a plateau as a function of T_R

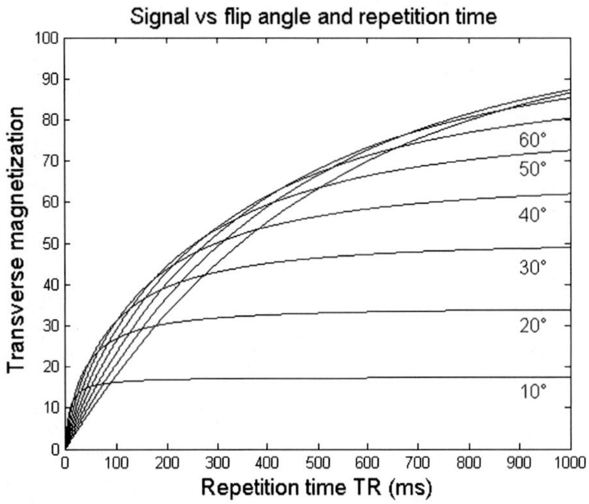

signal within a short acquisition time. This is very useful at high field as M_o is higher (and then part of the signal can be sacrificed), and also because $T1$ and then T_R increases with B_o.

More physics: the optimal flip angle. When the decay of the transverse magnetization is complete at the end of T_R, the steady state signal after several T_R intervals is written as

$$M_t(T_E) = \frac{M_o \sin\alpha[1-\exp(-T_R/T1)].\exp(-T_E/T2*)}{[1-\cos\alpha.\exp(-T_R/T1)]} \quad (1.39)$$

For given values of T_R and $T1$, M_t is maximal at the 'optimal flip angle' α (also named Ernst angle), given by the equation

$$\cos\alpha = \exp(-T_R/T1), \quad (1.40)$$

Figures 1.7.3 and 1.7.4 show the influence of the flip angle upon the signal at a given T1 value and upon an image.

1.7.2.2 *Fast imaging at shortest T_R values: recycling of transverse magnetization*

At very short T_R values ($T_R < 100$ ms) the transverse magnetization decay is not achieved at the end of the sequence, because T_R is shorter than $T2$ for some

Figure 1.7.4 Influence of the flip angle in gradient-echo imaging. (a) Transverse slice of a lemon done with $T_R = 600$ ms, $T_E = 5$ ms, flip angle $\alpha = 40°$. (b) Same acquisition done with $\alpha = 90°$. Fruit water $T1$ is about 1800 ms. At T_R 1000 ms, at flip angle 40° (the optimal value is 44°), the signal is 60% of the theoretical maximum available, whereas at flip angle 90°, it is only 28%. The contrast between water of the fruit pulp and lipids of the pip with shorter $T1$ is inverted between the two images

(a) (b)

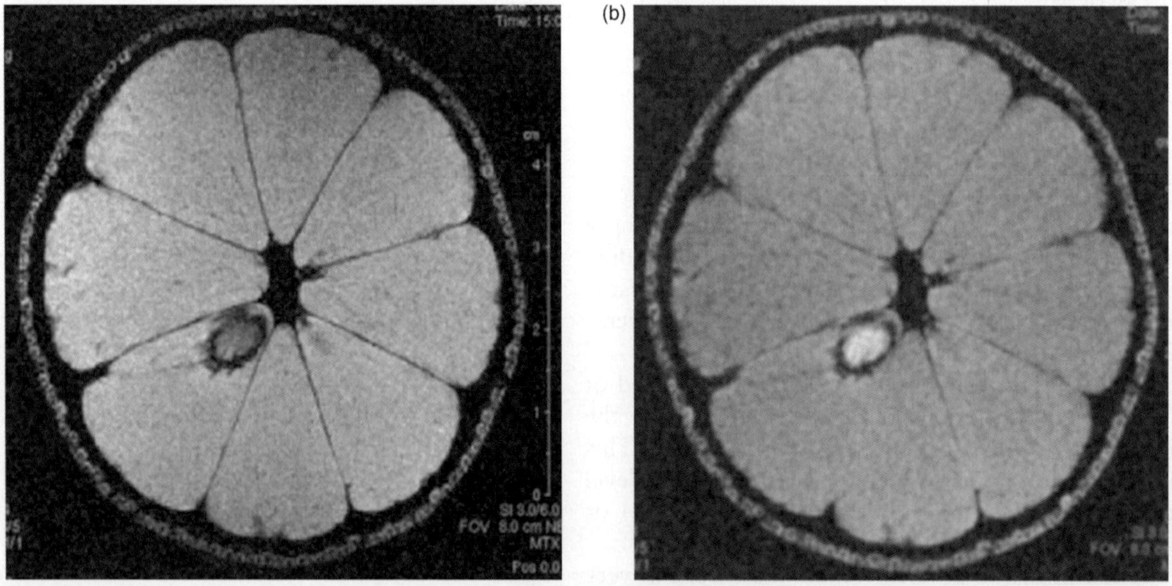

constituents of the sample. Then two distinct measurement strategies are either to spread the residual transverse magnetization by additional gradients or to recycle it.

The destruction (named spoiling) of the residual magnetization yields the contrast described in previous paragraph, depending on T_R, T_E and flip angle values.

The recycling of transverse magnetization yields, in adequate conditions, a signal proportional to M_o and to the ratio $T2/T1$, higher in liquids with long $T2$ values than in soft tissues with shorter $T2$. Then liquids such as blood in cardiac chambers, fluids in cysts, CSF in brain appear as bright. However, a very high homogeneity of the magnetic field is needed. The corresponding '*fully refocused*' sequences bear nicknames such as SSFP or FISP or FIESTA according to the commercial brand of the spectrometer.

1.7.3 More about relaxation times T1, T2, T2*

The relaxation times $T1$ and $T2$ are complex parameters, determined by molecular interactions, which depend on the composition and structure of tissues, and also at different degrees on the magnetic field B_o. $T1$ and $T2$ correspond to distinct physical mechanisms and their values are markedly different in most biological tissues. $T2$ is shorter or equal to $T1$, never longer. Every tissue has its own values of $T1$ and $T2$: This enables MRI to differentiate between different types of tissue.

The longitudinal relaxation time T1 is determined by several mechanisms that add their effects:

- Magnetic interactions between the water protons magnetic moments have weak efficiency because the water molecular motions are extremely fast. $T1$ is very long in pure water (3–5 s).
- Magnetic interactions between the water protons and the protons of macromolecules or proteins are more efficient: $T1$ is then around one second, depending on B_o.
- Magnetic interactions between water protons and paramagnetic substances (with unpaired electrons) are very efficient; these paramagnetic substances are used to shorten $T1$ and to increase contrast (cf paragraph 9).
- Magnetic interactions between fat protons are efficient because the lipid aliphatic chains do not move too fast. Fat has a high proton content and a very short $T1$. Its signal is very bright on $T1$-weighted images.

The transverse relaxation time T2 also conveys information on the tissue structure, and is very sensitive to some biological variations such as variations of water content and oxygenation of tissues.

T2 is long (~500 ms) in water, because water molecules rotate too fast and magnetic interactions between them are averaged by fast motions. It is long also in other biological fluids and in liquid tissues such as blood. It is shorter, in the range 10–100 ms, in tissues where water molecules have strong interactions with other bigger molecules and the water viscosity is high. *T2* is much shorter, less than 1 ms, in fibrous media such as tendons and bone, where macromolecules have high concentration and strongly interact with water.

In most, but not all, cases of pathology, *T1* and *T2* values are increased in comparison to their values in normal tissues. When the pathology causes the accumulation of iron, *T2* and sometimes *T1* values are decreased.

*The 'effective' transverse relaxation time T2** is similar to *T2* as a time measuring the decay of the NMR signal, but its causal mechanism is different: It is partly caused by the static magnetic field gradients. Since these local gradients do not fluctuate, the dephasing that they induce can be recovered by refocusing, as described in paragraph 6.4. Local variations of the magnetic field are more important when some parts of the object present a magnetic susceptibility (defined at paragraph 9.1) different from that of the bulk water. This is the case of air-filled structures, of bones, of erythrocytes and every biological structure containing iron. The influence of *T2** on NMR signals can be important, either as a parasitic effect (in spectroscopy; in cardiac imaging from the complex shape of lungs filled with air) or as a source of information (in bone structure studies; in brain studies using blood oxygen level dependent (BOLD) contrast introduced in paragraph 10.2).

Figure 1.7.5 Variation of the longitudinal relaxation time in brain structures, as a function of the magnetic field. To determine biologically relevant *T1* values, two juvenile C57/Bl6/J mice were scanned at three field strengths (4.7, 11.1 and 17.6 T) using a saturation recovery multislice spin-echo sequence in which the recovery time (T_R) was incremented to sample longitudinal relaxation. White matter (white symbols) in the corpus callosum, grey matter (black symbols) in the cortex and CSF (grey symbol) in the ventricles were segmented to provide a range of *T1* values. Region of interest analysis was utilized to provide mean signal intensities for each of the T_R times. These data were fit to a three component, single exponential model using a nonlinear least squares Levenburg-Marquadt algorithm to generate *T1* coefficients for each of the neuroanatomical structures (Padgett, Blackband and Grant, 2005) (*courtesy of Dr. K.R. Padgett, Dr. S.J. Blackband and Dr. S.C. Grant of the McKnight Brain Institute at the University of Florida, USA*)

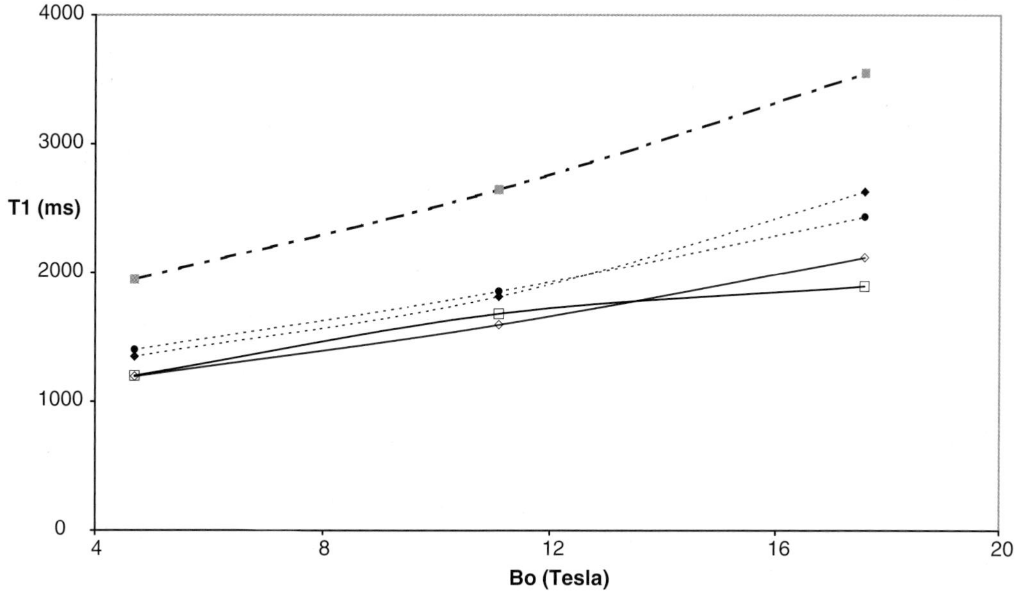

The values of the relaxation times in a given organ depend on many parameters: magnetic field intensity, temperature, species, age of animal and more interestingly modifications induced by pathology. Many data in the literature are obtained from ex vivo organs often studied at room temperature. In vivo measurements are not easily performed in organs like heart, requiring high quality synchronization of data acquisition.

*T*1 increases with the magnetic field *Bo* (Johnson, Herfkens and Brown, 1985); $T2^*$ and to a lesser degree *T*2 decrease with B_o. The increase of *T*1 values with B_o (Figure 1.7.5) results in longer imaging times; this counteracts with the signal increase at higher magnetic field.

1.8 Sensitivity, spatial resolution and temporal resolution

Key points

Sensitivity of MRI and MRS are measured by the signal-to-noise- ratio. The signal is determined by the number of nuclei detected in the volume of interest. Smaller voxels mean smaller signals, and then lower signal-to-noise ratio. To overcome the signal weakness at high spatial resolution, higher magnetic field B_o and/or more efficient receiver coil are needed. Selection of small voxels also needs strong magnetic field gradient intensity.

The spatial resolution, the signal-to-noise ratio and the acquisition time are three parameters strongly linked. For each MRI protocol, a compromise is negotiated between them.

The field of view is the spatial extent that can be mapped adequately without fold-over.

Sensitivity of MRS is further limited by the concentration of molecules studied and by the intrinsic sensitivity of the nucleus detected.

1.8.1 The spatial resolution

As MRI makes a correspondence between one volume element, the voxel, and one image element (or picture cell), the pixel, the in-plane resolution is given by the voxel dimensions δx and δy. (Often smaller pixel size is obtained by interpolation.) Usual values range from 1mm to 50 μm depending on resolution needed or affordable. Out-plane resolution is given by the slice thickness δz (usually larger, between 2 mm and 500 μm). 3D imaging, at the cost of a long acquisition time, allows to reach isotropic spatial resolution with thinner slices (δx, δy, δz between 100 μm and 40 μm).

The volume of one voxel is the product

$$\delta v = \delta x.\delta y.\delta z. \tag{1.41}$$

1.8.2 Determinants of the MRI signal

The signal of the voxel is related to its magnetization and to the volume of the voxel:

$$\delta M = M_o.\delta v. \tag{1.42}$$

At first a higher spatial resolution means a smaller voxel volume and a weaker elementary signal.

The magnetization M_o is proportional to the number of nuclei par volume unit and to their polarization *P* (see Section 1.2.4. and Table 1.2.1). The signal amplitude S is also proportional to the resonance frequency F_o. Then *S* increases roughly as the square of the magnetic field intensity B_o (or slightly less rapidly depending on instrumental factors and animal size).

The signal amplitude in a given acquisition sequence also depends on the values of T_R, T_E, *T*1, *T*2, as written in the Eq. (1.38) for the spin-echo sequence and α, $T2^*$ for the gradient-echo sequence.

At last, importantly, the signal amplitude is proportional to the receiver coil efficiency. A receiver coil is more efficient when a larger voltage is induced by the precession of a given magnetization. The voltage induced is higher when the receiver coil is very close to the sample. Choosing an efficient coil is the simplest and cheapest way to increase the signal-to-noise ratio of NMR measurements.

1.8.3 The noise

The electronic noise is a fluctuating voltage added to the NMR signal at reception (Webb, 2003). It neither depends on voxel size, nor on B1 intensity (i.e. on the flux of RF photons). The noise originates from the random electrical fluctuations in the coil and it is proportional to the resistance *R* of the receiver coil: A 'good' receiver coil has a low resistance. Some noise may also originate from the electric losses in the

Figure 1.8.1 Field of view and spatial resolution. Images of a fruit done at 1.5 T with a spin-echo sequence ($T_R = 1000$ ms, $T_E = 15$ ms). The acquisition time is 512 s; the matrix is 512×256. Left, FOV 10 cm; spatial resolution is $390\mu \times 195\mu \times 3000\mu$. Right, FOV 5 cm; spatial resolution is $195\mu \times 97\mu \times 3000\mu$. The spatial resolution is higher, from identical matrix and smaller field of view. Therefore the signal-to-noise ratio is lower (divided by four!). The fold over (also named aliasing) of an outer portion of the object is observed: The frequency spectrum corresponding to the width of the object is wider than the spectral range that can be analysed properly in the conditions of acquisition

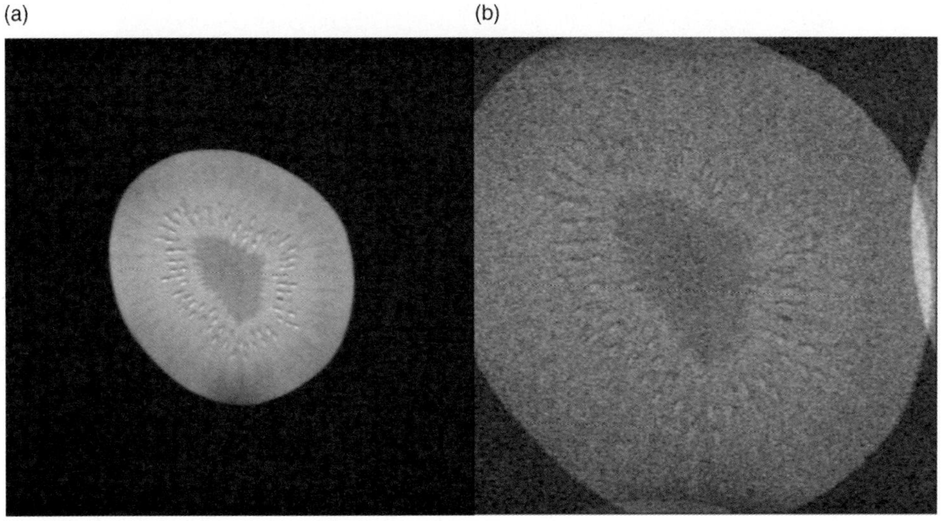

object; this is especially important for larger animals or those at very high magnetic field.

A small receiver coil has a lower resistance than a larger one, and it also picks noise (but also signal) from a smaller volume across the object.

The electronic noise is independent of the frequency. After Fourier transform, it is spread uniformly across the spectrum, and then across the image. It can be appreciated visually on images of a homogeneous structure, when signals of neighbouring pixels have different intensities, as represented on Figure 1.8.1. It can be measured easily from the fluctuation of signal in the background around an object, in the zones where signal would ideally be equal to zero, as may be seen in Figures 1.6.5 and 1.7.4.

The signal-to-noise ratio can be increased, at the cost of a longer acquisition time, by taking the average of several measurements, that is adding Na signals, thus multiplying signal – and also the measurement time – by N_a and the noise only by $(N_a)^{1/2}$.

The physiological noise, typical of living objects, results from motions that are not synchronized with the phase encoding: Signals from moving parts of the object are then depicted at improper location in the image, and they behave as some additional noise (see Figure 1.6.4).

1.8.4 The field of view and the spatial resolution

The spatial resolution is determined by the strength of the magnetic field gradient available and by the number of points sampled along the corresponding direction of the image.

The field of view is the maximal dimension of the object that can be represented accurately in given conditions of acquisition. There is a limitation of this dimension, because there is a limitation of the highest frequency that can be analysed by the spectrometer.

The spectral width available around the central frequency F_o is ΔF_o, related to parameters of acquisition.

The width of the spectrum of an object of length X, when its signal is read with a frequency encoding gradient G, is $\Delta F = (\gamma/2\pi).G.X$.

If $\Delta F > \Delta F_o$, the parts of the spectrum of the object that lie out of the interval $(-1/2\Delta F_o, +1/2\Delta F_o)$ are not correctly analyzed: They are 'undersampled'

Figure 1.8.2 Using a high temperature super-conductive coil. High resolution images of stems are done with voxel size $39 \times 39\,\mu m^2$ in plane and slice thickness $900\,\mu m$, at 1.5 T. (a) Detection with a small (12 mm diameter) circular copper coil. (b) Detection with a same sized high temperature super-conductive coil: The signal-to-noise ratio is improved by a factor of 5 (*courtesy of J.C. Ginefri, UMR CNRS 8081 and Université Paris-Sud, Orsay, France*)

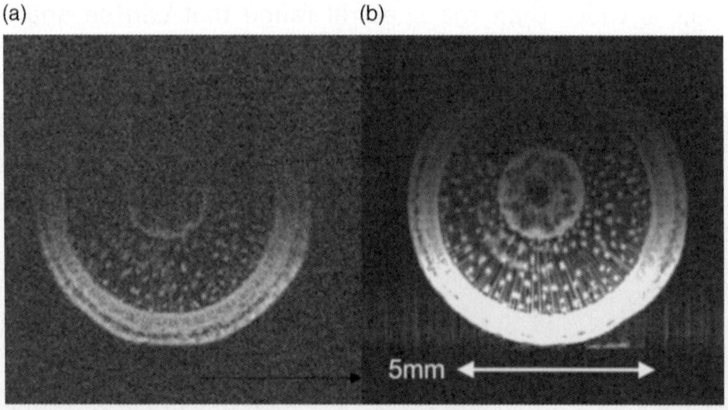

(Ness Aiver, 1997), and so they are translated into lower frequencies superimposed to the frequencies that represent the central part of the object. The corresponding regions of the object are wrongly localized, and are folded over the central zone. This artefact is named 'aliasing'; it is illustrated in Figure 1.8.1(b).

This makes it difficult to build the image of a small organ inside a large body; but technical recipes such as 'suppression' of the magnetization in parts of the object may solve this difficulty (Parzy *et al.*, 2003) as illustrated by figure 10.1.3.

1.8.5 More technology: High temperature supraconductive coils

The signal-to-noise ratio strongly depends on the efficiency of the RF receiver coil; the reason why coils are often optimised for a given experiment (Mispelter, Lupu and Briguet, 2006).

The signal-to-noise ratio (SNR) of images can be enhanced by using a receiver coil with extremely low resistance. The receiver coil efficiency can be further increased if the coil is built from a super-conducting material that has a resistance much lower than copper, at very low temperature. The coil is located near the object of study, itself kept at 'normal' temperature. This new technology needs complex and expensive cryogenic system, using liquid nitrogen (77 °K). Its implementation in pre-clinical or clinical environment is still in progress. However, the use of super-conducting coils opens a lot of investigations on different fields of biomedical applications. The SNR can be 4 to 15 times higher than that with a similar room-temperature copper coil, a function of imaged area (Ginefri *et al.*, 2005) as illustrated by Figure 1.8.2. Then current whole body MRI systems at 'moderate' field Bo can be used to examine small animals. The accessible spatial resolution at a given field is comparable to that usually obtained at a field two or three times higher. For several biomedical issues, imaging with super-conducting coils can offer a true alternative to higher magnetic field.

1.8.6 High resolution and Microscopy

Higher spatial resolution is needed for smaller animal imaging, mouse at first. Scaling from the human to the mouse corresponds to a decrease in linear dimension of approximately 15-fold. For example, if the voxel size is scaled down from $(1\,mm)^3$ to $(70\,\mu m)^3$, the voxel volume is divided by 3500 compared to usual clinical MRI, and the same figure is needed for sensitivity gain if equivalent signal-to-noise ratio is planned. Note that functional imaging is often done with coarser spatial resolution. Part of the sensitivity gain needed is obtained by increasing the field strength and by decreasing the receiver coil size (efficient and less expensive). The remaining sensitivity gain is obtained by averaging multiple acquisitions so that acquisition time can be several hours.

In vivo microscopy has been proposed initially as a tool to visualise the embryonic development, with the help of labelling some cells with a paramagnetic contrast agent (Jacobs and Fraser, 1994). *Ex vivo* 3D acquisitions with very high spatial resolution, though bringing no functional data, bear some advantages in regard to optical microscopy: The shape of organs and the relation between organs are clearly depicted, the field of observation is larger, 3D reconstruction of blood vessels is possible and the technique is non destructive, allowing other use of organs after MRI. Johnson *et al.* (2002) presented *ex vivo* high-resolution microscopy of mouse whole body or organs. In their study, acquisition times last 14 h, and fixation of tissue is done with a mixture of formaline and Gd-DTPA to shorten $T1$ of tissues and then T_R and the acquisition time.

1.9 Contrast agents for MRI

Key points

Contrast agents for MRI are either positive agents (increasing water signal) or negative agents (decreasing water signal). They contain strongly paramagnetic or ferromagnetic metallic compounds. They are not detected directly but modulate the signal of protons, a factor that increases their efficiency.

MRI natural contrast is easily made high by clever choice of T_R, T_E, depending at first on M_o, $T1$ and $T2$. Also MRI acquisitions can be sensitised to many other parameters (see paragraph 11). Yet contrast agents are widely used. They modify locally $T1$ or $T2$ or $T2^*$ values and help to get additional information such as the permeability of blood vessels, the vascularity of lesions, the extent of perfusion defects in brain or heart and the location of magnetically labelled cells.

Rinck (2001) wrote an excellent introduction to contrast agents for MRI. The field of application of contrast agents (CA) is wider for animal experimentation where the requirements of clinical safety are lower and then 'experimental' products built by research laboratories may be used.

1.9.1 What are diamagnetism, paramagnetism, ferromagnetism and superparamagnetism?

The magnetization of a sample depends on how magnetic moments inside are polarized (spontaneously or under action of the external magnetic field). The 'electronic' magnetization that derives from electronic magnetic moments is always much higher (about 10^4 times) than the nuclear magnetization.

Figure 1.9.1 Paramagnetism and ferromagnetism. The additional field ΔB, in the piece of magnetic material and around it, is proportional to the product $\chi . B_o$, where χ is the magnetic susceptibility of the material. This product is multiplicated by geometrical factors. χ is typically three orders of magnitude larger in the ferromagnetic material than in the paramagnetic material. Around the piece of ferromagnetic material a strong variation of the magnetic field takes place over a distance larger than the dimensions of the piece of material

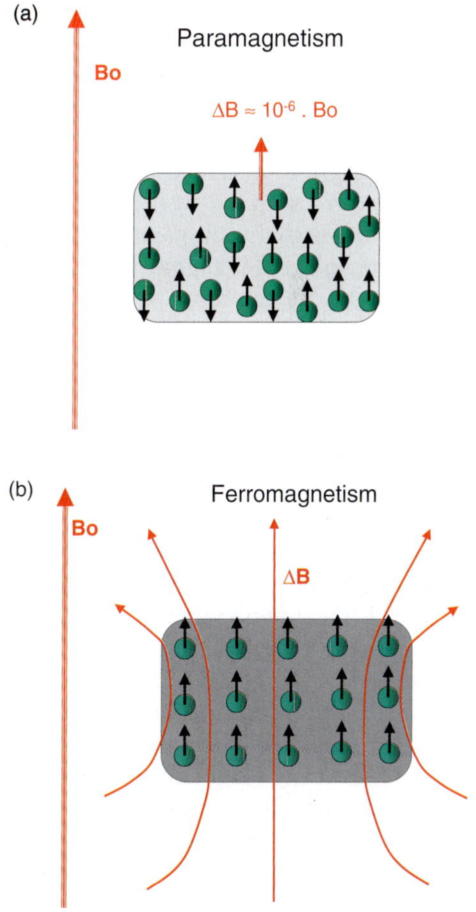

Table 1.9.1 Characteristics of some magnetic contrast agents at 0.47 Tesla

Contrast agent	k1 (s^{-1} . mM^{-1})	k2 (s^{-1} . mM^{-1})	Rat plasma half-life (min)	Size (nm)
Positives				
Gd-DTPA	4.5	5.9	20	1
Gd-DOTA	3.8	5.8	23	1
Gd-BOPTA	4.4	5.6	Biexponential (5.5/22)	1
Negatives				
AMI 225	23	107	10	80–120
AMI 227	22	53	118	30

Data given for the main contrast agents validated for human diagnostic are the relaxivities k1 and k2 that link T1 and T2 to the contrast agent concentration (see equations (1.46) and (1.47), the plasma half-life and the size of the molecule or particle.

The susceptibility is the ratio of the magnetization (defined at paragraph 1.2.3) to the external magnetic field:

$$\chi = M/Bo. \qquad (1.43)$$

In a sample such as drawn in fig 1.9.1 the additional magnetic field created by the internal magnetic moments is proportional to $\chi.B_o$. Around the sample, the influence of internal magnetic moments decays quickly across distance. If the surrounding material has a different susceptibility, the magnetic field outside has a different value, and there is a zone of inhomogeneous magnetic field at the interface.

Briefly, four main types of electronic magnetism are described:

Diamagnetism is observed in compounds with paired electrons such as water, fat and a vast majority of the organic compounds: The field inside a diamagnetic material is slightly lower than its value outside (the additional term is very weak: -10^{-8} times the external field).

Paramagnetism is observed in compounds with unpaired electrons such as iron, copper, nickel, manganese, rare earths as gadolinium and also dioxygen O^2. When the interactions between atoms are weak, magnetic moments are not oriented at zero external magnetic field, and they line parallel or anti-parallel to the magnetic field (see paragraph 1.2.3). The resulting effect is a slight increase of the magnetic field inside the material (the additional term is weak: $+10^{-6}$ to $+10^{-7}$ times the external field).

Ferromagnetism is observed in compounds with unpaired electrons as iron, cobalt, nickel, in metallic state, when the interactions between neighbouring atoms are strong. Magnetic moments are fully ordered under a weak magnetic field, and often stay ordered at zero external magnetic field. The magnetic field inside and around a piece of ferromagnetic material is very strong: The piece of ferromagnetic compound is a magnet.

Superparamagnetism is a phenomenon quite similar to ferromagnetism, in conditions where the total number of atoms is weak, such as in small solid particles smaller than 30 nm containing iron oxides. Then the magnetic moments are fully ordered in the presence of external magnetic field, but the magnetization does not persist in the absence of external magnetic field.

More physics: the relaxivity of contrast agents

The efficiency of a relaxation mechanism is expressed by the *relaxation rate* that is the inverse of the relaxation time: A high efficiency of relaxation corresponds to a high relaxation rate $R1$ (resp. $R2$) and then a short relaxation time $T1$ (resp. $T2$). When several mechanisms a, b,... contribute altogether to the relaxation, their contributions to the relaxation rate are additive. The relaxation of the longitudinal and transverse components of water nuclear magnetization are then written as

$$1/T1 = R1 = R1a + R1b + \dots, \qquad (1.44)$$

$$1/T2 = R2 = R2a + R2b + \dots. \qquad (1.45)$$

In the presence of a contrast agent, the relaxation rate of a tissue is the sum of its natural relaxation rate and of an additional term from the contrast agent. The relaxation rate from the contrast agent is the product of the concentration of the contrast agent, C, by its relaxivity k.

In a tissue with natural relaxation times $T1o$, $T2o$ (relaxation rates $R1o$, $R2o$), the relaxation rates are written as

$$R1 = R1o + k1.C \qquad (1.46)$$

also written

$$1/T1 = 1/T1o + k1.C,$$

$$R2 = R2o + k2.C, \qquad (1.47)$$

also written

$$1/T2 = 1/T2o + k2.C,$$

where $k1$ (resp. $k2$) are the longitudinal (resp. transverse) relaxivities of the contrast agent in the tissue and C is its concentration. $k1$ and $k2$, expressed in s^{-1}/mM^{-1}, measure how the CA at concentration 1 mM increases $R1$ or $R2$, and then shortens $T1$ or $T2$.

The positive contrast agents that contain gadolinium have similar efficiency for $T1$ and $T2$ shortening ($k1$ and $k2$ have similar values). As in most tissues, $T2$ is much shorter than $T1$, that is $R2o \gg R1o$, the shortening of $T2$ is negligible at low concentration of the agent.

With the negative contrast agents, $k2$ is larger than $k1$, corresponding to efficient $T2$ shortening; the shortening of $T2^*$ is still more effective, and better exploited by gradient-echo imaging. $T2^*$ shortening is not expressed by a relaxivity because it heavily depends on the measurement sequence. Table 1.9.1 shows the relaxivities of some contrast agents approved for diagnostic imaging.

1.9.2 Positive paramagnetic contrast agents

The positive contrast agents contain paramagnetic atoms with unpaired electrons, usually gadolinium. The gadolinium atom Gd has 7 unpaired electrons; its electronic magnetic moment is very large (10,000 times that of a proton). Furthermore, the kinetic of fluctuations of its magnetic moment is optimal, so that the relaxation of water molecules around this atom is strongly increased (just imagine it as a magnetic stirrer in a jug of water).

Positive contrast agents are molecules which contain one or several gadolinium atoms (or less frequently manganese atoms). As the free gadolinium ion Gd^{3+} is toxic, Gd is chelated to a complex with very high stability so that free Gd^{3+} release in tissues is negligible and the contrast agent clearance is complete, usually through renal elimination. Chelates of Gd are extremely safe products.

The shortening of $T1$ causes an increase of proton signal, as long as the acquisition is $T1$-weighted, which means that the repetition time T_R value is not too long (see Eq. (1.38)). Then these agents increase water signal of the zone inside which they are dispersed: They have to be exposed to water.

Conversely shortening of $T2$ causes a decrease of proton signal, more effective at long T_E values. As this effect is opposite to that induced by $T1$ shortening, it is better to handle the CA at a concentration low enough to shorten mostly $T1$ as illustrated in Figures 1.9.3 and 1.9.4.

Figure 1.9.3 Variation of water signal in the urinary bladder of a mouse with Gd concentration, during the urinary elimination of GdDTPA injected IP at a dose of 0.5 mM/kg; axial spin-echo image with TR 700 ms, TE 15 ms, at low spatial resolution (pixel size 273μ).Urine $T1$ and $T2$ are shortened by the contrast agent. The upper zone in urinary bladder contains light urine with lower concentration of Gd-DTPA. The water signal is strongly increased, from $T1$ shortening. The lower zone in urinary bladder contains heavier urine with higher concentration of Gd-DTPA. $T1$ and $T2$ are shortened under 20 ms so that the water signal is decreased, from $T2$ shortening. The urinary concentration of Gd-DTPA is about 10 mM at the transition between the bright and the dark zones

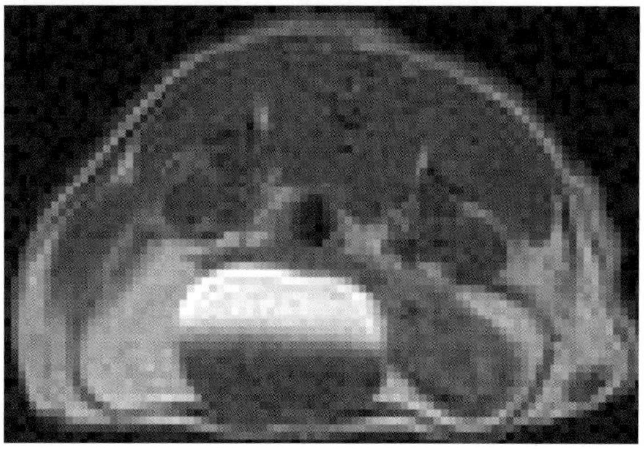

1.9.2.1 *How to use the positive contrast agents?*

They are a family of products with various molecular weights and biological destiny.

Small molecular weight gadolinium chelates (600 to 1000 Da) have extracellular diffusion.

Most of them (Gd-DTPA, Gd-DOTA, Gd-DTPA-BMA, Gd-HPDO3A) are rapidly eliminated by renal excretion. They diffuse quickly after injection in the extracellular compartment of all organs except brain. In brain, they diffuse only inside the lesions with blood brain barrier disruption. Thus $T1$-weighted images done after injection of the contrast agent enlighten the modification of the capillaries and of the extracellular space. When capillaries normally impermeable become leaky (in brain lesions such as tumours, abscess or inflammation) or in necrotic zones where the extracellular space volume is increased, strong signal increase is observed quickly after injection of the agent, as shown in Figure 1.9.3.

Figure 1.9.4. Myonecrosis induced by toxin injection in mouse leg. Images are obtained 24 h after injection of a snake venom toxin that induces necrosis of muscle fibres; the same slice is depicted with different contrasts that contribute to a better characterization of lesions. (a) *T1*-weighted spin-echo image with $T_R = 500$ ms, $T_E = 12$ ms. Fat signal is white. The treated leg (left of image) shows increased volume and slightly increased signal. (b) *T2*-weighted spin-echo image done with $T_R = 1500$ ms, $T_E = 102$ ms. Contrast depends heavily on *T2*. Oedema and fluids in the heterogeneous necrotic zone appear as bright. (c) *T1*-weighted spin-echo image done with the same parameters as (a), after IP injection of a positive CA (Gd-DTPA). The CA shortens *T1* in the extracellular space. In necrotic zones where the extracellular space is increased, the mean CA concentration is higher: Necrotic leg muscles in the treated leg exhibit a strong signal increase (*images from (Wishnia et al., 2001) with permission of Neuromuscular Disorders*)

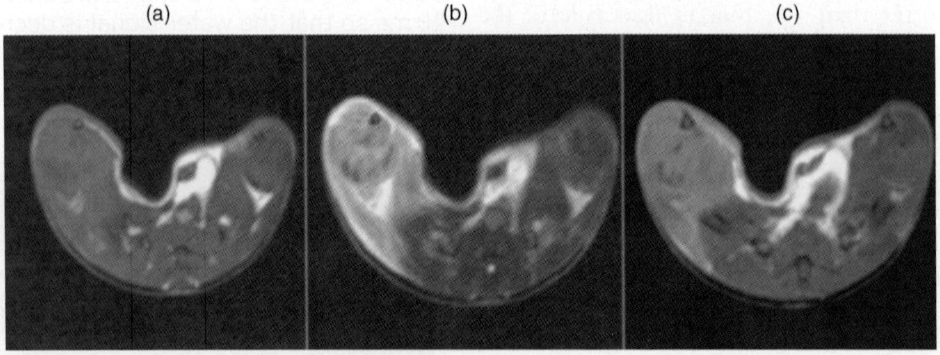

Other contrast agents (Gd-EOB-DTPA, Gd-BOPTA, MnDPP) that undergo hepatic captation, strongly increase liver signal and have hepatic elimination.

Macromolecular gadolinium chelates with molecular weight of more than 50 kDa stay in the vascular sector a longer time, making easier the visualisation of blood vessels by angiographic techniques, as shown in Figure 1.11.4, and the measurement of blood volume in an organ.

Gadolinium chelates can also be used to measure perfusion (see paragraph 1.11.7.1). This technique is named dynamic contrast enhancement (DCE). For brain studies, $T2^*$ decrease induced by a strong concentration of a positive agent around non-permeable brain capillaries is easily detected. For heart or kidney perfusion maps, *T*1 shortening is detected either with a positive agent or with a negative agent, which also shortens *T*1: Measurement of myocardial perfusion can be performed after injection of an ultra small particles of iron oxide (USPIO) at very low concentration using an imaging sequence extremely sensitive to *T*1 variations.

1.9.3 Negative contrast agents

Negative contrast agents are small particles of superparamagnetic material. The particle core made of iron oxides, of diameter 3–5 nm, is surrounded by a thick coating of a non-magnetic material such as dextran, starch, albumin or silicone. These particles are named MION (monocristalline iron oxyde nanoparticles) or more precisely SPIO (small particles of iron oxide) and USPIO depending on the thickness of coating that determine their biologic properties.

These particles behave as tiny magnets:They create a strongly heterogeneous magnetic field in their environment. The loss of homogeneity of the magnetic field causes weakening of tissue water signal. *T*2 is shortened by the diffusion of protons in the local magnetic field gradients. $T2^*$ is shortened much more efficiently: By using gradient-echo imaging, it is possible to detect these magnetic particles with a huge sensitivity. Over a distance of several microns around one particle, the signal of water is strongly diminished. This does not depend on interactions of the contrast agent with water molecules. If the particles are enclosed inside vessels or cells, they still destroy water signal at distance. Figure 14.2.1 in the report by Heryneck (Chapter 14) shows patterns of water signal destruction around 20 μg of iron inside brain.

Negative CA distribution and elimination vary with particle size: SPIO (particles with diameter 50–500 nm) are quickly cleared out of blood. Within liver and spleen, Kupffer cells selectively take these molecules up by phagocytosis. USPIO (particles with diameter under 50 nm) stay longer in circulating blood. Measurements of vascular parameters are then possible. USPIO are taken by macrophages and can be used to visualise lymph nodes or to detect inflammatory reaction inside an organ, as done in experimental models of arthritis, multiple sclerosis, diabetes (Billotey *et al.*, 2005).

Cell labelling with a negative CA before injection has been applied to visualise migration of stem cells at the vicinity of brain ischemic lesion (Hoehn *et al.*, 2002; Jendelova *et al.*, 2005). Single cell detection was demonstrated in optimised *ex vivo* conditions; it is possible if signal-to-noise and spatial resolution are high enough: with an isotropic spatial resolution equal to $(100\,\mu m)^3$, single cells loaded by iron mass as low as 2–10 pg can be detected as a signal void, depending on the signal-to-noise availability (Heyn *et al.*, 2005). However, this high sensitivity sets a limit to quantification if many loaded cells are close from one another. Also one has to check for viability of labelled cells, because after cell death iron in the tissue is ingested by macrophages and still detected *in situ*.

New kinds of magnetic nanoparticles are now developed for molecular imaging (Lanza *et al.*, 2004).

New NMR contrast agents are presented in Chapter 7.

1.10 Imaging of 'other' nuclei

Though hydrogen nucleus is the most favourable because of its concentration in the living tissue and its high resonance frequency, imaging nuclear magnetization of other nuclei is of interest: Hyperpolarized noble gases ^{3}He, ^{129}Xe and sodium ^{23}Na are used mostly for physiological studies.

1.10.1 Hyperpolarization and NMR of noble gases

In spite of the low gas density, roughly one thousand times lower than that of water, the nuclear magnetization of hyperpolarized gases can be directly measured by NMR since their nuclear polarization is increased by up to five orders of magnitude with techniques from atomic physics. So one can start with nearly all magnetic moments parallel to the magnetic field B_0. Two noble gases, the stable isotopes ^{3}He and ^{129}Xe, bear magnetic moments that can be polarized by optical pumping and then easily detected.

More physics: how to build hyperpolarization?

There are currently two efficient techniques. They both rely on an optical pumping process by a circularly polarized laser to prepare the atoms in a polarized electronic spin state. This polarization is then transferred to the nuclei. In the first technique, optical pumping is performed on alkali atoms (e.g. rubidium) in a high temperature and pressure cell. Spin exchange collisions with ^{129}Xe or ^{3}He transfer the polarization to their nuclei. It yields polarization up to 60%. In the second technique, optical pumping is performed directly on metastable ^{3}He gas excited in a radiofrequency discharge at low pressure. Metastability exchange collisions between ^{3}He atoms eventually prepare polarized ^{3}He nuclei up to 90%.

Nuclear polarization can reach values up to 90%. This 'hyperpolarization', prepared in a low field outside of the magnet, is up to 10^5 times higher than the thermal polarization obtained at equilibrium in the magnetic field of a standard MRI magnet. Hyperpolarized gases then yield high NMR signal. This makes possible to image the air spaces of lungs, which give no signal with normal MRI techniques because water is barely present and the nuclei in air, oxygen ^{16}O and nitrogen ^{14}N, bear no net magnetic moment. Special imaging techniques are to be used to make efficient use of the hyperpolarization, because the spontaneous longitudinal relaxation brings nuclei to the much lower equilibrium polarization within 15 seconds and each RF pulse contributes to the destruction of the huge initial longitudinal magnetization.

Two gases are used for lung imaging. Helium (the weakly abundant isotope ^{3}He) is expensive but well tolerated; it is mostly used for lung imaging as illustrated by Figure 1.10.1. High resolution rat lung images has been obtained with hyperpolarized helium (Dupuich *et al.*, 2003) and now are used to study lung function and respiratory diseases (Chen *et al.*, 2000). ^{129}Xe is abundant and cheap, and because it diffuses less rapidly it should ultimately yield sharper images. Moreover, it dissolves in blood (with possible side effect as an anaesthetic at high concentration). Although hyperpolarized xenon is stable for only tens of seconds in the blood, it is enough time to quickly image its transport to the brain and to distinguish white and grey matter

Figure 1.10.1. Rat lung imaging using hyperpolarized ^3He. The rat is ventilated with a respiratory cycle duration 1s and the acquisition of image is synchronized with ventilation. Measurements are done using a dedicated sequence, with $T_R = 10$ ms, $T_E = 30$ μs, 900 lines, according to a radial technique (Dupuich *et al.*, 2003). Spatial resolution is $156 \times 156\,\mu m^2$ in plane, without slice selection. Each elementary image explores 1/6th of the respiratory cycle. The image displayed here is the sum of three elementary images, showing bronchic tree and alveolae filled with helium (*courtesy of Y. Cremillieux, Laboratoire de RMN, UniversitéClaude Bernard, Lyon, France*)

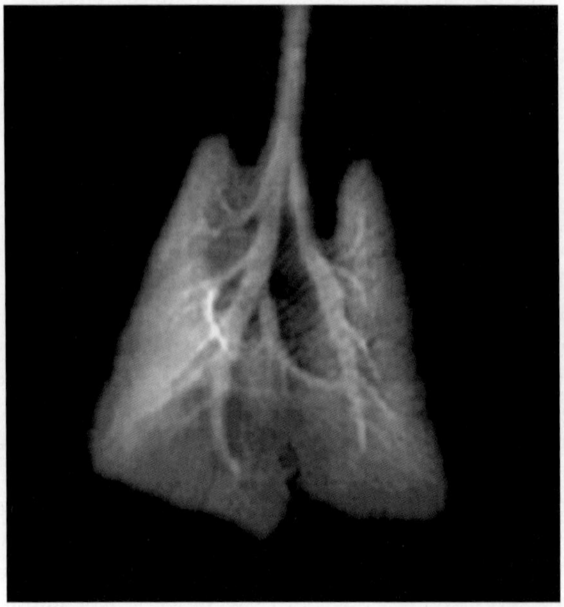

there. It can also be integrated into chemical carriers that allow to perform biochemical studies and angiography.

1.10.2 Sodium imaging

Sodium ^{23}Na is a 'difficult' but interesting nucleus: From the large difference between its concentration inside and outside cells, the quantification of the intracellular and extracellular sodium pools inside an organ is a sensitive index to detect and quantify ischemia and necrosis: When the intracellular ATP concentration is diminished, the function of the sodium ion pump is compromised resulting in increased intracellular sodium level. Intracellular and extracellular sodium signals are characterized by different *T2* values, typically ranging around 1 and 40 ms, respectively. Specific imaging techniques are needed to obtain extremely short echo times, less than 1 ms, so as to ensure that both intracellular and extracellular sodium ion pools contribute to the sodium MRI signal (Constantinides *et al.*, 2001). Applications to myocardial ischemia in animal modes have been published by Kim *et al.*,(1999) and Constantinides *et al.* (2001).

1.11 More parameters contributing to MRI contrast

NMR signal can be sensitised to many parameters other than Mo, T1 and T2; here we propose a list of those that are informative in the biological field. Some of them are presented in this paragraph and/or are illustrated in part II. Measurement of these parameters often requires to minimize the influence of local differences in the three major parameters, Mo, T1 and T2, a task not always easy.

Several measurement techniques are based upon modification of the water signal by magnetic agents; this concerns parameters such as

- Blood oxygenation (using haemoglobin as a contrast agent, with contrast dependent on $T2^*$),
- Blood volume (using an exogenous magnetic contrast agent),
- Vessel permeability (using an exogenous magnetic contrast agent),
- Macrophage activity (using an exogenous magnetic negative contrast agent),
- Quantification of endogenous iron accumulation,

Perfusion (using a bolus of an exogenous contrast agent).

Other measurement techniques rely upon the difference in resonance frequencies of water and other metabolites: this is the case of

- Magnetization transfer after irradiation of a proton pool other than water.
- Chemical composition of voxels, encoded by resonance frequency, in chemical shift imaging (CSI).

At last, other measurement techniques rely upon the sensitisation of the water signal to protons motion.

The measurements, which involve moving magnetic moments, rely either on *T*1 modifications or upon phase modification during the application of a magnetic field gradient. The displacement of protons is potentially a strong factor of degradation of measurements in anatomical MRI; the motion of organs creates artefacts, fought by cardiac and respiratory synchronization, or by single shot imaging. However motion of magnetic moments can be turned into a source of information. The applications where motion is detected and utilized to gain information are

- Angiography.
- Measurement of blood velocity.
- Measurement of myocardial contractility with tagging.
- Measurement of diffusion.

These parameters can be measured inside a single examination, either sequentially or sometimes interleaved with high simultaneity.

1.11.1 Quantification of iron storage

Iron is normally present in all tissues, bound to nitrogen atoms of heme in haemoglobin and myoglobin, bound to sulphur atoms in non-heme iron proteins such as aconitase, or bound to cytochroms. Storage of iron in excess is done by large proteins, hemosiderin and ferritin. In the inherited disorders of iron metabolism, as well as in chronic haemolysis, and in neurodegenerative diseases, iron accumulates in liver, kidney, heart and in specific brain zones.

Iron in solution or in a small protein may act as a paramagnetic agent that shortens the longitudinal relaxation time of blood water. Conversely, iron borne by a large storage protein, or contained in a small solid particle, has weaker influence upon *T*1 of water molecules at vicinity, but, behaving as a negative contrast agent, very efficiently 'kills' the NMR signal of water molecules at vicinity as illustrated by Figure 1.11.1.

Labelling of cells with a contrast agent made of iron oxide makes these cells detectable after transplantation.

MRI is very efficient for detection of endogenous or exogenous iron. The detection of iron, either loading a labelled cell or stored by a protein such as ferritin or hemosiderin, is based upon the strong local magnetic field around a particle of iron. This local magnetic field is inhomogeneous and large variation of water resonance frequency takes place around the particle. Thus the water signal is strongly weakened inside a volume much larger than the particle or the iron-loaded cell. The effective relaxation time $T2^*$ is the index that measures this water signal decay.

Gradient-echo imaging is very sensitive to small amounts of iron. Spin-echo imaging, less sensitive, can be used at best to evaluate more abundant iron deposits such as observed in liver with hemochromatosis. Many studies have been performed in order to quantify iron deposition in human or animal organs (Haacke *et al.*, 2005).

1.11.2 Blood oxygenation and BOLD contrast

Oxygen modulates the magnetic properties of the oxygen-binding proteins, haemoglobin and myoglobin. The deoxyhemoglobin molecule Hb, that contains one iron atom, is paramagnetic. The oxygenated haemoglobin HbO2, which combines Hb and O2 (both of them paramagnetic) is diamagnetic.

Deoxyhemoglobin in deoxygenated blood behaves as an endogenous magnetic contrast agent, contained in red blood cells. Each red cell containing Hb behaves as a small magnet. A vessel filled with deoxygenated blood can be described as a small rod filled with magnetic material, thus creating a small additional magnetic field outside the vessel. This additional magnetic field is proportional to the degree of deoxygenation of blood and to the magnetic field B_0. It also depends on the vessel size and on its orientation relative to B_0. The homogeneity of the magnetic field is degraded at short distance around the vessel leading to shortening of $T2^*$ in water surrounding the vessel. $T2^*$-weighted images, obtained by a gradient-echo sequence, are sensitive to local magnetic field inhomogeneity linked to blood deoxygenation. At last, variations of haemoglobin saturation in blood, resulting from physiological variations of oxygen supply or consumption, are translated into MR signal variations. Ogawa demonstrated on rat brain at $B_0 = 7\,\mathrm{T}$ that the visibility of blood vessels as dark lines was greatly increased by blood deoxygenation (Ogawa *et al.*, 1990). He named this effect the blood oxygen level dependent (BOLD) contrast, at the origin of so many brain functional imaging studies.

BOLD contrast is widely used to detect activated neurons in many functional imaging experiments,

Figure 1.11.1 Detection of iron. Axial slice done through liver at 1.5 T with a spin-echo sequence ($T_R = 500$ ms, 8 echoes with T_E ranging from 10 ms to 80 ms). Eight images of this slice are obtained at increasing values of the echo time. The measurement of mean signal in a region of interest pasted upon these images allows to fit the signal decay and to determine the local value of *T*2. Top: normal C57/Bl6 mouse, images at $T_E = 10$ ms and $T_E = 20$ ms. Liver (black arrow) yields signal comparable to that of muscle. Liver *T*2 is 33±1 ms. Bottom: C57/Bl6 mouse 2 h after IV injection of Endorem (Guerbet, Aulnay, France) at the dose of 20 μM iron per kg. The negative contrast agent is taken by Kuppfer cells and strongly decreases liver signal. Images of echoes at $T_E = 10$ ms and $T_E = 20$ ms. Liver (white arrow) is darker than muscle at $T_E = 10$ ms and still darker at $T_E = 20$ ms. Liver *T*2 is 20±1 ms (*courtesy of C.V. Denis and D. Geldwerth, INSERM U770 at Hopital Kremlin-Bicêtre, France*)

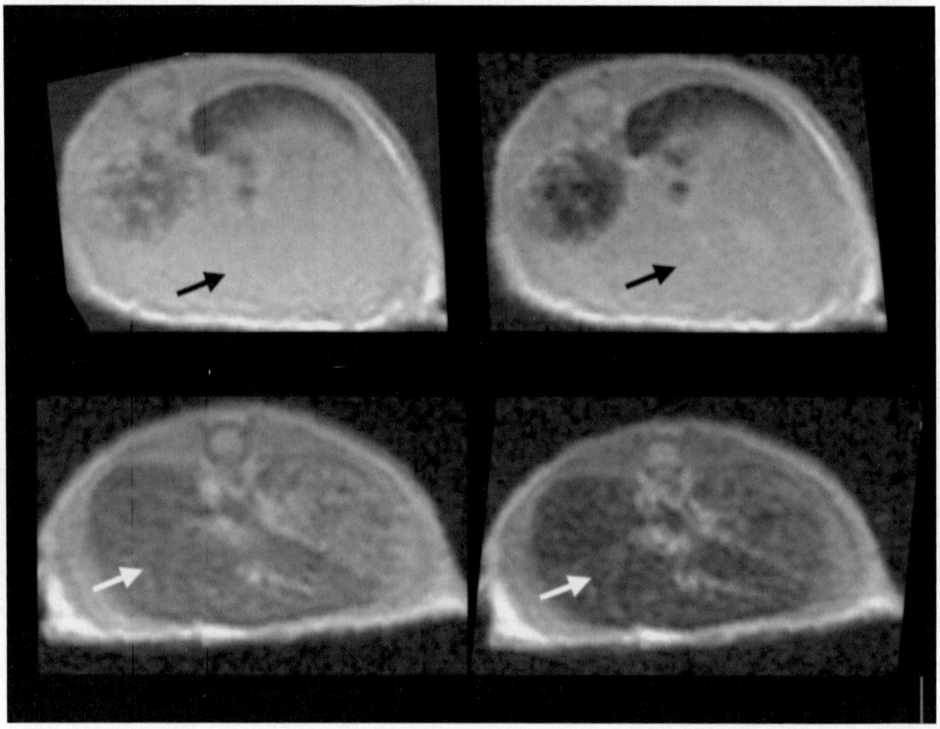

named by the acronym fMRI. Activation of brain neurons first induces an increase of oxygen consumption and a decrease of venous blood oxygenation, but it is followed a few seconds later by a strong increase of perfusion, such that venous oxygenation increases. In the activated zone, the blood oxygenation increase causes the decrease of the signal perturbation around capillaries and veins, so that the correlate of neuronal activation is NMR water signal increase (Figure 1.11.2).

This technique of localization of brain activation is fully atraumatic and offers a high spatial resolution; measurements can be done through the whole brain volume and repeated many times (as long as fatigue does not interfere with brain activity), so that complex protocols for brain stimulation can be performed. Often a repetitive stimulus is applied every 10–30 s, as perfusion adaptation typically needs 5–8 s. The coherence between the NMR signal variations and the stimulation protocol allows identification of activated voxels. The voxels where activation (or sometimes deactivation) is identified are often overlaid in colour upon an anatomical image of the same location. High-resolution BOLD studies, where voxels of 1 μl contain 600–800 neurons, can help to understand how neural networks are organized.

However, the signal variation depends on many parameters (perfusion, metabolic demand, shape of vascular tree), and its calibration in function of the

Figure 1.11.2 Detection of brain neuronal activation with an endogenous or exogenous magnetic agent. The intersection of a small vessel, filled with erythrocytes, with a brain voxel is schematised at rest (left column) and in activated state (right column). In the activated state, perfusion strongly increases; capillary oxygenation microvascular blood volume also increase. The signal of water around the vessel is displayed from white to dark grey.
Top: BOLD contrast from hemoglobin oxygenation. Water signal is weakened, from $T2^*$ shortening induced by the magnetization of deoxyhemoglobin in erythrocytes (represented as dark arrows). At rest, low blood oxygenation yields high extent of decreased signal around the vessel. In the activated state, higher blood oxygenation yields smaller extent of the decreased signal area and then higher signal of the corresponding voxel. Also from perfusion increase in activated brain, the capillary diameter increases.
Bottom: Exogenous contrast agent. Water signal is weakened, from $T2^*$ shortening induced by the magnetization of the intravascular negative contrast agent. When vessel volume increases, without significant change of the contrast agent concentration, the area of decreased signal around the vessel increases. The corresponding pixel is darker. Simultaneous variation of blood oxygenation induces much weaker signal variation

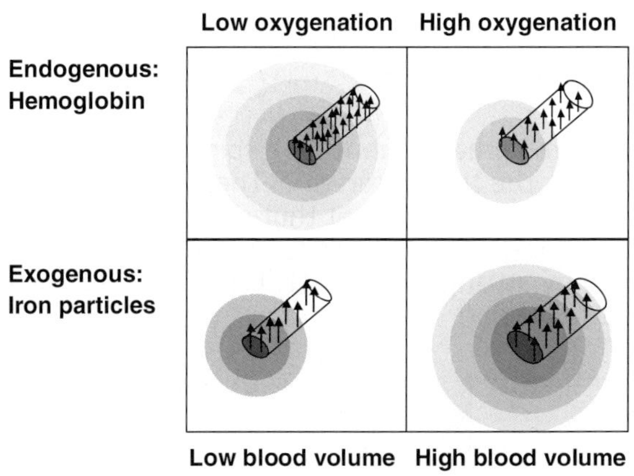

value of blood oxygen content and perfusion increase is difficult. Brain activation studies have been performed on many species from humans to rats and mice. Higher magnetic field is needed to explore smaller brains (Kim and Ogawa, 2002).

BOLD contrast detection is also applied to physiological studies of heart (Reeder *et al.*, 1999), kidney (Li *et al.*, 2003) and muscle (Jordan *et al.*, 2004).

1.11.3 Blood volume measurement using exogenous magnetic agents

After intravascular injection of a negative contrast agent with slow elimination, such as an USPIO (see 9.4), a modulation of microvascular volume can be detected from the induced variation of extravascular water signal (Figure 1.11.2). This is similar to BOLD contrast, but the magnetization of the intravascular contrast agent is much larger than that of deoxygenated hemoglobin in blood, so that the influence of oxygenation variation is negligible.

This technique is used to measure microvascular volume, and can also give information upon the distribution of vessel size in organs (Tropres *et al.*, 2004). Also the increase of sensitivity obtained by this technique allows detection of neuronal activation from the colocalized vascular volume increase. The signal variation during neuronal activation is typically 5 times higher than that detected with BOLD contrast at same B_0 value and is of opposite sign as shown in Figure 1.11.2. The increase in sensitivity also makes easier the detection of neuronal activation in the brain of awake animals (Vanduffel *et al.*, 2001).

1.11.4 Manganese-enhanced MRI (MEMRI)

Another way to study brain activation uses the shortening of water $T1$ by a positive contrast agent, manganese (Lin and Koretsky, 1997) and is named manganese-enhanced MRI (MEMRI). $MnCl_2$ is an efficient paramagnetic contrast agent (not used for human studies, because Mn^{2+} ion can interfere with many enzymatic mechanisms). Small doses, around 100 μM in water, efficiently shorten tissue water $T1$. The ion Mn^{2+}, an analogue of Ca^{2+}, is taken up through voltage gated Ca^{2+} channels, so that it can reflect the activity of cardiac and nervous cells. It is concentrated in active neurons and is washed out several days after local injections, then allows detection of neuronal activation during a long time period. Mn^{2+} is transported along axons and across synapses, and it has been shown to trace neuronal connections in the small animals central nervous system (Van der Linden *et al.*, 2004).

1.11.5 Water diffusion

Key points

Water diffusion derives from the random 3D motion of water molecules caused by thermal agitation. In tissues, it is limited by structural barriers (cell membrane, intracellular structures).

In the presence of intense magnetic field gradients, random displacements are converted into NMR signal attenuation that allows measuring the diffusion coefficient (see Figure 1.11.3).

Diffusion-weighted MRI is widely used for early detection of brain ischemia, being the first index that varies less than one hour after the ischemic event. It is also used to study connexions in the brain.

In biological tissues, water molecules are subject to the 'Brownian motion' caused by thermal agitation: Water molecules move randomly with velocities increasing with temperature and have frequent collisions with other molecules. This determines the viscosity of water in the tissue, and its diffusion coefficient.

A molecule with coefficient of diffusion D diffuses at a mean distance d from its initial location during a time interval t:

$$d^2 = 6Dt. \quad (1.48)$$

The value of the water diffusion coefficient is $2.4 \times 10^{-9}\,\mathrm{m^2\,s^{-1}}$ in pure water at 37°C, corresponding to a mean displacement of 15 μm during 50 ms. The diffusion coefficient of water is weaker in biological tissues. It is modulated by tissue structure and by cellular energetic status.

The water coefficient D may reflect the existence of several water pools with different viscosities and of intracellular structures that limit water displacement. When cell membranes restrict the displacement of water molecules, the value of D, called ADC for apparent diffusion coefficient, depends on the time interval allowed for displacement (LeBihan, 1995, Nicolay,van der Toorn and Dijkhuizen, 1995).

In an isotropic tissues, such as muscle or brain white matter tracts, the diffusion coefficient has different values along the direction of displacement. For example, a water molecule inside an axon can move freely along the axis of the fibre, whereas it has a low probability to cross the myelin sheath. If the fibres are parallel to axis z, water diffusion is more strongly limited along the x and y axes (small cell dimensions) than along z, and the diffusion coefficient is a set of three values $D_{xx} = D_{yy} < D_{zz}$. More generally the

Figure 1.11.3 Diffusion measurement. The diffusion-weighted sequence is a spin-echo sequence including two additional large gradient pulses separated by the time interval T. The 180° refocusing pulse invert phases at the middle of this time interval. A magnetic moment that stayed at the same position is dephased and then rephase by the two symmetrical gradient pulses. A magnetic moment that moved from position x to x + δx during T is dephased by an angle $\delta\phi = \gamma . G . T . \delta x$. The addition of elementary signals with random phases causes signal attenuation

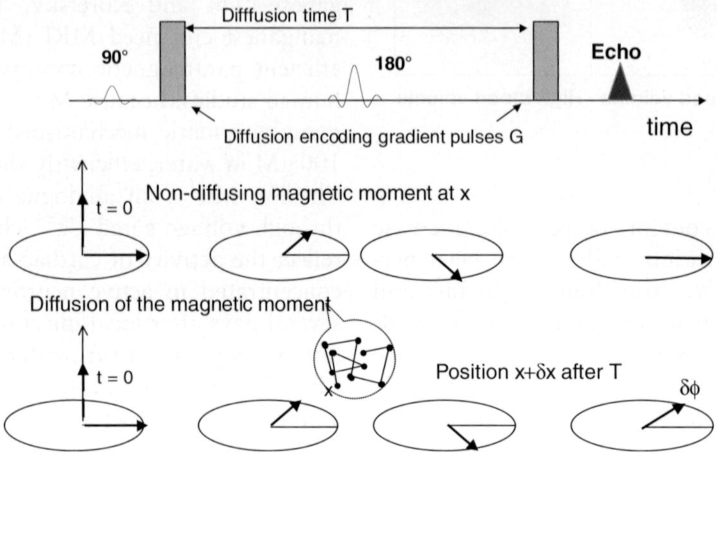

Figure 1.11.4 Multiparametric MRI study of rat cerebral ischemia induced with photochemical occlusion of proximal middle cerebral artery. Images are done at 1 h and 24 h after stroke induction using a 1.5 T clinical imaging system. In the *T2*-weighted high resolution images (T2WI) done with $T_R = 5680$ ms and $T_E = 100$ ms, no modification of signal is seen at 1 h (A1), whereas a large zone of increased signal is observed at 24 h (B1). Diffusion weighted images (DWI), obtained with a EPI spin-echo sequence, at lower spatial resolution, exhibit a zone of increased signal in the ischemic area at 1 h (A2) and 24 h (B2). The diffusion coefficient parametric image (ADC) is modified in the same area: The increased DWI signal is caused by the decreased attenuation due to the strong decrease of the water diffusion coefficient in the stroke area. The stroke is detected at 1 h after stroke induction (A2,A3) and has a wider extent at 24 h (B2,B3). Perfusion weighted images (PWI) are obtained by fast acquisition of *T2** weighted images during the first pass of a bolus of a positive contrast agent. Brain signal is diminished in the normally perfused tissue. Images at peak of contrast agent concentration show the extent of the perfusion defect at 1 h (A4) and at 24 h (B4). The various maps obtained at 24 h correspond well to the extent of the infarct on the histologic control (*images from (Chen et al., 2004) with permission of MAGMA*)

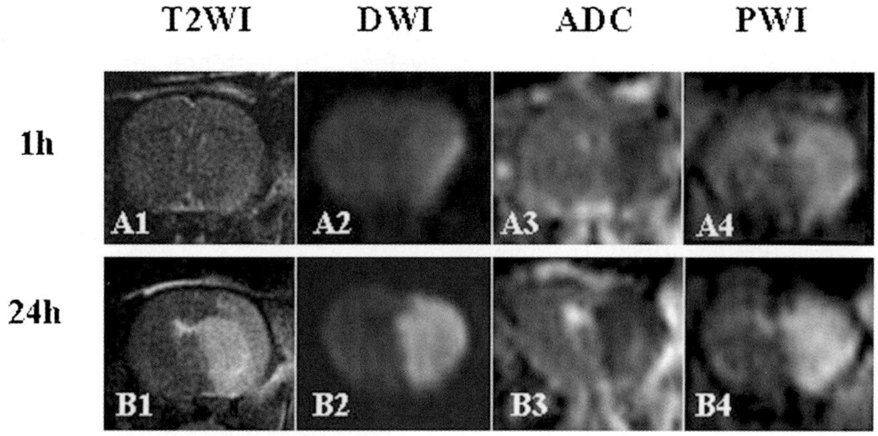

diffusion is described by a matrix of nine diffusion coefficients, the diffusion tensor. Diffusion tensor imaging (DTI) is the technique combining diffusion measurement under a combination of gradients directions that allows measuring this diffusion tensor (Zhang *et al.*, 2003).

Also, modification of cell size or content can induce large modifications of water diffusion coefficient. An important example is that of brain acute ischemia. In cat brain middle cerebral artery occlusion, the water apparent diffusion coefficient, measured by using diffusion weighted MRI, decreases strongly within 15 min following arterial occlusion (Moseley *et al.*, 1990). The decrease is around −40 to −60%, and the ADC stays at low values for several days. The stroke area, where water molecules displacements are weaker, appears as bright on diffusion-weighted images and dark on ADC maps as shown in Figure 1.11.4. This technique can be combined to other measurements such as perfusion, and it has gained much importance to measure the stroke volume and to evaluate the influence of therapies (Hoehn *et al.*, 2001).

More physics: How to measure the diffusion coefficient by using NMR

Conventional MRI utilizes contrast changes from *T1* and *T2* variations, reflecting the concentration of free water in tissues. Diffusion-weighted MRI utilizes the translation of water molecules during an interval of time *T*, in presence of a large magnetic field gradient: Random motion is converted into random dephasing of magnetic moments (Figure 1.11.3).

The classic diffusion-weighted sequence is a spin-echo sequence where intense gradient pulses of intensity *G* are added symmetrically before and after the 180° RF pulse. The effect of the first diffusion gradient pulse is to encode each proton with a given phase according its position, as is done during the phase encoding. If the proton does not move, the second diffusion gradient pulse brings the same phase and then (because of the 180° pulse) there is no net dephasing. Conversely if the proton has moved, there is no exact compensation of the first dephasing by the second gradient pulse: The magnetic moments that have moved randomly are dephased randomly. The resulting transverse magnetization is decreased without global net dephasing.

The signal collected is attenuated according to an exponential law:

$$S = S_o \exp(-b.\text{ADC}), \qquad (1.49)$$

where S_o is the signal in the absence of the gradient G, ADC is the 'apparent diffusion coefficient' of molecules for the applied gradient direction, b is proportional to G^2 and depends on the time interval T and the gradient G duration.

S is measured at several successive intensities of the gradient G to determine the ADC in each voxel. This makes it possible to build ADC images.

1.11.6 Magnetic resonance angiography

Magnetic resonance angiography (MRA) generates images of blood vessels (Bradley, 1992). The simplest principle for MR angiography, the 'time of flight' technique, is based upon the enhancement of magnetization caused by the flow inside a vessel, as illustrated in Figure 1.11.5. At a very short value of T_R, the signal of stationary water in the slice is weak, as $T_R \ll T1$: The time between successive measurements is too short for an efficient recovery of M_z. For blood water flowing in arteries or veins, if the velocity of blood is fast enough, that is its time of flight through the slice is shorter than T_R, in the slice under examination the water inside the vessel has not been submitted to previous selective irradiation because the circulation of blood brings fresh magnetization. Water in the vessel section has the maximal magnetization M_o. Then the vessel appears as a very bright structure on the image.

This technique requires acquisition with short values of T_R and T_E, at best obtained by 3D gradient-echo techniques. High spatial resolution is needed to visualise small diameter vessels without dilution in large voxels. Contrast agents can be injected in the vascular volume to shorten blood $T1$, in order to improve quality of the angiography (Miraux *et al.*, 2004).

Figure 1.11.5 Angiography. Magnetic resonance angiography (MRA) using the 'time of flight' method relies upon the difference in longitudinal magnetization between flowing and static water. Especially in short animals, shortening blood $T1$ with a positive contrast agent makes easier to perform 3D angiography in a given volume within a short time (Miraux *et al.*, 2004). 3D angiographic acquisition at $B_0 = 4.7$ T, after intravenous injection of Gd-DOTA that improved vessels visualisation during ten minutes. Acquisition is done with a gradient-echo sequence, the flip angle is 90°; $T_R = 12$ ms, $T_E = 3.1$ ms; the acquisition time is 1 min 38 s for one volume. The displayed volume is 65 × 38 × 32 mm (matrix acquisition 256 × 128 × 64). Three such volumes are needed to visualise vessels from aortic cross to top of brain. (a) Jugular veins; (b) vertebral arteries; (c) sub-clavian arteries; (d) aortal; (e) right ligated common carotid; (f) left common carotid. A: axial slice from the 3D acquisition. Vessel signal is much higher than the background tissue signal. The right ligated carotid artery appears with low signal and small section. B: construction of blood vessels from the 64 slices, using a maximum intensity projection (MIP) algorithm to reconstruct 3D views of vessels. *Images from Miraux et al., (2004) with permission of MAGMA*

(a)

(b)

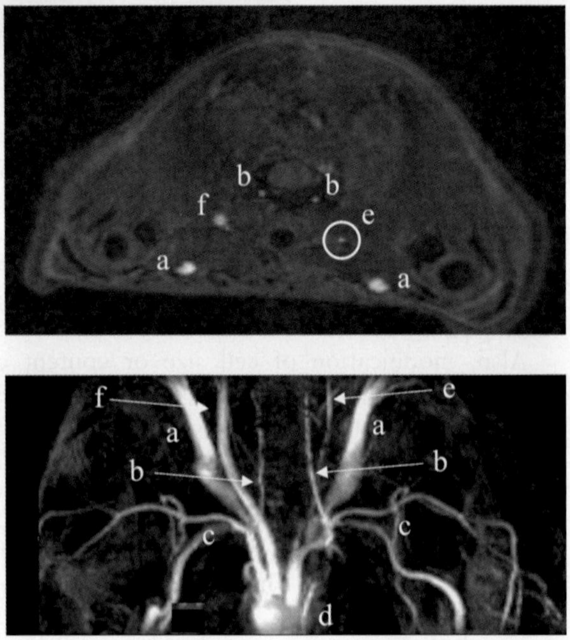

1.11.7 Perfusion measurement

Perfusion is the amount of blood flowing in the capillary bed of a given mass of tissue during a fixed period of time, delivering oxygen and nutrients. It is usually expressed in milliliters blood per 100 g tissue per min. Its value is of critical importance to maintain adequate energy status of the tissue. Perfusion measurements quantify the amount of blood flowing at low velocity in the smallest vessels. Two distinct MRI techniques allow perfusion measurements (Barbier, Lamalle and Decorps, 2001): The first one is based upon the follow-up of a contrast agent bolus after quick injection; the second one detects blood flow from the magnetic labelling of arterial water magnetization.

1.11.7.1 Dynamic contrast enhanced imaging (DCE-MRI)

After quick intravenous injection, a magnetic contrast agent (CA) can be detected during its first passage through capillaries of the organ under study. High-speed acquisition is needed (typ. one image every 0.5 s, depending on animal size).

The CA, before dilution in the extracellular space, modifies the blood relaxation times $T1$ and $T2$. Even though the CA is confined to the vascular space, it also modifies the tissue water signal around capillaries by two distinct mechanisms:

The first one is the shortening of tissue water $T1$ by exchange with the water containing the CA in capillaries. When using a gadolinium chelate, or an USPIO at very low concentration, $T1$ shortening can be detected by a very strongly $T1$-weighted sequence. This technique is widely used to detect ischemia and to evaluate myocardial perfusion.

The second one is the shortening of tissue water $T2^*$ around the capillaries, resulting from the high CA concentration in capillaries (Figure 1.11.2). Water $T2^*$ shortening around capillaries is easily detected by gradient-echo imaging as a strong signal drop during first pass of the CA (Caramia *et al.*, 1998). This technique is efficient to detect hypoperfused zones as illustrated by Figure 1.11.4, images A4 and B4. However accurate quantification of perfusion is difficult, and modification of capillary permeability can modify the curve of signal variation. Moreover there is no linear variation between the CA concentration and the signal variation. Also it is not possible to repeat perfusion measurement at short intervals because one has to wait the time needed for renal elimination after one injection.

DCE-MRI may also give information on vascular permeability. The size of the CA determines its issue out of the vascular sector, depending on capillary structure. The analysis of vascular and tissular signals by compartmental analysis of dynamic data allows to determine parameters that characterize the tissue (microvascular volume and permeability).

1.11.7.2 Arterial spin labelling

In arterial spin labelling (ASL) techniques, the blood water magnetization is used as an endogeneous contrast agent: The arterial water magnetization is modified upstream the organ by selective labelling of a thick slice containing the feeding artery (Detre *et al.*, 1994). Labelling is done either by a 180° RF pulse that inverts the longitudinal magnetization M_z inside the slice or by a 90° RF pulse that zeroes the longitudinal magnetization inside the slice (see paragraph 6.1). In the slice of the organ under study, located downstream, the effect of arterial labelling is related to the perfusion that brings some water with modified magnetization (here called 'spin') into the capillaries and then into the tissue. The resulting tissue magnetization is modified by a quantity related to perfusion, so that the perfusion is quantified from the difference between an image obtained with labelling and an image obtained without labelling.

ASL techniques, that rely upon a weak difference between two images, are more easily handled at high magnetic field and in organs highly perfused such as brain and kidney (Detre *et al.*, 1994), heart (Kober *et al.*, 2004) (Streif *et al.*, 2005) and muscle (Carlier and Bertoldi, 2005).

They allow absolute quantification of perfusion and do not require contrast agent injection, so that measurements can be repeated easily, and then allow dynamic studies under stress (Carlier and Bertoldi, 2005). With use of specific technological developments, ASL measurements can be coupled with simultaneous acquisitions of ^{1}H and ^{31}P NMR spectroscopy data. These protocols offer new possibilities whereby the microcirculatory control of cell oxygenation and high-energy phosphate metabolism can be explored (Reeder *et al.*, 1999).

1.12 More about applications

Multiparametric studies combining several NMR acquisitions, to probe sequentially or simultaneously several parameters, are of particular interest. MRI and MRS sequences offer a variety of contrast. Often a fast acquisition modulus, offering short acquisition time but weak contrast, is combined with a preparation modulus that sensitises magnetization to local value of a parameter such as $T1$, $T2$, $T2^*$, perfusion, diffusion, local tissue displacement.

Anatomy and physiology are the main grounds of application for NMR techniques. NMR applications to molecular imaging are less important, because optical and nuclear imaging techniques offer a much higher sensitivity for detection of compounds at low concentration. The only compound detected with very high sensitivity is the iron atom, detected through its influence upon water signal, so that applications of NMR in the field of cellular labelling and tracking are very promising.

NMR techniques (mostly spectroscopy) have been applied to cellular physiology (Balaban, 1984) since the 1970s and to *in vivo* studies of physiology on larger biological systems since the 1980s.

The following review papers give an extended view of the applications of MRI and MRS in specific domains of biomedical research such as brain diseases (Hoehn *et al.*, 2001; Dijkhuizen and Nicolay, 2003), drug discovery and development (Rudin *et al.*, 1999; Beckmann *et al.*, 2001), gene and cell therapies (Bell and Taylor-Robinson, 2000; Allport and Weissleder, 2001; Leroy-Willig *et al.*, 2003).

Many of these applications are illustrated in Part II of this book.

References

Allport, J. R., Weissleder, R., 2001. "*In vivo* imaging of gene and cell therapies." *Exp. Hematol.* **29**, 1237–1246.

Balaban, R. S., 1984. "The application of nuclear magnetic resonance to the study of cellular physiology." *Am. J. Physiol.* **246**, C10-9.

Barbier, E. L., Lamalle, L., Decorps, M., 2001. "Methodology of brain perfusion imaging." *J. Magn. Reson. Imaging* **13**, 496–520.

Beckmann, N., Mueggler, T., Allegrini, P. R., Laurent, D., Rudin, M., 2001. "From anatomy to the target: Contributions of magnetic resonance imaging to preclinical pharmaceutical research." *Anat. Rec.* **265**, 85–100.

Bell, J. D., Taylor-Robinson, S. D., 2000. "Assessing gene expression *in vivo*: Magnetic resonance imaging and spectroscopy." *Gene Therapy* **7**, 1259–1264.

Billotey, C., Aspord, C., Beuf, O., Piaggio, E., Gazeau, F., Janier, M. F., Thivolet, C., 2005. "T-cell homing to the pancreas in autoimmune mouse models of diabetes: *In vivo* MR imaging." *Radiology* **236**, 579–587.

Bolinger, L., Prammer, M., Leigh, J., 1988. "A multiple-frequency coil with a highly uniform *B*1 field." *J. Magn. Reson.* **81**, 162–166.

Bradley, W. G., 1992. "Recent advances in magnetic resonance angiography of the brain." *Curr. Opin. Neurol. Neurosurg.* **5**, 859–862.

Bushberg, J., Seibert, J., Leidholdt, E., Boone, J., 2001. *The Essential Physics of Medical Imaging*, Lippincott, Williams and Wilkins, Philadelphia.

Caramia, F., Yoshida, T., Hamberg, L. M., Huang, Z., Hunter, G., Wanke, I., Zaharchuk, G., Moskowitz, M. A., Rosen, B. R., 1998. "Measurement of changes in cerebral blood volume in spontaneously hypertensive rats following L-arginine infusion using dynamic susceptibility contrast MRI." *Magn. Reson. Med.* **39**, 160–163.

Carlier, P. G., Bertoldi, D., 2005. "*In vivo* functional NMR imaging of resistance artery control." *Am. J. Physiol. Heart Circ. Physiol.* **288**, H1028–H1036.

Chen, F., Suzuki, Y., Nagai, N., Peeters, R., Sun, X., Coudyzer, W., Marchal, G., Ni, Y., 2004. "Rat cerebral ischemia induced with photochemical occlusion of proximal middle cerebral artery: A stroke model for MR imaging research." *Magma* **17**, 103–108.

Chen, X. J., Hedlund, L. W., Moller, H. E., Chawla, M. S., Maronpot, R. R., Johnson, G. A., 2000. "Detection of emphysema in rat lungs by using magnetic resonance measurements of ^{3}He diffusion." *Proc. Natl. Acad. Sci. USA* **97**, 11478–11481.

Constantinides, C. D., Kraitchman, D. L., O'Brien, K. O., Boada, F. E., Gillen, J., Bottomley, P. A., 2001. "Non-invasive quantification of total sodium concentrations in acute reperfused myocardial infarction using ^{23}Na MRI." *Magn. Reson. Med.* **46**, 1144–1151.

Detre, J. A., Zhang, W., Roberts, D. A., Silva, A. C., Williams, D. S., Grandis, D. J., Koretsky, A. P., Leigh, J. S., 1994. "Tissue specific perfusion imaging using arterial spin labeling." *NMR Biomed.* **7**, 75–82.

Dijkhuizen, R. M., Nicolay, K., 2003. "Magnetic resonance imaging in experimental models of brain disorders." *J. Cereb. Blood Flow Metab.* **23**, 1383–1402.

Dupuich, D., Berthezene, Y., Clouet, P. L., Stupar, V., Canet, E., Cremillieux, Y., 2003. "Dynamic ^{3}He imaging for quantification of regional lung ventilation parameters." *Magn. Reson. Med.* **50**, 777–783.

Gadian, D., 1995. *Nuclear Magnetic Resonance and its Applications to Living Systems*, Oxford University Press, New York.

Ginefri, J. C., Poirier-Quinot, M., Robert, P., Darrasse, L., 2005. "Contrast-enhanced dynamic MRI protocol with improved spatial and time resolution for *in vivo* microimaging of the mouse with a 1.5-T body scanner and a superconducting surface coil." *Magn. Reson. Imaging* **23**, 239–243.

Haacke, E. M., Cheng, N. Y., House, M. J., Liu, Q., Neelavalli, J., Ogg, R. J., Khan, A., Ayaz, M., Kirsch, W., Obenaus, A., 2005. "Imaging iron stores in the brain using magnetic resonance imaging." *Magn. Reson. Imaging* **23**, 1–25.

Heyn, C., Bowen, C. V., Rutt, B. K., Foster, P. J., 2005. "Detection threshold of single SPIO-labelled cells with FIESTA." *Magn. Reson. Med.* **53**, 312–320.

Hoehn, M., Kustermann, E., Blunk, J., Wiedermann, D., Trapp, T., Wecker, S., Focking, M., Arnold, H., Hescheler, J., Fleischmann, B. K., Schwindt, W., Buhrle, C., 2002. "Monitoring of implanted stem cell migration in vivo: a highly resolved in vivo magnetic resonance imaging investigation of experimental stroke in rat." *Proc. Natl. Acad. Sci. USA* **99**, 16267–16272.

Hoehn, M., Nicolay, K., Franke, C., van der Sanden, B., 2001. "Application of magnetic resonance to animal models of cerebral ischemia." *J. Magn. Reson. Imaging***14**, 491–509.

Hornack, 2005. The basis of MRI, http://www.cis.rit.edu/htbooks/mri.

Jacobs, R. E., Fraser, S. E., 1994. "Magnetic resonance microscopy of embryonic cell lineages and movements." *Science* **263**, 681–684.

Jendelova, P., Herynek, V., Urdzikova, L., Glogarova, K., Rahmatova, S., Fales, I., Andersson, B., Prochazka, P., Zamecnik, J., Eckschlager, T., Kobylka, P., Hajek, M., Sykova, E., 2005. "Magnetic resonance tracking of human CD34+ progenitor cells separated by means of immunomagnetic selection and transplanted into injured rat brain." *Cell Transplant* **14**, 173–182.

Johnson, G. A., Cofer, G. P., Gewalt, S. L., Hedlund, L. W., 2002. "Morphologic Phenotyping with MR Microscopy: The Visible Mouse." *Radiology* **222**, 789–793.

Johnson, G. A., Herfkens, R. J., Brown, M. A., 1985. "Tissue relaxation time: *In vivo* field dependence." *Radiology* **156**, 805–810.

Jordan, B. F., Kimpalou, J. Z., Beghein, N., Dessy, C., Feron, O., Gallez, B., 2004. "Contribution of oxygenation to BOLD contrast in exercising muscle." *Magn. Reson. Med.* **52**, 391–396.

Kim, R. J., Judd, R. M., Chen, E. L., Fieno, D. S., Parrish, T. B., Lima, J. A., 1999. "Relationship of elevated ^{23}Na magnetic resonance image intensity to infarct size after acute reperfused myocardial infarction." *Circulation* **100**, 185–192.

Kim, S. G., Ogawa, S., 2002. "Insights into new techniques for high resolution functional MRI." *Curr. Opin. Neurobiol.* **12**, 607–615.

Kober, F., Iltis, I., Izquierdo, M., Desrois, M., Ibarrola, D., Cozzone, P. J., Bernard, M., 2004. "High-resolution myocardial perfusion mapping in small animals *in vivo* by spin-labelling gradient-echo imaging." *Magn. Reson. Med.* **51**, 62–67.

Lanza, G. M., Winter, P. M., Caruthers, S. D., Morawski, A. M., Schmieder, A. H., Crowder, K. C., Wickline, S. A., 2004. "Magnetic resonance molecular imaging with nanoparticles." *J. Nucl. Cardiol.* **11**, 733–743.

LeBihan, D., 1995. "Molecular diffusion, tissue microdynamics and microstructure." *NMR Biomed.* **8**, 375–386.

Leroy-Willig, A., Fromes, Y., Paturneau-Jouas, M., Carlier, P., 2003. "Assessing gene and cell therapies applied in striated skeletal and cardiac muscle: Is there a role for nuclear magnetic resonance?" *Neuromuscul. Disord.* **13**, 397–407.

Li, L., Storey, P., Kim, D., Li, W., Prasad, P., 2003. "Kidneys in hypertensive rats show reduced response to nitric oxide synthase inhibition as evaluated by BOLD MRI." *J. Magn. Reson. Imaging* **17**, 671–675.

Lin, Y. J., Koretsky, A. P., 1997. "Manganese ion enhances *T*1-weighted MRI during brain activation: An approach to direct imaging of brain function." *Magn. Reson. Med.* **38**, 378–388.

Miraux, S., Serres, S., Thiaudiere, E., Canioni, P., Merle, M., Franconi, J. M., 2004. "Gadolinium-enhanced small-animal TOF magnetic resonance angiography." *Magma* **17**, 348–352.

Mispelter, J., Lupu, M., Briguet, A., 2006. "NMR probeheads for biophysical and biomedical experiments." *Theoretical Principles and Practical Guidelines*, Imperial College Press, London.

Moon, R., Richards, J., 1973. "Determination of intracellular pH by 31P magnetic resonance." *J. Biol. Chem.* **248**, 7276–7278.

Moseley, M. E., Kucharczyk, J., Mintorovitch, J., Cohen, Y., Kurhanewicz, J., Derugin, N., Asgari, H., Norman, D., 1990. "Diffusion-weighted MR imaging of acute stroke: correlation with *T*2-weighted and magnetic susceptibility-enhanced MR imaging in cats." *AJNR Am. J. Neuroradiol.* **11**, 423–429.

Ness Aiver, M., 1997. *All You Really Need to Know About MRI Physics*, Simply Physics, Baltimore.

Nicolay, K., van der Toorn, A., Dijkhuizen, R. M., 1995. *In vivo* diffusion spectroscopy. *An overview: NMR in Biomedicine*, vol. 8, pp. 365–374.

Ogawa, S., Lee, T., Nayak, A., Glynn, P., 1990. "Oxygenation-sensitive contrast in magnetic resonance image of rodent brain at high magnetic fiels." *Magn. Reson. Med.* **14**, 68–78.

Padgett, K., Blackband, S., Grant, S., 2005.*Proceedings of the International Society for Magnetic Resonance in Medicine: 13th Scientific Meeting and Exhibition*, Miami, p. 2198.

Parzy, E., Fromes, Y., Wary, C., Vignaux, O., Giacomini, E., Leroy-Willig, A. and Carlier, P. G., 2003. "Ultrafast multiplanar determination of left ventricular hypertrophy in spontaneously hypertensive rats with single-shot spin-echo nuclear magnetic resonance imaging." *J. Hypertens.* **21**, 429–436.

Reeder, S. B., Holmes, A. A., McVeigh, E. R., Forder, J. R., 1999. "Simultaneous non-invasive determination of regional myocardial perfusion and oxygen content in rabbits: Toward direct measurement of myocardial oxygen consumption at MR imaging." *Radiology* **212**, 739–747.

Rinck, P., 2001. *Magnetic Resonance in Medicine: The Basic Textbook of the European Magnetic Resonance Forum*, 4th ed. Blackwell Science, Berlin, Germany.

Rudin, M., Beckmann, N., Porszasz, R., Reese, T., Bochelen, D., Sauter, A., 1999. "*In vivo* magnetic resonance methods in pharmaceutical research: Current status and perspectives." *NMR Biomed.* **12**, 69–97.

Stark, D., Bradley, W. J., 1992. *Magnetic Resonance Imaging*, Mosby-Year Books, St. Louis.

Streif, J. U., Nahrendorf, M., Hiller, K. H., Waller, C., Wiesmann, F., Rommel, E., Haase, A., Bauer, W. R., 2005. "*In vivo* assessment of absolute perfusion and intracapillary blood volume in the murine myocardium by spin labelling magnetic resonance imaging." *Magn. Reson. Med.* **53**, 584–592.

Tropres, I., Lamalle, L., Peoc'h, M., Farion, R., Usson, Y., Decorps, M., Remy, C., 2004. "*In vivo* assessment of tumoural angiogenesis." *Magn. Reson. Med.* **51**, 533–541.

Van der Linden, A., Van Meir, V., Tindemans, I., Verhoye, M., Balthazart, J., 2004. "Applications of manganese-enhanced magnetic resonance imaging (MEMRI) to image brain plasticity in song birds." *NMR Biomed.* **17**, 602–612.

Vanduffel, W., Fize, D., Mandeville, J. B., Nelissen, K., Van Hecke, P., Rosen, B. R., Tootell, R. B., Orban, G. A., 2001. "Visual motion processing investigated using contrast agent-enhanced fMRI in awake behaving monkeys." *Neuron* **32**, 565–577.

Webb, A., 2003. *Introduction to Biomedical Imaging*, John Wiley & Sons, Inc., Hoboken, New Jersey.

Webb, A., 2003. *Introduction to Biomedical Imaging*, John Wiley & Sons, Inc., Hoboken, New Jersey.

Wishnia, A., Alameddine, H., Tardif de Gery, S., Leroy-Willig, A., 2001. "Use of magnetic resonance imaging for non-invasive characterization and follow-up of an experimental injury to normal mouse muscles." *Neuromuscul. Disord.* **11**, 50–55.

Zhang, J., Richards, L. J., Yarowsky, P., Huang, H., van Zijl, P. C., Mori, S., 2003. "Three-dimensional anatomical characterization of the developing mouse brain by diffusion tensor microimaging." *Neuroimage* **20**, 1639–1648.

2 High Resolution X-Ray Microtomography: Applications in Biomedical Research

Nora De Clerck and **Andrei Postnov**

2.0 Introduction

Nowadays, microscopic imaging and advanced trends in molecular biology make it possible to analyse the smallest details in complex living structures. However physiologically, the question arises as to how these detailed structures are to be translated into the integrated function and spatial orientation of an organism. Therefore, broad interest has been growing in obtaining three-dimensional (3D) images in biological tissues (Postnov *et al.*, 2002a).

At present, several microscopic methods are available for biomedical research. Histology using optical and electronic microscopes requires ample sample preparation together with slicing of the object. Destruction of the specimen is a serious limitation especially when precious or unique objects have to be studied. When applying classical slicing techniques, information in the 3D space is lost in many cases. Moreover, it proved necessary to study samples in their natural surroundings preferably without preparation. These arguments clearly show the importance of non-invasive imaging techniques in biomedical research.

As discussed elsewhere in this book, a number of non-invasive tomography methods can be used. Nowadays, several 3D microscopic techniques are available. Some of the most frequently used are nuclear magnetic resonance imaging (MRI) and high-resolution X-ray microtomography (micro-CT). Both non-invasive techniques are complementary as MRI is most suitable to study soft tissues, whereas imaging by micro-CT will be preferred to analyse bones and calcified tissue.

In this chapter, we shall discuss the advantages and limitations of high-resolution *desktop X-ray micro-CT*. After an introduction of the physical *principles*, ample attention will be paid to the contribution of micro-CT in the field of biomedical imaging.

2.1 Principles of tomography

2.1.1 Introduction and definitions

Oxford dictionary defines tomography as a method of radiography displaying details in a selected plane within the body. The word tomography is derived from the Greek language where '*tomos*' means 'section'. Thus, X-ray tomography is a technique that visualizes the inner structure of samples by virtual cutting of the object by means of X-rays. We shall only consider X-ray shadow micro-CT, although tomographical methods are not restricted to shadow projections only. On the synchrotrons, X-ray phase contrast tomography is developing (Beckmann *et al.*, 1999). Tomography is also possible with electrons (electron tomography), visible light (optical tomography), positrons (positron tomography), acoustic waves (ultrasound tomography) and

Figure 2.1.1 Basic principles of imaging by micro-CT. (a): Shadow image of the skull of a fish; (b): reconstructed virtual cross-section through the skull; (c): 3D model built from an isotropic set of cross-sections

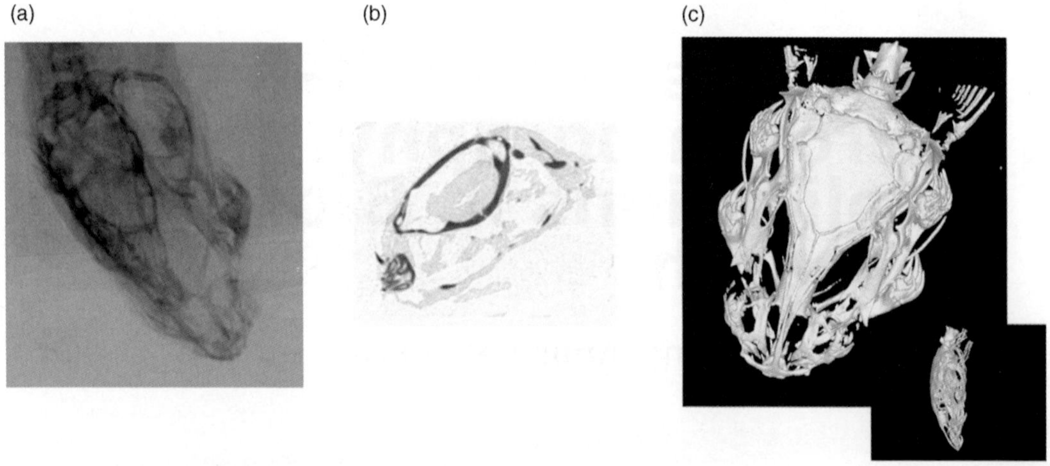

with the application of the nuclear magnetic resonance effect. The basic imaging principle of X-ray micro-CT is that images are reconstructed from X-ray projections or shadow images (Gilboy, 1995). The spatial distribution of the X-ray attenuation coefficients is measured in a plane as the X-ray beam is passing through an object from multiple orientations. Thanks to the tremendous developments in computing sciences, powerful computers are involved both in data storage and reconstruction of huge data sets. Therefore, the denomination of computed tomography abbreviated as CT was applied (Hounsfield, 1973).

X-ray computerized tomography (CT), in general, was introduced by Hounsfield about 30 years ago (Cormack, 1973; Hounsfield, 1973). X-ray CT soon found its applications in medicine as a useful diagnostic device (Hounsfield, 1980). Yet, despite of the enormous advantage, medical CT suffers from a relatively poor spatial resolution (typically 0.3–1.00 mm) due to the limitations in the radiation dose.

Recently, principles of X-ray computed tomography were implemented in desktop micro-CT instruments (Elliot and Dover, 1982; Rüegsegger, Koller and Müller, 1996; Sasov and Van Dyck, 1998) for application in fundamental research both in materials science and subsequently in biomedical research. Depending on the spatial resolution that can be obtained, a classification of the available CT techniques can be made (Davis and Wong, 1996; Ketcham and Carlson, 2001).

In the present chapter, we shall restrict ourselves to high-resolution X-ray micro-CT. The basic principles of imaging by micro-CT will be discussed below. X-rays that are generated by an X-ray source pass through an object resulting in a shadow image or scout view as illustrated in Figure 2.1.1(a). The X-rays that are attenuated by the object are captured by a detector. All measurements of the attenuation coefficients by the camera are stored in the computer as a floating-point matrix. Subsequent reconstruction will result in a virtual slice as shown in Figure 2.1.1(b). A typical CT image is called a virtual slice or cross-section. Scanning by micro-CT can be isotropic, that is the spatial resolution is the same in all three dimensions. Therefore, after acquisition of a stacked continuous series of CT images, data describing an entire volume becomes available which can be rendered in 3D space (3D reconstructions) as shown in panel C (Figure 2.1.1(c)). Afterwards it is possible to section these 3D models in any arbitrary orientation without any loss in quality.

2.1.2 Image acquisition

Figures 2.1.2 and 2.1.3 illustrate the configuration for data acquisition by micro-CT. Basically an X-ray source illuminates the object, which absorbs the X-rays. Attenuated X-rays are then captured by a detector. To collect enough information on shadow images, the investigated object should be illuminated from different orientations. There are two possibilities to achieve this either by rotating the object or by leaving it motionless. For *in vitro* scanning, the object can be placed on a rotating stage (Figure 2.1.2). In Figure 2.1.3, the camera and detector move around the object. This construction is more sophisticated and expensive and is used only for

Figure 2.1.2 Acquisition set-up for *in vitro* investigations. X-rays passing through an object are attenuated and captured by a detector. Source and detector remain motionless; the object (frog) is rotating

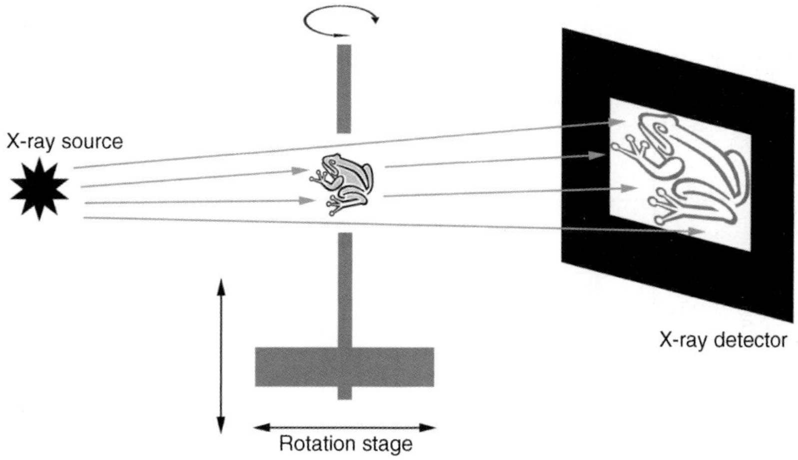

in vivo studies where the animal is laid out on a bed in the scanner.

In general, the following physical variables are defined for image acquisition.

Figure 2.1.3 Acquisition set-up for *in vivo* investigations. X-rays passing through an object are attenuated and captured by a detector. Object (frog) is laid out on an animal bed while source and detector are rotating around it

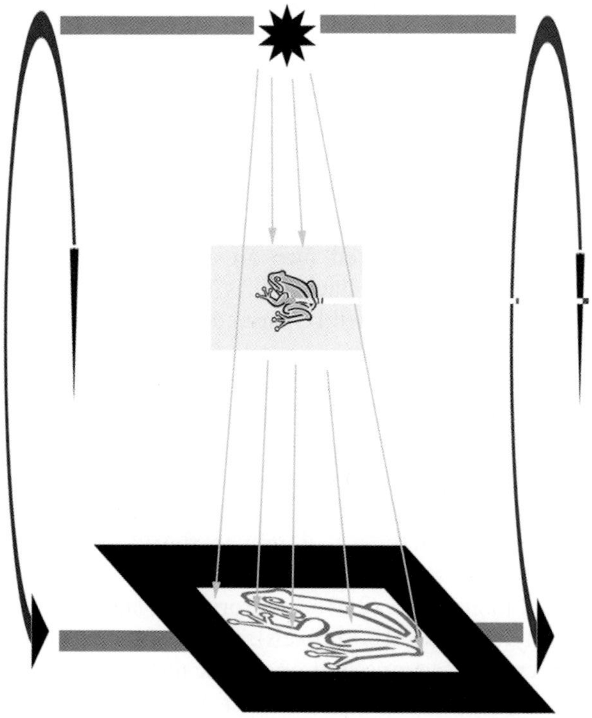

2.1.2.1 *Linear attenuation coefficient*

Absorption of X-rays has a statistical nature. An X-ray photon is either absorbed or passes through the material. Different materials have a different probability to absorb an X-ray photon. This probability also depends on the energy of the X-ray quantum.

The linear attenuation coefficient μ at a given energy E is defined as

$$dN/N = -\mu(E)dx, \qquad (1.1)$$

where N is the number of photons. An equivalent definition of $\mu(E)$ is

$$\mu(E) = -\ln\rho,$$

where ρ is the probability that an X-ray photon is absorbed on a given pathway through an object. Linear attenuation coefficients are measured in units of inverse length, for example for water it is 0.38 cm^{-1} at 30 keV (a photon energy typical for micro-CT).

2.1.2.2 *X-ray attenuation*

Let us define I_0 as an input energy generated by the X-ray source while I represents the transmitted energy; then Eq. (1.2) can be written as Beer's law derived from Eq. (1.1)

$$I = I_0 . e^{-\mu x}, \qquad (1.2)$$

where x represents the path length through the attenuating material. This formula describes only an ideal

Figure 2.1.4 X-ray emission spectrum representative for a Wolfram X-ray tube implemented in micro-CT. Spectrum is represented in terms of number of photons per energy interval, solid angle and tube charge (mAs) as a function of photon energy. From Hammersberg *et al.* (1998). Absolute energy spectra for an industrial micro-focal X-ray source under working conditions measured with a Compton scattering spectrometer – full spectra data. ***Linköping Electronic Articles in Mechanical Engineering***, No. 1, with permission by the authors

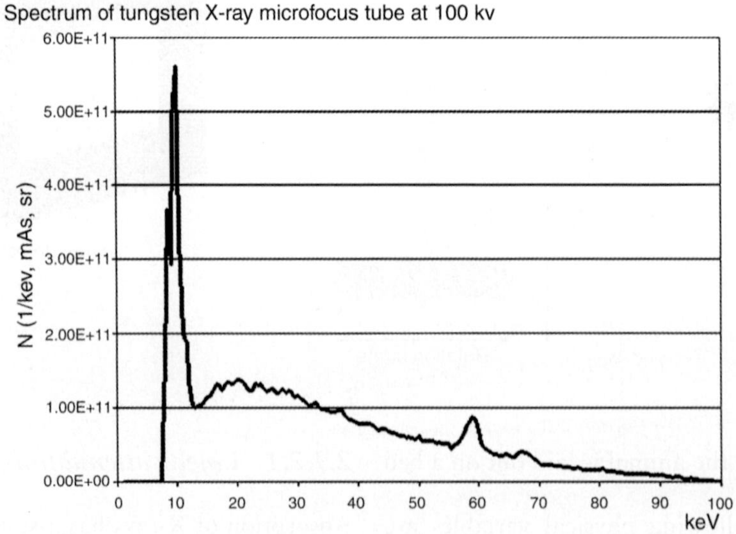

case for a monochromatic beam and for homogeneous material. In addition, to define the transmitted energy (I) more accurately, the emission spectrum of the X-ray source (X-ray tube in the case of micro-CT) should also be taken into account. A representative X-ray spectrum is shown in Figure 2.1.4 (Hammersberg *et al.*, 1998).

In general formula (1.2) can be written as

$$I = I_0 \int s(E) \, . \, e^{-\mu(E)x} dE, \tag{3}$$

where $s(E)$ is the spectrum of the X-ray source.

In contrast to MRI, X-ray absorption does not depend on the molecular structure (or the chemical formula) of the investigated material, but it is only affected by its elementary composition. When relating to biomedical applications, many organic compounds will not differ much in contrast as they are mainly constructed of low Z-number materials (C, O, H, N). With enhanced atomic numbers, linear attenuation coefficients increase dramatically. This is illustrated in Figure 2.1.5, where the linear attenuation coefficient of water and calcium are compared. Water and calcium were chosen because they both are representative constituents of tissue, water being the major component of living material, whereas calcium is present in all calcified tissue. The presence of calcium causes tissue to absorb so much that even slightly calcified tissues can be clearly separated from the rest of organic material. Physiologically, calcium is often associated with another dense element, phosphorus. Both are parts of calciumhydroxyapatite that is the key component in bone formation.

2.1.2.3 X-ray detection

In micro-CT scanners, a scintillator combined with a CCD (charge coupled device) camera is used for detection. Incident X-ray photons interact with the scintillator where they are absorbed and light photons are emitted. Light photons enter the fibre optic plate within the scintillator and are carried to the CCD where they are detected and converted into electrons. This CCD is a silicon wafer which is an electronic component segmented into an array of individual light-sensitive cells. Each cell is one element of the whole picture that will be formed. Consequently it is called a picture element or '*pixel*'. These picture elements are areas that detect X-rays independently and are separated in space. CCD line detectors, or so-called one-dimensional detectors (for example 1000 pixels located in one line), can be implemented. In this case, the investigated object needs to be moved and exposed several times to illuminate the whole volume. A lot of scanning time can be saved when two-dimensional (2D) detectors are applied (for example 1000 × 1000 pixels). In that case the whole object can be exposed at once. The modern trend for development of X-ray micro-CT detectors is to con-

Figure 2.1.5 Linear X-ray attenuation coefficients of water (black line) and calcium (grey line) as a function of energy

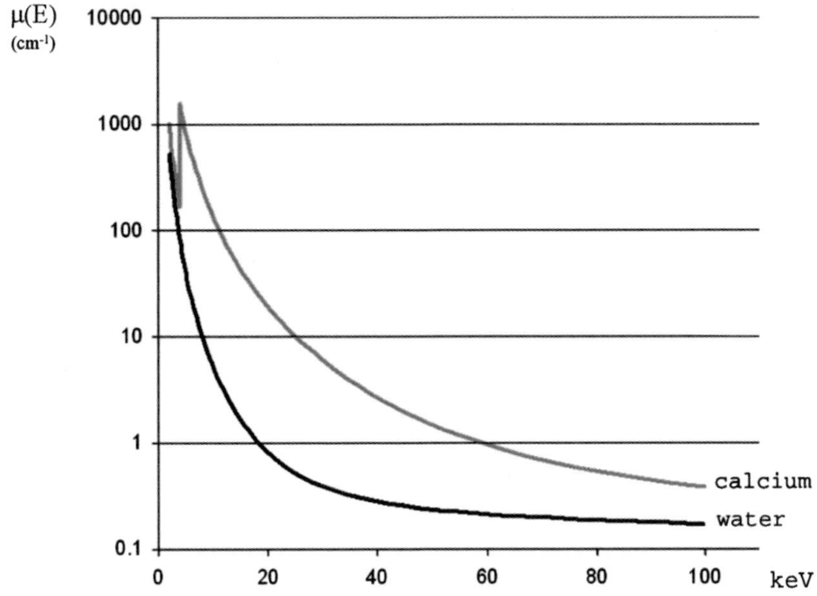

tain more and more pixels. Nowadays, up to 10 megapixel cameras are available (www.skyscan.be).

2.1.3 Image reconstruction

After acquisition, raw attenuation data are stored in the computer memory in a floating point matrix. A virtual slice is obtained by applying a reconstruction algorithm. The practical implementation of image reconstruction is always discrete. *Image reconstruction* from projections was originally described as the process of producing an image of a two-dimensional distribution (usually of some physical property) from estimates of its line integrals along a finite number of lines of known locations (Herman, 1980). Data from micro-CT analysis are composed of volume elements or *voxels*. A single virtual slice contains *pixels* or 2D image elements. Consequently, the process of image reconstruction is restricted to the definition of an array, that is a set of values that are associated with voxels of an image. To reconstruct a cross-section means to retrieve X-ray attenuation coefficients $\mu(x,y)$ from their integrals along the beam path.

An illustration of the simplest (but not the most accurate) reconstruction algorithm is the *back projection algorithm*. With this algorithm, it is sufficient to sum the intensities of all rays that pass through the reconstructed point. The shape of the object is defined by the intercept of the attenuation values that are back projected from the floating-point matrix and converted into a virtual slice. Thus, we obtain one single

Figure 2.1.6 Back projection of a point: Effect of increasing the number of projections. (a): 4 projections. (b): 8 projections. (c): 40 projections

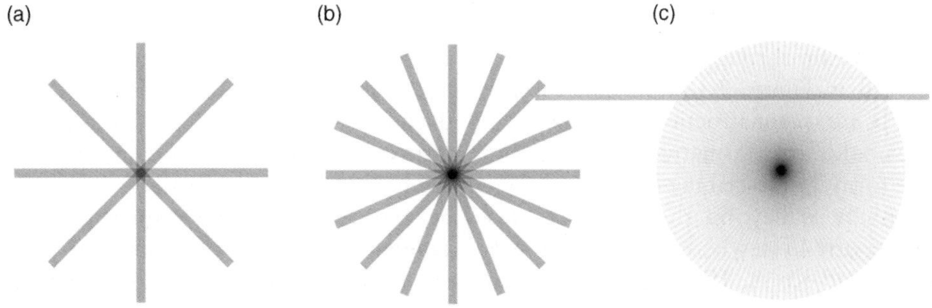

Figure 2.1.7 A diverging beam from a point source can mimic a parallel beam. For reconstruction, parallel rays from different positions of the source are taken into account (marked by the same colour)

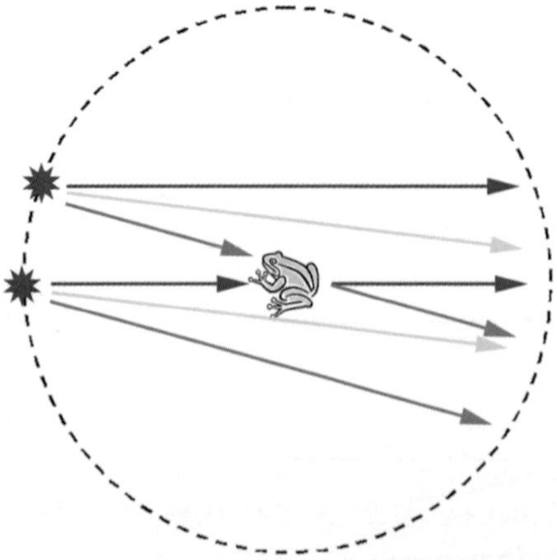

virtual cross-section. When repeating this, a series of cross-sections at different vertical levels through the object can be calculated. From Figure 2.1.6, it is easy to notice that the more the projections are taken into account for the calculation, the better is the contrast of the reconstructed point against the empty space. This can be achieved by rotating the specimen about its axis. Decreasing the rotation angle will increase the number of details that can be seen in the virtual cross-section. The discrete nature of CT limits the point size to the pixel size of the detector. However, the border of the reconstructed point is not sharp. The intensity of the beam of attenuated X-rays decays as an inverse ratio to the distance *(r); r* representing the distance the beam is elongated from the initial point where attenuation took place. This means that some traces of a point can be seen in places where physically they are not present. To reduce this artifact filtered back projection algorithm can be applied.

In our initial approach, the assumption was made that the X-rays emerged from the source as a parallel beam; however, most sources cannot generate parallel beams. In reality, a pointer source resulting in a fan beam through the object will be used as illustrated in Figure 2.1.7. In this case, a particular fan beam reconstruction algorithm is used. However, the exact mathematical background of these formulas describing different reconstruction algorithms is beyond the scope of this chapter and is reported elsewhere (Herman, 1980; Feldkamp, Davis and Kress, 1984; Gleason *et al.*, 1999).

2.2 Implementation

2.2.1 Analysis of virtual cross-sections

One of the most challenging questions for experimental micro-CT is a quantitative study of the reconstructed cross-sections. For a quantitative analysis of virtual cross-section, CT numbers are used. CT-number is defined as a value that is proportional to the average linear attenuation present in one voxel. CT numbers are displayed as grey values to be looked at in the visual field. In medical CT, an arbitrary scale is used routinely. In this case, CT numbers are compared to the attenuation value of water and displayed on a scale of arbitrary units called Hounsfield units (HU) named after Hounsfield. Water was assigned CT number 'zero', whereas air has number '−1000'. This scale appeared historically with the first medical scanner (Hounsfield, 1980). This choice could be easily explained. The human body predominantly has the density of water, and all organs (except bones and lungs) have small variations of several HU around the 'water background'. However, in the experimental situation the use of Hounsfield units can have its limitations as discussed elsewhere (Ketcham and Carlson, 2001).

2.2.2 Reconstruction artifacts

While reconstructing experimental datasets, there are many sources of distortions in the reconstructions originating from soft- and hardware, beam-hardening effect, partial volume effect, etc. These errors are called *artifacts*. Artifacts can dramatically affect quantification of the cross-sections once it is required to obtain information about intensities or shapes of an investigated sample. We shall briefly discuss sources of artifacts and methods that are used to correct them.

2.2.2.1 *Beam-hardening effect (BHE): Polychromatic sources*

Monochromatic (narrow energy band) sources are preferable for micro-CT but unfortunately they are not widely distributed. Synchrotron is an ideal source (Bonse *et al.*, 1992), but it remains expensive and with restricted access. In commercial micro-CT scanners, X-ray illumination is usually polychromatic. This is to be expected, as monochromatic X-rays cannot be generated separately: A monochromatic beam can only be obtained by cutting off all other energies. In this case even if strong characteristic lines are selected

Figure 2.2.1 Beam hardening effect during investigations of the human cochlea. (a): Beam-hardening is not corrected: The outer shape of the cochlea looks darker. (b): The same slice after the application of a polynomial correction

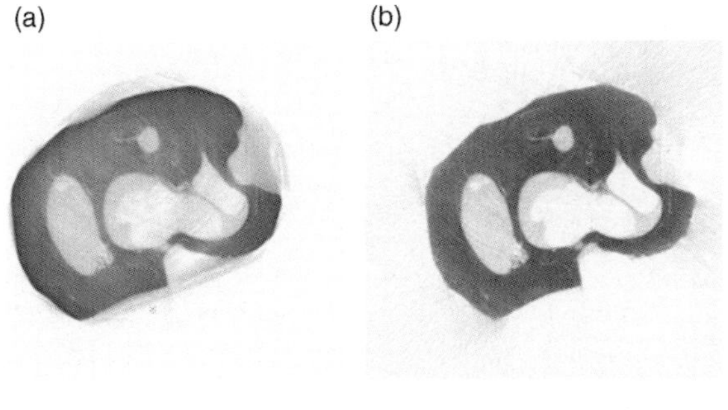

from the X-ray emission spectrum (cf. Figure 2.1.4), most of the power of the X-ray tube is not used. Some filters such as Al can remove the softest part of the spectrum, but normally, the beam remains polychromatic. Due to the polychromatic nature of the X-ray spectrum, beam-hardening effect (BHE) appears.

As a result of BHE, the outer surface of the sample usually seems denser than it really is, whereas the central part of the sample looks lighter as illustrated in Figure 2.2.1. This artifact can seriously affect quantitative measurements. To avoid BHE, a correction for the recorded signal is needed before reconstructing the image. Implementation of this correction requires different phantoms mimicking the composition and the density of the material studied.

Beam-hardening correction can be represented either as a table, which replaces the recorded CT number by another one or it can be defined as a polynomial function, the coefficients of which need to be determined for each particular situation as it is not possible to find one single correction for all different materials. An X-ray- spectrum that passes through a piece of metal will be attenuated differently than after a pathway through water. However, BHE correction can be accurate when a mixture of a particular material with a high linear absorption coefficient and another with a much lower absorption coefficient is used. As an example, we can study bone surrounded by tissue, considering bone as dense and tissue as a mixture of light materials (Postnov *et al.*, 2003).

2.2.2.2 Ring artifacts: X-ray detector under illumination

Another frequent artifact in virtual cross-sections is the presence of concentric rings that corrupt the quality of the picture and contain no useful information. Ring artifacts are clearly visible in Figure 2.2.2(a, b). These defects are always present being strong or negligible compared to the effective signal. The cause of ring artifacts is the detector. Every pixel should record the same signal in the beginning and at the end of the experiment if it is illuminated in the same way. In fact, the sensitivity of a pixel drifts with temperature. Even the shadow of investigated objects can affect the temperature of the matrix. All pixels in the X-ray camera initially have different sensitivity because it still is not possible to produce exactly identical pixels, although manufacturers try to make the sensitivity of each pixel as stable as possible. These irregularities in sensitivity are corrected by a procedure called *flat field correction* (FFC). However, when the temperature changes FFC does not work properly. Besides flat field correction, mathematical procedures have been described to eliminate remaining ring artifacts after data acquisition (Sijbers and Postnov, 2004).

Severe defects of CCD of the detector can lead to artifacts such as those presented in Figure 2.2.2 (c, d). These defects cannot be corrected because some pixels of the CCD (or optical waveguides to these pixels) are broken and do not register any X-ray photons or because their sensitivity is insufficient.

2.2.2.3 Partial volume effect

When a voxel represents more than one type of tissue, or a border between tissue and void, the CT number that is attributed to it represents some average of the attenuation properties of the tissues present in the voxel. This effect is referred to as the partial volume effect. As a consequence of this, the boundaries between different tissues can become blurred. What is even more important is that the values of the intensities in the borders separating tissues are not accurate: Half bone and half air in one pixel result in an intensity that is closer to air, but that is not identical to half of the sum of the intensities caused by both materials.

2.2.2.4 Geometrical distortions

Geometrical distortions usually appear if the CCD sensor and scintillator plate are separated by an optical waveguide. In that case, the detected shape can be skewed. This is illustrated in Figure 2.2.3(a). To correct for these distortions special grids were scanned to calculate corrections for every pixel (Figure 2.2.3(b)). This is of key importance for 2D X-ray CCD detectors.

Figure 2.2.2 Ring artifact correction illustrated in a cross-section through a human tooth. (a): Ring artifacts are not corrected. (b): Ring artifacts disappear after the implementation of the correction. (c): Ring artifacts caused by severe defects of the camera. (d): Correction methods cannot improve the picture

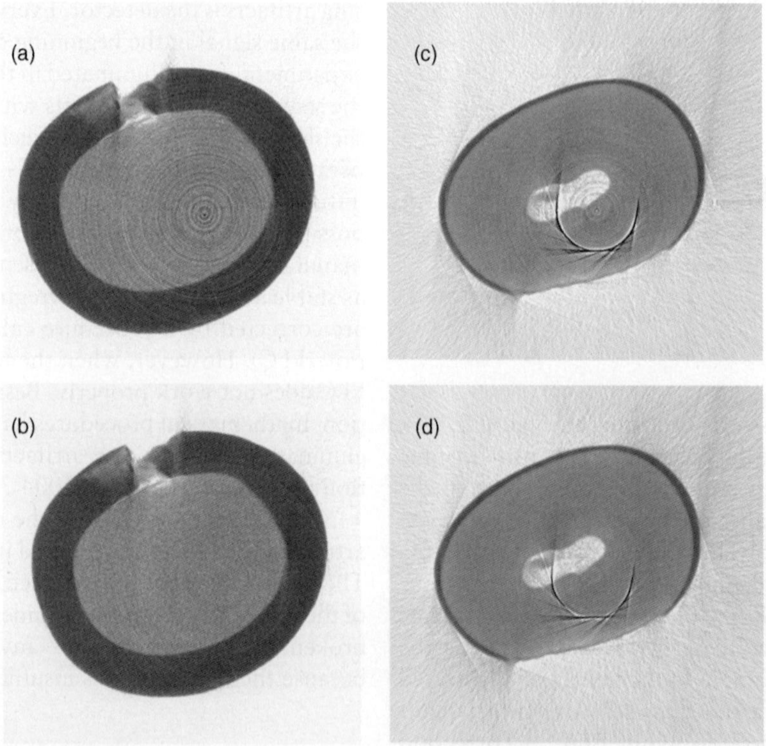

2.2.3 Segmentation

A major problem for the analysis of CT images is the process of segmentation. In the reconstructed virtual cross-sections, different tissues have to be separated from each other. For instance in bone scans, bony structures have to be distinguished from non-bone. This is not an easy task because it is complicated by noise, limitations in resolution and the beam-hardening effect. Inappropriate segmentation may induce artifacts in the interpretation of structural components. It is particularly important in the *in vivo* situation where resolution is lower and where the pictures possess much more noise. Different methods can be implemented to separate bone from non-bone. Recently, a method of local thresholding was applied successfully in the *in vivo* situation (Waarsing, Day and Weinans, 2004b).

2.2.4 Differences between micro-CT and medical CT

Although the same basic principles as in medical tomography are implemented, micro-CT is very different regarding the tasks that can be fulfilled. There are three

Figure 2.2.3 Geometrical distortions: Image of a grid. (a): Initial image of the grid without correction. (b): The grid after the implementation of a correction map for every pixel

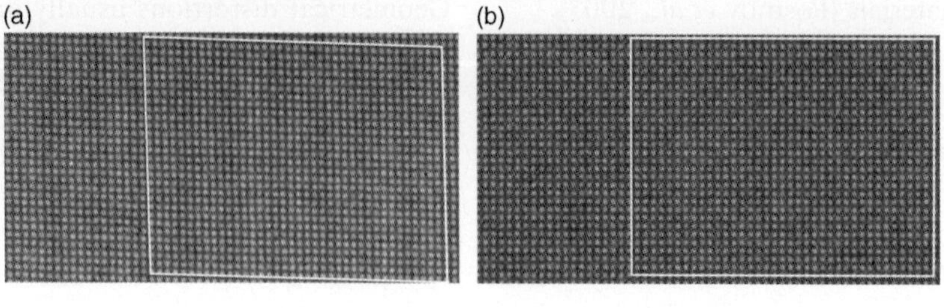

main differences: *Resolution*, *size* of the investigated object and *energy used*. All these key features are interrelated with each other. Further discussion will enable us to understand the advantages and new possibilities of micro-CT together with its limitations.

2.2.4.1 *Resolution*

Resolution of micro-CT, as can be understood from its denomination, is much higher than that of medical CT. In the newest desktop scanners the resolution can even be sub-micron: In that case they are referred to as nano-scanners (Sasov, 2004). However, typical resolution of micro-CT is about 10 μm being $100 \times 100 \times 100 = 1.000.000$ times better than in a medical scanner. In most micro-CT installations, resolution is isotropic. This is one of the key advantages allowing much more advanced quantitative analysis of obtained cross-sections and of 3D renderings.

Resolution is defined either by the source spot size (object is closer to the source than to the detector) or by the pixel size of the detector (radiography). The spot size of the microfocus X-ray tube used in laboratory scanners is of the level of 5–10 μm.

We should stress that the resolution of medical CT is restricted mostly by the X-ray dose limitations for the patient. To improve resolution, more X-ray photons should be absorbed in given volume, which can become a risk to the patient. Signal-to-noise ratio (SNR) is proportional to $\sim(N)^{1/2}$, where N is the detected signal. This implies that if you want to increase SNR two times, you need to absorb four times more photons. To gain a two times increase in spatial resolution while keeping the same image quality, the required X-ray dose should be enhanced $2^3 = 8$ times (two times in all three dimensions).

2.2.4.2 *Size*

Sizes of the samples are limited by the detector size. It is not possible to study an object that exceeds the dimensions of the detector because the basic principle of tomography, requiring that every part of an object must be illuminated from all directions, is not fulfilled. In high-resolution *in vitro* systems, typical dimensions are about 0.2–20 mm (Sasov and Van Dyck, 1998), but in some *in vivo* systems they can get up to 80–100 mm (Russo, 1998; Paulus *et al.*, 1999; Sasov, Dewaele and De Clerck, 2001; Lee *et al.*, 2003). Depending on the configuration of the scanner, resolution is determined by the size of the object. This implies that in order to obtain a large magnification the smallest possible object is required. An adequate combination between smaller sizes and high resolution opens new perspectives for innovative developments in micro-CT.

2.2.4.3 *Energy range*

Smaller resolved voxel sizes require sufficient signal to be absorbed in them. High-energy X-rays are well transmitted through a sample, and hence absorbed less. For micro-CT, the commonly used energies range between 40 and 100 keV.

Different energies result in various levels of contrast. Contrast can also be improved by implementing metal filters (Al, Ti, Cu) in front of the source removing softer X-rays. Soft X-rays are absorbed strongly and can always enhance contrast. Yet, the problem is that they can be totally absorbed, creating BHE instead of resolving fine details. Combination of the filter and the energy should be optimized for every object according to the requirements for the resulting contrast.

If the object is too thick or too dense for softer X-rays, then they should be removed in advance. For example, bones are too dense to be studied without Al filter. If no filter at all is used, strong BHE will appear. On the contrary, once ultra-soft material is studied such as dried lung tissue, the implementation of an Al filter will result in no signal at the detector, as all soft X-rays that could be absorbed in the carbon and nitrogen of soft organics were pre-filtered.

2.3 Contribution of microtomography to biomedical imaging

From the point of view of micro-CT imaging, biomedical preparations can be divided into different categories with specific challenges, depending on the X-ray density of the samples to be studied. The situation is summarized in Figure 2.3.1: The density of several biomedical applications is compared to the density of water and air. In the following discussion, we shall refer to this classification. For each field of interest, a distinction will be made between the *in vitro* and the *in vivo* situation, as there are major differences in data acquisition between both scanning conditions.

It is evident that micro-CT has ample applications in other fields than in biomedicine. However, this falls beyond the scope of the present chapter. The readers who are interested can find more information about

Figure 2.3.1 Classification of biological samples for imaging by micro-CT. Different densities are taken into account

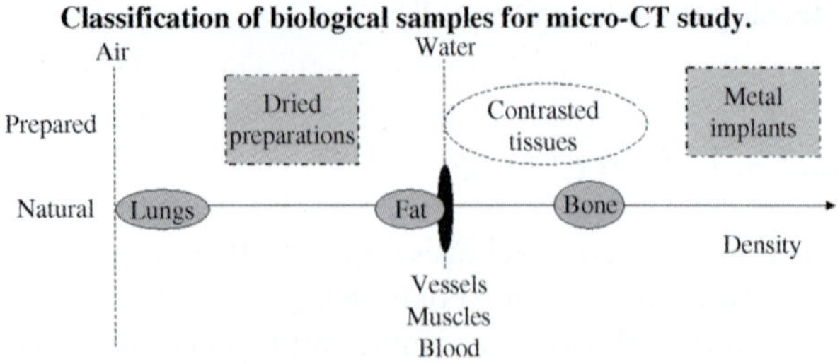

these subjects elsewhere (Sasov and Van Dyck, 1998; Ketcham and Carlson, 2001).

2.3.1 *In vivo* versus *in vitro* imaging by micro-CT

As mentioned in the introduction, most desktop micro-CT systems were developed initially for applications in materials science (Sasov and Van Dyck, 1998). Soon biologists and researchers in biomedical sciences became interested in the application of this imaging facility. The possibility to obtain 3D information during the lifetime of an animal opens wide perspectives for the longitudinal analysis of the evolution of several biomedical processes. It should be mentioned that the access to *in vivo* micro-CT is recent. Therefore, many efforts are made to obtain a validation of this *in vivo* imaging by *ex vivo* analysis.

One of the first *in vivo* scans was performed on live snails (Postnov *et al.*, 2002a) in a conventional *in vitro* micro-CT system. The snails were so small that they could easily be mounted in the system. A resolution of 10 μm was achieved. As discussed elsewhere (cf. section 3.2.1), resolution in this scanner is determined by the size of the object. Growth, development and regeneration as a function of time were studied in two species of snails that were known to grow fast. Their calcified shells developed simultaneously with their body. Comparison between 3D images of the animals at different moments during their lifetime showed how the shells grew and how they regenerated, when artificially damaged (Figure 2.3.2). These initial results clearly demonstrated that polychromatic desktop X-ray microtomography could be successfully applied to live animals.

It is obvious that the major interest of biomedical researchers goes to non-invasive imaging of small live laboratory animals such as rats and mice. As discussed previously (Davis and Wong, 1996), studies on small live animals can be very promising in therapeutic trials as the statistical uncertainty caused by inter-animal variation disappears. The accuracy of such studies can be improved by *in vivo* scanning. Non-invasive imaging would also induce an important reduction in the number of experimental animals that need to be sacrificed and studied. Therefore, a new series of typical *in vivo* scanners was developed with a special scanning geometry, and animal holders adapted to small laboratory animals (Russo, 1998; Paulus *et al.*, 1999; Sasov, Dewaele and De Clerck, 2001). Needless to say that new perspectives for *in vivo* scanning are opened in several animal models, especially in genetically manipulated mice (Nolan *et al.*, 2000). The fact that a spatial resolution up to 9 μm can be reached is an advantage to study tiny structures in mice.

Figure 2.3.2 *In vivo* investigations of growth and regeneration of a snail. (a): Normal growth of a snail. Dark part indicates the part that grew in 21 days. Two 3D models created at two different times were superimposed. (b): Regeneration of the damaged shell. From Postnov *et al.*, 3D *in vivo* X-ray microtomography of living snails *J. Microsc.* 205 (2), 201–205. With permission by Blackwell Publishing Company

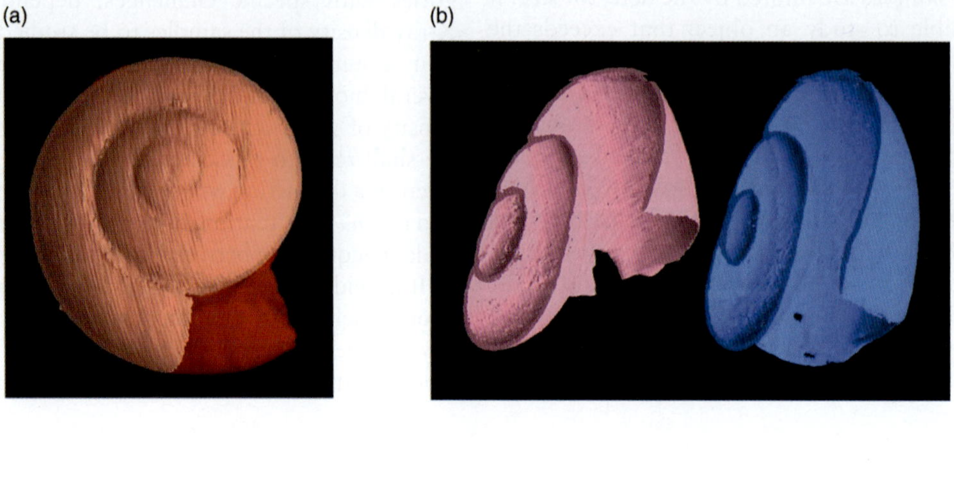

High-resolution *in vivo* micro-CT requires that the animal be restrained during acquisition. Global anaesthesia is required. Several procedures for anaesthesia have been described elsewhere (Flecknell, 1993). In summary, one can choose between injection method or gas anaesthesia. When gas anaesthesia is applied, forced respiration can be imposed on the animal as well as synchronization of the breathing cycle. It should be mentioned that intubation is an invasive procedure. Moreover, it should be kept in mind that general anaesthesia itself is a trauma to rats and mice, causing among other effects growth retardation (Salmon *et al.*, 2001).

Another harmful effect to the experimental animals is the application of X-rays (Dowseth, Kenny and Johnston, 1998) where the ionizing effect can result in immediate radiation damage and in long-term genetic damage.

The question of radiation dose during *in vivo* scanning is still under debate. Recently, the radiation dose to be delivered locally during a 20-min hindlimb scan of a rat in a desktop *in vivo* scanner was reported to be 400 mGy (Salmon and Sasov, 2005). Some authors (Spadaro *et al.*, 2003) suggested that in bone growth studies the administration of a radioprotectant drug might be useful. However, experience with *in vivo* scanning learns that single and repetitive scans after a time interval do not seem to have harmful effects. More research is required to answer these questions.

There still remains the possibility of scanning a whole animal *ex vivo*. Approximately one hour after sacrifice, this preparation becomes stable, and in this case movement artifacts can be avoided.

2.3.2 Bone

Bone growth retardation or age related changes in bone such as osteoporosis or other hormone mediated influences on the skeleton can seriously affect the mechanical stability of bone and eventually lead to bone fracture. These health problems will become more important as the current population is ageing, and bone fractions will represent a considerable economic burden as reviewed previously (Cumming, 1998). Osteoporosis in particular is a chronic, progressive bone disease caused by metabolic dysfunction. Current techniques for diagnosing osteoporosis and other bone diseases have been based on non-invasive measurements of bone mineral density (BMD) (Odgaard, 1997). BMD has a reasonable correlation to fracture risk as long as it refers to large populations but is less reliable for predicting fracture riks in individuals. The efficiency of treatment can only be evaluated on a statistical basis. Individual evaluation is required as reported elsewhere (Cumming, 1998). In addition, BMD alone is not sufficient as a measure for the overall bone quality as this is determined by its structural and material properties such as bone mass, geometry, architecture and composition of the bone (Einhorn, 1992). Therefore, both *in vitro* and *in vivo* imaging and 3D rendering of healthy and diseased bone became important.

2.3.2.1 *In vitro micro-CT analysis in bone*

As with synchrotron illumination (Nuzzo *et al.*, 2003), micro-CT using laboratory polychromatic X-ray sources already has several *in vitro* applications in bone research (Rüegsegger *et al.*, 1976; Rüegsegger, Koller and Müller, 1996; Elliott *et al.*, 1997; Postnov *et al.*, 2003). Due to the high X-ray attenuation coefficient of calcium, micro-CT is most suitable to study bone and calcified tissues (Postnov *et al.*, 2003).

Figure 2.3.3 Virtual cross-section through trabecular (a) and cortical bone (b) in a mouse femur

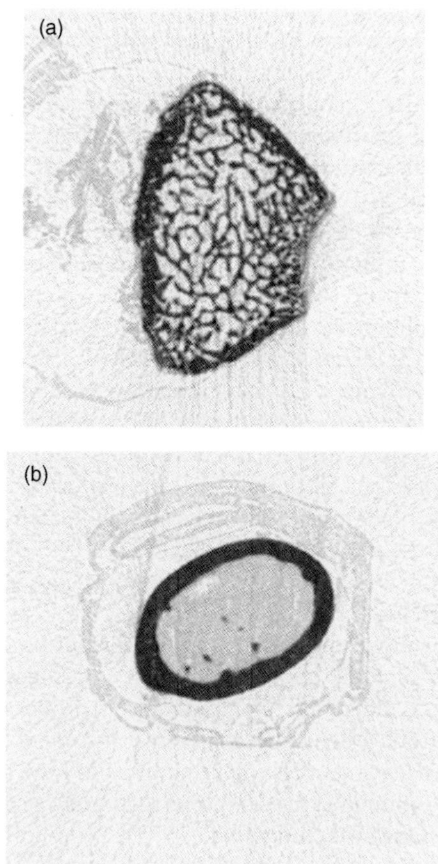

Figure 2.3.4 3D model of a mouse femur with a quarter opening. The model was cut after reconstruction. Trabecular structure is clearly visible

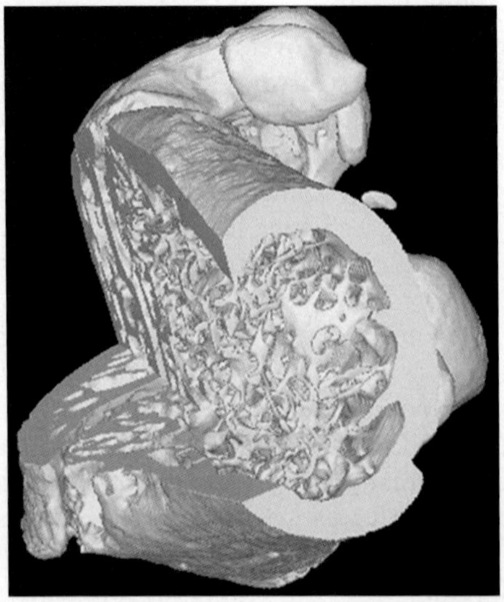

While explaining the principles of micro-CT, the possibilities were illustrated in a bone in Figure 2.1.1. A shadow picture, a virtual cross-section and a 3D model were shown. An advantage of micro-CT is that cortical bone can be distinguished from trabecular bone with its spongy structure as illustrated in Figure 2.3.3. Calculation allows quantification of several bone parameters based on the analysis of the virtual slices and the 3D model. These 3D models can be cut in any arbitrary direction, without loss of resolution when data acquisition was isotropic. This is shown in Figure 2.3.4. Besides a visual inspection of both virtual slices and 3D models, calculation of 3D morphometric parameters can be used as a quantification (Odgaard, 1997). In Table 2.3.1, the contribution of micro-CT to bone analysis is summarized.

To describe the architecture and condition of trabecular bone, volume fraction (bone volume/tissue volume) is an important parameter derived from micro-CT analysis. When the proper threshold (cf. Section 2.2.3.) was used to determine the 3D data, micro-CT was reported to be highly accurate (Ding, Odgaard and Hvid, 1999).

As mentioned above, it is a convenient practice to express mineral content as density of the sample. Therefore, micro-CT images require a proper density calibration. When expressing mineral content as physical density, the question remains as to how to measure the volume of the sample as accurately as possible (Ding, Odgaard and Hvid, 1999). Non-uniformity within the bone and porosity, especially in cancellous bone, remains a serious problem to determine volume (Odgaard, 1997). In addition, calibration of X-ray attenuation data is dependent on the type of scanner. In the *in vitro* situation quantitative analysis, in particular density measurements, became possible (Rüegsegger *et al.*, 1976; Davis and Wong, 1996; Postnov *et al.*, 2002b; Postnov *et al.*, 2003). Due to the polychromatic nature of the X-ray source, beam-hardening effect (cf. Section 2.2.2.1.) had to be corrected to calibrate the grey values in the virtual cross-sections. As their chemical composition strongly resembles bone, hydroxyapatite crystals (Ca_{10} $(PO_4)_6(OH)_2$) with different physical density were used as representative phantoms for bone. Besides measurement of overall density, it also became feasible to distinguish between different density windows in bone (Postnov *et al.*, 2003). This is illustrated in Figure 2.3.5. However, determination of local calcium density in every voxel still remains a challenge for micro-CT. Estimation of the standard deviation of the distribution of CT numbers in cross-sections of bones is needed taking into account partial volume effect, beam hardening, ring artifact correction, etc. in order to obtain local density in each pixel with an error distribution.

By comparison of micro-CT slices of bone biopsies with histological sections, it was shown that micro-CT measurements are representative for trabecular microstructures. In several studies of trabecular bone (Müller *et al.*, 1996; Ding, Odgaard and Hvid, 1999), micro-CT data were validated by conventional

Table 2.3.1 Contribution of micro-CT to bone analysis

Calcium balance	Bone architecture
Total Calcium amount (in grams)	Bone surface to volume ratio BS/BV
Calcium density (overall/local)	Trabecular thickness Tb.Th.
	Trabecular separation Tb.Sp.
	Trabecular number Tb.N.
	Trabecular pattern factor TBPf
	Degree of anisotropy
	Structure model index
	Relative bone volume BV/TV

Figure 2.3.5 Illustration of the application of a density window. (a): 3D model of a bone. (b): The same model is made semitransparent to visualize the densest part of a bone. (c): Model of the densest part of a bone is shown separately

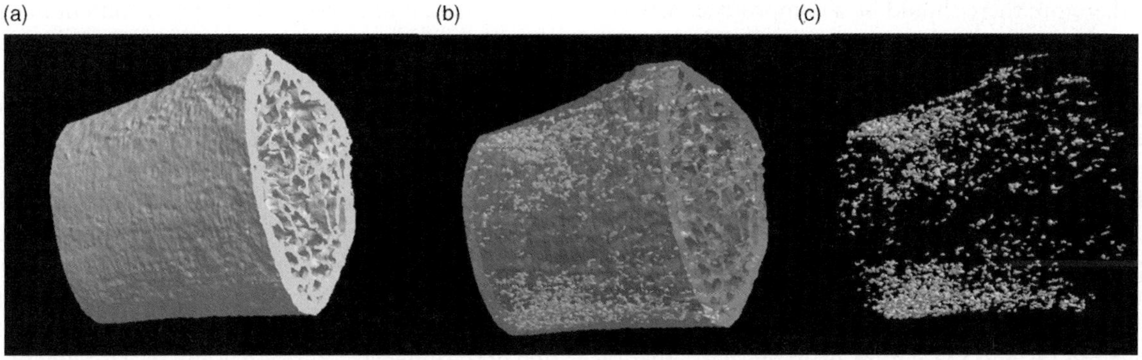

histology. Comparison between micro-CT and microradiography showed that micro-CT can be used for measuring human cortical bone porosity, although a higher resolution would improve this analysis (Cooper *et al.*, 2004). An important issue for the comparison of micro-CT with histology is the accurate correlation of the slices. Therefore, a correct alignment during scanning and actual cutting is required. Thickness of the knives is also a complicating factor. However, the actual resolution in histology is higher than in CT.

It is evident that bone analysis, both qualitatively and quantitatively, also reaches a higher resolution with synchrotron illumination (Nuzzo *et al.*, 2003). Yet, micro-CT offers a low-cost alternative solution in laboratory surroundings.

2.3.2.2 *In vivo scanning of bone*

Nowadays, imaging by high-resolution polychromatic desktop micro-CT has become feasible in live animals (Kennel *et al.*, 2000; De Clerck, Van Dyck and Postnov, 2003). Fast acquisition scanners even allow visualization of a whole mouse in approximately 1 min with a reduced spatial resolution (www.skyscan.be). Some *in vivo* experiments using synchrotron illumination were described previously (Kinney, Lane and Haupt, 1995; Lane *et al.*, 1998). It is important to conduct longitudinal *in vivo* experiments in animal models with different metabolic conditions affecting the quality of bone. Cortical and trabecular bone, together with its micro-architecture needed to be quantified as a function of time in live animals. In the *in vivo* situation each animal can serve as its own control. Such studies can open wide perspectives for the development of an appropriate pharmacological treatment for a specific bone disease as well as for the longitudinal follow-up of the therapy. They may also help to develop better diagnostics. A huge variety of animal models for several bone diseases is available (Wang *et al.*, 2001) including genetically manipulated mice (Nolan *et al.*, 2000). Recently, a longitudinal *in vivo* desktop scanning study (Waarsing *et al.*, 2004a) was performed in ovariectomized rats. In these experiments, scans of the proximal tibia of living rats were evaluated as a function of time by means of image registration.

The accuracy of imaging by micro-CT was validated by a number of studies: Comparison with histology could show deviations in 2D parameters depending on the resolution of scanning and depending on the segmentation method that was used (Laib and Rüegsegger, 1999; Kennel *et al.*, 2000; Waarsing, Day and Weinans, 2004b). However, in contrast to the large number of *in vitro* micro-CT studies in bone, scanning data in live animals remain rather scarse. This may be due to the fact that *in vivo* scanning has specific problems. Difficulties arise while thresholding the OI (object of interest) as the signal-to-noise ratio (SNR) is usually low. An improved segmentation method for *in vivo* micro-CT was developed (Waarsing, Day and Weinans, 2004b). To draw trabecular borders accurately and to reduce the influence of noise and partial volume effect, a local threshold segmentation algorithm was applied. In this case, local threshold has an advantage to global threshold in the sense that thin trabeculae with low attenuation can still be selected as bone. Also, statistical noise that is always present in the cross-sections is discarded even if the 'density' of this noise is high. The dynamical range of the object (or part of it) overlaps with the dynamic range of the background or the surrounding tissue. In most cases while scanning living material, the dose absorbed by the animal should be as small as possible, and therefore there should be a compromise between the quality of the sections (hence the SNR) and scanning time. Reduced quality of the cross-sections merely leads to qualitative analysis whereas quantitative information is required in trabecular bone in particular.

Especially in the *in vivo* situation where particular movement artifact cannot always be excluded (e.g. respiration and cardiac contractions), virtual cross-sections are often reconstructed with blurred borders. Although this provides some information about the OI and the nature of its borders, there is a strong need to draw a strict sharp border of trabecular bone. Accurate selection of the bone, including very thin trabeculae close to the resolution limit, is a tool of vital importance for further quantitative analysis of the trabecular bone parameters. A thorough and convenient algorithm can establish a 'gold standard' for micro-CT analysis of bone tissue. On top of mathematical problems of the reconstruction algorithm, scattering and reflection of X-ray photons contribute to blurring. A reduction of ring artifacts either on shadow images or on the virtual cross-sections should be implemented. Only images that are free of any trace of hardware instability can be analyzed quantitatively with reproducible results.

An additional challenge to further quantify bone scans, in the *in vivo* situaton should include a density calibration procedure together with the development of appropriate of hydroxyapatite phantoms for longitudinal *in vivo* scanning. These experiments should be further extended in the near future.

2.3.3 Calcified tissues other than bone

2.3.3.1 Micro-CT in dental research

As teeth are composed of calcified tissue, micro-CT became a suitable tool in dental research and education (Davis and Wong, 1996). Recently, 3D reconstructions of maxillary molars based on micro-CT images accurately showed the overall external and internal macromorphology of the teeth (Björndal *et al.*, 1999). These reconstructions were of high quality without destroying the teeth. Validation of micro-CT data with standard techniques proved that there was a good correlation between the number, positioning and cross-section of root canals as visualized by micro-CT. From this same study, it was also concluded that 3D models might improve pre-clinical training in endodontic procedures (Björndal *et al.*, 1999). Similar as in bone, the mineral content in teeth also has been analysed by micro-CT (Davis and Wong, 1996). In addition, micro-CT scanning was used for the early detection of hidden caries (Bottenberg *et al.*, 2003). Figure 2.3.6 shows an example of a micro-CT slice through a tooth with caries, together with the histological validation. Despite the fact that the tooth is affected, the enamel can remain smooth and undisturbed when inspected visually. An apparent intact surface enamel hides the slowly progressing lesion formation. Therefore, non-invasive visualization of the inner structure is a useful tool for diagnosis. Although histological sections are regarded as the 'gold standard' in caries research, micro-CT imaging is a viable alternative to histological validation (Bottenberg *et al.*, 2003). As mentioned before, histological slicing is a time-consuming procedure, involving disruption of tissue and loss of all information about the overall shape due to sectioning, as the teeth are irreversibly destroyed during preparation. Moreover, hard tissue sections need to have a minimal thickness to avoid fracture. In contrast to histology, micro-CT visualizes the distribution of calcium density with a linear correlation between CT numbers in the pixels.

Figure 2.3.6 Application of micro-CT in dental research. Comparison between micro-CT scanning and histology. (a): Tooth with caries. (b): Healthy tooth. Notice artificial cracks in enamel as a result of histological preparation

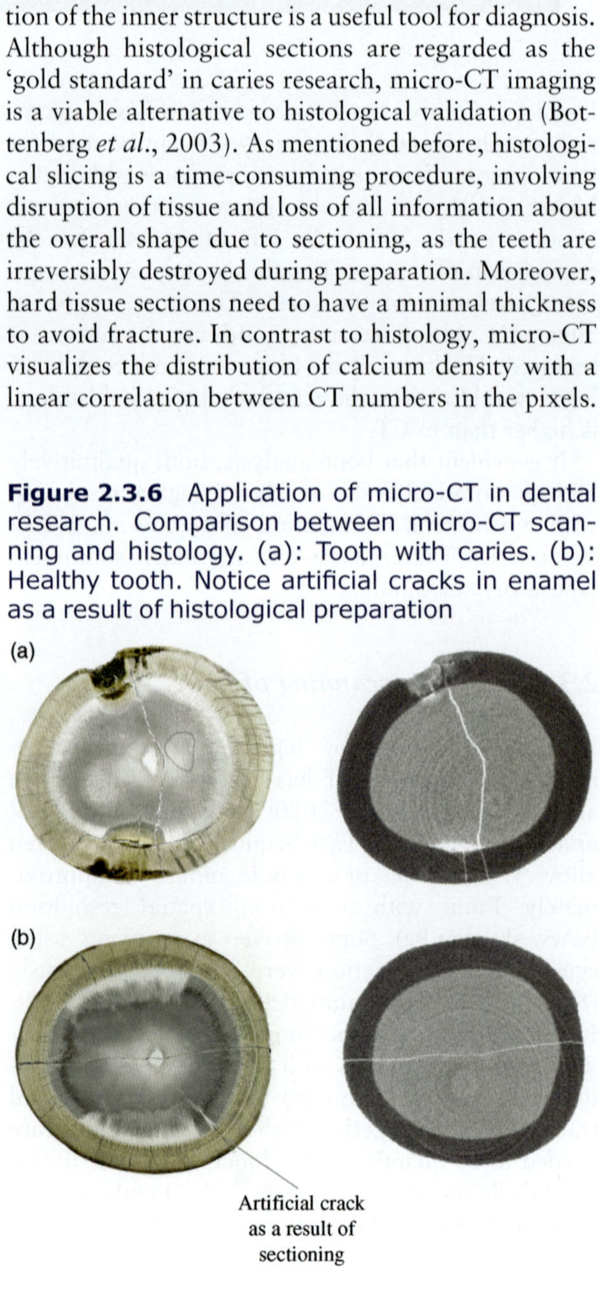

Real free-of-artifact tissue damage can be seen and studied. Up to now, micro-CT studies in dental research remain limited to the *ex vivo* situation.

2.3.3.2 *In vivo scanning of calcified tissues other than bone*

Another interesting biomedical application for *in vivo* micro-CT is the detection of calcified inclusions. Such calcifications may occur in blood vessels or in the heart in cardiovascular disease. In a preliminary report (Persy *et al.*, 2004), it was shown that in large blood vessels in live rats, calcifications could be detected by micro-CT. At present, further research is required to clarify the detection limit of these calcifications by micro-CT.

2.3.4 Imaging of soft tissues by micro-CT

2.3.4.1 *In vitro scanning of soft tissues*

As summarized in Figure 2.3.1, micro-CT is most suitable to visualize hard tissues. In contrast to these calcified structures, it is very difficult to differentiate by micro-CT imaging between different kinds of isolated soft tissues without sample preparation. At present, resolution is not high enough to visualize individual cells. As optical density of soft tissues and water are similar, isolated soft tissues cannot be scanned in saline or any watery solution.

However, it proved possible to build 3D models of a rat embryo (Boyde, De Clerck and Sasov, 2000). Micro-CT was also successfully used in dried lungs, where a distinction could be made between tumours and healthy lung tissue due to the increased density within the tumours. Figure 2.3.7 shows a virtual cross-section of a freeze-dried lung where tumours can be recognized as dense spots surrounded by normal tissue. In excised lungs, the airlines and bronchial tree could also be distinguished from the surrounding soft tissue. This is illustrated in Figure 2.3.8, which was obtained without any additional staining.

As expected, calcified inclusions can be detected in soft tissues. A representative example is shown in Figure 2.3.9, where the localization of calcifications in the human pineal gland was visualized (Postnov *et al.*, 2002b). The human pineal gland has been known as a mineralizing tissue, where the concretions are also composed of hydroxyapatite (Luke, 2001).

Figure 2.3.7 *In vitro* detection of lung tumours in freeze-dried lungs in a mouse. (a): Cross-section through a healthy lung. (b): Lung tumours can be identified as dense spots

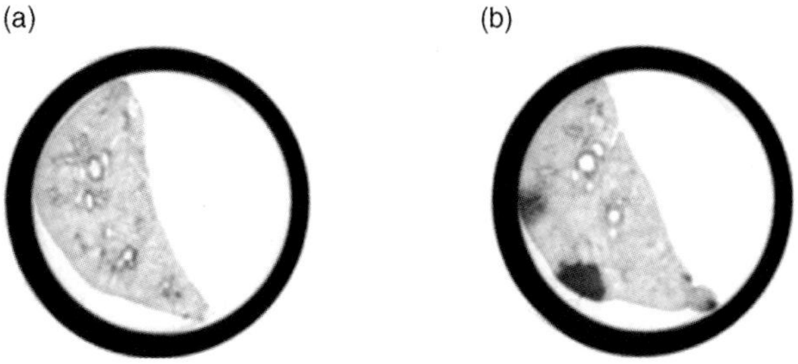

2.3.4.2 *In vitro analysis of soft tissue after sample preparation*

For *in vitro* imaging by micro-CT, several contrast agents or staining techniques were applied to visualize anatomical structures composed of soft tissues. Micro-CT allows to study their spatial distribution and architecture in the 3D space. In this context, myocardial muscle orientation, vasculature in skeletal muscle, coronary arteries and the renal vascular system have been visualized after staining. 3D rendering revealed their spatial orientation. PbO_4 suspended in a silicon polymer was used as an X-ray contrast agent (Jorgensen, Demirkaya and Ritman, 1998).

Karau *et al.* (2001) reported measurement of the arterial dimensions and location within the intact rat lung. In this report lungs were excised from rats and

Figure 2.3.8 3D representation of the bronchial tree of excised lungs of a mouse

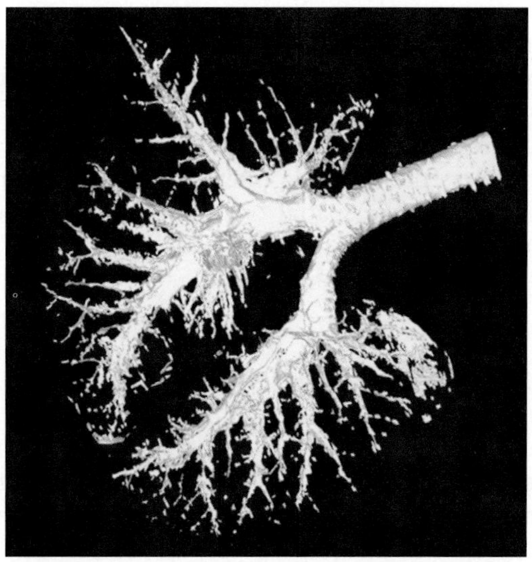

Figure 2.3.9 Detection of calcifications in excised human pineal gland tissue. (a): Virtual cross-section with calcifications as dark spots. (b): Spatial position of these calcification can be observed on a 3D model. Pineal glands were kindly provided by Prof. S. Saveliev, Institute of Human Morphology, Moscow

(a) (b)

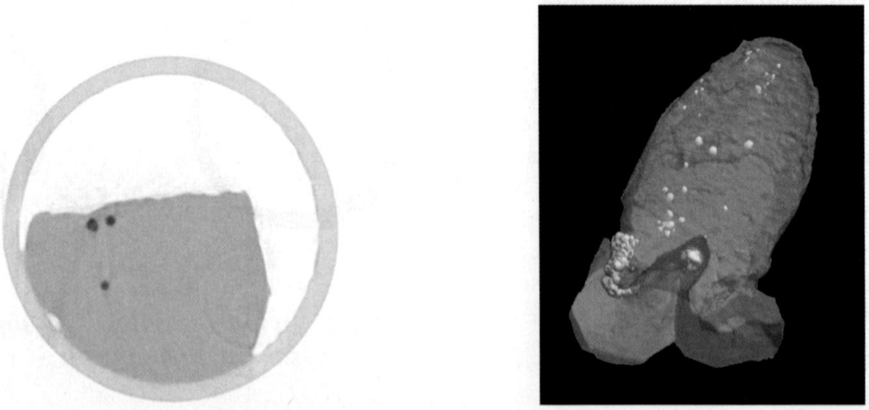

the pulmonary arterial trees were filled with an agent (perfluorooctyl bromide) as to enhance X-ray absorption. Fixation of lung tissue with formalin vapour and staining with silver nitrate can be a possibility to enhance X-ray contrast. High-resolution micro-CT was also used for accurate measurements of airway dimensions and airway narrowing in excised canine lungs (McNamara *et al.*, 1992).

Injections of a barium sulfate-gelatine-thymol mixture into the pulmonary microcirculation may be another example. In high-resolution micro-CT, osmium tetroxide may also improve the images as reported previously (Ritman, 2002). Recently, an overview has been published (Langheinrich *et al.*, 2004) summarizing several possibilities to visualize 3D models of the vessel wall and soft tissue architecture using different contrast perfusion and staining techniques to enhance X-ray contrast.

In these *in vitro* studies, the major advantage of micro-CT is the possibility to build 3D models. Needless to say that these experiments do not belong any longer to the category of non-invasive imaging, but nonetheless they provide interesting scientific information about the spatial organization of structures which may have an important impact on their integrated function and physiology.

2.3.4.3 *In vivo scanning of soft tissues*

The scanning situation of soft tissues in live animals becomes different from isolated preparations. *In vivo* scans of the whole body of small laboratory animals are possible (Russo, 1998; Sasov, Dewaele and De Clerck, 2001). Despite cardiac contraction and respiration, it is also feasible to scan the chest area of anaesthetized small laboratory animals. Intestinal movement may cause blurring of the virtual cross-section through that region.

As reported previously (Paulus *et al.*, 2000; De Clerck *et al.*, 2004), micro-CT is able to detect lung tumours in live animals. This will be discussed in a separate report. In contrast to the *in vitro* situation, an accurate estimate of the actual resolution of *in vivo* micro-CT is difficult as it is only possible for motionless objects. In the scans, the actual movement in every part of the cross-section is not known. In the heart region, tissue motions are much more pronounced than in areas closer to the vertebral column. As a result, some of the structures may become blurred and look larger than their actual size. As discussed above, movement artifacts may distort contours and shapes. That is the reason why it is very difficult to make a quantitative analysis of the virtual cross-sections obtained in live animals.

2.3.4.4 *In vivo scanning: contrast enhancement*

Due to the low X-ray contrast in soft tissues, the application of contrast agents is required for an improved imaging of organ systems composed of soft tissues. In many cases an important issue was to visualize the vascular bed of several organs. A number of contrast agents containing barium or iodine have been used (Ritman, 2002). Figure 2.3.10 illustrates the effect of injection of a contrast agent into a live

Figure 2.3.10 Contrast enhancement after injection of a contrast agent in a live mouse. (a): Shadow image of the whole mouse. (b): Contrast improvement in the heart area (top panel control animal, bottom panel mouse injected with contrast agent). (c): Contrast improvement in the kidney area (top panel control animal, bottom panel mouse injected with contrast agent)

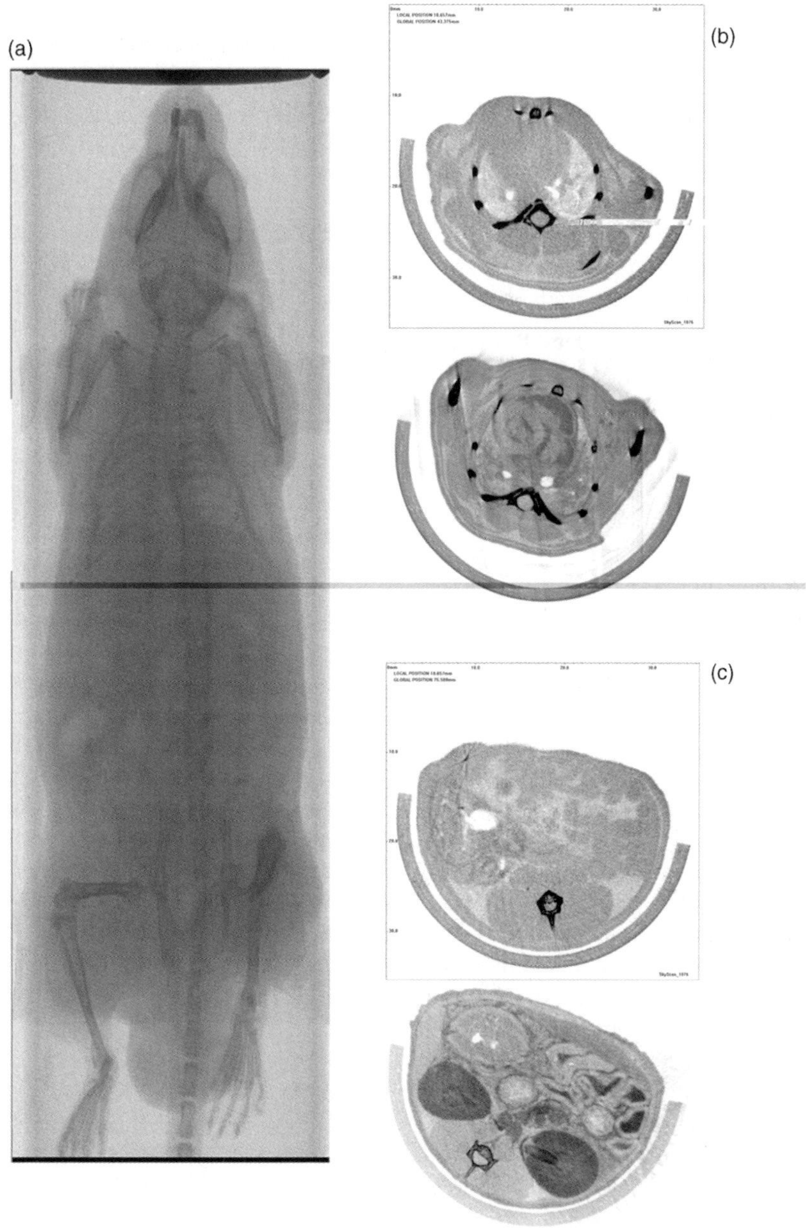

anaesthetized mouse in our laboratory. A solution containing iodine as a major radio-opaque material was injected through a catheter inserted in the tail vein. In the cardiovascular system, large blood vessels and the septum in the heart, as well as the kidneys (cortex and medulla) could be seen. However, injection of contrast agents did not improve the quality of the lung pictures. A compromise was found between duration of anaesthesia and radiation dose, and actual resolution and image quality. No lethal or visible damage was observed in the mice after scanning with the use of injection of clinically used contrast agents. Future research will still be required to refine the dose and application of contrast agent to be administered in small laboratory animals. A particular problem to be solved is the fast renal clearance of water-soluble contrast agents that are applied in human medicine. For imaging the hepatic circulation, injection of a lipid soluble agent (polyiodinated triglyceride) was reported previously (Bakan *et al.*, 2002). Application of contrast agents may also become promising particularly for studying angiogenesis (McDonald and Choyke, 2003).

However, in the field of molecular imaging a lot of research still needs to be done to fully exploit the possibilities of using different kinds of contrast agents for imaging done by micro-CT. In this respect, much development of radiopaque indicators together with the development of reconstruction and analysis software and scanner hardware is required (Ritman, 2002).

2.3.5 Applications of micro-CT in rare samples

Due to its non-invasive nature, micro-CT is an excellent tool to visualize paleontological samples and fossils on condition that they are isolated from the surrounding material.

Micro-CT has also proven useful in the description of skeletal morphology in rare samples, such as holotypes, and has helped to clarify the validity and phylogenetic position of certain species (Devaere *et al.*, 2005).

2.3.6 Bio-implants

One of the potential new applications of micro-CT is scanning bio-implants. The major challenge for micro-CT in this field is the visualization of different materials and tissues at the same time. In many bio-implants metal, parts such as Al or Ti are implemented causing X-ray scattering. A promising preliminary example is the experiment where it proved feasible to visualize the inner ear tissues together with the evaluation of the surgical aspects of newly developed cochlear implant electrodes relative to the intracochlear soft tissues (Postnov *et al.*, 2006). It opens further perspectives for the future visualization of other bio-implants.

Acknowledgement

The authors wish to thank Dr Kris Meurrens and Dr Piter Terpstra from Philip Morris research laboratories (Belgium and Germany) for stimulating discussions and for providing the mice models. Our thanks also go to Dr. Bart Truyen and Prof. P. Bottenberg, VUB. Financial support was obtained from the FWO (grant G. 0304.04). The authors wish to express their gratitude to Prof. Annemie Van der Linden (Bio-imaging Lab, University of Antwerp) and to Prof. Dirk Van Dyck and to all collaborators from VisionLab (University of Antwerp).

References

Bakan, D. A., Lee Jr., F. T., Weichert, J. P., Longino, M. A., Counsell, R. E., 2002. "Hepatobiliary imaging using a novel hepatocyte-selective CT contrast agent". *Acad. Radiol.* (Suppl. 1), S194–S199.

Beckmann, F., Heise, K., Kölsch, B., Bonse, U., Rajewsky, M. F., Bartscher, M., Biermann, T., 1999. "Three-dimensional imaging of nerve tissue by X-ray Phase contrast microtomography." *Biophys. J.* 76, 98–102.

Björndal, L., Carlsen, O., Thuesen, G., Darvann, T., Kreiborg, S., 1999. "External and internal macromorphology in 3D-reconstructed maxillary molars using computerized X-ray microtomography." *Int. Endodontic. J.* **32**, 3–9.

Bonse, U., Nusshardt, F., Busch, F., Kinney, J., Saroyan, R., Nichols, M., 1992. "X-Ray tomographic microscopy." In: Michette, A., Morrison, G., Buckley, C. (Eds.), *X-Ray Microscopy III.* Springer Series in Optical Sciences vol. 67. Springer-Verlag, Berlin, Heidelberg, pp. 167–176.

Bottenberg, P., Henin, L., Boca, C., Postnov, A., Wasek, A., De Clerck, N., Truyen B., 2003. "Application of desktop micro-CT imaging as gold standard in caries diagnosis." *J. Dental. Res.* **82**, B-386.

Boyde, A., De Clerck, N., Sasov, A., 2000. "Micro-CT of bones and soft tissues." *Microscopy and Analysis* vol 76 (UK ed) Wiley, 6, 70.

Cooper, D. M. L., Mathyas, J. R., Katzenberg, M. A., Hallgrimson, B., 2004. "Comparison of microcomputed tomographic and microtomographic measurements of cortical bone porosity." *Calcif. Tissue Int.* **74**, 437–447.

Cormack, A. M., 1973. "Reconstruction of densities from their projections with applications in radiological physics." *Phys. Med. Biol.* **18**, 195–207.

Cumming, S. R., 1998. "Review of the evidence for prevention, diagnosis and treatment and cost-effective analysis." *Osteoporosis Int.* **8** (Suppl. 4), S1–88.

Davis, G. R., Wong, F. S. L., 1996. "X-ray microtomography of bones and teeth." *Physiol. Meas.* **17**, 121–146.

De Clerck, N. M., Van Dyck, D., Postnov, A. A., 2003. "Non-invasive high-resolution μCT of the inner structure of living animals." *Microsc. Anal.* **81**, 13–15.

De Clerck, N. M., Meurrens, K., Weiler, H., Van Dyck, D., Vanhoutte, G., Terpstra, P., Postnov, A. A., 2004. "High resolution X-ray microtomography for the detection of lung tumours in living mice." *Neoplasia* **6**, 374–379.

Devaere, S., Adriaens, D., Teugels, G., De Clerck, N. M., Postnov, A. A., 2005. "Skeletal morphology of the holotype of *Gymnallabes nops* (Roberts and Stewart, 1976), using micro CT-scanning." *Cybium* **29**, 281–293.

Ding, M., Odgaard, S. Hvid, I., 1999. "Accuracy of cancellous bone volume fraction measured by micro-CT scanning." *J. Biomech.* **632**, 323–326.

Dowseth, D. J., Kenny, P. A., Johnston, R. E., 1998. *The Physics of Diagnostic Imaging.* Chapman and Hall Medical, London.

Einhorn, T., 1992. "A bone strength: The bottom line." *Calcif Tissue Int* **51**, 333–339.

Elliott, J. C., Dover, S. D., 1982. "X-ray microtomography." *J. Microsc.* **126**, 211–213.

Elliott, J., Davis, G. R., Anderson P., Wong F., Dowker, S., Mercier, C., 1997. "Application of laboratory microtomography to the study of mineralised tissues." *Anales. de Quimica. Int.* **93**, S77–S82.

Feldkamp, L. A., Davis, L. C., Kress, J. W., 1984. "Practical cone-beam algorithm." *J. Opt. Soc. Am. A.***1**, 612–619.

Flecknell, P. A., 1993. "Anaesthesia of animals for biochemical research." *Br. J. Anaesth.* **71**, 885–894.

Gilboy, W. B., 1995. "Microtomography with ionising radiations." *Appl. Radiat. Isot.* **46**, 689–699.

Gleason, S. S., Hari-Sarraf, H., Paulus, M. J., Johnson, D. K., Norton, S. J., Abidid, M. A., 1999. "Reconstruction of multi-energy X-ray computed tomography images of laboratory mice." *IEEE Trans. Nucl. Sci.* **46**, 1081–1086.

Hammersberg, P., Stenström, M., Hedtjärn, H., Måns Mångård, M., 1998. Absolute energy spectra for an industrial micro focal X-ray source under working conditions measured with a Compton scattering spectrometer – full spectra data. Linköping Electronic Articles in Mechanical Engineering vol. 1, http://www.ep.liu.se/ea/me/1998/001/.

Herman, G. T., 1980. *Image Reconstruction from Projections: The Fundamentals of Computerized Tomography.* Academic Press, New York, NY.

Hounsfield, G. N., 1973. "Computerized transverse axial scanning (tomography): Part I. Description of system." *Brit. J. Radiol.* **46**, 1016.

Hounsfield, G. N., 1980. "Computed medical imaging." *Med. Phys.* **7**(4), 283–290.

Jorgensen, S. M, Demirkaya, O., Ritman, E. L., 1998. "Three-dimensional imaging of vasculature and parenchyma in intact rodent organs with X-ray micro-CT." *Am. J. Physiol.* **275**, H1103–H1114.

Karau, K. L., Johnson, R. H., Molthen, R. C., Dhyani, A. H., Haworth, S. T., Hanger, C. C., Roerig, D. L., Dawson, C. A., 2001. "Microfocal X-ray CT imaging and pulmonary arterial distensibility in excised rat lungs." *Am. J. Physiol. Heart Circ. Physiol.* **281**, H1447–H1457.

Kennel, S. J., Davis, I. A., Branning, J., Pan, H., Kabalka, G. W., Paulus, M. J., 2000. "High resolution computed tomography and MRI for monitoring growth in mice undergoing radioimmunitherapy: Correlation with histology." *Med. Phys.* **27**, 1101–1107.

Ketcham, R. A., Carlson, W. D., 2001. "Acquisition, optimization and interpretation of X-ray computed tomographic imagery: Applications to the geosciences." *Comp Geosci*, **27**, 381–400.

Kinney, J. H., Lane, N. E., Haupt, D. L., 1995. "*In vivo*, three-dimensional microscopy of trabecular bone." *J. Bone Miner. Res.* **10**, 264–270.

Laib, A., Ruegsegger, P., 1999. "Comparison of structure extraction methods for *in vivo* trabecular bone measurements." *Comput. Med. Imaging Graph.* **23**, 69–74.

Lane, N. E., Thompson, J. M., Haupt, D., Kimmel, D. B., Modin, G., Kinney, J. H., 1998. "Acute changes in trabecular bone connectivity and osteoblast activity in the ovariectomized rat *in vivo*." *J. Bone Miner Res.* **13**, 229–236.

Langheinrich, A. C., Bohle, R. M., Breithecker, A., Lommel, D., Rau, W. S., 2004. "Micro-computed tomography of the vasculature in parenchymal organs and lung alveoli." *Fortschr. Rontgenstr.* **381**, 000A–000G.

Lee, S. C., Kim, H. K., Chun, I. K., Cho, M. H., Lee, S. Y., Cho, M. H., 2003. "A flat-panel detector based micro-CT: Performance evaluation for small animal imaging." *Phys. Med. Biol.* **48**, 4173–4185.

Luke, J., 2001. "Fluoride deposition in the aged pineal gland." *Caries Res.* **35**, 125–128.

McDonald, D. M., Choyke, P. L., 2003. "Imaging of angiogenesis: From microscope to clinic." *Nature Med.* **9**(6), 713–725.

McNamara, A., Muller, N. L., Okazawa, M., Arntorp, J., Wiggs, B., Pare, P. D., 1992. "Airway narrowing in excised canine lungs measured by high-resolution computed tomography." *J. Appl. Physiol.* **73**, 307–316.

Müller, R., Hahn, M., Vogel, M., Delling, G., Rüegsegger, P., 1996. "Morphometric Analysis of Noninavsively assessed bone biopsies; comparison of high-resolution computed tomography and histologic sections." *Bone* **18**, 215–220.

Nolan, P. M., Peters, J., Strivens, M., Rogers, D., Hagan, J., Spurr, N., Gray, I. C., Vizor, L., Brooker, D., Whitehill, E., 2000. "A systematic, genome-wide, phenotype-driven mutagenesis programme for gene function studies in the mouse." *Nat. Gen.* **25**, 440–443.

Nuzzo, S., Meneghini, C., Braillon, P., Bouvier, R., Mobillo, S., Peyrin, F., 2003. "Microarchitectural and physical changes during fetal growth in human vertebral bone." *J Bone Mineral Res*, **18**, 760–768.

Odgaard, A., 1997. "Three-dimensional methods for quantification of cancellous bone architecture." *Bone* **20**, 315–328.

Paulus, M. J., Sari-Sarraf, H., Gleason, S. S., Bobrek, M., Hicks, J. S., Johnson, D. K., Behel, J. K., Thompson, L. H., Allen, W. C., 1999. "A new X-ray computed tomography system for laboratory mouse imaging." *IEEE Trans. Nucl. Sci.* **46**, 558–564.

Paulus, M. J., Gleason, S. S., Kennel, S. J., Hunsicker, P. R., Johnson, D. K., 2000. "High resolution X-ray computed tomography: An emerging tool for small animal cancer research." *Neoplasia* **2**, 62–70.

Persy, V., Van Kuilenburg, J. T., Neven, E., Postnov, A., De Clerck, N., DHaese. P. C., De Broe, M. E., 2004. "Vascular calcification (VC) induced by high phosphate diet can be detected with *in vivo* micro-CT in chronic renal failure (CRF) rats." *J. Am. Soc. Nephrol.* **15**, 833A.

Postnov, A., De Clerck, N., Sasov, A., Van Dyck, D., 2002a. "3D *in vivo* X-ray microtomography of living snails." *J. Microsc.* **205**, 201–205.

Postnov, A., Van Dyck, D., Saveliev, S., Sasov, A., De Clerck, N. M., 2002b. "Definition of local density in biological calcified tissues using X-ray microtomography." *Prog. Biomed. Opt. Imaging (SPIE Meetings)* **3**(19), 749–755.

Postnov, A., Vinogradov, A., Van Dyck, D., Saveliev, S. V., De Clerck, N. M., 2003. "Quantitative analysis of bone mineral content by x-ray microtomography." *Physiol Meas.* **24**, 165–178.

Postnov, A., Zarowski, A., De Clerck, N. M., Vanpoucke, F., Offeciers, W., Van Dyck, D., Peeters, S., 2006. "High Resolution Micro-CT Scanning as Innovatory tool for Evaluation of the Surgical Positioning of Cochlear Implant Electrodes." *Acta Oto-Laryngologica*, **126**(5), 467–474.

Ritman, E. L., 2002. "Molecular imaging in small animals: Roles for micro-CT." *J. Cell Biochem. Supp.* **39**, 116–124.

Rüegsegger, P., Elsasser, U., Anliker, M., Gnehm, H., Kind, H., Prader, A., 1976. "Quantification of bone mineralization using computed tomography." *Radiology* **121**, 93–97.

Rüegsegger, P., Koller, B., Müller, R., 1996. "A microtomographic system for the nondestructive evaluation of bone architecture." *Calcif. Tissue Int* **58**, 24–29.

Russo, E., 1998. "Imaging devices to benefit both mouse and biologist." *The Scientist* **12**, 1–6.

Salmon, P. L., Collier, C. G., Hart, J. E., Halonen, K., 2001. "Influence of experimental procedures on the growth divergence of ovariectomised and sham operated rats." *J. Bone Miner. Res.* **16** (Suppl. 2), M442.

Salmon, P. L., Sasov, A.Y., 2005. "*In vivo* micro-CT: Optimising image quality, scan time and radiation dose: What can be realistically and safely achieved?" *Proc ECTS* **36** (Suppl. 2), S160–S161.

Sasov, A., 2004. "X-ray nanotomography." In: Bonse, U. (Ed.), *Proceedings of SPIE* vol. 5535. Developments in X-Ray Tomography IV. pp. 201–211.

Sasov, A., Van Dyck, D., 1998. "Desktop X–ray microscopy and microtomography." *J. Microsc.* **191**, 151–158.

Sasov, A., Dewaele, D., De Clerck, N., 2001. "Full-body *in vivo* Micro-CT system with 10 microns resolution." *Proceedings of the High Resolution Meeting*, Washington, DC, USA, September 9–11, vol. 196, pp. 18–20.

Sijbers, J., Postnov, A., 2004. "Reduction of ring artifacts in high resolution micro-CT." *Phys. Med. Biol.* **49**, N247–N253.

Spadaro, J. A., Baesl, M. T., Conta, A. C., Margulies, B. M., Damron, T. A., 2003. "Effects of irradiation on the appositional and longitudinal growth of the tibia and fibula of the rat with and without radioprotectant." *J. Pediatr. Orthop.* **23**, 35–40.

Waarsing, J. H., Day, J. S., van der Linden, J. C., Ederveen, A. G, Spanjers, C., De Clerck, N., Sasov, A., Weinans, H., 2004a. "Detecting and tracking local changes in the tibiae of individual rats: A novel method to analyse longitudinal *in vivo* micro-CT data." *Bone* **34** (1), 163–169.

Waarsing, J. H., Day, J. S., Weinans, H., 2004b. "An improved segmentation method for *in vivo* μCT Imaging." *J. Bone Miner Res.* **19**,1640–1650.

Wang, L., Banu, J., McMahan, C. A., Kalu, D. N., 2001. "Male rodent model of age-related bone loss in men." *Bone* **29**, 141–148.

3 Ultrasound Imaging

S. Lori Bridal, Jean-Michel Correas and **Geneviève Berger**

3.1 Principles of ultrasonic imaging and its adaptation to small laboratory animals

Ultrasound, like audible sound, is an elastic wave but with a frequency above the range of frequencies detected by the human ear (>20 kHz). Medical ultrasonic imaging typically uses frequencies between 1 and 15 MHz. Images formed from the 'echoes' returned at boundaries and scattered from small structures within tissues provide valuable anatomical information. Doppler ultrasound is used to assess the speed and direction of blood flow or heart wall movement based on the measurements of the shift in the ultrasonic wave's frequency as it returns from moving structures. Colour-coded Doppler information is often superimposed on the anatomical grey-scale image to combine functional and anatomical information in a single view.

Although ultrasonic imaging offers somewhat limited soft-tissue contrast and can be hindered if gas-filled or attenuating bone structures interfere with sound propagation, the advantages of this non-invasive technique are numerous. As it is easily portable, relatively low in cost, non-ionising and provides real-time imaging, diagnostic ultrasound has established its place as a modality of choice for the assessment of foetal health, blood flow, myocardial function, detection and characterisation of masses and guidance in needle biopsy. Currently, ultrasound is used in virtually every medical speciality, including obstetrics, cardiology, radiology, gastro-enterology, neurology, surgery and musculoskeletal applications.

Image spatial resolution is closely related to the ultrasonic frequency. Resolution limits in medical ultrasonography range from approximately 2 mm to 0.1 mm for frequencies of 1 MHz to 15 MHz, respectively. Doppler frequencies in clinical imaging systems typically provide the measurement resolution necessary to evaluate flow in vessels larger than 200 μm in diameter (Wells *et al.*, 1977). The energy in the ultrasonic wave lost during its propagation in biological tissue increases with frequency. That is why spatial resolution must generally be sacrificed to obtain images of deep tissues and organs. Broadband technology offers a compromise as it allows the combined use of higher frequency to study superficial tissues and lower frequency to visualise deeper structures.

Engineering advances such as enhanced bandwidth transducers, the introduction of digital technology and sophisticated image-formation routines have led to greatly improved image quality over the last decade. Higher frequency transducers have been developed and applied in the clinic for intravascular imaging of atherosclerotic plaque (Honda, Yock and Fitzgerald, 1999), dermatological applications (Foster *et al.*, 2000; Gupta *et al.*, 1996; Semple *et al.*, 1995) and visualisation of ocular structures (Pavlin *et al.*, 1991). In addition, new signal acquisition and analysis techniques have added considerably to the ultrasonic arsenal. Important advances have been made in Doppler analysis and display. Stabilised microbubbles, consisting of encapsulated low- solubility gases, have opened possibilities for quantitative capillary blood flow evaluation when used with innovative non-linear ultrasonic pulse sequencing.

As the capacities of ultrasound have continued to grow, so have the number of its contributions to studies in small animal models. The spatial resolutions and penetration depth offered by transducers, in the 20 – 60 MHz range, are well adapted for imaging of many structures of interest in the laboratory mouse. High frequency ultrasound has been used to study volumetric growth in a mouse melanoma model providing resolutions from 20 to 60 μm (Turnbull *et al.*, 1996; Goertz *et al.*, 2002; Cheung *et al.*, 2005). More recently, it has been applied to monitor tumour development in deeper organs such as the prostate, kidney and liver (Graham *et al.*, 2005; Jouannot *et al.*, 2006; Wirtzfeld *et al.*, 2005). *In utero* images of the embryonic mouse have been obtained as early as the 7th embryonic day (just after implantation in the uterus) up to the 17th day (near term). Images, obtained *in utero* at frequencies from 25 to 40 MHz, provide very sensitive evaluation of organogenesis (Turnbull *et al.*, 1995; Phoon, Aristizabal and Turnbull, 2000; Phoon *et al.*, 2004; Akirav *et al.*, 2005). High frequency pulsed Doppler measurements have been shown to provide information on flow in the mouse placenta (Zhou *et al.*, 2002). The feasibility of protocols for ultrasound contrast imaging has also been demonstrated in small animal models (Lucidarme *et al.*, 2003; Lucidarme *et al.*, 2004; Goertz *et al.*, 2005a; Goertz *et al.*, 2005b).

Commercial systems adapted to ultrasonic imaging of the mouse are becoming available. The system Vevo 770 for ultrasonic imaging in small animal research (VisualSonics, Toronto Canada) offers resolutions as fine as 30 μm, fields of view as large as 20 mm, three-dimensional imaging and high resolution Doppler modalities. Although the Diasus system (Dynamic Imaging, United Kingdom) is marketed for musculoskeletal, dermatological and breast imaging in humans, its large bandwidth probe (10–22 MHz) offers the advantages of an electronically scanned linear array system with resolutions on the order of 50 μm. Probes in the 12–15 MHz frequency range for superficial imaging with clinical systems offer very rapid frame rates that can contribute usefully to murine studies, if the spatial resolution needs are not too demanding. Non-linear imaging sequences available with certain clinical probes can also be of interest for functional contrast imaging in the mouse and rat.

This chapter presents a simple description of the physics and technology behind the most widespread modalities of ultrasonic imaging. Its goal is to provide the reader with the information needed to evaluate the capabilities and limitations of these modalities, to choose an ultrasound system and to develop ultrasonic imaging protocols for evaluation of laboratory animals. More detailed descriptions of ultrasonic imaging can be found in several books (Kremkau 1998; Webb 2003).

3.1.1 Ultrasonic waves

As opposed to radio waves and light waves, ultrasound is an elastic wave that requires a medium through which to travel. As an ultrasonic wave propagates through a tissue, the particles composing the tissue oscillate back and forth about a fixed mean position (on the order a few tenths of a nanometer). A very simple model for this process can be represented as a series of small particles connected by massless springs (Figure 3.1.1). Initially, the particles are equally spaced and rest at their equilibrium positions. The initiation of the wave can be imagined to be likened to the application of a repeated push and release at one end. This provokes a local compression and a local pressure change that will push on the adjacent segment of the medium. The next segment is compressed in turn, and the energy in the wave propagates away from the source. There is no global material displacement, only a local perturbation of the medium, which results in wave propagation.

Imagine that we look at this modelled medium at some instant after the wave has begun propagating (Figure 3.1.1(b)). Zones can be identified where the particles are closer together (*compression*) and where they are more widely spaced (*rarefaction*) than at equilibrium. These zones correspond to the positions of the maximum and minimum local acoustic pressures, respectively. The distance between two consecutive regions of peak compression (or peak rarefaction) represents the *wavelength*, λ. The number of complete oscillations that a particle makes about its equilibrium position per second is the *linear frequency*, ν, of the wave. The *wave propagation*

Figure 3.1.1 Longitudinal wave in a medium modelled as a series of small particles connected by massless springs. (a) Medium at equilibrium. (b) A sinusoidal push and release is applied to the medium at a linear frequency ν. At a fixed instant during the propagation of the resulting wave of wavelength λ, there are zones of compression (positive acoustic pressure) and rarefaction (negative acoustic pressure)

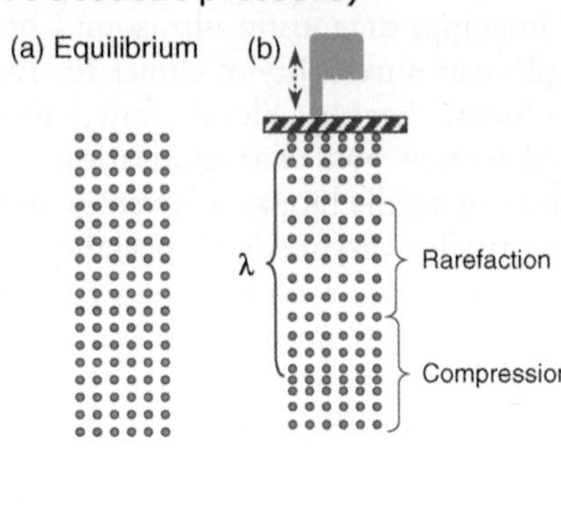

speed, c, is the product of the wavelength and the linear frequency,

$$c = \nu \cdot \lambda. \quad (3.1)$$

The propagation speed is inversely proportional to the square root of the mass per unit volume of a material, ρ, multiplied by its compressibility, κ. If two tissues have similar density, the more rigid of the two (the least compressible) will have a greater wave propagation speed.

As in this simple model, the back-and-forth particle motion in medical ultrasound is parallel to the direction of the wave's motion as it travels away from the source. Such a wave is referred to as *longitudinal* or *compressional*.

3.2 Pulse-echo transmission

3.2.0 Introduction

If you face a cliff and shout, some of the sound will be reflected from the barrier to return as an echo. This basic principle is at the heart of medical ultrasonic imaging. A single probe acts as source and receiver of ultrasound pulses in what is known as a 'pulse-echo' configuration. A pulse containing a few cycles of ultrasound is generated by an ultrasonic device and is transmitted into the body. Ultrasonic imaging pulses are typically 1 – 3 cycles long and Doppler pulses are typically 5 – 20 cycles long. Part of the energy in this pulse is reflected and scattered from structures within the body. Echoes are then returned to and detected by the same device. Obtaining a high quality ultrasonic image depends, in part, upon optimising each step along the path of the transmitted pulse and the returned echoes. The typical ultrasonic imaging configuration and some of the events that may occur along the pulse-echo path are illustrated in Figure 3.2.1.

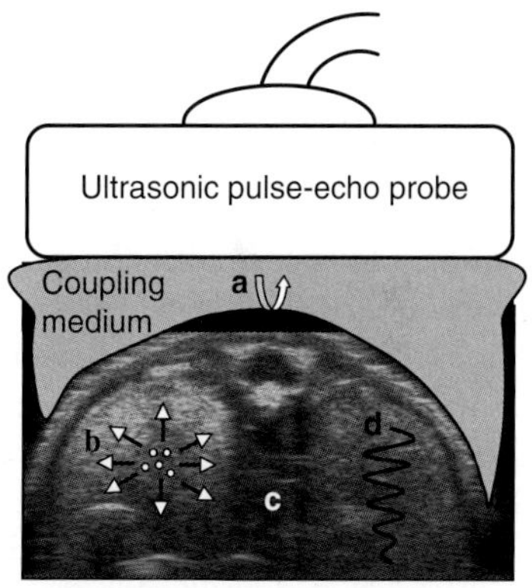

Figure 3.2.1 Illustration of pulse-echo ultrasound. The diagram superimposed on a transverse, dorsal ultrasonic image of mouse abdomen illustrates some of the events that may occur along the pulse-echo path. The pulses sent by the ultrasonic probe are transmitted into the body through a coupling medium (water or gel). (a) Interfaces between structures, such at that between the coupling-medium and the skin, give rise to strong reflections that are presented in the ultrasonic image as aligned groups of bright white pixels. (b) Within tissues and organs, structures that are smaller than the ultrasonic wavelength scatter ultrasound in all directions. Scattered energy that is redirected towards the ultrasonic probe is referred to as backscatter. The echoes from many non-resolved structures arrive at the probe simultaneously, creating an interference pattern. It is this interference pattern that gives the ultrasonic image its typical speckle texture containing a combination of bright and dark pixels. (c) Zones posterior to the backbone are in an acoustic shadow due to the strong reflection of the incident pulse by the bone. Gas-filled bodies will have a similar shadowing effect. (d) The average brightness of the ultrasonic speckle can be seen to decrease with depth. This is due to the attenuation of the energy in the ultrasonic pulse due to scattering and absorption processes

3.2.1 Reflection and refraction

Imagine that an ultrasonic wave propagating through a medium encounters an acoustic boundary that has lateral dimensions much greater than and surface roughness dimensions much smaller than the ultrasonic wavelength. Boundaries between tissues or organs and larger structures within organs often satisfy these conditions. At such an interface, a fraction of the intensity in the incident wave will be *reflected* back into the first medium (Figure 3.2.2). This fraction depends on the acoustic impedance difference between the media at the boundary. *Acoustic impedance, Z*, is defined as the density of a material multiplied by the speed with which the ultrasonic wave propagates through the medium. (The unit of acoustic impedance is the *rayl*, defined as a kg m^{-2} s^{-1}.)

For perpendicular incidence upon the interface, the *intensity reflection coefficient, R*, and *intensity transmission coefficient, T*, can be calculated if the acoustic impedance is known for each medium as

$$R = \frac{(Z_2 - Z_1)^2}{(Z_1 + Z_2)^2}, \quad (3.2)$$

$$T = \frac{4Z_1 Z_2}{(Z_1 + Z_2)^2}. \quad (3.3)$$

Figure 3.2.2 An ultrasonic wave of intensity I_0 arriving upon an acoustic boundary with impedance of Z_1 in the first medium and Z_2 in the second one. (a) Reflection and transmission paths for perpendicular incidence. (b) Reflection and transmission paths when the incident wave is at an angle of θ_i relative to the perpendicular axis of the interface

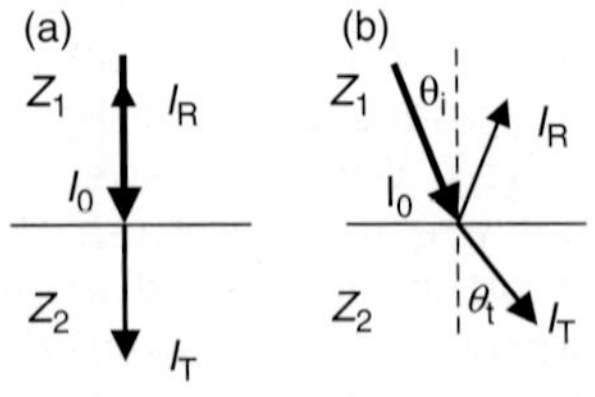

If the ultrasonic pulse arrives upon the interface at an angle of θ_i with respect to the perpendicular axis, the angular direction of the transmitted beam will be modified at the interface or *refracted*. The importance of this angular modification is proportional to the ratio of the speed of sound in the two materials:

$$\frac{\sin\theta_i}{\sin\theta_t} = \frac{c_1}{c_2}. \tag{3.4}$$

Table 3.2.1 summarises the typical acoustic impedance and the speed of sound values for biological media. If the impedance of the media on either side of the boundary are the same or *matched*, transmission is perfect ($T = 1$) and there is no reflection ($R = 0$). Because air and bone have very different acoustic impedance values than soft tissue, they will strongly reflect ultrasound and reduce ultrasonic transmission.

How important are the reflection and refraction effects for biological imaging? Let us first consider interfaces including bone or air. Estimates can be made based on the properties of biological tissues in Table 3.2.1 and the simple equations cited above. More than 40 per cent of the incident intensity in an ultrasonic wave in soft tissue will be reflected at an interface with bone. Angular refraction can also be very significant at an interface between soft tissue and bone. At an interface from soft tissue to gas in the lungs or bowel, more than 99 per cent of the incident intensity is reflected! Smaller trapped air bubbles in a tissue can seriously decrease signal transmission (depending on the concentration and size of the bubbles). These issues can be critical when imaging small animals. In human subjects, it is common to apply pressure by pushing the face of the transducer against the patient to displace air from bowel structures during abdominal imaging. Similar pressures applied in the small animal may damage organs, and thus only very light pressures can be used. An important strategy to improve ultrasonic transmission is to select a sound path that avoids gas-filled or bony structures. This is referred to as an *acoustic window*. For example, when imaging the heart, the ultrasonic source is positioned such that the ultrasonic beam passes through an acoustic window between the ribs. When an appropriate acoustic window can be found, ultrasonic transmission is rather good and, at typical soft tissue boundaries, less than 1 per cent of incident intensity is reflected. This means that a signal echo is obtained, allowing interface detection, while most of the energy is transmitted, allowing the investigation of tissues located behind the interface.

Table 3.2.1 Acoustic impedance and wave propagation speed of biological tissues and components (Kremkau, 1998; Webb, 2003)

	Characteristic acoustic impedance(Mrayl)	Speed of sound (mm/μs)
Soft tissue	1.58–1.63	1.54
Fat	1.38	1.45
Blood	1.61	1.55
Bone	7.8	3.5
Water	1.06	1.48
Air	0.0004	0.33

3.2.2 Scattering

Any surface roughness or particles with dimensions much smaller than or on the order of the ultrasonic wavelength will scatter rather than reflect. The *scattering cross section*, σ, refers to the total power scattered per unit of incident intensity. The angular distribution and intensity of scattering from a body depends on its precise geometry and acoustic properties as well as its size relative to the incident wavelength. Mathematical formula can be found to estimate the scattering cross section for specific scatterer geometry and acoustic properties (Faran, 1951; Hickling, 1962; Rose *et al*., 1995). For a scattering body much smaller than the wavelength, the angular distribution of the scattered energy about the body is approximately uniform, and the scattering cross section increases as the fourth power of frequency. As the size of the scattering body increases towards the wavelength dimension, scattering can become more complex with potentially strong angular dependence and strong variations as a function of frequency.

Clinical medical imaging in the frequency range from 1 to 15 MHz has wavelengths in soft tissue ranging from 1500 to 100 μm. For scattering from structures such as red blood cells (diameter ∼7 μm), the relatively simple descriptions limited to the long wavelength limit may generally be applied. As higher frequency ultrasound is applied (20 – 40 MHz with wavelengths in soft tissues from 75 μm to 38 μm) such approximations approach their limits for these structures. As ultrasonic frequency is increased, smaller structures in the tissue begin to contribute more to scattering. Significant scattering sites may include collagen and elastin structures or even cell nuclei and other cellular structures at very high ultrasonic frequencies (Insana *et al.*, 1990; Baddour *et al.* 2005).

3.2.3 Attenuation

As the ultrasound propagates through a medium both its pressure and intensity decrease exponentially as a function of the *propagation distance, z*. This energy loss can be described as

$$p(z) = p_0 \cdot \exp(-\alpha \cdot z) \tag{3.5}$$

or

$$I(z) = I_0 \cdot \exp(-2\alpha \cdot z), \tag{3.6}$$

where p_0 and I_0 represent the initial pressure and intensity, respectively, in the wave at $z = 0$, and α is the *amplitude (or pressure) attenuation coefficient.* The attenuation coefficient in the formula above is in units of cm^{-1}, if z is in centimetre. Most often, published values of the attenuation coefficient are provided in units of $\mathrm{d}B\,\mathrm{cm}^{-1}$. Conversion from one set of units to another follows:

$$\alpha(\mathrm{dB} \cdot \mathrm{cm}^{-1}) = 8.69 \cdot \alpha(\mathrm{cm}^{-1}). \tag{3.7}$$

The attenuation accounts for losses from scattering and absorption processes. Energy is lost because the beam is scattered away from the propagation direction. As particles are displaced by the ultrasonic wave, friction occurs, converting mechanical energy into heat (*absorption losses*). There are also *relaxation losses* related to energy lost as molecules return to their original configurations after ultrasonic displacement. Despite this complicated, multi-process nature, in most soft biological tissues, attenuation (in $\mathrm{dB\,cm}^{-1}$) increases linearly as a function of frequency within the range used in medical imaging. Specific measurements of attenuation made at higher ultrasonic frequencies, however, demonstrate levels that are superior to estimates made via simple linear extrapolation of the linear approximations used in the 1 – 15 MHz range. Figure 3.2.3 summarises attenuation estimates for several biological materials in the frequency range useful for small animal imaging. Because of the significant increase of attenuation at higher frequencies, penetration depths are limited to several millimetres.

Figure 3.2.3 Approximate values of the attenuation at high frequency in biological tissue. Approximate values of attenuation measured in *ex vivo* human tissue specimens. The small cross near 5 MHz represents the attenuation values typical of soft tissues in the clinical imaging range (∼1–5 $\mathrm{dB\,cm}^{-1}$) (Lockwood *et al.*, 1991; Pan, Zan and Foster, 1998)

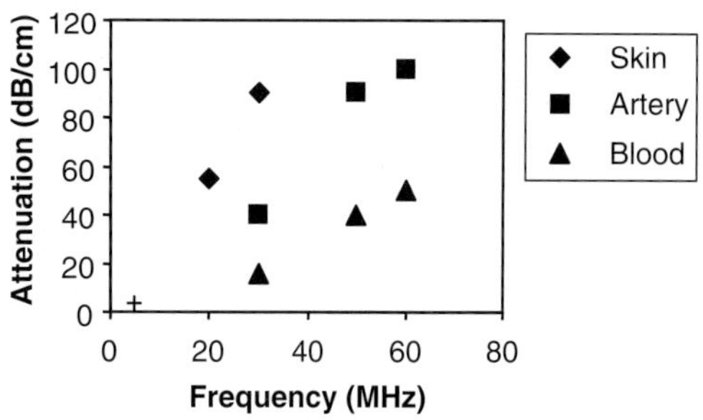

3.2.4 Coupling

Up to this point, we have only considered the propagation of the ultrasonic wave along the path within the body. One of the interfaces with maximum loss in energy to be encountered in the propagation path, however, is the interface between the ultrasonic device and the surface of the animal's body. Animal fur is filled with trapped air and impurities. In general, the fur is removed from the imaging site. Once dermatological tolerance is verified, a depilatory cream is applied to thoroughly remove the ultrasound-blocking hair. The site for ultrasonic wave transmission is then covered with an ultrasonic coupling gel to fill any air space between the ultrasonic device and the animal's body. For high frequency imaging, it is recommended to centrifuge the coupling gel prior to use to eliminate any dissolved gas in the gel, and for small animals it is important to warm the gel to body temperature. (Work has been reported for which hair has not been removed but has been fully saturated with ultrasonic coupling gel. This is clearly not optimal for imaging but may provide an alternative approach at lower frequencies if hair removal is not possible for a specific animal model.)

3.3 Ultrasonic transducers

The ultrasonic *transducer* that generates and detects the ultrasonic pulse is a critical component of the imaging system. The term transducer refers to the capacity to convert energy from one form to another. In the case of ultrasonic transducers, electrical energy in the form of an oscillating voltage is converted to a high frequency mechanical vibration for emission of the ultrasonic pulse. Conversely, ultrasonic echo vibrations returned to the transducer are converted into electrical signals and detected as voltage changes as a function of time. The transducer characteristics have a dominant impact on image resolution and the *signal-to-noise ratio* (SNR). The signal-to-noise ratio describes the ratio of voltage levels in a signal carrying information (for example, echoes from scattering structures) relative to voltage variations due to thermal and electronic noise sources. The diagram in Figure 3.3.1 presents the essential components of an ultrasonic transducer.

3.3.1 Piezoelectric materials

Materials that respond mechanically to an electric voltage potential are known as piezoelectric. Typically an alternating voltage is applied across opposite surfaces of a thin disk of piezoelectric having a thickness *d*. The thickness expands and contracts producing a motion with a linear frequency of ν. The linear frequency ν is related to the *angular frequency*, ω, by the simple relation, $\omega = 2\pi\nu$. The efficiency of transfer between the driving voltage and the piezoelectric response is optimised at the resonant frequency, ν_r, of the piezoelectric disk, determined by

$$\nu_r = \frac{c_{pz}}{2d}, \tag{3.8}$$

Figure 3.3.1 Diagram illustrating key components of a mono-element, cylindrical transducer

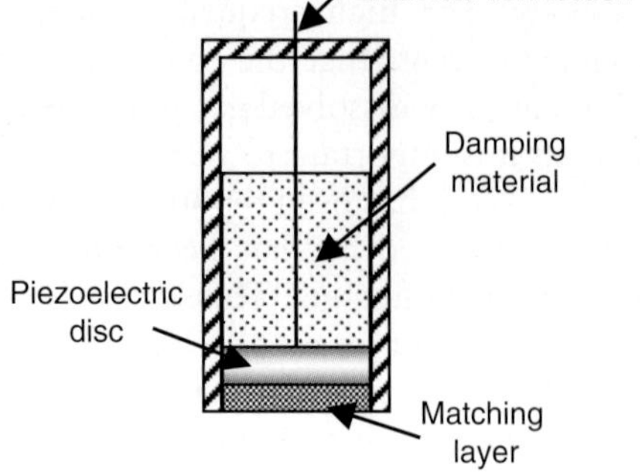

where c_{pz} is the speed of sound in the piezoelectric material. The conversion efficiency is also improved if the *electromechanical coupling coefficient* of the piezoelectric material is high.

For medical imaging, materials are sought which combine high electromechanical coupling coefficients with an acoustical impedance easily matched to that of soft tissue (~1.6 Mrayl). *Electrical impedance* of the materials is also an important parameter because it affects the efficiency obtained when matching the transducer element to the electronic circuitry. For high frequency imaging the material choices are further limited by the need to prepare the material in a thin-enough layer for high frequency resonance. Piezoelectric polymers such as polyvinylidene difluoride (PVDF) have been widely used in high frequency transducers. These materials offer the advantage of being producible in very thin layers and formed to a focusing lens shape. However, the electromechanical coupling is rather low, limiting transducer sensitivity. Lithium niobate piezoelectric crystals have a much higher electromechanical coupling and a relatively well-matched electrical impedance. Matching layers are necessary to overcome impedance mismatch between the crystal and biological tissue and a lens is necessary to focus the beam. The other principle families of piezoelectric materials include ceramics and piezo-composites. Ceramics have high electromechanical coupling but are difficult to match electrically. Piezo-composites consist of ceramic fibres in a polymer substrate. This preserves the high electromechanical coupling while improving acoustic and electrical impedance matching (Snook *et al.*, 2002).

3.3.2 Damping

For ultrasonic imaging, the excitation voltage is limited to only a few cycles to produce a short pulse centred at a frequency of ω. This temporal limitation is critical to insure good axial spatial resolution. However, as when you strike a bell hard and quickly, the bell's response rings-down over time, and the piezoelectric disk continues to vibrate after the excitation voltage has ceased with a gradual loss of vibration amplitude over time (Figure 3.3.2). To reduce the length of this 'ring-down', *damping material* (generally an acoustically-coupled plastic or epoxy containing small particles of metal powder to interfere with reverberations) is placed against the rear face of the piezoelectric disk. Effective damping will lead to the production of a shorter pulse containing a larger range of frequencies or *bandwidth* as illustrated in Figure 3.3.2.

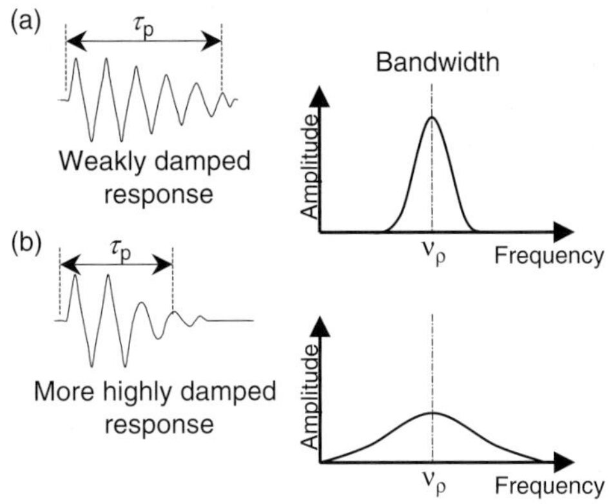

Figure 3.3.2 Relationship between damping, pulse duration and bandwidth. A brief electric impulsion is applied to excite a transducer with a resonant frequency of ν_r. (a) If the transducer is weakly damped, the transducer response will ring-down slowly resulting in an ultrasonic pulse with a long duration (several vibrational cycles). The frequency bandwidth of such a response is tightly centred about the resonant frequency of the transducer. (b) If the transducer is more highly damped, the ring-down is reduced. The pulse duration is shorter. The frequency bandwidth is larger. The overall response contains less total energy

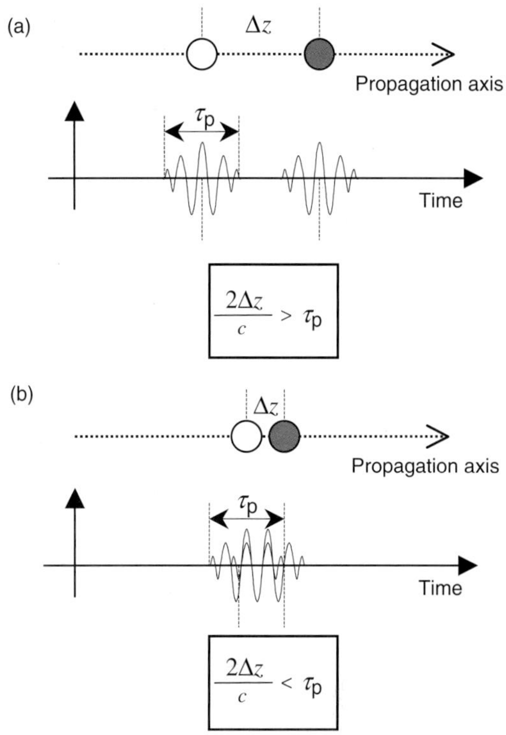

Figure 3.3.3 Axial resolution. The ultrasonic pulse has an effective duration of τ_p. (a) Two scattering structures on the propagation axis will be resolved if they are separated by a distance Δz greater than $\tau_p.c/2$. (b) If however, the axial distance between structures is less than $\tau_p.c/2$ the echoes of the pulses returned from the two structures will overlap in time. Such overlapping echoes will be summed constructively and destructively as a function of the interference between the two pulses. The two scatterers are not resolved

3.3.3 Axial resolution

The *axial resolution* is defined as the closest separation that two scattering or reflecting bodies may have while still being resolvable (Figure 3.3.3). Transducer damping is a key point in the optimisation of axial resolution. The more damped the transducer, the better the axial resolution will become. However, there is a price to pay. As the transducer is more and more strongly damped, the more the intensity of the transmitted pulse at the fundamental frequency of the transducer will be reduced. The theoretical limit for the axial resolution of the two scatterers with an ultrasonic wave is determined by the length of one wavelength of ultrasound in the medium as

$$\text{Axial resolution}_{\text{theor}} = \lambda = \frac{c}{\nu}, \tag{3.9}$$

where c is the speed of sound in the propagation medium. In practice, however, an ultrasonic pulse contains a few cycles of ultrasound and the total length of the effective pulse duration τ_p is also influenced by the quality of the system damping as described in the previous section. Two scatterers lying on the propagation axis will be resolvable if they are separated by a distance of at least one-half the length of the transmitted ultrasound pulse. The length of the pulse is easily related to its pulse duration for estimation of the axial resolution limit as

$$\text{Axial resolution} = \frac{1}{2}\tau_p \cdot c. \tag{3.10}$$

Structures separated by less than this distance along the ultrasonic line of site will not be distinguished as individual structures in the ultrasonic image. Axial resolutions on the order of 30 μm to 120 μm can be obtained for transducer centre-frequencies from 80 to 20 MHz, respectively.

3.3.4 Matching layer

Once a pulse has been produced, the challenge remains to efficiently couple its mechanical energy into the body. This is achieved by effective matching of the acoustic impedance of the piezoelectric crystal to that of the patient's skin and the coupling gel (Z_{sg} ~1–1.6 Mrayl). As the impedance is typically quite different for the piezoelectric and the biological material, a matching layer with an acoustic impedance of Z_m is interposed between the face of the transducer and the patient. The optimum acoustic impedance for the matching layer Z_m is related to the impedance of the piezoelectric Z_{pz} and the material into which the sound is transmitted Z_{sg} by

$$Z_m = \sqrt{Z_{pz} \cdot Z_{sg}}. \tag{3.11}$$

Optimum thickness for the matching later equals one-fourth the wavelength of the transmitted ultrasound.

3.3.5 Focusing and lateral resolution

Systems specifically designed for high-resolution imaging of the mouse most often use single-element high frequency transducers such as illustrated in the Figure 3.3.1. The piezoelectric disk (the active element) of the transducer has a radius *a* and produces a wave vibration with a linear frequency *ν*. The ultrasonic field produced by a vibrating source of limited extent has several characteristic properties. First, consider the case for a flat piezoelectric disk of radius *a*. Close to the transducer face in the region known as the *near-field* (or Fresnel zone), the field in the main beam is rather complex presenting rapid variations in the phase and amplitude of the ultrasonic wave front. The near field boundary is at a distance from the transducer face of approximately

$$\text{Near-field boundary} \cong \frac{a^2}{\lambda}. \tag{3.12}$$

Beyond this limit, in the far field (or Fraunhofer zone), the ultrasonic wave front behaves approximately as planar waves. The beam diverges laterally, and the intensity decreases smoothly as a function of the distance from the transducer. *Side lobes* may be produced by a diffraction grating effect at the transducer face. Side lobes are weaker beams of sound emitted from a transducer in directions other than that of the primary beam. The greater the ratio of the ultrasound wavelength to the transducer diameter, the fewer will be the number of side lobes. The angle between the main beam and the first side lobe decreases as the ultrasound wavelength decreases. Thus, for shorter wavelength (higher frequencies) side lobe effects, which sap the main beam of its energy and can lead to oddly timed echo artifacts, can become more bothersome.

Figure 3.3.4 Lateral resolution. (a) Two scattering objects at the focal distance *F* of the weakly focused transducer are not resolved because the lateral beam width is larger than the separation *D* between the two objects. (b) A more tightly focused transducer will receive the response from a single scattering object. By moving the transducer laterally, echoes can be received independently from the other scatterer. Thus, the image constructed from a series of scan lines will allow the resolution of the two separate scatterers. The shaded region of each field diagram represents the depth of field. As the lateral resolution of a transducer improves, the useful depth of field is reduced. (To maintain the same focal distance with improved lateral resolution as pictured in this diagram, either the radius or the ultrasonic frequency of the transducer in (b) must be increased relative to that of the transducer in (a).)

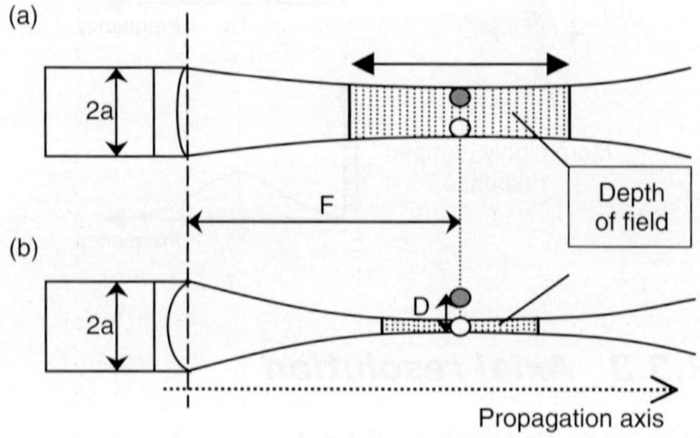

The *lateral resolution* is defined as the ability to resolve two scattering bodies at the same axial distance separated by a distance *D* (Figure 3.3.4). The narrower the lateral width of the ultrasonic beam, the finer will be the lateral resolution. At the near field boundary the field for a flat-disk transducer of radius, *a*, has a lateral beam width on the order of *2a*. Single-element, high frequency transducers typically have radii on the order of millimetres. Thus, to obtain lateral resolutions approaching the axial resolutions obtained at high frequencies, beam focusing is necessary. Most often, for high frequency transducers, this is accomplished by forming the active element of the transducer itself into a concave, curved form. The *focal distance F* is defined as the distance from the transducer to the region of the beam with the narrowest lateral width. For spherical focusing, the full width at half maximum (-6 dB beam width) in the focal region is related to the speed of sound *c*, the linear frequency of the wave ν, the focal distance F and the radius of the transducer a according to

$$\text{Lateral resolution} = \frac{c}{\nu} \cdot \frac{F}{2 \cdot a}. \tag{3.13}$$

As transducer focusing is tightened, the energy in the field is concentrated in a smaller and smaller zone about the focal distance, and the divergence of the field about the focus is more and more accentuated. Conventionally, we consider the useful region of the field to be that in which the intensity of the wave is greater than or equal to 50% of the maximum intensity at the focus. The total axial length of this region is called the *depth of field*. The depth of field is related to the speed of sound in the medium, the frequency of the wave, the focal distance and the radius of the transducer according to

$$\text{Depth of field}_{6\,\text{dB}} = 7.08 \cdot \frac{c}{\nu} \cdot \left(\frac{F}{2 \cdot a}\right)^2. \tag{3.14}$$

The ratio of the focal distance F of the transducer divided by the diameter of the transducer $(2 \cdot a)$ is called the *f-number*. A compromise must be made. Better lateral resolution is bought at the cost of a shorter useful working distance for imaging with a single element transducer. In the 20–80 MHz frequency range, lateral resolutions on the order of 200–50 μm are achievable. For a transducer with an f-number of 2.5, the depth of field is on the order of 3.4 and 0.8 mm at 20 MHz and 80 MHz, respectively.

Beam forming techniques can be applied to improve the useful depth of field and obtain a uniform lateral resolution throughout a more extensive depth if the transducer has multiple transmitting and receiving elements. The basic principles behind beam forming and the state of transducer array technology for high frequency ultrasound are discussed in the following section.

3.3.6 Multi-element transducers and beam forming

Nearly all medical-imaging transducers used for human imaging consist of arrays of many small piezoelectric elements each with its own electronics circuit for signal transmission and reception. Each element is physically and electrically isolated from its neighbours. A fixed focus can be created by curving the array or adding a lens. Arrays can be arranged in many configurations as illustrated in Figure 3.3.5. Better uniformity in the beam focusing can be obtained using techniques known as *beam forming*. Using digital technology, the effective focal length and aperture of a multi-element array transducer can be changed dynamically while the signal is being transmitted and received. The basic principle is illustrated in Figure 3.3.6. For a given focal point in the field, the time-of-flight to or from each transducer element is slightly different. The differences in these propagation delays can be compensated by varying the delays between transmitted or received signals in the electronic circuiting. In reception, by modifying the relative delays between elements progressively as echoes are returned from deeper and deeper regions, the reception focus can be optimised in real time for each depth.

Figure 3.3.5 Transducer array configurations. (a) Annular array with individual transducer elements in concentric rings. (b) Linear array with transducer elements aligned in a row. (c) Radial array (used for intraluminal imaging) with transducer elements oriented outwards from the circumference of a cylindrical support. (d) Multiple-dimensional array in two dimensions. Transducer elements are oriented in several adjacent rows. The lines extending from the transducer faces in (a), (b) and (d) demonstrate that timing of transmission and reception from individual elements and modification of the number of active elements in the effective aperture can be used to focus the ultrasonic beam at different positions

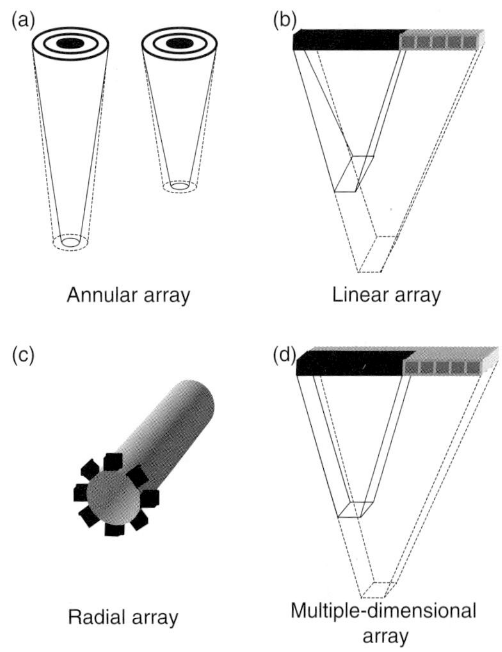

Figure 3.3.6 Beam forming. By adapting the relative delays between transmission and reception of signals at each element in a transducer array, the effective focal position can be varied

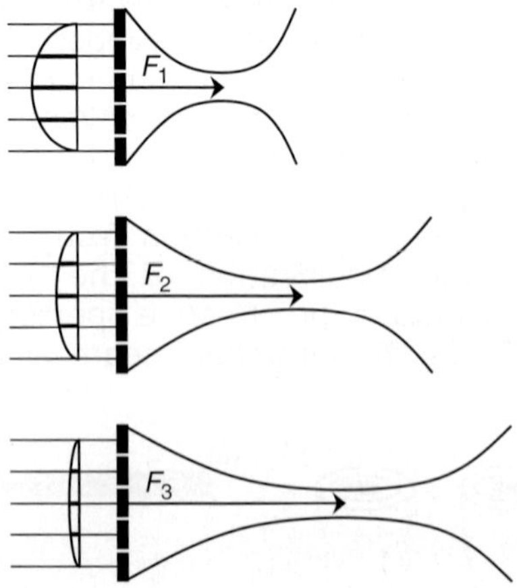

For an annular array, beam-forming techniques can be applied to obtain more uniform lateral focusing along the pulse-echo path and thus better depth of field. For linear arrays, the lateral resolution can be greatly improved for the lateral dimension parallel to the axis of the linear array. On the contrary, the lateral dimension of the field perpendicular to the axis of the linear array (the out-of scan-plane dimension) can only be focused at a fixed depth. This out-of-scan-plane dimension is known as the *slice thickness* or the *elevation dimension*. The fixed focus in the elevation dimension is generally determined by the geometry of a cylindrical lens fitted on the linear array. Multiple-dimensional arrays provide the best control of beam focusing characteristics (Figure 3.3.5(d)). For this array configuration, beam forming can be applied to optimise lateral resolution in both the in-scan-plane and elevation dimensions.

Prototype annular array systems have been presented with 40 MHz centre frequency (Ketterling *et al.*, 2005). Currently, work is also underway to develop 30 MHz linear arrays (Ritter *et al.* 2002). However, several technical difficulties must be confronted before multiple-element transducers become available above 20 MHz. These stem from the physical difficulties encountered in the miniaturisation of high frequency piezoelectric elements, the acoustic limitations related to side-lobe generation and lateral vibrational modes, and the high performance electronics needed to receive and record high frequency information rapidly along multiple channels.

3.4 From echoes to images

3.4.1 A-lines and envelope detection

The ultrasonic beam propagates through the medium and is partially reflected by interfaces and scattered by scattering bodies encountered along its path. This results in a series of echoes that are received by the transducer and converted to voltages. The stronger the echo, the greater is the absolute amplitude of the voltage that will be recorded. The resulting *radio frequency* signal or *A-line* presents negative and positive voltages as a function of time. If the signal is to be recorded digitally and if the frequency content in the signal is of interest, care must be taken to use a sufficient temporal sampling rate. At higher imaging frequencies, this condition implies the use of sampling rates on the order of 100s of samples per μs.

3.4.2 Time-of-flight and distance

The *time-of-flight*, Δt, is the time elapsed between the transmission of a pulse into the body and the return of an echo from a structure of interest. If the speed of sound, c, in the medium is known (approximately 1.54 mm/μs for soft tissue), the time-of-flight can be used to estimate the distance between the body's surface and the structure that produced the echo according to

$$d = \frac{c \cdot \Delta t}{2}. \tag{3.15}$$

If c is unknown, one can approximate using the averaged value of c in soft tissues. This value is close to the value of c in water (Table 3.2.1).

3.4.3 Time-gain compensation

To compensate for the reduction in signal amplitude with depth due to attenuation, time-gain compensation or TGC is applied by amplifying each segment of a received signal with an amplification factor that increases as a function of time (stronger amplification for greater distances between the transducer and the echo source.) The function describing gain versus time may be linear or non-linear and may be varied by the operator at the imaging system interface. After TGC application, the remaining signals are logarithmically compressed and amplified. Dynamic receiver filtering which narrows the received bandwidth for signals from deeper echo structure (attenuation will have decreased the useful bandwidth in these regions) may be applied to increase SNR. Once the signal has been received, digitised and amplified, it is processed via envelope detection, edge detection or another algorithm to detect echoes for display in an image format.

Figure 3.4.2 Scan line sweeping. (a) With a mono-element transducer, individual scan lines are acquired by mechanical displacement of the transducer between pulse-echo acquisitions. (b) Elements of multiple-element arrays can be fired in successive groups to rapidly form a series of scan lines with electronic scanning. (c) The relative delays between the transmission of signals from elements of a multiple-element array can be used to orient the beam

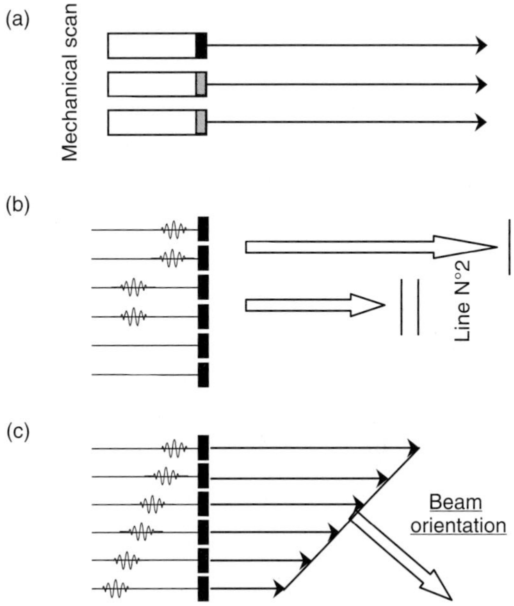

3.4.4 Scanning and image reconstruction

The envelope detected A-line can be presented as a line of dots with the brightness of each dot representing the echo strength at that location. The delay to each dot corresponds to the anatomic depth of the echo-generating structure. To form a cross-sectional anatomic image many pulses must be sent from different positions or different angles as illustrated in Figure 3.4.1. Each line thus acquired is known as a *scan line*. Scan lines can be produced via mechanical or electronic sweeping of the image plane (Figure 3.4.2). A grey-scale, cross-sectional image produced from dots representing echo strength received along many closely spaced scan lines is called a *B-scan*. The lateral dimension of a B-scan is determined by the lateral range of the beam sweeping, and its depth dimension is determined by the distance of propagation recorded for each scan line. The time necessary to sweep an image plane is limited roughly by the number of scan-lines times the time-of-flight to the deepest point in the image plane for multiple-element electronically scanning systems (on the order of several tens of milliseconds for conventional medical imaging leading to frame rates on the order of tens of hertz). For mechanically scanned systems (mono-element and annular array transducers), the imaging cadence is limited by the mechanical scanning speeds. Speeds can be obtained to permit real-time imaging (above 16 images per second).

Figure 3.4.1 Scan configurations. (a) Linear scan: Ultrasonic scan lines are acquired by moving the ultrasonic source laterally. (b) Sector scan: Ultrasonic scan lines are acquired by pivoting the ultrasonic source about a point. (c) Radial scan: The scan lines radiate from a central position (typically used for intraluminal imaging).

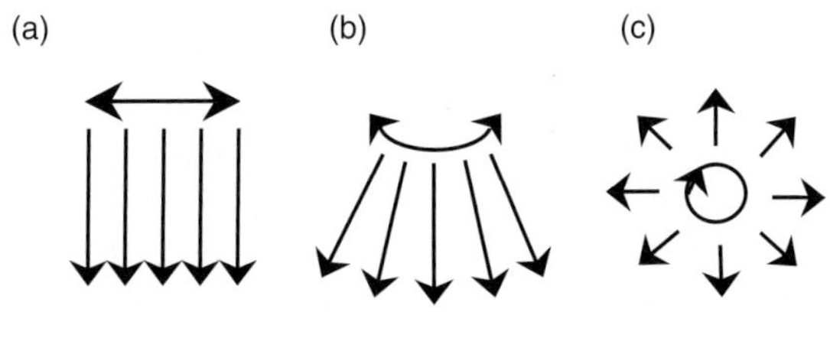

Figure 3.4.3 presents two transverse abdominal images of the mouse at 35 and 20 MHz. Shadowing in regions posterior to bony structures is evident in both images. The brightness of echoes from scattering structures in deeper regions due to attenuation is more clearly seen in the image at 35 MHz. Attenuation effects are less important at 20 MHz. The more sophisticated image formation techniques offered by the 20 MHz linear array system allow the lower frequency image to rival the quality of its higher frequency counterpart.

3.4.5 M-mode images

The term *M-mode* stands for motion-mode scanning. When examining a moving target within the body, instead of sweeping the transducer beam, a series of A-lines can be rapidly acquired along a fixed line of site and then displayed side by side in a grey-scale

Figure 3.4.3 B-mode images presenting a transverse abdominal plane of the mouse. (a) Scan lines were acquired by linear mechanical scanning of a 35-MHz centre-frequency broadband transducer. The resulting signals were envelope-detected and displayed in an image. Coupling between the transducer and the mouse was through a water path. (b) Image acquired with a 20 MHz Diasus linear array. This system permits rapid electronic scanning of the image plane and can take advantage of beam forming techniques. Coupling was achieved using gel. The compression of the linear array against the mouse leads to a compression of the abdominal cavity. The depth presented is approximately twice as deep in image (b) as in (a). Structures marked in the images: Vertebra (V), renal cortex (C), renal medula (M), liver (L). The contralateral kidney (K) and the abdominal aorta (A) are also marked with arrows in the 20 MHz image

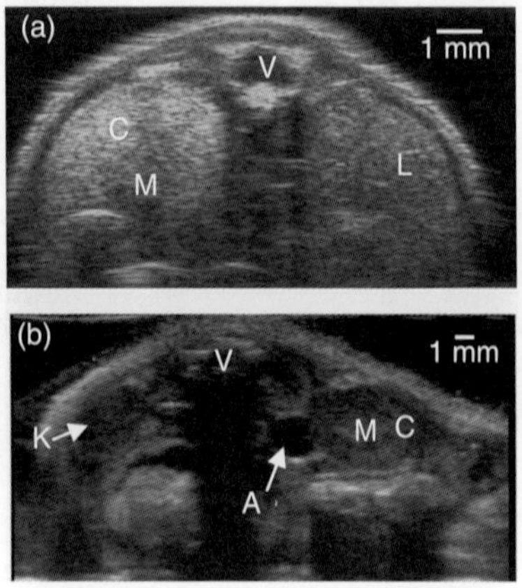

Figure 3.4.4 Liver tumour in a rat after diethyl nitrozamine ingestion. (a) Ultrasonic image of the liver in a rat. Imaging was performed with a 5–12 MHz linear array. (b) Compounding of the ultrasonic imaging yields an image with reduced speckle texture and better visualisation of a small hyperechoic mass with a diameter of 3 mm (arrow)

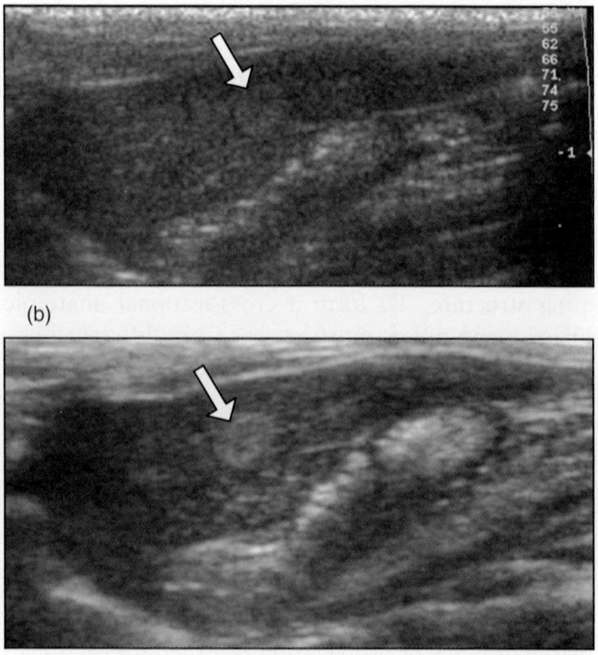

format. This format offers high temporal resolution limited only by the time-of-flight to the deepest point along an A-line. High frequency systems are often limited to mechanical scanning of mono-element transducers, and cardiac rhythm in small animals is very elevated compared to those in human patients (up to ~300 beats per minute for normal mice under anaesthesia). M-mode provides the most easily accessible solution for examining heart wall or valve motion in small laboratory animals.

3.4.6 Speckle, clutter and compound imaging

Ultrasonic *speckle* arises from coherent wave interference in tissue. This gives a granular appearance to what should appear as homogeneous tissue. The scattering structures giving rise to this are too small to be resolved individually, but the speckle pattern does vary as a function of the scatterer-size distributions. Clutter in the signal arises from side lobes, multi-path reverberations and tissue motion during imaging. Clutter and speckle reduce *contrast* between tissues with different echo responses. In other words, the ability to discriminate local changes in image brightness is limited by clutter and speckle noise.

Compound imaging uses scanning techniques to orient the beam from a multiple-element transducer as illustrated in Figure 3.4.2(c). Several B-mode images of the same scan plan are thus acquired with the slightly different scan-line angles. The images obtained from these slightly different viewpoints are then merged into a single compound image. Backscattered echoes from each view add coherently. Noise from clutter and speckle is reduced. The result provides a better visual definition of curved interfaces such as walls around arteries and better detection of small pathologies. Images with and without compound imaging are compared in Figure 3.4.4.

3.4.7 Three-dimensional imaging

To construct three-dimensional volume-rendering of structures or tissues, scans are typically acquired by controlled mechanical stepping of a two-dimensional array transducer with respect to the scanned object in the direction perpendicular to the scan plane. By pulling-back an intravascular ultrasound catheter (20 MHz) with an annular scanning mode as sequential images were acquired, three-dimensional information on atherosclerotic plaque in the coronary artery has been obtained (Schaar *et al.* 2004). In mice, high-resolution three-dimensional ultrasound has been used to evaluate tumour growth (Graham *et al.*, 2005; Wirtzfeld *et al.*, 2005).

3.5 Blood flow and tissue motion

Several different algorithms are used in ultrasound to analyse and display information related to blood flow or tissue displacement. Initially, the Doppler effect was exploited to measure blood velocity using a continuous wave Doppler approach (CW Doppler). Pulsed Doppler mode was developed to provide local estimates of flow and motion that could be overlaid on the B-mode anatomical image in what is known as Duplex imaging mode. Algorithms developed more recently use time-domain signal decorrelation to evaluate motion. Each technique is described briefly in the following sections.

3.5.1 Principle of Doppler measurement

The Doppler effect is perceived by a listener when the pitch of a moving sonic source such as a honking car horn or a police car siren is higher as the source approaches the listener and lower when it moves away. The Doppler effect occurs whenever there is relative motion between a source and a receiver of sound. To picture this more clearly, imagine a source sending sound at a frequency ν along a direct-line path to a receiver. If the receiver is stationary relative to the source, then the receiver will receive sound at the transmit frequency ν. This frequency is related to the sound speed in the medium and the wavelength according to the now-familiar equation

$$\nu_{r\ \text{stationary}} = \frac{c}{\lambda} = \nu. \tag{3.16}$$

If, however, the receiver is moving relative to the source with a velocity of v_R (this velocity is positive for a movement toward the source and negative for a movement away from the source), then the frequency detected by the receiver will be changed according to

$$\nu_{r\ \text{moving}} = \frac{c + v_R}{\lambda}. \tag{3.17}$$

The Doppler shift is the difference of the frequency perceived by a moving receiver compared to that perceived by a stationary receiver:

$$\Delta\nu_D = \frac{c + v_R}{\lambda} - \frac{c}{\lambda} = \frac{v_R}{\lambda} = \frac{v_R.\nu}{c}. \tag{3.18}$$

In the case of ultrasound being scattered from moving red blood cells, two successive Doppler shifts are involved. First, the sound from the transmitting transducer is received by the moving red blood cells. Second, the cells act as a moving source as they reradiate the ultrasound back to the transducer. This leads to a doubling of the Doppler frequency shift. Furthermore, only the component of the red blood cell velocity that is parallel to the ultrasound beam contributes to the Doppler shift. The resulting equation describing the Doppler shift for sound at a frequency of ν scattered by a moving red blood cell (or other structure) moving with a velocity of v_R is

$$\Delta\nu_{\text{Pulse-echo Doppler}} = \frac{2\nu \cdot v_R \cdot \cos\theta}{c}, \tag{3.19}$$

where θ is the angle between the direction of motion of the scattering body and the axis of the ultrasound beam and c is the speed of sound propagation in the medium (Figure 3.5.1). The angle θ (called the angle of incidence) can be estimated from simultaneously acquired B-mode scans and should be kept below 60°.

Figure 3.5.1 Ultrasonic measurement of the Doppler shift due to moving red blood cells. The ultrasonic beam arrives at an angle of θ with respect to the direction of the blood flow. The echoes scattered by the red blood cells moving at a velocity of v_R will be shifted in frequency due to the Doppler effect

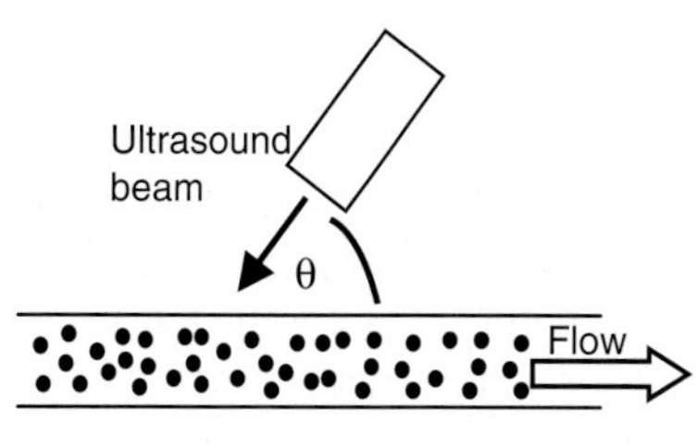

For clinical imaging frequencies, blood flow and tissue movement velocities and the propagation speed of sound in the body, the Doppler shifts fortuitously fall within the range of frequencies perceived by the human ear. Thus, visual display of Doppler information can be completed by actually listening to the audio frequencies of the Doppler shifts as they are being measured.

3.5.2 CW Doppler

For continuous wave Doppler, the transducer contains separate transmitting and receiving elements with beam overlap in a measurement volume. The transmitting element emits a continuous, uninterrupted ultrasonic wave of constant frequency and amplitude. The receiver continuously acquires echoes from the sensitive measurement volume. These echoes are amplified and compared to the frequency of the transmitted wave. A technique know as *phase quadrature detection* is used to calculate the Doppler shifts from moving structures, producing an output for shifts indicating positive flow towards the receiver and an output indicating negative flow away from the receiver. (The two outputs are distinguished by phase shift, and the definition of positive and negative directions is made by convention.) A low-pass filter limits the highest frequency shift of interest (fastest flow) and must be below the transmit frequency. A high-pass filter is also applied to remove high-intensity signals from slow moving structures that are not of interest such as the pulsatile movement of vessel walls (typical high-pass filter settings vary from 50–1000 Hz).

The simplest format for display of the Doppler measurement presents the amplitude of the Doppler shift on a vertical axis as a function of the time at which the measurement was made. Positive shifts indicate flow towards the transducer; negative shifts indicate flow away from the transducer (by convention). The higher the absolute value of the Doppler shift amplitude on the trace, the higher will be the velocity of the target. (Figure 3.5.2(a)). However, the Doppler signal originates from a number of scatterers moving differently with respect to the transducer, giving rise to a Doppler signal that contains a complex and changing set of frequencies. The relative power at the various frequency components present in the signal can be evaluated and displayed by presenting a grey-scale bar on the vertical axis (Figure 3.5.2(b)). The length of this bar indicates the total range of Doppler shift frequencies present and the relative brightness of each point along the bar is proportional to the amplitude of the signal (related to the number of moving targets) at that particular Doppler shift frequency. In general, the distribution of frequency shifts will be much larger for turbulent flow conditions. At any given time, the relative amount of power at each Doppler shift-frequency can be presented in spectral format as shown in Figure 3.5.2(c).

Figure 3.5.2 Display of the Doppler signal. (a) The Doppler shift is displayed as a function of time. By convention, positive values indicate movement toward the transducer and negative values indicate movement away from the transducer. (b) The full content of Doppler shift frequencies in the signal are displayed using different levels of brightness in a bar along the vertical axis. Pixel brightness is related to the relative number of scattering targets moving at the velocity giving a particular Doppler shift. (c) At a particular measurement time, the Doppler spectrum can be displayed. The spectrum diagrammed in part (c) of this figure represents the high frequency shifts measured at the time point $t = t_P$ in the Doppler chart in part (b) of this figure

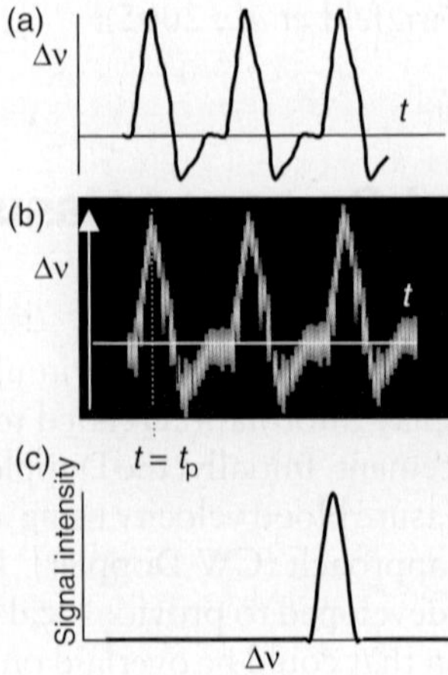

The chief advantage of CW Doppler is its high sensitivity. Also, CW Doppler does not suffer from 'aliasing' phenomenon caused by insufficient Doppler sampling described in the following section. CW Doppler is most used when it is not necessary to localise the moving structures. When flow from several vessels is superimposed in the measurement volume, the measurement becomes inaccurate. The use of sufficiently high frequencies, limits the measurement volume by attenuating the beams at deeper regions, and good CW Doppler evaluation of superficial vessels can be obtained.

3.5.3 Pulsed wave Doppler

Using a single transducer in pulse–echo mode, pulsed Doppler can evaluate movement and localise its position. Short pulses (typically a few cycles long) are transmitted at regular intervals. The time waited between transmission and the beginning of echo reception will determine the minimum depth of the pulsed-Doppler measurement window, and the total duration of echo acquisition (the time the receiver is effectively 'listening') defines the maximum range of depths in the measurement region. The lateral dimensions of the measurement volume are determined by transducer geometry and beam focusing as described in Sections 3.3.5 and 3.3.6.

The demodulator compares the phase of the received pulse with that of the oscillator (transmitted phase). The output waveform presents discrete estimates of Doppler shift as a function of time where the time between estimates equals the pulse repetition rate. Because the measurements of the Doppler shift frequency are sampled in time, the highest Doppler shift frequency that can be measured is limited by the Nyquist theorem. This states that the sampling frequency (pulse repetition rate) must be greater than or equal to twice the maximum measured Doppler shift frequency to determine that shift without aliasing. If the pulse repetition rate is too slow, the scattering target will already have moved by a distance too large on the scale of the wavelengths of the ultrasound cycles in the Doppler pulse before the next pulse reaches the target. The Doppler shift measured under such conditions is incorrect or aliased. Aliasing effects would appear on Doppler traces (such as those illustrated in Figure 3.5.2(a,b) without aliasing) as 'folding over' of sections of the curves between the positive and negative sections of the vertical scale. The maximum flow velocity that can be estimated with a pulsed Doppler system having a pulse repetition rate (PRR) and a transmit frequency of ν is

$$\text{max flow velocity} = \frac{PRR \cdot c}{4 \cdot \nu}. \tag{3.20}$$

As the pulse repetition rate must be longer to wait for a pulse to return from deeper regions, the maximum flow velocity will also be limited by the depth of the measurement region D_R as

$$\text{max flow velocity} = \frac{c^2}{8 \cdot \nu \cdot D_R}. \tag{3.21}$$

The combination of real-time imaging and Doppler techniques (most commonly pulsed-Doppler) is referred to as *duplex imaging*. Using the B-mode image, the angle θ between the direction of the ultrasonic beam and the principle axis of a blood vessel can be estimated and used to convert measurements of

Figure 3.5.3 Colour Duplex Doppler. Colour Doppler overlay on a B-mode image of heterogeneous liver in a woodchuck model with chronic hepatitis. Arrows mark a less echogenic zone on the B-mode image indicating an anterior mass corresponding to hepatocellular carcinoma. The Colour Doppler allows the detection of penetrating vessels

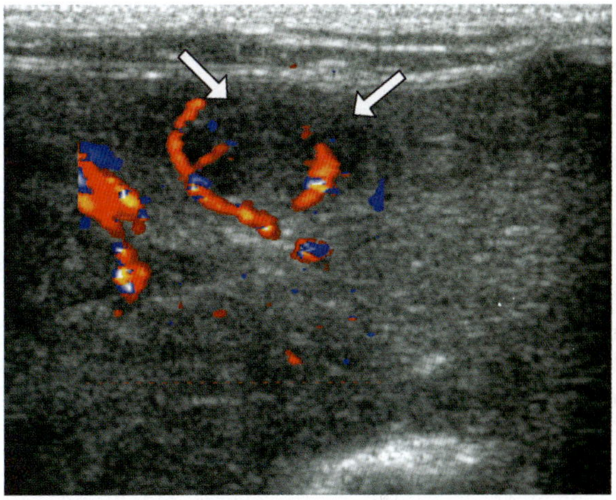

Doppler shift to measurements of mean blood velocity (see Eq. 3.19). Methods have been developed that allow the estimation of average Doppler shift frequency from the entire length of a scan line. An autocorrelation processor is applied to compare the quadrature-demodulated echoes received for two consecutive pulses. The output from the autocorrelation detector is zero for regions of echoes from stationary structures and non-zero where moving targets are indicated. Several pairs of pulses must be compared to improve signal to noise, and this system is very sensitive to clutter from large echoes at slow-moving but bright interfaces. The flow information is presented as a colour overlay on the B-mode image (Figure 3.5.3). The mean velocity, directional information (towards or away from the transducer) and the variance of the estimation are represented by the hue, the saturation and the luminance of the colour plot. The result is referred to as Colour Flow or Colour Doppler Imaging.

Power Doppler mode is an alternative display format that sacrifices velocity and directional information, displaying only a colour brightness linked to the total sum of power in the Doppler spectrum. As this mode simply integrates the total Doppler signal in the entire Doppler spectrum, regardless of relative Doppler shifts, the aliasing artifact linked to pulse repetition rate is no longer of importance. The greater the number of moving targets, the greater will be the total power and the brighter the display on the power

Doppler image. It should be noted that the number of moving targets is not the only factor effecting the measured Doppler power and that the relative scattering strength of an object will also have an effect on this parameter. This is why slow moving but highly reflective interfaces can introduce artifacts in power Doppler display manifested as bright flashes of colour.

3.5.4 Time-domain correlation

The last technique described in this section does not use the Doppler effect at all, but results in images with displays analogous to the Colour Doppler Imaging and Power Doppler.

First let us look at the nature of ultrasonic pulses returned from blood. The red blood cells (mean diameter of 7 μm) are the principal scattering element in blood. In general, their spacing is sufficiently sparse that each cell scatters the ultrasound independently. The wavelets scattered from each of these red blood cells combine according to their phase at the receiving transducer. If signals from several red blood cells arrive in phase, they will add constructively, and if they arrive out of phase they will cancel destructively. The ultrasonic echo pattern returned from a group of red blood cells results from this interference Figure 3.5.4. As the blood moves along the vessel a slight distance, the interference pattern acquired from a second pulse of ultrasound will be shifted by a time τ. Time-correlation methods are used to estimate the value of the shift in the position of the group of red blood cells between two consecutive ultrasonic pulses.

Figure 3.5.4 Time-domain correlation. (a) The ultrasonic echo acquired at a time t_0 presents the interference pattern produced by a group of red blood cells moving along a vessel at a velocity v_B. (b) At a time $t_0 + \Delta t$, a second pulse-echo acquisition is made along the same line of site. The group of red blood cells have moved by a distance $D = v_B \cdot \Delta T$. The limits of corresponding volumes of blood and ultrasonic signals are delimited with dotted vertical lines. The interference pattern in the ultrasonic signal is displaced in time by a time-shift of $\tau = 2D/c$ where c is the speed of sound in the medium

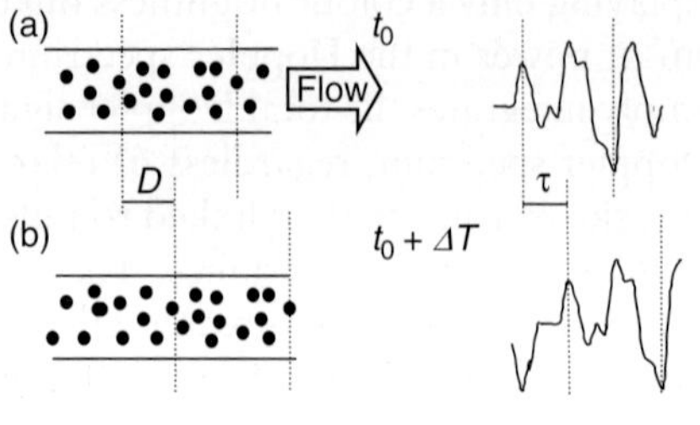

The more rapid the decorrelation of the signal (larger shift τ), the higher is the flow velocity. The red blood cell (RBC) velocity can be estimated from τ according to

$$\text{velocity RBC} = \frac{c \cdot \tau}{2 \cdot \Delta T}, \tag{3.22}$$

where ΔT is the time between consecutive pulses. Using two-dimensional cross-correlation methods, estimations of velocity can be obtained in two dimensions. Maximum measurable shifts are limited by two requirements. The group of tracked red blood cells must be maintained in the measurement volume for two consecutive pulses. The interference pattern must also remain recognisable. The limit placed on the maximum measurable velocity due to these constraints, is generally less severe than the limit imposed by the Nyquist frequency for pulsed-Doppler. Velocity information obtained by this method and overlaid on B-mode images is referred to as colour velocity imaging.

3.6 Non-linear and contrast imaging

For conventional ultrasonic imaging, the acoustic pressure in the transmitted pulse remains modest and wave propagation is approximately linear. This means that if an incident wave of amplitude A and angular frequency ω is sent into the body, the signal returned from scattering structures will consist of echoes with a lower amplitude and the same frequency as the incident wave. However, this simple relationship no longer holds when non-linear scattering structures, such as ultrasound contrast agents, are introduced in the medium or when the pressure of the incident wave becomes sufficiently high to provoke non-linear wave propagation effects. Great advantages can been taken using these non-linear effects for imaging and diagnosis as explained in the following sections.

3.6.1 Ultrasonic contrast agents

Ultrasound contrast agents consist of encapsulated gas microbubbles (Correas *et al.*, 2001). Because of the important differences between the density and compressibility of these microbubbles with respect to the blood and soft tissue surrounding them, they are excellent scatterers of ultrasound. Administered intravenously, these microbubbles must be smaller than 8–12 μm in diameter to traverse the pulmonary and capillary circulation. The majority of microbubbles giving rise to the contrast response from commercially available agents are typically between 2 and 5 μm in diameter. Following injection, they persist in the general circulation for several minutes to be ultimately eliminated by gas dissolution and metabolic processes. As these microbubbles are entirely contained in the vascular space and follow the blood velocity and flow, they can be used as blood flow tracers.

3.6.2 Ultrasonic response of contrast agents

When a contrast microbubble encounters an ultrasonic pulse, the bubble is cyclically compressed and dilated as the compression and rarefactional phases of the acoustic pressure wave pass. The oscillation is maximised if the frequency of the incident pulse (linear frequency ν, angular frequency ω) corresponds to the *resonant frequency* of the bubble (ν_{Br}, ω_{Br}). The resonant frequency is a specific property of a contrast microbubble, which depends on the nature of its gas, the composition of its encapsulating shell and its resting radius. The scattering cross section, σ, of a microbubble is maximum at its resonant frequency. The range of resonant frequencies of commercially available contrast agents is well adapted to coincide with the frequency range used in clinical imaging. As ultrasonic imaging systems use broadband pulses and as ultrasound contrast agents consist of a distribution of different bubble sizes, a range of bubbles will generally be excited at or near resonance. Resonance is a linear phenomenon. It implies that microbubble response will be optimal at a certain incident wave frequency, but for an excitation at a frequency ν, the returned response will be at that same frequency ν.

The amplitude A of the incident ultrasonic wave will determine whether the microbubble response is non-linear or not. At relatively weak acoustic pressures, a microbubble expands and contracts symmetrically about its equilibrium radius (Figure 3.6.1). The frequency of the wave scattered by the bubble in this low-pressure regime will be the same as the frequency of the incident wave (the incident or *fundamental* frequency ν). That is to say, the bubble behaves as a linear ultrasonic scatterer with a preference to responding most strongly at its resonant frequency. At more elevated incident pressures, the compression phase of the bubble is more limited than its expansion phase. The non-linear microbubble response introduces scattered frequencies that are harmonics, ultraharmonics and, sometimes, a subharmonic of the frequency of the incident pulse ($(N+1)\cdot\nu$, $((2N+1)/2)\cdot\nu$ and $(1/2)\cdot\nu$, respectively where N is an integer greater than 0). If the incident acoustic pressure is highly elevated, the microbubbles will be destroyed. This produces a very short but strong ultrasonic signal, containing a rich combination of ultrasonic frequencies (broadband). This also produces a rapid time decorrelation in the ultrasonic signal that is detected on Colour velocity imaging and displayed as a bright colour dot.

The precise amplitude levels at which microbubble behaviour will change from linear to non-linear and from non-linear to microbubble destruction depends on several factors. First, these threshold values will depend on the material properties (shell and gas) of

Figure 3.6.1 Illustration of a linear and non-linear microbubble response. Left: The first curve in the series of panels on the left demonstrates an incident acoustic pressure wave with a peak rarefactional pressure of A_1. The microbubble wall expands and contracts symmetrically about its equilibrium radius in response to this pressure wave. The resulting bubble radius changes are shown in the bottom curve on the left. The echo response from this symmetrically vibrating bubble is at the same frequency as the incident acoustic wave. Right: The first curve in the series of panels on the right demonstrates an incident acoustic pressure wave with a peak rarefactional pressure of A_2 ($A_2 > A_1$) The bubble response during expansion is greater than that during compression. The resulting bubble radius changes are shown in the bottom curve on the left. The echo response from this non-symmetrically vibrating bubble contains harmonic frequencies of the incident acoustic wave

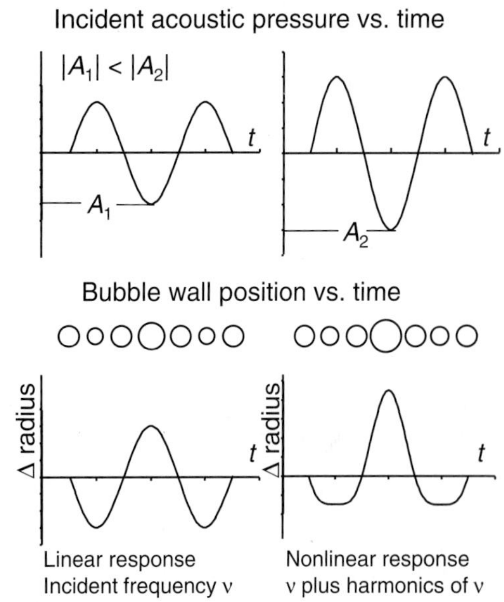

the microbubble. Thus, thresholds can be very different from one contrast agent to another. Second, thresholds depend indirectly on the size of microbubbles relative to the pulse frequency because the thresholds will be lower near resonance where the acoustic response of the microbubble is optimised. Finally, the viscosity of the medium around the microbubble, attenuation in the path between the transducer and the microbubble, and other environmental factors can modify the relationship between the transmit power and the microbubble acoustic response.

3.6.3 Ultrasonic intensity vs. contrast concentration

If the microbubbles in a cloud of contrast agent suspension are separated by at least an ultrasonic wavelength, the total effective scattering cross section of the ensemble of microbubbles can be estimated from the sum of the scattering cross sections of the individual microbubbles. This is important because it means that if a tissue is perfused with a suspension of microbubbles having a stable size distribution as a function of time, the additional backscattered intensity from a region of that tissue due to the contrast effect is linked to the concentration of microbubbles (number per unit volume). Techniques proposing quantitative functional evaluation of blood flow using ultrasonic contrast are based on this principle. However, when using such techniques, several effects that can alter the relationship between contrast concentration and acoustic intensity must be considered. Non-linear compression of intensity information displayed in an image format can skew relationships between apparent ultrasonic image brightness and contrast concentration. Attenuation in the sound path between the contrast-filled region and the transducer will reduce the apparent received amplitude of the contrast response. Modifications in the contrast size-distribution with time or between injections can modify the relationship between a fixed number of microbubbles and the ultrasonic intensity. To have comparable results with contrast ultrasound, therefore, (between subjects or for the same subject at different points in time) requires careful standardisation of image/signal analysis algorithms, region of interest placement and injection/measurement protocols.

3.6.4 Strategies for specific detection of contrast

In general, the contrast effect from clouds of microbubbles in the vascular cavities and vessels is detectable with B-mode imaging and standard Doppler techniques. This can be used to facilitate tasks such as endocardial border detection or Doppler detection of deeper vessels with low flow. It is more challenging to detect the presence of contrast microbubbles in the microvascular flow of the parenchyma or in weakly vascularised tumours. The relative blood volume (and thus the contrast concentration) is less than 25% in tissue parenchyma of highly vascularized organs such as the renal cortex and liver and can be as low as 10% in myocardial wall. Non-linear imaging techniques that separate contrast microbubble echoes from the linear response of tissue are key to the detection of microvascular flow.

Many specific ultrasonic pulse sequences have been developed to separate the non-linear contrast response from the linear response of the surrounding tissue. *Harmonic imaging* and *pulse inversion imaging* use non-linear signal processing techniques to select the signal due to the harmonic oscillatory response of microbubbles. Harmonic Imaging uses a large bandwidth transducer to transmit an ultrasonic pulse centred at the lower part of the transducer bandwidth into the medium (Figure 3.6.2). The received signal is filtered to retain only the echoes within the higher frequency range of the transducer. Thus, reception of the harmonic microbubble response is favoured rather than the linear response of the tissue. A disadvantage of this technique is that the bandwidth reduction achieved by filtering leads to a reduction in image spatial resolution. Pulse Inversion Imaging does not

Figure 3.6.2 Harmonic imaging. By using the lower part of the total transducer bandwidth in transmission and the higher part in reception, harmonic content resulting from the non-linear response of contrast microbubbles is preferentially received with respect to the linear echoes from tissue structures. The respective transmit and receive bandwidths can be narrowed further to minimise the potential to acquire linear response in the receiver bandwidth, but loss of spatial resolution will result

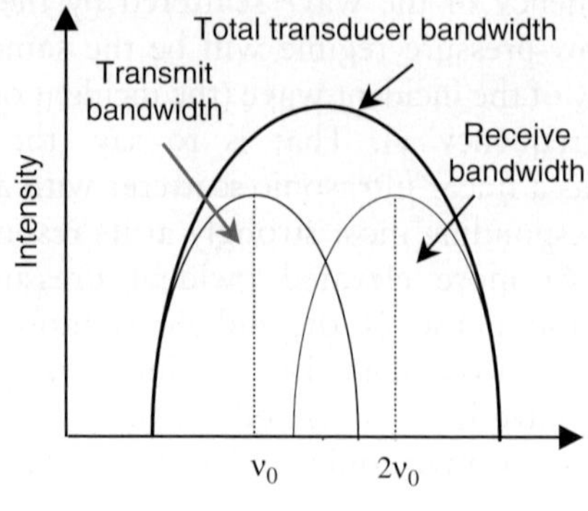

Figure 3.6.3 Principle of Pulse Inversion Imaging. A pulse sequence consisting of pulses of opposite phase is transmitted into the medium. The sum of the echoes returned for each of these pulses is summed. If the echoes are from non-linear scatterers, as illustrated, this sum will reveal the harmonic component

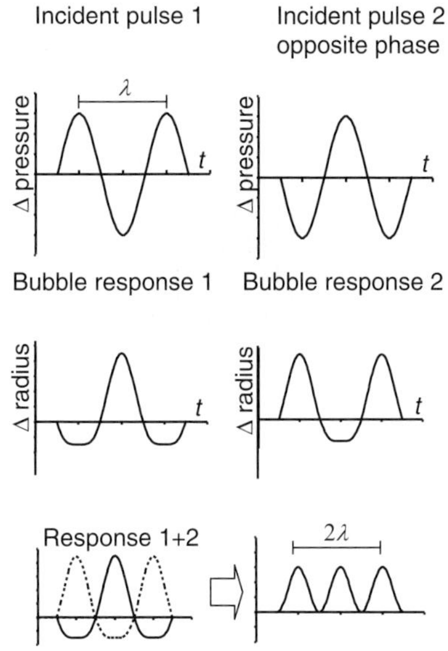

suffer from this filtering effect. Two successive ultrasonic pulses with opposite phases (Figure 3.6.3) are transmitted. In a linear medium, the echo response for the second transmitted pulse is the opposite of the echo response from the first, and adding the two received responses results in a signal with an amplitude of zero. However, when non-linear scatterers are present (or when non-linear wave propagation becomes important as described in Section 3.6.5) the sum of the returned echoes from the two pulses will not be zero. Instead, the non-linear content of the echoes is reinforced in the sum, and the linear content cancels out in the sum.

Harmonic power Doppler exploits the non-linear properties of the microbubbles to distinguish their Doppler response from that due to, for example, myocardial wall movement or movement of the ultrasonic probe in the hand of the operator. As described above, Power Doppler is highly sensitive to clutter signal that can be generated by motion of bright tissue interfaces. This clutter can be stronger even than the contrast enhanced signals from blood. By combing Doppler detection with harmonic filtering, the flash artifact can be reduced. Considering the second harmonic frequency as the transmit frequency in the Doppler equation, for example, the Doppler spectrum measured, based on the shift in this frequency, will essentially be generated from the movement of microbubbles. This technique is a powerful tool for detecting flow in small vessels of organs that are moving with cardiac pulsing or respiration. The technique has been applied most effectively, not by relying on the harmonic oscillation of the moving bubbles, but rather by destroying the microbubbles with a high-pressure acoustic wave. The bubble destruction results in a brief, broadband response from all bubbles (independent of their relative motion). Although very sensitive for the detection of weakly vascularised or slow flow zones, breaking the microbubbles to obtain their detection means that one must wait for refilling to obtain another look. If the flow being examined is capillary flow, the cost in data acquisition time is considerable.

Power pulse inversion is a hybrid technique between Doppler and pulse-inversion. Two phase-inverted pulses are transmitted at a time interval of T. Doppler analysis of the backscattered echoes reveals that the signal from linear scatterers are offset in the Doppler spectrum by a factor equal to half of the pulse repetition frequency ($1/2T$). The second harmonic response of non-linear scatterers lies between $-1/4T$ and $1/4T$ in the Doppler spectrum. By filtering the Doppler spectrum with a modified wall movement filter, the Doppler frequencies originating only from the moving non-linear scatterers can be selected. This technique is currently one of the most sensitive for flow detection in small vessels with contrast. Sensitive even at low incident acoustic pressure, this technique minimises microbubble destruction during imaging and non-linear wave propagation in the tissue.

3.6.5 Non-linear wave propagation

For linear wave propagation, the changes in pressure produced within the propagation medium (as illustrated in Figure 3.1.1(b)) are very small compared to the equilibrium atmospheric pressure. Thus, the particle displacements in the material and the local density variations are extremely small. However, ultrasonic imaging systems can deliver pulses with peak negative pressures in the range of several MPa (atmospheric pressure equals 101.3 kPa). Thus, within the range of conventional medical imaging, pressures can be attained at which the local changes of pressure and density in the medium will lead to local variations in the speed of sound.

Two parameters describing the level of non-linearity of a medium may be encountered. The parameter B/A is the thermodynamic parameter of non-linearity. The greater the value of B/A, the greater will be the importance of the first non-linear term in the equation describing the relationship between the density and pressure in the medium. Values of B/A are on the order of 5 for water, slightly higher in blood, 7 for liver and 10–11 for fatty tissue. The acoustic non-linear parameter calculates the additional sound speed at a point in the medium due to the thermodynamic non-linearity (B/A) and the local particle velocity during the wave's passage. It is a very small effect (on the order of 1 mm/s compared to a sound speed of 1540 m/s in soft biological tissue), but the effect on the wave will be accumulated along its entire propagation distance.

In the regions of the medium under local compression (higher densities) the local wave speed will be higher than in the regions of rarefaction. As the ultrasonic wave propagates, its waveform becomes more and more deformed. A pulse beginning with a sinusoidal form at a centre frequency of ν will become distorted. The spectral content of such a wave is full of harmonics at 2ν, 3ν, etc as demonstrated in the Figure 3.6.4.

The selective detection and display of harmonic energy generated by non-linear propagation is referred to as *native* or *tissue harmonic imaging*. Lateral resolution is enhanced and clutter is reduced compared to imaging at the fundamental transmit frequency. As higher frequencies are exploited, some loss in depth penetration can be anticipated. Microbubbles in the sound path will strongly increase the non-linearity of the medium and thus the production of non-linear propagation harmonics.

The non-linear propagation effect, however, can be problematic when the imaging goal is to specifically detect contrast microbubbles. If the amplitude of the incident ultrasonic pulse is well adapted for contrast imaging, the dominant non-linear response is due to the contrast microbubbles. However, as ultrasonic pressure levels applied are increased, the effect of non-linear propagation in tissue becomes more and more important. This means that the echoes from the tissue parenchyma will have more and more non-linear content making it more difficult to separate the contrast microbubbles form the surrounding tissue. Optimising the non-linear contrast response relative to tissue is part of the art of ultrasonic contrast imaging.

3.6.6 Contrast flow imaging

Direct measurement of microcirculatory flow with clinical Doppler is not possible because the flow speeds are too small. Contrast flow imaging allows an indirect evaluation of microvascular flow. Images comparing Doppler and contrast imaging of the vascularisation in a murine tumour model can be found in the application report by O Lucidarme paragraph 11.4.1, page 291.

Figure 3.6.4 Non-linear propagation. A waveform with high acoustic pressure detected after propagation across several centimetres of water. Initially the waveform consisted of several cycles of a sinusoid at a frequency of 4.6 MHz. (a) The waveform (peak rarefactional pressure of 3.5 MPa) is strongly distorted. (b) The spectrum contains considerable energy at the 2nd, 3rd and 4th harmonics

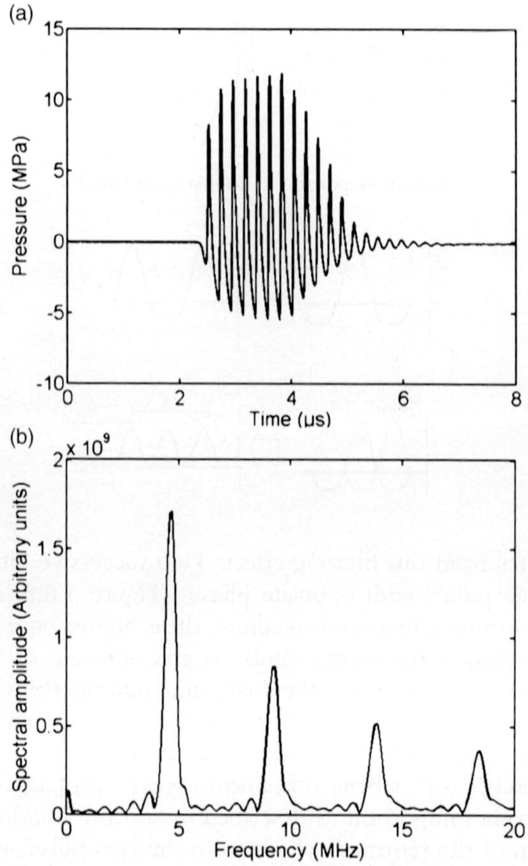

USCAs are used as blood pool tracers, in a similar manner to that used in nuclear medicine. Non-linear imaging sequences such as Pulse Inversion or Power Pulse Inversion are used to separate the contrast response from the surrounding parenchyma in real-time. One approach is to use a bolus injection of USCA, and then to acquire image time-intensity curves in regions of interest to follow the passage of the bolus. Functional indices, such as time to peak, peak intensity, wash-in slope, wash-out slope, duration of contrast enhancement and area under time-intensity curves can be calculated. Another approach

Figure 3.6.5 Perfusion imaging with contrast. Microbubbles are injected such that a uniform concentration of microbubbles circulates in the system. An image plane is selected containing the region in which flow is to be assessed. A series of high-pressure acoustic pulses are applied to clear the region of bubbles and then the progressive contrast enhancement produced by the return of the bubbles is monitored using non-linear imaging. The slope of the perfusion curve $1/\tau$ is related to the blood velocity and the plateau C_0 to the fractional blood volume in the region of interest

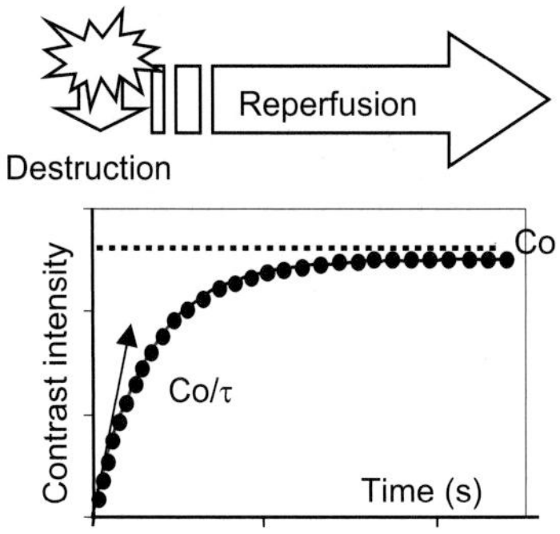

to functional contrast imaging is based on the destruction of microbubbles at high mechanical index followed by the study of the reperfusion curve observed at low mechanical index (Figure 3.6.5). The slope of the reperfusion curve is theoretically related to the blood velocity, and the level of enhancement to the local blood volume. Several variations of the above techniques exist, and techniques continue to be optimized to shorten the time needed for the functional imaging and improve the signal-to-noise ratio for contrast detection. Such flow assessment using USCAs is becoming an important technique for the assessment of tumor response to therapy (Eckersley *et al.*, 2002; Forsberg *et al.*, 2004; Iordanescu *et al.*, 2002).

3.6.7 Contrast agents and high frequency ultrasound

Ultrasound contrast agents have already considerably increased the imaging and diagnostic capacities of medical imaging in the 2 – 10 MHz frequency range. Although size distributions of existing USCAs were selected to favour resonant behaviour in this frequency range, subharmonic and ultraharmonic scattering by current contrast agents (Optison^{TM} and Definity^{TM}) has been demonstrated for transmit frequencies up to 26 MHz (Goertz *et al.* 2005a, Goertz *et al.* 2005b). Evidence of significant non-linear propagation and acoustic microbubble destruction has also be found at high frequencies (Goertz *et al.*, 2005a).

3.7 Discussion

A well-established, widely used and affordable technology, ultrasound imaging continues to grow and to expand its capabilities. Ultrasound sets itself apart from many other medical imaging modalities due to its real-time and non-invasive nature. Advances in high frequency ultrasound technology have been central in allowing the modality to enter the race for the creation of imaging systems adapted for small laboratory animals. Ultrasound contrast imaging has not only helped improve existing images but has also led to the modification of ultrasonic pulse sequences and signal analysis. The evaluation of the blood flow in tissues using ultrasonic contrast agents opens many new diagnostic opportunities. Contrast agent targeting for specific tissue discrimination or for therapeutic vectors is also under development.

Careful evaluation of imaging requirements (resolution, penetration depth, frame rate, field-of-view etc.) should help to choose the characteristics of the imaging system and the modality that is most well-adapted to a particular experimental protocol. Knowledge of the physical effects related to pulse-echo propagation and signal analysis should allow operators to avoid errors in image interpretation or quantification. High frequency ultrasound is the dominant contender for imaging protocols requiring hemodynamic evaluation or embryonic imaging. The real-time nature of high frequency ultrasound also makes it an ideal technique for image-guided injections. It offers unique capabilities for rapid phenotypic evaluation and tumour characterization. The non-invasive nature of ultrasound is highly favourable to serial *in vivo* studies. Ultrasound can be a powerful tool for the biologist. It is now up to the biologists to see that the tool is applied effectively.

References

Akirav, C., Lu, Y., Mu, J., Qu, D. W., Zhou, Y. Q., Slevin, J., Holmyard, D., Foster, F. S., Adamson S. L., 2005. "Ultrasonic detection and developmental changes in calcification of the placenta during normal pregnancy in mice." *Placenta* **26**, 129–137.

Baddour, R. E., Sherar, M. D., Hunt, J. W., Czamota, G. J., Kolios, M. C., 2005. "High-frequency ultrasound scattering from microspheres and single cells." *J. Acoust. Soc. Am.* **117**, 934–943.

Cheung, A. M., Brown, A. S., Hastie, L. A., Cucevic, V., Roy, M., Lacefield, J. C., Fenster, A., Foster, F. S., 2005. "Three-dimensional ultrasound biomicroscopy for xenograft growth analysis." *Ultrasound Med. Biol.* **31**, 865–870.

Correas, J-M., Bridal, L., Lesavre, A., Mejean, A., Claudon, M., Helenon, O., 2001. "Ultrasound contrast agents: properties, principles of action, tolerance, and artifacts." *Eur. Radiol.* **11**, 1316–1328.

Eckersley, R. J., Sedelaar, J. P., Blomley, M. J., Wijkstra, H., deSouza, N. M., Cosgrove, D. O., de la Rosette, J. J., 2002. "Quantitative microbubble enhanced transrectal ultrasound as a tool for monitoring hormonal treatment of prostate carcinoma." *Prostate* **51**, 256–267.

Faran Jr., J. J., 1951. "Sound scattering by solid cylinders and spheres." *J. Acoust. Soc. Am.* **23**, 405–418.

Forsberg, F., Ro, R. J., Potoczek, M., Liu, J. B., Merritt C. R., James K. M., Dicker A. P., Nazarian L. N., 2004. "Assessment of angiogenesis: implications for ultrasound imaging." *Ultrasonics* **42**, 325–330.

Foster, F. S., Pavlin, C. J., Harasiewicz, K. A., Christopher, D. A., Turnbull, D. H., 2000. "Advances in ultrasound biomicroscopy." *Ultrasound Med. Biol.* **26**, 1–27.

Goertz, D. E., Cherin, E., Needles, A., Karshafian, R., Brown, A. S., Burns, P. N., Foster, F. S., 2005a. "High-frequency, nonlinear B-scan imaging of microbubble contrast agents." *IEEE Trans. Ultrason. Ferroelectr. Freq. Control* **52**, 65–79.

Goertz, D. E., Needles, A., Burns, P. N., Foster, F. S., 2005b. "High-frequency, nonlinear flow imaging of microbubble contrast agents." *IEEE Trans. Ultrason. Ferroelectr. Freq. Control* **52**, 495–502.

Goertz, D. E., Yu, J. L., Kerbel, R. S., Burns, P. N. and Foster, F. S., 2002. "High-frequency Doppler ultrasound monitors the effects of antivascular therapy on tumor blood flow." *Cancer Res.* **15**, 6371–6375.

Graham, K. C. Wirtzfeld, L. A., MacKenzie, L. T., Postenka, C. O., Groom, A. C., MacDonald, I. C., Fenster, A., Lacefield, J. C., Chambers, A. F., 2005. "Three-dimensional high-frequency ultrasound imaging for longitudinal evaluation of liver metastases in preclinical models." *Cancer Res.* **65**, 5231–5237.

Gupta, A. K., Turnbull, D. H., Harasiewicz, K. A., Shum, D. T., Watteel, G. N., Foster, F. S., Sauder, D. N., 1996. "The use of high-frequency ultrasound as a method of assessing the severity of a plaque of psoriasis." *Arch. Dermatol.* **132**, 658–662.

Hickling, R., 1962. "Analysis of echoes from a solid elastic sphere in water." *J. Acoust. Soc. Am.* **34**, 1582–1592.

Honda, Y., Yock, P. G., Fitzgerald, P. J., 1999. "Impact of residual plaque burden on clinical outcomes of coronary interventions." *Catheter Cardiovasc. Interv.* **46**, 265–276.

Insana, M. F, Wagner, R. F., Brown, D. G., Hall, T. J., 1990. "Describing small-scale structure in random media using pulse-echo ultrasound." *J. Acoust. Soc. Am.* **87**, 179–192.

Iordanescu, I., Becker, C., Zetter, B., Dunning, P., Taylor, G. A., 2002. "Tumor vascularity: evaluation in a murine model with contrast-enhanced color Doppler US effect of angiogenesis inhibitors." *Radiology* **222**, 460–467.

Jouannot, E., Duong-Van-Huyen, J-P., Bourahla, K., Laugier, P., Lelievre-Pegorie, M, Bridal, S.L., 2006. "High freqeuncy ultrasound detection and follow-up of Wilms' tumor in the mouse." *Ultrasound. Med. Biol.* **32**, 183–190.

Ketterling, J. A., Aristizabal, O., Turnbull, D. H., Lizzi, F. L., 2005. "Design and fabrication of a 40-MHz annular array transducer." *IEEE Trans. Ultrason. Ferroelectr. Freq. Control* **52**, 672–681.

Kremkau, F. W., 1998. *Diagnostic Ultrasound: Principles and Instruments*,W. B. Saunders, London, ISBN 0721671438.

Lockwood, G. R., Ryan, L. K., Hunt, J. W., Foster, F. S., 1991. "Measurement of the ultrasonic properties of vascular tissues and blood from 35–65 MHz." *Ultrasound. Med. Biol.* **17**, 653–666.

Lucidarme, O., Franchi-Abella, S., Correas, J-M., Bridal, S. L., Kurtisovski, E. and Berger, G., 2003. "Blood flow quantification with contrast-enhanced US: 'entrance in the section' phenomenon – phantom and rabbit study." *Radiology* **228**, 473–479.

Lucidarme, O., Nguyen, T., Kono, Y., Corbeil, J., Choi, S. H., Varner, J., Mattrey, R. F., 2004. "Angiogenesis model for ultrasound contrast research: exploratory study." *Acad Radiol* **11**, 4–12.

Pan, L., Zan, L., Foster, F. S., 1998. "Ultrasonic and viscoelastic properties of skin under transverse mechnanical stress *in vitro*." *Ultrasound. Med. Biol.* **24**, 995–1007.

Pavlin, C. J., Harasiecwicz, K., Sherar, M. D., Foster, F. S., 1991. "Clinical use of ultrasound biomicroscopy." *Ophthalmology* **98**, 287–295.

Phoon, C. K., Aristizabal, O., Turnbull, D. H., 2000. "40 MHz Doppler characterization of umbilical and dorsal aortic blood flow in the early mouse embryo." *Ultrasound. Med. Biol.* **26**, 1275–1283.

Phoon, C. K., Ji, R. P., Aristizabal, O., Worrad, D. M., Zhou, B., Baldwin, H. S., Turnbull, D. H., 2004. "Embryonic heart failure in NFATc1-/- mice: novel mechanistic insights from *in utero* ultrasound biomicroscopy." *Circ. Res.* **95**, 92–99.

Ritter, T. A., Shrout, T. R., Tutwiler, R., Shung, K. K., 2002. "A 30-MHz piezo-composite ultrasound array for medical imaging applications." *IEEE Trans. Ultrason. Ferroelectr. Freq. Control* **49**, 217–230.

Rose, J. H., Kaufmann, M. R., Wickline, S. A., Hall, C. S., Miller, J. G., 1995. "A proposed microscopic elastic wave theory for ultrasonic backscatter from myocardial tissue." *J. Acoust. Soc. Am.* **97**, 656–668.

Schaar, J. A., Regar, E., Mastik, F., McFadden, E. P., Saia, F., Disco, C., de Korte, C.L., de Feyter, P. J., van der Steen, A. F., Serruys, P. W., 2004. "Incidence of high-strain patterns in human coronary arteries: assessment with three-dimensional intravascular palpography and correlation with clinical presentation." *Circulation* **109**, 2716–2719.

Semple, J. L., Gupta, A. K., From, L., Harasiewicz, K. A., Sauder, D. N., Foster, F. S., Turnbull, D. H., 1995. "Does high-frequency (40–60 MHz) ultrasound imaging play a role in the clinical management of cutaneous melanoma?" *Ann. Plast. Surg.* **34**, 599–605.

Snook, K. A., Zhao, J. Z., Alves, C. H., Cannata, J. M., Chen, W. H., Meyer Jr., R. J., Ritter, T. A., Shung, K. K., 2002. "Design, fabrication and evaluation of high frequency, single-element transducers incorporating different materials." *IEEE Trans. Ultrason. Ferroelectr. Freq. Control* **49**, 169–176.

Turnbull, D. H., Bloomfield, T. S., Baldwin, H. S., Foster, F. S., Joyner, A. L., 1995. "Ultrasound backscatter microscope analysis of early mouse embryonic brain development." *Proc. Natl. Acad. Sci. USA* **92**, 2239–2243.

Turnbull, D. H., Ramsay, J. A., Shivji, G. S., Bloomfield, T. S., From, L., Sauder, D. N., Foster, F. S., 1996. "Ultrasound backscatter microscope analysis of mouse melanoma progression." *Ultrasound. Med. Biol.* **22**, 845–853.

Webb, A. R., 2003. Ultrasonic Imaging. *Introduction to Biomedical Imaging*, John Wiley & Sons, Inc., Hoboken, New Jersey, pp. 107–151, ISBN 0471237663.

Wells, P. T., Halliwell, M., Skidmore, R., Webb, A. J., Woodcock, J. P., 1977. "Tumour detection by ultrasonic Doppler blood-flow signals." *Ultrasonics* **15**, 231–232.

Wirtzfeld, L. A., Wu, G., Bygrave, M., Yamasaki, Y., Sakai, H., Moussa, M., Izawa, J. I., Downey, D. B., Greenberg, N. M., Fenster, A., Xuan, J. W., Lacefield, J. C., 2005. "A new three-dimensional ultrasound microimaging technology for preclinical studies using a transgenic prostate cancer mouse model." *Cancer Res.* **65**, 6337–6345.

Zhou, Y. Q., Foster, F. S., Qu, D. W., Zhang, M., Harasiewicz, K. A., Adamson, S. L., 2002. "Applications for multifrequency ultrasound biomicroscopy in mice from implantation to adulthood." *Physiol. Genomics* **14**, 113–126.

4 *In Vivo* Radiotracer Imaging

Bertrand Tavitian, Régine Trébossen, Roberto Pasqualini and **Frédéric Dollé**

4.0 Introduction

The imaging methods presented in this chapter are based on the detection of radioactive nuclides (or radionuclides) that are not naturally present in animals (or humans) and once introduced inside, can be detected from outside their body. The image produced represents a map of the distribution of the radionuclide in the animal, which depends only on the manner in which its body handles the radionuclide, or the molecule it has been incorporated into. This in turn is dependent on (i) the nature of the radionuclide or radiochemical, (ii) its mode of administration, (iii) the physiological state of the animal, and (iv) the time delay between the introduction of the chemical and the acquisition of the image. Because they produce maps of the distribution of molecules, radiotracer-imaging techniques are essentially *molecular imaging* techniques and carry no direct anatomical information. Because of the very high sensitivity with which radionuclides can be detected, radioactively labelled compounds can be detected from outside the animal, even when they are present in negligible concentration (nanomolar or lower). And because '*life is essentially chemistry*' (Lord Kelvin), molecular imaging provides invaluable data for the exploration of life and disease from a biochemical point of view.

4.0.1 The radiotracer principle

The radiotracer principle, on which radiotracer imaging is based, is similar to the indicator principle by which chemists can determine the pH of a solution: A tiny amount of a chemical indicator is added to the solution, diffuses freely, and the change in colour reports on the concentration of H^+ ions *without substantially affecting the properties* of the solution. The use of radionuclides as indicators was invented just before WWI by the chemist, George de Hevesy, who considerably developed their applications, for which he was awarded the Nobel Prize in 1943. As he humorously reported in his Nobel lecture, the original idea came to de Hevesy from his frustration after failing in the task assigned to him by Rutherford, which was to separate Radium D from lead. As he could not separate the two compounds, he decided to use Radium D (actually, ^{210}Pb, a radioactive isotope of lead, as was discovered later) to trace the diffusion of lead, first in lead itself, then in plants and animals, and he went on to generalize the use of radiotracers with other isotopes, including artificial ones.

Radiotracers are basically radioactive indicators and share the same two basic properties with chemical indicators: (i) They can be used in very small amounts (i.e. *traces*) so that they do not disturb the system in which they are introduced; (ii) they report on the properties of the system. In contrast to indicators in solutions, radiotracers introduced into a living organism tend to concentrate in localized areas depending on their interactions with the heterogeneous chemical composition of the animal's tissues. The aim of molecular imaging with radiotracers is to provide maps of these interactions, which can be one or several of the following: Diffusion, metabolic degradation, incorporation into molecular complexes, affinity binding, enzymatic modification, sequestration and excretion. As all these events are time-dependent, radioactivity distribution maps evolve along time after radiotracer administration.

4.0.2 Radiotracers can be detected externally

In contrast to radioactivity counting in samples or tissue autoradiography, non-invasive imaging methods can follow the evolution of radiotracer distribution with time, providing additional information on biochemical events.

The high sensitivity of radiotracer imaging is linked to the relatively high levels of energy detected, with two immediate consequences: (i) Radiotracer imaging requires only very small amounts of labelled material because a significant fraction of the energy emitted by the radionuclide can travel through the body and reach the external detector; this allows application of the radiotracer principle, a condition which in turns permits quantification of molecular concentrations; however, it is at the expense of spatial resolution. (ii) Radiotracer imaging carries radio-security issues because part of the energy can be deposited inside the subject's or experimenter's body. This is detrimental, because both the necessity to radiolabel the imaging probe, and radioprotection concerns, limit the ease of use and increase the cost of the technique.

Nevertheless, radiotracer imaging allows comparing radiotracer distribution in the same animal at different times, in different physiological or pathological conditions, spares animal lives, and more importantly can be translated into humans. Accordingly, the three *in vivo* radiotracer-imaging techniques, scintigraphy, single photon emission computerized tomography (SPECT) and positron emission tomography (PET), were originally invented in clinical nuclear medicine departments. Their use is fully justified in pathologies threatening the life of patient such as cancer, cardiovascular or neurological diseases, and for the measurement of drug distribution or efficiency, which is made feasible by the capacity to introduce radionuclides in many different molecules of interest and quantify their distribution. Considerable pharmacokinetics and distribution data for hundreds of compounds of pharmacological interest has been gained in the last decades thanks to radiotracer imaging of labelled drugs, ligands, enzyme inhibitors, etc.

Only recently has the intrinsically low spatial resolution of these techniques progressed up to the point where they can be used in small laboratory animals for research purposes. The invention and commercialization of dedicated tomographs for small rodents has given to scientists an access to a myriad of small laboratory rodents' models for physiological and pathophysiological studies. This is rapidly changing the way in which *in vivo* biochemistry, molecular pathophysiology and pharmacological research are conducted, and increasing the importance of molecular imaging in biology and medicine.

This chapter deals with the physical principles and instruments of radiotracer imaging, whereas Chapter 7 describes some of the newest radiochemicals that can be used to assess biochemistry non-invasively. For further reading, refer to (Wagner *et al.*, 1995; Bushberg *et al.*, 2001; Webb, 2003; Bailey *et al.*, 2005).

4.1 Radioactivity

The nuclei of atoms contain two heavy particles of quasi-similar mass, the proton which has a positive electric charge of same value as the electron ($1\,\text{eV} = 1.6 \times 10^{-19}\,\text{C}$), and the neutron which bears no charge. An atom X is noted as $^{A}_{Z}X$, in which Z, the atomic number, is the number of protons, and A, the atomic mass, is the number of nucleons (protons + neutrons). Atoms with the same Z and different A are called isotopes and are chemically identical. Atoms with the same A and different Z have the same mass but differ chemically, and are called isomers.

Nuclei are stable when there is a balance between the number of protons Z and the number of neutrons they contain. Instable nuclei are those in which the relative amount of protons and neutrons is off-balance or which contain too many nucleons (high A). Instable nuclei tend to achieve stability by emitting radiation, a phenomenon called radioactivity discovered by Henri Becquerel in 1896. Among the naturally radioactive nuclei such as carbon-14, potassium-40 and tritium, some are normally present in living species; radioactive nuclei used for imaging are artificial.

A discussion on nuclear physics is out of the scope of this chapter, but one point should be stressed: The order of magnitude of the forces that bind nucleons together is several orders of magnitude higher than the energy that binds electrons to the nucleus.

A list of all known isotopes of natural and artificial elements, including the properties of those with radionuclides, can be found in Browne *et al.* (1978) or in (http://www.webelements.com/webelements/elements/text/periodic-table/radio.html). For safety data, one may consult for instance (http://www.nrc.gov/reading-rm/doccollections/cfr/part020/appb/).

4.1.1 Types of radioactivity

Radioactive transformation can lead to a change in the number of protons: In that case, that is in the process called disintegration, the atomic number and the chemical properties of the daughter and parent nuclei are different. Conversely, a nucleus can emit radiation without changing its atomic number, and this is called nuclear transition. The types of radioactivity depend on the nature of the radiation, which are

(i) Alpha radioactivity is the emission of a Helium nucleus (α particle, ${}^{4}_{2}\mathrm{He}$):

$$ {}^{A}_{Z}\mathrm{X} \rightarrow {}^{A-4}_{Z-2}\mathrm{X} + {}^{4}_{2}\mathrm{He} $$

Alpha radioactivity is found only in heavy atoms ($Z > 82$), just like fission, the splitting of a heavy nucleus into two smaller nuclei takes place, which is the basis for the use of radioactivity as a source of energy.

(ii) Beta radioactivity consists in the emission of either an electron, ${}^{0}_{-1}\mathrm{e}$ (β^{-} particle), or of a positron ${}^{0}_{+1}\mathrm{e}$ (β^{+} particle).

$$ \beta^{-}\text{disintegration}: {}^{A}_{Z}\mathrm{X} \rightarrow {}_{Z+1}^{\;\;A}\mathrm{X} + {}^{0}_{-1}\mathrm{e}\,\nu $$
$$ \beta^{+}\text{disintegration}: {}^{A}_{Z}\mathrm{X} \rightarrow {}_{Z-1}^{\;\;A}\mathrm{X} + {}^{0}_{+1}\mathrm{e}\,\nu $$

(iii) Gamma radioactivity is an isomeric transition in which there is emission of a γ photon which carries the difference in energy between the initial and final states of the nucleus, E_{i} and E_{f}:

$$ E_{\mathrm{i}} - E_{\mathrm{f}} = E_{\gamma} = h\nu, $$

where h is the Planck constant and ν is the radiation frequency.

Gamma rays of interest for imaging are in the $10^2 - 10^3$ keV energy range. As a rule, gamma radioactivity is an immediate phenomenon, but for some artificial radionuclides, called metastable, it can be delayed. This is the case of the metastable form of Technetium-99, noted ${}^{99\mathrm{m}}\mathrm{Tc}$, one of the most widely used radionuclides for biomedical imaging (see Section 4.6.4).

(iv) Electron capture: An electron from an orbital close to the nucleus (K or L shell) is captured by the nucleus. The gap in the orbital is filled by electrons from the outer shells with subsequent X or γ emission:

$$ {}^{A}_{Z}\mathrm{X} + {}^{0}_{-1}\mathrm{e} \rightarrow {}_{Z-1}^{\;\;A}\mathrm{X} $$

Alpha and beta radioactivity emit α and β particles which are stopped rapidly in matter and do not travel sufficient distances through biological tissues to be detected from outside the body. In contrast, gamma rays of sufficient energy (>100 keV) can travel several tens of cm in biological tissues and the emitting radiotracer can be detected from a distance. Radiotracer imaging is based on the detection of γ photon (or γ rays) from radiotracers labelled with nuclei emitting γ photons, either directly, or by electron capture (EC), or by annihilation of a positron. In that latter case, it is not the positron, which has a very short half-life ($\sim 10^{-7}$ s) and a short range, ($\sim$1 mm in water) that is detected, but its annihilation with an electron. Annihilation is the dematerialization of the positron–electron pair, or positronium, which creates two (rays of travelling in opposite directions (see Section 4.4.2.).
The law of conservation of mass and energy applies, and Einstein's equation

$$ E = mc^2, $$

indicates that the energy of each of the γ photons is 511 keV (i.e. 511 000 electron-volts).

4.1.2 Activity

Radioactivity is a physical, not a chemical process; therefore, it is independent from the chemical environment in which it occurs. Chemically speaking, a radioactive atom is essentially similar to the stable atom of same Z, and both isotopes interact similarly with their chemical environment. The radioisotope can therefore substitute the stable isotope in any chemical construction or biological organism with the substitution remaining unnoticed from this construction/organism.

Radioactivity is a probabilistic phenomenon, which means that in a given population of N unstable nuclei, a similar proportion dN will disintegrate in a given amount of time dt (first order law). The activity Q of a radionuclide is defined by

$$ Q = -\mathrm{d}N/\mathrm{d}t = \lambda N, \qquad (4.1) $$

where λ is the decay constant. The official unit for Q is the Becquerel (Bq), 1 Bq = 1 radioactive transformation per second; the old unit, the Curie (Ci) is still largely in use; 1 Ci = 3.7×10^{10} Bq.

A radionuclide is defined by its type of disintegration (nature of the emission), its energy (wavelength of the γ ray or kinetic energy of the particle) and the speed with which it decays (see Section 4.1.3).

4.1.3 Decay

Integrating Eq. (1) between times t and t_0, with $t_0 = 0$, yields

$$N_t = N_0 \exp(-\lambda t), \qquad (4.2)$$

which when written as

$$N_t/N_0 = \exp(-\lambda t),$$

shows that the relative amount of radioactivity depends only on time. In other words, the proportion of nuclei that undergoes radioactive transformation in a given time interval is a characteristic property of the radionuclide and is independent of the amount of radioactivity present. Therefore, it is convenient to express this property as a constant representing the time necessary for the transformation of one half (statistically) of the radioactive atoms present initially in a given collection, or radioactive half-life, $T_{1/2}$:

$$N_t/N_0 = 1/2 = \exp(-\lambda T_{1/2}).$$

Hence

$$T_{1/2} = \ln(2)/\lambda. \qquad (4.3)$$

$T_{1/2}$ is often noted as T or τ and sometimes called the radioactive half-life, which in biology is confusing because the actual half life of a radionuclide in a living organism is a combination of its radioactive period $T_{\text{radioactive}}$ and its half-life of elimination from the body, $T_{\text{elimination}}$. The effective biological half-life $T_{\text{effective}}$ is given by

$$1/T_{\text{effective}} = 1/T_{\text{radioactive}} + 1/T_{\text{elimination}} \qquad (4.4)$$

Knowing how much of a radioactive element is present at a given time in a given point, it is possible to predict, by a simple calculation, how much will remain at any time later or how much has been there at any time earlier. All that is needed apart from the knowledge of the isotope's period is a proper measurement of time.

After some time, the amount of radionuclide remaining will become negligible. The two numbers easy to memorize are

- After 10 periods, there is roughly one thousandth of the initial radioactivity left ($2^{10} = 1024$).
- After 20 periods, there is roughly one millionth of the initial radioactivity left ($2^{20} = 1\,048\,576$).

As a consequence, after some time, detection becomes impossible and irradiation negligible. Therefore, for biological applications it is advisable to use isotopes with the shortest period compatible with the desired observation time.

4.2 Interaction of gamma rays with matter

4.2.1 Ionization and excitation

Radioactive emissions are also called 'ionizing radiations' because they interact with matter in a way, which ultimately can result in its ionization. Interaction with matter is dependant on the type of emission and on its energy, and can occur anywhere away from the site of emission, that is inside the living organism, the detecting system (the tomograph) or elsewhere. Obviously, the ideal emission is one that results in minimal interactions inside the subject and maximal interactions in the detecting system, leading to minimal ionization of the living tissue and maximal sensitivity of detection.

Energetic radiations interact with matter mostly through ionization and excitation. Ionization is the process by which radiation expulses an electron from its orbital, leaving the atom with a positive charge. In contrast, excitation occurs when the electron remains on the atom but is transferred from its fundamental orbital to the one of a higher energy. The energy of binding of an orbital electron depends on the atom and the shell and ranges from approximately 10 eV (H) to a few thousands of keV (Pb). Naturally, an orbital electron that has been expulsed from the atom produces its own ionizations and excitations, and so does the return of an excited electron to its fundamental layer. One simple figure to remember is that a 1 MeV radiation (γ) or particle (electron) which transfers all its energy to water produces on average 30 000 ionizations and 100 000 excitations.

As stated above, in radiotracer imaging the interactions with matter concern essentially γ photons. For what concerns the positron, the time interval before its annihilation is very short, and thereafter the interactions with matter are mostly due to γ rays (see Section 4.2.2.2).

4.2.2 Gamma interactions

Ionization of tissue by gamma radiation is essentially an indirect consequence of the expulsion of

Figure 4.2.1 The photoelectric effect. The incident gamma photon transfers its energy $h\nu$ to a peripheral electron in orbit around the nucleus with orbital energy $E < h\nu$. The electron is expulsed, the incident photon disappears and the electronic gap is filled by the remaining atomic electrons with emission of X-ray

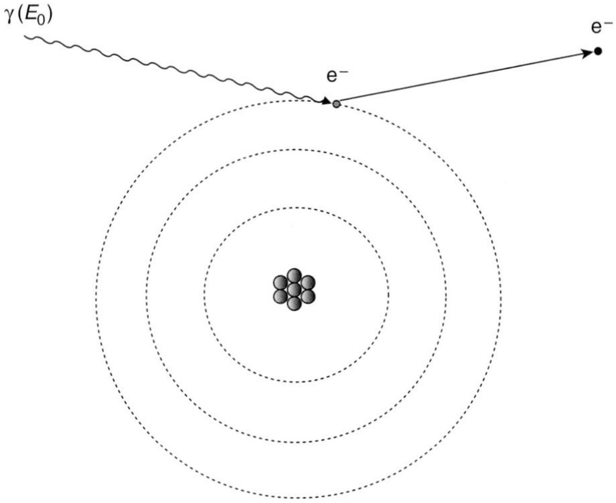

Figure 4.2.2 Compton scattering. The incident gamma photon transfers part of its energy to an orbital electron, which is expulsed, while the photon deviates from its incident trajectory and loses part of its energy

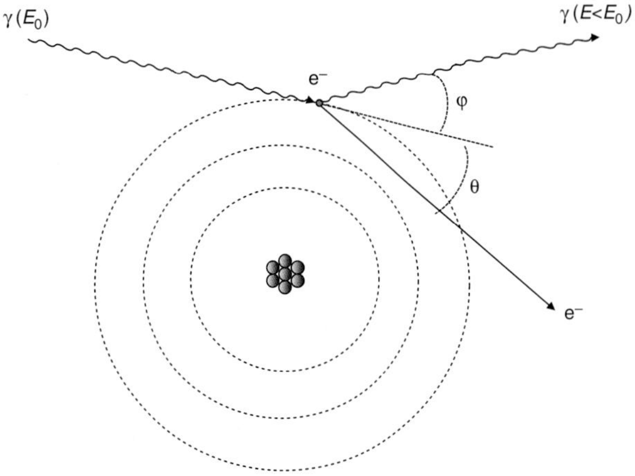

orbital electrons. The absorption of γ electromagnetic radiation by matter can occur in three major modes: Photoelectric effect, Compton diffusion and pair materialization.

(i) In the photoelectric effect (Figure 4.2.1), a photon transfers its energy hν to an orbital electron of energy $E < h\nu$. The electron is expulsed with a kinetic energy

$$\text{of } 1/2mv^2 = h\nu - E$$

The incident photon disappears and the electronic gap is filled by the remaining atomic electrons with emission of X-ray.

(ii) In the Compton effect or Compton scattering (Figure 4.2.2), the incident γ photon transfers only part of its energy to the orbital electron, which is expulsed, while the photon deviates from its incident trajectory and loses part of its energy: The scattered photon has the energy $h\nu' < h\nu$. The law of energy conservation implies

$$h\nu = h\nu' + 1/2mv^2 + E.$$

The difference of energy $h\nu - h\nu'$ depends on the value of the cosinus of the scattering angle φ; $h\nu - h\nu'$ is highest for backwards scattering, when $\varphi = 180°$ ($\cos(\varphi) = -1$)

(iii) Pair creation is the materialization of the photon by a mechanism, which is schematically the reverse of a positron–electron annihilation:

The law of mass-energy conservation implies that materialization occurs only with γ for which $h\nu > 1022$ keV.

The relative importance of photoelectric, Compton and materialization effects depends on (i) the density of the matter traversed and (ii) the energy of the incident γ photons. In living tissues and in the range of γ rays used in biology (from ~100 keV for γ emitters to 511 keV for the annihilation photons of positrons), Compton scattering is the predominant effect and accounts for 80–99% of the interactions, depending on the energy of the incident photon. The other interactions are essentially emission of photoelectrons; materialization occurs only for values of $E_\gamma > 1022$ keV.

4.2.3 Absorbed dose

In nuclear physics, absorption means that the radiation has transferred all its energy to the surrounding matter. Absorption depends on nature of the radiation, its energy and the density of the absorbing matter. In a homogeneous medium, such as water or lead for instance, the energy absorbed is related to the distance that the radiation travels to deposit all its energy:

$$E_{\text{absorbed}} = kl,$$

where l is the mean distance travelled by the radiation before being absorbed and k is the linear energy transfer to the tissue, expressed in keV per distance (usually cm). It is convenient to relate the proportion of γ photons absorbed to the thickness d of the material traversed, by introducing an attenuation coefficient which depends on the incident energy of photons E and the density of the material traversed:

$$Q_{\text{absorbed}} = Q_{\text{incident}} \exp(-\mu d) \qquad (4.5)$$

with μ in cm^{-1} and d in cm.

The ratio of Q_{absorbed} over Q_{incident} is called the absorption ratio, related to the transmission ratio; both are often expressed in percent:

$$\text{transmission ratio} = 1 - \text{absorption ratio}$$

Typical values of μ for 140 and 511 keV γ photons in different tissues and materials are shown in Table 4.2.1.

From these values and Eq. (4.5), it can be calculated that a thin (~1 mm) layer of lead will absorb or stop 99% of 140 keV, while over 2.6 cm are necessary to absorb 99% of 511 keV. Attenuation coefficients also explain why PET has a superior sensitivity to SPECT (transmission through tissues is much higher), and attenuation coefficient of the crystal detectors indicate that, for the same thickness, the BGO crystals will stop a larger fraction of 511 keV photons than the NaI(Tl) crystals, therefore being preferred for PET cameras. The ideal choice of the imaging system must take into account the γ rays emitted by the radionuclide from the double point of view of their detection (a high attenuation coefficient in the detection system is favoured) and the transmission through tissues (a low attenuation coefficient is preferable). The trade-off between the two depends on the radionuclide and will be detailed in the following paragraphs. Note that in this respect, smaller animals have the advantage over big ones that the thickness of absorbing tissue is reduced. The correspondence between percent of transmitted gamma ray and material thickness is shown in Figures 4.2.3 and 4.2.4 for 140 and 511 keV photons, respectively.

Table 4.2.1 Attenuation coefficients of different materials for 140 and 511 keV gamma rays

	μ at 140 keV (cm^{-1})	μ at 511 keV (cm^{-1})
Water	0.150	0.095
Muscle	0.155	0.101
Bone	0.284	0.179
NaI(Tl) crystal[a]	2.23	0.34
BGO crystal[b]	5.5	0.95
Lead	41	1.75

[a]Thallium-doped Sodium Iodide crystal used in SPECT cameras.
[b]Bismuth Germanate crystal used in PET cameras.

Figure 4.2.3 Attenuation curves for 140 keV gamma photons in different materials. The curves are percent of transmission for an incident gamma ray traversing a given material of varying thickness

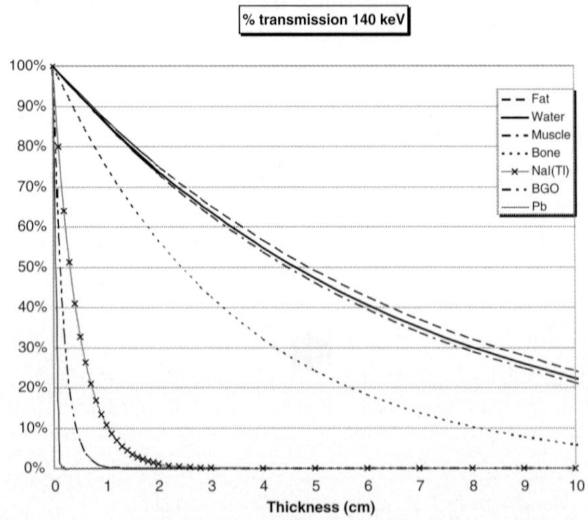

Figure 4.2.4 Attenuation curves for 511 keV gamma photons in different materials. The curves are percent of transmission for an incident gamma ray traversing a given material of varying thickness

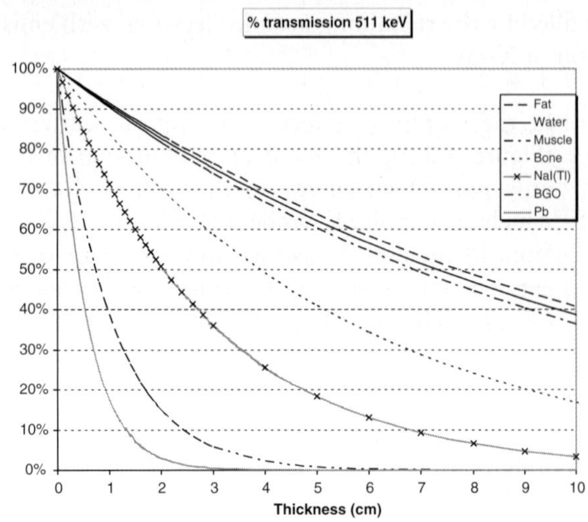

4.3 Radiotracer imaging with gamma emitters

4.3.1 Gamma-emitting radionuclides for radiotracer imaging

4.3.1.1 Type and energy of emitted radiation

Among the different nuclear properties, the type and the energy of radiation are crucial for imaging and in determining the safety rules for handling. Table 4.3.1 lists the main nuclear properties, half-life, route of decay and type and energy emitted of the major radionuclides routinely used in scintigraphy and SPECT.

The radioisotope most widely used in clinical imaging is technetium-99m (^{99m}Tc), employed in over half of all nuclear medicine procedures. This radionuclide displays almost ideal emitting radiation because decay occurs through a process called 'isomeric transition' that generates gamma rays and low energy electrons. As there is no high-energy beta emission, the radiation dose is low, and the gamma rays easily escape the tissues and are accurately detected by a gamma camera. In contrast, iodine-131 (^{131}I) can be used for imaging with conventional SPECT gamma camera, but its high energy γ-emissions are not optimally counted and its β-particle emissions increase the radiation dose delivered to the body.

Although the criteria for the selection of a suitable radionuclide for clinical use equally apply to animals, some radionuclides that are not useful for human studies due to unfavourable radiation or because they tend to deliver high doses and may be used only in animal experiments.

For instance, ^{125}I, a low-energy emitting radionuclide, can be used in mice, whereas it would be of no value in larger animals or humans due to the total attenuation of the photons through tissues. Also, for human examinations radionuclides giving low radiation doses to organs and tissues should be preferred; dosimetric considerations are generally less stringent in animal studies.

4.3.1.2 Half-life and decay

Half life is important not only because it determines the time during which imaging is possible after administration, but also from a practical point of view because it conditions the 'shelf' availability of the radionuclide and the time available for radiochemical labelling without important loss of radioactivity due to decay. With the relative exception of ^{99m}Tc, all the radionuclides listed in Table 4.2.1 display half-lives which are long enough to allow the use of long labelling syntheses.

With the exception of ^{131}I, all the radionuclides listed in Table 4.2.1 decay to stable or almost stable (half-life greater than 200 000 years) nuclides. ^{131}I mainly decays to stable ^{131}Xe, but a small fraction (0.8%) decays to the metastable ^{131m}Xe (Half-life = 12 days).

4.3.1.3 Production of gamma emitters

All gamma-emitting radionuclides are obtained artificially by one of the following methods:

- From decay of a radioactive parent obtained by neutron bombardment in nuclear reactors;
- By direct nuclear reactions induced by charged particles in a circular accelerator (a cyclotron), or after radioactive decay of a parent radionuclide;
- From fission products either as direct harvest or after radioactive decay of a fission product.

Table 4.3.1 Nuclear properties of gamma-emitting radionuclides routinely used

Radionuclide	Half-life	Route of decay	Main (γ or X-ray energy (keV) and probability (%)
^{67}Ga	3.56 Days	EC	93 (36%), 185 (20%), 300 (16%)
^{99m}Tc	6.02 h	IT	141 (89%)
^{111}In	2.83 Days	EC	171 (90%), 245 (94%)
^{123}I	13.3 h	EC	159 (83%)
^{125}I*	59.4 Days	EC	27 (73%), 27 (39%), 31 (25%), 35 (6%)
^{131}I	8.02 Days	(β^-	284 (6%), 344 (81%), 637 (7%)
^{201}Tl	3.04 Days	EC	69 (27%), 71 (46%), 80 (20%)

β−: electron emission; EC: electron capture; IT: isomeric transition
(*): for small animals only.

The parent–daughter decay pair represents a very common mode of production, provided that an efficient separation of the daughter radionuclide from its parent is possible. When the daughter and the parent display large differences in their chemical properties, the recovery of the daughter activity may occur by adopting simple separation techniques like liquid–solid extraction. In this favourable situation, the build-up and separation of the daughter radionuclide may occur in sterile and pyrogen-free conditions using a special apparatus, called generator.

Over 90% of radiopharmaceuticals used for clinical imaging applications are labelled with ^{99m}Tc. This is of course not only due to the interesting properties of this radionuclide described above, but also to the fact that it is easily available in a convenient generator that can be bought from a company and used on site at the clinician's demand.

^{99m}Tc is generated by decay of a longer-lived parent, ^{99}Mo (half life of 66 h or 2.8 days), generated from the fission of Uranium-235. [^{99}Mo] in the chemical form of molybdenate ammonium is adsorbed strongly to an alumina column and generates ^{99m}Tc, which can be easily eluted from the column by a simple wash with sterile saline. Washing the column yields pure, sterile ^{99m}Tc as sodium pertechnate $NaTcO_4$, which can be used for imaging as such or after further labelling of molecules (see Section 4.1.4). The elution procedure can be repeated several times and is so simple that it is called 'milking' by radiotechnologists. Milking is usually performed once or twice a day, for as long that the concentration of ^{99}Mo in the generator is sufficient, generally 2–3 weeks, after what the generator is replaced.

The mode of production for other gamma-emitting radionuclides is described in Section 4.6.1.2 and Table 4.6.1.

4.3.2 Instrumentation for detection of gamma emitters

4.3.2.1 *General scheme of a scintigraphic camera*

All imaging instruments, scintigraphic cameras, SPECT or PET tomographs, use the same basic set of elemental parts. Hence, the basic pieces of the simplest, the scintigraphic camera, may be taken as a relevant example for those of all imaging instruments (Figure 4.3.1).

Before detection, gamma rays are first collimated (i.e. geometrically 'selected' according to their direction) in a collimator. Detection occurs through absorption in a crystal, which transforms the gamma rays into photons in the visible range. Photons are then converted into electrons and amplified in a photomultiplier tube (PMT). The electric signals from the PMTs are spatially assigned by a logical circuit (Anger circuit) and the map of distribution is displayed on a computer.

Figure 4.3.1 Schematic representation of a gamma-camera head. Following the pathway of the gamma photons, the elements of a gamma camera head (Anger camera) are, from bottom to top, a collimator, a crystal scintillator, photomultiplier tubes or PMTs, a logical (Anger) circuit, and a computer for image reconstruction and display

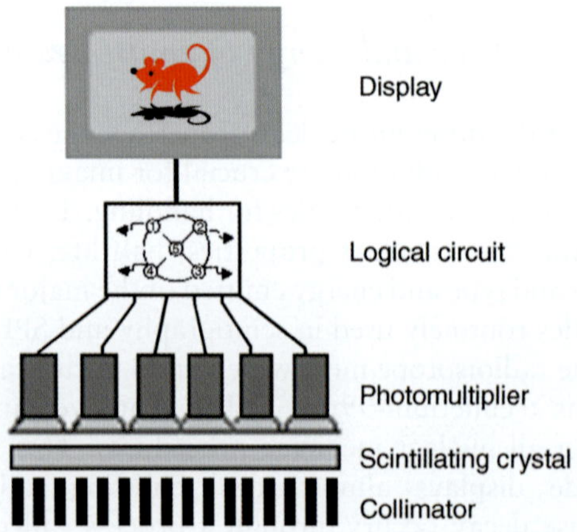

4.3.2.2 *The collimator*

Radionuclides emit energy randomly in all directions. Therefore, in order to link the point of emission and the point of detection, one requires a directional selection by a collimator, which selects the gamma rays according to their direction of propagation. The collimator consists of a network of holes drilled in a high-density material (lead or tungsten) placed before the detector so that only the gamma rays parallel to the holes will pass through. Its geometrical characteristics are optimized for the energy of the gamma rays and determine the final characteristics of the images. Selecting photons implies that only a fraction of incident photons is detected. The collimator is a key piece of gamma cameras; an example is depicted in Figure 4.3.2.

There are three types of collimators: Parallel collimators, cone beam collimators and fan beam collimators (Figure 4.3.3).

Figure 4.3.2 The beehive parallel collimator. (a) View from above (parallel to the large plane of the collimator) showing the hexagonal arrangement of the septa. (b) Side view of a virtual section of the collimator showing how the lead septa allow only the gamma photons parallel to the septa to reach the scintillation crystal

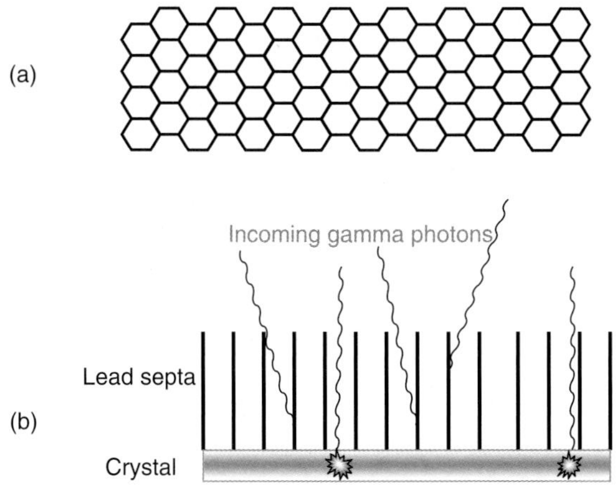

- Parallel collimators, the most common collimators in clinical imaging, consist of a network of parallel holes of adjusted diameter and height separated by lead. The angular aperture of the cone defined by each hole can be neglected, so that only gamma rays travelling at 90° angles to the collimator pass through, and the acquired data sets can be assumed as projections of the radioactivity along 2D quasi-parallel lines. The use of parallel collimators translates to a variable spatial resolution in the field of view: Spatial resolution depends on the distance of the source to the collimator and on the size of the hole. The sensitivity slightly varies with the distance of the source to the collimators and depends on the height of the holes (Figures 4.3.2 and 4.3.3(a)).
- Fan-beam collimators allow acquiring projections along a fan of the radioactivity. Along the fan, the lines of projections are no longer parallel. These collimators produce a zoom of the object along one direction and thus offer a good compromise between sensitivity and spatial resolution at the expense of a possible truncation of the object in one direction (Figure 4.3.3(b)).
- Cone beam collimators realize a zoom of the object in two directions, thus reducing the field of view of the detectors. The lines of projection are no longer parallel (Figure 4.3.3(c)). These collimators offer the best compromise between spatial resolution and sensitivity. They allow achieving spatial resolution lower than 1 mm and are well adapted to small animal imaging.

Figure 4.3.3 Some typical geometries for collimators. See text for explanations

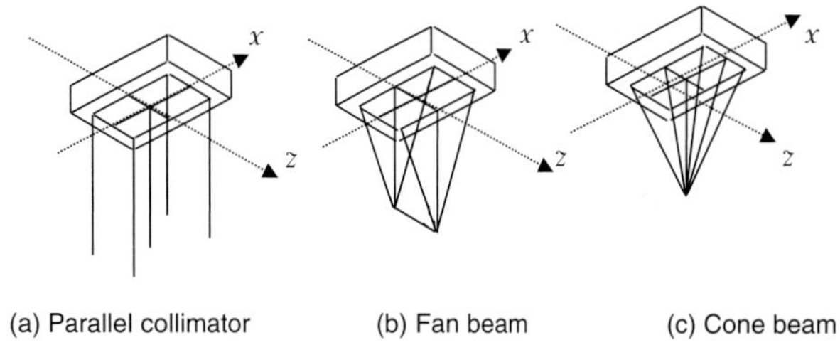

For fan beam and cone beam collimators, the sensitivity varies with the distance of the source to the collimator and the angulations of the lines of projection along one or two directions must be accounted for in the image reconstruction.

New collimator schemes such as sets of focusing cone beam pinhole collimators have been designed for high-resolution SPET (0.5 mm) while maintaining high efficiency (nearly 0.2%) for small animal imaging (Wirrwar *et al.*, 2001).

4.3.2.3 The detector

Scintillation crystal detectors are the most often used detectors for imaging. The principle of detection is to convert the energy of the incident photons into photons of low energy (~4 eV) in the visible light range (wavelengths around 400 nm) through photoelectric effect and Compton diffusion. The number of visible photons is proportional to the energy and the intensity of the incident gamma rays. As detectors for imaging, some relevant properties of scintillating crystals are

- the stopping power (given by the attenuation coefficient μ) that depends on the density of the crystal and the energy of the radiation;
- the scintillation decay time, that limits the rate of event acquisition;
- the light output (the ratio between light leaving the crystal and incident energy) that conditions the efficiency of gamma conversion.

Properties of the main scintillation crystals used for *in vivo* imaging are summarized in Table 4.3.2.

The most common scintillation material for gamma-emitting radiotracer imaging is NaI doped with thallium (NaI(Tl)) due to its good properties for the detection of photons of energies close to 140 keV (stopping power and energy resolution) and its low cost. Most of the PET cameras use LSO and BGO scintillation crystals.

Table 4.3.2 Main scintillating crystals used for imaging

	NaI(Tl)[a]	BGO[b]	LSO[c]	GSO[d]	LYSO[e]	LuAp[f]	YAP[g]
Density ($g.cm^{-3}$)	3.67	7.13	7.35	6.71	7.1	8.34	5.37
Effective atomic number	50	73	65	58	63	65	39
μ at 140 keV (cm^{-1})	2.2						
μ at 511 keV(cm^{-1})	0.34	0.9	0.8	0.67	0.83	0.91	
Scintillation decay (ns)	230	60/300	40	60/600	40	18	27
Relative light output (% NaI(Tl)	100	22	75	20	84	52	55
Comments	Hygroscopic		Auto-emission				
Application	Scintigraphy SPECT	PET	PET	PET	PET		PET and SPECT

[a]Thallium-doped sodium iodide.
[b]Bismuth germanate oxyde.
[c]Lutetium oxyorthosilicate.
[d]Gadolinium orthosilicate.
[e]Lutetium Yttrium orthosilicate.
[f]Lutetium aluminium perovskite.
[g]Yttrium aluminIum perovskite.

Small animal imaging represents extreme imaging conditions because it requires both high spatial resolution (typically lower than 1 mm) and high detection efficiency (typically more than 10%). This field of application has thus stimulated researches on new high-density materials and new detectors in order to improve both parameters simultaneously. One example is solid-state detectors, such as semi-conductors, which offer a direct conversion of the photon energy into an electrical signal. The main advantage of solid-state detectors over the other detectors is their good energy resolution: Values of 5% at 140 keV or less can be reached. These detectors appear very promising for small animal gamma imaging, while for the detection of 511 keV photons they have the disadvantage of a low stopping power.

4.3.2.4 *Photomultiplier tubes and Anger network*

In gamma cameras, one NaI(Tl) scintillation crystal is coupled to arrays of photomultiplier tubes (PMTs) or to photo-sensitive detectors that convert the low energy photon into electrical pulses. In order to localize the origin of the signal inside the crystal, the PMTs are coupled to the crystal in a well-defined geometrical organization, generally hexagonal, and the localization of the interaction of the gamma with the crystals is obtained by weighting the PMTs' position with the low energy gamma collected by each PMT. Hal Anger invented a logic circuit in which the PMTs in the array are multiplexed so as to optimize spatial localization, which is still in use today and gave his name to the first cameras (Anger camera). As indicated by their name, photomultiplier tubes amplify the signal and must be calibrated so that all PMT in the network amplify the signals from the scintillator with the same amplification factor.

For most detectors, spatial resolution and efficiency are inversely proportional. Hence, new improvements appear continuously, such as position-sensitive PMTs (PS-PMTs), or other technologies such as silicon avalanche photodiodes (APDs), in which individual APDs are coupled to scintillation crystals to provide energy and timing information without multiplexing of signal. APDs are also used in small animal PET cameras.

4.3.2.5 Planar scintigraphy or tomography

The image obtained in a planar camera is a projection of all the levels of emission from the lines parallel to the holes of the collimator. This technique, *planar scintigraphy*, is sufficient whenever the emission levels parallel to the plane of detection do not superimpose one on each other, or in other words when the plane of emission is sufficiently thin to be considered as having essentially two dimensions. It is still in use for clinical thyroid imaging. The projection image is similar in that case to a tissue autoradiography. However, animals and humans are 3D objects and most of the time it will be important to determine the depth of emission of the radiotracer by acquisition of 3D images.

A simple and unexpensive solution is to rotate the animal in front of a fixed planar scintigraphic camera. This can be done because of the small size of the animal, provided it is placed in a vertical position to avoid movement of its internal organs. Anatomy with an X-ray source can also be acquired simultaneously.

Alternatively, SPECT cameras are composed of one, two or three heads of planar detectors, which rotate around the subject. Each head consists of a collimator, a large crystal coupled to photomultipliers tubes via a light guide, and a logic circuit. The localization of the gamma rays interaction in the crystals is obtained using the same principle as that described above, with the addition of tomographic reconstruction, which will be described in Section 4.5.1. Clinical SPECT cameras used for nuclear medicine in humans are large cameras not adapted to small animals because their resolution is in the order of one cm. They can be used for large animals in veterinary medicine. SPECT cameras dedicated to small animals recently appeared on the market. In order to improve spatial resolution, the detector heads are most often equipped with cone-beam collimators and moved close to the centre of the tomograph. New collimator schemes such as sets of focusing cone beam pinhole collimators have been designed for high-resolution SPET (0.5 mm) while maintaining sufficient efficiency (nearly 0.2%) for small animal imaging (Wirrwar *et al.*, 2001). Some manufacturers propose lab SPECTS coupled with a CT scanner, offering the possibility to co-register the radiotracer distribution with an anatomical image of the animal. Until now these cameras have remained relatively expensive.

4.3.2.6 Data processing for SPECT

In order to obtain the radioactive distribution in the field of view of the collimators, the acquired data sets should be corrected for the detection of scattered events and for the attenuation of the gamma rays, and normalized prior to reconstruction.

Correction for scattered photons. The main interaction of gamma rays within the subject is Compton scattering. As a consequence, Compton scattering of the gamma in the object is associated with a false position of the emission of the gamma and degrades the contrast of the images. Energy based discrimination of the gamma rays by the detection system can be used to reject scattered gamma rays. As the detection crystals have a resolution in energy centred on the photopeak of typically 10% at 140 keV, that is 126–154 keV, only the scattered events which fall outside this energy window can be rejected. In addition, the scattered fraction varies in the different parts of the subject, and is highly dependant on the amount of matter traversed by the gamma ray. Therefore, a correction is needed to obtain reliable localization and quantification, even in small animals. Several methods have been designed to correct such phenomenon. One currently in use is to use a second energy window centred on a lower value (for instance, 121 keV for 140 keV gamma from ^{99m}Tc) which detects only the scattered events, and to subtract these from the total counts obtained at the photopeak window.

Correction for attenuation. Correction for attenuation of 100–200 keV gamma rays is important because, on average, less than 50% of 140 keV are transmitted through the body of an adult mouse without losing energy. Like scattering, attenuation depends on the thickness and geometry of the tissue, and must be corrected for. A map of the attenuation coefficients of the subject can be extrapolated when the geometry of the subject is known. Assuming the tissue homogeneous, the attenuation coefficient $\mu(x, y, z)$ (see Section 4.2.3) is considered identical in all directions and equal to μ. This map is then computed into the acquired data. Alternatively, a more precise attenuation map of the medium, $\mu(x, y, z)$, can be measured using a calibrated external radioactive source. ^{153}Gd, a long-lived emitter of 100 keV, is convenient for this purpose or using an anatomical image obtained from a CT-scanner.

4.4 Detection of positron emitters

4.4.1 Positron-emitting radionuclides

4.4.1.1 *Physical properties of major positron-emitting radionuclides*

The characteristics of the main positron emitters used for PET imaging are summarized in Table 4.4.1. In contrast to gamma emitters, many of the most interesting positron emitters are isotopes of low atomic mass elements found in organic biomolecules. This explains why PET is one of the only techniques that can image truly isotopic biomarkers, that is labelled radiotracers that have exactly the same chemical composition than that of the corresponding biomolecules. In addition, positron emitters of low atomic mass are pure (or almost pure) positron emitters with an extremely high specific radioactivity (much higher than ^{99m}Tc for instance). This allows their administration in very small amounts. The kinetic energy of the positron defines its maximal range of travel in the medium before annihilation, from under 1 mm to several mm and consequently the maximal spatial resolution, which could theoretically be obtained in the absence of all other limitations.

Other PET emitting radionuclides of higher atomic mass are available from some medical cyclotrons, such as ^{76}Br (half life 17 h), ^{124}I (4.2 days), etc. They can be useful for some studies for which longer half-lives are preferred; however, these radionuclides are not pure beta-+ emitters and have higher energies, meaning less favourable dosimetry and lower spatial resolution of the PET images.

4.4.1.2 *Production*

Positron emitters are produced in medical cyclotrons, circular accelerators of particles such as beams of protons (p) or deuterons (deuterium ions, d), which send these particles once they have reached very high energies to bombard pure elemental targets. The nuclear reactions that occur in the target are noted for instance ^{14}N(d,n)^{15}O, meaning that a ^{14}N target bombarded with a beam of deuterons releases ^{15}O and one neutron n. Other examples of common nuclear reactions that lead to positron emitting radionuclides are

- ^{14}N(p,α)^{11}C for carbon-11
- ^{16}O(p,α)^{13}N for nitrogen-13
- ^{18}O(p,n)^{18}F and (^{20}Ne(d,n)^{18}F) for fluorine-18, recovered as fluoride ion or fluorine gas, respectively.

The radionuclide is then rapidly incorporated into a molecule of interest by radiochemical procedures that will be described in Section 4.7.

A practical consequence of the short half-lives of positron emitting radionuclides is that they have to be produced in close proximity to their site of utilization: Just next to the PET camera for ^{15}O, or in a 2-h delivery time distance for ^{18}F. As a result of its shorter half-life (109.8 min) and its lower positron energy (maximum 635 keV), administration of fluorine-18-labelled radiopharmaceuticals gives a lowerradiation dose. Compared with the other conventional short-lived positron-emitting radionuclides carbon-11, nitrogen-13 and oxygen-15 with equally simple decay schemes, fluorine-18 has once more a relatively low positron energy and the shortest positron linear range in tissue (max 2.3 mm),

Table 4.4.1 Short-lived, positron-emitting radionuclides

Radionuclide	Half life (minutes)	β^+ Decay (%)	Max positron kinetic energy (keV)	Mean positron range in water (mm)	Energy of the detected gamma rays (keV)	Theoretical specific activity [a] ($Ci \cdot mol^{-1}$)
^{11}C	20.4	99	981	1.12	511	9215
^{13}N	10.0	99	1190	1.44	511	18430
^{15}O	2.07	100	1723	2.22	511	90960
^{18}F	109.8	97	634	0.6	511	1712

[a] The specific activity is defined as the radioactivity per unit mass in carrier-free radionuclide. Note that the specific radioactivity is inversely proportional to the half-life of the radionuclide, which is in the order of a few to a few tens of minutes.

resulting in the highest resolution in PET imaging. Even though these 2 h represent roughly one half-life, meaning that a little over half of the radioactivity will have vanished by the time the compound reaches the site of utilization, thanks to the very high specific radioactivity at the time of production, the decay during a few half-lives (between 3 and 5) will leave enough radioactivity for detection of the radiotracer by the PET cameras.

Accordingly, in developed countries, ^{18}F and even some ^{11}C labelled compounds are produced by facilities cyclotrons strategically located in or close to large cities and delivered daily to the nuclear medicine wards of their neighbourhood. Thus, centres that do not run an expensive cyclotron may benefit from the power of PET imaging.

A generator-based approach has also been proposed for delivery of a positron emitter, Gallium-68, a positron-emitter with a 68 min half-life. ^{68}Ga is eluted from its parent nuclide, Germanium-68 (^{68}Ge), similarly to ^{99m}Tc elution. Although radiolabelling with ^{68}Ga has until now been limited to its conjugation into chemical cages linked to biomolecules (much like ^{99m}Tc), the long shelf life of the ^{68}Ge generator (up to 1 year) is economical and may lead to further developments.

4.4.2 Positron annihilation

4.4.2.1 *Principle of positron annihilation*

A positron leaving the nucleus with a high kinetic energy is progressively slowed down by a cascade of successive interactions with the nuclei and electrons of the surrounding matter (Figure 4.4.1). Interactions are random in space, and the initial kinetic energy of the positron has a random value statistically distributed between 0 and E_{max}. Therefore, the direction and path length of the positron's travel in matter is random and statistically defined by the mean distance before annihilation, which is proportional to E_{max}, a characteristic of the radionuclide (see Table 4.4.1). When the kinetic energy of the positron is down to thermal levels (~0.025 eV) and it encounters an electron, these two anti-particles form a positronium (e^+/e^-), an evanescent particle which annihilates by dematerialization into two gamma photons. The conversion of matter into energy follows the Einstein relation or equation

$$E = mc^2,$$

Figure 4.4.1 Emission of a positron and its annihilation. The creation of a positron takes place in the nucleus by conversion of a proton into a neutron. The positron is emitted with a kinetic energy, which it loses in matter by random interactions, until it forms a positronium pair with an orbital electron. The positronium annihilates in two gamma rays of 511 keV emitted in opposite directions. *d* is the mean distance travelled by the positron from the nucleus to its site of annihilation

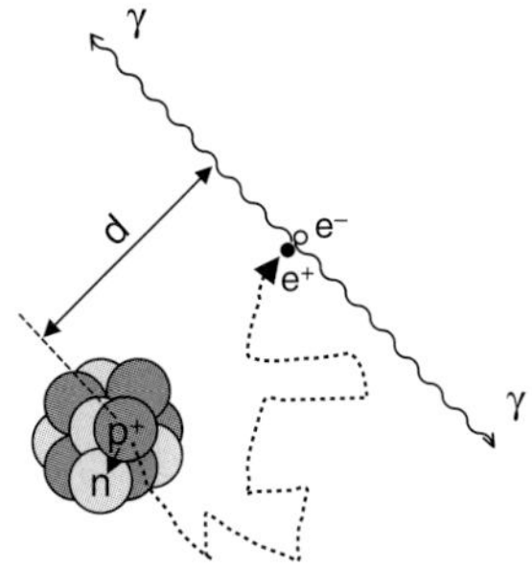

where m is the sum of the masses, and the two resulting gamma photons are emitted with the same energy, 511 keV, in opposite directions.

4.4.2.2 *Detection*

PET imaging is based on the simultaneous detection of two gamma rays by two opposite detectors and on the following model: The simultaneous detection of two gamma rays relates to the annihilation of one positron with an electron within the volume defined by the two detectors (Figure 4.4.2). Such an event is called a coincidence and the volume defined by the lines joining two opposite detectors is called a line of response (LOR). It follows that there is an intrinsic uncertainty in assigning the localization of the radiotracer, corresponding to the mean travel distance of the positron before annihilation. This uncertainty is in the order of 0.6 to several mm depending on the radionuclide.

The time interval during which two gamma rays are considered to originate from the annihilation of a unique positron is called the *time coincidence window*. In practice, an impulse is sent to a logical circuit at each event recorded in one of the detectors, and is assigned to the LOR if a second event is recorded in an opposite detector during the time coincidence window (Figure 4.4.2). Adjusting the width of the time coincidence window is important; it varies between 3 and 20 ns on current PET tomographs, depending on

Figure 4.4.2 Coincidence detection in a PET tomograph. Each event reaching detector 1 is assigned a short time interval during which an event reaching another detector (detector 2 shown is taken here as an example) will be registered as a coincidence corresponding to an annihilation event anywhere on the line of coincidence linking detectors 1 and 2

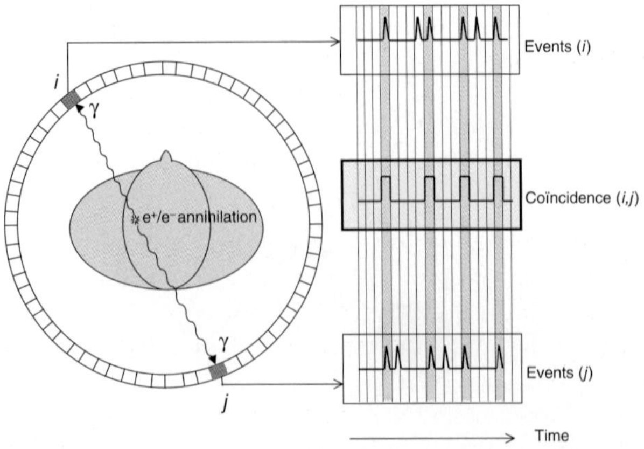

the properties of the detectors and the associated electronics (a 511 keV gamma ray travels 80 cm in nearly 3 ns). If the detectors and the electronics are sufficiently fast, the time delay between the two detected gamma rays can be measured. This measurement allows the localization of the positron annihilation along the line of propagation of the two gamma rays. Presently, tomographs allow measurement of the time of flight of the gamma rays with a precision of 650 psec at best, but current research focuses on the development of new detectors and electronics allowing this measurement. This translates to a 20 cm uncertainty on the localization of the annihilation.

4.4.2.3 Measurements and errors

Detection of two gammas in coincidence introduces a major difference between PET and SPECT. The flux I_i of gamma rays emitted in an attenuating medium of attenuation coefficient μ that reaches detector i situated at a distance d is related to the flux emitted I_0 by

$$I_i = I_0 e^{-\mu d} \tag{4.6}$$

The flux from the same source that reaches detector j in coincidence with i is

$$I_j = I_0 \; e^{-\mu(D-d)} \tag{4.7}$$

D being the distance between i and j, which is fixed by the diameter of the tomograph.

Therefore, the flux reaching simultaneously detectors i and j is the product

$$\begin{aligned} I_{ij} &= I_i \; I_j \\ &= KA \, e^{-\mu d} e^{-\mu(D-d)} \\ &= KA \; e^{-\mu D} \end{aligned} \tag{4.8}$$

where A is activity and k, a constant.

I_{ij}, the flux that is measured (i.e. the count rate), is independent of the place of annihilation along the LOR, and the level of attenuation depends only on the thickness of the tissue and not on the position of the source in the tissue. In other words, correction for attenuation in PET does not require, in contrast to SPECT, the knowledge of the depth of emission in the tissue.

Three types of coincidences are recorded (Figure 4.4.3):

Figure 4.4.3 Different types of coincidences. True coincidences correspond to the annihilation having taken place on the line of coincidence between the two detectors. Random coincidences correspond to two different annihilations, while scattered coincidences correspond to a deviation of at least one of the annihilation photons

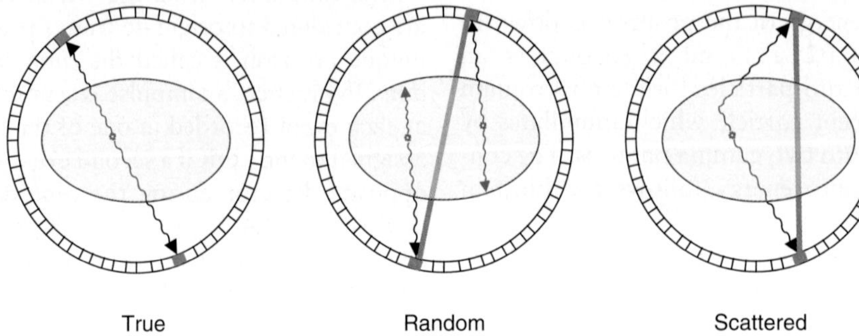

- *True coincidences*: Events in which the two annihilation gamma rays are not deviated before their detection.
- *Scattered coincidences:* Events in which at least one of the two annihilation gamma rays is scattered in the subject and deviates from its original direction of propagation. Scattered coincidences introduce a bias in the assignment of the LOR.
- *Random coincidences:* Events in which two gamma rays detected in the same time coincidence window correspond to the annihilation of two different positrons. The random coincidence rate is proportional to the width of the time coincidence window and to the square of the radioactive concentration between the two detectors. Random coincidences introduce homogenous noise in the measurements.

If the gamma ray is absorbed in the subject or slowed to the point that its energy when reaching the detector is far below 511 keV, it is not recorded and no coincidence is counted. This phenomenon, called *attenuation of the coincidences*, is important because only 17% of the coincidences can pass through 8 cm of tissue without losing energy. This figure falls to nearly 1% for the 511 keV gamma rays emitted at a depth of 15 cm, that is at the centre of the abdomen of a human subject. Figure 4.2.4, showing the percent of transmitted 511 keV as a function of tissue thickness, is an underestimation of the real attenuation because it does not consider all slowed gamma rays but only those which are totally absorbed by matter.

4.4.2.4 Corrections of the measurements

Correction for random coincidences. The random coincidence rate is estimated either using a measurement in a time coincidence window delayed from the on-line time coincidence window, or from the single photon count rate detected on each detector. In the latter case, the random coincidence are given by the following formulae:

$$R_{ij} = S_i, S_j, 2\tau, \qquad (4.9)$$

where R_{ij} is the rate of random coincidences along the LOR defined by detectors i and j, S_i is the rate of single photons detected by detector i, and τ is the width of the time coincidence window.

Correction for scattered coincidence detection. The main interaction of 511 keV with the subject is Compton scattering, in which the gamma ray loses energy and changes its direction of propagation. The characteristics of Compton scattering (i.e. angle of deviation and energy loss) have been established by Klein and Nishima. The rate of scattered coincidences is between 35 and 50% for abdominal imaging in human subjects and between 15 and 20% for cerebral imaging in human studies. The detection of such events mainly affects the low frequencies of the Fourier transform of the images. The presence of scattered coincidences in images thus has a small influence on visual inspection of the images, but it strongly affects the quantitative measurement of the radioactive concentration in various points of the organs.

Several methods have been designed to correct for such phenomenon. The most accurate method consists in computing the rate of scattered coincidences along lines of response from the Klein–Nishima equation. The computation requires the knowledge of the distribution of the attenuating medium of the subject, the estimation of the radioactive distribution in the medium and the modelling of the energy resolution of the detectors. Knowledge of the attenuating medium is obtained from the map of attenuation coefficients of the object $\mu(x, y, z)$ (see following paragraph). Knowledge of the spatial distribution of the radioactivity in the medium is obtained from a preliminary image reconstructed without scattered coincidence correction.

Correction for coincidence attenuation. The attenuation map of the medium, $\mu(x, y, z)$, is measured using an external radioactive source. The radioactive source can be a single photon-emitting source. The principle of the measurement is then similar to that of a computed tomography scans using an X-ray source. More and more frequently, PET scanners are coupled to a CT scanner. In this case, the attenuation coefficients are scaled to the appropriate energy (511 keV) using tabulated values. The attenuation map is then projected along each line of response to give the attenuation coefficient of the coincidences.

The radioactive source can also be a positron-emitting source. In this case, the attenuation coefficient of the coincidences along each line of response are obtained directly and used to correct the measurement of the true coincidences along the lines of response.

Normalization of the acquired data sets. Normalization allows correcting for the non-uniformity of detector efficiency. The normalization data set is obtained from a reference acquisition performed using a known radioactive source. This source can be

- positron emitting line sources inserted in the tomograph and rotating in front of the detectors thus simulating a perfect homogenous radioactive annular source
- a homogenous radioactive cylinder.

4.4.3 Instrumentation

Although the design of PET cameras is basically the same as that of SPECT cameras, there are some differences in the detectors used, the necessity of a coincidence circuit, and the fact that PET cameras can work without the need for a collimator.

4.4.3.1 Detectors

Three types of detectors are used for PET imaging: Multiwire proportional chambers, scintillation detectors coupled to photo-detectors, and more recently, semi-conductors.

Multiwire proportional chambers. Invented by Georges Charpak (Nobel prize 1992), they consist of an ionization chamber filled with an inert gas associated with a fine grid of metallic wires. The principle of gamma ray detection is as follows: A gamma ray produces an ionization of gas atoms as it passes through the gas; the ionization electrons are accelerated by an electric field and are collected at the wires. The advantages of proportional gas chambers are their high spatial resolution, which depends only on the distance between the wires. Its main disadvantage over the scintillation detectors is the low stopping power of gas for 511 keV photons, that requires using large volumes of detection in order to maintain the efficiency. This disadvantage has been partly overcome by using piles of wire grid and lead (Clark and Buckingham, 1975). Another disadvantage is the very poor energy resolution of the detectors which renders difficult correction of scattered events. This design of detectors has been adapted for small animal PET imaging (Jeavons *et al.*, 1983).

Crystals. Most PET cameras use LSO and BGO scintillation crystals (see Table 4.3.2). In order to increase the sensitivity of the PET cameras, new arrangements of crystal materials have been proposed: Two scintillating crystals with different timing properties have been coupled, allowing the measurement of the depth of interaction of the gamma rays in the crystals while increasing the volume of detection.

Packing. A particular problem posed by small animal PET cameras is the necessity to pack closely in a small volume a large number of detectors. The crystals and the photodetectors are arranged either into blocks of small crystals coupled to photomultipliers tubes or as small crystals coupled to a light guide, which is itself coupled to a large number of photomultipliers tubes, thus forming a semi-pixellated detector. Blocks and semi-pixellated detectors are generally assembled into a cylindrical geometry. Avalanche photodiodes are more and more often used in order to increase the packing fraction of the crystals thus reducing the gaps between detectors.

4.4.3.2 Coregistration of anatomy

The major recent improvement in clinical PET has been the capacity to include anatomical information in the molecular images through the addition of a CT scanner co-axial to the PET camera. This provides the benefit of both techniques, exquisite anatomical details from CT and sensitive molecular information from PET. However, it also considerably increases the cost because two instruments are necessary. Nevertheless, with the advent of small animal CT scanners (see Chapters 2 and 8), it can be expected that small animal PET-CT cameras will appear in the market in the near future.

4.5 Image properties and analysis

4.5.1 Tomographic image reconstruction

Image reconstruction requires a precise modelling of the link between the object to be reconstructed and the acquired data. In tomography, the link is used either to define the transition matrices between the measurements and the spatial distribution of radioactivity in the object, or to correct the measurements along each line of response in order to obtain the most accurate estimate of the integral of the radioactive distribution along each line of response. In the latter case, the tomographic image reconstruction reduces to the inversion of the transform.

Tomographic principles have been exposed in Chapter 2, and only the information specific to PET and SPECT imaging will be reported here; the major difference being that in contrast to CT, with radiotracer imaging, the source is inside the object under study and not focalized. Tomographic image reconstruction is an important step of the data processing because it determines the resulting image characteristics. Depending on the exploitation of the reconstructed images, visual inspection or quantitative analysis of the tracer kinetics, different reconstruction methods will be preferred. Therefore, a minimal knowledge of the different methods used for reconstruction is advisable.

After applying the corrections described in paragraphs in Sections 4.3.2.5 (SPECT) and 4.4.2.4 (PET), the number of events (coincidences for PET imaging and photons for SPECT imaging) for a given line of projection is an estimate of the integral of the radioactive concentration $f(x,y,z)$. It should be noted that projections under different angles of the radioactivity are acquired simultaneously on most PET cameras because the detectors are arranged in a cylindrical geometry. To the opposite, the acquisition of projections under various angles is most often performed sequentially on most SPECT cameras because the number of detectors is small. In that case, the projections acquired at various angles are not strictly equivalent. The tomographic image reconstruction consists in an inversion of this transform, named X-ray transform 2D, if the data acquisition is performed in 2D and X-ray transform 3D if the data acquisition is performed in 3D mode. The process can be performed either using analytic methods or using iterative methods.

4.5.1.1 *Reconstruction methods*

The most commonly used analytic reconstruction algorithm is the filtered back projection. It consists in backprojection on the image grid of the acquired and corrected projections. A filtering step is necessary in order to account for the response function of the backprojection operator. An apodization function is most often used with the filter to reduce the amplification of the noise at high frequencies.

The filtered backprojection algorithm is a linear reconstruction algorithm. This property ensures a better control of the trade off between signal-to-noise ratio and spatial resolution in images. Spatial resolution mainly depends on the choice of the apodization filter and is largely independent on counting statistics in projections. The main disadvantages of such methods are the following:

- The spatial sampling of the projections must be assumed linear;
- Signal-to-noise ratio in images is poor at low statistics;
- No *a priori* information on the object and data acquisition process can be included.

In contrast, iterative tomographic image reconstruction allows the incorporation of both an accurate description of the acquired data (geometry of the camera, statistical properties of the data, effects degrading the data themselves see) and *a priori* information on the object itself. *A priori* information on the object may be both spatial information and information on the tracer kinetics.

The more accurate the modelling, the more time consuming the reconstruction process is.

4.5.2 *Spatial resolution*

The spatial resolution is defined by the smallest distance separating two points of the subject that can be resolved separately on the resulting image. For tomographic imaging techniques, the spatial resolution can be isotropic in the three dimensions or anisotropic if resolution differs between the dimensions. Mathematical functions used to define the spatial resolution of a given tomograph and compare different tomographs are in the form

$$I(x, y, z) = S(x, y, z) \otimes F(x, y, z),$$

where $\otimes$ denotes a convolution of the function F applied to the object S in order to produce the image I.

Due to the heterogeneous and generally unpredictable nature of the object, it is generally impossible to define the true/real function, but approximations using regular geometrical forms can be made to approach F. One, which is largely used, is the Point Spread Function or PSF, which describes how a 'point source', that is a very small spherical object emitting gamma rays, will appear on the reconstructed image. Another useful function is the line spread function (LSF), which is the integration of the PSF along one axis. Calculation of the PSF of a complete tomograph can be extremely time consuming because it requires the calculation of all the individual PSFs, which represents several millions of individual functions for a PET camera with 8196 detectors.

Experimentally, when a point source is imaged in a camera, the resulting image can be fitted as a distribution of values centred on a maximal value corresponding to the centre of the object. The spatial resolution is defined as the width at half the maximal value (full width at half maximum (FWHM)).

The effective spatial resolution can be appreciated experimentally using objects of calibrated geometry called *phantoms*, such as the *Derenzo* phantom, which is a plastic cylinder drilled with calibrated holes of known diameters. Phantoms are very useful to evaluate tomographs and compare the quality of the images obtained in different acquisition modes or using different reconstruction paradigms. Shown in Figure 4.5.1 is a comparison of the images obtained with a Derenzo phantom (A) and in mice (B) acquired in three different PET cameras of decreasing spatial resolutions.

In SPECT, the spatial resolution depends on

- The intrinsic resolution of the crystal which is a function of its thickness;
- The Anger circuit;
- The geometry of the collimator. This is the main parameter on which one can act in order to improve the resolution;
- The degree of Compton scattering, which depends on the thickness of the tissue. Small animals have a very big advantage over larger ones in that respect.

In PET, the spatial resolution depends mainly on

- the spatial sampling of the lines of response, which is a characteristic of the camera;
- the positron range, which is a property of the radionuclide;
- the variation in the angle between the two annihilation gamma rays trajectories, which is statistically centred at $180° \pm 0.3°$. This later effect, which represents roughly one third of the degradation of spatial resolution in a clinical camera, can be accounted for in iterative reconstruction.

In addition, resolution is also degraded by post processing filtering in both imaging techniques.

Current clinical SPECT tomographs have spatial resolutions (FWHM) of 4–6 mm, and recent SPECTs for small animals have resolutions lower than 1 mm.

Current PET tomographs used for clinical imaging allows to obtain images with a spatial resolution between 2.5 and 5 mm (FWHM) in the three direc-

Figure 4.5.1 Reconstructed images of (a) a Derenzo phantom and (b) a mouse acquired in different PET cameras. (a) The phantom was filled with a positron emitter and the same number of counts acquired in (from left to right): a Concorde CTI-Siemens Focus 220, a Siemens-CTI HRRT, and a Siemens-CTI HR+. (b) the mouse was injected with 400 μCi of FDG and the same number of counts acquired in the three images. Identical reconstructions for all three acquisitions demonstrate the importance of spatial resolution on image quality

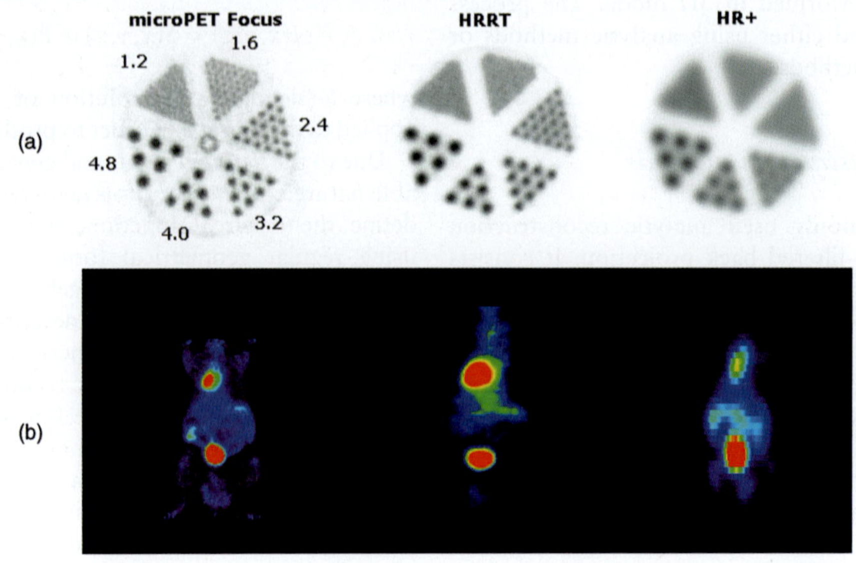

tions. For small animal PET imaging, spatial resolution is usually between 1 and 2 mm.

4.5.3 Sensitivity, signal-to-noise ratio and contrast ratio

The sensitivity is the ratio of the counts counted by the camera to those emitted by the object. It is best expressed in percent, but sometimes in Bq Bq^{-1}, or even Bq Ci^{-1}. Due to attenuation it is usually low in radiotracer imaging, and typical values range from 0.1–0.2% for small animal SPECT to 0.5–6% for small animal PET. Values given for a large volume placed in the field of view are more representative of the camera performance for real imaging than those which represent the counts recovered from a point source.

All imaging systems generate a significant amount of noise, and this is particularly important for radiotracer imaging because the absolute number of events detected per voxel is low. The signal-to-noise ratio (SNR) of an image is usually defined as the ratio of the mean voxel value to the standard deviation of the voxel values. Within certain limits, the noise is considered to follow a Poisson distribution. The SNR increases as the square root of the signal, and increasing the concentration of radionuclide within reasonable limits is beneficial to the SNR as long as the dead time of the system is low. However, even a very high SNR is not useful if there is no contrast between the regions of the subject. The contrast ratio is defined as the ratio of the highest to the lowest value that the system is capable of counting. High contrast ratio is a desired aspect of an imaging system; however, it depends on the radiotracer distribution as much as on the camera.

4.6 Radiochemistry of gamma-emitting radiotracers

4.6.1 General considerations

This chapter provides an overview of the nuclear and chemical properties of radionuclides to use for SPECT application in animal or in human beings. It is not intended as a textbook to prepare the reader to perform labelling by himself; rather it outlines the criteria, including those depending on cost and availability, that will allow a proper selection of the best suitable radionuclide for a given purpose. For further reading refer for instance to Adloff and Guillaumont (1993).

4.6.1.1 *Type and energy of emitted radiation*

Among the different nuclear properties, the type and the energy of radiation are crucial in assessing the feasibility of imaging and in determining the safety rules for handling.

The radioisotope most widely used in medicine is technetium-99m (^{99m}Tc), employed in over half of all nuclear medicine procedures. This radionuclide displays almost ideal emitting radiation because decay occurs through a process called 'isomeric transition' that generates gamma rays and low energy electrons. As there is no high-energy beta emission the radiation dose to the patient is low. In addition, low energy gamma rays easily escape the human body and are accurately detected by a gamma camera.

On the contrary, although iodine-131 (^{131}I) could be used for imaging, its high-energy γ-emissions are not optimally counted and its β^--particle emissions contribute to increase the radiation dose delivered to the body.

Although the criteria for the selection of a suitable radionuclide for clinical use equally apply to animals, some radionuclides that are not useful for human studies due to unfavourable radiation or because of delivering high doses may be used in animal experiments.

For instance, ^{125}I, a low-energy emitting radionuclide, can be used in small animals such as mice, whereas it would be of no value in imaging larger animals or humans due to the total attenuation of the photons through tissues. Similarly, whereas for human examinations radionuclides giving low radiation doses to organs and tissues should be preferred, dosimetric considerations will have a less stringent impact when applied to animal studies.

4.6.1.2 *Production and availability of gamma-emitting radionuclides*

The production of ^{99m}Tc from ^{99}Mo has been described in Section 4.3.1.3. The mode of production for other gamma-emitting radionuclides is reported in Table 4.6.1:

All the radionuclides listed in the above table are commercially available in a chemical form suitable for radiolabelling. In addition, those used routinely in humans are available as sterile and pyrogen-free preparations.

Table 4.6.1 Mode of production of gamma-emitting radionuclides

Radionuclide	Mode of production	Nuclear reactions	Type of irradiation
^{67}Ga	Direct	^{68}Zn(p, 2n)^{67}Ga	28 MeV protons (cyclotron)
^{99m}Tc	From ^{99}Mo decay (generator system)	^{235}U(n, fission) // → ^{99}Mo ^{99} Mo $\xrightarrow{\beta^-}$ ^{99}m Tc	Thermal neutrons (nuclear reactor)
^{111}In	Direct	^{112}Cd(p, 2n)^{111}In	26 MeV protons (cyclotron)
^{123}I	From ^{123}Xe decay	^{124}Xe(p, 2n)^{123}Cs → ^{123}Xe ^{123}Xe $\xrightarrow{EC}$ ^{123}I	30 MeV protons (cyclotron)
^{125}I	From ^{125}Xe decay	^{124}Xe(n, γ)^{125}Xe ^{125}Xe $\xrightarrow{EC}$ ^{125}I	Thermal neutrons (nuclear reactor)
	From ^{131}Te decay	^{130}Te (n, γ) ^{131}Te ^{131}Te $\xrightarrow{\beta^-}$ ^{131}I	Thermal neutrons (nuclear reactor)
^{131}I	From fission products	^{235}U(n, fission/ → ^{131}I	Thermal neutrons (nuclear reactor)
^{201}Tl	From ^{201}Pb decay	^{203}Tl(p, 3n)^{201}Pb ^{201}Pb $\xrightarrow{EC}$ ^{201}Tl	30 MeV protons (cyclotron)

4.6.1.3 Chemical differences between SPECT radionuclides

A rapid inspection of the radionuclides listed in the previous tables reveals that they belong to two distinct chemical groups, one including elements having metallic properties (gallium, technetium, indium and thallium), the other one including the halogen iodine.

The two groups display marked differences in their chemical properties so that completely different labelling strategies have to be elaborated.

The most noticeable difference between metals and iodine lies in the ability for the latter to form covalent bonds with carbon. Indeed, iodine, like the other halogens, can bind directly to an organic molecule through a direct C–I covalent bond. On the contrary, metal radionuclides would necessitate the presence of an appropriate set of coordinating atoms on the molecule, often called chelating ligands, to form a stable bond.

Despite their apparent large difference, the two binding modalities generate bonds of similar strength, but the introduction of a foreign atom induces some alterations of the original chemical and biological properties of the molecule. Therefore, the small radius of iodine compared with the size of the atoms necessary to co-ordinate a metal make it better suited for radiolabelling small molecules, especially for receptor binding studies. However, perturbations induced by labelling will be less and less pronounced as the size of the molecule increases, making metals suitable for labelling large molecules like peptides and proteins without a significant loss of the original properties of the molecule.

4.6.1.4 Macroscopic and tracer level chemistry

One important aspect encountered with the use of radioactive isotopes is that the element in question is present at an extremely low concentration. Different terms, sometimes intended as synonyms, are employed to describe this low concentration of a radioisotope: *Tracer level, carrier-free, no carrier-added.* As pointed out in a recent radiochemistry textbook[i] the above terms do not have a clear-cut signification and should therefore be carefully defined in order to avoid misleading. Generally, in analytical chemistry, *traces* cover the level from 100 to 1 ppm (i.e. from 100 to 1 μg/g).

- In radiochemistry the *tracer level* refers to amounts less than 10^{-8} g, which usually correspond to a concentration of about 10^{-10} mol L^{-1}.
- *Carrier-free* or, rather, *no carrier added* (the latter term should be preferred following the IUPAC recommendations) refers to a preparation of a

radioactive isotope which is essentially free from stable isotopes of the element in question (a subtle distinction between these two terms is nevertheless still in use to describe preparations of ^{99m}Tc – see below).

In radiopharmaceutical preparations the concentration is in the range of 10^{-5}–10^{-10} mol L^{-1}, because of the relatively high level of radioactivity handled.

It should be emphasized that the term *carrier-free* does not automatically imply that the concentration of the radionuclide is at tracer level; it only means that no stable isotope is present in the preparation (whether it has been intentionally added or not). This is easily explained by observing that the mass of a radionuclide associated with a given amount of radioactivity depends only on the value of the radioactive decay constant following the relationship:

$$A = \lambda N,$$

where A represent the radioactivity (decay rate) in becquerels (1 Bq = 1 desintegration per second or dps) and λ represents the decay constant in s^{-1}; N represents the number of radioactive nuclide responsible for the radioactivity A.

As the half-life ($T_{1/2}$) is linked to the decay constant by the relation $T_{1/2} = \ln(2)/\lambda$, the mass, in grams, of a given radionuclide is related to its half-life by the following relationship:

$$m(g) = \frac{A \times T_{1/2} \times Am}{\ln^{(2)} \times 6,02 - 10^{23}}$$

where A_m represent the atomic mass of the radioisotope and 6.02×10^{23} is the Avogadro number.

For relatively short-lived radionuclides, the mass associated with a given amount of radioactivity, let us say 10 MBq, is small (e.g. 0.015 μg for ^{125}I); indeed, carrier-free preparations of this radionuclide can be considered at tracer levels. This would not be the case for long-lived radionuclides ($T1/2 > 10\,000$ years), for which the mass associated with the same amount of activity will be in the order of milligrams, well above a tracer amount.

Practically, the theoretical specific activity calculated from $A = \lambda N$ is rarely achieved. Specific activity is often lowered by the unavoidable presence of minute amounts of the stable element contaminating the reagents used in the manufacturing process of the radionuclide. Contamination by elements different than the radionuclide (e.g. Fe^{3+} for $^{111}In^{3+}$) has a similar effect as that observed by reducing specfic activity because such elements may compete with the radionuclide for the same site of binding.

Theoretical and practical specific activities for gamma-emitting radionuclides are reported in Table 4.6.2:

Radiolabelling is merely a chemical reaction carried out with a radioactive isotope of an element. Therefore, one strategy could be to extend the results obtained with a stable isotope to preparations using the corresponding radioactive isotope. In fact, as pointed out by Baldwin[ii] for radioiodination, although isotopes (whether stable or radioactive) display very similar chemical properties, this is rarely successful.

Table 4.6.2 Specific activity of gamma-emitting radionuclides

		Theoretical specific activity		Practical specific activity	
Radionuclide	Half-life	(GBq/μg)	(GBq/μmol)	(GBq/μg)	(GBq/μmol)
^{67}Ga	3.56 Days	20.3	1357	0.02–0.04	1.3–2.7
$^{99m}Tc^{a}$	6.02 h	54.8^{b}	5425	20–50	2000–5000
^{111}In	2.83 Days	15.4	1707	>1.85	>68
^{123}I	13.3 h	70.9	8718	1–35^{c}	123–4300
^{125}I	59.4 Days	0.65	813	0.6	79
from ^{130}Te		4.6	602	0.74	97
^{131}I from ^{235}U	8.02 Days	4.6	602	1.7 – 2.5	222–327
^{201}Tl	3.04 Days	7.9	1589	0.004–0.04	0.8–8

[a]Due to the ^{99}Mo branching decay, ^{99m}Tc is always associated with ^{99}Tc.
[b]Value at the time of elution. Specific activity will then decrease by a factor of two every 6 h.
[c]Specific activity is strongly affected by the chemical purity of the reagent used in the manufacturing process.

Often, differences in chemical reactivity arise from accompanying impurities, which nature and amounts depend on the manufacturing process of the radionuclide. In addition, working with very low concentrations of one of the reagents brings about differences in the reaction kinetics. This effect is important for a reaction of second order with respect to the labelling atom.

4.6.1.5 Labelling strategies

Irrespectively of the radionucleide, strategies may be direct or indirect:

Direct labelling. In direct labelling, the radionuclide is bound directly to the target compound. Binding may occur by simple addition (e.g. addition of iodine to a C=C double bond, or addition of metal to a set of coordinating atoms) or by replacement of one atom of the target compound (e.g. replacement of an hydrogen by iodine).

The site of binding or of replacement may either pre-exist in the native target compound or has to be synthetically created before the labelling step. When labelling occurs with radioactive metals, the chemical entity that has to be added to the target molecule is called a '*bifunctional chelating agent*' or BCA. A BCA contains a functional group through which it can be covalently bound to the target compound, and a set of donor atoms able to form a stable complex with the metal. The two portions of the BCA may be separated by a spacer, the function of which is to minimize interactions between the site of the molecule responsible of its biological properties and the chelate. Extensive reviews cover the chemistry of BCA for labelling peptides and antibodies with technetium (Verbruggen, 1996), with diagnostic radiometals (Fichna and Janeka, 2003) and with therapeutic radiometals (Liu and Edwards, 2003).

Indirect labelling. In the indirect labelling the radionuclide is first bound to a BCA molecule, then the labelled BCA is conjugated to the target compound. For indirect iodination, the BCA entity is sometimes called a prosthetic group. Although the indirect labelling strategy has been proposed for ^{99m}Tc, its use is mainly adapted to longer-lived isotopes because the conjugation step, almost invariably followed by a separation step, is time consuming.

4.6.2 Labelling with metals

4.6.2.1 Basic concepts

Several excellent textbooks cover extensively the broad range of concepts related to the structure and reactivity of metal complexes (among others, Huheey *et al.*, 1993; Douglas *et al.*, 1983).

Two important properties should be taken into account in order to predict a reliable metal–ligand labelling:

- stability/instability, a thermodynamic concept;
- inertness/lability, a kinetic concept.

Thermodynamic stability/instability refers to concerned with the energetic aspects of metal–ligand association. Inertness/lability are kinetic terms referring to how quickly a reaction system reaches the equilibrium. Therefore, the thermodynamic stability gives information on the constant of formation of a given metal–ligand complex, whereas kinetics informs about the rate of replacement of ligands.

4.6.2.2 Thermodynamic stability

Thermodynamic stability is related to the difference of free energy (ΔG°) between the 'product' (metal–ligand complex) and reagents (metal and ligand taken separately).

Negative differences indicate high stability of the final product, whereas positive differences indicate instability. This is expressed by the following equation:

$$\Delta G = \Delta H - T\Delta S,$$

in which it can be seen that stability is related to enthalpic (ΔH) as well as to entropic ($T\Delta S$) factors. Note that the fact that a reaction has a negative ΔG does not necessarily imply that it will be completed in a reasonable period of time.

Some general rules can be applied to predict the formation of stable metal–ligand complexes. The enthalpic contribution can be positively affected by matching the electronic properties of the metal with those of the ligand (hard/soft acids and hard/soft bases concept).

Hard metals (small, non-polarizable, high charge density cations, e.g. Ca^{2+}, In^{3+}) preferentially bind to hard bases (small, non-polarizable high charge density anions, e.g. OH^-, $RCOO^-$, RNH_2). Soft metals (large, polarizable, low charge density cations, e.g. Cu^+, Tc^+, $Tc{\equiv}N^{2+}$) preferentially bind to soft bases (large, polarizable low charge density anions, e.g. P, R–NC, CO, R_2S, RS^-). Metals and ligands with intermediate hardness/softness (e.g. $Tc{=}O^{3+}$) bind to ligands with intermediate hardness/softness or to multidentate ligands with mixed hardness.

The entropic contribution can be positively affected by a right choice of the structure of the ligand. Therefore, the stability of the final metal–ligand complex will increase:

(a) by using polydentate in replacement of monodentate ligands

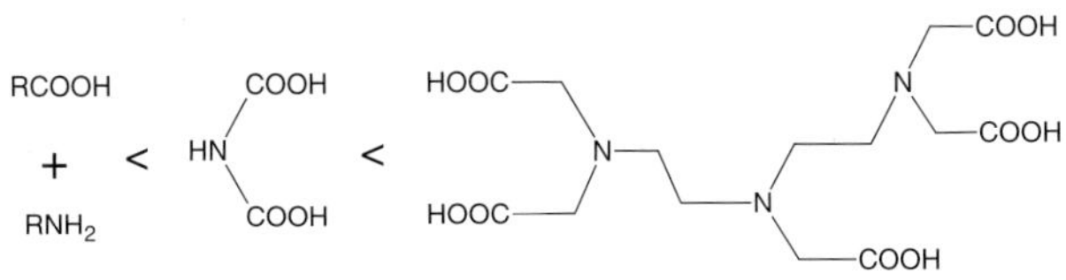

amine and carboxylic acid — aminod icarboxylate ligand (IDA) — polyamynocarboxylate ligand (DTPA)

(b) by increasing the number of cycles co-ordinating the metal

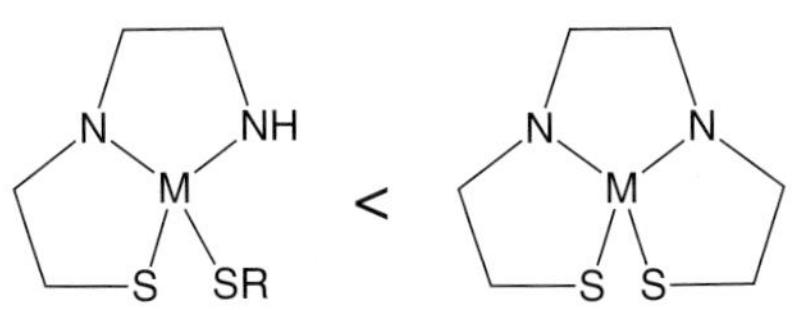

4.6.2.3 *Kinetic inertness*

The kinetic inertness represents the difference of energy between the activated state and the reagents (metal and ligand separately). It is related to the electronic status of the metal and to its difficulty to form activated intermediates (e.g. forming an heptaco-ordinated complex from an hexacoordinated complex) through a suitable low-energy pathway.

4.6.2.4 *Example of application*

These thermodynamic/kinetic concepts can be successfully applied to ^{111}In labelling of polyaminocarboxylate ligands (e.g. DTPA).

Radioactive indium (^{111}In) chloride is commercially available in 0.05 M HCl solution. Hydrochloric acid is used to prevent the precipitation of indium hydroxide both by keeping a low pH and by the formation of poorly stable ($K_{stab} \cong 10^3$) aqua-chloro complexes such as $^{111}[In(H_2O)_5Cl]^{2+}$. The Indium-DTPA complex, whose formation takes place at pH above 5, displays a very high stability constant ($K_{stab} \cong 10^{29}$). However, a direct reaction between indium aqua-chloro complexes and DTPA at pH $\geq$5 will yield only indium hydroxide. This occurs because the complexation of indium by DTPA proceeds too slowly to reach the equilibrium, allowing the much faster reaction of precipitation of indium hydroxide to succeed.

Indium-DTPA may nevertheless be obtained by an indirect reaction in which the pH of the indium solution is first raised in presence of acetate or citrate ions before adding DTPA as showed below:

$$[^{111}In(H_2O)_5Cl]^{2+} + AcO^- \xrightarrow{pH \geq 5} {}^{111}In(OAc)_3 \text{ (kinetically fast) } K_{stab} \cong 10^4$$

$$^{111}In(OAc)_3 + DTPA \longrightarrow {}^{111}In\text{-}DTPA \text{ (kinetically slow) } K_{stab} \cong 10^{29}$$

The kinetics of the formation of indium acetate or citrate complexes is very fast in comparison to that of the formation of indium hydroxide, thus preventing precipitation of indium.

DTPA is slowly substituted for the acetate (or citrate) ions, producing a compound whose stability is $\sim 10^{25}$ times higher than that of the intermediate compound.

4.6.3 *Chemical properties of gamma-emitting radiometals*

Table 4.6.3 lists the more relevant electronic and chemical properties of gallium, indium, thallium and technetium:

Technetium, a second-row transition metal, displays a very rich chemistry with documented examples of water stable complexes for five different oxidation states. On the contrary, the chemistry of the post-transition metals gallium, indium and thallium is restricted to that of the M^{3+} cation and, additionally, to the M^+ chemistry for thallium. Because of the marked difference between the two groups of metals, the chemistry of technetium will be discussed in a separate section.

4.6.3.1 *Chemical properties of gallium, indium and thallium*

Gallium(III) and indium(III) bind preferably to hard bases (carboxylate, hydroxyl and amine). Thallium(I) may bind to sulfur atoms. When macrocyclic ligands are used, the stability of the final chelate increases

Table 4.6.3 Electronic and chemical properties of radiometals for SPECT

	Gallium	Indium	Thallium	Technetium
Atomic number	31	49	81	43
Electronic configuration	$[Ar]3d^{10}4s^{2}4p^{1}$	$[Kr]4d^{10}5s^{2}5p^{1}$	$[Xe]4f^{14}3d^{10}4s^{2}4p^{1}$	$[Kr]4d^{5}5s^{2}$
Electronegativity	1.81 (III)	1.78 (III)	1.62 (I) 2.04 (III)	1.9
Radius M^{3+} (pm)	62	80	150 (Tl^{+})	136 (Tc°)
Oxidation state stable in water	III	III	I, III	I,III,IV,V,VII
Coordination number	6	6, 7, 8	6	5, 6, 7
Coordination geometry	Octahedral	Octahedral, square antiprismatic	Octahedral	Octahedral, square pyramidal, trigonal bipyramid

when the ionic radius of the metal matches the size of the cavity of the macrocycle (example, Ga-NOTA and Ga-DOTA complexes). However, due to their rigidity, the binding process of a metal to a macrocycle requires more energy than that needed to bind to a flexible acyclic ligand. Therefore, complexation reactions with macrocycles should be carried out at high temperatures for completeness. This may represent a serious drawback if the macrocyclic is used as a bifunctional chelating agent to label thermolabile substances.

4.6.3.2 *Labelling with ^{67}Ga and ^{111}In*

Compounds that have to be labelled with ^{67}Ga or with ^{111}In should contain a suitable chelating group, generally a derivative of an acyclic or cyclic polyaminocarboxylate ligand.

The general labelling procedure proceeds as follows:

(a) Addition of sodium acetate or sodium citrate to the radiometal in order to raise the pH of the solution to 4–5. A labile intermediate complex is obtained in few minutes.
(b) Addition of the compound to be labelled to the intermediate complex. The labelling reaction is generally completed in approximately 10–30 min at room temperature.

If binding is not complete, purification of the labelled compound from unreacted metal may be carried out by any suitable chromatographic technique (e.g. size-exclusion chromatography for large compounds).

4.6.3.3 *Examples of ^{67}Ga and ^{111}In labelled compounds*

There are several approved radiopharmaceuticals labelled with ^{111}In and only one labelled with ^{67}Ga.

^{67}Ga citrate

Preparations of ^{67}Ga citrate are used to localize sources of fever in patients with fever of unknown origin and to evaluate the evolution of inflammatory processes. The mechanism of ^{67}Ga uptake is not known but it is believed that the metal is trapped by cells after dissociation of the poorly stable citrate complex.

^{111}In-oxyquinoline

The neutral, lipid-soluble indium complex with hydroxyquinoleine (^{111}In-oxinate) is able to penetrate the cell membrane and to accumulate into the cell by dissociation mediated mechanism. Indium-111 labelled white cells are used in the detection of inflammatory tissues to which leukocytes migrate and accumulate. Labelling of leukocytes is performed by incubating a preparation of ^{111}In-oxyquinoline with isolated cells in a plasma-free medium.

^{111}In-pentetate (^{111}In-DTPA)

The negatively charged, high hydrophilic ^{111}In-DTPA complex is used for determining the flow pattern of cerebrospinal fluid (CSF) after intrathecal administration of the radiopharmaceutical.

^{111}In-pentetreotide (Octreoscan®)

Pentetreotide is a DTPA conjugate of the synthetic cyclic peptide octreotide, an analogue of the human hormone somatostatin. Once labelled, ^{111}In-pentetreotide localizes into primary and metastatic neuroendocrine tumours bearing somatostatin receptors.

^{111}In-labelled antibodies

Several IgG murine monoclonal antibodies labelled with Indium-111 have been approved by the European or by US authorities as diagnostic tools in the detection of benign or malignant disorders or as aid to patient management. They comprise antibodies for imaging prostate cancer (Capromab pendetide – ProstaScint®), for detecting metastatic diseases associated with colorectal and ovarian cancer (Satumomab pendetide – OncoScint®) and for the diagnosis of myocardial necrosis (Imciromab pentetate – Myoscint®). For all radiopharmaceuticals the native antibody has been chemically modified by introduction of a DTPA-like moiety to enable efficient indium chelation.

4.6.3.4 *^{201}Tl labelled compounds*

Thallium-201 is used only in the form of a simple chloride salt ($^{201}TlCl$). The mono-cationic thallium(I) accumulates in viable myocardium in a way analogous to that of potassium, allowing the detection of cardiac diseases of ischaemic origin. Because of the poor imaging quality of this radionuclide, very little work on thallium radiopharmaceuticals has been conducted in the last two decades.

4.6.4 Technetium

Technetium was the first element to be produced artificially. Currently, about 20 isotopes of technetium with a mass number from 90 to 110 have been documented. All of them are radioactive with half-lives ranging from 0.8 s (^{110}Tc) to 4.2 10^6 years (^{98}Tc). The radioisotope of interest in nuclear medicine is ^{99m}Tc, which is considered like the workhorse of the nuclear physician. The reasons are its optimal nuclear properties, its widespread availability and low cost and its chemical versatility. Several excellent reviews have witnessed the rapid evolution of new concepts in technetium chemistry (Deutsch *et al.*, 1983; Verbruggen, 1990; Schwochau, 1994; Banerjee *et al.*, 2005). In addition to these articles, an invaluable source of information related to the chemistry, radiochemistry and medical application of technetium compounds can be found in the proceedings of the six symposia devoted, since 1982, to technetium in chemistry and in medicine (Nicolini *et al.*, 1983, 1986, 1990, 1995, 1999, 2002). Only a small fraction of the enormous body of technetium literature will be commented in this section.

4.6.4.1 *Chemical properties of technetium*

Technetium is a second row transition element placed in group 7 of the Periodic Table, below manganese and above rhenium. As for other transition metals, the chemistry of technetium is much more similar to that heavier congener, rhenium, rather than to that of manganese. However, differences exist between technetium and rhenium. The main difference lies in the redox potential of the two elements: Tc(VII) is much easier to reduce than Re(VII). This is particularly important knowing that the starting chemical form for both radioactive nuclides is the ion $M(VII)O_4-$.

Technetium may be present in all oxidation states from −1 to +7, giving rise to compounds displaying coordination numbers ranging from 4 to 9. In aqueous or in hydro-alcoholic media, the most documented oxidation states are: +1, +3, +5 and +7 and, to a lesser extent, +4 and +6.

The most frequently encountered structures for oxidation state +5 are *trans*-dioxo core $(TcO_2)^+$, oxo core $(TcO)^{3+}$ and nitrido core $(TcN)^{2+}$.

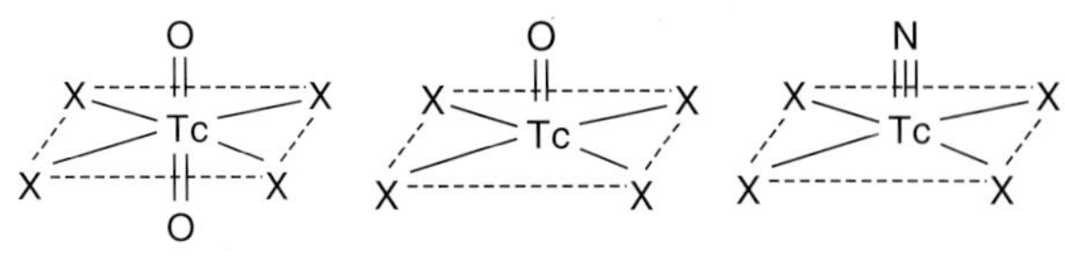

trans-dioxo core   oxo-core   nitrido core

Technetium chemistry in radiopharmaceuticals

Technetium-99m is obtained by decay of ^{99}Mo as shown in the following decay scheme:

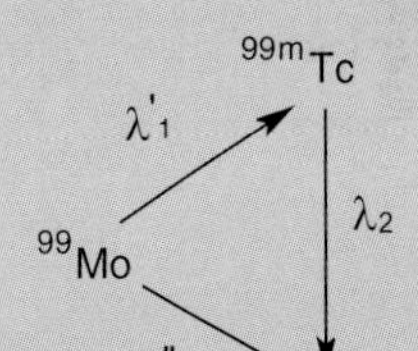

The separation of ^{99m}Tc from ^{99}Mo is obtained by eluting with saline a small chromatographic column filled with aluminium oxide. The eluate provides the chemical species $^{99m}TcO_4-$ in a sterile and pyrogen-free saline solution.

Due to the particular branching decay of ^{99}Mo, even fresh elutions from a generator always contain both ^{99m}Tc and ^{99}Tc nuclides. From the chemical point of view, the two isotopes are indistinguishable, that is they enter equally in the final radiopharmaceutical composition. The total amount of technetium depends on the total ^{99}Mo activity and on the time elapsed between two elutions. For a 3.7 GBq

^{99m}Tc generator eluted 24 h before, 37 MBq (1 mCi) of ^{99m}Tc are associated with approximately 7×10^{-7} mg of Tc ($^{99m}Tc + {}^{99}Tc$). This amount increases to approximately 4×10^{-6} mg if the generator has been eluted 96 h earlier.

Under the form of $^{99m}TcO_4^-$ ion, technetium does not bear any complex forming properties. For this reason, the metal must be reduced to oxidation states lower than +7 in order to bind various ligands.

Thus, the preparation of the technetium radiopharmaceuticals basically consists in a reduction reaction followed by formation of a co-ordination complex:

$$\mathbf{^{99m}TcO_4^- + reducing\ agent + ligand\ (L) \rightarrow [^{99m}TcL]}$$

The general procedure for preparing a kit for ^{99m}Tc labelling may be summarized as follows:

- All solutions should be made with Low Oxygen Content (LOC) water (i.e. water boiled and cooled under nitrogen/argon). All steps to be carried out in air tight vessels under nitrogen bubbling.
- Tin(II) chloride dihydrate is dissolved in LOC HCl (0.1–1 M). Solution is prepared in high concentrations in order to minimize their acid contents.
- If the molecule to be labelled is able to complex Sn^{2+}, a tin(II) chloride solution is directly added to the molecule dissolved in water. If the molecule does not complex tin(II), or if it is present at low concentration, a tin complexing anion is added first (e.g. glucoheptonate, tartrate, pyrophosphate, etc.)
- Other ingredients can be added to the solution (auxiliary reagents, stabilisers) and the pH is brought to the optimal value.
- The final solution is dispensed in vials and freeze-dried to obtain a product with a long shelf life.

Examples of ^{99m}Tc-cores obtained with a radiochemical yield greater than 90% in aqueous solvents are reported in Table 4.6.4.

4.6.4.2 ^{99m}Tc labelled compounds used in research and in clinical routine

There are many approved ^{99m}Tc radiopharmaceuticals, ranging from small molecules to large antibodies, each of them displaying very precise chemical and biological properties covering different diagnostic needs. In addition to approved radiopharmaceuticals, many other ^{99m}Tc-labelled compounds are being continuously studied in animals to find new clinical applications.

Examples of ^{99m}Tc radiopharmaceuticals and of research compounds used in animal studies are reported below. This list, which is intended to highlight the chemical and biological diversity existing between the different compounds, is far from being exhaustive. Excellent reviews discussing the structure–activity relationship of ^{99m}Tc radiopharmaceuticals have been published elsewhere (Clark and Podbielski, 1987; Dilworth and Parrots, 1998; see also Nicolini *et al.*, 1983, 1986, 1990, 1995, 1999, 2002).

4.6.4.3 Examples of ^{99m}Tc-radiopharmaceuticals

^{99m}Tc-sestamibi

Technetium-99m sestamibi is a lipophilic cationic complex which accumulates in viable myocardium tissue in manner analogous to that of $^{201}Tl^+$, giving comparable scintigraphic images. However, retention in the myocardial cell of the two cations is mediated by different mechanisms. Whereas, it has been shown that the uptake of $^{201}Tl^+$ correlates with the extent of sodium-potassium exchange, the myocardial uptake of ^{99m}Tc-sestamibi is not blocked when the sodium-potassium pump mechanism is inhibited. Studies in cultured cells have shown that retention of ^{99m}Tc-sestamibi occurs specifically within the mitochondria as a result of electrostatic interactions between the negative charges of the mitocondrial membrane and the positive charge of the lipophilic complex.

In addition to cardiology indications, this radiopharmaceutical has been also approved as aid in the diagnosis of malignancy in patients who are suspected of cancer in the breast combined with inconclusive mammography or palpable tumour and negative or inconclusive mammography.

^{99m}Tc-exametazime (^{99m}Tc-HMPAO)

Technetium-99m exametazime is a lipophilic neutral complex for use as an adjunct in the detection of altered regional cerebral perfusion and for radiolabelling of autologous leukocytes.

In this complex Tc(V), as Tc-oxo core, is bound to a ligand containing two amine and two oxime functions. The ligand behaves like a macrocycle by forming a hydrogen bond between the oxime functions.

Table 4.6.4 Examples of water stable ^{99m}Tc-cores

Tc cores	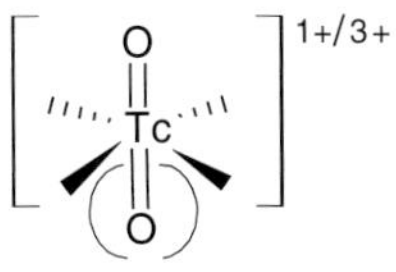	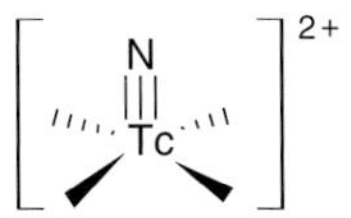	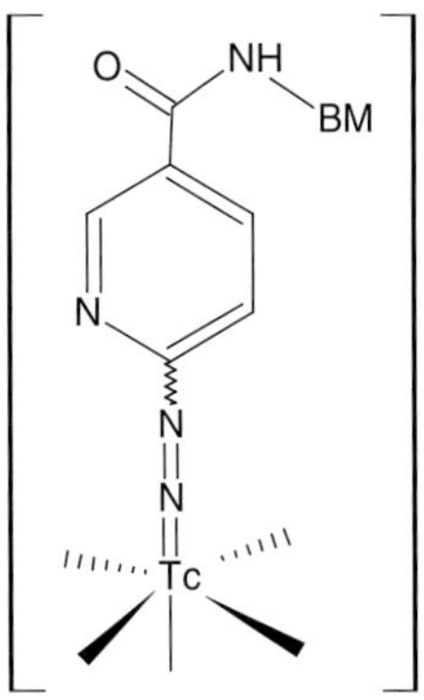	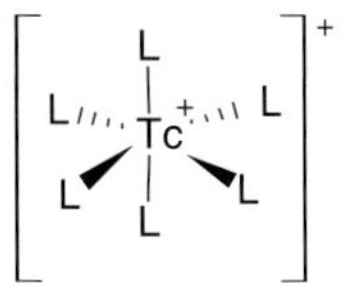	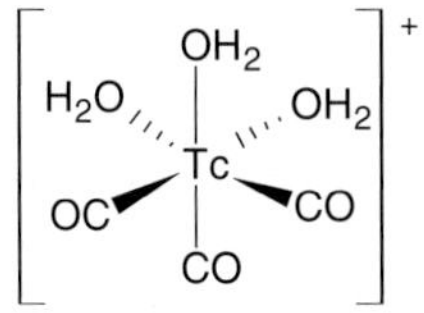
Reducing agent	Sn^{2+}	Sn^{2+}	Sn^{2+}	Sn^{2+}	$K_2[H_3BCO_2]$
Auxiliary reagents	None	Nitrido donor (>N-N<)	Hydrazinonicotinamide (HYNIC), Tricine, phosphines	L-Cysteine Citric acid	$K_2[H_3BCO_2]$
Coordinating atoms	N_4, $N_2O_2^-$, $N_2S_2^-$, N_3S^-, P_4, SS_3^- ... (N= amine/amide)	1,1-dithiolates, $N_2S_2^-$ N_3S^-, P_4, PS^- $P_2S_2^-$, $S_2S_2^-$...	R-N=N, O, O^-, P	Isonitriles (RN≡C)	N_2 (His), NO^-, N_3, N_2O^-, Cp^-
Ligand denticity	4, 3+1	4, 2+2, 3+1	1+5, 1+1+4	1	2, 2+1, 3
Charge of the complex	−1, 0, +1	0, +1	0, −3 (if phosphine is trinegative)	+1	−1, 0, +1
Comments	Syn/anti isomers may occur. Tc-oxo core can be oxidized to TcO_4^-	Syn/anti isomers may occur. Tc-nitrido core is extremely inert	Multiple linking isomers. HYNIC moiety should be part linked to the molecule	Complexes with high inertness	Tridentate ligands give complexes with higher inertness

The complex is unstable in the body where it is rapidly converted to a hydrophilic species. This conversion, which prohibits fast back-diffusion across cell membrane, is thought to be the reason for retention of the radioactivity in the brain and in other organs.

^{99m}Tc-bicisate (^{99m}Tc-ECD)

Technetium-99m bicisate is a lipophilic neutral complex for the localization of stroke in patients in whom stroke has already been diagnosed.

The neutral complex crosses intact cell membranes and intact blood brain barrier by passive diffusion. The ester groups may undergo hydrolysis leading to the negatively charged mono- and di-anionic species. In the brain cell, endogenous enzymes operate the hydrolysis of the ester group transforming the parent molecule into less polar metabolites, which are retained in the cell. As for ^{99m}Tc-exametazime, the stereochemistry of the ligand is important in determining the extent of the uptake in the brain.

^{99m}Tc-mertiatide (^{99m}Tc-MAG3)

Technetium-99m mertiatide is an anionic complex used as renal imaging agent in the diagnosis of renal failures, urinary tract obstructions and calculi. In this complex Tc(V), as Tc-oxo core, is bound to a ligand which can be viewed as a tripeptide (i.e. mercaptoacetylglycyl, –glycyl–glycine).

^{99m}Tc-depreotide

Technetium-99m depreotide is an example of peptide-based radiopharmaceutical labelled with ^{99m}Tc. The peptide is a synthetic somatostatin analogue comprising a binding domain for somatostatine receptors subtypes 2 and 3, and domain for ^{99m}Tc co-ordination.

The radiopharmaceutical is used in the detection of malignant solitary pulmonary lesions (small cell and non-small cell lung cancers) which over-express somatostastin receptors.

Unlike the bifunctional chelating agents, which can be viewed as a pendent arm added to the original molecule, in depreotide the ^{99m}Tc co-ordination site is integrated into the peptide structure as a three amino acid sequence. The three amino acids, (D-ap)-(L-Lys)-(L-Cys), where D-ap stands for D-aminopropionic residue, generate a N_3S set of co-ordinating atoms able to bind the ^{99m}Tc-oxo core in a neutral square pyramidal geometrical arrangement:

Many peptides, different for their size and their biological target, have been labelled with ^{99m}Tc either by co-ordination through an amino acid sequence or by the BCA approach (the latter mainly with the HYNIC moiety). The diagnostic potentiality of these labelled peptides has been checked in animal or, for some of them, in limited clinical studies. These peptides are not reported in this book but their description has been elsewhere discussed in literature (Weiner and Thakur, 2001).

^{99m}Tc-antibodies

Despite its relatively short half-life, there is potential advantage to use ^{99m}Tc- in replacement of ^{111}In for antibody labelling, because of the well-established superiority of its nuclear properties. Several ^{99m}Tc-labelled murine monoclonal antibodies were approved in Europe or in US so far as diagnostic tools in the detection of colorectal cancer (^{99m}Tc-arcitumab – CEA-Scan(®) and in imaging infection and inflammation foci (^{99m}Tc-besilesomab-Scintimun® Antigranulocyte, ^{99m}Tc-sulesomab -LeukoScan® and ^{99m}Tc-fanolesomab -NeutroSpec(®).

Approved antibodies may differ with regard to their isotype, for example belonging to the IgG class (^{99m}Tc-arcitumab, ^{99m}Tc-besilesomab, ^{99m}Tc-sulesomab) or to the IgM class (^{99m}Tc-fanolesomab). For a

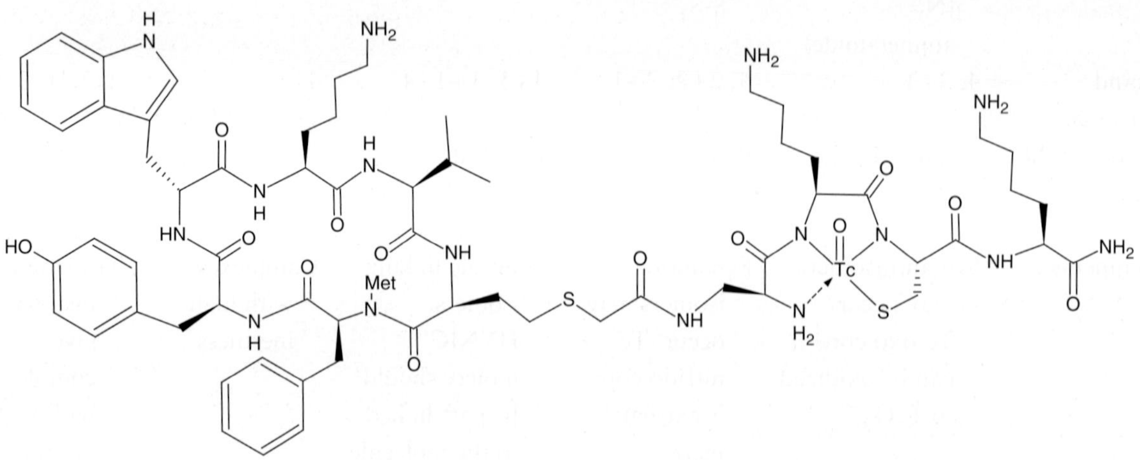

^{99m}Tc-Depreotide

The internal sequence (D-ap)-(L-Lys)-(L-Cys) found in depreotide is not the only one which can work efficiently. Other amino acid sequences have been used successfully, such as terminal Gly–Gly–Cys (N_3S set), terminal Me_2Gly–Ser–Cys (N_3S set), internal Cys–Gly–Cys (N_2S_2 set). The advantage of the use of an amino acid sequence over the BCA approach is that insertion of amino acids in the peptide skeleton may occur through a fast and automated solid phase synthesis.

same isotype, antibodies may be used in different size, for example as intact IgG (^{99m}Tc-besilesomab) or as Fab' fragments (^{99m}Tc-arcitumab and ^{99m}Tc-sulesomab). All the approved ^{99m}Tc-antibodies are labelled with ^{99m}Tc using the so-called 'direct method'.

^{99m}Tc-antibodies: direct labelling

In principle, reduced forms of ^{99m}Tc can bind to the amino acid residues of a non-modified antibody. In practice, the probability of ^{99m}Tc to interact with an accessible set of donor atoms (e.g. N_3S) displaying the appropriate geometry for a stable co-ordination is relatively low. Indeed, when non-modified antibodies directly react with reduced ^{99m}Tc, a large fraction of the radioactivity non-specifically bound to the macromolecule is lost very rapidly. Although the role of the thiol (sulfhydryl) groups in the direct labelling of proteins was described in 1975 (Steigman *et al.*, 1975), systematic studies on the importance of the thiol group began only in the 80s. It became apparent that there was a correlation between the number of free thiols present on the antibody and the amount of radioactivity firmly bound on it (Paik *et al.*, 1985). For antibodies or fragments carrying few or no accessible thiol groups (e.g. intact IgG, F(ab)'$_2$ fragments), the generation of free thiols from disulphide groups is the key step for successful labelling. Several reducing agents (stannous ions, thiols, phosphines) have been proposed to generate sulfhydryls from disulphide groups. It has been shown that thiol-based compounds, such as 2-mercaptoethanol (Mather and Ellison 1990) and aminoethanethiol (Garron *et al.*, 1991) are efficient and reliable reagents, capable of reducing intra-chain or inter-chain disulphide bond without structural deterioration or loss of the immunobiological activity of the antibody. Industrial processes capable of manufacturing gram-size batches of pharmaceutical grade antibodies are now routinely used for the preparation of commercial kits.

An antibody in which free thiol groups have been generated is termed 'activated' (i.e. prone to be labelled). Because the chemical activation takes place on endogenous disulphide groups, this process is rarely perceived as a true chemical modification of the native antibody. The labelling method is thus often incorrectly reported as 'direct labelling'. Usually, as the disulfide groups in IgG antibodies are far from the hypervariable region, the immunoreactivity of the activated antibody is unlikely to be altered by the reduction process. Nevertheless, the effect of the parameters governing disulphide reduction (molar ratio between antibody and reducing agent, pH, time of reaction) on the immunoreactivity of the final activated product has to be checked for each new antibody.

^{99m}Tc-antibodies: indirect labelling

This refers to the attachment of an exogenous BCA to the antibody. Generally the BCA contains an activated ester, which is able to bind to the ε-NH_2 groups of lysine residues of the antibody. Different chelating systems for ^{99m}Tc antibody labelling have been proposed and tested and the reader will found an exhaustive presentation and discussion of the chemical properties of these chelates in (Verbruggen, 1996).

^{99m}Tc-antibodies: preformed-chelate technique

This technique was elaborated during the first attempts of antibody labelling with BCA, in order to overcome the low labelling yields and the poor stability of ^{99m}Tc-antibody binding. The following labelling strategy was thus defined (Fritzberg *et al.*, 1988):

- Reaction of a small BCA (a diamide dithioltetetradentate ligand -N_2S_2- bearing a carboxylate function) with reduced ^{99m}Tc. Formation of a stable Tc-oxo N_2S_2 chelate.
- Activation of the carboxylate function with 2,3,5,6-tetrafluorophenol (transformation of carboxylate into the more reactive tetrafluorophenyl ester).
- Reaction of the activated ester of Tc-oxo N_2S_2 chelate with the ε-NH_2 groups of lysine on the antibody.

The advantage of this method is to provide antibodies with highly stable labelling, because the non-specific bonding to low-affinity sites is avoided. The main disadvantage is the low overall yield of labelling (about 30–50%), which necessitates lengthy purification processes after each chemical step.

4.6.5 Labelling with iodine

4.6.5.1 *Basic concepts*

From the point of view of organic chemistry, labelling molecules with a radioactive isotope of iodine is merely the formation of a C–I bond. Therefore, the basic concepts useful for understanding radiolabelling with iodine are easily available in organic chemistry textbooks (March and Smith, 2001).

With the exception of astatine, iodine is the least reactive halogen with regard to the formation of a C–I bond. It is also the halogen that forms the weakest carbon–halogen bond.

The stability of the C–I bond is strongly dependent of the nature of carbon to which iodine is bound (Table 4.6.5).

Table 4.6.5 Bond energies of different carbon-halogen moieties

	C–X bond energy (kJ mol^{-1})		
X	Phenyl-X	Vinyl-X	Alkyl-X
H	460	452	398
F	523	492	444
Cl	398	381	339
Br	335	343	285
I	268	297	222

4.6.5.2 *Chemistry of iodine labelling*

According to the organic chemistry terminology, introduction of a radioactive iodine atom into a molecule may occur through one of the following reactions:

(i) **Aliphatic/aromatic nucleophilic substitution**

Halogen exchange reactions are the most common nucleophilic method for the introduction of radioiodine into small molecules. Iodinated molecules represent an excellent substrate because the C–I bond (the least energetic C–halogen bond) can be easily broken and almost quantitatively replaced by radioiodine.

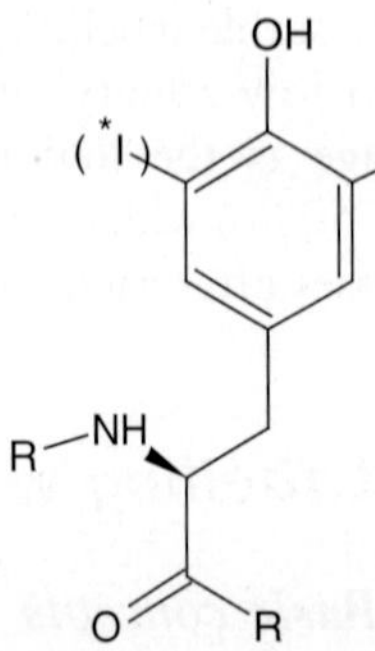

Tyrosine

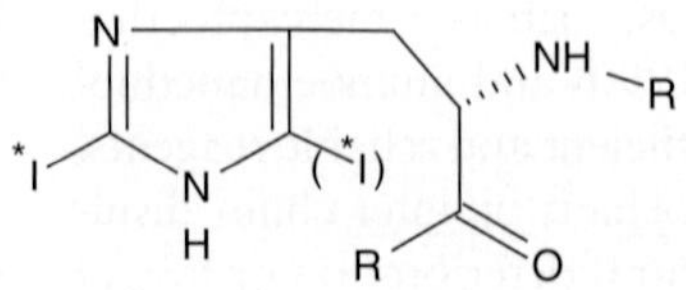

Histidine

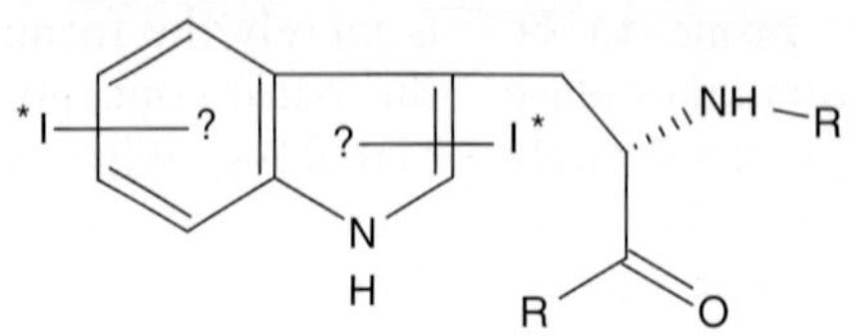

Tryptophane

However, as most of the iodine atoms in the final molecule will be stable ^{127}I nuclides, it is not possible, by this method, to prepare compounds in a high specific activity, as demanded for ligand–receptor investigations.

Bromine containing compounds may also be used as substrates for iodine exchange, provided that the reaction is carried out in more drastic conditions. Theoretically, the radioiodinated compound may be separated from the large excess of bromine precursor by HPLC obtaining a high-specific activity compound. In practice, very often the chromatographic characteristics of the bromo- and iodo-derivatives are so close that an efficient separation of the two compounds is not efficiently achieved.

(ii) **Aliphatic/aromatic electrophilic substitution**

Electrophilic substitution labelling with radioiodine is the most used chemical route to label small molecules and proteins, mainly because the final radioiodinated compound can be obtained in a high specific activity using mild conditions.

The reaction is carried out by the electrophilic species $^*I^+$ on electron-rich substrates like aromatic rings or organo-metallic compounds. Therefore, oxidation of iodide can be considered the key step in the preparation of radioiodinated compounds.

(iii) **Addition to carbon–carbon multiple bonds.**

4.6.5.3 *Radioiodination of proteins*

Proteins and peptides may react with electrophilic iodine in a single-pot reaction through the following residual groups:

The three amino acid residues display a different degree of reactivity towards iodination. Tyrosine, which bears the most reactive group, is (mono) iodinated at first. Histidine reacts only when the labelling reagents are in large excess or when the reaction time is greatly increased. Tryptophane seldom reacts. For long reaction times and in the absence of steric hindrance, mono-iodotyrosine may undergo a second attack to form di-iodo tyrosine.

In order to avoid multiple radioiodination sites and limit the extent of side reactions on sensitive residues, the proteins may be labelled by an indirect approach. Following this strategy, radioiodine is first bound to a prosthetic group, which is then attached to the macromolecule, generally by conjugation on a ε-NH_2 lysine residue.

One of the most used reagent, N-succimidyl-3 (4-hydroxyphenyl)proponiate, known as 'Bolton and Hunter' reagent, is commercially available. A sketch of reactions involved in protein labelling with such reagent is depicted below:

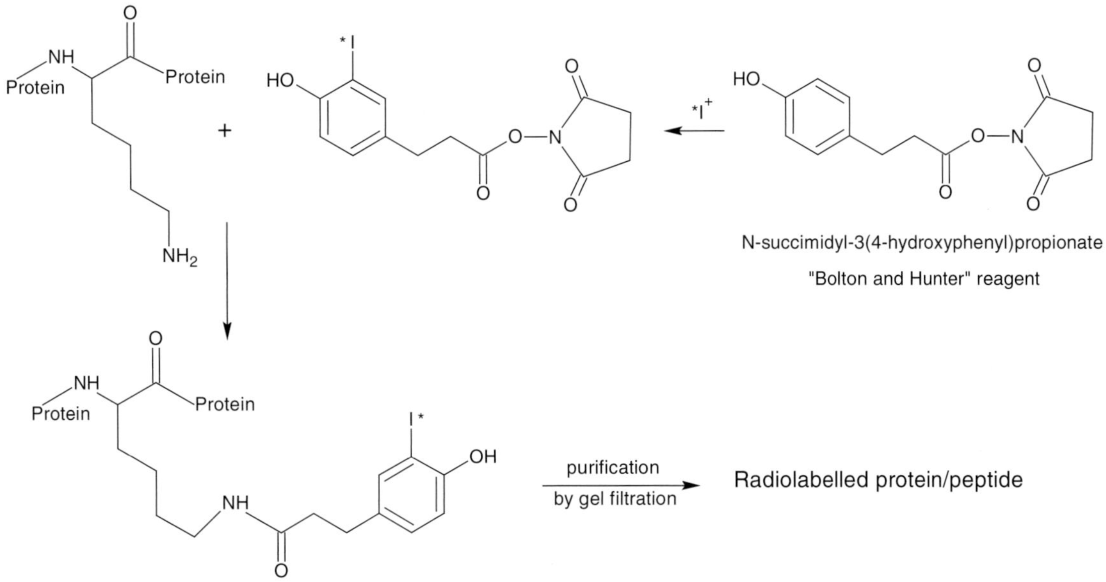

The conjugation of the prosthetic group to the protein is rarely a quantitative process. Therefore, a purification step, usually carried out by gel filtration, is necessary to obtain a pure product.

4.6.5.4 Examples of radioiodinated gamma-emitting radiopharmaceuticals

Although from the clinical point of view (imaging quality and dosimetry), ^{123}I can be considered the preferred radioisotope for iodine-containing radiopharmaceuticals; its high manufacturing cost has prevented ^{123}I-radiopharmaceuticals from a large industrial development.

Therefore, besides sodium iodide itself, which is used in the diagnosis of thyroid function, the number of worldwide commercially available ^{123}I-radiopharmaceuticals is modest and limited to specific countries or regions. The following is a list of the most used ^{123}I-radiopharmaceuticals in clinical practice.

^{123}I-Iobenguane

Due to its structural resemblance with norepinephrine, ^{123}I iobenguane (*m*-iodobenzyl guanidine or MIBG) enters neuroendocrine cells by an active mechanism, and it is stored in the neurosecretory granules. Therefore MIBG scintigraphy is used to image tumours of neuroendocrine origins, particularly those of sympathoadrenal system (pheochromocytomas, paragangliomas and neuroblastomas), although other neuroendocrine tumours (carcinoids, medullar thyroid carcinoma, etc.) are also visualized (Bombardieri *et al.*, 2003). In cardiology, MIBG is used in the functional studies of the myocardium (sympathetic innervations).

^{123}I-Ioflupane

Like (^{123}I) β-CIT, ^{123}I-ioflupane (N-ω-fluoropropyl-2β-carbomethoxy-3β-(4-iodophenyl) nortropane or (^{123}I) β-FP-CIT) is a cocaine analogue, which has shown to bind with high affinity to the dopamine transporter (DAT) located in the membrane of the presynaptic nigrostrial nerve terminals.

These tropane derivatives are used mainly in the early diagnosis of Parkinson's disease (and other degenerative Parkinsonian disorders), in the differential diagnosis between degenerative and non-degenerative Parkinsonism and tremor disorders including non-classical essential tremor and psychogenic tremor and in the differential diagnosis between dementia with Lewy bodies and Alzheimer's disease (Tatsch *et al.*, 2002).

The main difference between the (^{123}I)β-CIT and (^{123}I)FP-CIT, lies in their different binding kinetics to dopamine transporters. (^{123}I)-β-CIT reaches its maximum of striatal binding approximately 24 h after bolus injection, whereas (^{123})I-FP-CIT reaches this maximum only 3–6 h after injection. Consequently, (^{123}I)FP-CIT SPECT imaging can be performed on the day of tracer injection, whereas (^{123}I)-β-CIT necessitates imaging on the day after injection, when 70% of the initially administered activity has been lost by decay.

^{123}I-IBZM

Although not formally approved, (^{123}I) IBZM, (*(S)*-2-hydroxy-3-iodo-6-methoxy-(1-ethyl-2-pyrrolidinylmethyl)-benzamide), is used in clinical research as an imaging agent capable to bind both *in vitro* and *in vivo* to the post synaptic dopamine D2 and D3 receptors. Because of the presence of two activating groups, radioiodination of BZM is performed directly on the benzene ring.

^{123}I-Fatty acids

In fasting patients, free fatty acid metabolism represents the source of energy for normal myocardium. This metabolic pathway is suppressed in ischemic myocardium. Therefore, several radioiodinated fatty acid tracers have been developed for the detection of metabolically altered myocardium. Among them, the β-methylated ^{123}I-15-(p-iodophenyl)-3-(R,S)-methyl pentadecanoic acid (^{123}I-BMIPP) has been approved in Japan. The metabolic oxidative mechanism of fatty acids is inhibited by the presence of a methyl group in the beta position of BMIPP leading to a longer retention time of the radioactivity in the myocardium.

4.7 Radiochemistry of positron-emitting radiotracers

4.7.0 Introduction

The design of radiotracers or radiopharmaceuticals labelled with short-lived positron-emitters requires, beside the inescapable selection of a chemical structure of interest to be labelled, the choice of the radionuclide to be used.

The selection of a chemical structure is generally based on already existing and available information. For example, the pharmacological characterization of the target molecule is crucial for receptor studies. Qualitative data, such as selectivity, specificity, antagonist or agonist character, and quantitative data, such as affinity (K_D, K_i, IC_{50}), number and localization of binding sites (B_{max}), permit to classify potential candidates. Other biological information, such as efficiency (ED_{50}), toxicology (LD_{50}), metabolism and pharmacokinetics as well as physico-chemical information, such as partition coefficient, can also influence the final selection. When known (or foreseeable), the metabolic parameters of the target molecule also come into play. Other considerations, such as availability of the radionuclide, dosimetry of the radiotracer (and its possible metabolites) as well as radiotoxicity of the isotope used, may have their say.

The choice of the radionuclide and the position of the labelling are generally determined by the chemical structure of the target molecule to label. The physical half-life of the chosen radionuclide should however match the biological half-life of the studied process. For example, in repeated blood flow measurements, oxygen-15 is ideal using [^{15}O]water, while carbon-11 but especially fluorine-18 are preferable in the study of slower processes. The ease of introduction of the radionuclide, from a chemical point of view, determines in most cases both the radionuclide to be used and the associated radiochemical pathway.

4.7.1 Short-lived positron-emitter radiochemistry: general considerations

Most of the challenges associated with the handling of these radionuclides are direct consequences of their physical properties: Half-life, decay mode and specific radioactivities.

An evident consequence of the short half-lives is that these radioisotopes normally have to be produced (with the possible exception of fluorine-18) on the site of their use by dedicated biomedical cyclotrons.

Of primordial importance is the development of methods and techniques for the synthesis of short-lived positron-emitting radiotracers and radiopharmaceuticals, with time being the most critical parameter (Långström *et al.*, 1999). It is a rule of thumb that the preparation of a radiopharmaceutical (radiosynthesis including purification and formulation, usually for intravenous injection) should be achieved within three half-lives of the radioisotope. Minimizing the radiosynthesis time will increase both the overall radiochemical yield and the final specific radioactivity, the latter being of great importance for the pharmacological studies of high-affinity receptors present in very low concentrations. Therefore, inclusion of the radionuclide as late as possible in the synthetic sequence is an important aspect to consider when evaluating synthetic strategies.

As another consequence of the time constraint, the number of chemical steps involved in the preparation of the radiopharmaceutical, even if classified and reported as rapid reactions, must be kept to a minimum. An additional limitation is that the yield must be maximized with the shortest possible synthesis time for each radiochemical step. As the radiochemical yield is a function of both the chemical yield and the radioactive decay, the maximum radiochemical yield is attained before the reaction has proceeded to completion. Synthetic methods are often modified when compared to those applied in conventional organic syntheses with the corresponding stable isotopes. For example, drastic reaction conditions may be used in labelling procedures, despite the fact that the chemical yield is lower, if the increase in reaction rate is large enough. Time optimization also influences the type of procedures used for the synthesis, work-up and final purification. Examples are the use of one-pot procedures in order to reduce preparation time by simplifying the technical handling, microwave technologies in order to accelerate reaction rates and reducing heating times, and Sep-Pak® cartridges in order to shorten the purification and formulation step. The choice of the solvent of reaction is also important, from the point of view of both the kinetic parameters of the reaction and the purification process.

Finally, a consequence of the short half-lives of these radioisotopes is that a very high quantity of starting radioactivity is engaged in the preparation of a radiopharmaceutical labelled with carbon-11 or fluorine-18: Typically, at the Curie level or more (>37 GBq).

Positron-emission gives rise to secondary highly penetrating 511 keV gamma radiation; therefore, radiosyntheses are performed in lead-shielded hotcells, with a typical wall thickness of 50–75 mm, closed and ventilated, in order to minimize the radiation dose delivered to the chemist. As a consequence, a radiopharmaceutical preparation imposes an exhaustive and advanced automation. It is noteworthy that the gamma rays facilitate process monitoring by providing a simple and sensitive means of radioactivity detection. The susceptibility of the chemical reaction to automation should be taken into account at the early design stage of the radiochemical pathway. Certain manipulations of classical chemistry, such as liquid–liquid extraction or precipitation, cannot be envisaged.

The specific radioactivity of a radiopharmaceutical labelled with a short-lived positron-emitter such as carbon-11 or fluorine-18, even though in practice much lower than the theoretical ones (see Table 4.4.1), is usually very high: At least one Curie per micromole (1 Ci/μmol = 37 GBq/μmol), and up to tens of Ci/μmol (hundreds of GBq/μmol). Given the specific radioactivity and the level of radioactivity currently used in the radiosyntheses, the quantity of starting radionuclide in terms of associated mass engaged is extremely low (less than 1 μmole). These small amounts of starting radionuclides are another opportunity to reduce the time of reactions, transfers and purifications. Owing to the low masses involved, the stoechiometrical ratio between the non-radioactive starting material (often called precursor for labelling) and the radioactive one (starting radionuclides or labelled intermediate reagents) is usually in the order of 1000–1. As a consequence, the starting radionuclide (or intermediate labelled reagent) is consumed fast because of pseudo-first-order reaction kinetics. Typically, the concentrations involved are sub-millimolar or even in the low micromolar range, implying that these radiosyntheses should remain compatible with high-dilution conditions. This radiochemistry also demands high-purity reagents that preserve the high specific activities and that introduce minimal competing reactive impurities that can compromise this trace chemistry. The use of such small amount of reagent is also beneficial for the technical handling and offers various possibilities of miniaturization in order to facilitate automation and to speed up handling. This is exemplified by the convenient application of high performance liquid chromatography (HPLC) at the semi-preparative scale (often with direct injection of the crude reaction mixture) as the purification process. Beside its intrinsic efficiency, this separation method is very well adapted to low mass quantities and is fast, thus compatible with short half-lives. The choice of the HPLC-solvents in the purification is important in order to facilitate the final formulation step (usually for intravenous injection).

Finally, in a large number of cases, the radiopharmaceuticals are often used on a daily basis within the framework of studies, which are often long (several months or years), implying a viable and reproducible production chain, leading to a sterile and pyrogen-free radiopharmaceutical of high radiochemical purity. Therefore, microprocessor-controlled automated synthetic devices are developed in order to fulfil routine pharmaceutical production and are becoming mandatory in order to meet the demands related to good laboratory practice (GLP) and good manufacturing practice (GMP).

4.7.2 Oxygen-15 radiotracers and radiopharmaceuticals

Oxygen-15 has a half-life of 2.07 min and decays by the emission of a positron (100%) having a maximum energy of 1.7 MeV, to the stable nuclide nitrogen-15. Due to its short half-life, the variety of compounds that have been labelled with oxygen-15 is limited to O_2, CO_2, and H_2O. Molecular [^{15}O]oxygen and [^{15}O]carbon dioxide are usually produced in-target by irradiation of nitrogen containing 0.20–0.50% of oxygen and 0.2–0.25% of carbon dioxide, respectively. [^{15}O]carbon monoxide is often produced by reacting molecular [^{15}O]oxygen with activated charcoal at 900°C

[^{15}O]water ($H_2[^{15}O]O$) is the most popular oxygen-15-labelled radiopharmaceutical currently used. It is readily produced by the palladium- or platinum-catalysed reaction of hydrogen with molecular [^{15}O]oxygen:

$$[^{15}\mathrm{O}]\mathrm{O}_2 \xrightarrow[\substack{\mathrm{Pd,150^\circ C}\\ 30-60\,\mathrm{s}}]{\mathrm{H_2}} \mathrm{H}_2[^{15}\mathrm{O}]\mathrm{O}$$

The short half-life of oxygen-15 is advantageous in minimizing the radiation dose to the patient and when making serial measurements (rapid sequential administrations of the radiopharmaceutical), but can be a disadvantage when prolonged individual studies are necessary. The short half-life makes it essential that imaging facilities for the use of this radionuclide are close to the point of production, and the production and processing equipment require careful design to minimize any unnecessary loss of activity by decay. When possible, 'on-line' production systems are preferred. In such systems, the radioactive oxygen-15-labelled gases are continuously produced and processed, in order to remain available throughout the whole of a clinical session. When continuous production is not possible or uneconomic, batch-wise production is used either 'on-line' or in conjunction with rapid transport facilities.

4.7.3 Nitrogen-13 radiotracers and radiopharmaceuticals

Nitrogen-13 has a half-life of 10.0 min and decays by the emission of a positron (100%) with a maximal energy of 1.19 MeV to the stable nuclide carbon-13 (Clark and Buckingham, 1975). Nitrogen-13 was one of the earliest positron-emitting radionuclides produced. It was first prepared by the irradiation of boron nitride with alpha-particles using the ^{10}B(αn)^{13}N nuclear reaction (Joliot and Curie, 1934) and was then used to prepare the first, and still today the most important nitrogen-13-labelled radiotracer, [^{13}N]ammonia. At present, nitrogen-13 is mostly cyclotron-produced by irradiation of a liquid natural-water target via the ^{16}O(p,α)^{13}N nuclear reaction.

Beside its use as a radiopharmaceutical, [^{13}N]ammonia is also a key reagent for the introduction of nitrogen-13 into more complex chemical structures (Welch, 1977). An extensive range of nitrogen-13-labelled α-amino acids (e.g. [^{13}N]alanine, [^{13}N]leucine, [^{13}N]methionine, [^{13}N]phenylalanine, [^{13}N]valine, [^{13}N]tyrosine, [^{13}N]aspartic acid and [^{13}N]glutamic acid) have been prepared, mainly using biosynthetic methods involving enzymes, leading therefore to regio- and enantiospecific aminations. The enzymes are often immobilised on a support which prevents the contamination of the final products by antigenic and pyrogenic macromolecules and enables the enzyme to be reused for several productions.

The preparation of other nitrogen-13-labelled amino acids, such as [^{13}N]-γ-aminobutyric acid, [^{13}N]asparagine and [^{13}N]glutamine, as well as [^{13}N]-β-phenethylamine, [^{13}N]amphetamine and [^{13}N]dopamine, have also been reported (Welch and Redvanly, 2003).

4.7.4 Carbon-11 radiotracers and radiopharmaceuticals

Carbon-11 has a half-life of 20.4 min and decays by the emission of a positron (99%) with a maximum energy of 0.960 MeV to the stable nuclide boron-11 (Browne *et al.*, 1978). Carbon-11 is mostly cyclotron-produced by the irradiation of a natural molecular-nitrogen gas target using the ^{14}N(p,α)^{11}C nuclear reaction. Carbon-11-labelled carbon dioxide ([^{11}C]CO_2) at high specific radioactivity is the most commonly prepared primary precursor (i.e. in-target produced). Normally the target contains small trace amounts of added molecular oxygen (0.1–2.0%) to facilitate [^{11}C]carbon dioxide formation. Some carbon monoxide ([^{11}C]CO) is also formed simultaneously, but can be easily removed cryogenically. Carbon-11-labelled methane ([^{11}C]CH_4) is the other principal primary precursor, produced by irradiation of a nitrogen/hydrogen gas mixture (typically 95/5) via the same nuclear reaction. It is also often produced alternatively by reduction of [^{11}C]CO_2 with hydrogen over hot nickel (400 °C).

The successful use of carbon-11 in the labelling of an impressive variety of chemical structures is undoubtedly related to the exhaustive development of the so-called secondary precursors. These usually one-carbon small reactive molecules are prepared from a primary precursor ([^{11}C]CO_2 or [^{11}C]CH_4), often by 'on-line' or one-pot procedures and are used as alternative building blocks for the labelling of different chemical functions in the target radiopharmaceutical (Welch, 2003).

The most frequently used secondary precursors are the carbon-11-labelled methylation agents methyl iodide ([^{11}C]CH_3I) and methyl triflate ([^{11}C]CH_3OTf) (Crouzel *et al.*, 1987). As shown below in Scheme A, [^{11}C]Methyl iodide is usually prepared from [^{11}C]carbon dioxide using the well-known two step protocol, consisting of [^{11}C]carbon dioxide trapping and conversion into [^{11}C]methoxide with lithium aluminium hydride ($LiAlH_4$) in tetrahydrofuran (THF) followed by iodination using aqueous hydriodic acid (aq. HI). Even though this process is very reliable, it suffers from the drawback that the lithium aluminium hydride is inevitably contaminated with carrier carbon that reduces specific radioactivity. Alternative processes, in which [^{11}C]methyl iodide is prepared from [^{11}C]methane, usually provide higher specific radioactivities. In the latter method, [^{11}C]methane undergoes a free radical iodination in a circulating gas phase while the [^{11}C]methyl iodide formed is continuously trapped to prevent further iodination (Reaction B below). [^{11}C]Methyl triflate is prepared from [^{11}C]methyl iodide using silver triflate (Reaction C below).

$$\text{A. } {}^{11}CO_2 \xrightarrow[\text{then } H_2O]{LiAlH_4/THF} {}^{11}CH_3OH \xrightarrow{aq.HI} {}^{11}CH_3I$$

$$\text{B. } {}^{11}CH_4 \xrightarrow[>650^\circ C]{I_2} {}^{11}CH_3I$$

$$\text{C. } {}^{11}CH_3I \xrightarrow[200^\circ C]{AgOTf} {}^{11}CH_3OTf$$

Production of [^{11}C]methyl iodide and [^{11}C]methyl triflate from [^{11}C]carbon dioxide or [^{11}C]methane.

[^{11}C]Methyl iodide is also used for the preparation of other valuable secondary precursors. [^{11}C]Carbon monoxide ([^{11}C]CO), normally obtained from [^{11}C]carbon dioxide, is more and more often used for the synthesis of labelled carbonyl compounds such as aldehydes and ketones. Other useful precursors include hydrogen [^{11}C]cyanide (H[^{11}C]CN), [^{11}C]phosgene ([^{11}C]$COCl_2$) and [^{11}C]diazomethane ([^{11}C]CH_3N_2), usually obtained on-line from [^{11}C]methane. Hydrogen [^{11}C]cyanide is used for the synthesis of [^{11}C]nitriles which can then be converted to [^{11}C]amines and [^{11}C]carboxylic acids. [^{11}C]Phosgene is used in the synthesis of often cyclic [^{11}C]ureas and [^{11}C]carbamates. Finally, [^{11}C]diazomethane is used in the synthesis of [^{11}C]methyl esters, when chemo-selectivity in the structure to be labelled is required relatively to nucleophilic amine and alcohol functions.

From a practical radiochemical point of view, the chemistry used for the preparation of carbon-11-labelled radiotracers and radiopharmaceuticals is dominated by [^{11}C]methyl iodide and [^{11}C]methyl triflate alkylations of heteroatoms (sulfur, oxygen and nitrogen). The first application of [^{11}C]methyl iodide was the synthesis of [^{11}C]methionine (*S*-methylation). Later the utilization of [^{11}C]methyl iodide (and [^{11}C]methyl triflate) in alkylations of oxygen- and nitrogen-nucleophiles such as phenolates, carboxylates, amines or amides, became the most common way for introducing carbon-11 into a molecule. A far-from-exhaustive list of carbon-11-labelled radiotracers and radiopharmaceuticals includes [^{11}C]raclopride and [^{11}C]PE2I (*O*-methylation), (R)-[^{11}C]deprenyl, [^{11}C]DASB, (R,S)-[^{11}C]MQNB, (R,S)-[^{11}C]PK-11195 and [^{11}C]flumazenil (*N*-methylation). Amide formation from amines have also been used, as for example in the preparation of [^{11}C]WAY-100635 via the intermediate secondary precursor cyclohexyl[^{11}C]carbonyl chloride. Insertion of a carbonyl function between two heteroatoms using [^{11}C]phosgene led to the successful preparation of the urea (S)-(-)-[^{11}C]CGP-12177 and the carbamate [^{11}C]befloxatone.

Although today a substantial number of carbon-11-labelled compounds used as radiotracers and radiopharmaceuticals contain a *N*- or *O*-methyl group, and may thus be labelled by [^{11}C]alkylating agents ([^{11}C]methyl iodide for example), new synthetic strategies giving access to other labelling positions have been developed, including [^{11}C]carbon-carbon bond forming reactions. The latter have been successfully applied in radiotracer chemistry, using for example, (a) alkylation on a stabilized carbanion using [^{11}C]methyl iodide and other [^{11}C]alkyl halides, (b) palladium-mediated cross-coupling using [^{11}C]methyl iodide (Stille- and Suzuki reactions, involving an organotin or organoboron reagent respectively as well as the Heck reaction, involving alkenes), (c) reaction of [^{11}C]cyanide with electrophilic carbons and (d) palladium-mediated carbonylation insertion reaction using [^{11}C]carbon monoxide.

4.7.5 Fluorine-18 radiotracers and radiopharmaceuticals

Fluorine-18 has a half-life of 109.8 min and decays by the emission of a positron (97%) having a maximum energy of 0.635 MeV, to the stable nuclide oxygen-18 (Browne *et al.*, 1978). Fluorine-18 is mostly cyclotron-produced and the most widely used process is the irradiation of a >95% oxygen-18-enriched liquid water target via the $^{18}O(p,n)^{18}F$ nuclear reaction. Fluorine-18 is then recovered from the target as [^{18}F]fluoride anion in an aqueous solution at high specific radioactivity (>185 GBq/μmol). Fluorine-18 can also be produced by irradiation of neon gas containing 2% of carrier F_2 ($^{20}Ne(d,n)^{18}F$) and is recovered from the target as molecular [^{18}F]fluorine gas at low specific radioactivity (<750 MBq/μmol).

Fluorine-18 appears nowadays as the most attractive positron-emitting radioisotope for radiopharmaceutical chemistry and PET imaging, part of this continuous growing interest probably due to the successful use in clinical oncology of 2-[^{18}F]fluoro-2-deoxy-D-glucose ([^{18}F]FDG), actually the most widely used PET-radiopharmaceutical. Compared with other positron-emitting radiohalogens used in PET such as bromine-76 (half-life: 16.1 h) or iodine-124 (half-life: 4.18 days), fluorine-18 displays simpler decay- and emission properties with high positron abundance (97%) (Browne *et al.*, 1978). Its half-life is long enough to give access to relatively extended imaging protocols compared to what is possible with carbon-11. Therefore, this facilitates kinetic studies and high-quality metabolite- and plasma analysis because of higher count rates and better statistics over a longer time. On the contrary, the half-life is too long for repeated imaging with the same or different radiotracers, while it is possible with carbon-11, nitrogen-13 and oxygen-15. From a chemical point of view, the half-life of fluorine-18 allows multi-step synthetic approaches that can be extended over hours, and fluorine-18 has therefore been effectively used for the labelling of numerous relatively simple or complex bioactive chemical structures, including high-molecular-weight macromolecules such as peptides and proteins (de Bruin *et al.*, 2005 and references therein) and oligonucleotides (Kühnast *et al.*, 2004 and references therein).

Fluorine-18 can be reliably and routinely produced in high quantities (several Cor/tens of GBq) on widely implemented biomedical cyclotrons of relatively low-energy proton beam (e.g. 18 MeV). This distinctive advantage, combined with its favourable half-life, permits the transport and the use of fluorine-18-labelled radiopharmaceuticals (such as the archetype [^{18}F]FDG) at 'satellite' PET units that do not have access to an on-site cyclotron facility.

Fluorine-18 also presents some drawbacks. Isotopic labelling is limited to chemical structures already containing a fluorine atom and most of the molecules of interest, for biological applications do not originally contain fluorine, and compounds containing C–F bonds are rare in living nature. Another disadvantage of fluorine-18 is the restricted versatility of the possible labelling strategies, especially when compared with carbon-11. Radiofluorinations comprise electrophilic fluorinations and nucleophilic substitutions.

Electrophilic radiofluorinations involve low specific radioactivity fluorine-18-labelled reagents; the simplest one being molecular [^{18}F]fluorine ([^{18}F]F_2). Other electrophilic reagents, less reactive and therefore also less destructive, have also been reported, such as xenon di[^{18}F]fluoride (Xe[^{18}F]F_2) and acetyl[^{18}F]hypofluorite (CH_3COO[^{18}F]F). Being prepared from [^{18}F]F_2 which is produced in a carrier-added way, electrophilic radiofluorinations are thus necessarily carrier-added and result in final products with low at best moderate specific radioactivities. This has limited the usefulness of electrophilic fluorinations to the radiosynthesis of compounds for which there is no need for high specific radioactivity, and where the chemical species in question is not toxic. A second drawback of electrophilic fluorinations is that the maximum achievable radiochemical yield is limited to 50% (for reactions with [^{18}F]F_2, only one of the two fluorine atoms is incorporated into the product). Electrophilic radiofluorinations require electron-rich structures, and usually involve [^{18}F]F_2 in direct (often poorly regioselective) fluorination of vinylic and aromatic derivatives as well as fluorodemetallation with tin derivatives. Despite these severe limitations, electrophilic radiofluorination was the method initially used to prepare [^{18}F]FDG and is still used for the preparation of [^{18}F]Fluoro-L-DOPA.

In the present decade, there has been a dramatic increase in the fluorine-18 literature, both in the sheer number of new radiotracers published, and in the number of those reaching at least preliminary human evaluation. Most of these fluorine-18-labelled radiotracers and radiopharmaceuticals have been synthesized using no-carrier-added (and therefore with high specific radioactivity, usually >185 GBq/μmol) nucleophilic substitutions mainly in *homo*aromatic and aliphatic series (Kilbourn, 1990).

*Homo*aromatic nucleophilic substitutions with [^{18}F]fluoride usually require activated aromatic rings, bearing both a good leaving group (e.g. a halogen, a nitro- or a trimethylammonium group) and a strong electron-withdrawing substituent (e.g. a nitro-, cyano- or acyl group) preferably placed *para* to the leaving group, whereas aliphatic nucleophilic substitutions only require a good leaving group usually a halogen or a sulphonic acid derivative such as mesylate, tosylate or triflate. Labelling procedures involve pre-activation of cyclotron-produced, no-carrier-added, aqueous [^{18}F]fluoride by evaporation to dryness from an added base (typically K_2CO_3) and, usually, the added kryptand Kryptofix-222, in order to form the so-called naked fluoride anion as its K[^{18}F]F-K_{222} complex. Nucleophilic substitutions are then performed in an aprotic polar solvent under alkaline conditions, either on a suitable direct precursor of the target molecule (one step procedure) or on a indirect precursor followed by one or more chemical steps leading to the target radiotracer. [^{18}F]FDG and [^{18}F]fluorothymidine ([^{18}F]FLT) on the one hand, and [^{18}F]setoperone on the other hand, are selected examples of radiopharmaceuticals prepared by aliphatic nucleophilic substitution and *homo*aromatic nucleophilic substitution, respectively.

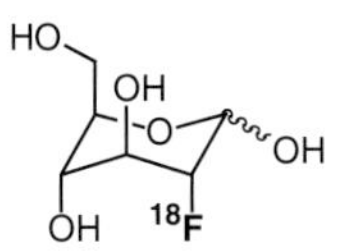

[^{18}F]FDG

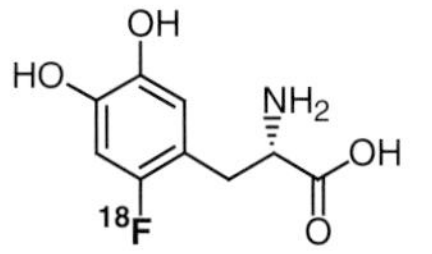

[^{18}F]Fluoro-L-DOPA

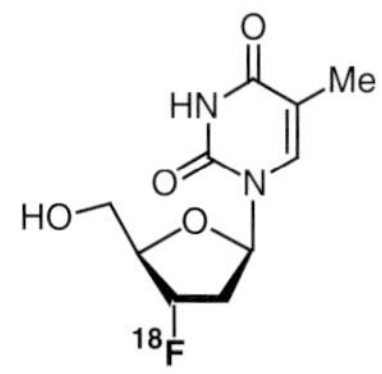

[^{18}F]FLT

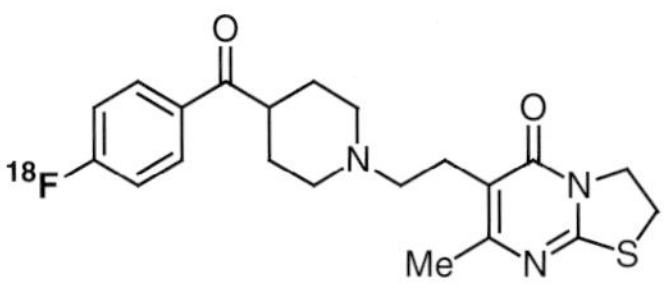

[^{18}F]setoperone

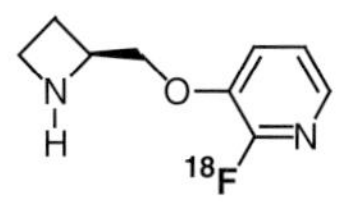

[^{18}F]F-A-85380

Chemical structures of [^{18}F]FDG, [^{18}F]Fluoro-L-DOPA, [^{18}F]FLT, [^{18}F]Setoperone and [^{18}F]F-A-85380.

More recently, the field of aromatic nucleophilic substitutions with [^{18}F]fluoride has been extended to *hetero*cyclic chemical structures bearing a fluoropyridinyl moiety (Dolci *et al.*, 1999). Similar to the aliphatic nucleophilic radiofluorinations, only a good leaving group is required (a halogen, or better a nitro- or a trimethylammonium group), and, except if one considers *meta*-fluorination, there is no need for an additional strong electron-withdrawing substituent for activation of the aromatic ring such as in the *homo*aromatic nucleophilic radiofluorinations. Nucleophilic *hetero*aromatic substitution and consequent fluorine-18 incorporation are generally performed in DMSO using K[^{18}F]F-K_{222} complex and conventional heating at a moderately high temperature (120–150 °C) or using microwave irradiation (100 W) for a short period of time (1–2 min) and often lead to high radiochemical yields. [^{18}F]F-A-85380 (Dollé *et al.*, 1998) is one selected example of radiopharmaceuticals prepared by *hetero*aromatic nucleophilic substitution.

4.8 Major radiotracers and imaging applications

All radiopharmaceuticals used in nuclear medicine are suitable for animal imaging. Regulations on radiopharmaceuticals intended for clinical use also require that candidate radiopharmaceuticals be first tested in animals. In addition, a large number of fundamental studies can benefit from radiotracer imaging in laboratory animals, especially considering the recent possibility to image in mice and rats. These two species constitute a considerable number of models for human pathologies, notwithstanding the usual cautions when translating from rodent to man. The repeatability of molecular imaging is particularly useful in diseases with slow progression such as cancer, neurodegeneration, or inflammatory disorders. Transgenic animals give access to fundamental studies of monogenic diseases and genetically related disorders. Immunodeficient mice, and now, rats, are precious as cancer models and to study graft-host relationships. Rat models of cardiovascular and neurological diseases are well established and have been used for years to test drug candidates, etc.

For the biochemist and the molecular pharmacologist, the recent access to small animal radiotracer imaging is a benediction. It is the first time in the history of biomedical research that precise biochemistry and even genetic molecular studies can be performed in organs *non-invasively*, or to put it directly, without sacrificing the animal. This is a major progress because it limits considerably the sacrifice of animals for research, but it also opens new research avenues that could not be possible without these techniques. Indeed, a main characteristic of living species is the organized and concerted manner in which they respond to environmental cues and adapt to physical and chemical modifications of this environment. At the individual level, this adaptation can only be studied by a non-destructive process, meaning molecular imaging for the biochemical aspects. Conversely, molecular changes induced by a pathological process can be examined at different stages. Pharmacokinetics, distribution and even activity of drugs can be tested directly, and compared under different physiological or pathological conditions, and gene expression analysed at the level of a whole organism.

Accordingly, the field of experimental radiotracer imaging has virtually exploded in the last years, and attempting to summarize all its developments is an incommensurable task. Even more, it is a task bound to become obsolete by the time this chapter is published. Imagination is the limit, and as mentioned already, hundreds of radiotracers are available already, and thousands potentially. Therefore, this section presents an overview of some radiotracers in use and of their applications.

4.8.1 Basics of radiotracer distribution

When a radiotracer is introduced into the bloodstream of an animal by a bolus intravenous injection, its concentration becomes homogeneous throughout all the arterial blood pool in a few seconds. Thereafter, the biodistribution of the radiotracer will depend essentially on two mechanisms that have opposite effects:

1. The tissue barriers that oppose to the diffusion of the radiotracer. The first barrier is naturally the endothelium of the vascular system; its passage by the radiotracer is called extravasation.
2. The affinity of the radiotracer for endogenous molecules, such as receptors, enzymes, membranes, etc. The major distinction between specific and unspecific sites is that the former are saturable in contrast to the latter.

In addition, the radiotracer can be metabolized to other radioactive compounds and the knowledge of its metabolism is necessary in order to interpret the biodistribution images. This is generally done by sequential arterial sampling and analysis of the radiolabelled compounds present in blood and plasma.

The evolution of the distribution over time is the resultant of the different interactions of the radiotracer, and is analysed on images by measuring in a region of interest (ROI) that has been identified usually on anatomical bases, through the use of time-activity curves (TAC). Interpretation of TACs in different ROIs indicates the movements of the tracer in the different body compartments. The term compartment designates a biochemical pool of the radiotracer that is not necessarily coincident with an organ or substructures of an organ visible on the images. For instance, a tracer located nearby a cell is present in two compartments, whether it is bound to a membrane receptor or unbound (free) in its close vicinity. Rate equations can be applied to a compartmental model describing the time course of the concentration of the radiotracer in the different compartments. For modeling, the number of compartments must be limited to 2 or 3; for instance:

- one plasmatic compartment and one tissular compartment, or
- one plasmatic and two tissular compartments: One for the free radiotracer and one for the radiotracer bound to its receptor.

Although it may be tedious to implement, compartmental analysis of radiotracer imaging studies provides reliable measurements of the concentration of receptors in the brain and heart.

One thing to keep in mind though is that the final objective of radiotracer imaging is to produce a biomarker, that is, an agent that produces image contrast based on biochemistry or biochemistry-based mechanisms with a relevant physiological interpretation. Test–retest of the same radiotracer in identical imaging conditions defines its accuracy as a biomarker. Many radiotracers available have accuracies (Coefficient of variation = standard deviation divided by the mean of a group of values) under 20%, which is comparable to most analytical dosages on blood samples. Moreover, provided that metabolism is accounted for, radiotracer imaging performs better than most sampling techniques for the study of organ pharmacokinetics of a compound because all time points are analysed on the same samples.

4.8.1.1 Diffusion versus retention

Laws that govern the diffusion of compounds in solution were established by Fick?

$$\Phi_{1,2} = (\mathrm{d}n/\mathrm{d}t) = K_n(\mathrm{C}_1 - \mathrm{C}_2), \tag{10}$$

where $\Phi_{1,2}$ is the flux from compartment 1 to compartment 2 for compound n, K_n is the permeability coefficient of the membrane separating compartements 1 and 2 for compound n, and C_1 and C_2 are the concentrations in compartments 1 and 2.

Compounds with $K = 0$ do not extravasate from the circulation and can be used as indicators of the plasmatic pool.

For other compounds, it is possible to represent their relative passage from plasma to tissue by measuring their relative concentrations in both compartments at equilibrium. At equilibrium, the volume of distribution DV is the volume that would be occupied by plasma if its radiotracer concentration were the same as that in the tissue.

The relative distribution volumes is often used by pharmacologists to compare different organs, and is also of value in imaging to compare uptake in regions with different affinities for a radiotracer.

4.8.1.2 Ligand–receptor interaction

In the case where binding to the receptor is reversible, the radiotracer is called a radioligand, L^*, and its interaction with an endogenous receptor R (a protein on or within a target cell) is a ligand–receptor interaction to which the classical equation of *in vitro* binding assays applies:

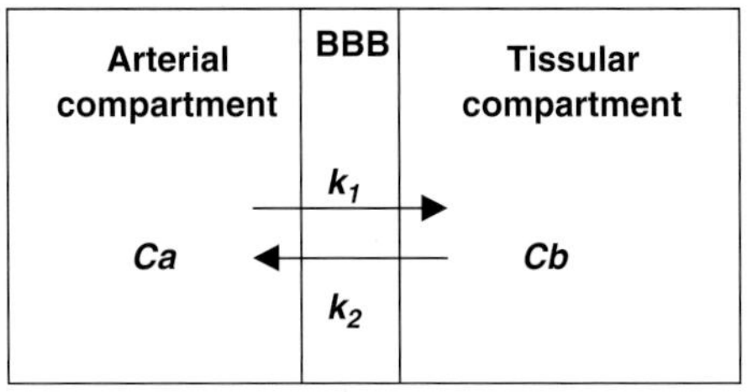

Figure 4.8.1 Two-compartment model of tracer kinetics. The radiotracer distributes in a plasmatic pool with concentration *Ca* and a tissue pool with concentration *Cb*. k_1 is the influx (from plasma to tissue) and k_2 the efflux rate constant. The concentration in tissue pool *Cb* varies over time as $dCb(t)/dt = k_1Ca(t) - k_2Cb(t)$

$$L^* + R \underset{K_{\mathrm{off}}}{\overset{K_{\mathrm{on}}}{\rightleftharpoons}} L^*R$$

At equilibrium, the law of mass reaction implies

$$K_{\mathrm{on}}[R][L^*] = K_{\mathrm{off}}[L^*R] = K_{\mathrm{D}} \tag{4.11}$$

The dissociation rate constant $K_{\mathrm{D}} = K_{\mathrm{off}}/K_{\mathrm{on}}$ is the inverse of the affinity of L^* for R. The total number of binding sites B_{max}, is

$$B_{\mathrm{max}} = [R] + [L^*R] \tag{4.12}$$

Substitution of Eq. (12) in Eq. (11) gives

$$[L^*R] = B_{\mathrm{max}}[L^*]/([L^*] + K_{\mathrm{D}}) \tag{4.13}$$

In *in vitro* binding assays the binding of L^* to the receptors is competed by the addition of large amounts of unlabelled ligand L. As $[L]$ approximates $([L] + [L^*])$, plotting $[L^*R]$ as a function of $[L]$ yields B_{max} and K_{D}.

In vivo, $[L^*R]$ and $[L^*]$ are not measured separately, and the total tissue radioactivity concentration, $[RA] = [L^*R] + [L^*]$. In addition, the pharmacological effects of the ligand preclude its administration in significant amounts; therefore, direct measurement of B_{max} and K_{D} is problematic. However, as the concentration of radiotracer bound to the receptor is theoretically negligible in respect to the total concentration of receptors[1], $[L^*R]$ is very small compared to $[R]$, and Eq. (12) can be approximated to

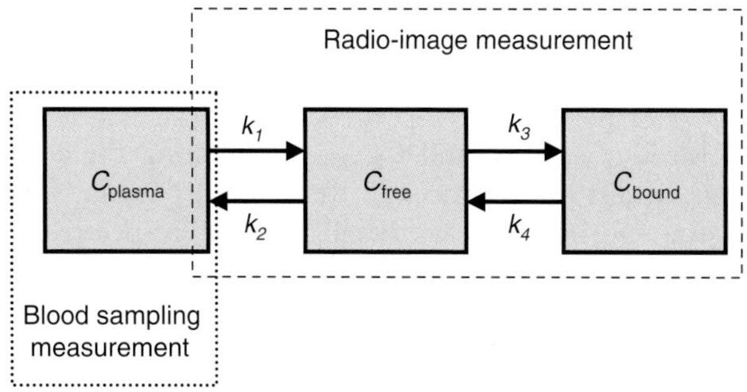

Figure 4.8.2 Three-compartment model of tracer kinetics. The radiotracer distributes in a plasmatic pool and two tissue pools, one in which the tracer is free and one in which it is associated with its receptor. See text for the relevant equations

$$B_{\mathrm{max}} = [R] \tag{4.14}$$

Therefore, Eq. (13) simplifies into

$$B_{\mathrm{max}}/K_{\mathrm{D}} = [L^*R]/[L^*] = \mathrm{BP} \tag{4.15}$$

BP, the *binding potential*, is the ratio of bound to free radiotracer concentrations. If the affinity of the radiotracer for the receptor is similar in all organs (K_{D} constant), this ratio is directly proportional to B_{max}, the concentration of receptor sites in the tissue.

4.8.1.3 Kinetic modelling

Two-compartment models. For a two-compartment model such as the one representing the passage of the blood–brain barrier separating plasma from cerebral parenchyma (Figure 4.8.1):

$$dCb(t)/dt = k_1Ca(t) - k_2Cb(t) \tag{4.17}$$

with
Ca: arterial concentration, measured by blood sampling;

Cb: cerebral concentration, measured by imaging;
k_1: influx rate constant;
k_2: efflux rate constant.

If blood flow F is known, the permeability × surface coefficient PS can be deduced from:

$$\mathrm{PS} = -\ln(1 - (k_1/F)) \times F \tag{4.18}$$

[1]This hypothesis must be verified knowing the specific radioactivity and the estimated receptor concentration.

Three-compartment models. Three-compartment models such as the one represented in Figure 4.8.2 are largely used for neuropharmacological studies; they usually include:

- an arterial plasma fraction, in exchange k_1/k_2 with
- a free tissue fraction, itself in exchange k_3/k_4 with
- a receptor-bound tissue fraction.

With C_{plasma}, C_{free} and C_{bound} representing the concentrations of the tracer in the arterial plasma, free tissue and bound tissue compartments, respectively, the following set of equations apply

$$dC_{bound}(t)/dt = k_3C_{free}(t) - k_4C_{bound}(t) \quad (4.19)$$

$$dC_{free}(t)/dt = k_1C_{plasma}(t) - k_2C_{free}(t) - k_3C_{free}(t) + k_4C_{bound}(t) \quad (4.20)$$

At equilibrium, there is no net transfer of radiotracer from one compartment to another; therefore

$$k_3C_{free}(t) - k_4C_{bound}(t) = 0$$
$$C_{bound}/C_{free} = k_3/k_4 = BP \quad (4.21)$$

Substituting Eq. (10) in Eq. (9):

$$C_{bound}/C_{plasma} = k_1k_3(k_2k_4)^{-1} = BP^* \quad (4.22)$$

BP^* is easier to apprehend than BP because C_{plasma} can be measured in arterial samples. However, BP^* differs from BP by a factor k_1/k_2, meaning that passage from blood to brain must be similar when comparing BP^* in different regions, individuals, or conditions.

For a more detailed description of the models for quantification of brain receptors, the reader may refer to (Ichise *et al.*, 2001) and references therein.

Trapping of radiotracer metabolites. In the case, where the radiotracer is trapped definitively in some chemical form in one compartment, radioactivity accumulates in that compartment. The rate of accumulation can be deduced from the curve of radioactivity over time in the tissue measured by a radiotracer imaging technique. The accumulation model developed by Louis Sokolov for brain autoradiography with 2-[^{14}C]-2-deoxy-glucose has been applied with great success to the PET tracer 2-[^{18}F]-2-deoxy-2-fluoro-glucose (better known under the name FDG) for the measurement of glucose utilization by tissues.

In the FDG model (Figure 4.8.3), glucose and FDG are considered to follow the same first steps of their metabolic route in the body:

Figure 4.8.3 The FDG model (schematic). In this simplified representation of the FDG model developed by Louis Sokolov and others, FDG and glucose use the same membrane transporter to enter the cell in which they are both phosphorylated on Carbon 6 by hexokinase. Glucose-6-phosphate is converted to phosphorylated fructose while FDG-6-phsophate is trapped. Accumulation of radioactivity reflects the actvitity of the transporter and hexokinase

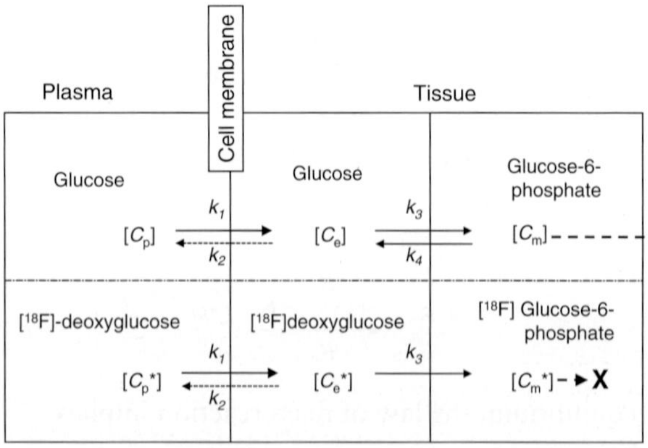

1. uptake by a membrane glucose transporter of the GLUT family;
2. phosphorylation by hexokinase into glucose-6-phosphate or FDG-6-phosphate;
3. after that, glucose-6-phosphate is converted into fructose-6-phosphate and further metabolized in the glycolytic pathway,
4. whereas FDG-6-phosphate is unable to isomerize because isomerization concerns the –OH group at the two position which is lacking,
5. and is trapped because phosphorylation prevents retrotransport by the GLUT transport system.

Louis Sokolov and others developed the equations that measure glucose utilization (transport plus phosphorylation) in tissues through a simple measurement of the radioactivity over time curve. Because glucose utilization is highly correlated with local energy metabolism, it is a reliable index of metabolic activity in the brain and can be used to detect local changes in cerebral regions during normal or pathological activation of brain functions. The FDG became a best seller of radiotracers when it was proven that tumours with increased utilization of glucose can be detected with FDG imaging. Today, FDG imaging in oncology is the major clinical application of PET.

Other models. In addition to the ones briefly introduced above, many other mathematical models have been applied to radiotracer imaging data. Some have been designed to simplify the interpretation of data by the use of graphic representations, such as the Logan plot (Logan *et al.*, 1996) or the Patlack plot (Patlak *et al.*, 1983). The use of a reference region in which $[R]$ is assumed to be nil (no receptor) is particularly interesting because it allows measuring the binding potential without measurements of the radiotracer in the blood (Lammertsma and Hume, 1996), reducing invasive procedures. For a full discussion of mathematical modelling of PET and SPECT data, the reader is referred to (Ichise *et al.*, 2001).

4.8.2 Some examples of radiotracers and their applications

The most important applications of non-invasive methods are imaging of deep organs in which sampling for biochemical analysis is dangerous, deleterious, or irrelevant. These include the brain, heart, kidney, lungs, and to a lesser extent the liver. In addition, radiotracer imaging allows scanning the whole body, therefore screening for local molecular anomalies without *a priori* knowledge of their localization, and this is precious in disseminated disease such as cancer or infections.

4.8.2.1 *Cerebral imaging*

Although presently supplanted by other techniques such as MRI and PET, the measurement of cerebral blood perfusion with ^{99m}Tc radiotracers is a classical procedure in nuclear medical imaging. It uses blood perfusion agents such as ^{99m}Tc-HMPAO, a lipophilic compound which crosses the blood-brain barrier rapidly, enters the brain parenchyma where it is converted to hydrophilic metabolites which remain trapped. Radioactivity distribution after ^{99m}Tc-HMPAO is considered to reflect the relative regional cerebral blood flow (rCBF), although other mechanisms could explain the brain retention of this compound. It is used in studies that affect rCBF such as epilepsy, brain tumours and dementia. Other radiotracers for the imaging of rCBF are available, such as Neurolite (^{99m}Tc-ethylcysteine dimer), ^{99m}Tc-DTPA or ^{99m}Tc-glucoheptonate.

PET imaging with Oxygen-15-labelled water ($H_2{}^{15}O$) is a very efficient and elegant way to measure blood flow. It has been used for years in brain activation studies, until it was supplanted by functional MRI, which does not require the injection of radioactivity.

The most active field of radiotracer imaging is the imaging of neurotransmission, which can be studied with radiotracers specific for neurotransmitter receptors, enzymes or uptake systems. Table 4.8.1 gives a list of PET and SPECT radiotracers used to image the distribution of specific neuroreceptors in brain regions. In cases where a model to quantify the receptor density with the radiotracer has been validated, the occupancy of receptors by CNS system drugs can be indirectly measured by competition with the radiotracer with cerebral binding sites. This is a major application of SPECT and PET for CNS drug development in the pre-clinical and clinical phase.

Reproduced with permission from (Brooks, 2005).

4.8.2.2 *Cardiac imaging*

Receptor mapping has many applications in the heart as well as in the brain, and is easier to model because of the absence of the blood-brain-barrier for access to the tissue compartment. One of the major applications of cardiac radiotracers is the assessment of myocardial viability in coronary artery disease and left ventricular dysfunction, and radiotracer imaging plays an important clinical role in this field. In particular, PET imaging using [^{18}F]-fluorodeoxyglucose is regarded as the metabolic gold standard of tissue viability. Myocardial PET perfusion can also be imaged with rubidium-82. A favoured PET tracer for cardiac metabolism is [^{13}N]ammonium.

Two major SPECT radiotracers, [^{99m}Tc]-sestamibi and [^{99m}Tc]-tetrofosmin are available at lower cost than PET tracers and are widely used in clinical studies.

Cardiac imaging in small animals requires gating of the signal to limit image blurring because of the rapid heart movements. In order to do so, an ECG is recorded simultaneously with the images and the image frames are separated according to their time of acquisition in the cardiac cycle.

4.8.2.3 *Other organs*

Lungs can be explored with ^{133}Xe; the macrophagic function in liver is explored with [^{99m}Tc]-labelled sulphur colloid. Renal excretion is measured with ^{99m}Tc-DTPA or ^{99m}Tc-glucoheptonate.

4.8.2.4 *Cancer and metastasis*

Today, [^{18}F]-fluorodeoxyglucose (FDG) is the gold standard in cancer imaging. Tumours have a higher metabolic rate of glucose utilization than normal tissue and take up FDG with avidity.

Table 4.8.1 Radiotracers available for studying neuropharmacology

Biological application	Tracers
Blood flow	H2 ^{15}O, ^{15}O-butanol. ^{99m}Tc-HMPAO, ^{133}Xe
Oxygen metabolism	^{15}O2
Glucose metabolism	^{18}F-2-fluoro-2-deoxyglucose (FDG)
Dopamine storage/DDC activity	^{18}F-6-fluorodopa (F-dopa), β-^{11}C-dopa
Monoamine vesicle transporters	^{11}C-dihydrotetrabenazine (DTBZ)
Dopamine transporters (DAT)	^{11}C-CFT, ^{11}C-RTI 32, ^{18}F-CFT, ^{123}I-β-CIT ^{123}I-FP-CIT, ^{123}I-IPT, ^{123}I-altropane
Dopamine D1 type sites	^{11}C-SCH23390
Dopamine D2 type sites	^{11}C-raclopride, ^{11}C-FLB456, ^{11}C-methylspiperone, ^{18}F-spiperone,
Noradrenaline transporters	^{18}F-fluorethylspiperone, ^{76}Br-bromospiperone, ^{123}I-epidepride,
Noradrenaline _2	^{123}I-iodobenzamide (IBZM)
Serotonin storage	^{11}C-BATA
Serotonin transporters	^{18}F-2-fluorethoxyidazoxan
Serotonin HT1a	^{11}C-methyltryptophan
Serotonin HT2a	^{11}C-DASB, ^{123}I-β-CIT
Acetylcholinesterase activity	^{11}C-WAY100635
Cholinergic vesicle transporters	^{11}C-MDL100907, ^{18}F-altanserin, ^{18}F-setoperone
Muscarinic M1 sites	^{11}C-MP4A, ^{11}C-physostigmine
Nicotinic sites	^{18}F-fluoroethoxybenzovesamicol, ^{11}C-vesamicol, ^{123}I
Histamine H1 sites	-benzovesamicol ^{11}C-tropanylbenzylate, ^{11}C-NMPB,
Opioid μ sites	^{18}F-FP-TZTP, ^{123}I-QNB
Opioid μ, κ, and δ sites	^{11}C-MPA, ^{11}C-A-85380, ^{18}F-A-85380, ^{123}I-A-85380
Central benzodiazepine sites	^{11}C-dothiepin
Central benzodiazepine sites (α5 subunits)	^{11}C-carfentenil, ^{18}F-cyclofoxy ^{11}C-diprenorphine
Peripheral benzodiazepine sites	^{11}C-flumazenil
Substance P/NK1 sites	^{11}C-RO15-4513
Adenosine A2A sites	^{11}C-PK11195, ^{18}F-PK11195, ^{123}I-PK11195
NMDA voltage channels	^{18}F-SPARQ, ^{11}C-GR205171
Amyloid	^{11}C-SCH 442416
Phosphoglycoprotein activity	^{11}C-CNS 5161 ^{18}F-FDDNP, ^{11}C-PIB, ^{11}C-SB13
	^{123}I-IMPY
	^{11}C-carfentenil

FDG imaging can easily be applied to animal models of tumours, such as immunodeficient mice bearing xenografts of human tumours or tumoural cell lines. In addition, in the recent years, several transgenic models of mice carrying activated forms of oncogenic proteins (such as *ras, myc,* different forms of mutated Receptor Tyrosine Kinases, p53, etc.) have been created with the aim to mimic the development of the human disease.

Among other PET tracers for oncology are

- [^{18}F]-DOPA for neuroendocrine tumours;
- [^{18}F]-FLT, an analog of thymidine to measure proliferation;
- [^{11}C]-citrate for prostate cancer;
- [^{11}C]- and [^{18}F]-labelled estrogen derivatives for breast cancer;
- [^{18}F]-MISO for tumour hypoxia.

Among other SPECT tracers for oncology are

- ^{99m}Tc]-phosphonates (MDP, HMDP, osteoscan) to explore tumour-induced bone remodelling;
- [^{67}Ga]-citrate for Hodgkin and other lymphomas, leukemias and lung cancer;
- ^{201}Tl to image viable tumoural tissue;
- [^{111}In]-octreotide which shows uptake in tumours over expressing somatostatin receptors.

These lists are limited and much more compounds are in use or in trials. Concerning small animals, radiotracer imaging has two major applications in the field of experimental oncology:

1. Follow-up of the time course of the disease progression, and detection of metastases non-invasively, in order to define and validate the animal model, and/or the radiotracer before its clinical use.
2. Measurement of the efficacy of a drug treatment. Once the correlation between imaging and disease course has been made, it is possible to use eventual modifications in the images as a surrogate biomarker of the therapeutic action of drug candidates in order to assess experimentally their efficacy. Another application is to label the drug directly and image its pharmacokinetics and distribution. This can be performed preferentially wit PET for small molecules, or with either PET or SPECT for large molecules such as antibodies or aptamers.

4.8.2.5 Other applications

Many other applications of radiotracer imaging in experimental animals exist. Indeed, radiotracer development and validation is a very active field of research. Imaging of gene expression through specific reporters, of drug delivery by vectors, of stem cell migration, of gene and cellular therapy, of membrane remodelling, etc. is becoming increasingly accessible thanks to new radiotracers. Radiochemistry, the first step of radiotracer imaging, is making considerable progress constantly, and imaging applications of research radiotracers cannot be listed here. The reader should refer to chapter 7 of this book for significant examples.

One important characteristic of radiotracer imaging techniques and procedures developed in small animals is that they can easily be translated into clinical applications because the imaging techniques are basically the same. Hence, the future of this imaging technology is bright.

References

Adloff, J. P., Guillaumont, R., 1993. *Fundamentals of Radiochemistry*, CRC Press Inc., Boca Raton, FL.

Bailey, D. L., Townsend, D. W., Valk, P. E., Maisley, M. N., (Eds.), 2005. *Positron Emission Tomography*, Springer, London.

Banerjee, S. R., Maresca K. P., Francesconi L., Valliant J., *et al.*, 2005. "New directions in the coordination chemistry of ^{99m}Tc: A reflection on technetium core structures and a strategy for new chelate design." *Nucl. Med. Biol.* **32**, 1–20.

Bombardieri, E., Aktolun, C., Baum, R. P., Bishof-Delaloye, A., Buscombe, J., Chatal, J. F., Maffioli L., *et al.*, 2003. "^{131}I/^{123}I-Metaiodobenzylguanidine (MIBG) scintigraphy: Procedure Guidelines for Tumor Imaging." *Eur. J. Nucl. Med. Mol. Imaging* **30**, B132–B139.

Brooks, D., 2005. "PET and SPECT in CNS drug development." *NeuroRx* **2**, 226.

Browne, E., Dairiki, J. M., Doebler, R. E., 1978. In: Lederer, C. M., Shirley, V.S. (Eds), *Table of Isotopes*, Wiley-Interscience Pub., New York, Chichester, Brisbane, Toronto, pp. 1–1523.

Bushberg, J., Seibert, J., Leidholdt, E., Boone, J., 2001. *The Essential Physics of Medical Imaging*, Lippincott, Williams and Wilkins, Philadelphia.

Clark, J. C., Buckingham, P. D., 1975. In: Clark, J.,C., Buckingham, P.D. (Ed.), *Short-Lived Radioactive Gases for Clinical Use*, Butterworths, London, Boston, pp. 1–353.

Clark, M. J., Podbielski, L., 1987. "Medical diagnostic imaging with complexes of ^{99m}Tc."*Coord. Chem. Rev.* **78**, 253–331.

Crouzel, C., Långström, B., Pike, V., Coenen, H. H., 1987. "Recommendations for a practical production of [^{11}C]methyl iodide." *Appl. Rad. Isot.* **38**, 601–603.

de Bruin, B., Kühnast, B., Hinnen, F., Yaouancq, L., Amessou, M., Johannes, L., Samson, A., Boisgard, R., Tavitian, B., Dollé, F., 2005. "1-[3-(2-[^{18}F]Fluoropyridin-3-yloxy)propyl]pyrrole-2,5-dione: Design, synthesis and radiosynthesis of a new [^{18}F]fluoropyridine-based maleimide reagent for the labeling of peptides and proteins." *Bioconj. Chem.* **16**, 406–420.

Deutsch, E., Libson, K., Jurisson, S., 1983. "Technetium chemistry and technetium radiopharmaceuticals." *Prog. Inorg. Chem.* **30**, 73–139.

Dilworth, J. R., Parrots, S. J., 1998. "The biomedical chemistry of technetium and rhenium." *Chem. Soc. Rev.* **27**, 43–55.

Dolci, L., Dollé, F., Jubeau, S., Vaufrey, F., Crouzel, C., 1999. "2-[^{18}F]Fluoropyridines by no-carrier-added nucleophilic aromatic substitution with K[^{18}F]F-K_{222} – a comparative study." *J. Label Coépds. Radiopharm.* **42**, 975–985.

Doll , F., Valette, H., Bottlaender, M., Hinnen, F., Vaufrey, F., Guenther, I., Crouzel, C., 1998. "Synthesis of 2-[^{18}F]fluoro-3-[2(S)-2-azetidinylmethoxy]pyridine, a highly potent radioligand for *in vivo* imaging central nicotinic acetylcholine receptors." *J. Label Compds. Radiopharm.* **41**, 451–463.

Douglas, B., McDaniel, D. H., Alexander, J. J., 1983. *Concept and Models of Inorganic Chemistry*, second ed. John Wiley & Sons, Inc., New-York (NY).

Fichna, J., and Janecka, A., 2003. "Synthesis of target specific radiolabeled peptides for diagnostic imaging." *Bioconjug. Chem.* **14**, 3–17.

Fritzberg, A. R., Abrams, P.G, Beaumier, P. L., Kasina, P. L., Morgan, A. C., *et al.*, 1988. "Specific and stable labelling of antibodies with technetium – 99m with a diamide dithiolate chelating agent." *Proc. Natl. Acad. Sci. USA* 85, 4025–4029.

http: // www . webelements . com / webelements / elements/text/periodic-table/radio.html

http://www.nrc.gov/reading-rm/doc-collections/cfr/part020/appb/

Garron, J.-Y., Moinereau, M., Pasqualini, R., Saccavini, J.-C., 1991. "Direct ^{99m}Tc labelling of monoclonal antibodies: Radiolabeling and *in vitro* stability." *Int. J. Rad. Appl. Instrum.* **B 18**, 695–703.

Huheey, J. E., Keiter, E. A., Keither, R.L., 1993. *Inorganic Chemistry: Principles of Structure and Reactivity*, fourth ed. Harper Collins College Publisher, New York (NY).

Ichise, M., Meyer, J. H., Yonekura, Y., 2001. "An introduction to PET and SPECT neuroreceptor quantification models." *J. Nucl. Med.* **42** (5), 755–763.

Jeavons, A., Parkman, C., Donath, A., Frey, P., Herlin, G., Hood, K., Magnanini, R., Townsend, D., 1983. "The high-density avalanche chamber for positron emission tomography." *IEEE Trans. Nucl. Sci.* **NS-30**, 640–645.

Joliot, F., Curie, I., 1934. "Artificial production of a new kind of radioelement." *Nature* **133**, 201–202.

Kilbourn, M. R., 1990. In: Kilbourn, M.R. (ed). *Fluorine-18 Labelling of Radiopharmaceuticals (Nuclear Science Series)*, National Academy Press, Washington, pp. 1–149.

Kühnast, B., Lagnel, - de Bruin, B., Hinnen, F., Tavitian, B., Dollé, F., 2004. "Design and synthesis of a new [^{18}F]fluoropyridine-based haloacetamide reagent for the labeling of oligonucleotides: 2-Bromo-N-[3-(2-[^{18}F]fluoro-pyridin-3-yloxy)-propyl]-acetamide." *Bioconj. Chem.* **15**, 617– 627.

Lammertsma, A. A., Hume, S. P., 1996. Simplified reference tissue model for PET receptor studies. *Neuroimage* **4**(3 Pt. 1), 153–158.

Långström, B., Kihlberg, T., Bergstrom, M., Antoni, G., Bjorkman, M., Forngren, B. H., Forngren, T., Hartvig, P., Markides, K., Yngve, U., Ogren, M. 1999. "Compounds labelled with short-lived beta-plus-emitting radionuclides and some applications in life sciences. The importance of time as a parameter." *Acta. Chem. Scand.* **53**, 651–669.

Logan, J., Fowler, J. S., Volkow, N. D., Wang, G. J., Ding, Y. S., Alexoff, D. L., 1996. Distribution volume ratios without blood sampling from graphical analysis of PET data. *J. Cereb. Blood Flow Metab.* **16**(5), 834–840.

Liu, S., Edwards, D. S., 2003. "Bifunctional chelators for therapeutic lanthanide radiopharmaceuticals." *Bioconjugate Chem* **14**, 7–34.

March, J. A., Smith, M. B., 2001. *Advanced Organic Chemistry. Reactions, Mechanisms and Structure*, fifth ed. John Wiley & Sons, New York.

Mather, S. J., Ellison, D., 1990. "Reduction-mediated technetium-99m labelling of monoclonal antibodies." *J. Nucl. Med.* **31**, 692–697.

Nicolini, M., Bandoli, G., Mazzi, U., (Eds.), 1983, 1986, 1990, 1995, 1999, 2002. *Technetium and Rhenium in Chemistry and Nuclear Medicine*, Vol. I to VI. Verona: Ed. Cortina International, Raven Press, New York.

Paik, C. H., Phan, L. N., Hong, J. J., Sahami, M. S., Reba, R. C., Steigman, J., Eckelman, W. C., 1985. "The labelling of high affinity sites of antibodies with ^{99m}Tc." *Int. J. Nucl. Med. Biol.* **2**, 3–8.

Patlak, C. S., Blasberg, R. G., Fenstermacher, J. D., 1983. Graphical evaluation of blood-to-brain transfer constants from multiple-time uptake data. *J. Cereb. Blood Flow Metab.* **3**(1), 1–7.

Schwochau, K., 1994. "Technetium radiopharmaceuticals: fundamentals, synthesis, structure and development." *Angew. Chem. Int. Ed. Engl.* **33**, 2258–2267.

Steigman, J., Williams, H. P., Soloman, N. A., 1975. "The importance of the protein sulfhydryl groups in HSA labeling with technetium-99m." *J. Nucl. Med.* **16**, 573 (abs.).

Tatsch, K., Asenbaum, S., Bartenstein, P., Catafau, A., Halldin, C., Pilowsky, L. S., Pupi, A., 2002. "EANM procedure guidelines for brain neurotransmission SPET using ^{123}I-labelled dopamine transporter ligands." *Eur. J. Nucl. Med. Mol. Imaging* **29**, B23–B29.

Verbruggen, A., 1996. Bifunctional chelators for technetium-99m. In: Mather, S.J. (Ed.), *Current Directions in Radiopharmaceutical Research and Development*, Kluwer Academic Publisher, Dordrecht (NL), pp. 31–46.

Verbruggen, A.M.,1990. "Radiopharmaceuticals: State of the art." *Eur. J. Nucl. Med.* **17**, 346–364.

Wagner, H. N., Szabo, Z., Buchanan, J. W., (Eds.), 1995. *Principles of Nuclear Medicine*, Saunders, Philadelphia.

Webb, A., 2003. *Introduction to Biomedical Imaging*, John Wiley & Sons, Inc., Hoboken, New Jersey.

Weiner, R. E., Thakur, M. L., 2001. "Radiolabeled peptides in diagnosis and therapy." *Semin. Nucl. Med.* **31**, 296–311.

Welch, M. J., 1977. In: Welch, M. J. (Ed). *Radiopharmaceuticals and other Compounds Labeled with Short-Lived Radionuclides*, Pergamon Press, Oxford, New York, Toronto, Sydney, Paris, Frankfurt, pp. 1–246.

Welch, M. J., Redvanly, C. S., 2003. In: Welch, M. J., Redvanly, C.S. (Eds.). *Handbook of Radiopharmaceuticals – Radiochemistry and Applications*, Wiley-Interscience Pub., New York, Chichester, Brisbane, Toronto, pp. 1–848.

Wirrwar, A., Schramm, N., Vosberg, H., Muller-Gartner, H. W., 2001. "High resolution SPECT in small animal research." *Rev. Neurosci.* **12**(2), 187–93.

5 Optical Imaging and Tomography

Antoine Soubret and **Vasilis Ntziachristos**

5.0 Introduction

Optical imaging is a fundamental biomedical tool. It spans many different applications; from the generic visual inspection and photography of assays and gels, to advanced microscopy techniques and deep tissue imaging. This chapter is particularly concerned with optical imaging of tissues *in vivo*, a field of the imaging sciences that encompass a large variety of methods and approaches because different light source and detector combinations and different photon-tissue interactions can be utilized to form images. Contrast mechanisms include light absorption, scattering, tissue auto-fluorescence and spectral information. In addition, changes in the properties of light propagating through tissue, for example polarization or interference changes due to structural or functional tissue characteristics, can be further used to form images. For example, Optical Coherence Tomography (OCT) (Fujimoto, 2003) capitalizes on detecting coherence between a low coherence light beam incident on tissue and the corresponding back-reflected beam indicative of the presence of a scatterer at some depth in the range of 0.5–2 mm in dense tissues. In another example, hyper-spectral imaging can capture colour differences associated with changes in the concentrations of tissue chromophores such as oxy- and deoxy- haemoglobin or melanins and offers useful functional information even if it used at coarser resolution (Vo-Dinh *et al.*, 2004).

While endogenous tissue contrast (i.e. absorption, scattering) offers significant insights into tissue anatomy, function and disease-related structural and physiological alterations, the ability to visualize molecular events and gene-expression can be facilitated by the use of fluorescence and bioluminescence reporter technologies. There are two major approaches in visualizing molecular contrast in tissues. *Direct imaging* involves the administration of a fluorescence probe that specifically localizes on an intended target and reports on the presence and concentration of this target to the optical system. *Indirect imaging* involves the use of transgenic biomedical methods, which construct cells and tissues that intrinsically express fluorescence or bioluminescence. In this way, gene regulation, transcription, translation and protein function can be indirectly studied by using fluorescence or bio-luminescence light as reporters of a specific function. Similarly, cell traffic can be studied by employing cells stably expressing fluorescent proteins or bio-luminescence, used to report on their location *in vivo*. Fluorochromes used in direct imaging methods are described in detail in Chapter 7. Fluorescence and bioluminescence reporter technologies used for *in vivo* imaging are described in Chapters 7.3 and 7.5, respectively.

These fundamental contrast mechanisms can be either visualized microscopically or macroscopically. Standard microscopy can be used for observing cell mono-layers and thin (5–10 μm) tissue slices with high resolution (0.2 μm) (Murphy, 2001; Hoppert, 2003). Thicker tissue specimens can be observed with confocal and multi-photon microscopy at depths of a few hundred microns. Confocal and multi-photon microscopies are well suited to image fluorescence in tissues and have also gained wide acceptance in biomedical research and small animal imaging for *in vivo* tissue microscopy, also termed as *intravital*

Figure 5.0.1 Absorption spectra of tissue and spectral emission range of common fluorescent probes and proteins

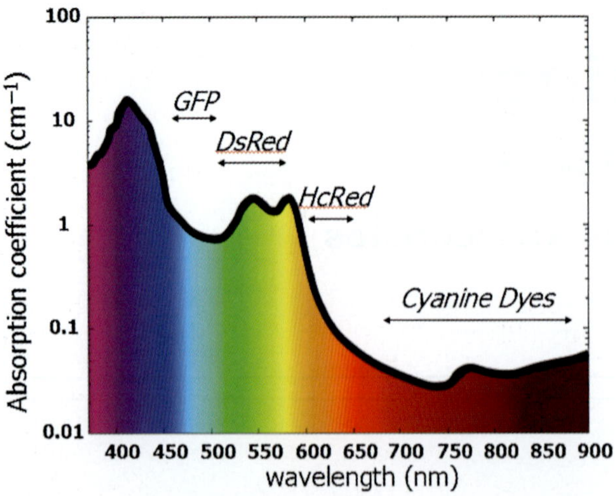

microscopy. The principle of operation and the performance of these technologies are analytically described in Chapter 6.

For imaging deeper than a few hundred microns in tissues, macroscopic optical imaging is necessary. Optical imaging of tissues is possible through several centimetres, especially in the near-infrared (NIR) region. Tissue exhibits low light attenuation in this spectral window as shown by Figure 5.0.1. This is easily observed in everyday life by noting red light propagating through one's finger, for example, after being illuminated by a red laser pointer. One of the complications, however, in optical imaging is that photons in the visible and infrared wavelengths are highly scattered by tissue organelles and membranes resulting in 'photon diffusion', that is photons do not propagate along straight lines when injected in tissue but follow diffusive patterns that limits the quantification ability and maximum resolution achieved. Correspondingly, advances in mathematical models of photon propagation and data processing schemes, combined with appropriate detection schemes, restore the ability to quantify in tissues although microscopic resolution is lost because no coherent light is available for tissue thickness or depth larger than 0.5–1 mm.

In contrast therefore to microscopy, macroscopic imaging of tissues is possible at greater depths but with reduced resolution. There are two main classes of macroscopic optical imaging. A first class of methods using simple photographic imaging yields qualitative two-dimensional images. The most common planar imaging method is epi-illumination imaging. In this technique a light source illuminates the animal or tissue of interest, and the changes of the surface light attenuation or the fluorescence back-emitted from the animal is recorded with a CCD camera using appropriate filters (see Ntziachristos *et al.*, 2003 and Figure 5.0.2). This is a simple, easy to implement

Figure 5.0.2 Epi-illumination imaging. (a) Light is expanded on the tissue surface and light emitted back from the surface is captured by a CCD camera through appropriate filters. (b) Images can be captured at the same wavelength(s) as the incident light, in order to record preferential tissue attenuation. (c) In addition, using a different filter, images can be captured at a longer wavelength, to capture possible fluorescence excited by the incident light. Fluorescence can be endogenous tissue fluorescence or due to the administration of a fluorescent dye or probe. This approach can typically image optical contrast at depths of a few millimeters (2–3) and is surface weighted, that is contrasts at the surface are recorded with higher sensitivity than the deeper-seated contrast

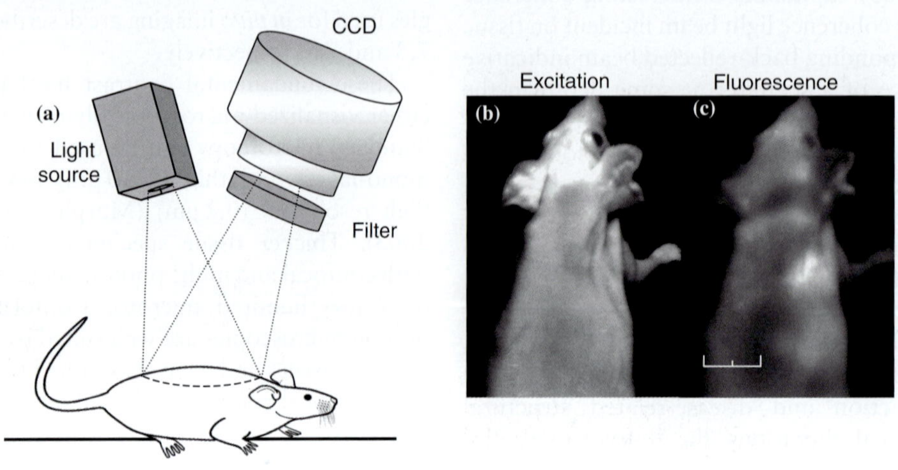

technique but also comes with some significant limitations, as it does not account for the non-linear dependence of signal intensity with depth and optical properties. In addition, the method is 'surface weighted', that is it images superficial activity with higher sensitivity compared to deep-seated activity.

An alternative two-dimensional planar imaging approach is transillumination. In this method the source and detector are placed on the opposite sides of the tissue or animal of investigation as shown in Figure 5.0.3. Figure 5.0.3(b–e) demonstrates that transillumination can be used to image contrast that are deeper in tissues than what is commonly achieved by epi-illumination. This example entails fluorescence contrast and further utilizes appropriate normalization methods that correct fluorescence signals for optical contrast variations by dividing fluorescence images by corresponding images captured at the excitation wavelengths as described in Ntziachristos *et al.* (2005).

A second class of optical imaging methods is the use of advanced theoretical models that accurately model photon propagation in tissues. Generally, these methods are combined with tomographic principles (i.e. multi-projection approaches) to offer three dimensional quantitative imaging at improved resolutions compared to epi- or trans-illumination of two-dimensional imaging. Common implementations are shown in Figure 5.0.4(a–c). Figure 5.0.4(d) depicts a

Figure 5.0.3 Transillumination imaging. (a) Principle of transillumination imaging where the light source and light detector are placed on opposite sides of the tissue illuminated. (b–d) Images obtained from a euthanized nude mouse implanted with a fluorescent tube, inserted through the oesophagus. (b) Mouse epi-illumination image (photograph) obtained at the excitation wavelength. (c) Fluorescence epi-illumination image obtained with the set-up shown in Figure 5.0.2. The fluorescent tube is not evident on this image. (d) Transillumination fluorescence image of the animal showing contrast consistent with the presence of the tube. (e) Normalized transillumination image; here the fluorescence image of (d) is corrected by a corresponding transillumination image at the excitation wavelength (not shown) that captures difference of light attenuation as it propagates through the tissue. This normalized transillumination image demonstrates significantly better imaging characteristics and accurately detects the implanted tube compared to all other images

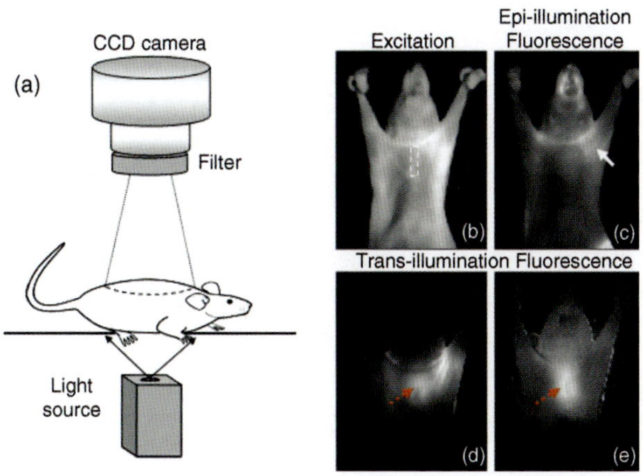

Figure 5.0.4 Optical tomography implementations utilize multiple point-source and point-detector pairs at different geometrical arrangements. (a) In the *reflectance* arrangement, sources and detectors are placed on the same side of tissue. This set-up is similar to the epi-illumination imaging shown in Figure 5.0.2. The most distinct difference is that it time-shares point sources rather than utilizing a single expanded beam as in Figure 5.0.2(a). (b) In the *limited angle projection* arrangement, sources and detectors are placed at a transillumination geometry. Similar to (a), the difference with transillumination is that only one point source is on at the time, effectively implementing multiple source-detector par projections at limited projection angles through the medium. (c) In *complete projection*, point sources and detectors are assumed to be placed around the medium of investigation (here shown at a 180° arrangement for simplicity). (d) A single source detector projection through a medium, also called *sensitivity function*. White indicates high density of photons; dark indicates the opposite. Multiple source-detector pairs create multiple such sensitivity functions that can cover the entire volume

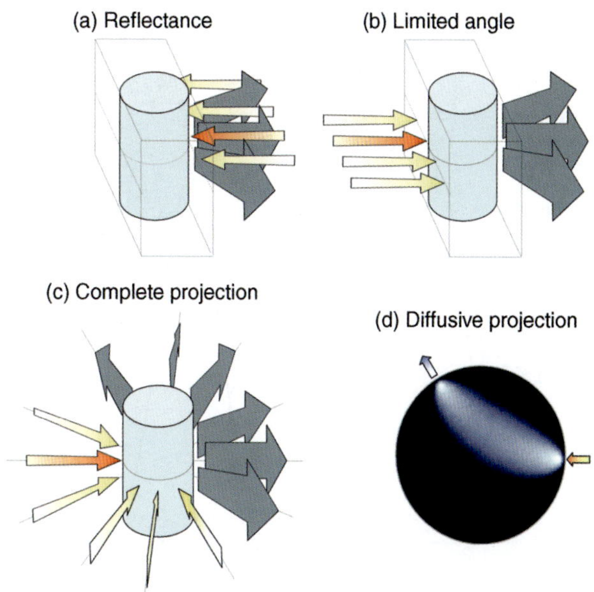

characteristic photon propagation pattern through a diffusive medium for a given source–detector pair. This pattern describes the most probable photon paths that photons follow for the given source and detector pair. Optical tomography uses such photons that propagated through tissues assuming several different source detector pairs, following similar diffusive patterns, so that entire volumes can be sampled. Imaging performance depends on the geometrical arrangement of sources and detectors employed, the tissue geometry and optical properties, and the theoretical and inversion model employed. This imaging approach offers the method of choice for quantitative optical imaging and is analytically described in Sections 5.2–5.4.

Tomographic methods can be categorized by the illumination domain employed. The simpler form is to employ a source of constant intensity, generally termed as 'CW', where CW stands for 'continuous wave'. CW tomography records the light attenuation as it propagates through tissue, or the resulting emitted fluorescence intensity, at multiple projections and offers robust imaging and excellent signal-to-noise characteristics. Conversely, certain advantages can be achieved when intensity modulated light or ultra-short photon pulses at the femptosecond to picosecond range and corresponding advanced detection techniques are used instead of CW light. Intensity modulation tomography typically operates at the 100 MHz–1 GHz modulation frequency and resolves the amplitude and phase shifts of the propagating photon wave through tissue, instead of just amplitude as in CW imaging. Time-resolved methods detect the arrival of photons as a function of time and build a histogram of photon number at different time-gates, within a time scales of a few nanoseconds. These advanced illumination methods are summarized in Figure 5.0.5 for a single source-detector pair. Their use, over CW imaging, can explicitly resolve absorption from scattering or fluorescence concentration and lifetime. Such measurements become important for quantitatively differentiating tissue optical properties, which is useful when intrinsic tissue chromophores are visualized. Furthermore, time-resolved methods offer the option of selecting early arriving photons, that is the photons that travelled along the shortest routes from the source to the detector. This approach can select photons that track less diffusive patterns than the bulk of photons and, therefore, can yield better spatial resolution at the expense of lower sensitivity (Das, Liu and Alfano, 1997).

By using multiple source-detector pairs and appropriate illumination schemes, macroscopic tissue imaging can yield three-dimensional imaging and quantification and resolution that cannot be achieved

Figure 5.0.5 Advanced illumination methods. (a) Intensity modulated light is used to illuminate tissue and establishes a photon wave in tissue with the same frequency. Appropriate demodulating detectors can resolve the *amplitude change* and *phase shift* of the wave and utilize this information to separate the tissue absorption from tissue scattering or separately calculate fluorescence concentration and lifetime. (b) Time resolved method: An ultra-short pulse of light is used to illuminate the tissue. A detector resolves the arrival of photons as a function of time and builds a curve of the characteristic shape shown that depends on the absorption and scattering of tissue. A similar technique can be used to resolve fluorescence lifetime

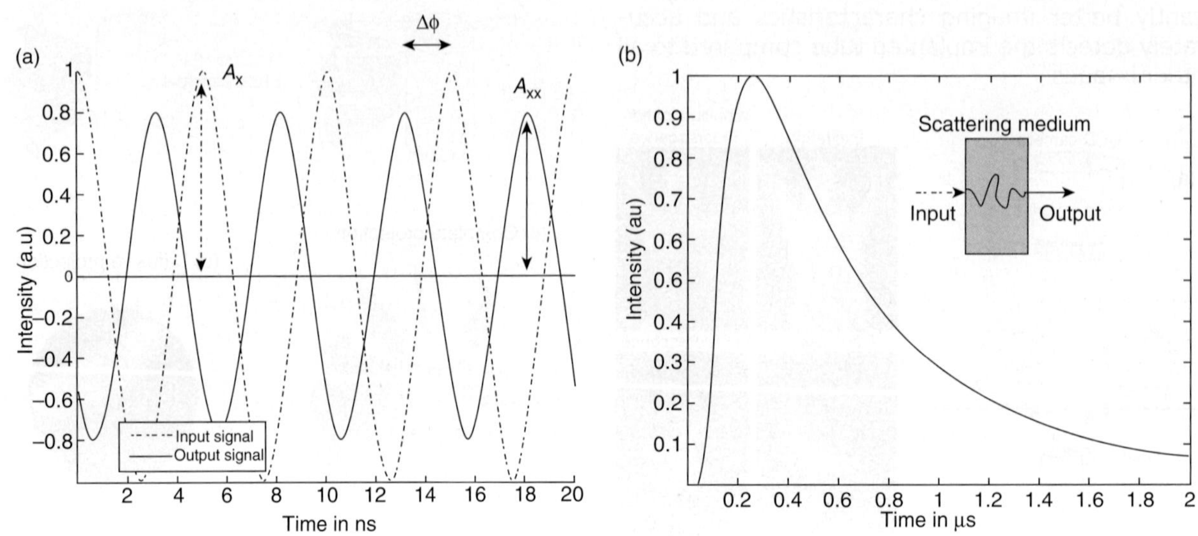

by planar imaging methods. Such methods can be used with different contrast mechanisms. For example, resolving the spatial distribution of absorbers at multiple wavelengths can yield images of oxy- and deoxy-haemoglobin, water, lipids and organelle concentrations. Similarly, the bio-distribution of various exogenous fluorochromes marking molecular function or gene-expression can be similarly resolved. This chapter summarizes the basic characteristics of photons interacting with tissues at the microscopic and macroscopic level to illustrate various contrast mechanisms that exist for optical imaging of tissues. It describes the basic light characteristics and separately observes interactions that are better described using the photon nature of light (Section 5.1.1) from interactions better described by considering light as a wave (Section 5.1.2). Then, it presents basic measures of optical characteristics of tissues and launches into the description of light propagation in tissues for macroscopic observations (Section 5.2). Finally, the formation of quantitative modelling for optical tomography is presented, and key aspects of planar imaging and Fluorescence Molecular Tomography are presented. The fundamental light properties presented in Section 5.1 are also useful to explain microscopic *in vivo* imaging, which is presented in Chapter 6.

5.1 Light – tissue interactions

Light is electromagnetic radiation with wavelengths that generally range from the ultra-violet to the near-infrared (i.e. wavelengths that are visible or almost visible by the human eye). Its prominent characteristic is that it exhibits properties of both waves and particles (Loudon, 1983; Hecht, 2002). When it propagates though tissue, light undergoes different interactions associated with the structural arrangement and physical properties of the micro-environment. These interactions can be described at the atomic and molecular level when considering the particle nature of light (photons) or at the macroscopic scale where wave properties and bulk optical coefficients are used to characterize average interaction features.

For optical imaging applications, the selection of a description approach of light as a wave or a sum of particles largely depends on the ratio of the light wavelength to the size of the object considered (i.e. molecules, cells, tissues). For example, reflection, refraction and diffraction are best described by electromagnetic wave theory (Griffiths, 1999; Jackson, 1999; Hecht, 2002) as they typically consider interfaces with dimensions much larger than the wavelength of light. Conversely, interactions involving the atomic structure of the object illuminated, such as absorption and emission are better described by considering the particle nature of light (Loudon, 1983; Fox, 2001). Similarly, the principle of operation of different optical imaging techniques is best described by utilizing the wave or particle nature of light, depending on the interaction or technology used. Interferometric methods, for example, utilize the wave description, whereas charge coupled devices utilize the particle nature of light.

The following sections describe the most pertinent light-matter interactions, with focusing on these interactions that are related to generating contrast or explaining the characteristics underlying optical imaging of tissues. Section 5.1.1 describes the interaction of light with tissue matter by considering the particle nature of light. Section 5.1.2 discusses the interaction of the wave nature of light and tissue matter. Finally, the optical description of tissues as associated with these interactions is summarized in Section 5.1.3.

5.1.1 Photon (particle) description of light–matter interaction

The *particle* or *photon* description of light propagation in tissue looks at the different physical or photochemical phenomena that individual photons experience in tissues. Photons are elementary particles having zero mass, no electric charge, and an indefinitely long lifetime. They are the quanta of electromagnetic energy. Each photon carries energy E and momentum $\vec{p}$. In vacuum, photons move in straight lines at the speed of light c_0 and are conserved in number, except if emitted or absorbed by an atom or a molecule. As shown in Figure 5.1.1 , the two fundamental photon-matter interaction associated with optical imaging of tissues are scattering and absorption(Loudon, 1983; Fox, 2001; Hecht, 2002; Valeur, 2002). The absorption phenomenon can lead to secondary processes like the emission of new photons or the transfer of the photon's energy to matter through different non-radiative decay pathways.

At the molecular level, these interactions are graphically captured by the Jablonski energy diagram shown in Figure 5.1.2. This diagram represents the orbital energy levels of the molecules and explores how electrons in the molecules are excited from the

Figure 5.1.1 Photon-matter interaction at the molecular level

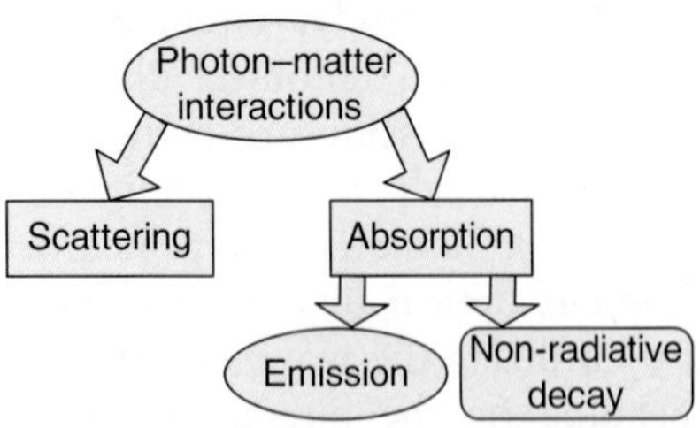

ground state into higher electronic energy states by the incident light, and the events that occur as these excited molecules fall back into lower energy states.

5.1.1.1 *Scattering*

Scattering is the redirection of the radiation of a photon after interaction with matter. If during the process there is no energy transfer between the photon and matter, this interaction is called *elastic* or *Rayleigh scattering*, and the scattered photon has same wavelength as the incident photon. In contrast, if energy transfer from the photon to the matter occurs then the process is called *inelastic* or *Raman scattering*, and the scattered photon is of lesser energy (longer wavelength) compared to the incident photon. In tissues, Rayleigh scattering is dominant and Raman scattering is not common and requires high incident light intensity to produce statistically important samples that can be practically detected. These scattering processes are usually very fast ($\sim 10^{-15}$s).

5.1.1.2 *Absorption*

When a photon incident on a molecule has an energy corresponding to the energy difference of two molecular orbital levels, the radiation field transfers

Figure 5.1.2 Jablonski diagram, which illustrates the different possible interactions of light with the different energy level of matter. The ground, first and second singlet electronic states (which correspond to states having only paired electron spins and a total electron spin quantum number equal to 0) are respectively depicted by S0, S1 and S2, and T1 is the first electronic triplet state (which corresponds to a state having one set of electron spins unpaired and a total electron spin quantum number of 1). The others horizontal lines correspond to the different vibrational and rotational energy levels of the molecules

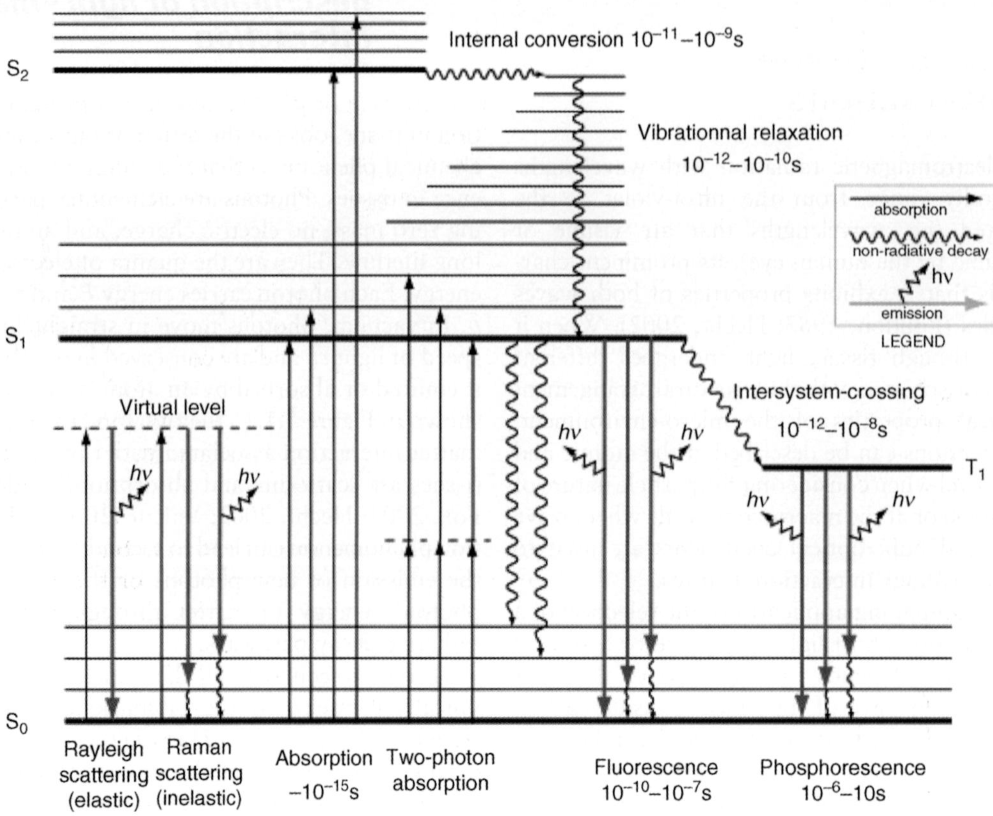

its energy to the molecule and a transition from the ground level to a higher energy singlet state (S1, S2...) occurs. This phenomenon is called *absorption* and results in the loss of the incident photon. The type of transition that the molecules undergo depends on the wavelength of the light. The electrons of a molecule are excited to higher orbital by ultraviolet, visible or near-infrared light; the vibration modes of the molecules are excited by infrared light, and the rotation modes are excited by microwave radiations (Lakowicz, 1999; Valeur, 2002; Hollas, 2004). Excitation of high-energy orbitals is also possible by the simultaneous absorption of two (or multiple) photons of lower energy, via a short-lived virtual state, such that the total energy of the two (or multiple) photons corresponds to an energy transition of the molecule. This phenomenon is the basis of two- or multi-photon microscopy and it takes place at areas of very high light intensity (i.e. around the focus point of a laser beam) (Murphy, 2001; Tuchin, 2002; Hoppert, 2003; Vo-Dinh, 2003; Goldman and Spector 2005;). Following the absorption of a photon, the energy transferred to the molecules by the photon can be either converted to another energy form by the molecules or partly re-emitted through the following photoluminescence processes:

- *Non-radiative decay*: Three non-radiative deactivation processes (i.e. processes that do not emit any photons) are involved in the interaction of light with matter: *Vibrational relaxation*, *intersystem crossing*, and *internal conversion* (Loudon, 1983; Fox, 2001; Valeur, 2002). The vibrational relaxation facilitates energy transfer through collisions to the surrounding environment (e.g. the solvent surrounding the excited molecule). This is the most common of the three non-radiative decay processes and occurs very quickly ($<1 \times 10^{-12}$ s). This means that most excited state molecules in liquid never emit any energy because the solvent absorbs the energy before other deactivation processes can occur. *Intersystem crossing* is a radiation-less transition between two different spin states (i.e. between a singlet and a triplet state) of the molecule, that is, the spin of an excited electron is reversed during the process. *Internal conversion* is a radiation-less transition between two energy states of the molecule having the same spin angular momentum (i.e. singlet to singlet state). After this conversion, the molecule is placed in an extremely high vibrational level. This excess of vibrational energy is usually lost by collision with other molecules through the vibrational relaxation process.

- *Fluorescence and phosphorescence emission*:

Molecules that are excited to high energy levels by light can decay to lower levels by emitting radiation (photoluminescence effects). Fluorescence is emission light from singlet-excited states, in which the electron in the excited orbital is paired (of opposite sign) to the second electron in the ground-state orbital (Lakowicz, 1999; Valeur, 2002). Return to the ground state is usually fast ($10^{-10} - 10^{-7}$s), so that typical fluorescence lifetimes for organic fluorescent molecules are near 1– 10 ns.

Phosphorescence is emission of light from triplet-excited states, in which the electron in the excited orbital has the same spin orientation as the ground-state electron. Transition rates are slow ($10^{-3} - 10$ s), so phosphorescence lifetimes are typically milliseconds to seconds. The spectra of these emission processes vary widely and are dependent on the chemical structure of the fluorophore and the solvent in which it is dissolved.

5.1.2 Wave nature of light–matter interaction

Section 5.1 described basic interactions of photons with matter, as it pertains to applications to optical imaging of tissues. This section describes in greater detail the macroscopic appearance of common light interactions with matter when considering the wave nature of light. This section begins with a description of the general properties of electromagnetic light waves in order to remind of the basic terminology that is used throughout the book. Then, the major interactions of light with matter at the macrocopic level are considered, as summarized in Figure 5.1.3. Where appropriate, the connection to the microscopic observation is established.

5.1.2.1 Electromagnetic light waves

Herein we briefly review the basic principles of electromagnetic light waves as they pertain to phenomena and theories closely associated with optical imaging (Griffiths, 1999; Jackson, 1999).

Frequency, wavelength of an electromagnetic wave. The classical description of light considers light as the linear combination of plane waves propagating in different directions and carrying an electric and a magnetic components $(\vec{E}, \vec{B})$. By definition, every wavefront of a plane wave is a parallel plane normal to the direction of propagation $\hat{s}$ of the wave as shown in Figure 5.1.4. A plane wave is characterized by its frequency ν (i.e. a monochromatic wave) and a

Figure 5.1.3 Wave–matter interaction at the macroscopic level

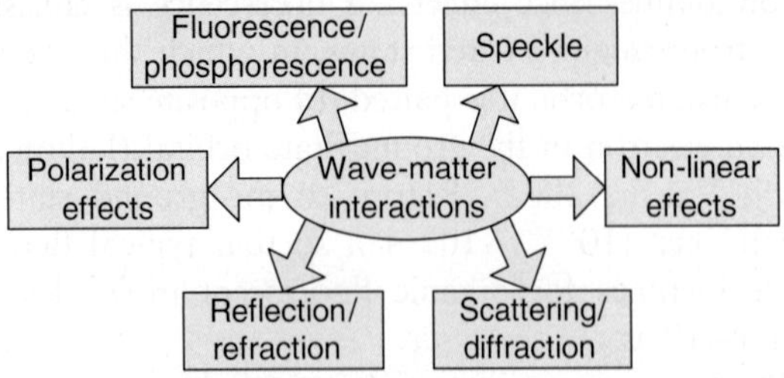

Figure 5.1.4 Electromagnetic wave

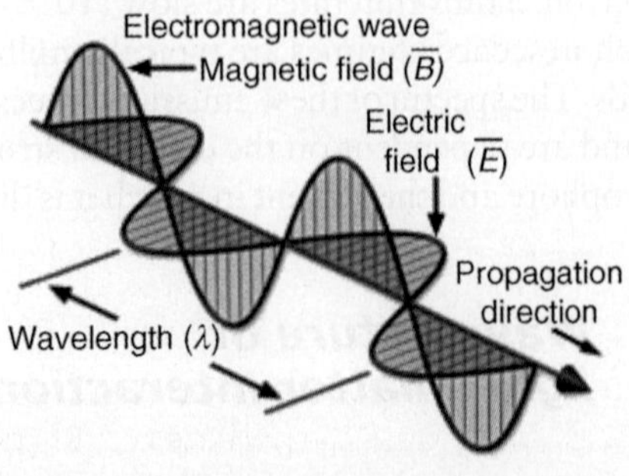

wavelength that can be determined as $\lambda_0 = c_0/\nu$, where c_0 is the speed of light in vacuum (i.e. $c_0 \approx 3 \times 10^{10}$cm/s). From a microscopic point of view, an electromagnetic wave can be seen as the carrier of a high number of photons, each of which has an energy E related to the frequency ν of the wave by the Planck constant h:

$$E = h\nu. \tag{5.1}$$

Index of refraction in dense media. It is generally observed that the speed of a light wave is slower when light propagates inside a non-scattering material compared to the speed of light in vacuum. The ratio of the speed of propagation in vaccum over that inside the medium is the refractive index of this medium:

$$n = \frac{c_0}{c}. \tag{5.2}$$

Consequently, the wave induced inside the material propagates with a different wavelength $\lambda = c/\nu$ compared to the vacuum wavelength $\lambda_0 = c_0/\nu$. It is important to notice that only the spatial characteristic of the wave (i.e. the wavelength λ) is modified but not the frequency ν (and thus the energy) of the wave.

Electric vector, magnetic vector and polarization. The electromagnetic nature and polarization of light are properties that are used in several optical imaging applications including microscopic applications and in imaging through dense tissues. To explain these properties, it is useful to review first the electric and magnetic vectors of the electromagnetic (EM) wave. These vectors are used to describe the electric $\vec{E}(\vec{r},t)$ and the magnetic $\vec{B}(\vec{r},t)$ fields of light at each point in space $\vec{r}$ and time t . For a monochromatic plane wave of frequency ν , we have

$$\vec{E}(\vec{r},t) = E_0 \mathrm{Re}\left[e^{\mathrm{i}(\vec{k}\cdot\vec{r}-\omega t)}\right]\hat{u},$$
$$\vec{B}(\vec{r},t) = \frac{E_0}{c_0}\mathrm{Re}\left[e^{\mathrm{i}(\vec{k}\cdot\vec{r}-\omega t)}\right]\hat{s}\times\hat{u}, \tag{5.3}$$

where Re[z] is the real part of the complex number z, $\omega = 2\pi\nu$ is the angular frequency of the wave and E_0 is the amplitude of the electric field. The wave vector $\vec{k}$ is defined by

$$\vec{k} = n\frac{\omega}{c_0}\hat{s}, \tag{5.4}$$

with c_0 the speed of light in vacuum, n the index of refraction of the medium and $\hat{s}$ the direction of propagation of the wave. As the vector $\vec{k}$ has only real component, the equation describes a pure oscillating wave which is not attenuated as it propagates.

For a spherical monochromatic wave produced at the origin of the coordinate system, that is, from a point source, the expressions for the electromagnetic field are similar to the equation where a $1/r$ dependence has to be added to ensure that the energy is conserved, that is

$$\vec{E}(\vec{r},t) = \frac{E_0}{r}\mathrm{Re}\left[e^{\mathrm{i}(k\,r-\omega t)}\right]\hat{u},$$
$$\vec{B}(\vec{r},t) = \frac{E_0}{c_0 r}\mathrm{Re}\left[e^{\mathrm{i}(k\,r-\omega t)}\right]\hat{s}\times\hat{u}, \tag{5.5}$$

with $k = n\omega/c_0$ and $r = \sqrt{x^2+y^2+z^2}$ the distance from the point $\vec{r} = (x, y, z)$ to the origin of the coordinate where the source is located.

Coming now to the polarization of light $\hat{u}$, this is then a property that describes the direction that the electric field of the EM wave points toward to. As light is a transverse electromagnetic wave, the polarization direction $\hat{u}$ is in a direction perpendicular to the direction of wave propagation. Ordinarily, a ray of light can be described as consisting of a mixture of waves vibrating along different directions perpendicular to its line of propagation. If all the directions of polarization are equally probable, then light is said to be unpolarized, otherwise the light is said to be polarized. It is found, for example, that reflected light is always polarized to some extent. If two orthogonal axes, assumed as the *horizontal* and the *vertical* direction (and orthogonal to the direction of propagation) are defined, then any state of polarization can be decomposed as

$$\hat{u} = \begin{pmatrix} u_H \\ u_V \end{pmatrix}, \tag{5.6}$$

where u_H and u_V are the horizontal and the vertical component of the polarized light. If u_H and u_V are real constant numbers, the light is said to be *linearly polarized*, that is have a polarization vector $\hat{u}$ which is constant as the wave propagates. Correspondingly, the *circularly polarized* light is defined as

$$\hat{u}_{\pm} = \begin{pmatrix} u_H \\ e^{\pm i\pi/2} u_V \end{pmatrix}, \tag{5.7}$$

where the plus and minus sign corresponds, respectively, to left and right circular polarization. Because of the $\pi/2$ phase difference between the horizontal and vertical component, this wave corresponds to a polarization vector $\hat{u}$ rotating clockwise or counterclockwise at the frequency ν of the wave.

Irradiance and energy of an electromagnetic wave. Often in optical imaging, only the light *strength* or *intensity* is necessary. In this case an appropriate quantity used is the light irradiance $I(\vec{r}, t)$, which is defined as the optical power of light per unit area (unit of Watt/cm^2) incident on a surface (or leaving a surface). The irradiance is proportional to the square of the amplitude E_0 of the electromagnetic field and is obtained by the following equation:

$$I(\vec{r}, t) = \frac{1}{2\mu_0} \hat{n} \cdot \left(\vec{E}(\vec{r}, t) \times \vec{B}^*(\vec{r}, t) \right), \tag{5.8}$$

where $\hat{n}$ is the vector normal to the surface and μ_0 is the permeability of vacuum.

Interference. Interference is an important property of light and is exploited as a source of contrast in many optical imaging application including microscopy (Murphy, 2001; Hoppert, 2003) and optical coherence tomography (Fujimoto, 2003; Boppart, 2004). Interference is the ability of two waves to add constructively or destructively depending on the relative phase of the two waves. Based on such phenomena, an observation at a given point illuminated by two identical light beams may yield intensity levels ranging from double the light intensity of each individual beam to complete darkness (zero intensity), depending on the relative phase of the two waves. Quantitatively we can express the measured irradiance due to the interference of two monochromatic plane waves of irradiance $I_1(\vec{r}, t)$ and $I_2(\vec{r}, t)$ as

$$I(\vec{r}, t) = I_1(\vec{r}, t) + I_2(\vec{r}, t) + 2\sqrt{I_1(\vec{r}, t) I_2(\vec{r}, t)} \cos \alpha(\vec{r}). \tag{5.9}$$

The third term of this equation is the interference term and depends on the relative phase $\alpha(\vec{r}) = \vec{k}_1 \cdot \vec{r} - \vec{k}_2 \cdot \vec{r}$ of the two interfering waves. Interference is possible only if the waves have the same wavelength. The interference pattern observed depends on certain parameters of the incident light field, the underlying medium and light source used. An important factor is the *coherence* of the light source employed, which is a property that describes the length and the time scale where the incident field can maintain a determined phase at different points in time and space. Generally, lasers are highly coherent source of light in time and space due to the well-synchronized process of stimulated emission; therefore, interference effects can be observed over large distances and long travelling times. Conversely, light bulbs virtually produce incoherent light due to the random emission along the filament, and interference effects can be observed at very short time and space scales.

Interference effects in tissues can be spoiled due to the random micro-movement of particles (e.g. Brownian motion, particle diffusion and flow, muscle movement, thermal agitation, etc). These phenomena induce random temporal fluctuations in the propagating light and result in loss of coherence within very short periods of time (typically within a few tenths of nanoseconds (Tuchin, 2002; Vo-Dinh, 2003)). In this case a typical detection processes that is slower than a few tenths of nanoseconds will be unable to detect interference effects and the measured intensity $\bar{I}_{\text{meas}}(\vec{r}, t) = \langle I(\vec{r}, t) \rangle_{\text{average}}$ will be an averaged signal of the different interference patterns. In this case interference effects are greatly reduced and in most cases we have

$$\bar{I}(\vec{r}, t) = \bar{I}_1(\vec{r}, t) + \bar{I}_2(\vec{r}, t), \tag{5.10}$$

for two waves interacting with each others (to be compared with Eq. (5.9)).

Interference (and loss of it) is exploited in many optical imaging methods and, in particular, in microscopy. Perhaps the most widely known method is the differential interference contrast (DIC) microscopy for observing living specimens, which generally have little inherent contrast. This is because interference measurements are very sensitive to small differences in thickness and refraction of non-stained tissue slices. In another popular imaging technique, Optical Coherence Tomography (Fujimoto, 2003; Boppart, 2004), interference is used to detect and time-analyzed, back-scattered coherent light to gain depth information. Reduction of coherence effects is also indicative of particle movement and has been exploited as a contrast mechanism to image blood flow (Cheung *et al.*, 2001; Tuchin, 2002; Ayata *et al.*, 2003; Vo-Dinh, 2003; Yuan *et al.*, 2005).

5.1.2.2 *Reflection/refraction*

A light wave incident at the interface of two media attaining refractive indexes of n_0 and n_1 , respectively will be partly propagated in both media after reflection and refraction processes occur as shown in Figure 5.1.5. The reflected wave is commonly called the specular wave because the angle of incidence and reflection are the same. The angle of reflection θ_0 and refraction θ_1 follow the well-known Snell's law of refraction:

$$n_0 \sin(\theta_0) = n_1 \sin(\theta_1). \tag{5.11}$$

Microscopically, the reflected and transmitted plane waves are due to the Rayleigh scattering interference pattern produced by each atom and induced by the incident plane wave. Because of the regular arrangement of atoms at the microscopic level, the constructive/destructive arrangement produces a very regular pattern which can be decomposed as a reflected and refracted wave (Born, Wolf and Bhatia, 1999; Hecht, 2002). When $n_0 > n_1$, there is an angle of incidence called the critical angle $\theta_c = \sin^{-1}(n_1/n_0)$, which when exceeded, no refraction occurs, but all light is reflected back into the n_0 medium; a phenomenon called total internal reflection. However, close to the boundary, in the n_1 medium, a rapidly decaying wave is established, termed as the evanescence wave, which is exploited in total internal reflection microscopy applications.

Figure 5.1.5 Snell reflection and refraction

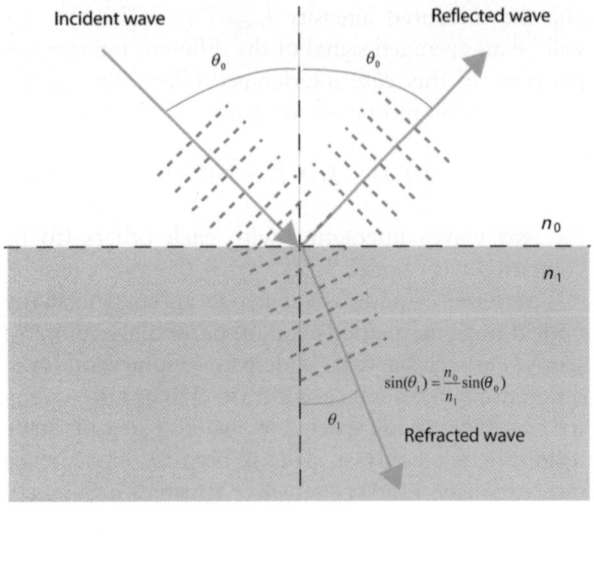

5.1.2.3 *Absorption*

The cumulative macroscopic effect of the microscopic absorption described in Section 5.1.1 is customarily described in non-scattering fluids by using the measure of *absorbance* A (also known as the optical density) of the sample as (Bass and Optical Society of America, 1995):

$$A = -\log_{10}\frac{I}{I_0}, \tag{5.12}$$

where I_0 is the incident plane wave irradiance and I is the irradiance exiting the sample as shown in Figure 5.1.6. The quantity A can be easily measured with a spectrophotometer.

Experimentally, it has been observed that the absorbance coefficient A is proportional to the thickness of the sample. Consequently, in the optical community it is customary to characterize the absorption of sample not with the absorbance A but with an absorption coefficient per unit length $\mu_a[\text{cm}^{-1}]$ defined as (Bohren and Huffman, 1983; Bass and

Figure 5.1.6 Geometry used to define the absorption of a sample

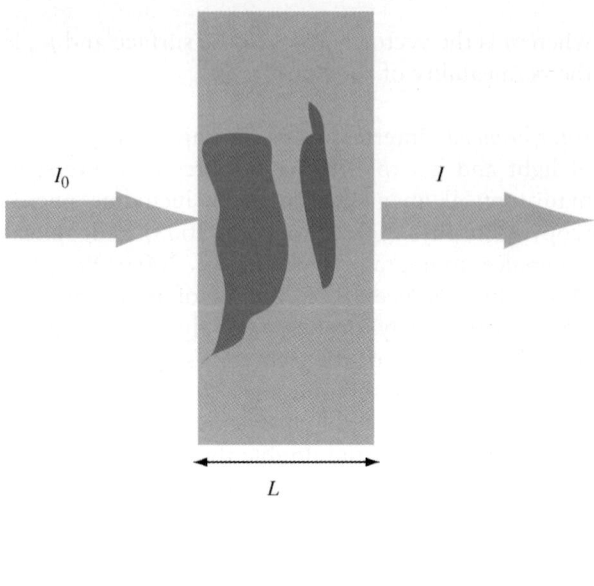

Optical Society of America, 1995; Born, Wolf and Bhatia, 1999)

$$\mu_a = -\frac{1}{L}\ln\frac{I}{I_0}. \quad (5.13)$$

The absorbance is expressed in function of the absorption coefficient as

$$A = \mu_a L/\ln(10), \quad (5.14)$$

where the factor ln(10) is necessary to account for the decimal logarithm used in the definition (5.12) in comparison with the definition (5.13) based on the Neperian logarithm.

Furthermore, as the absorbance A and the absorption coefficient μ_a are proportional to the molar concentration $C[M]$ of the absorbing species in the sample, the molar absorptivity (also known as molar extinction coefficient) ε [$cm^{-1}M^{-1}$] can also be used to characterize the absorption by using the following definition:

$$\varepsilon = \frac{A}{CL}, \quad (5.15)$$

which is related to the absorption coefficient through the following relationship:

$$\mu_a = \varepsilon C\ln(10). \quad (5.16)$$

Equation (5.15), rewritten in the form $A = \varepsilon CL$, denotes a linear relationship between absorbance and concentration, and is known as the Beer–Lambert law, or more commonly, as Beer's law.

It is important to note that all the definitions introduced in this section for characterizing the absorption of a medium are valid only if the medium is *non-scattering*. Scattering and absorbing media, such as tissues, are considered in the next section.

The absorption processes can be considered by the electromagnetic wave theory when a generalized complex refractive index is introduced, that is

$$n = n' + in''. \quad (5.17)$$

Assuming a plane wave, as described by Eq.(5.3), the wave vector $\vec{k} = n\frac{\omega}{c_0}\hat{s}$ becomes a complex number, that is $\vec{k} = \vec{k}' + i\vec{k}''$, and the electromagnetic fields are exponentially attenuated in agreement with experiments, that is

$$\vec{E}(\vec{r},t) = E_0\,\mathrm{Re}[e^{i(\vec{k}'\cdot\vec{r}-\omega t)}]e^{-\vec{k}''\cdot\vec{r}}\hat{u}. \quad (5.18)$$

Employing the definition of irradiance in Eq. (5.8), a relationship between the absorption coefficient μ_a and the imaginary part of the complex refractive index n can be obtained:

$$\begin{aligned}\mu_a &= 2\,\vec{k}''\cdot\hat{s}L \\ &= 2\omega/c_0\,n'',\end{aligned} \quad (5.19)$$

with $\omega = 2\pi\nu$ as the angular frequency of the wave and c_0 the speed of light in vacuum.

This complex refractive index is defined for a homogeneous medium but can be generalized (Bohren and Huffman, 1983; Tsang, Kong and Shin, 1985; Sheng, 1990 Sihvola and Institution of Electrical Engineers, 1999) when introducing an averaging procedure for an inhomogeneous media (i.e. a medium made of materials having different complex refractive indexes). Several expressions have been proposed to calculate this generalized refractive index in function of the refractive indexes of its homogeneous constituents. The Maxwell–Garnett and Bruggemann formula are two of the most popular theories used for this purpose. However, for biological specimens where significant complexity and variation exists in the different structures, implementation of such theoretical approaches becomes challenging, and experimental measurements are often the only way to assess the optical attenuation of a sample.

5.1.2.4 *Diffraction and Scattering*

The term diffraction is usually used to describe the deviation of light from rectilinear propagation occurring whenever a portion of the light wavefront is obstructed in some way by an object (Hecht, 2002). The term scattering is usually used when light is interacting with dispersed particles like in the atmosphere or with a rough surface or random arrangement of particles like in suspension (Chandashekhar, 1960; Kerker, 1969; Van de Hulst, 1981; Bohren and Huffman, 1983; Born, Wolf and Bhatia, 1999)(See Figure 5.1.7).

However, as mentioned by Van de Hulst (1981) and emphasized by Bohren's article in (Bass, 1995) the definition of diffraction and scattering is so broad that it is almost meaningless. Actually, in terms of light-matter interaction, there is no practical difference between scattering and diffraction. Quoting Bohren in (Bass, 1995), 'A distinction must be made between a physical process and the superficially different theories used to describe it. There is no fundamental difference between specular reflection and refraction by films, diffraction by slits, and scattering by particles. All are consequences of light interacting with matter. They differ only in their geometries and the approximate theories that are

Figure 5.1.7 Scattering processes observed at different scales. (Left) A multiple scattering medium can be seen as composed of multiple scattering subunits, collectively scattering the incident light and the light produced by the other subunits. (Centre) Each of the scattering subunits can be isolated and characterized by the angular scattering distribution it produces after illumination by a plane wave. Different subunits, for example different cells or organelles, may have different angular scattering distribution patterns. (Right) At the molecular level, the scattering profile produced by the subunit is due to the combination of the Rayleigh scattering of each molecule constituting the subunit

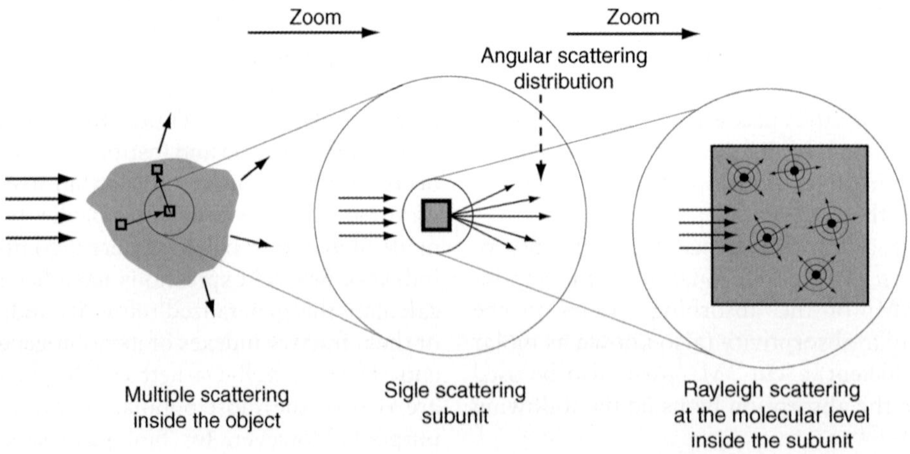

sufficient for their quantitative description. The different terms used to describe them are encrustations deposited during the slow evolution of our understanding of light and matter'. Thus in the following we will consider scattering as a general process of changing the direction of the incident light by interaction with matter.

We can understand the point of view of Van de Hulst and Bohren in realizing that, as in the reflection/refraction effect, diffraction and scattering can also be seen microscopically as the bulk effect of the Rayleigh scattering by matter molecules on the incident light field as also explained in Figure 5.1.7. Generally, scattering effects are mostly observable when the wavelength of the incident wave λ is comparable with the characteristic size of the scatterer (or subunit as shown in Figure 5.1.7).

To macroscopically characterize scattering, it is a custom to consider a small part of the scattering medium as a 'particle' or subunit (but still macroscopic in comparison with the atomic scale; Figure 5.1.7, centre). The angular scattering profile produced by this particle illuminated by a plane wave is described by a phase function $p(\hat{s}, \hat{s}_0)$, such that the scattered irradiance $I_{sca}(\vec{r})$ on the detector located at $\vec{r}$ produced by a plane wave of irradiance $I_{inc}(\vec{r}_0)$ incident in the direction $\hat{s}_0$ is

$$I_{sca}(\vec{r}) = \frac{\sigma_s}{4\pi R^2}\, p(\hat{s}, \hat{s}_0)\, I_{inc}(\vec{r}_0), \qquad (5.20)$$

where $R = ||\vec{r} - \vec{r}_0||$ is the distance between the particle and the detectors, $\hat{s}$ is the scattered direction given by the unit vector source-detector $\hat{s} = (\vec{r} - \vec{r}_0)/R$, with $\vec{r}_0$ the location of the diffuser considered as shown in Figure 5.1.8. The phase function is defined such that the integration over all angles is unity:

$$\int \frac{d^2\hat{s}}{4\pi} p(\hat{s}, \hat{s}_0) = 1. \qquad (5.21)$$

In general, the exact expression of the phase function in biological media is not very well known and varies from one type of tissue to another. It is custom to use the Henyey–Greenstein function to describe the light scattering pattern (Ishimaru, 1997; Tuchin and

Figure 5.1.8 Definition of the phase function of a 'small' scatterer element isolated from a tissue like structure

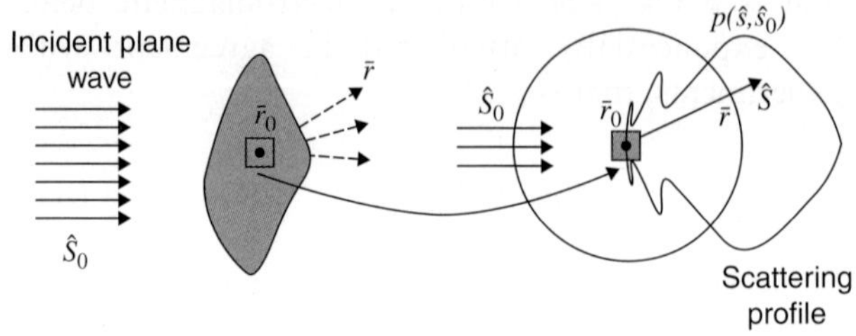

Society of Photo-optical Instrumentation Engineers, 2000; Tuchin, 2002; Vo-dinh, 2003):

$$p(\hat{s}, \hat{s}_0) = \frac{1}{4\pi} \frac{1 - g^2}{(1 + g^2 - 2g\cos\theta)^{3/2}}, \qquad (5.22)$$

where the angle θ is defined by $\hat{s} \cdot \hat{s}_0 = \cos\theta$. The Henyey–Greenstein is a very simple function only parameterized by its anisotropy factor g defined as

$$g = \int \hat{s} \cdot \hat{s}_0 p(\hat{s}, \hat{s}_0)\, d^2\hat{s}. \qquad (5.23)$$

However, it allows describing highly forward scattering medium for $g \simeq 1$ and isotropic scattering for $g \simeq 0$ as shown in Figure 5.1.9.

The level (or 'strength') of the scattering element in Eq. 5.20 is defined by the scattering cross-section σ_s (cm^2), which is the hypothetical area normal to the incident radiation that would geometrically intercept the total amount of radiation actually scattered by a scattering particle.

Depending on the ratio of the light wavelength over the geometrical cross-section σ_{geom} (e.g. for a sphere of radius a, we have $\sigma_{geom} = \pi a^2$), three scattering regimes are identified, that is the Rayleigh regime ($\sigma_{geom} \ll \lambda^2$), the Mie regime ($\sigma_{geom} \approx \lambda^2$) and the geometrical optics ($\sigma_{geom} \gg \lambda^2$) regime. Rayleigh scattering is considered isotropic (when the polarization is not taken into account), whereas the Mie and geometrical scattering regime produce highly forward scattering patterns.

Similar to the scattering cross-section, we define the absorption cross-section σ_a (cm^2) of the scatterer element as the hypothetical area normal to the incident radiation that would geometrically intercept the total amount of radiation actually absorbed by a scattering particle.

The scattering and absorption cross-section are directly dependent on the ratio of the index of refraction of the scatterer over the index of refraction of the surrounding medium. This relationship is usually unknown for tissue because there is typically a mixture of different scattering elements. However, for a scatterer having a very simple geometry, like a spherical shape, an analytical expression can be obtained for the phase function and the cross-sections in function of the refractive indexes and the geometrical properties of the diffuser. The mathematical expressions detailed are quite involved and uses decompositions in term of spherical harmonics (Kerker, 1969; Van de Hulst, 1981; Bohren and Huffman, 1983; Ishimaru, 1997).

Naively, if we consider a tissue-like structure made of multiple identical diffusers as previously described in Figure 5.1.7, we could suppose that the irradiance measured at the detectors will be the sum of the light irradiance scattered by each element. However, due to the multiple interactions between each element (i.e.

Figure 5.1.9 Henyey–Greenstein phase function for $g = 0$, 0.3 and 0.6

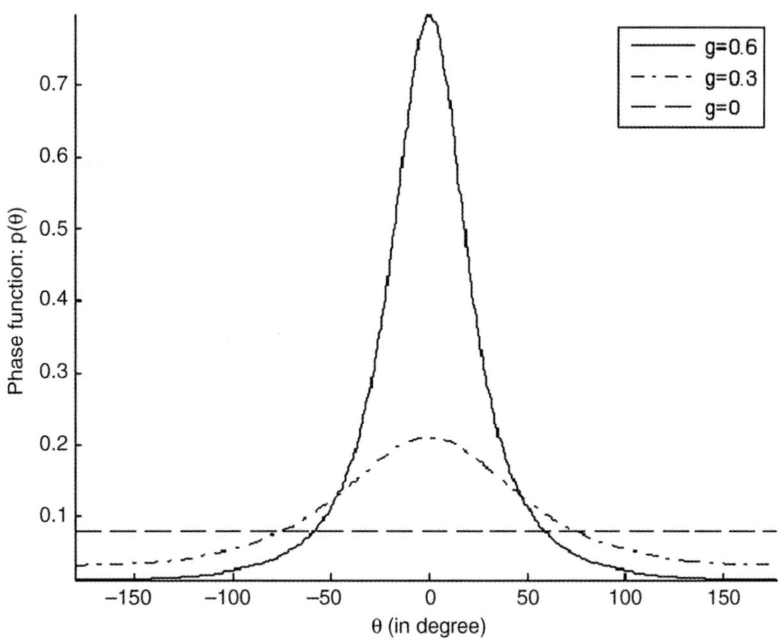

the light scattered by one scattering subunit will then be scattered by the other scattering subunits), the resulting scattering pattern cannot be seen as simply the sum of the scattering pattern of each subunit. In Section 5.2, we describe the physical theory describing the light propagation in such multiple-scattering medium. However, we can already introduce the scattering coefficient μ_s and absorption coefficient μ_a as a measure of the level of scattering and absorption for a multiple scattering medium, which are defined as

$$\mu_s = C\sigma_s[\text{cm}^{-1}],\ \mu_a = C\sigma_a[\text{cm}^{-1}], \tag{5.24}$$

where C [cm^{-3}] is the concentration of the scattering particle subunit in the tissue like structure. In particular, it can be demonstrated that the scattering length $l_s = 1/\mu_s$ corresponds to the average distance between two scattering events, and the absorption length $l_a = 1/\mu_a$ is the average length covered by the photons before being absorbed (Davison, 1957; Case and Zweifel, 1967; Duderstadt and Martin, 1979; Tsang, Kong and Shin, 1985; Sheng, 1990; Tsang and Kong, 2001).

5.1.2.5 *Speckle and coherence effects*

Scattering processes contribute to light diffusion and loss of image resolution, as it becomes easily apparent when shining light through tissue for example. However, despite this effect, *macroscopic coherent effects* can be observed at random arrangement of scatterers, such as cellular distributions in tissue or sphere suspensions in gel. Such phenomena can be observed especially when the positions of the scatterers are fixed, or the measurement is done over time spans when there is no movement of the observed scattering particles . In such cases the scattered waves can interfere and produce a characteristic interference pattern called *speckle pattern*, as shown in Figure 5.1.10. This pattern is the direct result of the interference property of light and the random nature of the medium. Such interference pattern can be used for characterizing movement because moving scatterers spoil such interference patterns (Sheng, 1990; Tuchin, 2002; Ayata *et al.*, 2003;Vo-Dinh, 2003).

Other coherent effects such as the backscattering enhancement (Sheng, 1990; Kim *et al.*, 2005) or other interference effects of scattered wave can be used to characterize matter and biological media. Optical Coherence Tomography (OCT) is certainly one of the best examples of the use of coherent effect in biological imaging. In OCT, the photons which are not multiple scattered are selected through an

Figure 5.1.10 Speckle pattern

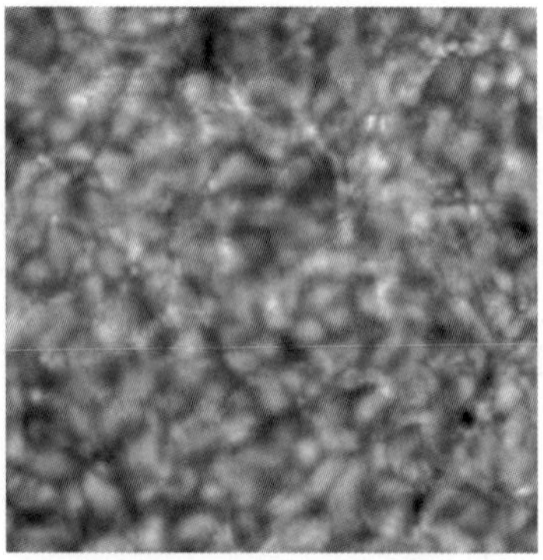

interferometer based on the properties, that is, only photons which are scattered only one time can interfere with the incident light (Tuchin, 2002; Vo-Dinh, 2003; Boppart, 2004).

5.1.2.6 *Fluorescence/Phosphorescence*

The atomic and molecular fluorescence and phosphorescence processes are directly observable macroscopically. Usually, these processes are characterized by three quantities (Das, Liu and Alfano, 1997; Lakowicz, 1999; Valeur, 2002; Vo-Dinh, 2003):

- The quantum yield (or quantum efficiency) of the fluorochrome, that is the ratio between the number of fluorescence photons emitted and the number of photons absorbed: $\eta_{\text{fluo}} = \frac{N_{\text{emitted}}}{N_{\text{absorbed}}}$.
- The absorption coefficient of the fluorochrome, that is $\mu_a^{\text{fluo},\lambda_{\text{exc}}} = \varepsilon\ \ln(10)\, C_{\text{fluo}}$ [cm^{-1}] at the excitation wavelength λ^{exc} , where ε [cm^{-1}M^{-1}] is the molar absorptivity of the fluorochrome and C_{fluo} is the molar fluorochrome concentration.
- The fluorescence lifetime of the photoluminescence process τ_{fluo}[s] , which is the mean time spent in the excited state, such that if $I(t_0)$ is the irradiance produced by the fluorochrome just after being excited by a short pulse of light at time t_0, then the irradiance at a later time t is

$$I(t) = \frac{\exp(-(t-t_0)/\tau_{\text{fluo}})}{\tau_{\text{fluo}}} I(t_0). \tag{5.25}$$

The factor τ_{fluo} in the denominator is necessary to insure the conservation of energy, that is

$$\int_{t_0}^{+\infty} I(t)dt = I(t_0). \tag{5.26}$$

Table 5.1.1 Characteristic values of the Cy5.5 fluorochrome

Quantum yield	$\eta_{\text{fluo}} = 0.23$
Fluorescence lifetime	$\tau_{\text{fluo}} = 1\,ns$
Decadic absorption	$\varepsilon = 190\,000$

More complicated multi-exponential time decays could have also been incorporated depending on the complexity of the time decay of the dye.

The higher the value of the product $\eta_{\text{fluo}}\mu_a^{\text{fluo},\lambda_{\text{ex}}}$, the greater fluorescence will be observed. Both of these factors are important in the overall brightness of the dye and both must be considered when comparing different fluorophore. For example, the dye Cy5.5 from Amersham Biosciences has the properties listed in Table 5.1.1, and the spectrum is shown in Figure 5.1.11.

5.1.2.7 Polarization effects

Polarization effects can be used for imparting contrast in optical imaging, because the state of light polarization may change when the light interacts with matter, that is, is reflected, refracted or scattered. The amount and nature of the change depends on geometrical and optical features of the matter that light interacts with. One of the major applications of polarization effects in biology, for example, is the ability to discriminate between photons which are multiply scattered and photons which experience only few scattering events. This is because photons that are scattered only a few times keep their polarization, whereas multiple scattering photons loose macroscopic polarization. Therefore, polarization can be used to differentiate between properties of superficial and sub-surface tissues (Tuchin and Society of Photo-optical Instrumentation Engineers, 2000; Vo-Dinh, 2003). To understand this dependence, we can assume a tissue structure that is illuminated by a linearly polarized light and the co-polarized $I_{//}$ and cross-polarized $I_{\perp}$ components of the backscattered light are recorded; then the multiple scattering can be significantly reduced by subtracting off the depolarized portion of the total signal. This can be achieved by subtracting the cross-polarized component from the co-polarized component, that is $I_{//} - I_{\perp}$. A normalized quantity known as the degree of polarization

$$P = \frac{|I_{//} - I_{\perp}|}{I_{//} + I_{\perp}}, \tag{5.27}$$

which ranges from 0% for unpolarized light to 100% for completely polarized radiation, is also very useful to select single or multiple scattered light. Similarly, the use of polarized light can be employed to image birefringence effects in structures and tissues.

Figure 5.1.11 Excitation and fluorescence emission for the Cy5.5 NHS ester dye from Amersham Biosciences

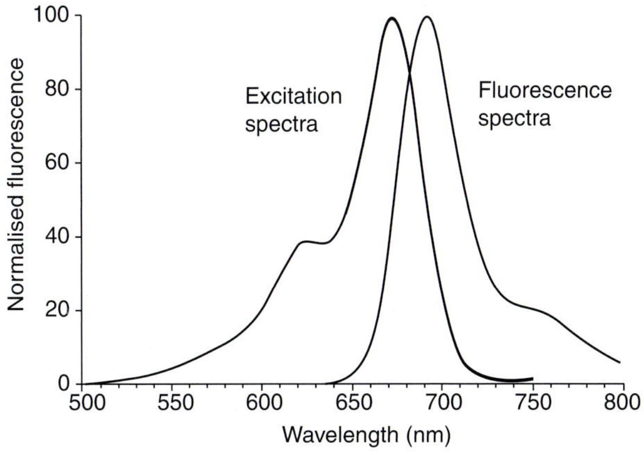

5.1.3 Optical description of tissue

The interaction of light with living tissue is a combination of elementary interactions of light with matter described in the previous sections. Optical microscopy, presented in Chapter 6, capitalizes on some of these basic interactions to impart tissue contrast at the micron to sub-micron scales. Conversely, macroscopic observations tend to summarize the average effects of microscopic interactions using appropriate coefficients such as a mean index of refraction, a mean absorption coefficient, a mean scattering coefficient, a mean phase function etc. Similarly, *in vivo* imaging of thick tissue does not generally resolve microstructure but is rather sensitive to variation of these macroscopic optical parameters describing the interaction of light with tissue. Figure 5.1.12 illustrates some of the interactions primarily responsible for the scattering pattern observed.

The average index of refraction, absorption coefficient, scattering coefficient, phase function and fluorescence properties and the most important cells constituents involved in the light tissue interaction are described in the following sections.

Figure 5.1.12 Example of light-cell interactions

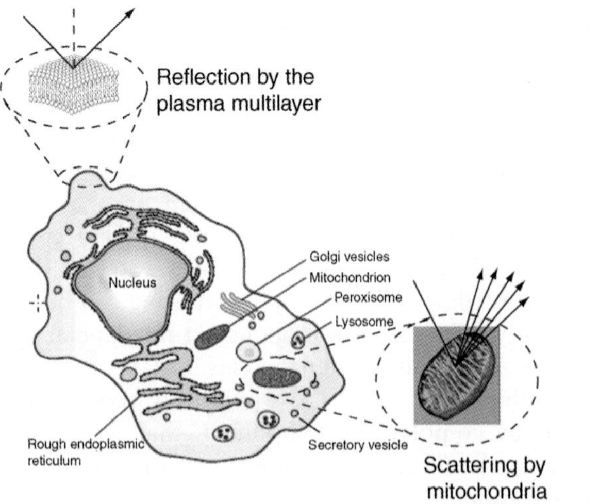

5.1.3.1 *Refractive indices*

Changes of the index of refraction lead to light reflection, refraction and scattering, as described in Sections 5.1 and 5.2. As tissues are heterogeneous, we can define an average optical index that accounts for the optical indexes of different tissue constituents. Several approximations exist to calculate this effective index depending on the geometry and the arrangement of the different constituents (Bohren and Huffman, 1983; Sihvola and Institution of Electrical Engineers, 1999). The simplest one is the volume weighted average of the optical index constituents. This effective coefficient characterized the reflected and refracted part of the diffuse wave scattered by the tissue.

As water is the major constituent of tissue and attains the lowest index of refraction ($n = 1.33$) compared to structural tissue components, solvents and other bio-polymers, it also represents the minimum value for the average index of refraction expected to be found in tissue. Approximate values for the index of refraction of various organs are given in the Table 5.1.2.

5.1.3.2 *Scattering coefficients and phase functions*

Tissue structures of the order of magnitude of the light wavelength are mostly responsible for the strong light scattering experienced by photons propagating in tissue. In the visible and near-infrared range (400–1000 nm), tissue scattering is a mixture of 'Rayleigh' and 'Mie' scattering, as described in the section 5.1.2.

Table 5.1.2 Typical values of the optical indices of human organs

Tissue constituents	Optical index
Water, soft tissues	1.33
Extracellular fluids, intracellular cytoplasm	1.35–1.38
Brain, aorta, lung, stomach, kidney, bladder	1.36–1.40
Fatty tissues	1.45
Skin epidermal tissue	~1.6

Rayleigh scattering is due to supporting tissues like elastin and collagen, which are small-scale variation of the visible wavelength. The Mie scattering is produced by organelles and subcomponents of organelles like mitochondria, cell nucleus, endoplasmic reticulum, Golgi apparatus and smaller structures like lysosomes and perioxisomes (see Figure 5.1.13) (Tuchin and Society of Photo-optical Instrumentation Engineers, 2000; Tuchin, 2002; Vo-Dinh, 2003). In blood, the disk-shaped red cells are the strongest scatterers, and the scattering properties of blood depend on the volume fraction of the red cells and their degree of agglomeration. As a Henyey–Greenstein phase function is usually used to describe the scattering angular variation, only the anisotropy factor g is necessary to define this function. As the parameter g is difficult to measure *in vivo*, a value between 0.8 and 0.99 is usually assumed. Furthermore, it is also customary to introduce the reduced scattering coefficient combining the scattering coefficient and the factor g by

$$\mu_s^{'} = \mu_s(1 - g), \quad (5.28)$$

because this parameter can easily be measured by time-resolved or modulated frequency system. Extensive list of *ex vivo* and *in vivo* scattering properties can be found in the following references (Cheong, Prahl and Welch, 1990; Tuchin, 2002; Vo-dinh, 2003) and some *in vivo* values are summarized in Table 5.1.3.

Figure 5.1.13 Biological constituents of tissue responsible for the scattering of light in the visible region

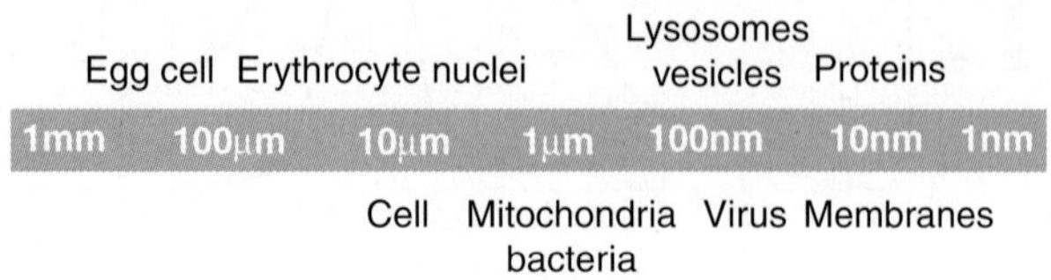

Table 5.1.3 Typical optical properties for some human tissue constituents obtained *in vivo*

Tissue	Absorption coefficient $\mu_a(\mathrm{cm}^{-1})$	Reduced scattering coefficient $\mu_s'(\mathrm{cm}^{-1})$
Breast (Tromberg *et al.*, 1997) $\lambda = 674\,\mathrm{nm}$	0.04	11
Muscle (Torricelli *et al.*, 2001) (human) $\lambda = 780\,\mathrm{nm}$	0.15	9
Brain (Ntziachristos *et al.*, 1999) (human) $\lambda = 780\,\mathrm{nm}$	0.15	12
Lung (Beek *et al.*, 1997) (human) $\lambda = 780\,\mathrm{nm}$	0.02	20
Upper torso (Niedre, private communication) (mouse) $\lambda = 750\,\mathrm{nm}$	0.25	20
Lower torso (Niedre, private communication) (mouse) $\lambda = 750\,\mathrm{nm}$	0.5	15

5.1.3.3 Absorption coefficient

Several tissue chromophores contribute to the absorption of light. The most important absorbers in the visible and the near-infrared are oxy- and deoxy-haemoglobin, but other tissue chromophores may contribute to visible light absorption, such as various metabolites, melanins etc., whereas lipids and water may contribute in the absorption of near-infrared light. As shown in Figure 5.0.1, the near-infrared offers significantly less attenuation to light and is preferred for optical imaging applications for achieving high detection sensitivity and depth penetration of several centimeters in tissues. Figure 5.1.14 depicts the absorption spectra of the most important tissue absorbers in this spectral window.

Figure 5.1.14 Absorption spectra of the major absorbers in the near-infrared window, reproduced from Taroni *et al.* © copyright, 2003, Royal Society of Chemistry. Main tissue constituents absorbing in the 600–1000 nm spectral range. The curves to 100% water (▲), and lard (△), and to 100 μM of oxy- (□) and deoxy- (■) haemoglobin.

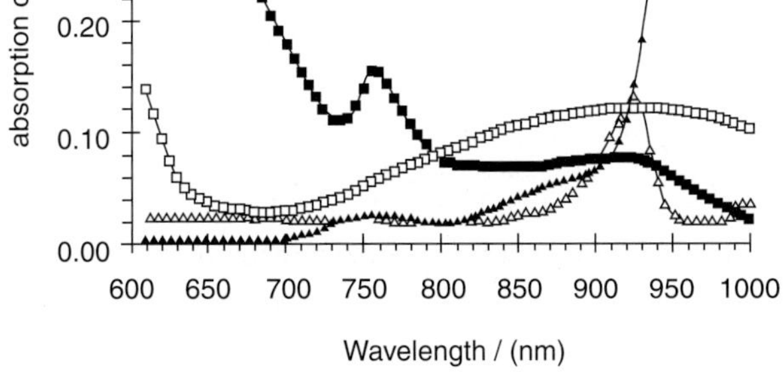

In the infrared window (>900 nM), water is the dominant absorber due the absorption by the vibrational energy states of the water molecules. At shorter wavelength, the haemoglobin absorption dominates. Due to the strong dependence of light absorption in tissue by haemoglobins, its use has been considered for monitoring blood saturation as implemented by the pulse oxi-meter and in recording tissue function or angiogenesis (Tuchin and Society of Photo-optical Instrumentation Engineers, 2000; Boas *et al.*, 2001; Hielscher *et al.*, 2002; Tuchin, 2002; Vo-Dinh, 2003)

5.1.3.4 Tissue fluorescence emission

Autofluorescence is the natural fluorescence of tissues due to the fluorescence of endogenously produced biofluorophores such as keratin, porphyrins, NAD(P)H, collagen and elastin. In certain cases fluorescence imaging or spectroscopy can resolve the presence and concentration of these tissue constituents to characterize metabolic processes or as markers of disease for diagnostic purposes. Figure 5.1.15 depicts the auto-fluorescence spectra of common tissue constituents assuming a 337 nM excitation.

In addition to the naturally occurring fluorescence in tissues, extrinsic fluorescence can be introduced for optical imaging applications by using transgenic technology to introduce fluorescent protein expressing genes in tissues or after the preferential bio-distribution of fluorescent molecules (fluorescent probes), extrinsically administered for diagnostic or therapeutic

Figure 5.1.15 Autofluorescence spectra from different tissue constituents following a 337 nm excitation

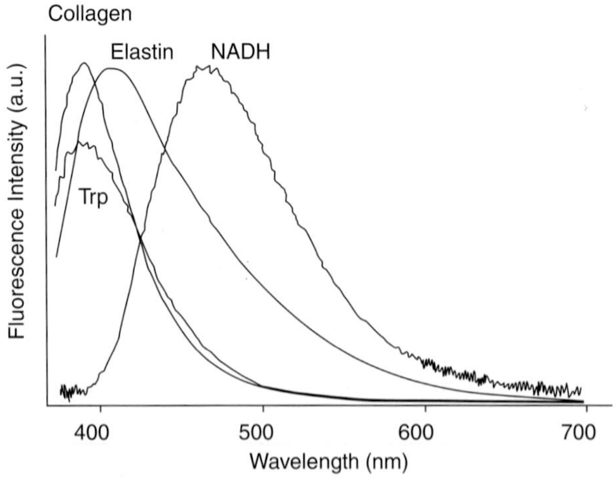

reasons, for example receptor targeted fluorescent dyes. Fluorescent probes can further impart molecular specificity and the ability to image *in vivo* the function of proteins and enzymes, follow cell movement and migration and study gene expression and regulation. As such, these technologies find an increasing application to biology and potentially to clinical practice as well (Weissleder and Mahmood, 2001; Cherry, 2004).

5.2 Light propagation in tissues

In the preceding section, we have introduced the different elements necessary to describe phenomena associated with light propagation in tissues. Based on this knowledge, physical photon propagation models can be derived for quantitative macroscopic tissue observation. Although much of today's optical imaging is performed with simple photographic methods that do not make use of such models, photography fails as an accurate biomedical imaging method because it does not account for the non-linear dependence of light propagation in tissues as a function of propagation distance and optical properties. Herein, we summarize key quantities, methods and models that can be employed for quantitative optical imaging of tissues.

5.2.1 Radiometry

Before embarking on the description of physical models of photon propagation in tissues, it is necessary to define basic photometric quantities associated with

Figure 5.2.1 Definition of the radiance

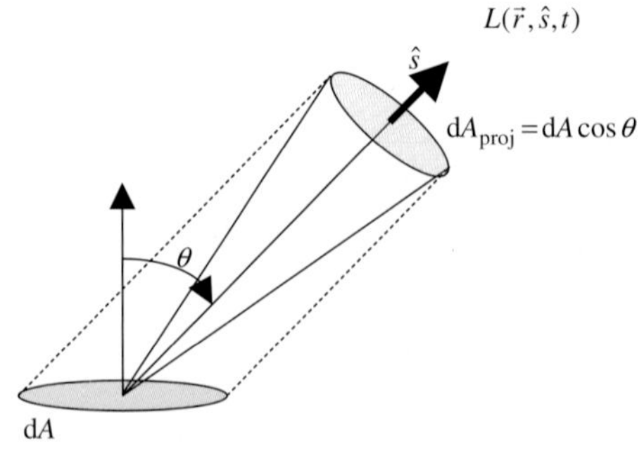

radiometry in general and used to describe an electromagnetic wave at a point $\vec{r}$ and time t inside the tissue, as summarized in Table 5.2.1.

From all the quantities defined in the Table 5.2.1, the radiance carries the most information on the light propagation. This is because the radiance provides the rate of energy flowing in the direction $\hat{s}$ through a surface dA on each point on tissue, as illustrated in Figure 5.2.1.

Compared with the other quantities defined in the Table 5.2.1, the radiance is the only one taking into account the directivity of the light along the direction $\hat{s}$ and can be generally used to describe light propagation through clear, moderately scattering and highly scattering (diffusive) media, that is through a lens, a semi-transparent tissue cavity or a highly diffusive tissue volume, assuming that macroscopic coherent interference effects can be neglected. The different regimes of photon propagation and the corresponding mathematical formulations that model this propagation are described in the following paragraphs.

5.2.2 Models of photon propagation

Macroscopic measurements of photons passing through media of different consistency often require different theoretical models for quantitatively describing their propagation. The different propagation regimes associated with optical imaging are summarized in Figure 5.2.2 that shows a progressive change from the *ballistic regime* where light propagation is described by the laws of geometrical optics as found in standard optical textbook (Born, Wolf and Bhatia, 1999; Hecht, 2002) to the *diffusive regime* where the directivity of the incident light is lost due the multiple scattering processes. In media with moderate scattering (or of small dimensions, i.e. 1–10 mm), light can propagate without totally loosing the memory of the incident light direction.

The radiance $L(\vec{r}, \hat{s}, t)$, described in the previous paragraph, can be generally used to describe light

Table 5.2.1 Most important photometric quantities used in optical imaging. (Adapted from Moseley and Sliney (1997).)

Quantity	Symbol	Unit	Definition
Energy	$E(\vec{r},t)$	Joule (J)	The total energy in a radiation field or the total energy delivered by such a field.
Power or radiant flux	$P(\vec{r},t)$	Watt (W)	The rate at which energy is transferred from one region to another by the radiation field. $P = \mathrm{d}E/\mathrm{d}t$.
Irradiance	$I(\vec{r},t)$	W cm^{-2}	The flux per unit area received by a real or imaginary surface. $I = \mathrm{d}P/\mathrm{d}A$, where $\mathrm{d}A$ is the area of the surface element.
Fluence rate or spherical irradiance	$\phi(\vec{r},t)$	W cm^{-2}	At a given point in space, the power incident on a small sphere divided by the cross-sectional area of that sphere.
Energy density	$w(\vec{r},t)$	J cm^{-3}	The energy per unit volume of the radiation field. $w = \mathrm{d}E/\mathrm{d}V$, where V is the volume.
Radiance	$L(\vec{r},\hat{s},t)$	W cm^{-2}sr^{-1}	The flux per unit projected area per unit solid angle leaving a source or a reference surface. $L = \mathrm{d}P/\mathrm{d}\hat{s}\,\mathrm{d}A_{\mathrm{proj}}$, where $\mathrm{d}A_{\mathrm{proj}} = \mathrm{d}A\cos\theta$ is the projected area, θ the angle between the outward surface normal of the area element $\mathrm{d}A$ and the direction of observation $\hat{s}$ (see Figure 5.2.1)

Figure 5.2.2 Different regimes of light propagation and theories describing them

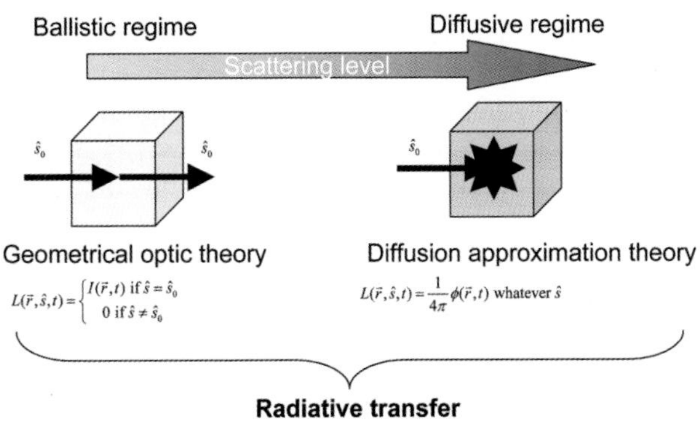

propagation as it passes through a medium and possibly progressively loosing directivity of the incident light. Theoretically, phenomena associated with this propagation can be modeled by the *radiative transfer* theory (Chandrashekhar, 1960; Sobolev, 1963,1975; Case and Zweifel, 1967; Van de Hulst, 1980; Bohren and Huffman, 1983; Kong and Shin, 1985; Rytov, Kravtsov and Tatarskiĭ, 1987; Ishimaru, 1997; Tsang, Tsang, Kong and Ding, 2000), which can serve as a generic model of photon propagation in tissues. However, due to the complexity and computational burden associated with the calculations of the radiance and the use of the radiative transfer model, such solutions are rarely used for bio-optical imaging. Instead, either ballistic solutions are employed in special cases where reduced scattering or early arriving photons are utilized based on geometrical optics or else diffusive solutions as described in the following.

5.2.3 Diffusion model

The diffusion model is based on the general principle of conservation of energy and the approximate Fick's law.

5.2.3.1 *Conservation of energy*

The conservation energy principle applied on a finite volume of the diffusive medium can be formulated as

Variation over time of the density of energy
= −**power density leaving the volume**
− **power density absorbed inside the volume**
+ **power density produced by the sources**

Mathematically, this principle has the following form:

$$\frac{1}{c}\frac{\partial\phi(\vec{r},t)}{\partial t} = -\nabla\cdot\vec{j}(\vec{r},t) - \mu_a\phi(\vec{r},t) + S(\vec{r},t), \quad (5.29)$$

where c [cm s^{-1}] is the speed of light in the medium, $\phi(\vec{r},t)$ is the fluence rate [W cm^{-2}] defined in Table 5.2.1, and which can be written as the integral over all the direction of the radiance $L(\vec{r},\hat{s},t)$:

$$\phi(\vec{r},t) = \iint_{4\pi} L(\vec{r},\hat{s},t)\mathrm{d}\hat{s}. \quad (5.30)$$

$\vec{j}(\vec{r},t)$ is the flux density vector [W cm^{-2}] defined by the first moment of the radiance $L(\vec{r},\hat{s},t)$:

$$\vec{j}(\vec{r},t) = \iint_{4\pi} L(\vec{r},\hat{s},t)\,\hat{s}\,\mathrm{d}\hat{s}, \quad (5.31)$$

and $S(\vec{r},t)$ [W cm^{-3}] is the power per unit volume produce by the sources.

We note that to obtain Eq. (5.29), we have used the following relationship between the fluence rate $\phi(\vec{r},t)$ and the density of energy $w(\vec{r},t)$:

$$w(\vec{r},t) = \phi(\vec{r},t)/c, \quad (5.32)$$

to obtain the first left-hand side of the equation (Ishimaru, 1997). We also note that the flux density vector $\vec{j}(\vec{r},t)$ is closely related to the irradiance by

$$I(\vec{r},t) = \vec{j}(\vec{r},t)\cdot\hat{n}. \quad (5.33)$$

This last relation is particularly useful to calculate the irradiance leaving a diffusion medium knowing the flux density vector $\vec{j}(\vec{r},t)$ and the orientation of the surface $\hat{n}$.

5.2.3.2 *Fick's law*

To solve Eq. (5.29), another relationship between the fluence rate $\phi(\vec{r},t)$ and the flux density vector $\vec{j}(\vec{r},t)$ is necessary. For highly scattering medium, the photons follow random path trajectories. After a few scattering events, the photons have a diffusive behaviour. The mathematical theory of diffusion process is well established, and it has been established that the Fick's law of the form

$$\vec{j}(\vec{r},t) \approx -D\vec{\nabla}\phi(\vec{r},t), \quad (5.34)$$

provides a good description of the diffusion processes (Patterson, Chance and Wilson, 1989; Ishimaru, 1997), whereas the diffusion coefficient D [cm] is introduced. To calculate D, we can consider the diffusion approximation of the radiance $L(\vec{r},\hat{s},t)$ dependence on fluence and flux (i.e. Davison, 1957; Ishimaru, 1997),

$$L(\vec{r},\hat{s},t) = \frac{1}{4\pi}\phi(\vec{r},t) + \frac{3}{4\pi}\vec{j}(\vec{r},t)\cdot\hat{s}, \quad (5.35)$$

which states that the radiance is mostly isotropic with a corrective factor provided by the flux density vector $\vec{j}(\vec{r},t)$. Combination of Eqs. (5.35) and (5.29) can lead (for derivation see Ishimaru, 1997; Davison, 1957; Duderstadt and Martin, 1979; Case and Zweifel, 1967; Aronson and Corngold, 1999) to the well-known dependence of D on the absorption and scattering coefficients (i.e. Case and Zweifel, 1967):

$$D = \frac{1}{3[\mu_s' + a\,\mu_a]}, \quad (5.36)$$

where μ_s' is the reduced scattering coefficient defined by $\mu_s' = \mu_s(1-g)$. This coefficient μ_s' is a construction that approximates the diffusion of photons as an isotropic scattering phenomenon, even though each individual scattering event is anisotropic. Hence, a medium scattering mostly in the forward direction (i.e. $g \approx 1$) will be equivalent to an isotropic scattering medium having a smaller scattering coefficient given by μ_s' . For typical optical properties in the NIR window, $a \approx 0.2$ seems to provide a good order of magnitude of this coefficient (Case and Zweifel, 1967; Aronson and Corngold, 1999). However, in the near-infrared window, $\mu_s' \gg \mu_a$, and it is customary to define the diffusion coefficient by

$$D = \frac{1}{3\mu_s'}. \quad (5.37).$$

5.2.4 *Diffusion equation*

5.2.4.1 *Homogeneous medium*

Direct substitution of Eq. (5.34) in the conservation of energy principle Eq. (5.29) yields the *time-domain diffusion equation*, that is

$$\frac{1}{c}\frac{\partial\phi(\vec{r},t)}{\partial t} - D\vec{\nabla}^2\phi(\vec{r},t) + \mu_a\phi(\vec{r},t) = S(\vec{r},t). \quad (5.38)$$

When intensity modulated sources of angular frequency ϖ are utilized, the source term can be written as $S(\vec{r},t) = S(\vec{r})\cos(\overline{\omega}t)$, and the fluence rate will be also modulated in time, that is $\phi(\vec{r},t) = \phi(\vec{r})\cos(\overline{\omega}t)$. In this case, to simplify the calculations, it is customary to introduce complex time dependence, that is

$$S(\vec{r},t) = S(\vec{r})\exp(-\mathrm{i}\varpi t),$$
$$\phi(\vec{r},t) = \phi(\vec{r})\exp(-\mathrm{i}\overline{\omega}t). \quad (5.39)$$

Introducing Eqs. 5.39 in Eq. 5.38 yields the *frequency domain diffusion* equation:

$$D\vec{\nabla}^2\phi(\vec{r}) + k(\overline{\omega})^2\phi(\vec{r}) = -\frac{S(\vec{r})}{D}, \quad (5.40)$$

where we have introduced

$$k(\overline{\omega})^2 = \frac{-\mu_a}{D} + \frac{\mathrm{i}\overline{\omega}}{cD}. \quad (5.41)$$

We note that the angular frequency ϖ indicates the frequency of modulation of the light intensity and is different from the natural angular frequency ω of the photons emitted by the light source at wavelength λ so that $\omega = 2\pi c/\lambda$.

Equations (5.38) and (5.40) are general equations valid for homogeneous media. Equation (5.40) hold for sources of constant intensity, that is constant wave (CW) sources, assuming $\overline{\omega} = 0$.

5.2.4.2 Heterogeneous medium

For a medium where the optical properties μ_a and D can be spatially dependent, $\mu_a = \mu_a(\vec{r}), D = D(\vec{r})$, as is the case for *in vivo* imaging, the substitution of Eq. (5.34) with a spatially diffusion coefficient in the conservation of energy principle Eq. (5.29) yields the following diffusion equation:

$$\frac{1}{c}\frac{\partial\phi(\vec{r},t)}{\partial t} - \vec{\nabla}D(\vec{r})\vec{\nabla}\phi(\vec{r},t) + \mu_a(\vec{r})\phi(\vec{r},t) = S(\vec{r},t), \quad (5.42)$$

which is similar to the diffusion equation for a homogeneous medium excepted that the diffusion coefficient cannot be taken out of the gradient operator $\vec{\nabla}$ due to the spatial dependence of $D(\vec{r})$. Similar to the previous section, a time frequency equation can be derived from Eq. (5.42). Equation (5.42) is the starting point of most tomographic reconstruction algorithms. It can be used to derive a *forward model*, where the fluence rate inside a diffusive medium can be calculated as a function of the optical property variation of the medium and the source distribution, and can be subsequently utilized to solve for the *inverse problem* where measurements at the tissue boundary are used to extract the optical properties of the medium.

5.2.5 Solutions of the diffusion equation

The analytical derivation of solutions of the diffusion equation for the different possible cases present in *in vivo* optical imaging and tomography is beyond the scope of this chapter. However, some representative and commonly utilized solutions are herein described because they illustrate how such solutions can be used for image formation as appropriate for bio-optical imaging and tomography. The solutions considered are written for light sources whose intensity is sinusoidaly modulated with a pulsation $\overline{\omega}$. Constant intensity (CW) sources are then a sub-category of these solutions where the pulsation $\overline{\omega}$ is zero.

Solutions for homogenous media can be easily derived for the diffusion equation and are summarized in Table 5.2.2, in the frequency domain for an infinite homogeneous diffusive medium and assuming a point source (i.e. a focused laser spot). For comparison between the scattering and the non-scattering cases, we recapitulated the corresponding solutions obtained for non-scattering media.

In the frequency domain, the generic propagation of a photon field in an infinite homogeneous medium has well-known solutions (Morse and Feshbach, 1953) $\phi_{\overline{\omega}}(r,t)$ to the diffusion equation (5.40) for a point source located at the origin of coordinate. For a non-bounded scattering medium, this function has a well-known spherical distribution solution:, that is

$$\phi_{\overline{\omega}}(\vec{r},t) = \frac{A}{D}\,\mathrm{Re}\left[\frac{e^{\mathrm{i}(k(\overline{\omega})r - \overline{\omega}t)}}{r}\right], \quad (5.43)$$

where $r = \|\vec{r}\|$, $k(\overline{\omega})^2 = \frac{-\mu_a}{D} + \frac{\mathrm{i}\overline{\omega}}{cD}$ and A is the amplitude of the source of light.

Equation (5.43) is in many respects a fundamental equation as it describes the basic characteristics of propagation of a photon field generated by a point source and propagating inside a diffuse medium with characteristics described by the wave number $k(\overline{\omega})$. Equation (5.43) reveals the basic non-linear dependence of the field strength (and phase because $k(\overline{\omega})$ has a real part at non-zero angular frequencies) to depth or propagation distance, r, and optical

Table 5.2.2 Analytical formula describing the propagation of light in non-scattering and scattering media for a frequency modulated intensity source of angular frequency ϖ in function of the distance $r = \|\vec{r}\|$.

Time modulated intensity of pulsation ϖ	Electric field: $\vec{E}$	Fluence rate: ϕ
Non-scattering, absorbing media	$\vec{E}_\omega(\vec{r},t) = \mathrm{Re}\left[E_0(t)\frac{e^{i(k(\omega)r-\omega t)}}{r}\right]\hat{u}$ $E_0(t) = E_0\cos(\varpi t)^{1/2}$ $k(\omega) = (n' + in'')\omega/c_0$ $= n'\omega/c_0 + i\mu_a/2$	$\phi_{\varpi}(\vec{r},t) = A\,\mathrm{Re}\left[\frac{e^{-i\varpi t}e^{-\mu_a r}}{r^2}\right]$ $A = E_0^2$
Scattering and absorbing media: Ballistic component	$\vec{E}_\omega(\vec{r},t) = \mathrm{Re}\left[E_0(t)\frac{e^{i(k(\omega)r-\omega t)}}{r}\right]\hat{u}$ $k(\omega) = n'\omega/c_0 + i(\mu_s + \mu_a)/2$	$\phi_{\varpi}(\vec{r},t) = A\,\mathrm{Re}\left[\frac{e^{-i\varpi t}e^{-(\mu_s+\mu_a)r}}{r^2}\right]$ $A = E_0^2$
Highly scattering and absorbing media: Diffuse component DPDW wave	No elementary analytical formula available(Tsang, Kong and Shin, 1985; Tsang and Kong, 2001).	$\phi_{\varpi}(\vec{r},t) = \frac{A}{D}\,\mathrm{Re}\left[\frac{e^{i(k(\varpi)r-\varpi t)}}{r}\right]$ $k(\varpi) = \sqrt{-\frac{\mu_a}{D} + i\frac{\varpi}{cD}}$

properties, μ_a, D . Mathematically, these modulated diffuse photon density wave (DPDW) are very similar to electromagnetic waves; for a non-scattering medium and a non modulated source (i.e. $\varpi = 0$), the electric vector for a point source is given by (see Table 5.2.2)

$$\vec{E}_\omega(r,t) = E_0\mathrm{Re}\left[\frac{e^{i(k(\omega)r-\omega t)}}{r}\right]\hat{u}, \qquad (5.44)$$

where the wave number k is $k(\omega) = n\omega/c_0$ (n can be complex for an absorbing medium).

Except for the polarization component $\hat{u}$ and the definition of the wave number k, the expressions (5.43) and (5.44) are mathematically identical. (However, it is important to notice that the equivalent of the natural pulsation ω of the electromagnetic wave for the diffusive wave is the modulated pulsation ϖ and not ω). Because of this similarity, it has been demonstrated that these diffusive waves described by Eq.(5.43) behave like electromagnetic waves, that is they can be reflected/refracted, and scattered at the interface of two *diffusing media* having different optical properties (usually different absorption and/or scattering coefficients) (Oleary *et al.*, 1992, 1993; Boas *et al.*, 1993, 1994, 1997; Tuchin, and Society of Photo-optical Instrumentation Engineers, 2000; Ripoll, Nieto-Vesperinas and Arridge, 2001; Tuchin, 2002; Vo-Dinh, 2003). However, we emphasize that even though mathematically very similar, these waves are physically very different. For example, the interference effect of the electromagnetic wave can be observed in vacuum for CW sources, whereas the diffusive wave interferences are only observed in a scattering medium for a source with time intensity modulation $\varpi \neq 0$.

Solutions in the time-domain have also been derived. However, they typically offer more complex solutions for tomographic imaging (Arridge and Hebden, 1997; Tuchin, 2002; Vo-Dinh, 2003). Therefore, solutions in the frequency domain are generally preferred. Photon fields that are produced in the time-domain, that is using photon pulses, can be easily treated with frequency domain solutions by converting the time-domain data to frequency domain data using the Fourier transform. This generally leads to more manageable tomographic problems, without loss of generality or performance of the approach.

5.2.6 Solutions of the diffusion equation for heterogeneous media

The discussion in the previous paragraph focused on homogeneous diffuse media, namely media where the diffusion coefficient was spatially invariant. These solutions are fundamental for deriving methods for tomography of inhomogeneous media as is the case for *in vivo* tomography. Here, we will focus on analytical solutions derived in the frequency domain for optically heterogeneous media. It is then straightforward to obtain solutions for the CW case (where $\varpi = 0$) or the time domain via the Fourier transform.

5.2.6.1 *Absorptive heterogeneity*

Due to the change in concentration of various tissue chromophores, and in particular haemoglobin, there is significant spatial variation of optical properties in living tissue. The goal of finding appropriate solutions for tomography of this absorption variation is to express the fluence rate produced by the variation of absorption in the medium as a function of the fluence rate of a reference state. Tissues typically exhibit a reduced scattering coefficient that is significantly higher than the absorption coefficient, that is $\mu_s' \gg \mu_a$. We will also assume that $D \approx 1/3\mu_s'$, that is the diffusion coefficient is independent of absorption. If the absorption reference state is $\mu_{a0}(\vec{r})$ (which can be inhomogeneous), we a looking for the effect of an absorption variation $\delta\mu_a(\vec{r})$ to the fluence rate of the photon field established in the medium. For example we would like to be able to calculate how much the photon field will change, at different positions in the medium and the boundary of the medium if there is a local change in the concentration of a chromophore. The total, spatial variant absorption is of the medium is obviously given by

$$\mu_a(\vec{r}) = \mu_{a0}(\vec{r}) + \delta\mu_a(\vec{r}). \quad (5.45)$$

Assuming a unit point source, it can be shown that the field dependence on the perturbation $\delta\mu_a(\vec{r})$ is given by (Morse and Feshbach, 1953; Kak and Slaney, 2001; Tuchin, 2002; Vo-Dinh, 2003)

$$\phi^{\varpi}(\vec{r},\vec{r}_s) = \phi_0^{\varpi}(\vec{r},\vec{r}_s) - \int d^3\vec{r}_1 G_0^{\varpi}(\vec{r},\vec{r}_1)\frac{\delta\mu_a(\vec{r}_1)}{D}\phi^{\varpi}(\vec{r}_1,\vec{r}_s), \quad (5.46)$$

where $G_0^{\varpi}(\vec{r},\vec{r}_s)$ is the Green function (Morse and Feshbach, 1953) of the diffusion equation, Eq.(5.40), that is

$$G_0^{\varpi}(\vec{r},\vec{r}_s) = \frac{\exp(ik(\varpi)||\vec{r}-\vec{r}_s||)}{4\pi||\vec{r}-\vec{r}_s||}, \quad (5.47)$$

and $\phi_0^{\varpi}(\vec{r},\vec{r}_s)$ is the fluence rate in a reference medium of absorption $\mu_{a0}(\vec{r})$. Therefore, if we can calculate or measure $\phi_0^{\varpi}(\vec{r},\vec{r}_s)$ for a reference medium, we can then express the fluence rate for a different state of absorption in the medium using Eq. (5.46). Commonly, the reference state is assumed in a homogeneous medium, which can then utilize the solutions described in Section 5.2.5 derived with the appropriate boundary conditions.

Equation (5.46) is a non-linear equation on the parameter $\delta\mu_a(\vec{r})$; the fluence rate ϕ^{ϖ} appears on both sides of the equation. Appropriate numerical techniques (Atkinson, 1976; Colton and Kress, 1983; Press and Numerical Recipes Software, 1997) can be used to calculate this equation in an iterative manner. Another common approach is to assume a first-order approximation where it is postulated that $\phi^{\varpi}(\vec{r}_1,\vec{r}_s) \approx \phi_0^{\varpi}(\vec{r}_1,\vec{r}_s)$. This approximation is known as the Born approximation which is generally valid when $\delta\mu_a(\vec{r}) \ll \mu_{a0}(\vec{r})$ (Messiah, 1976; Kak and Slaney, 2001). Then we reach a popular equation in problems of diffuse optical tomography, that is

$$\phi^{\varpi}(\vec{r},\vec{r}_s) = \phi_0^{\varpi}(\vec{r},\vec{r}_s) - \int d^3\vec{r}_1 G_0^{\varpi}(\vec{r},\vec{r}_1)\frac{\delta\mu_a(\vec{r}_1)}{D}\phi_0^{\varpi}(\vec{r}_1,\vec{r}_s). \quad (5.48)$$

In this case, we are in the *perturbation* or linear regime, and the fluence rate ϕ^{ϖ} is only a function of ϕ_0^{ϖ} . Experimentally, $\phi^{\varpi}(\vec{r},\vec{r}_s)$ is the quantity measured for light propagating through tissues. Equation (5.48) therefore relates this measurement to the unknown distribution of absorption heterogeneity $\delta\mu_a(\vec{r})$. Solving (inverting) Equation (5.48) can therefore give three dimensional absorption images. Inversion is described in Section 5.3.

5.2.6.2 *Scattering heterogeneity*

The derivation of solutions for heterogeneous scattering is similar to the derivation of solutions for absorbing heterogeneity. The goal here is to express the fluence rate detected from a medium with a spatially varying diffusion coefficient $D(\vec{r})$. This derivation assumes a known fluence rate from a reference medium with a diffusion coefficient $D_0(\vec{r})$ and an unknown perturbation of this unknown coefficient $\delta D(\vec{r})$, so that $D(\vec{r}) = D_0(\vec{r}) + \delta D(\vec{r})$. As with the absorptive heterogeneity, we will suppose that the diffusion coefficient is independent of the absorbing coefficient, that is $\mu_{s0}'(\vec{r}) \gg \mu_a$ and $\mu_s'(\vec{r}) \gg \mu_a$ such that

$$D(\vec{r}) = \frac{1}{3\mu_s'(\vec{r})}, \quad D_0(\vec{r}) = \frac{1}{3\mu_{s0}'(\vec{r})}. \quad (5.49)$$

Using this approximation, we can derive, similar to the previous section, an equation linking the fluence rate ϕ^{ϖ} for the distribution $D(\vec{r})$ with the fluence rate ϕ_0^{ϖ} for the distribution $D_0(\vec{r})$:

$$\phi^{\varpi}(\vec{r},\vec{r}_s) = \phi_0^{\varpi}(\vec{r},\vec{r}_s) + \int_V \vec{\nabla} G_0(\vec{r},\vec{r}_1)\cdot\frac{\delta D(\vec{r}_1)}{D_0(\vec{r}_1)}\vec{\nabla}\phi^{\varpi}(\vec{r}_1,\vec{r}_s)d^3\vec{r}_1. \quad (5.50)$$

For small variation of the diffusion coefficient $\delta D(\vec{r}) \ll 1$, a first Born approximation can be used to calculate $G(\vec{r}, \vec{r}_s)$:

$$\phi^{\overline{\omega}}(\vec{r}, \vec{r}_s) = \phi_0^{\overline{\omega}}(\vec{r}, \vec{r}_s) + \int_V \vec{\nabla} G_0(\vec{r}, \vec{r}_1) \cdot \frac{\delta D(\vec{r}_1)}{D_0(\vec{r}_1)} \vec{\nabla} \phi_0^{\overline{\omega}}(\vec{r}_1, \vec{r}_s) d^3 \vec{r}_1. \quad (5.51)$$

If both the absorption and the scattering coefficient change, a linear combination of Eqs. (5.48) and (5.51) can be used to calculate the total fluence rate detected from the medium.

5.2.6.3 *Fluorescence heterogeneity*

Similar solutions can be reached for describing the photon field detected at the emission wavelength due to the distribution of fluorochromes inside diffusive media. This derivation assumes that fluorophores attain a single lifetime (Das, Liu and Alfano, 1997; Oleary *et al.*, 1996) and that the fluorescent radiation is well separated in energy from that of incident photons, so that the possibility of the excitation of fluorophores by the fluorescent re-emission can be safely ignored. Here, fluorophore and chromophore absorptions are treated separately and fluorophore-induced scattering is assumed to be negligible.

As described in Section 5.1.3, if $\phi_{\lambda_{ex}}^{\text{pulse}}(\vec{r}_1, t_1)$ is the fluence rate produced by a short *pulse* of light at the wavelength λ_{ex} , then the fluence rate over time at the location of the fluorochrome $\vec{r}_1$ is

$$\mu_{\text{a}}^{\text{fluo},\lambda_{ex}}(\vec{r}_1)\eta_{\text{fluo}} \frac{\exp(-(t_2 - t_1)/\tau_{\text{fluo}})}{\tau_{\text{fluo}}} \phi_{\lambda_{ex}}^{\text{pulse}}(\vec{r}_1, t_1). \quad (5.52)$$

Here, η_{fluo} and τ_{fluo} are the quantum efficiency and lifetime of the fluorochrome (assumed here spatially invariant), and $\mu_a^{\text{fluo},\lambda_{ex}}(\vec{r})$ is the fluorochrome absorption which is directly proportional to fluorochrome concentration C.

For a modulated source of pulsation $\overline{\omega}$, if $\phi_{\lambda_{ex}}^{\overline{\omega}}(\vec{r}_1, t_1)$ is the fluence rate produced by this source, then the fluence rate produced by the fluorochrome is given by the Laplace transform of Eq. (5.52) at the point $i\overline{\omega}$, that is

$$\frac{\mu_a^{\text{fluo},\lambda_{ex}}(\vec{r}_1)\eta_{\text{fluo}}}{1 - i\overline{\omega}\tau_{\text{fluo}}} \phi_{\lambda_{ex}}^{\text{pulse}}(\vec{r}_1, t_1). \quad (5.53)$$

An expression for the fluence rate at the boundary of the scattering medium can be derived in integrating Eq. (5.53) overall the scattering medium volume and in propagating the fluence rate from the fluorochrome to the boundary using the Green function Eq.(5.47), and we obtain

$$\phi^{\overline{\omega}}(\vec{r}, \vec{r}_s) = \int d^3\vec{r}_1 G_{0,\lambda_{fl}}^{\overline{\omega}}(\vec{r}, \vec{r}_1) \frac{\eta_{\text{fluo}}\mu_a^{\text{fluo},\lambda_{ex}}(\vec{r}_1)}{1 - i\overline{\omega}\tau_{\text{fluo}}} \phi_{\lambda_{ex}}^{\overline{\omega}}(\vec{r}_1, \vec{r}_s) . \quad (5.54)$$

To calculate the Green function $G_{0,\lambda_{fl}}^{\overline{\omega}}(\vec{r}, \vec{r}_1)$, the absorption and the scattering coefficient appearing in the factor $k(\overline{\omega})$ of Eq. (5.41) are evaluated at the emission wavelength λ_{fl} . In fact, the optical properties at the excitation λ_{ex} and the emission λ_{fl} wavelength can be different if the excitation and the emission peak of the fluorochrome are spectrally well apart.

A solution that has proven particularly useful for fluorescence tomography of living tissues is the normalized Born approximation, which divides the left part of Eq. (5.54) with a measurement at the excitation wavelength and the right part with the corresponding analytical function solution (Ntziachristos, Weissleder and Mahmood, 2001), that is

$$\phi^{n\text{Born}}(\vec{r}_d, \vec{r}_s) = \left[\frac{\phi_{\lambda_{fl}}^{\overline{\omega}}(\vec{r}_d, \vec{r}_s)}{\phi_{\lambda_{ex}}^{\overline{\omega}}(\vec{r}_d, \vec{r}_s)}\right]_{\text{measured}} = \frac{\eta_{\text{fluo}}}{\phi_{\lambda_{ex}}^{\overline{\omega}}(\vec{r}_d, \vec{r}_s)} \int d^3\vec{r}_1 G_{0,\lambda_{fl}}^{\overline{\omega}}(\vec{r}_d, \vec{r}_1) \frac{\mu_a^{\text{fluo},\lambda_{ex}}(\vec{r}_1)}{1 - i\overline{\omega}\tau_{fl}} \phi_{\lambda_{ex}}^{\overline{\omega}}(\vec{r}_1, \vec{r}_s) . \quad (5.55)$$

The advantage of Eq. (5.55) is that it is insensitive to several spatially dependent experimental factors and theoretical assumptions, in particular non-uniform illumination, light coupling issues on tissue, theoretical inaccuracies in modelling the boundary conditions or the presence of tissue optical heterogeneity. For example, it has been shown that Eq. (5.55) can offer better accuracy compared to Eq. (5.54) in imaging the fluorescence distribution in diffusive media when there is high variation in background absorption and scattering (Soubret *et al.*, 2005).

5.2.7 Boundary conditions and analytical versus numerical solutions

The derivations presented in the above section contain functions to describe the fluence rate of diffuse photon waves in diffuse media. A simple form of this fluence rate is given by Eq. (5.43) derived for an infinite homogenous medium. However, tissues are bounded by air-tissue interfaces that change the photon propagation characteristics compared to the ones predicted by solutions calculated for infinite media.

The presence of boundaries can be accounted for by calculating Green's functions that incorporate their effect, without the loss of generality of the solutions in Eqs. (5.46), (5.48), (5.50), (5.51) and (5.54)–(5.55). There are several approaches to calculate these improved Greens functions. The description of light propagation in diffusing media presented in the previous chapters is usually termed as 'analytical' because explicit equations are derived for relating the photon measurements at the tissue boundary to the internal distribution of optical contrast, such as the absorption, scattering or fluorescence variation compared to that of a reference medium. The analytical expressions are usually computationally efficient and can be easily implemented. In these cases the Kirchoff approximation or higher order methods can be used to incorporate the effects of a surface into the photon propagation in tissues (Ripoll *et al.*, 2001a,b; Schulz, Ripoll and Ntziachristos, 2003, 2004).

Alternatively, numerical solutions of the diffusion equation can be used to incorporate the effects of boundaries. Typical methods for these solutions include finite difference and finite element methods implementing the diffusion equation, with the finite element method, generally offering better implementation flexibility and performance (Arridge, 1999; Arridge *et al.*, 2000a,b; Arridge and Schweiger, 1995; Bluestone *et al.*, 2004a,b; Dehghani *et al.*, 1999, 2000; Eppstein *et al.*, 2001; Hielscher, 2005; Pogue *et al.*, 1999; Schweiger and Arridge, 1997; Schweiger *et al.*, 1995). Numerical methods are known not only to offer significant computational burden but also to attain superior flexibility in iteratively calculating the non-linear problem of Eqs. (5.46) or (5.50) (instead of the linear approximation offered by Eqs. (5.48) and (5.51) as well). In this respect limitations of the use of Green's functions calculated for infinite media can be addressed, although the computational aspects of such implementations are challenging. Numerical solutions are also used to overcome the known limitations of the diffusion approximation, by solving the radiative transfer equation (Abdoulaev and Hielscher, 2003; Klose and Hielscher, 2002; Klose *et al.*, 2002), or by using Monte Carlo models (Wang, Jacques and Zheng, 1995) although these methods tend to further increase the computation burden which limits the applicability of such methods to large scale problems.

5.3 Reconstruction and inverse problem

Section 5.2 presented theoretical solutions to the diffusion equation that can describe light propagation in diffusing tissue. Therefore, it provides the *forward model*, that is it can predict the measurements for a known optical contrast distribution. Tomography however presents the *inverse problem* where the photon distribution on the boundary is known, and this data is used to *reconstruct* the unknown distribution of optical properties inside the tissue. The use of physical models for the description of photon propagation yields a quantitative inverse problem for diffuse optical tomography.

5.3.1 Linear inversion

The typical steps of an inversion of the linear equations described by Eqs. (5.48), (5.51), (5.54), and (5.55) are summarized in Figure 5.3.1. Data collection is followed by a pre-processing step to reject noise and formulate the appropriate data vector. For example, in the normalized Born approximation, it is used to create a ratio of measurements at emission and excitation wavelengths. Then, after calculating a forward model F, an iterative process estimates the fluorescence distribution or the absorption/scattering coefficient $\vec{x}$ inside the medium, and compares these predictions to the experimental measurements. The iteration process ends when the predicted measurements are close enough to the experimental values.

The calculation of the forward model is typically based on the discritization of the volume integrals presented in Eqs. (5.48), (5.51), (5.54), and (5.55). Discritization segments the volume of interest into voxel elements (voxels) and calculates the contribution of some optical contrast (absorption, scattering, fluorescence) in each voxel to the photon measurement for each source-detector pair employed. This process effectively calculates *weights* that estimate the importance of each voxel in the volume for each particular measurement. Mathematically, this process converts the volume integrals in Eqs. (5.48), (5.51), (5.54), and (5.55) to summations of weights, multiplying each voxel value by a corresponding weight. For example, Eq. (5.55) after discritization can be written as

$$\phi^{n\text{Born}}(\vec{r}_d,\vec{r}_s) = \frac{1}{\phi^{\overline{\omega}}_{\lambda_{\text{ex}}}(\vec{r}_d,\vec{r}_s)} \sum_{n=1}^{N} G^{\overline{\omega}}_{0,\lambda_{fl}}(\vec{r},\vec{r}_n) \times \frac{\eta_{\text{fluo}}\mu_a^{\text{fluo},\lambda_{\text{ex}}}(\vec{r}_n)}{1 - i\overline{\omega}\tau_{\text{fluo}}} \phi^{\overline{\omega}}_{\lambda_{\text{ex}}}(\vec{r}_n,\vec{r}_s)\cdot h_v, \quad (5.56)$$

Figure 5.3.1 Different steps involved in a linear reconstruction algorithm

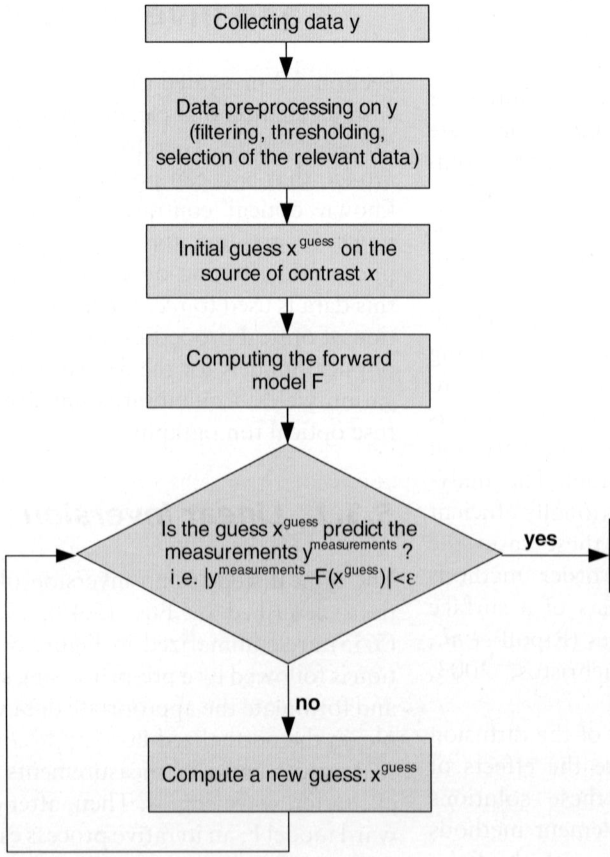

where the time decay τ_{fluo} of the fluorochrome is supposed to be constant, h_v is the volume of each voxel, N is the total number of voxels and $\vec{r}_n$ is now a vector pointing to discrete locations denoting the centre of each voxel. For M measurements, M equations like in Eq. (5.56) can be written which in matrix form is simply given as

$$\vec{y} = W\vec{x}, \qquad (5.57)$$

W is an $M \times N$ matrix of weights, the *weight matrix*, $\vec{x} = (x_i)_{i=1...N}$ is a vector of the optical property in each voxel and $\vec{y}$ is the vector of measurements for M source-detector pairs. In the example of Eq. (5.56), the fluorescence distribution $\vec{x}$ is expressed as a function of the fluorochrome absorption:

$$x_j = \frac{\eta_{\text{fluo}} \mu_a^{\text{fluo},\lambda_{\text{ex}}}(\vec{r}_j)}{1 - i\overline{\omega}\tau_{\text{fluo}}}, \qquad (5.58)$$

and the weight matrix can be defined as a function of the Green's functions of the system such that

$$W_{ij} = h_v \cdot G_{0,\lambda_{\text{fl}}}^{\overline{\omega}}(\vec{r}_i^{\,d}, \vec{r}_j)\phi_{\lambda_{\text{ex}}}^{\overline{\omega}}(\vec{r}_j, \vec{r}_i^{\,s}) / \phi_{\lambda_{\text{ex}}}^{\overline{\omega}}(\vec{r}_i^{\,d}, \vec{r}_i^{\,s}), \quad (5.59)$$

where $G_{0,\lambda_{\text{fl}}}^{\overline{\omega}}(\vec{r}_i^{\,d}, \vec{r}_j)$ describes the propagation of light through the mesh point located at $\vec{r}_j$ to the detector located at r_i^d in the diffusion approximation model, and $\phi_{\lambda_{\text{ex}}}^{\overline{\omega}}(\vec{r}_j, \vec{r}_i^{\,s})$ describes the propagation of light through the source point located at $\vec{r}_i^{\,s}$ to the mesh point located at $\vec{r}_j$. Figure 5.0.4(d) depicts the relative weights for one source and one detector, demonstrating the bulk propagation of photons through a diffuse circular medium. Brighter areas correspond to voxels that can significantly affect the measurement, whereas darker areas do not contribute much to this measurement. These patterns are known as 'banana patterns' and are characteristic of the volume sampled from each source detector pair in optical tomography problems. In contrast to profiles seen when using high-energy photons, for example X-rays, these diffuse optical profiles 'bend', depending on the exact location of source and detector, possibly covering the entire volume, even with a single source, if sufficient detectors are placed around the boundary. In addition, they appear wide, and the width progressively increases when moving away from the source and detector due to the characteristics of diffusive propagation in media. This variable width contributes to the characteristic depth-dependent resolution common to optical tomography images.

In Eq. (5.57), the unknown vector of voxels $\vec{x}$ is the unknown distribution of the optical property imaged. Solution (inversion) of this system therefore yields the three-dimensional image of optical property distribution. The inverse problem can be seen as an optimization problem (Bertero and Boccacci, 1998; Hansen, 1998; Kaipio and Somersalo, 2005; Vogel, 2002) where the object $\hat{x}$ to be reconstructed is found as the minimum of a 'cost function' C, that is

$$\hat{x} = \arg \min_{\vec{x}^{gs} \in \Omega} C(\vec{x}^{gs}), \tag{5.60}$$

where the vector $\vec{x}^{gs} = (x_i)_{i=1...M}$ contains an estimation (guess) of the value. One of the simpler and useful cost functions is based on a least square formulation, written as

$$C(\vec{x}^{gs}) = ||\vec{y} - F(\vec{x}^{gs})||^2. \tag{5.61}$$

In this case the function, $F(\vec{x}^{gs}) = W\vec{x}^{gs}$ is the forward calculation (prediction) of measurement assuming an estimate of optical parameters $\vec{x}^{gs}$. Therefore the estimated values $F(\vec{x}^{gs})$ for an assumed distribution $\vec{x}^{gs}$ are compared to the actual measurements $\vec{y}$, and this difference is minimized by updating the vector $\vec{x}^{gs}$ until it reaches a minimum.

The inverse problem in diffuse optical tomography is an ill-posed problem. This means, in the general sense, that the problem is very sensitive to numerical error and needs to be treated with a regularization process for efficient inversion. Tikhonov regularization is one of the best-known regularization approaches where a constraint on the fluctuation of the norm $||\vec{x}^{gs}||$ is added to the cost function:

$$C(\vec{x}^{gs}) = ||\vec{y} - F(\vec{x}^{gs})||^2 + \lambda||\vec{x}^{gs}||^2. \tag{5.62}$$

Minimizing the function Eq. (5.62), which depends quadratically on the unknowns $\vec{x}$, is a linear algebra problem because it can be formulated as finding the zeros of a system equation given by the gradient of Eq. (5.62). This system can be solved by a large set of iterative method such as the Algebraic Reconstruction Technique (ART) (Herman, 1997), the Krylov subspace methods (Hansen, 1998; Saad, 2003) like the Generalized Minimum Residual method (GMRES) and conjugate type gradient methods (BICGSTAB, LSQR, etc.). Direct inversion of the system can also be achieved with Singular Value Decomposition methods (Hansen, 1992; Golub, 1989), which are particularly efficient for small-scale problems (for weight matrix smaller than 2000 × 2000 elements). A trade-off for the value of λ has to be found between a very noisy reconstructed solution for low λ values and an over smoothed solution for high values of λ . Automatic numerical methods λ , such as the L curve analysis (Hansen, 1992), have been developed to find the best value of λ but in increasing the computation burden and without guarantying that the optimal value of λ will be found.

The Tikhonov regularization in Eq. (5.62) is a particular case of a penalty method which can be more generally written as

$$C(\vec{x}^{gs}) = ||\vec{y} - F(\vec{x}^{gs})||^2 + Q(\vec{x}^{gs}), \tag{5.63}$$

where $Q(\vec{x}^{gs})$ can herein include some more generic constraints also including prior information that may be available, such as distribution constraints based for example on anatomical images of the volume imaged optically (Arridge, 1999; Pogue *et al.*, 1999; Bdoulaev and Hielscher, 2003; Klose and Hielscher, 2002; Arridge and Schweiger, 1998; Arridge and Simmons, 1997; Barbour *et al.*, 1995; Hielscher and Bartel, 2001; Li *et al.*, 2003; Hielscher, Klose and Hanson, 1999; Klose and Hielscher, 1999; Schweiger and Arridge, 1999; Schweiger, Gibson and Arridge, 2003; Milstein *et al.*, 2002, 2003; Roy and Sevick-Muraca, 1999, 2001). Using this penalty function, the constraint optimization problem is transformed to an unconstrained optimization problem that can be efficiently inverted (Bertero and Boccacci, 1998; Kaipio and Somersalo, 2005; Vogel, 2002; Fletcher, 1987; Nocedal and Wright, 1999).

5.3.2 Non-linear inversion

The description of Section 5.3.1 described a generic inversion method for the linearized problem of Eqs. (5.48), (5.51), (5.54), and (5.55). However, when reconstructing the absorption or scattering contrast, inversion based on the non-linear equations (5.46) and (5.50) is sometimes considered for improved imaging accuracy. Typically, an inversion procedure in this case considers an additional iterative step, that is it first iteratively calculates a vector $\vec{x}^{gs}$ as discussed in Section 5.3.1, and then uses this solution to compute a new forward model F. The updated forward model is calculated by incorporating changes that are due to the effects of the contrast seen in $\vec{x}^{gs}$. Then a new vector $\vec{x}^{gs}_{\text{new}}$ is calculated based on the updated forward model, and this process is repeated until a convergence criterion is satisfied. The forward model update is generally performed by a finite difference or finite-element procedures but could also be performed based on Eqs. (5.46) and (5.50).

5.4 Fluorescence molecular tomography (FMT)

Of importance to imaging of vertebrates, in particular small animal imaging is the use of fluorescence technologies for imparting molecular contrast *in vivo*. As however discussed in the introduction, simple photographic imaging offers a single view imaging and cannot resolve depth. This has two major consequences. First, superficial activity may block or reduce contrast compared to deep-seated activity. Second, as was shown by Eq. (5.54), the fluorescence intensity recorded depends linearly on fluorochrome concentration (or fluorochrome amount present in a lesion), but it has a strong non-linear dependence to lesion depth and to the optical properties of the lesion and the surrounding tissue. Therefore, the resulting image may be significantly distorted compared to the true underlying activity. Fluorescence Molecular Tomography has therefore evolved as a tomographic method combining the theoretical mainframe described in Sections 5.2 and 5.3 with advanced instrumentation in order to overcome many of the limitations of planar imaging (epi-illumination and transillumination) and yield a robust and quantitative modality for imaging fluorescent reporters *in vivo*. The method illuminates and collects photon signals at multiple projections and offers three-dimensional, quantitative fluorescence images *in vivo*. Some of the key-building elements are discussed in the following sections.

5.4.1 Free-space and non-contact approaches

Original tomographic systems and methods were based on the use of fibres to couple light to and from tissue and the use of matching fluids to improve fibre coupling or simplify the boundary conditions used in the forward problem. As the technology evolved, it moved away from the use of fibres and instead it implemented flying spot illumination and CCD-based detection using *non-contact* technology and multi-view imaging. In addition, while several data collection implementations utilized matching fluids to simplify experimental and theoretical requirements, evolution of theories and methodologies also allowed collection assuming only air-tissue interfaces (i.e. at the absence of matching fluids), a technology denoted as *free-space*. Depending on the application, fibre or free-space systems can be used for data collection. Generally however, free-space systems collect higher information content measurements leading to improved imaging performance.

Free space imaging requires the collection of the surface geometry which can be achieved using photogrammetry or contour mapping techniques. This information is combined with the appropriate theoretical models in order to obtain accurate description of the forward model of photon propagation in diffuse media and air (Ripoll *et al.*, 2003; Schultz, Ripoll and Ntziachristos, 2003) as described in Sections 5.2 and 5.3. These techniques are essential for offering experimental simplicity while allowing for multi-projection viewing and high-spatial sampling of photon fields. An imaging example is shown in Figure 5.4.1.

5.4.2 Complete projection tomography

To achieve superior imaging performance, it is important to illuminate tissue using a large number of projections and detect signal around the tissue boundary, similar to other tomographic techniques such as X-ray CT, PET or SPECT. Typical geometries employed for tomography are shown in Figure 5.0.4. Reflectance (Figure 5.0.4(a)) and limited angle projection (Figure 5.0.4(b)) approaches can be easily implemented using simple theoretical models for modelling the boundary conditions. Non-contact and free-space technologies however facilitate the theoretical mainframe in order to practically implement a significantly larger number of projections (Figure 5.0.4(c)) for fluorescence and more generally for diffuse optical tomography. Cylindrical geometries have been implemented in the past using fibres for illumination and detection. However, the combination of CCD-cameras and non-contact sources yield a superior data set and improved imaging capacity.

A typical complete projection tomography free-space scanner is shown in Figure 5.4.1. The animal is rotated in front of the illumination path based on a beam scanner device that spatially scans the position of a collimated or focused laser beam on the animal surface. Photon detection is based on a CCD camera that collects light transilluminating the animal at different projections over 360°. Obviously the optics can be also mounted on a rotational gantry so that a mouse placed on the horizontal position could be imaged.

Combination of modalities with complementing features is a very attractive strategy in fluorescence

Figure 5.4.1 Complete projection free-space tomography. (a) A typical scanner based on CCD capture technology and laser beam scanner. (b) Superposition of a tomographic image with the capture mouse surface. In this case fluorescence contrast is due to a fluorescent tube inserted post-mortem through the esophagues (blue arrow) and an additional smaller tube implanted sub-surface (red arrow)

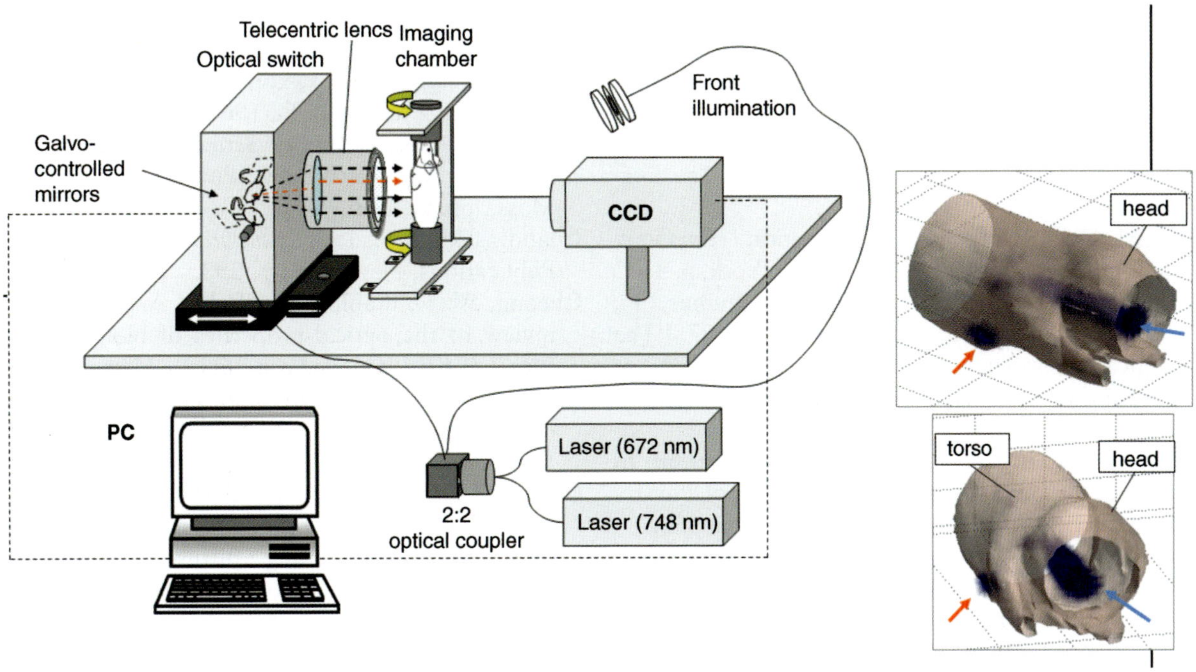

tomography. Typically, fluorescence tomography offers depth-dependent resolutions of several hundred microns to millimeters for small animal imaging (Graves *et al.*, 2003.), and this number worsens when imaging larger volumes. In addition, fluorescence contrast is mainly functional or molecular. Therefore, co-registration with other modalities that reveal anatomy can be helpful in better understanding the source of contrast. Examples of multi-modality imaging using FMT are shown in Chapter 8.

References

Abdoulaev, G. S., Hielscher, A. H., 2003. "Three-dimensional optical tomography with the equation of radiative transfer." *J. Elect. Imaging* **12**(4), 594–601.

Aronson, R., Corngold, N., 1999. "Photon diffusion coefficient in an absorbing medium." J. Opt. Soc. Am. A-Opt. *Image Sci. Vision* **16**(5), 1066–1071.

Arridge, S. R., 1999. "Optical tomography in medical imaging." *Inverse Problems* **15**(2), R41–R93.

Arridge, S. R., Dehghani, H., Schweiger, M., Okada, E., 2000a. "The finite element model for the propagation of light in scattering media: A direct method for domains with nonscattering regions." *Med. Phys.* **27**(1), 252–264.

Arridge, S. R., Hebden, J. C., 1997. "Optical imaging in medicine. 2. Modelling and reconstruction." *Phys. Med. Biol.* **42**(5), 841–853.

Arridge, S. R., Hebden, J. C., Schwinger, M., Schmidt, F. E. W., Fry, M. E., Hillman, E. M. C., Dehghani, H., Delpy, D. T., 2000b. "A method for three-dimensional time-resolved optical tomography." *Int. J. Imaging Syst. Technol.* **11**(1), 2–11.

Arridge, S. R., Schweiger, M., 1995. "Photon-measurement density-functions. 2. Finite-element-method calculations." *Appl. Opt.* **34**(34), 8026–8037.

Arridge, S. R., Schweiger, M., 1998. "A gradient-based optimisation scheme for optical tomography." *Opt. Express* **2**(6), 213–226.

Arridge, S. R., Simmons, A., 1997. "Multi-spectral probabilistic diffusion using Bayesian classification." *Scale-Space theory in Computer Vision*, vol. 33. Springer-Verlag, Berlin, pp. 224–235.

Atkinson, K. E., 1976. "A survey of numerical methods for the solution of Fredholm integral equations of the second kind." *Society for Industrial and Applied Mathematics*, Philadelphia, pp. vii, 230.

Ayata, C., Ozdemir, Y., Dunn, A., Atochin, D. N., Huang, P. L., Muzykantov, V. R., Murciano, J. C., Boas, D. A., Moskowitz, M. A., 2003. "Laser speckle-flowmetry: A novel two-dimensional technique for the study of cerebral blood flow in normal and ischemic mouse brain, in vivo." *Stroke* **34**(1), 251.

Barbour, R. L., Graber, H. L., Chang, J. W., Barbour, S. L. S., Koo, P. C., Aronson, R., 1995. "MRI-guided optical tomography: Prospects and computation for a new imaging method." *IEEE Comput. Sci. Eng.* **2**(4), 63–77.

Bass, M., 1995. "Optical society of America." *Handbook of Optics*, 2nd ed. McGraw-Hill, New York, p. 4 v.

Beek, J. F., Vanstaveren, H. J., Posthumus, P., Sterenborg, H., Vangemert, M. J. C., 1997. "The optical properties of lung as a function of respiration." *Phys. Med. Biol.* **42**(11), 2263–2272.

Bertero, M., Boccacci, P., 1998. *Introduction to Inverse Problems in Imaging*, Institute of Physics Pub., Bristol, UK/Philadelphia, PA, pp. xi, 351.

Bluestone, A. Y., Stewart, M., Lasker, J., Abdoulaev, G. S., Hielscher, A. H., 2004a. "Three-dimensional optical tomographic brain imaging in small animals, part 1: Hypercapnia." *J. Biomed. Opt.* **9**(5), 1046–1062.

Bluestone, A. Y., Stewart, M., Lei, B., Kass, I. S., Lasker, J., Abdoulaev, G. S., Hielscher, A. H., 2004b. "Three-dimensional optical tomographic brain imaging in small animals, part 2: Unilateral carotid occlusion." *J. Biomed. Opt.* **9**(5), 1063–1073.

Boas, D. A., Brooks, D. H., Miller, E. L., Dimarzio, C. A., Kilmer, M., Gaudette, R. J., Zhang, Q., 2001. "Imaging the body with diffuse optical tomography." *IEEE Signal Proc. Mag.* **18**(6), 57–75.

Boas, D. A., Oleary, M. A., Chance, B., Yodh, A. G., 1993. "Interference, diffraction, and imaging with diffuse photon density waves." *Biophys. J.* **64**(2), A357–A357.

Boas, D. A., Oleary, M. A., Chance, B., Yodh, A. G., 1994. "Scattering of diffuse photon density waves by spherical inhomogeneities within turbid media – analytic solution and applications." *Proc. Nat. Acad. Sci. USA* **91**(11), 4887–4891.

Boas, D. A., Oleary, M. A., Chance, B., Yodh, A. G., 1997. "Detection and characterization of optical inhomogeneities with diffuse photon density waves: A signal-to-noise analysis." *Appl. Opt.* **36**(1), 75–92.

Bohren, C. F., Huffman, D. R., 1983. *Absorption and Scattering of Light by Small Particles*, John Wiley & Sons, Inc., New York pp. xiv, 530.

Boppart, S. A., 2004. "Optical coherence tomography - Principles applications and advances." *Minerva Biotechnol* **16**(4), 211–237.

Born, M., Wolf, E., Bhatia, A. B., 1999. *Principles of Optics: Electromagnetic Theory of Propagation, Interference and Diffraction of Light*, 7th (expanded) ed. Cambridge University Press, Cambridge, England, pp. xxxiii, 952.

Case, K. M., Zweifel, P. F., 1967. "Linear transport theory." *Addison-Wesley Series in Nuclear Engineering*. Addison-Wesley Pub. Co., Reading, Mass., pp. ix, 342.

Chandrasekhar, S., 1960. *Radiative Transfer*, Dover Publications, New York, p. 393.

Cheong, W. F., Prahl, S. A., Welch, A. J., 1990. "A review of the optical-properties of biological tissues." *IEEE J. Q. Electron* **26**(12), 2166–2185.

Cherry, S. R., 2004. In vivo molecular and genomic imaging: new challenges for imaging physics. *Phys. Med. Biol.* **49**(3), R13–R48.

Cheung, C., Culver, J. P., Takahashi, K., Greenberg, J. H., Yodh, A. G., 2001. "In vivo cerebrovascular measurement combining diffuse near-infrared absorption and correlation spectroscopies." *Phys. Med. Biol.* **46**(8), 2053–2065.

Colton, D. L., Kress, R., 1983. *Integral Equation Methods in Scattering Theory*. John Wiley & Sons, Inc., New York, pp. xii, 271.

Das, B. B., Liu, F., Alfano, R. R., 1997. "Time-resolved fluorescence and photon migration studies in biomedical and model random media." *Rep. Prog. Phys.* **60**(2), 227–292.

Davison, B., 1957. Neutron transport theory. *International Series of Monographs on Physics*, Clarendon Press, Oxford, p. 450.

Dehghani, H., Arridge, S. R., Schweiger, M., Delpy, D. T., 2000. "Optical tomography in the presence of void regions." *J. Opt. Soc. Am. A-Opt. Image Sci. Vision* **17**(9), 1659–1670.

Dehghani, H., Delpy, D. T., Arridge, S. R., 1999. "Photon migration in non-scattering tissue and the effects on image reconstruction." *Phys. Med. Biol.* **44**(12), 2897–2906.

Duderstadt, J. J., Martin, W. R., 1979. *Transport Theory*, John Wiley & Sons, Inc., New York, pp. x, 613.

Eppstein, M. J., Dougherty, D. E., Hawrysz, D. J., Sevick-Muraca, E. M., 2001. "Three-dimensional Bayesian optical image reconstruction with domain decomposition." *IEEE Trans. Med. Imaging* **20**(3), 147–163.

Fletcher, R., 1987. *Practical Methods of Optimization*, 2nd ed. John Wiley & Sons, Ltd., Chichester, pp. xiv, 436.

Fox, M., 2001. Optical properties of solids. *Oxford Master Series in Condensed Matter Physics*, Oxford University Press, Oxford, New York, pp. xii, 305.

Fujimoto, J. G., 2003. "Optical coherence tomography for ultrahigh resolution in vivo imaging." *Nat. Biotechnol.* **21**(11),1361–1367.

Goldman, R. D., Spector, D. L., 2005. In: D. S. Robert, D. S., Goldman, D., eds. *Live Cell Imaging: A Laboratory Manual*, Cold Spring Harbor Laboratory Press, Cold Spring Harbor, NY, pp. xvi, 631.

Golub, G. H., Van Loan, C. F., 1989. "Matrix computations." *Johns Hopkins Series in the Mathematical Sciences*, vol. 3, 2nd ed. Johns Hopkins University Press, Baltimore, pp. xix, 642.

Graves, E., Ripoll, J., Weissleder, R., Ntziachristos, V., 2003. "A sub-millimeter resolution fluorescence molecular imaging system for small animal imaging." *Med. Phys.* **30**(5), 901–911.

Griffiths, D. J., 1999. *Introduction to Electrodynamics*, 3rd ed. Prentice Hall, Upper Saddler River, NJ, pp. xv, 576.

Hansen, P. C., 1992. "*Analysis of Discrete Ill-Posed Problems by Means of the L-Curve*," Siam Rev **34**(4), 561–580.

Hansen, P. C., 1998. "Rank-deficient and discrete ill-posed problems: numerical aspects of linear inversion." *SIAM Monographs on Mathematical Modeling and Computation. Siam*, Philadelphia, pp. xvi, 247.

Hecht, E., 2002. *Optics*, 4th ed. Addison-Wesley, Reading, Mass., pp. vi, 698.

Herman, G. T., 1997. In: Leondes, C. T., ed. *Algebraic Reconstruction Techniques in Medical Imaging in Medical Imaging Systems Techniques and Applications*, Gordon and Breach Science Publishers, Amsterdam, pp. 1–42.

Hielscher, A. H., 2005. "Optical tomographic imaging of small animals." *Curr. Opin. Biotechnol.* **16**(1), 79–88.

Hielscher, A. H., Bartel, S., 2001. "Use of penalty terms in gradient-based iterative reconstruction schemes for optical tomography." *J. Biomed. Opt.* **6**(2), 183–192.

Hielscher, A. H., Bluestone, A. Y., Abdoulaev, G. S., Klose, A. D., Lasker, J., Stewart, M., Netz, U., Beuthan, J., 2002. "Near-infrared diffuse optical tomography." *Disease Markers* **18**(5–6), 313–337.

Hielscher, A. H., Klose, A. D., Hanson, K. M., 1999. "Gradient-based iterative image reconstruction scheme for time-resolved optical tomography." *IEEE Trans. Med. Imaging* **18**(3), 262–271.

Hollas, J. M., 2004. *Modern Spectroscopy*, 4th ed. John Wiley & Sons, Ltd, Chichester, Hoboken, pp. xxvii, 452.

Hoppert, M., 2003. *Microscopic Techniques in Biotechnology*, Wiley-VCH, Weinheim, Great Britain, pp. xii, 330.

Hulst, H. C.v.d., 1980. *Multiple Light Scattering: Tables, Formulas, and Applications*, Academic Press, New York, p. 2 v.

Hulst, H. C.v.d., 1981. *Light Scattering by Small Particles*, Dover Publications, New York, p. 470.

Ishimaru, A., 1997. "Wave propagation and scattering in random media." *IEEE OUP Series on Electromagnetic Wave Theory*. IEEE Press/Oxford University Press, Oxford, New York, pp. xxv, 574.

Jackson, J. D., 1999. *Classical Electrodynamics*, 3rd ed. John Wiley & Sons, Inc., New York, pp. xxi, 808.

Kaipio, J., Somersalo, E., 2005. *Statistical and Computational Inverse Problems*, Springer, New York, pp. xvi, 339.

Kak, A. C., Slaney, M., 2001. "Principles of computerized tomographic imaging." *Classics in Applied Mathematics*, vol. 33. Society for Industrial and Applied Mathematics, Philadelphia, pp. xiv, 327.

Kerker, M., 1969. *The Scattering of Light and Other Electromagnetic Radiation*. Academic Press, New York, pp. xv, 666.

Kim, Y. L., Liu, Y., Turzhitsky, V. M., Wali, R. K., Roy, H. K., Backman, V., 2005. Depth-resolved low-coherence enhanced backscattering. *Optics Lett.* **30**(7), 741–743.

Klose, A. D., Hielscher, A. H., 1999. "Iterative reconstruction scheme for optical tomography based on the equation of radiative transfer." *Med. Phys.* **26**(8), 1698–1707.

Klose, A. D., Hielscher, A. H., 2002. "Optical tomography using the time-independent equation of radiative transfer – Part 2: Inverse model." *J. Q. Spect. Rad. Transfer* (5), 715–732.

Klose, A. D., Netz, U., Beuthan, J., Hielscher, A. H., 2002. "Optical tomography using the time-independent equation of radiative transfer – Part 1: Forward model." *J. Q. Spect. Rad. Transfer* **72**(5), 691–713.

Lakowicz, J. R., 1999. *Principles of Fluorescence Spectroscopy*, 2nd ed. Kluwer Academic/Plenum, New York, pp. xxiii, 698.

Li, A., Miller, E. L., Kilmer, M. E., Brukilacchio, T. J., Chaves, T., Stott, J., Zhang, Q., Wu, T., Chorlton, M., Moore, R. H., Kopans, D. B., Boas, D. A., 2003. "Tomographic optical breast imaging guided by three-dimensional mammography." *Appl. Opt.* **42**(25), 5181–5190.

Loudon, R., 1983. *The Quantum Theory of Light*, 2nd ed. Oxford science publications. Clarendon Press, Oxford University Press, pp. xiv, 393.

Messiah, A., 1976. *Quantum Mechanics*, North-Holland Pub. Co./John Wiley & Sons, Inc., Amsterdam/New York, p. 2 v.

Milstein, A. B., OH, S., Reynolds, J. S., Webb, K. J., Bouman, C. A., Millane, R. P., 2002. "Three-dimensional Bayesian optical diffusion tomography with experimental data." *Opt. Lett.* **27**(2), 95–97.

Milstein, A. B., Oh, S., Webb, K. J., Bouman, C. A., Zhang, Q., Boas, D. A., Millane, R. P., 2003. "Fluorescence optical diffusion tomography." *Appl. Opt.* 42(16), 3081–3094.

Morse, P. M., Feshbach, H., 1953. "Methods of theoretical physics." *International Series in Pure and Applied Physics*, McGraw-Hill, New York, P. 2 v.

Moseley, H., Sliney, D. H., 1997. "Radiometric quantities and units." *Phys. Med. Biol.* **42**(5), U5–U6.

Murphy, D. B., 2001. *Fundamentals of Light Microscopy and Electronic Imaging*, Wiley-Liss, New York, pp. xii, 368.

Niedre, M., private communication.

Nocedal, J., Wright, S. J., 1999. Numerical Optimization. *Springer Series in Operations Research*, Springer, New York, pp. xx, 636.

Ntziachristos, V., Bremer, C., Weissleder, R., 2003. "Fluorescence imaging with near-infrared light: new technological advances that enable in vivo molecular imaging." *Eur. Radiol.* **13**(1), 195–208.

Ntziachristos, V., Ma, X. H., Yodh, A. G., Chance, B., 1999. "Multichannel photon counting instrument for spatially resolved near infrared spectroscopy." *Rev. Sci. Instrum.* 70(1), 193–201.

Ntziachristos, V., Turner, G., Dunham, J., Windsor, S., Soubret, A., Ripoll, J., Shih, H. A., 2005. "Planar fluorescence imaging using normalized data." *J. Biomed. Optics* **10**(6), 64007.

Ntziachristos, V., Weissleder, R., 2001. "Experimental three-dimensional fluorescence reconstruction of diffuse media by use of a normalized Born approximation." *Opt. Lett.* **26**(12), 893–895.

Oleary, M. A., Boas, D. A., Chance, B., Yodh, A. G., 1992. "Refraction of diffuse photon density waves." *Phys. Rev. Lett.* **69**(18), 2658–2661.

Oleary, M. A., Boas, D. A., Chance, B., Yodh, A. G., 1993. "Refraction of diffuse photon density waves." *Biophys. J.* **64**(2), A357.

Oleary, M. A., Boas, D. A., Li, X. D., Chance, B., Yodh, A. G., 1996. "Fluorescence lifetime imaging in turbid media." *Opt. Lett.* **21**(2), 158–160.

Patterson, M. S., Chance, B., Wilson, B. C., 1989. "Time resolved reflectance and transmittance for the noninvasive measurement of tissue optical-properties." *Appl. Optics* **28**(12), 2331–2336.

Pogue, B. W., Mcbride, T. O., Prewitt, J., Osterberg, U. L., Paulsen, K. D., 1999. "Spatially variant regularization improves diffuse optical tomography." *Appl. Opt.* **38**(13), 2950–2961.

Press, W. H., Numerical Recipes Software (Firm), 1997. *Numerical Recipes in C*, 2nd v2.08 ed. Cambridge University Press, Cambridge, England.

Ripoll, J., Nieto-Vesperinas, M., Arridge, S. R., 2001a. "Effect of roughness in nondiffusive regions within diffusive media." J. Opt. Soc. Am. A-Opt. Image Sci. *Vision* **18**(4), 940–947.

Ripoll, J., Ntziachristos, V., Carminati, R., Nieto-Vesperinas, M., 2001b. "Kirchhoff approximation for diffusive waves." *Phys. Rev. E* **6405**(5).

Ripoll, J., Schultz, R., Ntziachristos, V., 2003. "Free-space propagation of diffuse light: Theory and experiments." *Phys. Rev. Lett.* **91**(10), 103901-1-4.

Roy, R., Sevick-Muraca, E. M., 1999. "Truncated Newton's optimization scheme for absorption and fluorescence optical tomography: Part I theory and formulation." *Opt. Express* **4**(10), 353–371.

Roy, R., Sevick-Muraca, E. M., 2001. "Three-dimensional unconstrained and constrained image-reconstruction techniques applied to fluorescence, frequency-domain photon migration." *Appl. Opt.* **40**(13), 2206–2215.

Rytov, S. M., Kravtsov, Y. A., Tatarskii, V. I., 1987. *Principles of Statistical Radiophysics*, 2. rev. and enl. ed. Springer-Verlag, Berlin; New York, p. v.

Saad, Y., 2003. *Iterative Methods for Sparse Linear Systems*, 2nd ed. Siam, Philadelphia, pp. xviii, 528.

Schultz R, Ripoll, J., Ntziachristos, V., 2004. "Experimental fluorescence tomography of arbitrarily shaped diffuse objects using non-contact measurements." *IEEE Trans. Med. Imaging* **23**(4), 1701–1703.

Schulz, R.B., Ripoll, J., Ntziachristos, V., 2003. "Noncontact optical tomography of turbid media." *Opt. Lett.* **28**(18), 1701–1703.

Schulz, R. B., Ripoll, J., Ntziachristos, V., 2004. "Experimental fluorescence tomography of tissues with noncontact measurements." *IEEE Trans. Med. Imaging* **23**(4), 492–500.

Schweiger, M., Arridge, S. R., 1997. "The finite-element method for the propagation of light in scattering media: Frequency domain case." *Med. Phys.* **24**(6), 895–902.

Schweiger, M., Arridge, S. R., 1999. "Optical tomographic reconstruction in a complex head model using a priori region boundary information." *Phys. Med. Biol.* **44**(11), 2703–2721.

Schweiger, M., Arridge, S. R., Hiraoka, M., Delpy, D. T., 1995. "The finite-element method for the propagation of light in scattering media – Boundary and source conditions." *Med. Phys.* **22**(11), 1779–1792.

Schweiger, M., Gibson, A., Arridge, S. R., 2003. "Computational aspects of diffuse optical tomography." *Comput. Sci. Eng.* **5**(6), 33–41.

Sheng, P., 1990. "Scattering and localization of classical waves in random media." *World Scientific Series on Directions in Condensed Matter Physics*, vol. 8. World Scientific, Singapore/Teaneck, NJ, pp. viii, 635.

Sihvola, A. H., Institution of Electrical Engineers, 1999. "Electromagnetic mixing formulas and applications." *IEE Electromagnetic Waves Series*, vol. 47. Institution of Electrical Engineers, London, pp. xii, 284.

Sobolev, V. V., 1963. *A Treatise on Radiative Transfer*. Van Nostrand, Princeton, N.J., p. 319.

Sobolev, V. V., 1975. "Light scattering in planetary atmospheres." *International Series of Monographs in Natural Philosophy*, vol. 76, 1st ed. Pergamon Press, Oxford, New York, pp. xvii, 256.

Soubret, A., Ripoll, J., Ntziachristos, V., 2005. "Accuracy of fluorescent tomography in the presence of heterogeneities: Study of the normalized born ratio." *IEEE Trans. Med. Imaging* **24**(10), 1377–1386.

Taroni, P., Pifferi, A., Torricelli, A., Comelli, D., Cubeddu, R., 2003. "In vivo absorption and scattering spectroscopy of biological tissues." *Photochem. Photobiol. Sci.* **2**(2),124–129.

Torricelli, A., Pifferi, A., Taroni, P., Giambattistelli, E., Cubeddu, R., 2001. "In vivo optical characterization of human tissues from 610 to 1010 nm by time-resolved reflectance spectroscopy." *Phys. Med. Biol.* **46**(8), 2227–2237.

Tromberg, B. J., Coquoz, O., Fishkin, J., Pham, T., Anderson, E. R., Butler, J., Cahn, M., Gross, J. D., Venugopalan, V., Pham, D., 1997. Non-invasive measurements of breast tissue optical properties using frequency-domain photon migration. *Philosophical Transactions of the Royal Society of London Series B-Biological Sciences*, 1997. 352(1354), 661–668.

Tsang, L., Kong, J. A., 2001. "Scattering of electromagnetic waves." *Wiley Series in Remote Sensing*, John Wiley & Sons, Inc., New York, pp. xvi, 413.

Tsang, L., Kong, J. A., Ding, K.-H., 2000. "Scattering of electromagnetic waves." *Wiley Series in Remote Sensing*, Jhon Wiley & Sons, Inc., New York, pp. xiv, 426.

Tsang, L., Kong, J. A., Shin, R. T., 1985. "Theory of microwave remote sensing." *Wiley Series in Remote Sensing*, John Wiley & Sons, Inc., New York, pp. xiii, 613.

Tuchin, V. V., 2002. *Handbook of Optical Biomedical Diagnostics*, SPIE Press, Bellingham, Wash, pp. xv, 1093.

Tuchin, V. V., Society of Photo-optical Instrumentation Engineers, 2000. Tissue optics: light scattering methods and instruments for medical diagnosis. *Tutorial Texts in Optical Engineering TT 38*. SPIE Press, Bellingham, Wash, pp. xxv, 352.

Valeur, B., 2002. *Molecular Fluorescence: Principles and Applications*, Wiley-VCH, Weinheim, New York, pp. xiv, 387.

Vo-Dinh, T., Stokes, D. L., Wabuyele, M. B., Martin, M. E., Song, J. M., Jagannathan, R., Michaud, E., Lee, R. J., Pan, X. G., 2004. "A hyperspectral imaging system for in vivo optical diagnosis." *IEEE Eng. Med. Biol. Mag.* **23**(5), 40–49.

Vo-Dinh, T., 2003. *Biomedical Photonics Handbook*, CRC Press, Boca Raton, Fla. (1 v. (various pagings), [32] of plates.)

Vogel, C. R., 2002. "Computational methods for inverse problems. Frontiers in Applied Mathematics." *Society for Industrial and Applied Mathematics*, Philadelphia, pp. xvi, 183.

Wang, L. H., Jacques, S. L., Zheng, L. Q., 1995. "Mcml - Monte-Carlo modeling of light transport in multilayered tissues." *Comput. Method Prog. Biomed.* **47**(2), 131–146.

Weissleder, R., Mahmood, U., 2001. "Molecular imaging." *Radiology* **219**(2), 316–333.

Yuan, S., Devor, A., Boas, D. A., Dunn, A. K., 2005. "Determination of optimal exposure time for imaging of blood flow changes with laser speckle contrast imaging." *Appl. Opt.* **44**(10), 1823–1830.

6 Optical Microscopy in Small Animal Research

Rakesh K. Jain, Dai Fukumura, Lance Munn
and **Edward Brown**

6.0 Introduction

Of all the techniques suitable for imaging living tissues, optical microscopy has the highest spatial resolution and is able to distinguish objects to less than 1 μm (Jain *et al.*, 2002). This compares favourably to other *in vivo* imaging techniques, such as PET, CT and MRI, with resolutions of >10 μm (Weissleder, 2002). Unfortunately, the superior resolution of light microscopy has historically come at the expense of diminished depth penetration, with typical epifluorescence and confocal laser-scanning microscopes (CLSM) able to penetrate ~100 μm into tissue (Jain *et al.*, 2002). The recent introduction of the multiphoton laser-scanning microscope (MPLSM) to *in vivo* imaging has extended this depth penetration to over half a millimetre. Although this is a vast improvement, it is still insufficient to access most of the tissues and organs in the mouse without surgical intervention. This chapter will provide an introduction to the technology of confocal and multiphoton microscopy and then outline some important applications in small animal research, focusing specifically on tumour biology and on the least invasive applications of MPLSM and CLSM: those that utilize chronically implanted windows and hence do not require surgical intervention at the time of imaging.

6.1 Confocal laser scanning microscopy

A typical widefield epifluorescence microscope has excellent lateral resolution but essentially no depth resolution: the amount of light reaching the eyepiece from the plane of interest is the same as the amount of light reaching the eyepiece from planes above and below (Bradbury and Bracegirdle, 1998). To address this lack of depth resolution, Minsky patented the confocal microscope in 1957 (Minsky, 1988), and it evolved into its modern form using laser excitation in the late 1980s (White *et al.*, 1987). A typical confocal microscope (see Figure 6.1.1(a)) builds up an image point by point as follows: A laser beam is focused into the tissue by an objective lens, and fluorescence photons are generated when excitation light interacts with fluorophores (i.e. fluorescent dyes or intrinsically fluorescent molecules in the tissue). The focused beam excites these fluorescence photons in a double cone within the tissue, and some of these photons are collected by the objective lens and directed to a photomultiplier tube, which converts the light to electrons. The number of detected electrons is plotted on a television monitor as brightness of a pixel on the screen. The beam is then moved slightly to an adjacent location, and the process is repeated until a complete image is built up on the monitor. Up to this point in our description, the technique still does not have depth resolution: the number of photons collected from the plane of interest (the 'focal plane', i.e. the plane perpendicular to the optical axis at the focus of the beam) is the same as the number collected from planes above and below. However, if a small aperture is positioned in the detection pathway of the system such that light from the plane of interest is focused through the pinhole, light from above or below the plane of interest will not be focused through the pinhole and will be attenuated (see Figure 6.1.1(b)).

Figure 6.1.1 (a) Instrumentation diagram of a confocal laser-scanning microscope. The excitation laser (blue) is directed into the objective lens through a pair of movable scanning mirrors. Fluorescence is generated in the sample (green), collected by the objective lens, and directed by the scanning mirrors onto a stationary aperture. Fluorescence from the plane of interest passes through the aperture and is detected. (b) Details of the confocal aperture. Left: fluorescence from the plane of interest (cone defined by solid lines) is focused through the aperture, while fluorescence from above (dashed) and below (dotted) the plane of interest is focused below and above the aperture, respectively, and hence is attenuated. Right: scattering of fluorescence photons from the plane of interest (solid arrow) can prevent signal from reaching the detector, while scattering of florescence photons from outside the plane of interest (dashed arrow) can increase noise

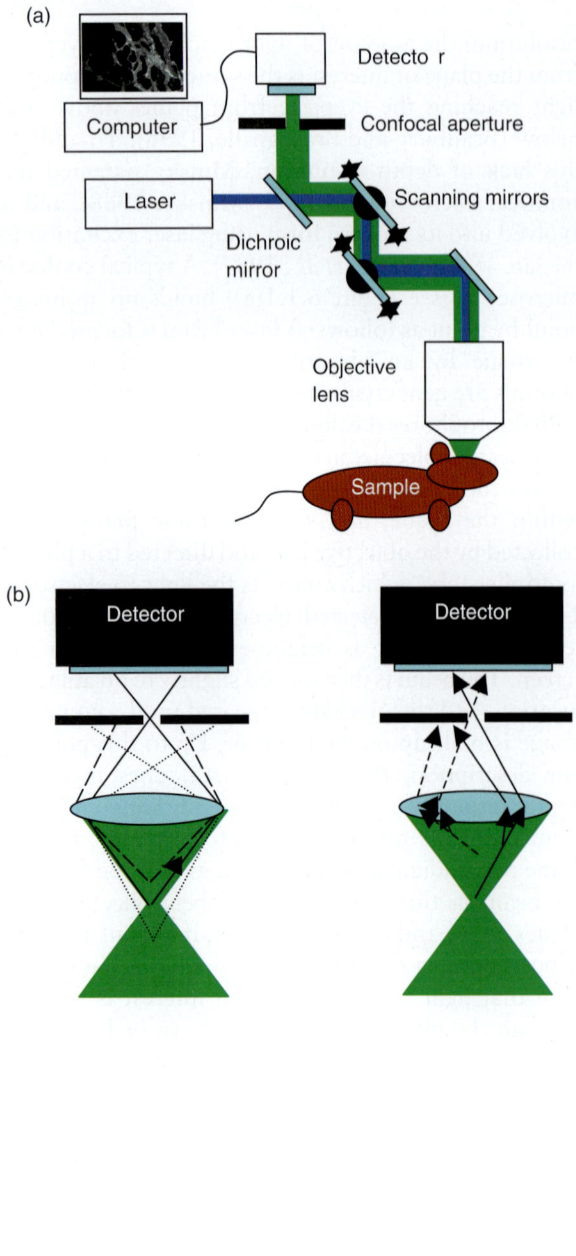

This provides imaging depth resolution of approximately 1 μm, accompanying a lateral resolution of less than 1 μm (Sheppard and Shotton, 1997).

The overall depth penetration of the system *in vivo* is a function of scattering and absorption in the tissue as well as photobleaching and photodamage. As the depth of interest increases, the amount of scattering that fluorescence photons experience as they travel from the focal plane to the objective lens increases. Signal photons (from the focal plane) are scattered out of the pinhole, decreasing signal, while noise photons (from above or below the focal plane) are scattered into the pinhole, increasing noise. To overcome the attenuation of signal photons by scattering, it is tempting to increase the excitation laser power. Unfortunately, this also increases photobleaching and photodamage. Photobleaching occurs because a given fluorophore can only be excited a finite number of times before if converts into a nonfluorescent molecule, a process often accompanied by the generation of singlet oxygen or some other toxic product. This process not only decreases signal by decreasing the pool of usable fluorophores but also damages nearby cells and tissue. As a result of these various limitations, the confocal laser-scanning microscope can generate three-dimensionally resolved images down to ~100 μm in scattering tissue (Jain *et al.*, 2002).

6.2 Multiphoton laser scanning microscopy

In 1990, Watt Webb and co-workers introduced the multiphoton laser scanning microscope (MPLSM) (Denk *et al.*, 1990). Like a confocal microscope, the MPLSM builds an image point by point (see Figure 6.2.1(a)). A laser beam is focused into the tissue by an objective lens, and the focused beam excites fluorescence within the tissue. However, the MPLSM uses a long-wavelength laser which emits photons of half the energy required to excite typical dye molecules. Therefore, fluorescence is only generated where lower energy excitation photons are packed so densely together that there is a chance for two excitation photons to interact with a dye molecule approximately at the same time. The resultant ‘multiphoton excitation’ is therefore confined to a small (~1 μm^3) volume located at the focus of the laser beam (see Figure 6.2.1(b)). The probability of fluorescence excitation is also enhanced by concentrating excitation photons in time as well as in space, using pulsed lasers (a typical laser used for multiphoton excitation emits light pulses

Figure 6.2.1 (a) Instrumentation diagram of a multiphoton laser-scanning microscope. The excitation laser (red) is directed into the objective lens through a pair of movable scanning mirrors. Fluorescence is generated in the sample (green) collected by the objective lens and directed into detectors. (b) Detail of the multiphoton focal volume. Compare this diagram to Figure 6.1.1(b). As there is no fluorescence generated outside the plane of interest, there is no need to pass through a confocal aperture. Consequently, fluorescence signal from the plane of interest can be somewhat scattered and still reach the detectors. Conversely, there is no fluorescence from outside of the plane of interest that can be scattered into the detector to increase noise

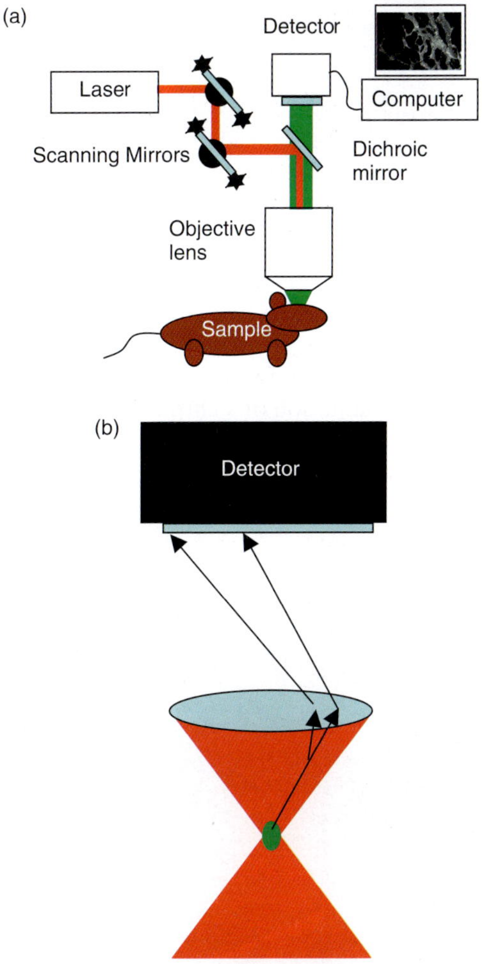

that are $\sim 1 \times 10^{-13}$ s long and occur at a rate of $\sim 1 \times 10^{8}$ times per second).

A typical MPLSM has imaging resolution of less than 1 μm laterally and ~1 μm in depth similar to the confocal microscope. However, the MPLSM achieves this resolution without the use of a confocal aperture in the detection system. Consequently, the MPLSM is far less sensitive to loss of signal photons by scattering, and the scattering of noise (i.e. out-of-plane) photons into the detector is not a significant obstacle, since essentially no such photons are generated in the first place. This allows the MPLSM to image at far greater depths than the CLSM, with imaging at over 500 μm in living brain tissue being reported (Kleinfeld *et al.*, 1998).

6.3 Variants for *in vivo* imaging

The vast majority of confocal and multiphoton laser-scanning microscopy has been performed on *in vitro* systems, i.e. cells in a dish, or on *ex vivo* systems, such as acutely prepared slices of tissue. However, the application of these microscopes to *in vivo* preparations does not require any significant modification of the instrument (Brown *et al.*, 2001). Some care must be taken, however, in objective lens selection. The quality of image deteriorates within just a few microns of the surface of the imaged material if the index of refraction of the immersion medium (oil, glycerol, water or air) does not closely match with that of the material being imaged. Because tissue has an index of refraction essentially equal to water, water immersion lenses should be used to achieve any significant imaging depth. Care should also be taken in the selection of fluorophores. Longer wavelengths of light travel farther through tissue than the shorter ones; therefore, longer wavelength dyes (i.e. TRITC vs. FITC, dsRED vs. GFP) should be chosen whenever possible.

6.4 Surgical preparations

Although the imaging depth of the MPLSM is a significant improvement over confocal and epifluorescence microscopy, it is still extremely challenging to image through skin into an underlying organ or tumour. Therefore, most chronic *in vivo* imaging with the MPLSM and other light microscopies is performed using chronic transparent window preparations, in which the tissue of interest is located under a surgically implanted window (the tissue of interest can also be acutely surgically accessed for imaging purposes, reviewed by Jain *et al.*, 2003). Two of the most common chronic window preparations for light microscopy are the cranial window (Yuan *et al.*, 1994) and dorsal skinfold chamber (Leunig *et al.*, 1992) (see Figures 6.4.1 and 6.4.2).

Figure 6.4.1 Mouse cranial window

The cranial window, as its name suggests, is the surgical removal of a circle of skull and its replacement by a window. This preparation can last for a year or more and is frequently used in neuroscience (see Figure 6.4.3), often in a variation in which the skull is thinned down to form the window itself (Zuo *et al.*, 2005). The cranial window is also often used to study tumour biology, wherein a tumour cell line is implanted on the surface of the brain before the window is glued in place. The dorsal skinfold chamber is almost exclusively used for tumour biology studies and consists of implantation of a chronic window in a fold of skin on a laboratory animal. This provides optical access to the underside of the skin, and tumours can be

Figure 6.4.2 Mouse dorsal skinfold chamber

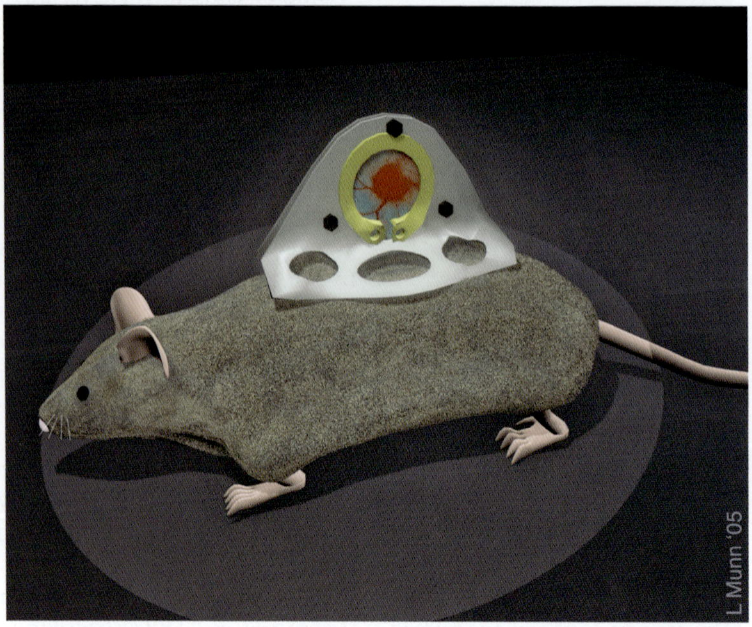

Figure 6.4.3 *In vivo* MPLSM imaging of neurons. A cranial window was implanted in a transgenic mouse that expressed green fluorescence protein in a subset of neurons. Image is 400 μm across and is a maximum intensity projection of 120 optical sections spaced 3 μm apart. Courtesy of Dr Ania Majewska at University of Rochester Medical Center (*with the permission of Dr Ania Majewska*)

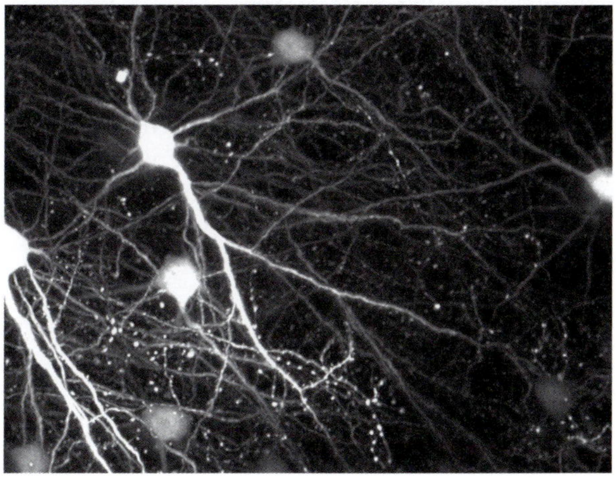

grown in the subcutaneous space. For other tissue preparations, see Jain *et al.* (2002).

6.5 Applications

6.5.1 Imaging cells in living tumours

The exceptionally high spatial resolution of light microscopy allows us to image individual cells and even subcellular organelles. Furthermore, the low phototoxicity of MPLSM (Squirrell *et al.*, 1999) coupled with its high spatial resolution allows repeated imaging of fluorescently labelled cells, often several hundreds of microns into living tissue. The earliest application of the MPLSM to the *in vivo* imaging of tumours (Brown *et al.*, 2001) investigated the spatial distribution of vascular endothelial growth factor (VEGF)-expressing stromal cells within a tumour. A dorsal skinfold chamber was implanted on a transgenic mouse which expressed green fluorescent protein (GFP) under the control of VEGF promoter ($VEGF^P$-GFP) and nonfluorescent tumour cells were implanted in it. The MPLSM revealed that VEGF-expressing (and hence expressing GFP reporter) cells from the host animal penetrate deep in the tumour and form sleeve-like structures around tumour blood vessels (Brown *et al.*, 2001) (see Figure 6.5.1). Transgenic

Figure 6.5.1 *In vivo* MPLSM imaging of host cells within a tumour. A nonfluorescent breast tumour cell line was grown in the dorsal skin chamber of a $VEGF^P$-GFP transgenic mouse. The tumour blood vessels were highlighted by intravenous injection of TRITC-dextran (molecular weight 2×10^6) (red) while VEGF-expressing host cells were visualized by GFP (green). Image is 250 μm across

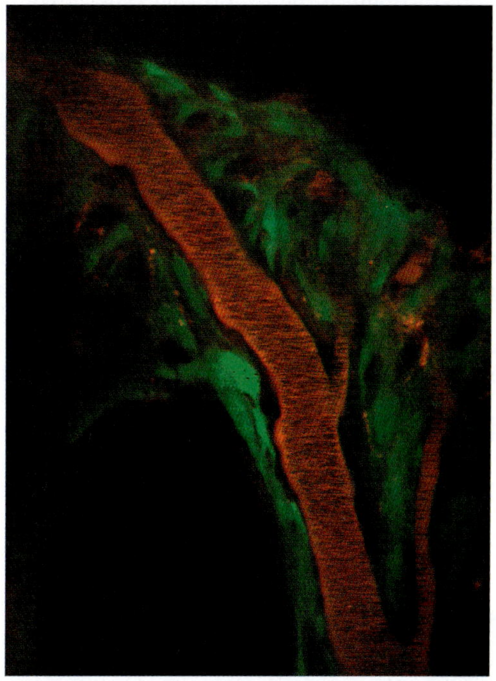

GFP reporter mice combined with *in vivo* MPLSM has also revealed transplantability of host stromal cells within transplanted tumour tissue (Duda *et al.*, 2004). These stromal cells survive and proliferate after transplantation and participate in initial growth of transplanted tumours. In another study, a specially constructed high-speed MPLSM was used to generate time-lapse images of labelled host immune cells within tumour vessels. This allowed the quantification of parameters describing the host/tumour immune response such as the total leukocyte flux, rolling fraction and adhering density (Padera *et al.*, 2002b).

MPLSM and CLSM have also been used *in vivo* to image tumour cells themselves. MPLSM of GFP-labelled gliosarcoma cells, in conjunction with near-infrared CLSM, has localized the proteolytic dequenching of cathepsin-sensitive near-infrared fluorescent probes in the tumours grown in mouse ears (Bogdanov *et al.*, 2002). A high-speed CLSM has been used to quantify circulating fluorescent tumour cells in the blood vessels in rodent ears and skull after intravenous injection (Georgakoudi *et al.*, 2004; Sipkins *et al.*, 2005).

6.5.2 Imaging and quantification of blood and lymphatic vessels

A complete understanding of the angiogenic and lymphangiogenic process during cancer progression requires the imaging of tumour-associated blood and lymphatic vessels, *in vivo*, with sufficient spatial resolution to distinguish the endothelial cells surrounding the smallest vessels. Consequently, one useful application of the MPLSM is the imaging of blood and lymphatic vessels *in vivo* (see Figure 6.5.2). Brown *et al.* used intravenous injection of 2M MW FITC-dextran to highlight the vasculature of a tumour growing in the dorsal skin chamber and imaged the tumour vessels with the MPLSM (Brown *et al.*, 2001). MPLSM tumour angiography has also revealed normalization of chaotic tumour vasculature during an anti-angiogenic treatment (Tong *et al.*, 2004; Winkler *et al.*, 2004) as well as the angiogenic properties of a tumour suppressor protein (Garkavtsev *et al.*, 2004). MPLSM was applied to the lymphatic system by Padera *et al.* (2000a) who utilized an injection of FITC-dextran to highlight lymphatics and thereby reveal an absence of functional lymphatics inside the tumour but lymphatic hypertrophy at the tumour periphery. MPLSM lymphangiography has more recently revealed the effects of compressive forces on the function of blood and lymphatic vessels in tumours (Padera *et al.*, 2004) and has shed light onto the abnormal function of peritumour lymphatics induced by VEGF-C (Isaka *et al.*, 2004). CLSM imaging of shallow vessels within mouse skull bone marrow has revealed the presence of unique endothelial microdomains which is home to circulating tumour cells (Sipkins *et al.*, 2005).

Figure 6.5.2 *In vivo* MPLSM imaging of tumour blood vessels. A colon adenocarcinoma-derived cell line was grown in the cranial window of a SCID mouse. The tumour blood vessels were highlighted by intravenous injection of FITC-dextran (molecular weight 2×10^6). Fifteen optical sections were generated spaced 5 μm apart and merged together in a maximum intensity projection. Image is 370 μm across

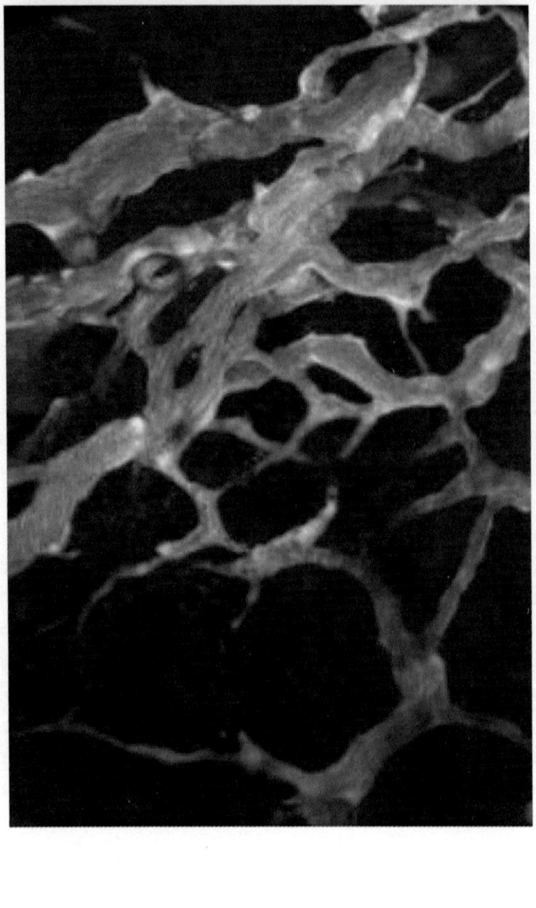

6.5.3 Future directions

As discussed above, chronic, minimally invasive CLSM and MPLSM have been performed on the exposed skin or in chronically implanted window preparations, such as the cranial window and dorsal skinfold chamber. The development of miniaturized endoscopic devices will greatly enhance chronic, minimally invasive CLSM and MPLSM by allowing optical imaging of many interior surfaces of the animal without surgical intervention and will allow imaging of many other organs and tissues via minimally invasive acute laparoscopy. These endoscopic devices are already being developed. A side-viewing micromachined CLSM that is only ~2.5 mm wide has been demonstrated (Dickensheets and Kino, 1996). A forward-viewing fibre-optic CLSM has also been developed and used to image brain, bladder and other organs (D'Hallewin *et al.*, 2005). Forward-viewing GRIN lens-based microscopes have also been demonstrated (Jung *et al.*, 2004; Levene *et al.*, 2004) which are only a few millimetres in diameter. In widefield, confocal and multiphoton modes they are already able to laparoscopically image brain tissue, blood vessels, etc. In the future, such microendoscopes will become commonplace and the regions of patients and experimental animals accessible to optical microscopy will greatly increase.

Acknowledgements

The work described here was supported by grants from the NIH (P01CA80134 and R24 CA85140 to

R.K.J., R01 HL64240 and R01 CA96915 to L.M. and D.F., respectively), the Department of Defense (Era of Hope Scholar Award to E.B.) and the Whitaker Foundation (Biomedical Engineering Research Grant to E.B.).

References

Bogdanov Jr., A. A., Lin, C. P., Simonova, M. *et al.*, 2002. "Cellular activation of the self-quenched fluorescent reporter probe in tumor microenvironment." *Neoplasia* **4**(3), 228–236.

Bradbury, S., Bracegirdle, B., 1998. *Introduction to Light Microscopy.* Springer-Verlag, New York.

Brown, E., Campbell, R., Tsuzuki, Y. *et al.*, 2001. "In vivo measurement of gene expression, angiogenesis, and physiological function in tumors using multiphoton laser scanning microscopy." *Nat. Med.* **7**(7), 864–868.

D'Hallewin, M. A., Khatib, S. El, Leroux, A. *et al.*, 2005. "Endoscopic confocal fluorescence microscopy of normal and tumor bearing rat bladder." *J. Urol.* **174**(2), 736–740.

Denk, W., Strickler, J. H., Webb, W. W. *et al.*, 1990. "Two-photon laser scanning fluorescence microscopy." *Science* **248**(4951), 73–76.

Dickensheets, D., Kino, G., 1996. "Micromachined scanning confocal optical microscope." *Opt. Lett.* **21**(10), 764–766.

Duda, D. G., Fukumura, D., Munn, L. L. *et al.*, 2004. "Differential transplantability of tumor-associated stromal cells." *Cancer Res.* **64**(17), 5920–5924.

Garkavtsev, I., Kozin, S. V., Chernova, O. *et al.*, 2004. "The candidate tumour suppressor protein ING4 regulates brain tumour growth and angiogenesis." *Nature* **428**(6980), 328–332.

Georgakoudi, I., Solban, N., Novak, J. *et al.*, 2004. "In vivo flow cytometry: a new method for enumerating circulating cancer cells." *Cancer Res.* **64**(15), 5044–5047.

Isaka, N., Padera, T. P., Hagendoorn, J. *et al.*, 2004. "Peritumor lymphatics induced by vascular endothelial growth factor-C exhibit abnormal function." *Cancer Res.* **64**(13), 4400–4404.

Jain, R. K., Brown, E., Munn, L. *et al.*, 2003. In: Spector D., G. R. (Ed.), *Intravital Microscopy of Normal and Diseased Tissue in the Mouse. Live Cell Imaging: A Laboratory Manual*, Cold Spring Harbor Press, Cold Spring Harbor, New York.

Jain, R. K., Munn, L. L., Fukumura, D. *et al.*, 2002. "Dissecting tumour pathophysiology using intravital microscopy." *Nat. Rev. Cancer* **2**(4), 266–276.

Jung, J. C., Mehta, A. D., Aksay, E. *et al.*, 2004. "*In vivo* mammalian brain imaging using one- and two-photon fluorescence microendoscopy." *J. Neurophysiol.* **92**(5), 3121–3133.

Kleinfeld, D., Mitra, P. P., Helmchen, F. *et al.*, 1998. "Fluctuations and stimulus-induced changes in blood flow observed in individual capillaries in layers 2 through 4 of rat neocortex." *Proc. Natl. Acad. Sci. USA* **95**(26), 15741–15746.

Leunig, M., Yuan, F., Menger, M. D. *et al.*, 1992. "Angiogenesis, microvascular architecture, microhemodynamics, and interstitial fluid pressure during early growth of human adenocarcinoma LS174T in SCID mice." *Cancer Res.* **52**(23), 6553–6560.

Levene, M. J., Dombeck, D. A., Kasischke, K. A. *et al.*, 2004. "*In vivo* multiphoton microscopy of deep brain tissue." *J. Neurophysiol.* **91**(4), 1908–1912.

Minsky, M. (1988). "Memoir on inventing the confocal scanning microscope." *Scanning* **10**, 128–138.

Padera, T. P., Kadambi, A., Tomaso, E. di. *et al.*, 2002a. "Lymphatic metastasis in the absence of functional intratumor lymphatics." *Science* **296**(5574), 1883–1886.

Padera, T. P., Stoll, B., So, P. *et al.*, 2002b. "Conventional and high-speed intravital multiphoton laser scanning microscopy of microvasculature, lymphatics, and leukocyte-endothelial interactions." *Molecular Imaging* **1**, 9–15.

Padera, T. P., Stoll, B. R., Tooredman, J. B. *et al.*, 2004. "Pathology: cancer cells compress intratumour vessels." *Nature* **427**(6976), 695.

Sheppard, C. J. R., Shotton, D. M., 1997. *Confocal Laser-Scanning Microscopy*, Springer-Verlag, New York.

Sipkins, D. A., Wei, X., Wu, J. W. *et al.*, 2005. "*In vivo* imaging of specialized bone marrow endothelial microdomains for tumour engraftment." *Nature* **435**(7044), 969–973.

Squirrell, J. M., Wokosin, D. L., White, J. G. *et al.*, 1999. "Long-term two-photon fluorescence imaging of mammalian embryos without compromising viability." *Nat. Biotechnol.* **17**(8), 763–767.

Tong, R. T., Boucher, Y., Kozin, S. V. *et al.*, 2004. "Vascular normalization by vascular endothelial growth factor receptor 2 blockade induces a pressure gradient across the vasculature and improves drug penetration in tumors." *Cancer Res.* **64**(11), 3731–3736.

Weissleder, R., 2002. "Scaling down imaging: molecular mapping of cancer in mice." *Nat. Rev. Cancer* **2**(1), 11–18.

White, J. G., Amos, W. B., Fordham, M. *et al.*, 1987. "An evaluation of confocal versus conventional imaging of biological structures by fluorescence light microscopy." *J. Cell. Biol.* **105**(1), 41–48.

Winkler, F., Kozin, S. V., Tong, R. T. *et al.*, 2004. "Kinetics of vascular normalization by VEGFR2 blockade governs brain tumor response to radiation: role of oxygenation, angiopoietin-1, and matrix metalloproteinases." *Cancer Cell.* **6**(6), 553–563.

Yuan, F., Salehi, H. A., Boucher, Y. *et al.*, 1994. "Vascular permeability and microcirculation of gliomas and mammary carcinomas transplanted in rat and mouse cranial windows." *Cancer Res.* **54**(17), 4564–4568.

Zuo, Y., Yang, G., Kwon, E. *et al.*, 2005. "Long-term sensory deprivation prevents dendritic spine loss in primary somatosensory cortex." *Nature* **436**(7048), 261–265.

7 New Radiotracers, Reporter Probes and Contrast Agents

Coordinated by Bertrand Tavitian

7.0 Introduction

Bertrand Tavitian

For our pre-historical ancestors, hunters–gatherers, *to trace* food and foes, that is to recognize the traces left by comestible or dangerous animals in the primitive forests, was an absolute obligation for survival. Hence, they developed tracing skills, which ultimately led to our present species, *Homo sapiens*. Tracing is indeed the essential means by which we recognize what has been there when we were not present, and elaborate rules that tend to organize the chaos surrounding us. Our brain is designed to reckon tiny changes in our surroundings, which we can indirectly attribute to something or somebody else whom we do not see, and the large part of our brain devoted to the treatment of visual information attests that rendering visual traces meaningful is a fundamental activity in Humans. Tracing is also a fundamental activity for *in vivo* imaging scientists. Similar to the hunters in the forest, scientists must learn to recognize meaningful biological characteristics through the indirect traces that these leave on images: calcifications on X-rays, proton density changes in MRI, ultrasound deflexions, etc.

The word 'trace' derives from the supine of the Latin verb 'trahere', to pull, and, very early on (ca. 14th century in English), it was used in the fields of travel, broidery, drawing and writing, and also with the metaphoric senses of sketching, elaborating, and so on. As a noun, 'trace' covers two senses: 'a trace that is followed (by the hunter)' and a trace that (the animal) leaves behind'. Coming back to imaging, these two senses are mixed in the derivative *tracer*, which is a compound acting as a beacon for a biological activity.

Non-tracer imaging is essentially a physics–chemistry interaction, based on the interaction of the type of energy (electromagnetic wave, see Book Introduction), used by the imaging instrument, with the endogenous molecules. This makes the system simple to describe because the physics of detection is, by definition, invariant with time. Conversely, it limits imaging to a limited set of endogenous molecules which have a special (with respect to the bulk of molecules found in the living organism) type of interaction with the imaging system. Using a tracer adds a chemical dimension to imaging and extends considerably the number of biochemical compounds or activities which can be imaged. Image contrast is now also based on the interaction of the tracer, a chemical compound in the broad sense, with the endogenous molecules. This has two consequences: First, tracers are artificial indicators of the biological activity, and caution must be exerted as to their accuracy; like all chemicals, they are prone to transformation by living organisms. Second, tracers generally introduce a temporal dimension into imaging because the way they interact with endogenous molecules, and therefore the contrast they produce, is time dependant. Tracer imaging is becoming a working horse for pharmacokinetics and biodistribution studies, and, in particular, pre-clinical tracer imaging is invading the field of drug development. Moreover, tracers represent the domain of excellence of molecular imaging. As imaging becomes molecular, the imaging scientists are now endowed with the capacity, not only to interpret traces on the images provided by the different instruments at their disposal but also to use chemistry, cellular, molecular and animal biology in order to build artificial traces that document a biological phenomenon.

The inventiveness in tracer design is extraordinary, and the present chapter on 'new tracers' can only give a taste of that burgeoning field by presenting some recent achievements in that direction. The first section presents the sophisticated chemistry presently explored to ameliorate paramagnetic tracers. This line of research is based in particular on extremely careful analyses of the interaction of the tracers with the local environment they encounter in living individuals, not only water but also subtle differences in physiological parameters such as temperature, oxygen pressure, enzymes, etc. In addition, the targeting of magnetic contrast agents by functionalization is a promising field with applications in other imaging modalities. The second section follows that direction and describes multimodal agents with the capacity to be detected by several imaging modalities, that is MRI and optical or nuclear techniques. Ideally, these agents could be used to benefit from the advantages of each of the imaging techniques. The third section describes briefly some of the new radiotracers developed for nuclear imaging techniques, a technique which is based only on the tracer principle and which has a long history of tracer development. In contrast, the use of fluorescent tracers described in the fourth section is pretty recent, as optical imaging has only recently reached the capacity to image living animals. Finally, Section 7.5 describes one of most spectacular achievements from the transfer of molecular biology techniques towards *in vivo* imaging, the use of genetically encoded reporter genes specially designed for imaging purposes. Just as much as the introduction of the beta-galactosidase reporter gene into bacteria and later into vertebrates completely transformed molecular genetics up to the point that it is now a routine technique taught in basic biology courses, the encoding of genes engineered for reporter imaging of gene expression is likely to become one of the major tools of non-invasive biology. With that technique, artificiality hits both the chemistry and the biology of the imaging system, opening a future vertiginously extendable.

7.1 New radiotracers

Bertrand Tavitian, Roberto Pasqualini and Frédéric Dollé

7.1.1 General considerations

Radiochemistry for *in vivo* imaging applications is a very active field of research, and descriptions of new radiotracers appear every week in specialized journals (among others, *The Journal of Labelled Compounds and Radiopharmaceuticals, The Journal of Applied Radiation and Isotopes, The Journal of Nuclear Medicine, The European Journal of Nuclear Medicine and Molecular Imaging, Bioconjugate Chemistry, ...*). However, a very small proportion of these new radiotracers make it to pre-clinical *in vivo* imaging in animals and even less so to clinical applications. Hence, the purpose of this section is rather to indicate gross lines of recent research which have opened new fields of applications to radiotracer imaging. The reader should be well aware that the examples shown here represent personal choices of the authors and are by no means exhaustive, nor do they constitute a generally accepted view of the radiochemistry community.

Initially, the radiotracer principle was developed using crude radionuclides, such as phosphorous, calcium, iodine, etc. Progress in radionuclide tracer production is supported by progress in basic radiochemistry, that is the introduction of radioactive atoms into chemical structures of interest. This progress has been a permanent process, and advances in technetium and fluorine chemistries have led to many new ^{99m}Tc and ^{18}F labelled radiotracers. Some recent examples of these advances will be presented below. A general trend over years has been the introduction of radionuclides into more and more complex molecules. Modern radiochemistry now tackles complex macromolecules issued from biotechnological advances, interestingly, the increase in the size and complexity of the radiotracers parallels that of the type of biological information which can be addressed: The more complex the molecule, the more complex and subtle the type of information it carries (Table 7.1.1). The major recent fact in radionuclide imaging is the attempt to address more complex biological questions by labelling peptides, proteins, and nucleic acids.

However, the increased sizes of the radiotracers goes along with increased difficulties in their delivery to their target sites, due to a lower diffusion into tissues and a higher number of non-specific interactions. This has led to attempts to address imaging agents by vectorization agents, designed to favour their biodistribution and reduce their non-specific interactions in the body. In addition, natural macromolecules are naturally recognized by any organism's chemical and cellular sentinels as exogenous, and therefore, are prone to attack by degradation enzymes and immune recognition. Therefore, chemical modifications of the natural peptides, or oligonucleotides are most often mandatory for *in vivo* applications such as imaging.

A final remark is that proteins and nucleic acids are polymers of amino acids and nucleotides,

Table 7.1.1 Biochemical signalling of biological information

Type of information	Molecules	Tracers
• **Basic information is carried by small molecules**		
Food and metabolism	Glucose, amino-acids	[^{18}F]FDG, [^{11}C]methionine, [^{11}C]acetate, etc
Sensory and motor	Transmitters	[^{11}C]flumazenil, [^{11}C]raclopride, [^{18}F]DOPA, etc
• **Complex information and processing is mediated by macromolecules**		
Self and recognition	Poly-osides, proteins, peptides	Under development RGD-peptides
Cell growth and division	Peptides, proteins	Annexin, somatostatine, antibodies
Heredity	Nucleic acids	Under development
Regulation and integration	Proteins and nucleic acids	Under development

respectively. This consideration has two important consequences. First, a high level of chemical diversity can be obtained using a few elementary building blocks: The number of different polymers formed out of a given number of units is in the power of 4 for natural oligonucleotides, and in the power of 20 for natural peptides. Second, the same radiochemical procedures can be applied to any peptide or oligonucleotides, allowing to build a diversity of radiotracers and reducing the amount of work necessary for labelling. Radiochemistry has access to more and more combinatorial libraries of small or large polymeric molecules, and this may well represent the major trend of future developments in the radiotracer field.

7.1.2 Examples of research ^{99m}Tc-radiotracers

7.1.2.1 *Technetium-nitride dithiocarbamate complexes*

The technetium-nitride $[Tc{\equiv}(N)]^{2+}$ core constitutes a functional moiety in which the Tc^{+5} ion is multiply bound to a nitride ion (N^{3-}). This core exhibits a very high chemical stability towards both oxidation and reduction reactions and towards pH variations. In addition, this soft acid core displays a marked selective reactivity in forming bonds with ligands containing soft donor atoms, such as S and P. The chemistry of technetium nitride (TcN) compounds intended for radiopharmaceuticals was first developed by Baldas and Bonnyman (1985). However, the proposed labelling method was complex. A new efficient method was proposed later and applied to the preparation of neutral dithiocarbamate complexes for cardiac imaging (Pasqualini *et al.*, 1994).

The basic chemistry is represented by the following two chemical reactions:

(i) Preparation of a labile intermediate species containing the ^{99m}TcN core:

$$^{99m}TcO_4^- + \text{Reducing agent} + {>}N{-}N{<} \rightarrow [^{99m}TcN]_{interm}$$

(ii) Ligand exchange reaction by a suitable ligand on the $[^{99m}TcN]_{interm}$ species:

$$[^{99m}TcN]_{interm} + \text{ligand} \rightarrow [^{99m}Tc(N)L].$$

Reaction (i) is quite general. In principle, any chemical compound derived from hydrazine, that is containing the >N–N< moiety, can generate the TcN core when reacting with $^{99m}TcO_4^-$ in presence of a reducing agent. Practically however, only a few compounds display the chemical properties allowing quantitative formation of $[^{99m}TcN]_{interm}$ species in a short time (<10 min) at room temperature. Among them, succinohydrazide (SDH) proved to be a very efficient nitride donor. Pharmaceutical kits formulated with SDH have been used in the preparation of the neutral, square pyramidal complex bis(*N*-ethoxy, *N*-ethyl dithiocarbamato)nitrido technetium (V) [^{99m}Tc] or ^{99m}TcN-NOET, a perfusion imaging agent investigated in pre-clinical and clinical studies as replacement of (^{201}Tl) thallium chloride (Jeetly *et al.*, 2004).

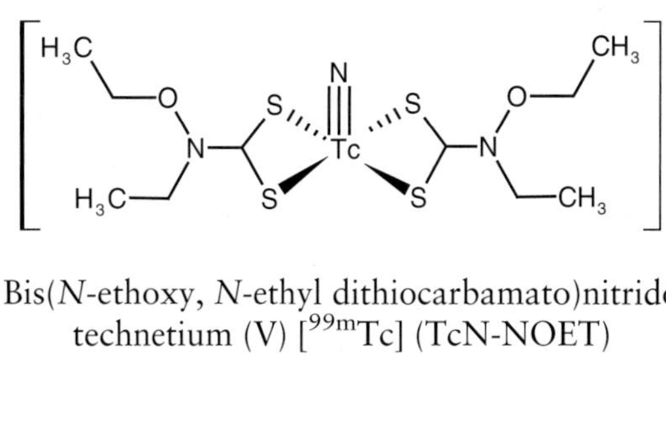

Bis(*N*-ethoxy, *N*-ethyl dithiocarbamato)nitrido technetium (V) [^{99m}Tc] (TcN-NOET)

7.1.2.2 *Asymmetrical Tc-nitride complexes*

Due to its particular electronic properties, the technetium-nitride core offers the peculiar possibility of obtaining pure asymmetrical heterocomplexes when reacting with two different bidentate ligands of a specific structure. The advantage of asymmetrical over symmetrical complexes is easily perceived when only one large bioactive molecule has to be bound to the $[^{99m}TcN]^{2+}$ core.

It has been shown that reaction of $[^{99m}TcN]^{2+}$ with a suitable bidentate phosphine yields the stable metal nitride fragment $[^{99m}Tc(N)(PXP)]^{2+}$. This fragment reacts selectively with electron-rich soft (donor ligands containing S, O and N as coordinating atoms (Bolzati *et al.*, 2000). In contrast to the geometry of symmetrical bis-dithiocarbamate complexes, TcN heterocomplexes adopt a trigonal bipyramidal (tbp) geometry, in which the two *trans* positions are occupied by the two phosphorous atoms. A sketch of the structure of a dithiocarbamate derivative of the $[^{99m}Tc(N)(PXP)]^{2+}$ fragment, named ^{99m}Tc-DBODC5, which has shown high cardiac uptake in animals with very low lung and liver uptake (Boschi *et al.*, 2003), is reported below:

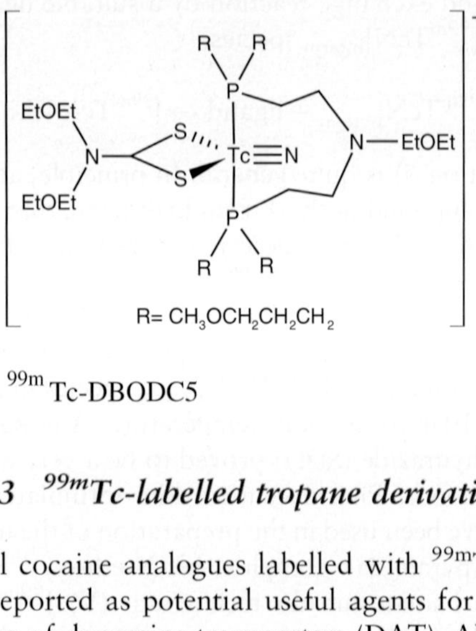

^{99m}Tc-DBODC5

7.1.2.3 *^{99m}Tc-labelled tropane derivatives*

Several cocaine analogues labelled with ^{99m}Tc have been reported as potential useful agents for *in vivo* imaging of dopamine transporters (DAT). All ^{99m}Tc tracers are derived from phenyl tropane-2β-carboxylate, a cocaine derivative with high affinity for DAT. The tropane structure is modified to allow binding of ^{99m}Tc, by a chelating moiety, as for TRODAT-1 (Kung *et al.*, 1997) and for TECHNEPINE (Meltzer *et al.*, 1997), by the use of the so-called 3 + 1 ligand approach and by insertion of the $Tc(I)(CO)_3$ core (TROTEC-1) (Hoepping *et al.*, 1998). For all these compounds, in order to avoid difficulties in crossing the blood brain barrier, the chemistry of the bonding atoms is tailored so that the ^{99m}Tc chelate has a zero net charge.

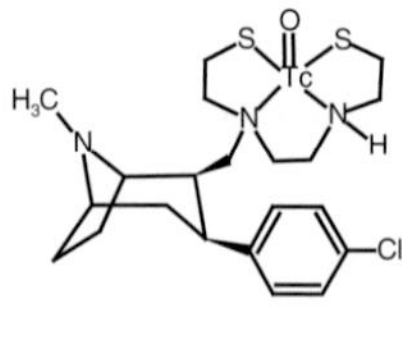

TRODAT-1

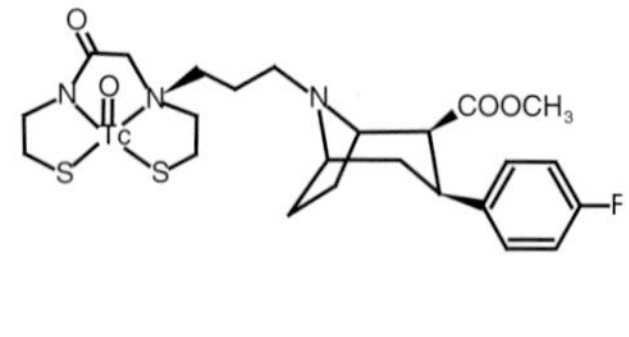

TECHNEPINE

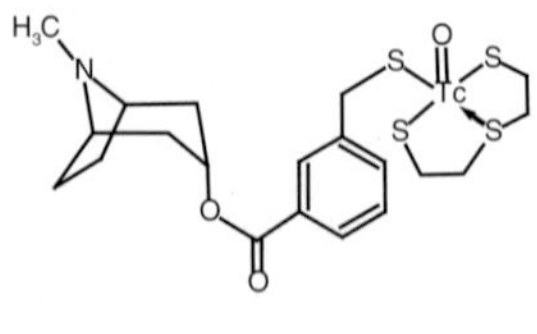

3+1 Tropane derivative

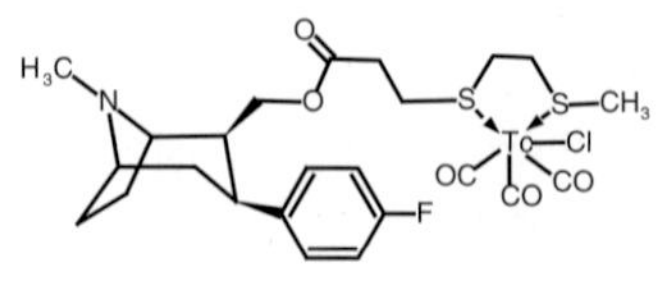

TROTEC-1

Among the several ^{99m}Tc-labelled tropane derivatives reported in literature, TRODAT-1 has showed encouraging results in humans, being able to image the basal ganglia in a distribution that was consistent with selective dopamine transporter binding (Mozley *et al.*, 1998).

7.1.3 Examples of research $^{123/125}$I-labelled compounds

The potential utility of radioiodinated tracers imaging agents has been explored in many pathological models of almost all organs, although receptor imaging in central nervous system and in brain remains the most prolific field of interest for scientist working with radioiodinated tracers. Several reviews focused on ^{123}I-labelled radiopharmaceuticals (Bourguignon *et al.*, 1997; Ross and Seibyl, 2004), or aimed at more general subjects like molecular targeting (Britz-Cunningham and Adelstein, 2003), brain receptor (Heiss and Herholz, 2006) and cardiac nervous system imaging with radiotracers are available (Langer and Halldin, 2002).

7.1.3.1 *Dopamine transporter (DAT) imaging*

To obviate the lack of specificity and the sub-optimal kinetic properties of the commercially available $[(^{123}I)]\beta$-CIT and $[(^{123}I)]$FP-CIT, other DAT tracers have been developed. One of them, PE2I, has showed enhanced specificity (high DAT, low 5HT affinity) and very favourable kinetics properties in rat models (Guilloteau *et al.*, 1998).

As for the other tropane derivative, labelling is performed by an electrophilic attack of $^{123/125}I^+$ on a tributyltin group which, for this particular tracer, is located on a vinylic carbon:

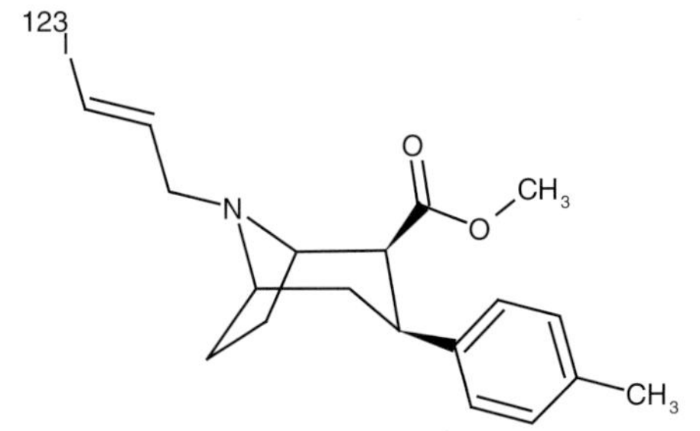

[(^{123}I)]-PE21

7.1.3.2 *Serotonin transporters*

Although the DAT tropane derivative tracers can be used for imaging the central nervous system serotonin transporters (5-HTT), there has been a considerable interest in the development of selective tracers for imaging 5HTT with SPECT. The iodinated compound, 2-((2-((dimethylamino)methyl)phenyl)thio)-5-^{123}I-iodophenylamine (^{123}I-ADAM), has been selected as a good candidate for imaging 5-HTT in non-human primates and in normal volunteers (Newberg *et al.*, 2004).

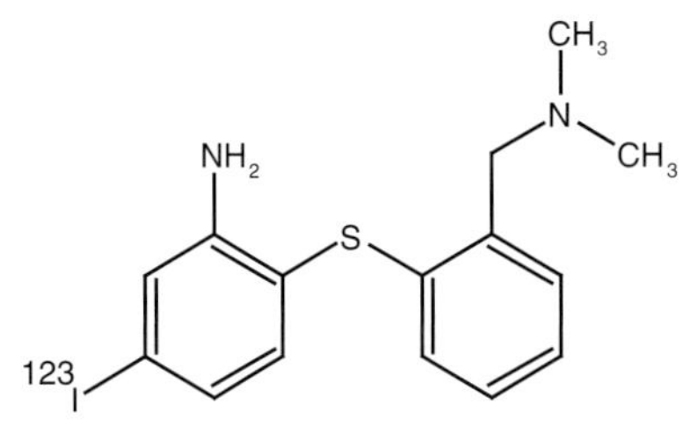

[(^{123}I)]-ADAM

7.1.3.3 *Benzodiazepine receptors*

The molecular targets of benzodiazepines are inhibitory neurotransmitter receptors directly activated by (γ-aminobutiric acid (GABA). GABA transmission is altered in epilepsy, in anxiety and in other psychiatric disorders. Radioiodinated iomazenil (ethyl-5,6-dihydro-7-iodo-5-methyl-6-oxo-4H-imidazo[1,5-a] [1,4]-benzodiazepine-3-carboxylate) was proposed as a potential marker of benzodiazepine receptors after *in vitro* studies and biodistribution in rats (Beer *et al.*, 1990). In contrast to the previous radioiodinated neuroreceptor ligands obtained by electrophilic substitution on a trialkytyltin precursor, radiolabelling of ^{123}I-iomazenil is performed by a nucleophilic exchange on the bromo analogue followed by a HPLC purification that isolates pure ^{123}I-iomazenil with a specific activity better than 74 GBq/μmol.

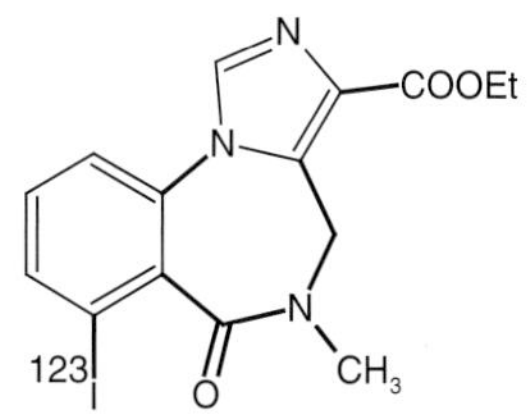

[(^{123}I)]-iomazenil

7.1.3.4 *Radioiodinated tumour agents*

Besides radioiodinated antibodies (see Chapter 4), only very few radioiodinated small molecules have been used with success as receptor-specific tumour agents in clinical research. Indeed, MIBG remains the only example of an approved radioiodinated pharmaceutical in detecting neuroendocrine tumours.

The following is a brief description of radioiodinated compounds that have been extensively used in animal and clinical research.

L-[(^{123}I)] iodo-α-methyl tyrosine. L-^{123}I-α-methyl tyrosine generated much interest after the demonstration that its uptake specifically reflects the increased amino acid transport in gliomas (Langen *et al.*, 1991). Although in humans the tumour-to-background ratios in brain tumours are generally between 1.5 and 2.5, SPECT studies suffer from low count density which can be remedied by a long acquisition time and by the injection of a relatively large dose of radioactivity.

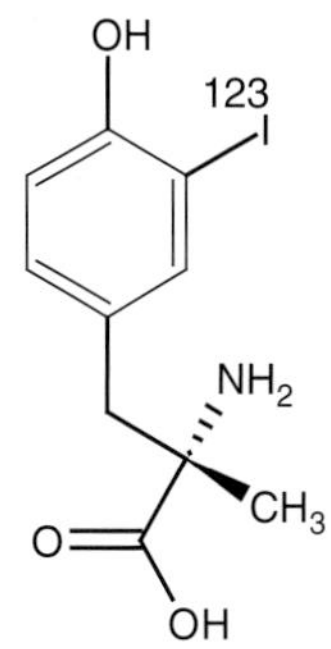

L-[(^{123}I)] Iodo-α-methyl tyrosine

5-[($^{123/125}$I)] Iodo-2′-deoxyuridine. 5-Iodo-2′-deoxyuridine (IUdR) labelled with ^{125}I or ^{123}I has attracted attention because of its structural analogy with the DNA base thymidine. The van der Waals radius of iodine is similar in size to the 5-methyl group of thymidine and may replace this group without altering the overall size of the native molecule. After uptake by the cell, IUdR undergoes biochemical phosphorylation and is incorporated in DNA. Although this mechanism has been proven efficient in cell studies, there is a lack of efficacy in animal and human studies because IUdR suffers of a low uptake and extensive *in vivo* dehalogenation when injected intravenously (Kassis, Adelstein and Mariani 1996).

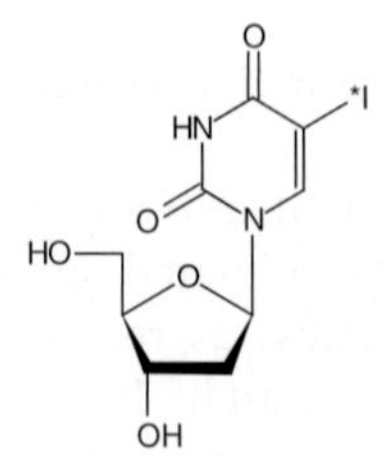

($^{123/125}$I) IUdR

Iodobenzamides derivatives for melanome imaging. Radioiodinated benzamides may accumulate in melanotic tissues, probably by a non-specific interaction mechanism with melanine components. A low-specific activity radioiodinated *N*-alkyl benzamide, namely *N*-(2-diethylaminoethyl)-4-iodobenzamide (I-BZA) showed, 60 min after IV injection to melanoma bearing mice, an uptake of 6.5% and 4% ID/g in murine and human melanotic cells, respectively (Michelot *et al.*, 1991). A Phase II clinical trial evaluating ^{123}I-BZA as an imaging agent of primary melanomas and metastases showed a diagnostic sensitivity of 81%, accuracy of 87% and specificity of 100% (Michelot *et al.*, 1993). The melanoma/non-target tissue ratio has been improved by modifying the pharmacokinetic properties, by replacement of 4-iodo in I-BZA benzene ring by 3-iodo-4-methoxy substituents to yield *N*-(2-diethylaminoethyl)-3-iodo-4-methoxybenzamide (I-MBA; Nicholl *et al.*, 1997). This compound, now under clinical evaluation, is a good example of the development of a radiotracer for clinical purposes by animal studies.

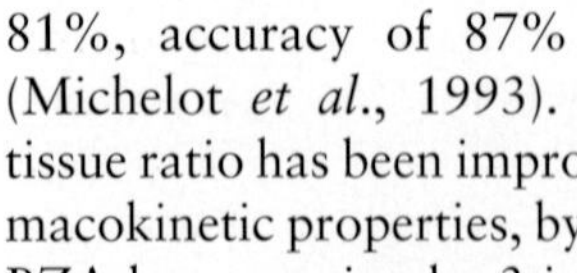

[(^{123}I)] I-BZA

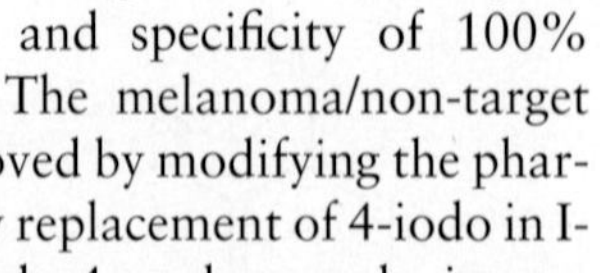

[(^{123}I)] I-MBA

7.1.4 Examples of research positron-emitting radiotracers

7.1.4.1 *Fluoropyridines*

As reported in Chapter 4, Section 8, positron-emitting radiochemistry requires efficient methods in order to introduce the isotope into the molecules of interest. Among others, one recent advance has been the optimization of efficient methods for the fluorine-18 labelling of pyridine rings, which led to the radiosynthesis of [^{18}F]F-A-85380 (2-[^{18}F]fluoro-3-[2(S)-2-azetidinylmethoxy]pyridine), a selective and high-affinity radioligand for the evaluation and study of $\alpha_4\beta_2$ nicotinic acetylcholine receptors (Dollé *et al.*, 1999). [^{18}F]F-A-85380 is synthesized by no-carrier-added nucleophilic aromatic substitution by K[^{18}F]F-K$_{222}$ complex with (3-[2(S)-*N*-(tert-butoxycarbonyl)-2-azetidinylmethoxy]pyridin-2-yl)trimethylammonium trifluoromethanesulfonate as a highly efficient labelling precursor, followed by TFA removal of the Boc protective group and HPLC purification. Total synthesis time is 50–53 min from the end of cyclotron fluorine-18 production and radiochemical yields, with respect to initial [^{18}F]fluoride ion radioactivity, are 68–72% (decay-corrected) and 49–52% (non-decay-corrected). Specific radioactivities at the end of the synthesis reach 2–5 Ci/μmol (74–185 GBq/μmol).

Interestingly, [^{18}F]F-A-85380 and its pyrrolidinyl analogues have been used as precursors for the preparation of ten *N*-substituted-closely related [^{18}F] fluoropyridinyl derivatives. For example after preparation of [^{18}F]F-A-85380, a condensation reaction of the cyclic amine with the appropriate commercially available isothiocyanate, isocyanate and acylhalide was conducted and final HPLC purification was done to give the expected corresponding *N*-aromatic/aliphatic]-thioureas, -ureas, and -amides (Josserand *et al.*, 2006). This approach could well prefigure a type of 'combinatorial radiochemistry' of potential considerable appeal.

7.1.4.2 *Macromolecules*

Complex high-molecular-weight bioactive chemical structures, such as single-stranded oligonucleotides, peptides and proteins are increasingly proposed as radiopharmaceuticals and their applications are rapidly gaining importance in nuclear medicine. Methods for radiolabelling these macromolecules have been of interest for several decades. The direct labelling of these macromolecules with fluorine-18 is generally not feasible, and labelling is usually performed by conjugation of a prosthetic group carrying the radioisotope with a reactive function of the macromolecule. This strategy has the advantage of offering a wide choice of chemical routes, including drastic chemical conditions, for the preparation of the labelled prosthetic group entity, followed by the conjugation of the latter with a macromolecule using mild conditions needed to preserve the latter's integrity (Wilbur, 1992; Okarvi, 2001).

The structures of some fluorine-18-labelled reagents which have already been described in the literature are shown below. Most of them, such as for example the activated ester *N*-succinimidyl 4-[^{18}F]fluorobenzoate ([^{18}F]SFB) (Wester, Hamacher and Stöcklin, 1996), were designed for coupling to the peptide or protein *via* an amino function borne by an amino acid residue (*N*-terminus α-NH_2 or internal lysine ε-NH_2) or *via* an alkylamine linker.

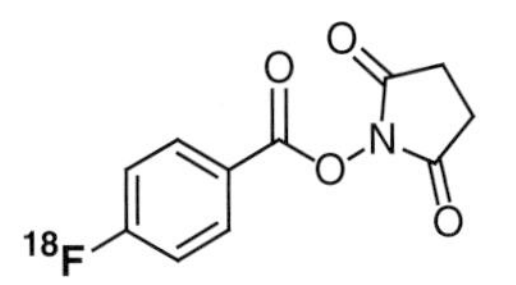

[^{18}F]SFB

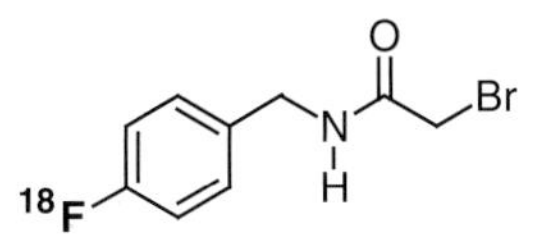

[^{18}F]FBnBrA

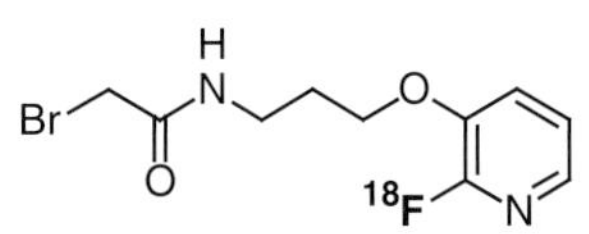

[^{18}F]FPyBrA

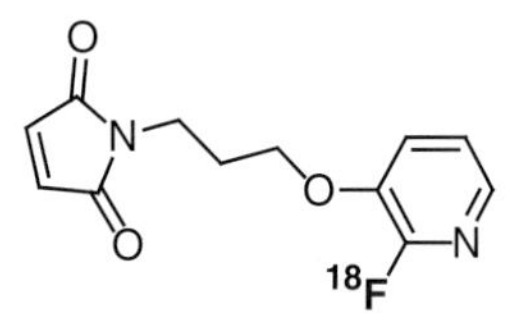

[^{18}F]FPyME

Chemical structures of [^{18}F]SFB, [^{18}F]FBnBrA, [^{18}F]FPyBrA, [^{18}F]FPyME.

N-(4-[^{18}F]fluorobenzyl)-2-bromoacetamide ([^{18}F]-FBnBrA) is another fluorine-18-labelled reagent, designed for coupling to oligonucleotides *via* phosphorothioate monoester functions (Dollé *et al.*, 1997). This reagent has been reliably and routinely applied to the fluorine-18-labelling of natural phosphodiester DNA oligodeoxyribonucleotides. The methodology has also been applied to many popular chemical modifications of oligonucleotides, such as full-length phosphorothioate diester internucleosidic-bond deoxyribonucleotides, hybrid methylphosphonate/phosphodiester internucleosidic-bond deoxyribonucleotides and 2-O-Methyl-modified oligoribonucleotides (see Tavitian, 2003, for review), nucleic acids with a peptidic backbone or PNAs (Hamzavi *et al.*, 2003) and Spiegelmers (L-RNA or L-DNA), (Kuhnast *et al.*, 2003).

Based on the nucleophilic *hetero*aromatic *ortho*-radiofluorination technique, two distinctive 2-[^{18}F]fluoropyridinyl-based reagents ([^{18}F]FPyBrA and [^{18}F]FPyME) were recently designed for the prosthetic labelling of peptides, proteins and oligonucleotides *via* selective conjugation with sulphur-containing functions, and can therefore advantageously be used for the design and development of new peptide- and protein-based radiopharmaceuticals for PET (de Bruin *et al.*, 2005).

7.1.4.3 Generators of positron-emitting radionuclides

In the early days of PET, the availability of positron-emitting radionuclides was limited by the availability of cyclotron facilities. As a consequence, generators of positron-emitting isotopes based on the technetium generator principle (i.e. a daughter positron-emitter generated by decay of a long lived parent isotope) were proposed by several groups (Ehrhardt and Welch, 1978; Loc'h, Maziere and Comar, 1980), but their uneasiness of use, relatively low yield and the fact that the positron-emitters generated are heavy metals which require indirect conjugation techniques to the radiotracer led to little success. Recently, there has been a regain of interest in positron-emitter generators, due to the interest in macromolecular radiotracers, such as peptides, in which addition of a metal conjugate leads to relatively minor modifications of the molecule in comparison to smaller radiotracers. Generator-based PET tracers are cheap to produce, offer long shelf lives, and radiolabelling can be performed in the absence of an on-site cyclotron facility; y a t il d autres noyaux que Ga concernés?

Gallium-68 ($T_{1/2}$ 68 min) is a generator-based positron-emitting radionuclide obtained from its parent ^{68}Ge ($T_{1/2}$ 271 days). ^{68}Ga may be eluted once or twice a day from commercially available ^{68}Ge/^{68}Ga generators with a shelf life of approximately one year, and this metal easily coordinates to the commonly used conjugate 1,4,7,10-tetraazacyclododecane-1,4,7,10-tetraacetic acid (DOTA). Recently, automated kits have been proposed for the production of several ^{68}Ga-labelled PET radiopharmaceuticals, such as peptides (Maecke, Hofmann and Haberkorn, 2005), oligonucleotides (Roivainen *et al.*, 2004), antibodies (Schuhmacher *et al.*, 2001), or diverse bioconjugates (Velikyan, Beyer and Langstrom, 2004). In the eyes of their promoters, positron-emitting generators may well represent the future of PET research radiotracers.

References

Baldas, J., Bonnyman, J., 1985. "Substitution reaction of 99mTcNCl4- -a new route to a new class of 99mTc-radiopharmaceuticals." *Int. J. Appl. Radiat. Isot.* **36**, 133–139.

Beer H. F., Blauenstein P. A., Hasler P. H., Delaloye B., Riccabona G., Bangerl I., Hunkeler W. *et al.*, 1990. "In vitro and in vivo evaluation of iodine-123-Ro 16-0154: A new imaging agent for SPECT investigations of benzodiazepine receptors." *J. Nucl. Med.* **31**, 1007–1014.

Bolzati, C., Boschi, A., Duatti, A., Refosco, F., Tisato, F., Bandoli, G., 2000. "Geometrically controlled selective formation of nitrido(V) asymmetrical heterocomplexes with bidentate ligands." *J. Am. Chem. Soc.* **122**, 4510–4511.

Boschi, A., Uccelli, L., Bolzati, C., Duatti, A., Sabba, N. Moretti, E., Di Domenico, G. *et al.*, 2003. "Synthesis and biologic evaluation of monocationic asymmetric 99mTc-nitride heterocomplexes showing high heart uptake and improved imaging properties." *J. Nucl. Med.* **44**, 806–814.

Bourguignon, M. H., Pauwels, E. K. J., Loc'h, C., Mazière, B., 1997. "Iodine-123 labelled radiopharmaceuticals and single photon emission tomography: A natural liaison." *Eur J. Nucl. Med.* **24**, 331–344.

Britz-Cunningham, S. H., Adelstein, S. J., 2003. "Molecular targeting with radionuclides: state of the science." *J. Nucl. Med.* **44**, 1945–1961.

de Bruin, B., Kühnast, B., Hinnen, F., Yaouancq, L., Amessou, M., Johannes, L., Samson, A., Boisgard, R., Tavitian, B., Dollé, F., 2005. "1-[3-(2-[^{18}F]Fluoropyridin-3-yloxy)propyl]pyrrole-2,5-dione: Design, synthesis and radiosynthesis of a new [^{18}F]fluoropyridine-based maleimide reagent for the labeling of peptides and proteins." *Bioconjugate. Chem.* **16**, 406–420.

Dollé, F., Hinnen, F., Vaufrey, F., Tavitian, B., Crouzel, C., 1997. "A general method for labeling oligodeoxynucleotides with ^{18}F for *in vivo* PET imaging." *J. Label. Compds. Radiopharm.* **39**, 319–330.

Dollé, F., Dolci, L., Valette, H., Hinnen, F., Vaufrey, F., Guenther, I., Fuseau, C., Coulon, C., Bottlaender, M., Crouzel, C., 1999. "Synthesis and nicotinic acetylcholine receptor in vivo binding properties of 2-fluoro-3-[2(S)-2-azetidinylmethoxy]pyridine: A new positron emission tomography ligand for nicotinic receptors." *J. Med. Chem.* 42, 2251–2259.

Ehrhardt, G. J., Welch, M. J., 1978. "A new germanium-63/gallium-68 generator." *J. Nucl. Med.* **19** (8), 925–929.

Guilloteau, D., Emond, P., Baulieu, J-L., Garreau, L., Frangin, Y., Pourcelot, L., Mauclaire, L. *et al.*, 1998. "Exploration of the dopamine transporter: In vitro and in vivo characterization of a high-affinity and high-specificity iodinated tropane derivative (E)-*N*-(3-iodoprop-2-enyl)-2β-carbomethoxy-3β-(4′-methylphenyl)nortropane (PE2I)." *Nucl. Med. Biol.* **25**, 331–337.

Hamzavi, R., Dollé, F., Tavitian, B., Dahl, O., Nielsen, P., 2003. "Modulation of the pharmacokinetic properties of PNA: Preparation of galactosyl, mannosyl, fucosyl, *N*-acetylgalactosaminyl, and *N*-acetylglucosaminyl derivatives of aminoethylglycine peptide nucleic acid monomers and their incorporation into PNA oligomers." *Bioconjugate. Chem.* **14**:941–954.

Heiss, W-D., Herholz, K., 2006. "Brain receptor imaging." *J. Nucl. Med.* **47**, 302–312.

Hoepping, A., Brust, P., Berger, R., Seifert, S., Spies, H., Alberto, R., Johannsen, B., 1998. "TROTEC-1: A new high-affinity ligand for labeling of the dopamine transporter." *J. Med. Chem.* **41**, 4429–4432.

Jeetly, P., Sabharwal, N. K., Soman, P., Kinsey, C., Raval, U., Bhonsle, U., Lahiri, A., 2004. "Comparison between Tc-^{99m}N-NOEt and Tl-201 in the assessment of patients with known or suspected coronary artery disease." *J. Nucl. Cardiol.* **11**, 664–672.

Josserand, V., Pélerin, H., deBruin, B., Jego, B., Kühnast, B., Hinnen, F., Ducongé, F., Boisgard, R., Beuvon, F., Chassoux, F., Daumas-Duport, C., Ezan, E., Dollé, F., Mabondzo, A., Tavitian, B., 2006. "Evaluation of drug penetration into the brain: A double study by *in vivo* imaging with positron emission tomography and using an *in vitro* model of the human blood-brain barrier." *J. Pharmacol. Exp. Therap.* **316** (1), 79–86.

Kassis, A. I., Adelstein, S. J., Mariani, G., 1996. "Radiolabeled nucleoside analogs in cancer diagnosis and therapy." *Q. J. Nucl. Med.* **40**, 301–319.

Kung, M. P., Stevenson, D. A., Plossl, K., Meegalla, S. K., Becwith, A., Essman, W. D., Mu, M. et al. 1997. "[99mTc]TRODAT-1: A novel technetium-99m complex as a dopamine transporter imaging agent." *Eur. J. Nucl. Med.* **24**, 372–380.

Kühnast, B., Klussmann, S., Hinnen, F., Boisgard, R., Rousseau, B., Fürste, J. P., Tavitian, B., Dollé, F., 2003. "Fluorine-18- and iodine-125 labelling of Spiegelmers." *J. Label. Compds. Radiopharm.* **46**, 1205–1219.

Langen, K. J., Roosen, N., Coenen, H. H., Kuikka, J. T., Kuwert, T., Herzog, H., Stocklin, G. *et al.*, 1991. "Brain and brain tumor uptake of

L-3-[^{123}I]iodo-alpha-methyl tyrosine: Competition with natural L-amino acids." *J. Nucl. Med.* **32**, 1225–1229.

Langer, O., Halldin, C., 2002. "PET and SPET tracers for mapping the cardiac nervous system." *Eur. J. Nucl. Med. Mol. Imaging* **29**, 416–434.

Loc'h, C., Maziere, B., Comar, D., 1980. "A new generator for ionic gallium-68." *J. Nucl. Med.* **21**(2),171–173.

Maecke, H. R., Hofmann, M., Haberkorn, U., 2005 "(68)Ga-labeled peptides in tumor imaging." *J. Nucl. Med.* **46**(Suppl 1), 172S–178S.

Meltzer, P., Blundell, P., Jones, A. G., Mahmood, A., Garada, B., Zimmerman, R. E., Davison, A. *et al.*, 1997. "A technetium-99m SPECT imaging agent which targets the dopamine transporter in primate brain." *J. Med. Chem.* **40**, 1835–1844.

Michelot, J. M., Moreau, M. F., Labarre, P. G., Madelmont, J. C., Veyre, A. J., Papon, J. M., Parry, D. F., 1991. "Synthesis and evaluation of new iodine-125 radiopharmaceuticals as potential tracers for malignant melanoma." *J. Nucl. Med.* 32, 1573–1580.

Michelot, J. M., Moreau, M. F., Veyre, A. J., Bonafous, J. F., Bacin, F. J., Madelmont, J. C., Bussiere, F. *et al.*, 1993. "Phase II scintigraphic clinical trial of malignant melanoma and metastases with iodine-123-*N*-(2-diethylaminoethyl-4-iodobenzamide)." *J. Nucl. Med.* **34**, 1260–1266.

Mozley, P. D., Stubbs, J. B., Plossl, K., Dresel, S. H., Barraclough, E. D., Alavi, A., Araujo, L. I. *et al.*, 1998. "Biodistribution and dosimetry of TRODAT-1: A technetium-99m tropane for imaging dopamine transporters." *J. Nucl. Med.* **39**, 2069–2076.

Newberg, A. B., Plössl, K., Mozley, P. D., Stubbs, J. B., Wintering, N., Udeshi, M., Alavi, A. *et al.*, 2004. "Biodistribution and imaging with ^{123}I-ADAM: A serotonin transporter imaging agent. *J. Nucl. Med.* **45**, 834–841.

Nicholl, C., Mohammed, A., Hull, W. E., Bubeck, B., Eisenhut, M., 1997. "Pharmacokinetics of iodine-123-IMBA for melanoma imaging." *J. Nucl. Med.* **38**, 127–133.

Okarvi, S. M., 2001. "Recent progress in fluorine-18 labelled peptide radiopharmaceuticals." *Eur. J. Nucl. Med.* **28**, 929–938.

Pasqualini, R., Duatti, A., Bellande, E., Comazzi, V., Brucato, V., Hoffschire, D., Fagret, D. *et al.* 1994. "Bis(dithocarbamato) nitrido technetium-99m radiopharmaceuticals: A class of neutral myocardial imaging agents." *J. Nucl. Med.* **35**, 334–341.

Roivainen, A., Tolvanen, T., Salomaki, S., Lendvai, G., Velikyan, I., Numminen, P., Valila, M., Sipila, H., Bergstrom, M., Harkonen, P., Lonnberg, H., Langstrom, B., 2004. "68Ga-labeled oligonucleotides for *in vivo* imaging with PET." *J. Nucl. Med.* **45**(2), 347–355.

Ross, S. A., Seibyl, J. P., 2004. "Research applications of selected 123I-labeled neuroreceptor SPECT imaging ligands." *J. Nucl. Med. Technol.* **32**, 209–214.

Schuhmacher, J., Kaul, S., Klivenyi, G., Junkermann, H., Magener, A., Henze, M., Doll, J., Haberkorn, U., Amelung, F., Bastert, G., 2001. "Immunoscintigraphy with positron emission tomography: Gallium-68 chelate imaging of breast cancer pretargeted with bispecific anti-MUC1/anti-Ga chelate antibodies." *Cancer Res.* **61**(9), 3712–3717.

Tavitian, B., 2003. "*In vivo* imaging with oligonucleotides for diagnosis and drug development." *Gut* **52**, iv40– iv47.

Velikyan, I., Beyer, G. J., Langstrom, B., 2004. "Microwave-supported preparation of (68)Ga bioconjugates with high specific radioactivity." *Bioconjugate Chem.* **15**(3), 554–560.

Wester HJ, Hamacher K, Stöcklin, G., 1996. "A comparative study of N.C.A. fluorine-18 labeling of proteins via acylation and photoactivation." *Nucl. Med. Biol.* **23**, 365–372.

Wilbur, D.S., 1992. "Radiohalogenation of proteins: An overview of radionuclides, labeling methods and reagents for conjugate labeling." *Bioconjugate Chem.* **3**, 433–470.

7.2 Multimodal constructs for magnetic resonance imaging

Willem J.M. Mulder, Gustav J. Strijkers and Klaas Nicolay

Many pathological processes are accompanied by (I) an increased vascular permeability, (II) the up-regulation of cell surface receptors and (III) a massive infiltration of cells of the immune system into diseased tissue. Visualization of these pathophysiological parameters with magnetic resonance imaging (MRI) has gained much interest in recent years and is dependent on the combination of contrast enhanced MRI with an appropriate contrast agent. Recent developments in chemistry have led to the development of many different probes for MRI. In this chapter, examples of innovative multimodal MRI probes will be given.

Multimodal MRI probes have been demonstrated to be useful when combined with a label for a different, preferably complementary, imaging modality.

One of the first examples of such a bimodal probe was reported by Huber *et al.* (1998). They describe bifunctional contrast-enhancing agents for optical detection and MRI. The main advantage of this combination is that it integrates optical properties for high-resolution fluorescence microscopy and imaging and magnetic properties for *in vivo* MRI visualization of intact and opaque organisms. More recently, MRI probes not only for the combination with other imaging modalities (such as ultrasound, CT and nuclear methods) but also for the combination with drug targeting have been developed (Lanza *et al.*, 2004).

For the assessment of increased vascular and tissue permeability, long circulating, macromolecular contrast agents, of which the clearance can preferably be controlled, are required. Dafni *et al.* (2003) developed albumin triply labelled with Gd-DTPA, a fluorescent label and biotin. The paramagnetic and fluorescent properties of this interstitial macromolecular agent allow the investigation of angiogenesis associated vascular hyperpermeability with MRI. Active clearance of this contrast material was achieved by rapidly removing it from the circulation with an avidin-chase, thus also allowing early experimental differentiation between vascular leak and lymphatic drain. Furthermore, with histology and confocal microscopy, this contrast agent could also be visualized in the tumour and tissue sections. The methodology used is depicted in Figure 7.2.1.

Figure 7.2.1 Triply labelled albumin consists of BSA labelled with Gd-DTPA, a fluorescent marker and biotin. (a) The biotin group allows the active removal of this contrast material from the circulation by a chase with avidin. The difference MIPs (Maximal intensity projections) clearly demonstrate that upon injection of avidin the MR signal enhancement becomes less. (b) Albumin with different fluorescent labels allows the differentiation between lymphatic uptake and blood vessel. (1) Carboxyfluorescein (FAM)-labelled biotin6-BSA-GdDTPA injected 60 min before tissue excision. (2) Carboxyrhodamine (ROX)-labelled BSA injected 3 min before tissue retrieval. (3) Overlay image demonstrates the lymphatic uptake in green. (*Taken from Van den Burg et al., 2003 with permission from Wiley Interscience*)

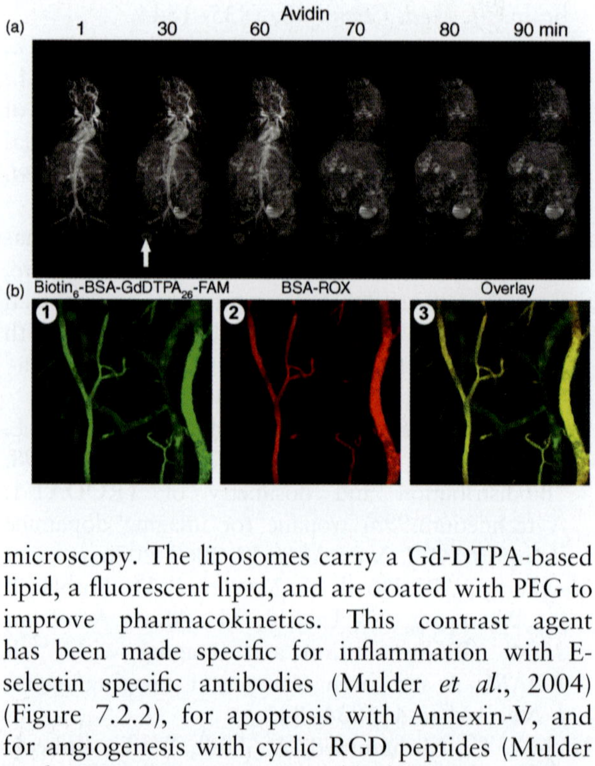

Visualizing the up-regulation of cell surface receptors is an attractive way for characterizing and staging disease progression. Antibodies and peptides, conjugated to a radiolabel, have been used extensively to image such receptors with nuclear methods like SPECT, PET and scintigraphic imaging. To deal with the low inherent sensitivity of MRI, compared to these nuclear methods, very potent, usually nanoparticulate, MRI contrast agents conjugated to targeting ligands are required. The size of these conjugates poses a limit to the receptors that can be reached. Cell surface receptors expressed at the vasculature of pathological tissue, therefore, are especially attractive because they can be reached directly from the blood circulation and do not require the contrast agents to leave the vascular compartment and penetrate into the tissue. An early study on such an approach was reported by Sipkins *et al.* (1998). They made use of paramagnetic, polymerized liposomes conjugated with $\alpha v\beta 3$-specific antibodies via an avidin-biotin linkage, for the detection of $\alpha v\beta 3$ integrin expressed at angiogenic tumour blood vessels with MRI.

Mulder *et al.* (2004) have developed a bimodal targeted liposomal contrast agent, for the detection of molecular markers with both MRI and fluorescence microscopy. The liposomes carry a Gd-DTPA-based lipid, a fluorescent lipid, and are coated with PEG to improve pharmacokinetics. This contrast agent has been made specific for inflammation with E-selectin specific antibodies (Mulder *et al.*, 2004) (Figure 7.2.2), for apoptosis with Annexin-V, and for angiogenesis with cyclic RGD peptides (Mulder *et al.*, 2005) (Figure 7.2.3(a)). The latter material was used to identify the angiogenic endothelium in tumour bearing mice with *in vivo* MRI and *ex vivo* fluorescence microscopy (Mulder *et al.*, 2005). The cyclic RGD-peptide has high affinity for the $\alpha v\beta 3$-integrin, which is upregulated at endothelial cells of angiogenic blood vessels. MRI revealed that upon intravenous injection of the contrast agent (Figure 7.2.3(a)), the RGD-liposomes localized to a large extent in the tumour rim, which is known to have the highest angiogenic activity (Figure 7.2.3(b)). It was established with fluorescence microscopy that RGD-liposomes were exclusively associated with tumour blood vessels (Figure 7.2.3(c)). This study illustrates the

Figure 7.2.2 *In vitro* fluorescence microscopy and MRI of E-selectin expression. (a) Endothelial cells were stimulated to express E-selectin and incubated with E-selectin specific bimodal liposomes (sample 2). Samples (1), (3) and (4) served as controls. (b) Fluorescence microscopy revealed the association of the bimodal liposomes with the cells in (2) and not in case of the control incubations. (c) The association of the liposomes with the cells in sample (2) resulted in an increased signal intensity of the cell pellet on T_1-weighted images as compared to the control samples

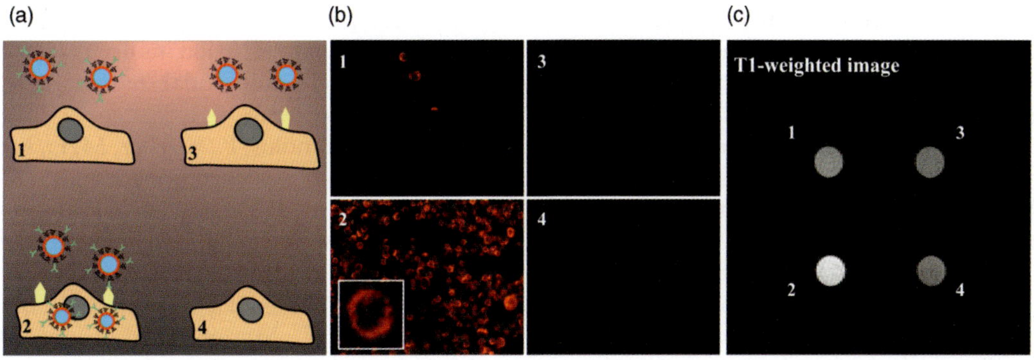

strengths of the use of complementary visualization techniques and multimodal contrast agents.

Lanza and Wickline have jointly developed a nanoparticulate contrast agent that consists of a perfluorocarbon core, which is covered with a monolayer of lipids (Lanza *et al.*, 2004). Initially, they used their technology in combination with ultrasound, exploiting the acoustic properties of the contrast agent. Their technology was further extended and made suitable for MRI by incorporating paramagnetic amphiphilic chelates in the lipid monolayer. Furthermore, because of the perfluorocarbon core of the nanoparticles, they can be imaged using ^{19}F MRI. The main advantage of this technique is that it allows imaging of the contrast material without any background, so-called hot spot imaging. More recently, they have modified these nanoparticles for CT and SPECT imaging. In addition, the perfluorocarbon nanoparticles have been used for

Figure 7.2.3 (a) Tumour bearing mice were injected with $\alpha v\beta 3$-specific bimodal RGD-liposomes. (b) T_1-weighted images measured before (left) and 35 min after (middle, right) the injection of the RGD-conjugated liposomes. The arrow (middle image) indicates a bright region appearing at the periphery of the tumour. Signal enhancement in pixels in the tumour of at least three times the noise level is colour coded according to the pseudo-colour scale on the right. (c) Fluorescence microscopy of DAPI coloured ten-μm sections from dissected tumours revealed circular (left) and longitudinal (middle) distribution patterns of the rhodamine fluorescence from RGD-liposomes and these were invariably associated with blood vessels. A slice through the middle of the tumour (right) showed no rhodamine fluorescence

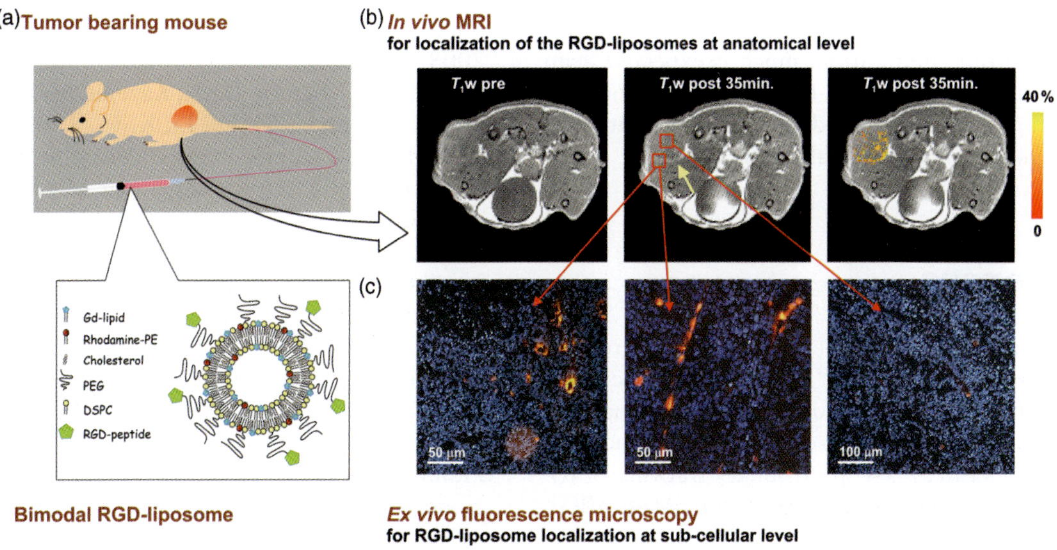

therapeutic purposes in animal models of cardiovascular disease and cancer, allowing image-guided therapy.

Magneto-optical probes (CLIO-Cy5.5), based on superparamagnetic iron oxide nanoparticles with a cross-linked dextran coating (CLIO) labelled with the near-infrared fluorochrome Cy5.5 have been used for VCAM-1 detection in a mouse model of atherosclerosis (Kelly *et al.*, 2005) and for the visualization of cardiomyocyte apoptosis (Sosnovik *et al.*, 2005). In both studies, the combination of MR and optical imaging was shown to be very useful, and the correlation between both techniques was good.

The MR visualization of the infiltration of cells of the immune system, like T-lymphocytes or monocytes at pathological sites and the migration of stem cells has recently received much attention. Cells are magnetically labelled *ex vivo* or *in vivo*, with either superparamagnetic nanoparticles or paramagnetic agents to allow the detection of the cells with MRI. The use of multimodal contrast agents for cell labelling is also extremely useful for improved understanding of the fate of the contrast agent, to assess the viability, and to validate the MRI findings. Examples of bimodal paramagnetic cell labelling agents are gadolinium rhodamine dextran (GRID) (Modo *et al.*, 2002) and gadolinium-rhodamine nanoparticles (Vuu *et al.*, 2005). Tat(FITC)-Cy3.5-CLIO is an example of an iron oxide based cell labelling agent with two different optical labels for detailed understanding of the fate of this nanoparticulate conjugate upon internalization by cells (Koch *et al.*, 2003).

In conclusion, bimodal contrast agents for combined MR and optical imaging are excellent tools for studying pathological processes. The implementation of other tracers, for example SPECT or ultrasound, within the same probe will enhance the applicability of the probe and may facilitate the choice for the optimal detection or imaging method with the same probe. In addition, much is expected from the combination of a probe for imaging and therapy, which may lead to improved and faster evaluation of therapy and a more personalized treatment.

References

Dafni, H., Gilead, A., Nevo, N., Eilam, R., Harmelin, A., Neeman, M., 2003. "Modulation of the pharmacokinetics of macromolecular contrast material by avidin chase: MRI, optical, and inductively coupled plasma mass spectrometry tracking of triply labeled albumin." *Magn. Reson. Med.* **50**(5), 904–914.

Huber, M. M., Staubli, A. B., Kustedjo, K., Gray, M. H., Shih, J., Fraser, S. E., Jacobs, R. E., Meade, T. J., 1998. "Fluorescently detectable magnetic resonance imaging agents." *Bioconjugate Chem.* **9**(2), 242–249.

Kelly, K. A., Allport, J. R., Tsourkas, A., Shinde-Patil, V. R., Josephson, L., Weissleder, R. 2005. "Detection of vascular adhesion molecule-1 expression using a novel multimodal nanoparticle." *Circ. Res.*, **96**(3), 327–336.

Koch, A. M., Reynolds, F., Kircher, M. F., Merkle, H. P., Weissleder, R., Josephson, L., 2003. "Uptake and metabolism of a dual fluorochrome Tat-nanoparticle in HeLa cells." *Bioconjugate Chem.* **14**(6), 1115–1121.

Lanza, G. M., Winter, P., Caruthers, S., Schmeider, A., Crowder, K., Morawski, A., Zhang, H., Scott, M. J., Wickline, S. A., 2004. "Novel paramagnetic contrast agents for molecular imaging and targeted drug delivery." *Curr. Pharm. Biotechnol* 5(6), 495–507.

Modo, M., Cash, D., Mellodew, K., Williams, S. C., Fraser, S. E., Meade, T. J., Price, J., Hodges, H., 2002. "Tracking transplanted stem cell migration using bifunctional, contrast agent-enhanced, magnetic resonance imaging." *Neuroimage* **17**(2), 803–811.

Mulder, W. J., Strijkers, G. J., Griffioen, A. W., van Bloois, L., Molema, G., Storm, G., Koning, G. A., Nicolay, K. 2004. "A liposomal system for contrast-enhanced magnetic resonance imaging of molecular targets." *Bioconjugate Chem.* **15**(4), 799–806.

Mulder, W. J., Strijkers, G. J., Habets, J. W., Bleeker, E. J., van der Schaft, D. W., Storm, G., Koning, G. A., Griffioen, A. W., Nicolay, K., 2005. "MR molecular imaging and fluorescence microscopy for identification of activated tumor endothelium using a bimodal lipidic nanoparticle." *FASEB J.* **19**(14), 2008–2010.

Sipkins, D. A., Cheresh, D. A., Kazemi, M. R., Nevin, L. M., Bednarski, M. D., Li, K. C., 1998. "Detection of tumor angiogenesis *in vivo* by αVβ3-targeted magnetic resonance imaging." *Nat. Med.* **4**(5), 623–626.

Sosnovik, D. E., Schellenberger, E. A., Nahrendorf, M., Novikov, M. S., Matsui, T., Dai, G., Reynolds, F., Grazette, L., Rosenzweig, A., Weissleder, R., Josephson, L., 2005. "Magnetic resonance imaging of cardiomyocyte apoptosis with a novel magneto-optical nanoparticle." *Magn. Reson. Med.* **54**(3), 718–724.

Vuu, K., Xie, J., McDonald, M. A., Bernardo, M., Hunter, F., Zhang, Y., Li, K., Bednarski, M., Guccione, S., 2005. "Gadolinium-rhodamine nanoparticles for cell labeling and tracking via magnetic resonance and optical imaging." *Bioconjugate Chem.* **16**(4), 995–999.

7.3 Fluorescence reporters for biomedical imaging

Benedict Law and Ching-Hsuan Tung

Current advances in optical technologies may facilitate the development of biomedical imaging. Approaches based on fluorescence are economical and also potentially safer than the existing magnetic resonance (MR) and radiological techniques. Biomolecules such as water, lipid, and haemoglobin and deoxyhaemoglobin have minimal absorption coefficients at the near-infrared (NIR) window (700 – 1000 nm) (Ntziachristos, Bremer and Weissleder; 2003). For these reasons, fluorescent reporters that absorb and emit in this region have deeper tissue penetration and are ideally suited for *in vivo* imaging. *De novo* fluorescence probe designs are aimed to identify anatomic features as well as provide accurate and reliable information. Recently, fluorochromes targeted probes, and fluorogenic enzyme-mediated reporters derived from polymers or nanoparticles have been developed to improve the visualization of tissues, organs, physiological processes and disease status. When these compounds are employed for animal research, they allow a direct and real-time observation of physiological events *in vivo*. In this chapter, we will summarize the recent advance in fluorophores and optical contrast agents used in animal experiments.

7.3.1 *Fluorochromes*

The ideal fluorochrome for *in vivo* imaging should have excitation and emission in the NIR window. It should also be biocompatible, have a high extinction coefficient, and good quantum yield, such that the fluorescence output from very low concentrations of reporters can be detected. Organic cyanine dyes are the most commonly employed fluorochromes for animal imaging. However, a few inorganic quantum dots (qdots) have also been used for *in vivo* applications.

7.3.1.1 *Indocyanine dyes*

During the last decade, indocyanine green (ICG) derivatives have been synthesized to improve the photophysical and chemical properties of the dyes (Mishra *et al.*, 2000). The core structure of cyanine dyes consists of two heterocyclic rings, which are linked by a polymethine bridge (Figure 7.3.1). The addition of multiple charged groups onto the fluorophore can reduce the tendency of the dyes to aggregate (Lin, Weissleder and Tung 2002), thus increasing their solubility in aqueous media. The introduction of extra aromatic rings at the heterocycles end can be used to tune the absorption and emission bands to longer wavelengths. Extending the length of the polymethine bridge with an extra –CH=CH– unit creates an additional red shift of about 100 nm. Furthermore, the chemical stability can be improved *via* the introduction of moieties such as cyclohexenyl group, to increase the rigidity of the polymethine linker (Oswald *et al.*, 1999; Pham *et al.*, 2003).

Figure 7.3.1 Chemical structures of representative indocyanine derivatives used for animal imaging

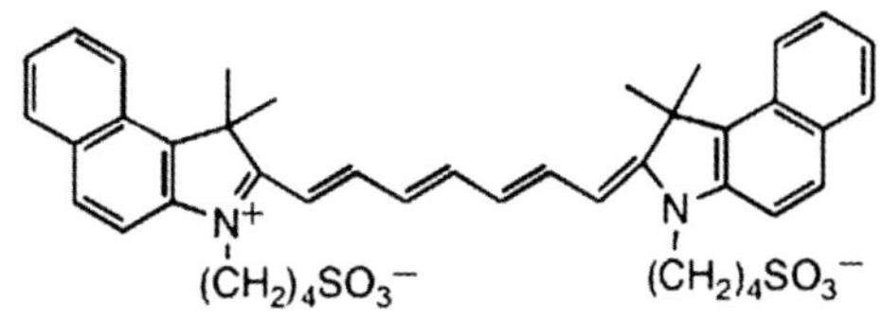

ICG. Ex/Em = 775/831 nm

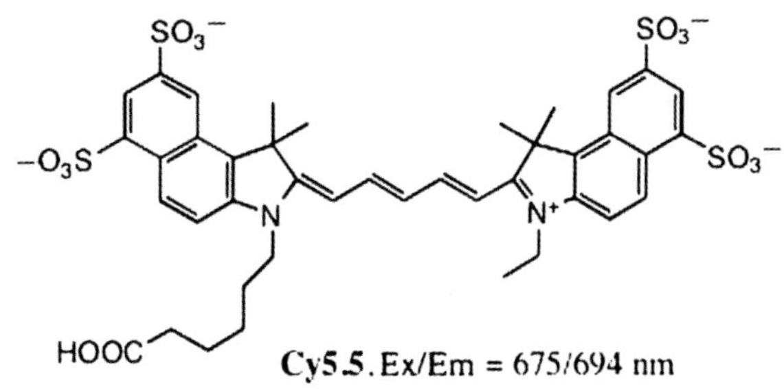

Cy5.5. Ex/Em = 675/694 nm

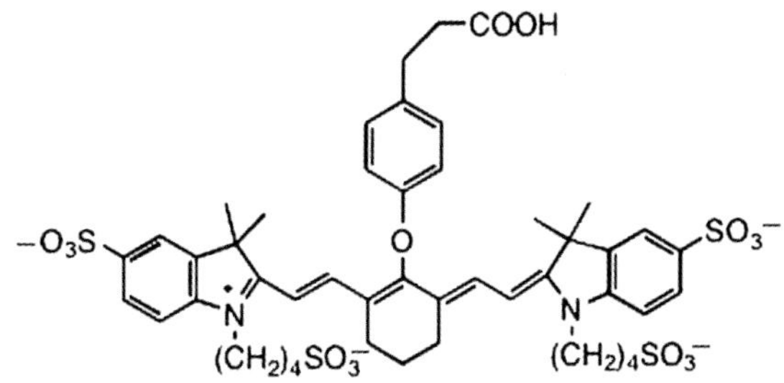

IRDye78. Ex/Em = 768/796 nm

Biological applications of ICG are limited because it lacks a synthetic handle for further chemical conjugation. For labelling purposes, different

Figure 7.3.2 A schematic illustration of multi-functional qdot. Coatings can increase the aqueous solubility and at the same time, reduce the toxicity. Ligands such as peptides, antibodies and proteins are conjugated on the surface of qdots for specific site targeting

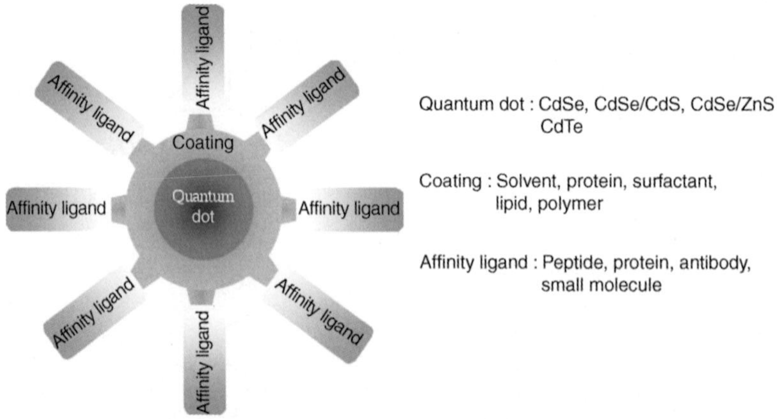

activated mono- or bi-functional groups including hydoxysuccinimide, iodoacetyl, hyrazide, maleimide and phosphoramidite esters have been introduced at the distal end of various derivatives (Mujumdar *et al.*, 1996; Licha *et al.*, 2002; Lin, Weissleder and Tung, 2003). These functional groups allow for labelling of different bio-molecules such as peptide, proteins, antibodies and DNA (Southwick *et al.*, 1990; Neblock *et al.*, 1992; Mujumdar *et al.*, 1993; Yu *et al.*, 1994; Randolph and Waggoner, 1997; Gruber *et al.*, 2000). One commonly used indocyanine derivative is Cy5.5^{TM}, which has a high absorption coefficient (190 000 M^{-1} cm^{-1}) and good quantum yield (0.23) (Figure 7.3.1) (Wessendorf and Brelje, 1992).

7.3.1.2 *Quantum dots*

Quantum dots (qdots) are nanometer-sized crystalline particles engineered from groups II–VI elements (e.g. CdSe, CdTe, CdS, and ZnSe), group III–V elements (e.g. InP and InAs) or IV–VI (e.g. PbSe, PbTe) and have been investigated for the last two decades as an alternative to organic fluorophores (Bruchez *et al.*, 1998; Chan *et al.*, 2002; Larson *et al.*, 2003; Smith, Gao and Nie, 2004; Gao *et al.*, 2005). This technology offers several advantages over the traditional dyes. The absorbance onset and emission maximum (400 – 2000 nm) can be easily tuned by varying the size of the nanocrystals, and they typically give narrow emission and broad absorption spectra (Bruchez *et al.*, 1998). Different discrete nanometer-sized crystals can be excited by a single source at a specific wavelength and give rise to multicoloured emission spectra (Jaiswal *et al.*, 2003). In addition, due to their inorganic composition, qdots are less susceptible to photobeaching and, therefore, can be used for long term monitoring of physiological events (Jaiswal *et al.*, 2003). For biological labelling (Figure 7.3.2), it is necessary to modify qdots to increase aqueous solubility (Michalet *et al.*, 2005) so that biomolecules such as peptides (Akerman *et al.*, 2002; Pinaud *et al.*, 2004), proteins (Gao, Chan and Nie, 2002; Wu *et al.*, 2003), antibodies (Jaiswal *et al.*, 2003; Chan and Nie, 1998; Goldman *et al.*, 2002a,b) and oligonucleotides (Lakowicz *et al.*, 2000; Dubertret *et al.*, 2002) can be conjugated by either covalent or electrostatic interactions (Michalet *et al.*, 2005).

Because qdots commonly contain elements such as cadmium and selenium, there are important toxicity issues that need to be addressed when used for *in vivo* imaging (Akerman *et al.*, 2002; Wu *et al.*, 2003; Ballou *et al.*, 2004; Hoshino *et al.*, 2004; Kim *et al.*, 2004; Morgan *et al.*, 2005). For example, CdSe qdots can be oxidized by either air or UV irradiation to produce free cadmium cations. Low level of cadmium (100 μM) are known to have cytotoxic effects in heptocytes (Santone, Acosta and Bruckner, 1982). By coating the qdots with ZnS, dihydrolipoic acid (DHLA) (Jaiswal *et al.*, 2003), bovine serum albumin (BSA) (Gao, Chan and Nie, 2002) or polyacrylate (Wu *et al.*, 2003), the stability of the nanoparticles can be increased, thus reducing their toxicity.

7.3.2 Designs of imaging contrast agents

ICG is a clinically approved contrast agent for monitoring liver function (Rowell *et al.*, 1965), cardiac

Figure 7.3.3 Different approaches for *in vivo* imaging. (a) Simple molecules such as organic fluorophores or qdots are commonly used to image anatomical changes in tumours. This includes the 'enhanced permeability and retention' effect in angiogenesis. (b) Fluorophores can be coupled to ligands including drug inhibitors, peptides, proteins or antibodies for site specific targeting. (c) Probes can be designed to image the changes of functional events in different disease status. Signals can be detected only after specific activation by targeted enzymes

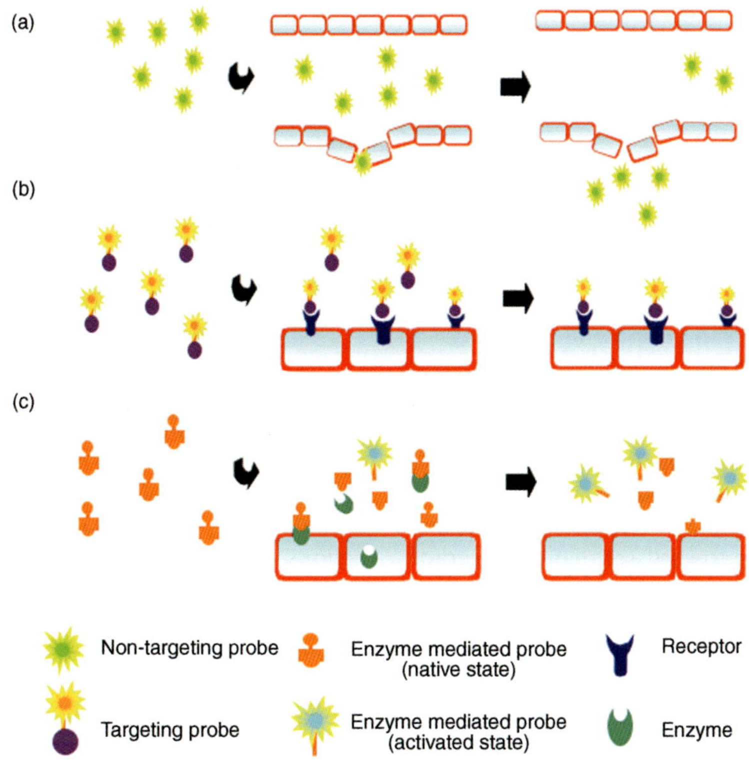

output (Krovetz and Gessner, 1965) and ophthalmic retinal angiography (Horiguchi *et al.*, 2003). Recently, ICG has been evaluated for breast tumour imaging (Intes *et al.*, 2003; Ntziachristos *et al.*, 2000). Other indocyanine derivatives have been used in animals to study myocardial infarction (Nakayama *et al.*, 2002) and arthritis (Hansch *et al.*, 2004a,b). Similarly, fluorescent qdots have been tested as lymph node tracers in pig (Parungo *et al.*, 2005). In general, information obtained from these studies using unfunctionalized fluorochromes can only provide data on the anatomic nature of the diseases (Figure 7.3.3(a)). To expand the capability of optical imaging techniques, probes reporting specific physiological events could provide a better understanding of the disease at a molecular level. The more detailed picture of the disease state may ultimately be beneficial, especially for clinical diagnosis and for development of appropriate treatment regimens.

There are two major classes of molecular probes for *in vivo* imaging based on targeted and the enzyme-mediated strategies. The former are aimed to locate surface proteins, receptors and extracellular targets, whereas the latter are used to determine enzymatic activities. The targeted approach is normally more straightforward. It could be as simple as fluorochromes conjugated to targeting moieties, such as peptides, proteins, antibodies, ligands and even drug inhibitors (Figure 7.3.3(b)). The enzyme-mediated probes (EMP), however, require a more sophisticated design strategy (Figure 7.3.3(c)).

7.3.3 Targeted probes

Proteins and antibodies are classes of targeting ligands that are frequently used for nuclear imaging. Optical imaging technology has made the use of non-ionizing fluorochrome-labelled contrast agents more attractive. For example, fluorescence-labelled anti-EGFR has been applied to image human squamous cell carcinoma in nude mice (Folli *et al.*, 1994). A single-chain fluorescent antibody fragment (NovoMab-G2-scFv-Cy5) derived from human anti-tumour MAb was injected subcutaneously to athymic mice with human melanoma tumour cells (Ramjiawan *et al.*,

2000). Tumour fluorescence peaked 2-h post injection. Bio-distribution data suggested that nearly 70% of fluorescent probe accumulated in the kidneys. The antibody-targeting strategy has also been applied for imaging angiogenesis (Neri *et al.*, 1997) and arthritis (Hansch *et al.*, 2004b). The advantage of using antibodies for optical imaging is their high affinities, despite the fact that the pharmacokinetic and pharmacodynamic properties can be more complex than other traditional small organic molecules (Lobo, Hansen and Balthasar, 2004).

Small synthetic peptides can also be used as targeting molecules. They are generally non-immunogenic and can be conjugated easily to a variety of fluorophores by standard protocols. Advances in peptide synthesis and phage display technologies (Landon and Deutscher, 2003) can generate high affinity and selective ligands for receptors. Moreover, the plasma half-life of peptide is generally shorter than that of the larger sized proteins and, thus, improves the signal-to-background ratio during imaging. Combinatorial synthesis and cell binding assay have been used to identify vasoactive intestinal peptide (VIP) ligands for improved stability towards proteolytic degradation (Bhargava *et al.*, 2002). A structurally optimized indotricarbocyanine dye (ITCC) conjugated to VIP analogue ([Arg8]-VIP-ITCC) was administrated to RIN38($VPAC_1$) tumour bearing mice for NIR imaging. In another study, indodicarbocyanine (IDCC) conjugated to octreotide (Becker *et al.*, 2000) was injected intravenously to mice bearing somatostain 2 (SST2) receptor expressing tumors. Both studies demonstrated a rapid increase in fluorescence at the tumour sites within 30 min. The feasibility of using peptides as targeting molecules has been further demonstrated with endostatin-Cy5.5 for angiogenesis imaging (Citrin *et al.*, 2004), epidermal growth factor-Cy5.5 for breast cancer (Ke *et al.*, 2003) and RGD-Cy5.5 for integrin expressing cancer cells (Chen, Conti and Moats, 2004; Achilefu *et al.* 2005).

Small ligands such as folic acid or bisphosphonate can also be used to direct specific binding *in vivo*. Folic acid labelled with a NIR fluorescence tag has been used to image folate receptor expressing tumour and inflamed arthritic joints (Moon *et al.*, 2003; Chen *et al.*, 2005 a,b). A bisphosphonate analog specific for hydroxyapatite has been synthesized for image osteoblast activity and microcalcification in breast cancer (Zaheer *et al.*, 2001; Lenkinski *et al.*, 2003).

Several unmodified qdots have also been tested in animals for tumour imaging (Larson *et al.*, 2003; Michalet *et al.*, 2005; Ballou *et al.*, 2004; Kim *et al.*, 2004; Gao *et al.*, 2004). The enhanced permeability and retention effect in tumours (Duncan, 2003) suggest that qdots may possibly target tumours. Further modification of qdots with peptides or antibodies can improve their selectivity. For example, vasculature recognition of normal lungs and tumours has been achieved by various peptide-coated qdots (Akerman *et al.*, 2002). In another study, qdots were conjugated to a prostate-specific membrane antigen (PSMA) monoclonal antibody containing an amphiphilic triblock co-polymer for imaging prostate cancer (Gao *et al.*, 2004). In both studies, significant contrast was observed at the target sites. However, the conjugated qdots also demonstrated a non-specific accumulation in liver and spleen.

7.3.4 Enzyme-mediated probes (EMPs)

EMPs are molecular probes designed to monitor enzyme activity *in vivo*. These probes are assembled as molecular switches to report functional events and can be further subdivided into bond degrading and bond forming probes, based on their mechanism of actions.

The bond-degrading EMP usually contains enzyme degradable linkages that bring multiple fluorochromes, or a fluorochrome and a quencher into close proximity of each other. Under these conditions, strong fluorescence quenching occurs. Once the cleavable linkers are recognized and degraded/hydrolyzed by target enzymes, the quenched fluorescence signal is recovered from the librated fluorochromes. Such designs are highly sensitive for the detection of small amount of enzymes because each enzyme can activate multiple probes. This results in a continuous flow of amplified fluorescencent signal, thus increasing the detection sensitivity and reducing the background fluorescence. One of the examples is the early development of a broad protease activatable probe (Figure 7.3.4(a)) (Weissleder *et al.*, 1999). The probe is built on a non-immunologic co-polymer consisting of poly-L-lysine partially protected by multiple methoxypolyethylene glycol moieties for tumoural delivery (Bogdanov *et al.*, 1997). Loading the polymer with a large number of Cy5.5 fluorochromes results in quenched fluorescence. Certain trypsin-like proteases, such as cathepsin B and cathepsin L can digest the polymer *via* the free lysine residues on the backbone. The generation of fluorochrome containing polymer fragments after enzymatic cleavage gives rise to amplified fluorescence emission. This broad-spectrum proteases sensitive probe has been employed to image diseases that overexpress cathepsins. These include the detection of cancer (Bremer *et al.*, 2002; Marten *et al.*,

Figure 7.3.4 *De novo* design of enzyme-mediated probes. (a) The bond-degrading EMP is activated *via* the degradation of the backbone template. Fluorophores are subsequently released to give fluorescence signal. (b) The design of the second type of bond-degrading EMP is similar to (a) except enzyme degradable peptide substrates are conjugated to the template. The selectivity of the probes is determined by the peptide sequences. (c) The bond-formation EMP. Unlike (a) and (b), this probe is not optically silent in its native state. A labelled peptide substrate, recognized by specific cross-linking enzyme such as transglutaminase, could be linked to target sites covalently, leading to increased accumulation of fluorophores

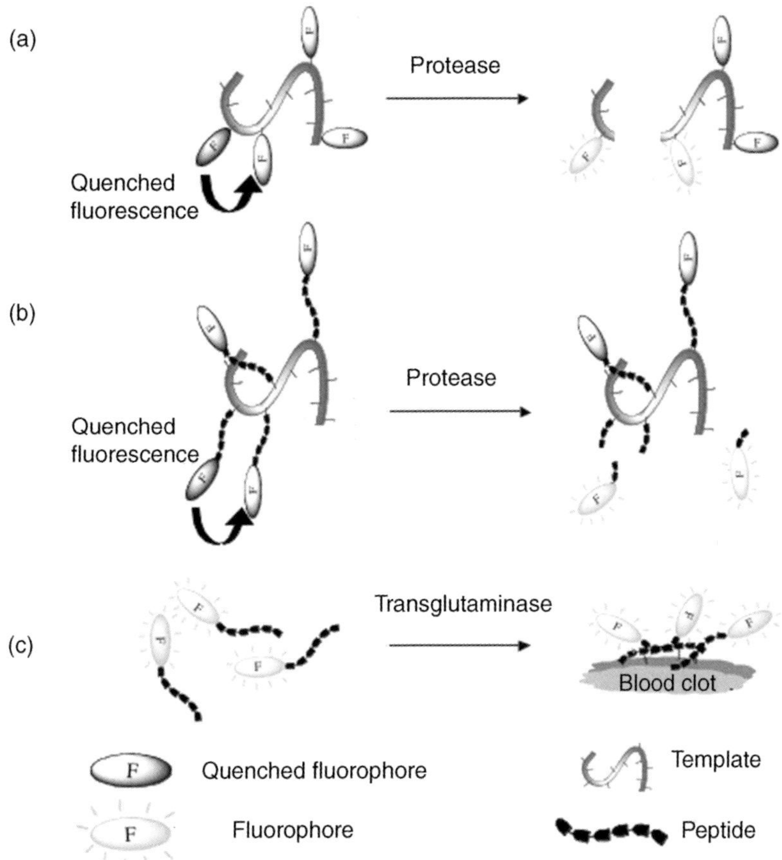

2002), arthritis (Lai *et al.*, 2004; Ji *et al.*, 2002; Wunder *et al.*, 2004) and atherosclerosis (Chen *et al.*, 2002).

Imaging probes with selectivity for other proteases have also been developed (Figure 7.3.4(b)). The major difference between the selective and broad-spectrum probes is the site of attachment for the fluorochrome. In protease selective probe, fluorochrome (Cy5.5) is attached to the graft-copolymer backbone through specific peptide substrates instead of directly to the poly-*L*-lysine backbone. Specificity and sensitivity is dependent on the introduced peptide sequences. This approach has been employed to image several disease-associated proteases activities, such as cathepsin D (Tung *et al.*, 2000), metalloproteinases (Bremer, Tung and Weissleder, 2001; Bremer *et al.*, 2001), caspases (Messerli *et al.*, 2004), thrombin (Jaffer *et al.*, 2002) and HIV proteases (Shah *et al.*, 2004).

To image non-proteolytic enzymes, a sugar based EMP was developed to measure enzymatic activity of beta-galactosidase, one of the most used reporter genes in biomedical research (Tung *et al.*, 2004). The intact probe has short fluorescence emission, while the emission from the galactosidase-mediated hydrolysis product is red-shifted for approximately 100 nm. This imaging probe provides a useful tool to study enzymatic response in real time.

The second category of EMP involves the formation of new chemical bonds. The probe is designed to have an enzyme-dependent reactive group and a fluorochrome. The reactive group on the reporter will react specifically *via* formation of a stable covalent bond (Figure 7.3.4(c)). Background fluorescence signal is extremely low because unreacted probe is washed away from the site of reaction. Recently, a NIR fluorochrome was anchored to a peptide substrate sequence extracted from the amino-terminus of alpha-2 antiplasmin (Tung *et al.*, 2003). When employed *in vivo*, the transglutaminase activity efficiently cross-links the probe to newly formed blood clots (Jaffer *et al.*, 2004), which has been applied to study cerebral thrombi and blood-brain barrier disruption (Kim *et al.*, 2005).

7.3.5 Prospective

Optical imaging is a relatively new technology involving multi-disciplinary collaboration between scientists. Probe design is one of the fundamental research areas. A better knowledge of human biology can expedite the probe development process and enable development of more sophisticated imaging strategies. These fluorochrome-based probes can offer real-time insight of physiological events and provide unique information about the disease *in vivo*. The clinical needs for specific molecular imaging tools are obvious, despite the fact that current fluorescence imaging agents are primarily limited to animal research. As discussed in this book, many researches are working to improve the sensitivity of imaging equipment for human uses. With the continuing discovery and recognition of novel molecular markers for different diseases, we expect more novel imaging probes and strategies will be developed in the near future.

References

Achilefu, S., Bloch, S., Markiewicz, M. A., Zhong, T., Ye, Y., Dorshow, R. B., Chance, B., Liang, K., 2005. "Synergistic effects of light-emitting probes and peptides for targeting and monitoring integrin expression." *Proc. Natl. Acad. Sci. USA* **102**, 7976–7981.

Akerman, M. E., Chan, W. C., Laakkonen, P., Bhatia, S. N., Ruoslahti, E., 2002. "Nanocrystal targeting *in vivo*." *Proc. Natl. Acad. Sci. USA* **99**, 12617–12621.

Ballou, B., Lagerholm, B. C., Ernst, L. A., Bruchez, M. P., Waggoner, A. S., 2004. "Noninvasive imaging of quantum dots in mice." *Bioconjugate Chem.* **15**, 79–86.

Becker, A., Hessenius, C., Bhargava, S., Grotzinger, C., Licha, K., Schneider-Mergener, J., Wiedenmann, B., Semmler, W., 2000. "Cyanine dye labeled vasoactive intestinal peptide and somatostatin analog for optical detection of gastroenteropancreatic tumors." *Ann. N. Y. Acad. Sci.* **921**, 275–278.

Bhargava, S., Licha, K., Knaute, T., Ebert, B., Becker, A., Grotzinger, C., Hessenius, C., Wiedenmann, B., Schneider-Mergener, J., Volkmer-Engert, R., 2002. "A complete substitutional analysis of VIP for better tumor imaging properties." *J. Mol. Recognit.* **15**, 145–153.

Bogdanov Jr., A., Wright, S. C., Marecos, E. M., Bogdanova, A., Martin, C., Petherick, P., Weissleder, R., 1997. "A long-circulating co-polymer in "passive targeting" to solid tumors." *J. Drug. Target.* **4**, 321–330.

Bremer, C., Bredow, S., Mahmood, U., Weissleder, R., Tung, C. H., 2001. "Optical imaging of matrix metalloproteinase-2 activity in tumors: feasibility study in a mouse model." *Radiology* **221**, 523–529.

Bremer, C., Tung, C. H., Bogdanov Jr., A., Weissleder, R., 2002. "Imaging of differential protease expression in breast cancers for detection of aggressive tumor phenotypes." *Radiology* **222**, 814–818.

Bremer, C., Tung, C. H., Weissleder, R., 2001. "*In vivo* molecular target assessment of matrix metalloproteinase inhibition." *Nat. Med.* **7**, 743–748.

Bruchez Jr., M., Moronne, M., Gin, P., Weiss, S., Alivisatos, A. P., 1998. "Semiconductor nanocrystals as fluorescent biological labels." *Science* **281**, 2013–2016.

Chan, W. C., Maxwell, D. J., Gao, X., Bailey, R. E., Han, M., Nie, S., 2002. "Luminescent quantum dots for multiplexed biological detection and imaging." *Curr. Opin. Biotechnol* **13**, 40–46.

Chan, W. C., Nie, S., 1998. "Quantum dot bioconjugates for ultrasensitive nonisotopic detection." *Science* **281**, 2016–2018.

Chen, J., Tung, C. H., Mahmood, U., Ntziachristos, V., Gyurko, R., Fishman, M. C., Huang, P. L., Weissleder, R., 2002. "*In vivo* imaging of proteolytic activity in atherosclerosis." *Circulation* **105**, 2766–2771.

Chen, W. T., Khazaie, K., Zhang, G., Weissleder, R., Tung, C. H., 2005b. "Detection of dysplastic intestinal adenomas using a fluorescent folate imaging probe." *Mol. Imaging* **4**, 67–74.

Chen, W. T., Mahmood, U., Weissleder, R., Tung, C. H., 2005a. "Arthritis imaging using a near-infrared fluorescence folate-targeted probe." *Arthritis. Res. Ther.* **7**, R310–R317.

Chen, X., Conti, P. S., Moats, R. A., 2004. "*In vivo* near-infrared fluorescence imaging of integrin alphavbeta3 in brain tumor xenografts." *Cancer Res.* **64**, 8009–8014.

Citrin, D., Scott, T., Sproull, M., Menard, C., Tofilon, P. J., Camphausen, K., 2004. "*In vivo* tumor imaging using a near-infrared-labeled endostatin molecule." *Int. J. Radiat. Oncol. Biol. Phys.* **58**, 536–541.

Dubertret, B., Skourides, P., Norris, D. J., Noireaux, V., Brivanlou, A. H., Libchaber, A., 2002. "*In vivo* imaging of quantum dots encapsulated in phospholipid micelles." *Science* **298**, 1759–1762.

Duncan, R., 2003. "The dawning era of polymer therapeutics." *Nat. Rev. Drug. Discov.* **2**, 347–360.

Folli, S., Westermann, P., Braichotte, D., Pelegrin, A., Wagnieres, G., van den Bergh, H., Mach, J. P., 1994. "Antibody-indocyanin conjugates for immunophotodetection of human squamous cell carcinoma in nude mice." *Cancer Res.* **54**, 2643–2649.

Gao, X., Chan, W. C., Nie, S., 2002. "Quantum-dot nanocrystals for ultrasensitive biological labeling and multicolor optical encoding." *J. Biomed Opt.* **7**, 532–537.

Gao, X., Cui, Y., Levenson, R. M., Chung, L. W., Nie, S., 2004. "*In vivo* cancer targeting and imaging with semiconductor quantum dots." *Nat. Biotechnol* **22**, 969–976.

Gao, X., Yang, L., Petros, J. A., Marshall, F. F., Simons, J. W., Nie, S., 2005. "*In vivo* molecular and cellular imaging with quantum dots." *Curr. Opin. Biotechnol* **16**, 63–72.

Goldman, E. R., Anderson, G. P., Tran, P. T., Mattoussi, H., Charles, P. T., Mauro, J. M., 2002b. "Conjugation of luminescent quantum dots with antibodies using an engineered adaptor protein to provide new reagents for fluoroimmunoassays." *Anal. Chem.* **74**, 841–847.

Goldman, E. R., Balighian, E. D., Mattoussi, H., Kuno, M. K., Mauro, J. M., Tran, P. T., Anderson, G.P., 2002a. "Avidin: A natural bridge for quantum dot-antibody conjugates." *J. Am. Chem. Soc.* **124**, 6378–6382.

Gruber, H. J., Hahn, C. D., Kada, G., Riener, C. K., Harms, G. S., Ahrer, W., Dax, T. G., Knaus, H. G., 2000. "Anomalous fluorescence enhancement of Cy3 and cy3.5 versus anomalous fluorescence loss of Cy5 and Cy7 upon covalent linking to IgG and noncovalent binding to avidin." *Bioconjugate Chem.* **11**, 696–704.

Hansch, A., Frey, O., Hilger, I., Sauner, D., Haas, M., Schmidt, D., Kurrat, C., Gajda, M., Malich, A., Brauer, R., Kaiser, W. A., 2004a. "Diagnosis of arthritis using near-infrared fluorochrome Cy5.5." *Invest. Radiol.* **39**, 626–632.

Hansch, A., Frey, O., Sauner, D., Hilger, I., Haas, M., Malich, A., Brauer, R., Kaiser, W. A., 2004b. "*In vivo* imaging of experimental arthritis with near-infrared fluorescence." *Arthritis. Rheum.* **50**, 961–967.

Horiguchi, M., Nagata, S., Yamamoto, N., Kojima, Y., Shimada, Y., 2003. "Kinetics of indocyanine green dye after intraocular surgeries using indocyanine green staining." *Arch. Ophthalmol* **121**, 327–331.

Hoshino, A., Hanaki, K., Suzuki, K., Yamamoto, K., 2004. "Applications of T-lymphoma labeled with fluorescent quantum dots to cell tracing markers in mouse body." *Biochem. Biophys. Res. Commun.* **314**, 46–53.

Intes, X., Ripoll, J., Chen, Y., Nioka, S., Yodh, A. G., Chance, B., 2003. "*In vivo* continuous-wave optical breast imaging enhanced with Indocyanine Green." *Med. Phys.* **30**, 1039–1047.

Jaffer, F. A., Tung, C. H., Gerszten, R. E., Weissleder, R., 2002. "*In vivo* imaging of thrombin activity in experimental thrombi with thrombin-sensitive near-infrared molecular probe." *Arterioscler Thromb. Vasc. Biol.* **22**, 1929–1935.

Jaffer, F.A., Tung, C.H., Wykrzykowska, J. J., Ho, N. H., Houng, A. K., Reed, G. L., Weissleder, R., 2004. "Molecular imaging of factor XIIIa activity in thrombosis using a novel, near-infrared fluorescent contrast agent that covalently links to thrombi." *Circulation* **110**, 170–176.

Jaiswal, J. K., Mattoussi, H., Mauro, J. M., Simon, S. M., 2003. "Long-term multiple color imaging of live cells using quantum dot bioconjugates." *Nat. Biotechnol* **21**, 47–51.

Ji, H., Ohmura, K., Mahmood, U., Lee, D. M., Hofhuis, F. M., Boackle, S. A., Takahashi, K., Holers, V. M., Walport, M., Gerard, C., Ezekowitz, A., Carroll, M. C., Brenner, M., Weissleder, R., Verbeek, J. S., Duchatelle, V., Degott, C., Benoist, C., Mathis, D., 2002. "Arthritis critically dependent on innate immune system players." *Immunity* **16**, 157–168.

Ke, S., Wen, X., Gurfinkel, M., Charnsangavej, C., Wallace, S., Sevick-Muraca, E.M., Li, C., 2003. "Near-infrared optical imaging of epidermal growth factor receptor in breast cancer xenografts." *Cancer Res.* **63**, 7870–7875.

Kim, D. E., Jaffer, F. A., Weissleder, R., Schellingerhout, D., Tung, C. H., 2005. "Near-infrared fluorescent imaging of cerebral thrombi and blood-brain barrier disruption in a mouse model of cerebral venous sinus thrombosis." *J. Cereb. Blood. Flow. Metab.* **25**, 226–233.

Kim, S., Lim, Y. T., Soltesz, E. G., De Grand, A. M., Lee, J., Nakayama, A., Parker, J. A., Mihaljevic, T., Laurence, R. G., Dor, D. M., Cohn, L. H., Bawendi, M. G., Frangioni, J. V., 2004. "Near-infrared fluorescent type II quantum dots for sentinel lymph node mapping." *Nat. Biotechnol* **22**, 93–97.

Krovetz, L. J., Gessner, I.H., 1965. "A new method utilizing indicator-dilution technics for estimation of left-to-right shunts in infants." *Circulation* **32**, 772–777.

Lai, W. F., Chang, C. H., Tang, Y., Bronson, R., Tung, C. H., 2004. "Early diagnosis of osteoarthritis using cathepsin B sensitive near-infrared fluorescent probes." *Osteoarthritis Cartilage.* **12**, 239–244.

Lakowicz, J. R., Gryczynski, I., Gryczynski, Z., Nowaczyk,K., Murphy, C. J., 2000. "Time-resolved spectral observations of cadmium-enriched cadmium sulfide nanoparticles and the effects of DNA oligomer binding." *Anal. Biochem.* **280**, 128–136.

Landon, L. A., Deutscher, S. L., 2003. "Combinatorial discovery of tumor targeting peptides using phage display." *J. Cell. Biochem.* **90**, 509–517.

Larson, D. R., Zipfel, W.R., Williams, R. M., Clark, S. W., Bruchez, M. P., Wise, F. W., Webb, W. W., 2003. "Water-soluble quantum dots for multiphoton fluorescence imaging *in vivo*." *Science* **300**, 1434–1436.

Lenkinski, R. E., Ahmed, M., Zaheer, A., Frangioni, J.V., Goldberg, S. N., 2003. "Near-infrared fluorescence imaging of microcalcification in an animal model of breast cancer." *Acad. Radiol.* **10**, 1159–1164.

Licha, K., Riefke, B., Ebert, B., Grotzinger, C., 2002. "Cyanine dyes as contrast agents in biomedical optical imaging." *Acad. Radiol.* **9**(2), S320–S322.

Lin, Y., Weissleder, R., Tung, C. H., 2002. "Novel near-infrared cyanine fluorochromes: Synthesis, properties, and bioconjugation." *Bioconjugate Chem.* **13**, 605–610.

Lin, Y., Weissleder, R., Tung, C.H., 2003. "Synthesis and properties of sulfhydryl-reactive near-infrared cyanine fluorochromes for fluorescence imaging." *Mol. Imaging* **2**, 87–92.

Lobo, E. D., Hansen, R. J., Balthasar, J. P., 2004. "Antibody pharmacokinetics and pharmacodynamics." *J. Pharm. Sci.* **93**, 2645–2668.

Marten, K., Bremer, C., Khazaie, K., Sameni, M., Sloane, B., Tung, C. H., Weissleder, R., 2002. "Detection of dysplastic intestinal adenomas using enzyme-sensing molecular beacons in mice." *Gastroenterology* **122**, 406–414.

Messerli, S. M., Prabhakar, S., Tang, Y., Shah, K., Cortes, M. L., Murthy, V., Weissleder, R., Breakefield, X. O., Tung, C. H., 2004. "A novel method for imaging apoptosis using a caspase-1 near-infrared fluorescent probe." *Neoplasia.* **6**, 95–105.

Michalet, X., Pinaud, F. F., Bentolila, L. A., Tsay, J. M., Doose, S., Li, J. J., Sundaresan, G., Wu, A. M., Gambhir, S. S., Weiss, S., 2005. "Quantum dots for live cells, *in vivo* imaging, and diagnostics." *Science* **307**, 538–544.

Mishra, A., Behera, R. K., Behera, P. K., Mishra, B. K., Behera, G. B., 2000. "Cyanines during the 1990s: A review." *Chem. Rev.* **100**, 1973–2012.

Moon, W. K,; Lin, Y., O'Loughlin, T., Tang, Y., Kim, D. E., Weissleder, R., Tung, C. H., 2003. "Enhanced tumor detection using a folate receptor-targeted near-infrared fluorochrome conjugate." *Bioconjugate Chem.* **14**, 539–5345.

Morgan, N. Y., English, S., Chen W., Chernomordik V., Russo, A., Smith, P. D., Gandjbakhche, A., 2005 "Real time *in vivo* non-invasive optical imaging using near-infrared fluorescent quantum dots." *Acad. Radiol.* **12**, 313–323.

Mujumdar, R.B., Ernst, L.A., Mujumdar, S.R., Lewis, C.J., Waggoner, A.S., 1993. "Cyanine dye labeling reagents: Sulfoindocyanine succinimidyl esters." *Bioconjugate Chem.* **4**, 105–111.

Mujumdar, S. R., Mujumdar, R. B., Grant, C. M., Waggoner, A. S., 1996. "Cyanine-labeling reagents: sulfobenzindocyanine succinimidyl esters." *Bioconjugate. Chem.* **7**, 356–362.

Nakayama, A., del Monte, F., Hajjar, R. J., Frangioni, J. V., 2002. "Functional near-infrared fluorescence imaging for cardiac surgery and targeted gene therapy." *Mol. Imaging* **1**, 365–377.

Neblock, D. S., Chang, C. H., Mascelli, M. A., Fleek, M., Stumpo, L., Cullen, M. M., Daddona, P.E., 1992. "Conjugation and evaluation of 7E3 x P4B6, a chemically cross-linked bispecific F(ab')2 antibody which inhibits platelet aggregation and localizes tissue plasminogen activator to the platelet surface." *Bioconjugate Chem.* **3**, 126–131.

Neri, D., Carnemolla, B., Nissim, A., Leprini, A., Querze, G., Balza, E., Pini, A., Tarli, L., Halin, C., Neri, P., Zardi, L., Winter, G., 1997. "Targeting by affinity-matured recombinant antibody fragments of an angiogenesis associated fibronectin isoform." *Nat. Biotechnol* **15**, 1271–1275.

Ntziachristos, V., Bremer, C., Weissleder, R., 2003. "Fluorescence imaging with near-infrared light: new technological advances that enable *in vivo* molecular imaging." *Eur. Radiol.* **13**, 195–208.

Ntziachristos, V., Yodh, A. G., Schnall, M., Chance, B., 2000. "Concurrent MRI and diffuse optical tomography of breast after indocyanine green enhancement." *Proc. Natl. Acad. Sci. USA* **97**, 2767–2772.

Oswald, B., Patsenker, L., Duschl, J., Szmacinski, H., Wolfbeis, O.S., Terpetschnig, E., 1999. "Synthesis, spectral properties, and detection limits of reactive squaraine dyes, a new class of diode laser compatible fluorescent protein labels." *Bioconjugate Chem.* **10**, 925–931.

Parungo, C. P., Ohnishi, S., Kim, S. W., Kim, S., Laurence, R. G., Soltesz, E. G., Chen, F. Y., Colson, Y. L., Cohn, L. H., Bawendi, M. G., Frangioni, J. V., 2005. "Intraoperative identification of esophageal sentinel lymph nodes with near-infrared fluorescence imaging." *J. Thorac. Cardiovasc. Surg.* **129**, 844–850.

Pham, W., Lai, W. F., Weissleder, R., Tung, C. H., 2003. "High efficiency synthesis of a bioconjugatable near-infrared fluorochrome." *Bioconjugate Chem.* **14**, 1048–1051.

Pinaud, F., King, D., Moore, H. P., Weiss, S., 2004. "Bioactivation and cell targeting of semiconductor CdSe/ZnS nanocrystals with phytochelatin-related peptides." *J. Am. Chem. Soc.* **126**, 6115–6123.

Ramjiawan, B., Maiti, P., Aftanas, A., Kaplan, H., Fast, D., Mantsch, H. H., Jackson, M., 2000. "Noninvasive localization of tumors by immunofluorescence imaging using a single chain Fv fragment of a human monoclonal antibody with broad cancer specificity." *Cancer* **89**, 1134– 1144.

Randolph, J. B.; Waggoner, A. S., 1997. "Stability, specificity and fluorescence brightness of multiply-labeled fluorescent DNA probes." *Nucl. Acid. Res.* **25**, 2923–2929.

Rowell, L. B., Blackmon, J. R., Martin, R. H., Mazzarella, J.A., Bruce, R.A., 1965. "Hepatic clearance of indocyanine green in man under thermal and exercise stresses." *J. Appl. Physiol.* **20**, 384–394.

Santone, K. S., Acosta, D., Bruckner, J. V., 1982. "Cadmium toxicity in primary cultures of rat hepatocytes." *J. Toxicol. Environ. Health* **10**, 169–177.

Shah, K., Tung, C. H., Chang, C. H., Slootweg, E., O'Loughlin, T., Breakefield, X. O., Weissleder, R., 2004. "*In vivo* imaging of HIV protease activity in amplicon vector-transduced gliomas." *Cancer Res.* **64**, 273–278.

Smith, A. M., Gao, X., Nie, S., 2004. "Quantum dot nanocrystals for *in vivo* molecular and cellular imaging." *Photochem. Photobiol.* **80**, 377–385.

Southwick, P. L., Ernst, L. A., Tauriello, E. W., Parker, S. R., Mujumdar, R. B., Mujumdar, S. R., Clever, H. A., Waggoner, A. S., 1990. "Cyanine dye labeling reagents–carboxymethylindocyanine succinimidyl esters." *Cytometry* **11**, 418–430.

Tung, C. H., Ho, N. H., Zeng, Q., Tang, Y., Jaffer, F. A., Reed, G. L., Weissleder, R., 2003. "Novel Factor XIII Probes for Blood Coagulation Imaging." *Chembiochem.* **4**, 897–899.

Tung, C.H., Mahmood, U., Bredow, S., Weissleder, R., 2000. "*In vivo* imaging of proteolytic enzyme activity using a novel molecular reporter." Cancer Res. **60**, 4953–4958.

Tung, C. H., Zeng, Q., Shah, K., Kim, D. E., Schellingerhout, D., Weissleder, R., 2004. "*In vivo* imaging of beta-galactosidase activity using far red fluorescent switch." *Cancer Res.* **64**, 1579–1583.

Weissleder, R., Tung, C. H., Mahmood, U., Bogdanov Jr., A., 1999. "*In vivo* imaging of tumors with protease-activated near-infrared fluorescent probes." *Nat. Biotechnol.* **17**, 375–378.

Wessendorf, M. W., Brelje, T. C., 1992. "Which fluorophore is brightest? A comparison of the staining obtained using fluorescein, tetramethylrhodamine, lissamine rhodamine, Texas red, and cyanine 3.18." *Histochemistry* **98**, 81–85.

Wu, X., Liu, H., Liu, J., Haley, K. N., Treadway, J. A., Larson, J. P., Ge, N., Peale, F., Bruchez, M. P., 2003. "Immunofluorescent labeling of cancer marker Her2 and other cellular targets with semiconductor quantum dots." *Nat. Biotechnol* **21**, 41–46.

Wunder, A., Tung, C. H., Muller-Ladner, U., Weissleder, R., Mahmood, U., 2004. "*In vivo* imaging of protease activity in arthritis: a novel approach for monitoring treatment response." *Arthritis. Rheum.* **50**, 2459–2465.

Yu, H., Chao, J., Patek, D., Mujumdar, R., Mujumdar, S., Waggoner, A. S., 1994. "Cyanine dye dUTP analogs for enzymatic labeling of DNA probes." *Nucl. Acid. Res.* **22**, 3226–3232.

Zaheer, A., Lenkinski, R. E., Mahmood, A., Jones, A. G., Cantley, L.C., Frangioni, J.V., 2001. "*In vivo* near-infrared fluorescence imaging of osteoblastic activity." *Nat. Biotechnol.* **19**, 1148–1154.

7.4 New contrast agents for NMR

Silvio Aime

The contrast in MR images is the result of a complex interplay between instrumental and endogeneous parameters, mainly T_1 and T_2 of water protons. In tissues water protons, relaxation is mainly determined by their interactions with macromolecular structures like proteins, sugars and lipids eventually enhanced when such systems include paramagnetic ions like iron, copper and manganese. Therefore, the search for contrast agents (CAs) has been addressed to

Figure 7.4.1 Chemical structure of $[Gd\text{-}DTPA]^{2-}$

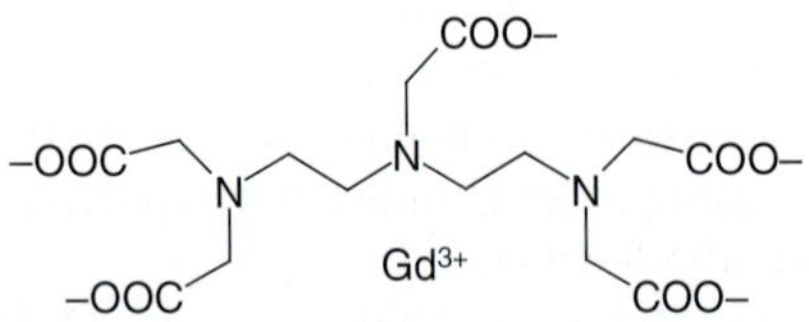

chemicals that markedly alter the relaxation times of water protons in the tissues where they distribute. Generally, the purpose is to reduce T_1 in order to obtain an intense signal in the MR image in short times and a better S/N ratio with the acquisition of a higher number of measurements. CAs that reduce either T_1 or T_2 are called *positive*, whereas those that affect predominantly T_2 are called *negative*. Positive agents are represented by paramagnetic metal complexes, mostly based on Gd(III) ion because it contains seven unpaired electrons (high effective magnetic moment), and a long electronic relaxation time, and it can be wrapped by multidentate ligands to yield thermodynamically stable complexes. The prototype of this class of CAs is $[Gd\text{-}DTPA]^{2-}$ (Figure 7.4.1) widely used in the clinical practice as extracellular agent, reporter of Blood Brain Barrier leakage, organ perfusion, flow measurement and so on (Aime *et al.*, 1998).

Negative agents are represented by iron oxide particles surrounded by coating materials like dextran, carboxydextran or other hydrophilic materials. The diameter of the iron oxide core is just 3–5 nm, whereas the overall particle may range from few tens to few hundreds of nanometre in size (Roch, Muller and Gillis, 1999) (Figure 7.4.2).

These agents provide excellent (negative) contrast when administered at iron doses as low as 8–15 μmol/kg body weight. The pharmacodynamic properties of these CAs depend upon the size, the characteristics of the hydrophilic cover and the overall electric charge. They have found applications in angiography, liver and spleen imaging as well as reporters of macrophages activity at linphonodes.

The need of having CAs useful for Molecular Imaging applications has been addressed through the design of systems characterized by higher sensitivity and endowed with proper targeting capabilities in order to accumulate at the sites of interest. In principle, either paramagnetic Gd(III) complexes or iron oxide particles can be delivered to the target sites by linking, on their surface, the vector that recognizes the given biological structure. Iron oxide particles own enough sensitivity and have been used to acquire the proof of concept in several MR-Molecular Imaging protocols. For instance, when bound to transferrin, it has been possible to visualize tumour cells in mouse as these cells overexpress the transferrin receptors to accomplish the increased requirement of iron need for cellular proliferation (Osterloh and Aisen, 1989).

Figure 7.4.2 Schematic representation of an iron-oxide particle

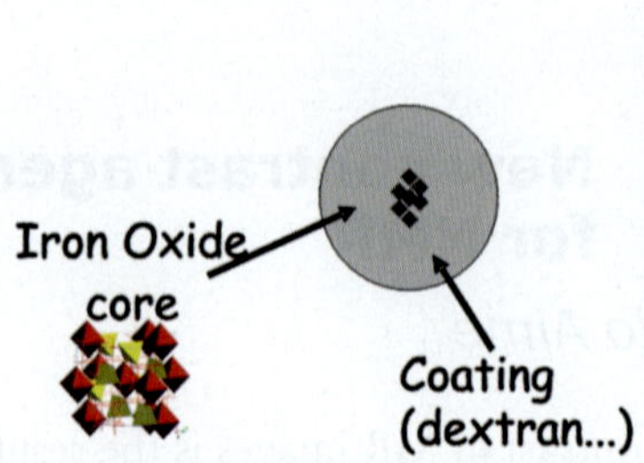

SPIO (Superparamagnetic Iron Oxides)
-Median diameter > 50 nm
-Accumulate in the RES of the liver and spleen
- Effective on T2 relaxation

Es. of Commercial agent:
Endorem®
Internal diameter = 4.3-4.8 nm
Particle diameter = 200 nm

USPIO (Ultrasmall Superparamagnetic Iron Oxides)
Median diameter < 50 nm
- Do not accumulate in the RES system as fast as larger particles
- longer plasma half-life
-Effective on T2 relaxation

Es. of Commercial agent:
Sinerem®
Internal diameter = 4.3-4.9 nm
Particle diameter = 50 nm

Figure 7.4.3 Schematic representations of three eptadentate ligands able to form stable Gd(III)-chelates with two coordinated water molecules

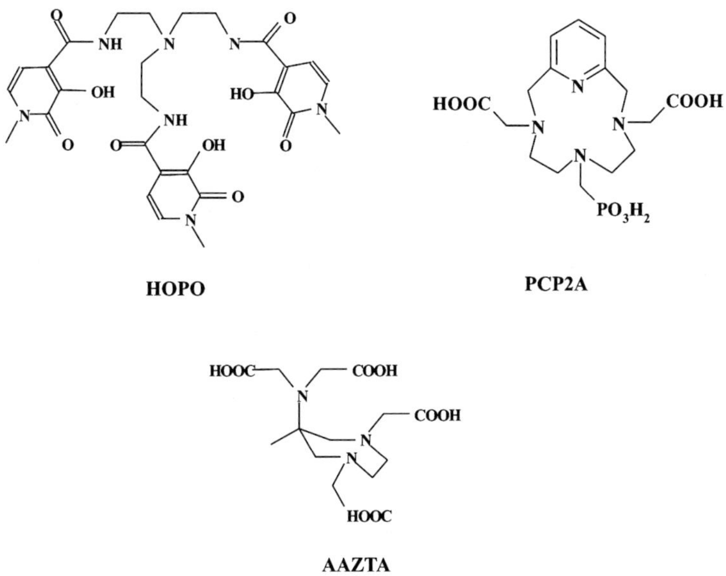

For Gd(III) based agents, it is necessary to set up strategies aimed at allowing the accumulation of enough paramagnetic units at the target. Different approaches have been tackled to pursue this goal that have involved the synthesis of Gd-containing macromolecular systems based on polymers, micelles, dendrimers as well as supramolecular adducts (Aime *et al.*, 2001; Andre *et al.*, 1999; Aime *et al.*, 2000; Bligh *et al.*, 1991; Wiener *et al.*, 1994). To attain higher sensitivity, it is also useful to improve the relaxivity of each Gd(III) centre. In the last two decades much work has been devoted to elucidate the relationships between structure, dynamics and relaxivity of Gd(III) chelates (relaxivity, r_1 = enhancement of water proton relaxation rate at 1 mM concentration of the paramagnetic agent). Most promising systems are those that conjugate the occurrence of two water molecules in the inner co-ordination sphere and a high termodynamic stability. Examples of ligands for these types of Gd(III)-chelates are shown in Figure 7.4.3.

7.4.1 Responsive agents

An advantage of Gd(III) based systems in respect to iron-oxide particles relies on the fact that their relaxivity can be made dependent on a specific parameter of the micro-environment in which the CA distributes through a proper design of the coordinating ligand. For instance, a pH-responsive agent has been designed by introducing an arm that contains a sulfonamide moiety in a DOTA derived structure. At acidic pH, the latter cannot coordinate the Gd ion, whereas at basic pH it deprotonates and can replace two coordinated water molecules, thus causing a large change in the observed relaxivity (Lowe *et al.*, 2001). Other examples deal with responsiveness to pO_2, temperature, specific enzymatic activity, etc. Among the latter class, the seminal system is represented by the β-galactosidase reporting complex developed some years ago by T. Meade and coworkers (Moats, Fraser and Meade, 1997). They synthesised a derivative of GdHPDO3A containing a β-galactose moiety (EGad) which acts as a lid for the access of water to the inner sphere coordination cage (Figure 7.4.4). The action of the β-galactosidase enzyme causes the hydrolysis of the glucosidic linkage to yield GdHPDO3A-like complex which contains one exchanging water. Thus, the removal of the lid causes a 20% increase of the observed relaxivity. An interesting application as reporter of cell transfection has been suggested for this responsive system. In fact, the gene expression of the β-galactosidase enzyme is often used as a marker to report about the successful cell transfection. The gene is simply included into the plasmid used for gene transfection.

Figure 7.4.4 Structure of EGad complex and schematic representation of the cleavage of glucosidic moiety

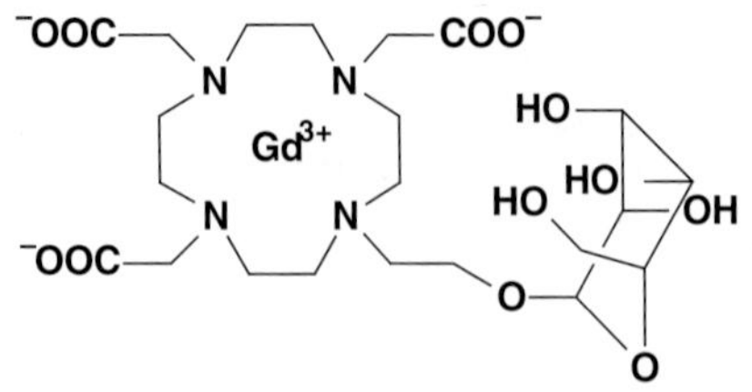

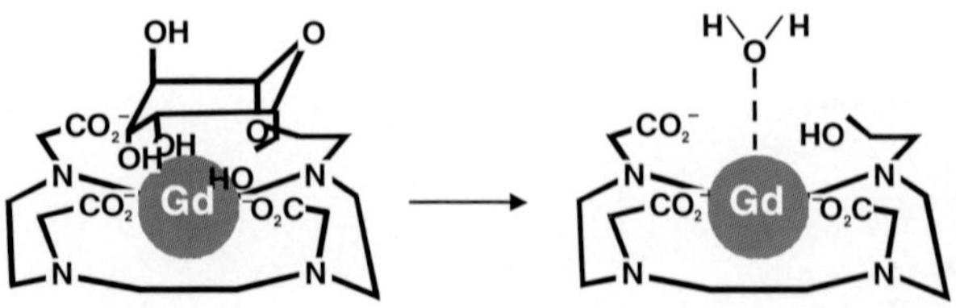

7.4.2 Cellular labelling

The rapidly growing field of cellular therapy can receive an important support from the availability of efficient procedure of cell tracking *in vivo*. Up to now most attention has been devoted to the use of iron oxide particles (Bulte and Kraichman, 2004) as this kind of labelling is very efficient and, in principle, up to a single cell can be detected by this methodology. Iron-oxide particles added to incubation media enter the cells by phagocytosis. Alternatively, it has been shown that the commercially available Ferridex can be easily internalized into cells by applying a short, low intensity electroporating pulse (Walczak *et al.*, 2005).

Incubation of cells in the presence of Gd(III) chelates leads to their entrapment into endosomes by fluid endocytosis (pinocytosis) that consists of a spontaneous process involving the progressive invagination of the cellular membrane which ends up in the formation of small vescicles ($\leq$150 nm diametres). The commercial Gd-HPDO3A (Prohance®) has proved useful to label Endhotelial Progenitor Cells that, once implanted subcutaneously in a mouse model, can be detected up to 2 weeks after implantation (Geninatti Crich *et al.*, 2004). This labelling procedure appears of general applicability. It has been tested in several tumour cell lines and also in pancreatic islets maintaining an excellent cell viability. In the case of soluble paramagnetic chelates, the use of electroporation allows the internalization of the contrast agent in the cytoplasm rather then in the endosomes. These results are particularly useful for the attainment of high cellular relaxation rates in respect to the pinocytosis mediated labelling. For cells endowed with phagocytotic activity, cell labelling can be pursued with biodegradable particles made of Gd-chelates. The insolubility of the Gd-chelates is given by the conjugation to the ligand surface of long aliphatic chains. Once internalized in endosomes, the particles are degraded with the release of the soluble Gd(III)-complexes. The release of the insolubilizing moiety can be made responsive to a specific enzymatic activity through the introduction of a suitable spacer between the chelate and the aliphatic chain (Aime *et al.*, 2002a).

7.4.3 Targeting cells with suitably functionalized CAs

A number of investigations have been carried out either with iron-oxide and Gd(III) based systems. Targets may be on the endothelial cells, in the extracellular matrix, on the cellular membrane and intracellular. A well-studied endothelial receptor is represented by $\alpha_v\beta_3$ Integrin receptor as reporters of neoangiogenesis in tumors and reumatoid arthritis. The vector to this target is represented by the tripeptidic RGD sequence. Being present in a relatively small number (10^3–10^5 per cell) their visualization requires the use of a construct containing ca. 10^5 Gd-chelates. Alternatively, Integrins receptors have been visualized by a three-step procedure that consists of targeting the receptor with a suitably biotinilated Ab, followed by the administration of Streptavidin (that has a very high binding affinity to biotin) and, finally, of a biotinilated liposome whose payload is made of a huge number of Gd(III) chelates.

A representative target in the extravascular matrix is represented by Matrix Metallo Proteases that are highly expressed in metastatic tumours and in other diseases involving extensive tissue re-modeling.

Cellular internalization *via* receptor, that is the route of choice in Nuclear Medicine assay, often implies a shift to receptor-mediated endocytosis when the native ligand is structurally modified upon conjugation to the Imaging reporter. An interesting system that is internalized into hepatocytes by the same route of the endogeneous species is represented by Gd-loaded Apoferritin (Aime *et al.*,

2002a). It has been shown that it is possible to host up to 8–10 Gd(III) chelates in the inner cavity of the protein. The exterior of such Gd(III)-loaded Apoferritin is exactly the same of the parent Ferritin and then, once administered intravenously, it is quickly cleared-up by the specific receptors on the hepatocytes membrane. Interestingly, the relaxivity shown by each Gd(III) complex entrapped in the apoferritin is very high (ca.80 $mM^{-1} s^{-1}$), that is, a value ca 20 times higher then the r_1 value of the free chelate. This reduces the number of Gd necessary for the MR-visualization to ca 5×10^7/cell. Targeting cells through transporters of nutrients (e.g. aminoacids) and pseudonutrients (e.g. polyamines) is another, particularly attractive, approach (Aime *et al.*, 2002a).

7.4.4 New emerging agents

Few years ago, a novel class of MRI CAs, the so-called CEST (Chemical Exchange Saturation Transfer) agents, has been proposed. The CEST agent is a molecule containing labile protons, whose exchange rate with the solvent water protons is smaller than the separation of their resonance frequencies (Ward, Aletras and Balaban, 2000). When the mobile protons' absorption is selectively saturated by a proper rf irradiation field, the chemical exchange process transfers saturated magnetization from the CA to the bulk water signal, whose intensity will then decrease accordingly. The extent of contrast then relies on the efficiency of the saturation transfer which, in turn, is dependent on the number of exchangeable protons and on their exchange rate. Particularly, interesting systems have been found in the class of Lanthanide Paramagnetic Complexes (Para-CEST agents) (Aime *et al.*, 2002b; Zhang *et al.*, 2001).

References

Aime, S., M. B., Fasano, M., Terreno, E., 1998. *Chem. Soc. Rev.* **27**, 19.

Aime, S., Barge, A., DelliCastelli, D., Fedeli, F., Mortillaro, A., Nielsen, F.U., Terreno, E., 2002b. *Magn. Res. Med.* **47**, 639.

Aime, S., Botta, M., Fasano, M., Terreno, E., 2001. In: Merbach, A.E., Toth, E., (Eds.) *The Chemistry of Contrast Agents in Medical Magnetic Resonance Imaging*. John Wiley & Sons, Ltd, Chichester, p. 193.

Aime, S., Botta, M., Garino, E. *et al.*, 2000. *Chem. Eur. J.* **6**, 2609.

Aime, S., Cabella, C., Colombatto, S., Geninatti Crich, S., Gianolio, E., Maggioni, F., 2002a. *JMRI* **16**(4), 394–406.

Andre, J. P., Toth, E., Fischer, H., Seelig, A., Macke, H. R., Merbach, A. E., 1999. *Chem. Eur. J.* **5**, 2977.

Bligh, S.W., Harding, C. T., Sadler, P. J. *et al.*, 1991. *Magn. Res. Med.* **17**, 516.

Bulte, J. W. M., Kraichman, D. L., 2004. *Curr. Pharm. Biotech.* **5**, 567.

Geninatti Crich, S., Biancone, L., Cantaluppi, V., Esposito, G., Russo, S., Camussi, G, Aime, S., 2004. *Magn. Res. Med.* **51**, 938.

Lowe, M. P., Parker, D., Reany, O., Botta, M., Aime, S., Pagliarin, R., 2001. *J. Am. Chem. Soc.* **123**, 7601.

Moats, R. A., Fraser, S. E., Meade, T. J., 1997. *Angew. Chem. Int. Ed. Engl.* **36**, 726.

Osterloh, K., Aisen, P., 1989. *Biochim. Biophys. Acta.* **1011**, 40–45.

Roch, Muller, R. N., Gillis, P., 1999. *J. Chem. Phys* **110**, 5403.

Wiener, E. C., Brechbiel, M. W., Brothers, H. *et al.*, *Magn. Res. Med.* **31**, 1.

Walczak, P., Kedziorek, D. A., Gilad, A. A., Lin, S., Bulte, J. W. M., 2005. *Magn. Res. Med.* **54**, 769.

Ward, K. M., Aletras, A. H., Balaban, R. S., 2000. *J. Magn. Res.* **143**, 79.

Zhang, S., Winter, P., Wu, K.,.Sherry, A. D., 2001. *J. Am. Chem. Soc.* **123**, 1517.

7.5 Imaging techniques – reporter gene imaging agents

Huongfeng Li and Andreas H. Jacobs

7.5.0 Introduction

Advances in genetic engineering and molecular biology promote the development of gene therapy which holds significant promise in the treatment of many human diseases. In order to achieve a controlled and effective target-specific expression of gene for application of gene therapy, the ability to image the locations(s), magnitude and time variation of therapeutic gene expression in animal models as well as in patients has been significantly improved over the last years. Under different imaging techniques, radiotracer-imaging approaches can non-invasively and highly sensitively image the reporter gene for *in vivo* use in animals and humans. The development and application of radionuclide imaging with positron emission tomography (PET) and single-photo emission computed tomography (SPECT) have been extensively reviewed over the past 5 years (Gambhir *et al.*, 2000; Weissleder and Mahmood, 2001; Nichol and Kim, 2001; Blasberg and Tujvajev, 2002; Min and Gambhir, 2004). In the past decade most of the molecular imaging in gene therapy focuses on oncologic applications, but further studies include cardiovascular and neurological applications. In the following, marker gene/marker substrate combination enabling imaging of gene expression and gene therapy *in vivo* are summarized.

7.5.1 Reporter gene imaging agents

7.5.1.1 *E.coli cytosine deaminase*

The *E. coli* cytosine deaminase gene (*cd*), which was one of the first reporter genes to be studied for imaging reporter gene expression, had been used as a PET marker gene with fluorinated 5-fuorocytosine (5-FC) (Haberkorn *et al.*, 1996). However, because of the relatively slow uptake of 5-FC in *cd*-transduced cells, the metabolization of 5-FC to toxic 5-fluorouracil (5-FC), and the rapid efflux of 5-FU from transduced cells, the CD/5-FC combination does not meet some of the critical requirements for an ideal marker gene/marker substrate combination for radionuclide imaging (Tjuvajev *et al.*, 1996). On the contrary, the ability of MRS to distinguish signals from chemically distinct compounds provides the advantage over PET methods to measure gene expression. Conversion of the non-toxic prodrug 5-fluorocytosine (5-FC) to the chemotherapeutic agent 5-fluorouracil (5-FU) by yeast cytosine deaminase (yCD) could be observed and quantitated in colorectal tumour xenograft in living subjects using ^{19}F MRS (Brix *et al.*, 1996; Stegman *et al.*, 2000). These studies demonstrate the feasibility of using MRS to non-invasively monitor therapeutic transgene expression in tumours. Morever, indirect assessment of CD activity has been performed using a *cdtk* fusion gene and [^{124}I]FIAU-PET imaging (Hackman *et al.*, 2002).

7.5.1.2 *Aromatic L-amino acid decarboxylase*

Recently, Bankiewicz *et al.* (2002) delivered the aromatic-amino-dopadecarboxylase (AADC) gene to the striatum of experimental animals as a therapeutic gene via an adeno-associate viral vector (AAV) to study Parkinson's disease. Because of decarboxylation and storing in the striatal neurons of 6-[^{18}F]fluoro-l-m-tyrosine ([^{18}F]FMT) as tracer for AADC expression, *in vivo* visualization of gene expression in the monkey brain could be observed by PET (Bankiewicz *et al.*, 2000). The possibility of non-invasive assessment of vector-mediated gene expression after CED-mediated vector delivery into the central nervous system had been shown through monitoring the CED-dependent magnitude of AADC expression by co-registering MRI and [^{18}F]FMT-PET with tyrosine hydroxylase (TH) and AADC immunohistochemical sections (Bankiewicz *et al.*, 2000).

7.5.1.3 *Reporter gene for optical imaging*

For optical imaging of reporter gene expression in living animals, the firefly and Renilla luciferase gene have been used as marker genes with D-luciferin and coelenterazine as specific marker substrates, respectively. Different luciferases catalyze the energy-dependent reactions which emits visible photos at specific wavelengths which can be detected by use of a cooled charged couple detector (CCD) camera or photon-counting cameras (Bhaumik and Gambhir, 2002; Contag *et al.*, 2000). This method has been used for the determination of adenoviral-mediated luciferase gene transduction into the rat myocardium (Wu *et al.*, 2002a) and skeletal muscle of mice (Wu *et al.*, 2001). Most importantly, the study, using this method to non-invasively quantitate the growth and therapy-induced changes in tumour burden in the intracranial rat 9L gliosarcoma model (Rehemtulla *et al.*, 2000a,b), showed that this imaging system is rapid, easy to handle and relatively cheap compared to the PET and MRI, although the last two methods can offer a more detailed 3D analysis. This imaging system has been further developed to study protein-protein interactions *in vivo* by a modification of the yeast two-hybrid system adapted for mammalian cells (Ray *et al.*, 2002), to detect the luciferase enzyme as a transcriptional reporter in order to monitor both the presence and the function of transgenes in transgenic animals (Zhang *et al.*, 2001). The kinetics of gene expression after administration of lentiviral vector encoding FL could be easily determined (Tsui *et al.*, 2002). Another reporter gene is adenovirus-encoding GFP and is used for non-invasive, whole-body, real-time fluorescence optical imaging of transgene expression in the major organs of nude mice including the brain and liver (Yang *et al.*, 2000). Imaging of the transduced lentivirus in non-dividing hepatocytes in living nude mice with GFP has also been investigated (Pfeifer *et al.*, 2001). Both fluorescence and bioluminescence imaging have been used for tracking the migration of cells in living small animals (Hardy *et al.*, 2001; Weninger, Manjunath and Von Andrian, 2002). A further method of using prostate-targeted gene expression has been developed to assess the Cathepsin-B expression in breast cancer-bearing mice, the matrix metalloproteinase-2 (MMP-2) activity by near-infrared fluorescence imaging (NIRFI) (Min and Gambhir, 2004; Bremer *et al.*, 2002; Mahmood *et al.*, 1999).

7.5.1.4 β-Galactosidase

Using a contrast agent which is enzymatically processed by β-galactosidase, a histochemical marker gene, the vector-mediated gene expression *in vivo,* can be also monitored by MRI (Louie *et al.*, 2000). The specific contrast agent, (1-(2-(β-galacto- pyranosyloxy) propyl)-4,7,10-tris(carboxymethyl)-1,4,7,10-tetra- azacyclododecane) gadolinium(III) (EgadMe), was designed and synthesized according to the clinical contrast agent, Gd(HPDO3A), that has been modified with a carbohydrate 'cap' that blocks the access of water to the gadolinium. When access of water to gadolinium is blocked, signal enhancement by the contrast agent is turned 'off'. The cap is attached to the contrast agent through a ß-galactosi- dase-cleavable linker. Enzyme cleavage releases the cap and opens water access to the gadolinium ion, turning the contrast agent 'on' (Louie *et al.*, 2000). In a study employing Xenopus embryos, the ß-galactosidase expression to cellular resolution through MRI demonstrated the ability of MRI to detect gene expression in living animals (Louie *et al.*, 2000).

7.5.1.5 Dopamine type-2 receptor

Dopamine-2-receptor (D2R) is being used as endogenous as well as exogenous PET reporter gene for applications in the central nervous system. By using the specific D2-receptor ligand [^{11}C]raclopride (Ogawa *et al.*, 2000; Umegaki *et al.*, 2002) after the administration of *d2r*-carrying vectors in rats, the different levels of D2R expression could be differentiated by PET. Furthermore, a mutant gene has been studied by using 3-[2′-[^{18}F]fluoroethyl]-spiperone (FESP) (Maclaren *et al.*, 1999; Liang *et al.*, 2001), and shows uncoupling of signal transduction but preservation of the affinity of receptor for tracer ligand (Liang *et al.*, 2001). By placing the D2R gene and a mutated HSV-1-*tk* gene under transcriptional control of a bidirectional, tetracycline-responsive element, Sun *et al.* (2001) could demonstrate that various levels of PET marker gene expression can be differentiated by PET in cell lines stably expressing these constructs depending on the state of induction.

7.5.1.6 Somatostatin and other receptors or membrane transporters

Receptors for human type 2 somatostatin receptor (hSSTr2) (Zinn and Chaudhuri, 2002) and the sodium iodide symporter (NIS) (Chung, 2002; Spitzweg and Morris, 2002) are other examples of receptor-based imaging reporters beside D2R which can be imaged by PET/SPECT. One of the problems with receptor-based imaging might be that quantification of gene expression may be difficult due to saturation effects. The NIS gene was applied in radioiodide gene therapy due to its ability to facilitate the accumulation of iodide by thyroid follicular cells (Chung, 2002; Haberkorn *et al.*, 2003). Moreover, conventional imaging agents such as radioiodide and [^{99m}Tc]pertechnetate can be used to monitor NIS expression with SPECT (Cho *et al.*, 2002; Min *et al.*, 2002; Groot-Wassink *et al.*, 2002).

7.5.1.7 Synthetic receptors

Imaging of novel cell-surface expressed fusion proteins typically not present on eukaryotic cells has been studied for investigating phenotypic changes in genetically manipulated cells (Bogdanov, Simonova and Weissleder, 2000). The proposed artificial receptor comprises a peptide-based chelate that binds the [^{99m}Tc] and a membrane-anchoring domain. The most promising fusion protein to date consists of a metallothionein (MT)-derived C-terminal peptide fused to a type II membrane protein containing the N-terminal membrane-anchoring domain of neural endopeptidase (PEP). This leads to cell-surface expression of ^{99m}Tc-binding sites and enabling 'tagging' of transduced cells with ^{99m}Tc for indirect assessment of the level of gene expression (Bogdanov, Simonova and Weissleder, 2000).

A human transferring receptor (ETR) has been engineered as a potential transporter for accumulation of targeted MR contrast agents to non-invasively follow the uptake of a sterically protected iron-containing probe (Weissleder *et al.*, 2000) and monocrystalline iron oxide nanoparticles (Tf-MION; Moore *et al.*, 1998, 2001)) or crosslinked iron oxide (Tf-CLIO) in ETR-expressing cells (Ichikawa *et al.*, 2002). The transferrin is recognized by the receptor, and the entire particle is endocytosed by the cell, bringing in iron that acts as a contrast agent by affecting the T2 rate (Marten *et al.*, 2002). *In vivo*, ETR-positive tumours accumulated more Tf-MION and had higher signal intensities on T1-weighted MR images and lower signal intensities on T2-weighted images than negative control tumours suggesting that the presence or the absence of marker gene expression can be detected by MR using this MR-marker receptor; however, quantification of gene expression by the change in MR signal intensity is still difficult (Moore *et al.*, 2001). A recent study using an HSV-1-based amplicon vector system for transgene delivery demonstrate that MRI of ETR expression can serve as a surrogate for measuring therapeutic transgene expression (Ichikawa *et al.*, 2002).

7.5.1.8 *Herpes simplex virus type 1 thymidine kinase (HSV-1-tk)*

HSV-1-*tk* was originally suggested as a marker enzyme for detection of viral encephalitis by autoradiography (Saito *et al.*, 1984), and then was used as a suicide gene first by Moolten (Moolten, 1986). The viral thymidine kinase gene has broad substrate specificity and can phosphorylate nucleoside analogs like pyrimidine and purine derivatives as well as deoxythymidine (De Clercq, 1993). Thus, the HSV-1-*tk* viral gene has been applied as suicide gene in clinical studies and is the most widely exploited enzyme-based reporter gene whose expression is detected by radionuclide product sequestration (Borrelli *et al.*, 1988; Moolten and Wells, 1990). Under the two major classes of radiolabelled substrates (acycloguanosine and pyrimidine derivatives) which are used for HSV-1-*tk* reporter gene imaging, 9-(4-[^{18}F]-fluoro-3-hydroxy- mehtylbutyl)guanine (FHBG) (Alauddin and Conti, 1998) and (2-fluoro-2′-deoxy-β- D-arabino-furanosyl-5-iodo-uracil) ([^{18}F]FIAU (Alauddin, Conti and Fissekis, 2003; Mangner *et al.*, 2003), [*I]FIAU (Tovell *et al.*, 1988)) are currently the best substrates for HSV-1-*tk*. They have been used for PET imaging of HSV-1-*tk* reporter gene expression in living animals (Tjuvajev *et al.*, 2002; Min, Iyer and Gambhir, 2003) and in a phase I/II clinical trial of liposomal vector-mediated gene therapy of recurrent glioblastoma (Jacobs *et al.*, 2001) to determine the biologically active target tissue. HSV-1-*tk* together with D2R as PET imaging reporter genes has been used for reporting transcriptional regulation at the level of induction (Sun *et al.*, 2001). Tjuvajev *et al.* (1995) were the first to use a recombinant replication-deficient STK retrovirus containing HSV-1-*tk* to transduce RG2 glioma cells *in culture* and image the TK expression in experimental rat model *in vivo*. With [^{124}I]FIAU-PET, a half-log difference between viral doses could be distinguished, so that the proliferation in xenograft tumour infected by oncolytic HSV-1 mutant vectors could be determined (Bennett *et al.*, 2001; Jacobs *et al.*, 2001). Rat myocardial tissue has been studied through transduction of adenoviral-mediated HSV-1-*tk* reporter gene (Bengel *et al.*, 2000) and HSV-1-*sr*39TK (Wu *et al.*, 2002 b) protein and by measuring the uptake of the radiolabelled [^{125}I]FIAU and [^{18}F]FHBG, respectively. The application of HSV-1-*tk*/HSV-1-*sr39tk* for central nervous system directed research with PET is not optimal because all the reporter probes for HSV-1-*tk*/HSV-1-*sr39tk* have poor permeability to cross the blood-brain barrier (BBB). Further development of more lipophilic radionuclide probes has been performed, for example the radioiodinated 5-(E)-(2-[^{125}I]iodovinyl)-2′-fluoro 2′-deoxyuridine (IVFRU), 5-(E)-(2-[^{125}I] iodovinyl)-2′-fluoro-2′-deoxyarabinouridine (IVFAU) (Morin *et al.*, 1997) and 2-fluoro-2-deoxy-ß-D-arabino- furanosyl-5-ethyluracil ([^{18}F]FEAU) (Allauddin *et al.*, 2005), to enable imaging HSV-1-*tk*/HSV-1-*sr39tk* gene expression in neurological applications.

7.5.2 Conclusion

With the development of human gene therapy and molecular medicine, optimal imaging techniques and marker gene/marker substrate combinations are required and expected to map gene expression noninvasively *in vivo*. Moreover, genetic engineering requires the generation of more robust gene-transfer vectors, both viral and non-viral. Further improvement of reporter gene imaging agents, especially the attempt to develop probes which freely pass the BBB, is required because most reporter gene imaging agents currently available, especially those which are targets for HSV-1-TK, do not penetrate the intact BBB. The advancements of multimodality reporter gene approaches employing various imaging technologies will be helpful for both gene therapy investigations and molecular testing of pre-clinical animal models and eventually for successful safe and efficient gene therapy in patients.

References

Alauddin, M. M., Conti, P. S., 1998. "Evaluation of 9-(4-[^{18}F]-fluoro-3-hydroxy-mehtylbutyl) guanine (FHBG): A new potential imaging agent for viral infection and gene therapy using PET." *Nucl. Med. Biol.* **25**(3),175–180.

Alauddin, M. M., Conti, P. S., Fissekis, J. D., 2003. "A general synthesis of 2'-deoxy-2'-[^{18}F]fluoro- 1-ß-D-arabinofuranosyluracil and its 5-substituted nucleosides." *J. Label Compd. Radiopharm.* **46**, 285–289.

Allauddin, M. M., Hajitou, A., Soghomonyan, S., Ozawa, M., Balatoni, J., Pasqualini, R., Arap, W., Gelovani, J., 2005. "Micro-PET Imaging of HSV-tk gene delivery and expression with [^{18}F]FEAU in tumor-bearing mice." *Mol. Imaging* **4**(3), 282.

Bankiewicz, K. S., Eberling, J. L., Kohutnicka, M., Jagust, W., Pivirotto, P., Bringas, J., Cunningham, J., Budinger, T. F., Harvey-White, J. 2000. "Convection-enhanced delivery of AAV vector in parkinsonian monkeys; in vivo detection of gene expression and restoration of dopaminergic function using prodrug approach." *Exp. Neurol.* **164**, 2–14.

Bengel, F. M., Anton, M., Avril, N., Brill, T., Nguyen, N., Haubner, R., Gleiter, E., Gansbacher, B., Schwaiger, M., 2000. "Uptake of radiolabeled 2′-fluoro-2′-deoxy-5-iodo-1-beta-D-arabinofurano- syluracil in cardiac cells after adenoviral transfer of the herpes virus thymidine kinase gene - the cellular basis for cardiac gene imaging." *Circulation* **102**, 948–950.

Bennett, J. J., Tjuvajev, J., Johnson, P., Doubrovin, M., Akhurst, T., Malholtra, S., Hackman, T., Balatoni, J., Finn, R., Larson, S. M., Federoff, H., Blasberg, R., Fong, Y. M. et al., 2001. "Positron emission tomography imaging for herpes virus infection: implications for oncolytic viral treatments of cancer." *Nat. Med.* **7**, 859–863.

Bhaumik, S., Gambhir, S. S., 2002. "Optical imaging of Renilla luciferase reporter gene expression in living mice." *Proc. Natl. Acad. Sci. USA* **99**, 377–382.

Blasberg, R. G., Tujvajev, J. G., 2002. "Molecular-genetic imaging: A nuclear medicine-based perspective." *Mol. Imaging* **1**, 280–300.

Bogdanov Jr., A. A., Simonova, M., Weissleder, R., 2000. "Engineering membrane proteins for nuclear medicine: Applications for gene gehrapy and cell tracking." *Q. J. Nucl. Med.* **44**, 224–235.

Borrelli, E., Heyman, R., Hsi, M., Evans, R. M., 1988. "Targeting of an inducible toxic phenotype in animal cells." *Proc. Natl. Acad. Sci. USA* **85**, 7572–7576.

Bremer, C., Tung, C. H., Bogdanov Jr., A., Weissleder, R., 2002. "Imaging of differential protease expression in brease cancers for detection of aggressive tumor phenotypes." *Radiology* **222**, 814–818.

Brix, G., Bellemann, M. E., Haberkorn, U., Gerlach, L., Lorenz, W. J., 1996. "Assessment of the biodistribution and metabolism of 5-fluorouracil as monitored by 18F PET and 19F MRI: A comparative animal study." *Nucl. Med. Biol.* **23**, 897–906.

Cho, J. Y., Shen, D. H.Y., Yang, W., Williams, B., Buckwalter, T. L.F., La Perle, K. M.D., Hinkle, G., Pozderac, R., Kloos, R., Nagaraja, H. N., Barth, R. F., Jhiang, S. M., 2002. "*In vivo* imaging and radioiodine therapy following sodium iodide symporter gene transfer in animal model of intracerebral gliomas." *Gene. Therapy.* **9**, 1139–1145.

Chung, J. K., 2002. "Sodium iodide symporter: Its role in nuclear medicine." *J. Nucl. Med.* **43**, 1188–1200.

Contag C H, Jenkins D, Contag PR, Negrin RS (2000). "Use of reporter genes for optical measurements of neoplastic disease *in vivo*." *Neoplasia* **2**, 41–52.

De Clercq, E., 1993. "Antivirals for the treatment of herpes virus infections." *J. Antimicrob. Chemoth.* **3**(Suppl. A), 1–2132.

Gambhir, S. S., Herschman, H. R., Cherry, S. R., Barrio, J. R., Satyamurthy, N., Toyokuni, T, Phelps, M. E., Larson, S. M., Balatoni, J., Finn, R., Sadelain, M., Tjuvajev, J., Blasberg, R., 2000. "Imaging transgene expression with radionuclide imaging technologies." *Neoplasia* **2**, 118–138.

Groot-Wassink, T., Aboagye, E. O., Glaser, M., Lemoine, N. R., Vassaux, G., 2002. "Adenovirus biodistribution and non-invasive imaging of gene expression in vivo by positron emission tomography using human sodium/iodide symporter as reporter gene." *Hum. Gene. Ther.* **13**, 1723–1735.

Haberkorn, U., Kinscherf, R., Kissel, M., Kübler, W., Mahmut, M., Sieger, S., Eisenhut, M., Peschke, P., Altmann, A., 2003. "Enhanced iodide transport after transfer of the human sodium iodide symporter gene is associated with lack of retention and low absorbed dose." *Gene. Therapy* **10**, 774–780.

Haberkorn, U., Oberdorfer, F., Gebert, J., Morr, I., Haack, K., Weber, K., Lindauer, M., vn Kaick, G., Schackert, H. K., 1996. "Monitoring gene therapy with cytosine deaminase: *In vitro* studies using tritiated 5-fluorocytosine." *J. Nucl. Med.* **37**, 87–94.

Hackman, T., Doubrovin, M., Balatoni, J., Beresten, T., Ponomarev, V., Beattie, B., Finn, R., Bornmann, W., Blasberg, R., Tjuvajev, J., 2002. "Imaging expression of cytosine deaminase-herpes virus thymidine kinase fusion gene (CD/TK) expression with [^{124}I]FIAU and PET. Mol." *Imaging* **1**, 36–42.

Hardy, J., Edinger, M., Bachmann, M. H., Negrin, R. S., Fathman, C. G., Contag, C. H., 2001. "Bioluminescence imaging of lymphocyte trafficking *in vivo*." *Exp. Hematol.* **29**, 1353–1360.

Ichikawa, T., Högemann, D., Saeki, Y., Tyminski, E., Terada, K., Weissleder, R., Chiocca, E. A., Basilion, J. P., 2002. "MRI of transgene expression: correlation to therapeutic gene expression." *Neoplasia* **4**, 523–530.

Jacobs, A., Tjuvajev, J. G., Dubrovin, M., Akhurst, T., Balatoni, J., Beattie, B., Joshi, R., Finn, R., Larson, S. M., Herrlinger, U., Pechan, P. A., Choacca, E. A., Breakefield, X. O., Blasberg, R. G., 2001. "Positrion emission tomography-based imaging of transgene expression mediated by replication-conditional, oncolytic herpes simplex virus type 1 mutant vectors in vivo." *Cancer Res.* **61**, 2983–2995.

Jacobs, A., Voges, J., Reszka, R., Lercher, M., Gossmann, A., Kracht, I., Kaestle, C., Wagner, R., Wienhard, K., Heiss, W. D., 2001. "Positron-emission tomography of vector-mediated gene expression in gene therapy for gliomas." *Lancet* **358**, 727–729.

Liang, Q., Satyamurthy, N., Barrio, J. R., Toyokuni, T., Phelps, M. P., Gambhir, S. S., Herschman, H. R., 2001. "Non-invasive, quantititative imaging in living animals of a mutant dopamine D2 receptor reporter gene in which ligand binding is uncoupled from signal transduction." *Gene. Ther.* **8**, 1490–1498.

Louie, A. Y., Huber, M. M., Ahrens, E. T., Rothbacher, U., Moars, R., Jacobs, R. E., Fraser, S. E., Mead, T. J., 2000. "In vivo visualization of gene expression using magnetic resonance imaging." *Nat. Biotechnol.* **18**, 321–325.

Maclaren, D. C., Gambhir, S. S., Satyamurthy, N., Barrio, J. R., Sharfstein, S., Toyokuni, T., Wu, L., Berk, A. J., Cherry, S. R., Phelps, M. E., Herschman, H. R., 1999. "Repetitive, non-invasive imaging of the diopamine D2 receptor as a reporter gene in living animals." *Gene. Ther.* **6**, 785–791.

Mahmood, U., Tung, C. H., Bogdanov Jr., A., Wsissleder, R., 1999. "Near-infrared optical imaging of protease activity for tumor detection." *Radiology* **213**, 866–870.

Mangner, T. J., Klecker, R. W., Anderson ,L., Shields, A. F., 2003. "Synthesis of 2′-deoxy-2′- [^{18}F]fluoro-ß-arabinofuranosyl nucleosides, [^{18}F]FAU, [^{18}F]FMAU, [^{18}F]FBAU and [^{18}F]FIAU, as potential PET agents for imaging cellular proliferation." *Nucl. Med. Biol.* **30**, 215–224.

Marten, K., Bremer, C., Khazaie, K., Sameni, M., Sloane, B., Tung, C. H., Weissleder, R., 2002. "Detection of dysplastic intestinal adenomas using enzyme-sensing molecular beacons in mice." *Gastroenterology* **122**, 406–414.

Min, J. J., Chung, J. K., Lee, Y. J., Shin, J. H., Yeo, J. S., Jeong, J. M., Bom, D. S.L., Lee MC (2002). In vitro and in vivo characteristics of a human colon cancer cell line, SNU-C5N, expressing sodium-iodide symporter." *Nucl. Med. Biol.* **29**, 537–545.

Min, J. J., Gambhir, S. S., 2004. "Gene therapy progress and prospects: non-invasive imaging of gene therapy in living subjects." *Gene. Ther.* **11**, 115–125.

Min, J. J., Gambhir, S. S., 2004. "Gene. Therapy progress and prospects: Non-invasive imaging of gene therapy in living subjects." *Gene. Ther.* **11**, 115–125.

Min, J. J., Iyer, M., Gambhir, S. S., 2003. "Comparison of FHBG and FIAU for imaging of HSV1-tk reporter gene expression: Adenoviral infection versus stable transfection." *Eur. J. Nucl. Med. Mol. Imaging.* **30**, 1547–1560.

Moolten, F. L., 1986. "Tumor chemoosensitvity conferred by inserted herpes thymidine kinase gnes: paradigm for prospective cancer control strategy." *Cancer. Res.* **46**, 5276–5281.

Moolten, F. L., Wells, J. M., 1990. "Curability of tumors bearing herpes thymidine kinase genes transfected by retroviral vectors." *J. Natl. Cancer Inst.* **82**, 297–300.

Moore, A., Basilion, J. P., Chiocca, E. A., Weissleder, R., 1998. "Measuing transferrin receptor gene expression by NMR imaging." *Biochim. Biophys. Acta.* **1402**, 239–249.

Moore, A., Josephson, L., Bhorade, R. M., Basilion, J. P., Weissleder, R., 2001. "Human transferrin receptor gene as a marker gene for MR imaging." *Radiology* **221**, 244–250.

Morin, K. W., Atrazheva, E. D., Knaus, E. E., Wiebe, L. I., 1997. "Synthesis and cellular uptake of 2'-substituted analogues of (E)-5-(2-[^{125}I]Iodovinyl)-2′-deoxyuridine in tumor cells transduced with the herpes simplex type-1 thymidine kinase gene. Evaluation as probes for monitoring gene therapy." *J. Med. Chem.* **40**, 2184–2190.

Nichol, C., Kim, E. E., 2001. "Molecular imaging and gene therapy." *J. Nucl. Med.* **42**, 1368–1374.

Ogawa, O., Umegaki, H., Ishiwata, K., Asai, Z., Ikari, H., Oda, K., Toyama, H., Ingram, D. K., Roth, G. S., Iguchi, A., Senda, M., 2000. "*In vivo* imaging of adenovirus-mediated over-expression of dopamine D2 receptors in rat striatum by positron emission tomography." *Neuro. Report.* **11**, 743–748.

Pfeifer, A., Kessler, T., Yang, M., Baranov, E., Kootstra, N., Cheresh, D. A., Hoffman, R. M., Verma, I. M., 2001. "Transduction of liver cells by lentiviral vectors: analysis in living animals by fluorescence imaging." *Mol. Ther.* **3**, 319–322.

Ray, P., Pimenta, H., Paulmurugan, R., Berger, F., Phelps, I,M., Gambhir, S. S., 2002. "Non-invasive quantitative imaging of protein-protein interactions in living subjects." *Proc. Natl. Acad. Sci. USA* **99**, 3105–3110.

Rehemtulla, A., Hall, D. E., Stegman, L. D., Chen, G., Bhojani, M. S., Chenevert, T. L., Ross, B. D.,

2000a. "Molecular imaging of gene expression and efficacy following adenoviral-mediated brain tumor gene therapy." *Mol. Imaging* **1**, 43–55.

Rehemtulla, A., Stegman, L. D., Cardozo, S. J., Gupta, S., Hall, D. E., Contag, C. H., Ross, B. D., 2000b. "Rapid and quantitative assessment of cancer treatment response using in vivo bioluminescence imaging." *Neoplasia* **1**, 491–495.

Saito, Y., Rubenstein, R., Price, R. W., Fox, J. J., Watanabe, K. A., 1984. "Diagnostic imaging of herpes simplex virus encephalitis using a radiolabeled antiviral drug: Autoradiographic assessment in an animal model." *Ann. Neurol.* **15**, 548–558.

Spitzweg, C., Morris, J. C., 2002. "The sodium iodide symporter: its pathophysiological and therapeutic implications." *Clin. Endocrinol. (Oxf)* **57**, 559–574.

Stegman, L. D., Rehemtulla, A., Hamstra, D. A., Rice, D. J., Jonas, S. J., Stout, K. L., Chenevert, T. L., Ross, B. D., 2000. "Diffusion MRI detects early events in the response of a glioma model to the yeast cytosine deaminase gene therapy strategy." *Gene. Ther.* **7**, 1005–1010.

Sun, X., Annala, A. J., Yaghoubi, S. S., Barrio, J. R., Nguyen, K. N., Toyokuni, T., Satyamurthy, N., Namavari, M., Phelps, M. E., Herschman, H. R., Gambhir, S. S., 2001. "Quantitative imaging of gene induction in living animals." *Gene. Ther.* **8**, 1572–1579.

Sun, X., Annala, A. J., Yaghoubi, S. S., Barrio, J. R., Nguyen, K. N., Toyokuni, T., Satyamurthy, N., Namavari, M., Phelps, M. E., Herschman, H. R., Gambhir, S. S., 2001. "Quantitative imaging of gene induction in living animals." *Gene. Ther.* **8**, 1572–1579.

Tjuvajev, J. G., Doubrovin, M., Akhurst, T., Cai, S., Balatoni, J., Alauddin, M. M., Finn, R., Bornmann, W., Thaler, H., Conti, P., Blasberg, R. G., 2002. "Comparison of radiolabeled nucleoside probes (FIAU, FHBG, and FHPG) for PET imaging of HSV1-tk gene expression." *J. Nucl. Med.* **43**, 1072–1083.

Tjuvajev, J. G., Finn, R., Watanabe, K., Joshi, R., Oku, T., Kennedy, J., Beatie, B., Koutcher, J., Larson, S., Blasberg, R. G., 1996. "Non-invasive imaging of herpes virus thymidine kinase gene transfer and expression: A potential method for monitoring clinical gene therapy." *Cancer. Res.* **56**, 4087–4095.

Tjuvajev, J. G., Stockhammer, G., Desai, R., Uehara, H., Watanabe, H., Gansbacher, B., Blasberg, R. G., 1995. "Imaging the expression of transfected genes in vivo." *Cancer. Res* **55**, 6126–6132.

Tovell, D. R., Samuel, J., Mercer, J. R., Misra, H. K., Xu, L., Wieber, L. I., Tyrrell, D. L., Knaus, E. E., 1988. "The in vitro evaluation of nucleoside analogues as probes for use in the non-invasive diagnosis of herpes simplex encephalitis." *Drug. Des. Del.* **3**, 213–221.

Tsui, L. V., Kelly, M., Zayek, N., Rojas, V., Ho, K., Ge, Y., Moskalenko, M., Mondesire, J., Davis, J., Van, R. M., Dull, T., McArthur, J. G., 2002. "Production of human clotting Factor IX without toxicity in mice after vascular delivery of a lentiviral vector." *Nat. Biotechnol.* **20**, 53–57.

Umegaki, H., Ishiwata, K., Ogawa, O., Ingram, D. K., Roth, G. S., Zoshimura, J., Oda, K., Matsui-Hirai, H., Ikari, H., Iguchi, A., Senda, M., 2002. "*In vivo* assessment of adenoviral vector-meditated gene expression of dopamine D2 receptors in the rat striatum by positron emission tomography." *Synapse* **43**, 195–200.

Weissleder, R., Mahmood, U., 2001. "Molecular imaging." *Radiology* **219**, 316–333.

Weissleder, R., Moore, A., Mahmood, U., Bhorade, R., Benveniste, H., Chiocca, E. A., Basilion, J. P., 2000. "*In vivo* magnetic resonance imaging of transgene expression." *Nat. Med.* **6**, 351–355.

Weninger, W., Manjunath, N., Von Andrian, U. H., 2002. "Migration and differentiation of CD8+ T cells." *Immunol. Rev.* **186**, 221–233.

Wu, J. C., Inubushi, M., Sundaresan, G., Schelbert, H. R., Gambhir, S. S., 2002a. Optical imaging of cardiac reporter gene expression in living rats." *Circulation* **105**, 1631–1634.

Wu, J. C., Inubushi, M., Sundaresan, G., Schelbert, H. R., Gambhir, S. S., 2002b. "Positron emission tomography imaging of cardiac reporter gene expression in living rats." *Circulation* **106**, 180–183.

Wu, J. C., Sundaresan, G., SIyer, M., Gambhir, S. S., 2001. "Non-invasive optical imaging of firefly luciferase reporter gene expression in skeletal muscles of living mice." *Mol. Ther.* **4**, 297–306.

Yang, M., Baranov, E., Moossa, A. R., Penman, S., Hoffman, R. M., 2000. "Visualizing gene expression by whole-body fluorescence imaging." *Proc. Natl. Acad. Sci. USA* **97**, 12278–12282.

Zhang, W., Feng, J. Q., Harris, S. E., Contag, P. R., Srevenson, D. K., Contag, C. H., 2001. "Rapid *in vivo* functional analysis of transgenes in mice using whole body imaging of luciferase expression." *Transgenic Res.* **10**, 423–434.

Zinn, K. R., Chaudhuri, T. R., 2002. "The type 2 human somatostatin receptor as a platform for reporter gene imaging." *Eur. J. Nucl. Med. Mol. Imaging* **29**, 388–399.

8 Multi-Modality Imaging

Coordinated by Vasilis Ntziachristos

8.0 Introduction

Vasilis Ntziachristos

Multi-modality imaging has emerged as a strategy that combines the strengths of different modalities and yields a hybrid imaging platform with characteristics superior to those of any of its constituents considered alone. This imaging trend has been intensified in recent years due to the emergence of molecular imaging, which brings a new dimension in the imaging sciences and requires highly synergistic visualization approaches. This is because molecular imaging, by definition, provides information that was not available in conventional radiological imaging, but regardless needs to be superimposed on more traditional images of anatomy and function to be fully understood.

Collecting images from different modalities to improve diagnostic information has been always performed at some level, for example the use of ultrasound as a follow-up procedure to mammography to improve the specificity of the detection. What distinguishes current trends is the accuracy achieved in the co-registration of the images collected from different modalities. This is primarily achieved by 1) the development of true hybrid systems that collect images under the same or identical placement conditions, as seen by the recent evolution of clinical X-ray-CT/PET systems and 2) by the development of sophisticated image registration algorithms as elaborated in 8.2. An important feature of multi-modality approaches is that besides the straightforward superposition of images, it is possible to utilize the high-resolution information coming from an anatomical or functional image, in order to improve the imaging performance of the low resolution modality. This can be achieved for example by constructing attenuation maps based on anatomical or functional images which are then used to compute more accurate theoretical models of energy propagation in tissue (forward model) and therefore improve the fidelity and quantification of the reconstructed images.

These trends have also propagated into the small animal imaging arena with dedicated multi-modality systems being increasingly developed for improving the *in-vivo* visualization capacity in pre-clinical research. This is a relatively young field but it is expected to grow significantly in the future to serve a similarly advancing field of interrogating biology and new potent drugs on transgenic animal models using novel reporter strategies developed for *in-vivo* biomedical research. This chapter presents the key elements of multimodality imaging in 8.1 and showcases multi-modality small animal imaging utilizing micro-CT/ micro-SPECT approaches in 8.2 and FMT – micro-MRI approaches in 8.3. These two hybrid approaches are given as implementation and utility examples of multi-modal imaging, whereas many more combinations of different synergistic modalities can improve the accuracy in imaging living systems.

8.1 Concurrent imaging versus computer-assisted registration

Fred S. Azar

Multi-modality imaging involves the integration of imaging data, which are generally complementary in nature. For example, multi-modality imaging enables

Figure 8.1.1 Multi-modality imaging spectrum: Concurrent imaging versus computer-assisted registration

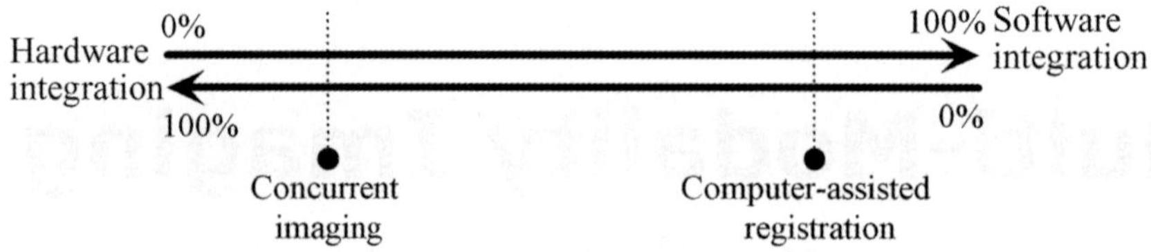

the combination of anatomical, functional and molecular information, such that biochemical activities can be detected, quantified and registered to a specific location in the animal's body. One may want to combine images of an animal from different modalities taken at the same time point, or images from the same modality taken at different time points.

Anatomical modalities, primarily describe morphology and include computed tomography (CT) and magnetic resonance imaging (MRI). Functional modalities primarily provide information on the body metabolism and include positron emission tomography (PET), single photon emission computed tomography (SPECT), various MRI implementations including functional MRI (fMRI) and diffuse optical tomography (DOT). Furthermore, several of these modalities have been put to use in molecular imaging, that is reporting on cellular and sub-cellular activity, especially for small animal imaging, and are further complemented by fluorescence molecular tomography (FMT).

Multi-modality imaging requires two important steps:

a. *Image registration*: That is bring the modalities involved into spatial alignment. This represents the main step in multi-modality imaging, and usually involves a mix of hardware and software components.
b. *Image fusion*: Once the data have been spatially aligned, combine the resulting data into the same display. This step involves only software operations.

The main challenge of multi-modality imaging is the ability to keep track of animal body position and movement (external and internal), during and between image acquisitions. There are two major trends in achieving registration. The hardware approach attempts to use concurrent acquisition methods or image under identical placement conditions at two separate time points. The software approach attempts to identify convenient markers on images acquired from two different modalities or at two different time points and register one image on the other based on a set of rigid measurements (i.e. fiducials) or models that take into account the physics of possible deformation between images.

One may imagine concurrent imaging and computer-assisted registration as two opposite ends of the same multi-modality imaging spectrum as shown in Figure 8.1.1.

The process of multi-modal imaging can be divided into four major steps. In each of these steps special care is taken to minimize and track animal movement (depending on the body part imaged and the modalities involved), and thus facilitate image registration without compromising the imaging endpoint:

1. *Animal preparation*: The animal can be anaesthetized to minimize movement during and between image acquisitions; imaging contrast agents can be injected into the animal, which may enhance structure recognition in all modalities involved and thus facilitate image registration through software post-processing techniques. In some cases such as studies of animal heart motion, markers can be implanted into the cardiac wall.
2. *Imaging setup*: The animal can be placed or restrained on the animal bed so as to minimize its movements. In some cases the animal has to be moved from one imaging device to another. The setup may then involve placing the animal in a custom-made cradle, which helps preserve the spatial alignment between the image acquisitions. Other solutions involve attaching external imaging markers to the animal; these markers are designed to be traceable and accurately detectable in the modalities involved.
3. *Image acquisition*: Most of the time, *in vivo* imaging is performed on animals; therefore, organ movement and local deformations can occur inside the animal during the process of image acquisition. The major reasons for organ deformation during image acquisition are due to animal breathing and heart beat. In some cases, it is possible to track or model those deformations for a better image registration outcome.
4. *Image post-processing*: Software image registration is performed after the imaging data has been

acquired, and it can rely on the information collected during all previous steps.

In the world of small animal imaging today, it is technically straightforward to physically combine imaging devices, due to the smaller dimensions compared with clinical systems. Thus most multi-modality pre-clinical imaging devices enable concurrent imaging of small animals, or co-linear imaging (sliding the animal from one device to the other as both devices are rigidly fixed to each other). For example, such devices in the market today enable PET/CT and SPECT/CT imaging (Gamma Medica Inc., Northridge CA; Siemens CTI Concorde Inc., Knoxville TN), as well as MRI/PET (Bruker Biospin GmbH, Germany & Siemens CTI Concorde) and planar X-ray/fluorescence/luminescence imaging (Kodak Molecular Imaging, Rochester, NY). Other companies are developing animal beds compatible with different modalities, which can be used to facilitate multi-modality image registration. However there are many instances, especially in research environments using prototype imaging systems where concurrent imaging is not possible. Therefore, in general, computer-assisted registration methods are necessary in multi-modality imaging.

To explain the basic steps involved in software registration, let us suppose that two images (2D or 3D) from two imaging modalities must be registered, and then fused. One image must be selected as the reference, and the other image becomes the secondary or moving image. The moving image is registered to the reference image through a set of applied transformations. Generally, when registration of anatomical to functional modalities is required, the anatomical image (usually the high-resolution image) is chosen as the reference, and the functional or molecular image (usually the low-resolution image) as the moving image.

Most often, computer-assisted registration requires four consecutive steps:

1. Computation of a similarity measure and/or a difference measure quantifying a metric for comparing the images (Wells *et al.*, 1996; Huesman *et al.*, 2003),
2. Use of an optimization scheme, which searches through the parameter space (e.g. rigid body motion) in order to maximize the similarity measure or minimize the difference measure (Maes *et al.*, 1996; Eberl *et al.*, 1996; Zuo *et al.*, 1996),
3. Use of an image warping (or interpolation) algorithm, which applies the latest computed set of parameters to the moving image and brings the moving image into alignment with the reference image,
4. Fusion of the moving image with the reference image, which allows displaying the combined information from the two images in the same space.

Computer-assisted image-based registration methods used in clinical imaging may be directly used in small animal imaging with minor modifications. Examples of clinical multi-modal image-based registration include MR-PET (Studholme, *et al.*, 1995; Mäkelä *et al.*, 2001), CT-PET (Pietrzyk *et al.*, 1997; Van Dalen *et al.*, 2004) and MR-DOT (Azar *et al.*, 2004, 2005).

Registration methods can be based on two types of information, external objects placed into the imaged space (during imaging setup for example) or image information generated by the animal:

Image registration based on external objects: These objects are usually fiducial markers designed to be detectable in all modalities involved. The markers can be invasive (Ellis *et al.*, 1996) or non-invasive (Wang, *et al.*, 1995) (such as markers glued to the skin). Since external objects cannot include patient-related image information, this type of image registration often leads to rigid body motion (i.e. translations and rotations). Therefore, local deformations within the animal body cannot be taken into account.

Image registration based on image information generated by the animal: This type of registration allows for non-rigid transformations. Registration can be based on a finite set of landmark points (user-defined or computer-generated), on segmented structures (lines or surfaces), or directly on measures derived from image voxel (or pixel) values. Landmark-based methods can be applied to any type of images and are mostly used to find rigid body transformations (Besl and McKay, 1992). The segmentation-based methods can be rigid-model based (Feldmar and Ayache, 1996) (where the same structures are identified in both images and used in the registration procedure), or deformable-model based (Thirion, 1996) (where the structure identified in the moving image is elastically deformed to fit the structure identified in the reference image). Voxel-based registration methods are the most flexible of all methods (Maintz *et al.*, 1996; Wells *et al.*, 1996), however they must be used with a good understanding of the multi-modal registration problem.

Independently of the image registration method used, it is important in software registration approaches to study and confirm the accuracy of the technique. The result achieved by a registration method may look 'reasonably good' but could represent the underlying truth with varying accuracy.

Therefore, controlled phantom studies, simulations or comparisons against gold standards (i.e. results obtained by an accepted and previously validated technique) become important in the development of such tools. Generally, therefore, it is a good practice to compare results from a new computer-assisted registration method to those obtained from a concurrent imaging device.

References

Azar, F. S., ElBawab, M., *et al.*, 2005. "Multimodal 3D registration of non-concurrent diffuse optical tomography with mri of breast cancer." In: *Proceedings of the Fourth Conference of International Society of Molecular Imaging*, Cologne, Germany.

Azar, F. S., Lee, K., *et al.*, 2004. "A software platform for multi-modal information integration and visualization: Diffuse optical tomography and mri of breast cancer." In: *Proceedings of the Third Conference of International Society of Molecular Imaging*, St. Louis, MO.

Besl, P. J., McKay, N. D., 1992. "A method for registration of 3D shapes." *IEEE Trans. Pattern Anal. Machine Intell.* **14**(2), 239–256.

Eberl, S., Kanno, I., *et al.*, 1996. "Automated interstudy image registration technique for SPECT and PET." *J. Nucl. Med.* **37**, 137–145.

Ellis, R. E., Toksvig-Larsen, S., *et al.*, 1996. "A biocompatible fiducial marker for evaluating the accuracy of CT image registration." In: Lemke, H. U., Vannier, M. W., Inamura, K., Farman, A.G. (Eds.), *Computer Assisted Radiology*, Vol. 1124 of Excerpta Medica – International Congress Series, Elsevier, Amsterdam, pp. 693–698.

Feldmar, J., Ayache, N., 1996. "Rigid, affine and locally affine registration of free-form surfaces." *Int. J. Comput. Vis.* **18**(2), 99–119.

Huesman R.H., Klein G., *et al.*, 2003. "Deformable registration of multi-modal data including rigid structures." *IEEE Trans. Nucl. Sci.* **50**(3).

Maes, F., Collignon, A., *et al.*, 1996. "Multi-modality image registration by maximization of mutual information." In: *Mathematical Methods in Biomedical Image Analysis*, IEEE Computer Society Press, Los Alamitos, CA, pp. 14–22.

Maintz, J. B. A., van den Elsen, P. A., *et al.*, 1996. "Comparison of edge-based and ridge-based registration of CT and MR brain images." *Med. Image Anal.* **1**(2).

Mäkelä, T. J., Clarysse, P., *et al.*, 2001. "A new method for the registration of cardiac PET and MR images using deformable model based segmentation of the main thorax structures." In: Niessen, W., Viergever, M. A. (Eds.), *Proceedings of the Fourth International Conference on Medical Image Computing and Computer Assisted Intervention (MICCAI'01)*, Lecture Notes in Computer Science 2208, Springer, pp. 557–564.

Press, W. H., Teukolsky, S. A., Vetterling, W. T., Flannery, B. P., 1992. *Numerical Recipes in C*, second ed., Cambridge University Press, Cambridge, UK.

Pietrzyk, U., Herholz, K., *et al.*, 1994. "An interactive technique for three-dimensional image registration: Validation for PET, SPECT, MRI and CT brain studies." *J. Nucl. Med.* **35**, 2011–2018.

Studholme, C., Hill, D. L., Hawkes, D. J., 1995. "Automated 3D MR and PET brain image registration." In: Lemke, H. U., Imamura, K., Jaffe, C.C and Vannier, M. W. (Eds.), *Computer Assisted Radiology*, pp. 248–253.

Thirion, J., 1996. "Non-rigid matching using demons." In: *Computer Vision and Pattern Recognition*, IEEE Computer Society press, Los Alamitos, CA, pp. 245–251.

Van Dalen, J. A., Vogel, W., *et al.*, 2004. "Accuracy of rigid CT-FDG-PET image registration of the liver." *Phys. Med. Biol.* **49**, 5393–5405.

Wang, M. Y., Fitzpatrick, J. M., Maurer, C. R., 1995. "Design of fiducials for accurate registration of CT and MR volume images." In: Loew, M. (Ed.), *Medical Imaging*, SPIE, Vol. 2434, pp. 96–108.

Wells, W. M., III, Viola, P., Atsumi, H., Nakajima, S., Kikinis, R., *et al.*, 1996. "Multi-modal volume registration by maximization of mutual information." *Med. Image Anal.* **1**(1), 35–51.

Zuo, C. S., Jiang, A., *et al.*, 1996. "Automatic motion correction for breast MR imaging." *Radiology* **198**(3), 903–906.

8.2 Combination of SPECT and CT

Jan Grimm

The use of radioactive tracers for molecular imaging is a highly appealing approach due to the very high sensitivity and the negligible internal background of nuclear imaging techniques. Biologically active molecules can be labelled with a single radiotracer molecule, which is usually sufficient for adequate imaging as described in Chapter 4. This makes nuclear imaging especially suitable to detect interactions with molecules in low abundance (Blankenberg

Figure 8.2.1 Combination of SPECT with CT: SPECT (a), CT (b) and overlay of both (c). A mouse was injected with 99mTc-MDP for a bone scan. The SPECT dataset alone provides very little information on the anatomical location of the signal (a). After combination with the CT dataset (b) the combined overlay (c) provides accurate information on the anatomical localization of the signal. The intense signal in the upper left on (a) can now be attributed to the bones in the paw

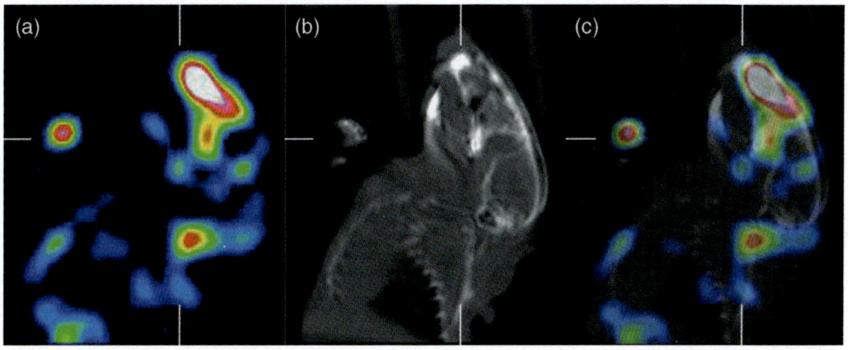

and Strauss, 2002; Mountz *et al.*, 2002). Methods to label a large variety of molecules with radionuclides have been developed and are used clinically today around the world. For SPECT imaging, gamma-emitting nuclides (e.g. 99mTc, 111In, 67Ga or 123I) are widely used. Compared to the rapidly decaying positron-emitting isotopes used for PET imaging, nuclides used for SPECT imaging provide longer half-lives (several hours to days versus seconds or minutes) and often do not require a cyclotron for their generation. This makes SPECT tracers relatively cheap and easy to handle. Furthermore, since nuclides used for SPECT imaging differ in their individual spectra, more than one nuclide can be detected at the same time, allowing for the simultaneous imaging of two different molecules (Meoli *et al.*, 2004; Zhou *et al.*, 2005), which is not feasible using PET tracers.

Therefore, SPECT-scanners are an important modality for small animal imaging, however, similarly to other molecular imaging modalities they offer resolutions of the order of 1 mm and lack background anatomical information. By combining the high sensitivity of SPECT for the radiolabel with the high anatomical resolution of CT, an accurate localization of the radiotracer can be obtained in three dimensions at high resolution (Figure 8.2.1). Combined SPECT/CT scanners accommodate the SPECT and the CT system within the same shielded housing, usually mounted onto the same aligned gantry for easy co-registration. The animal is scanned twice, once with SPECT and once with CT, and both datasets are combined and can be displayed as overlays (Figure 8.2.1) in all orientations. The CT dataset can also be used for attenuation correction of the SPECT data.

Most animal CT scanners are equipped with lower voltage X-ray tubes, and the soft tissue contrast is worse than in clinical scanners due to the small differences in attenuation coefficients between soft tissues. Therefore, contrast-enhancing agents are frequently used to adjust for this shortcoming (Figure 8.2.2). The average dose from a single CT scan for a mouse is around 0.2 Gray, the dose for the SPECT scan depends on the injected amount of tracer.

Figure 8.2.2 Application of iodinated contrast agent: CT scan without contrast agent (a) and after intravenous injection of iodinated contrast agent (b). Within the kidneys the cortex, medulla and pelvis can be distinguished after contrast agent application. The inferior vena cava can be seen between the kidneys as well

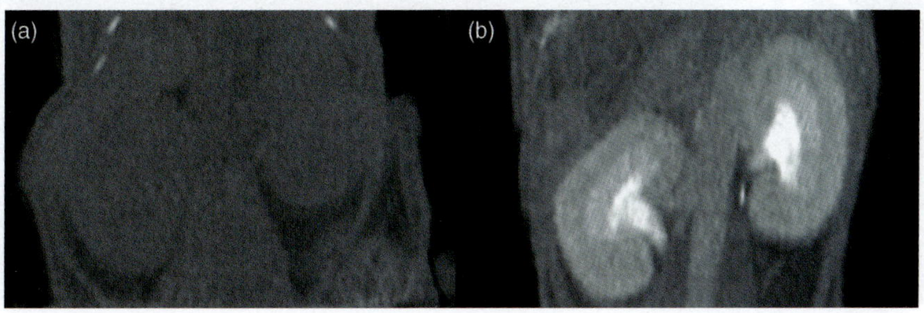

8.2.1 Applications

Labelling peptides or proteins with nuclides allows their distribution to be followed *in vivo*, provided there is a favourable combination of the biological half-life of the molecule, the physical half-life of the nuclide and the sensitivity of the detector (Blankenberg and Strauss, 2002; Acton and Kung, 2003). The expression of receptors can be evaluated using radiolabelled ligands. Already in clinical use are somatostatin analogues such as 111In-D-Phe-DPA-octreotide (OctreoScan^{TM}, Mallinkrodt Medical Inc, St. Louis, MO) (Breeman *et al.*, 2001) or 99mTc-depreotide (Neotect^{TM}, Berlex Laboratories, Richmond, CA) (Virgolini *et al.*, 2001) to detect the somatostatin receptor, which is over-expressed on many tumour cells. Radio labelled peptides targeted against αvβ3-integrin or proteins targeted against the VEGF-receptor can be used to image angiogenesis in infarcted tissues (Meoli *et al.*, 2004) or tumours (Haubner *et al.*, 1999; Blankenberg *et al.*, 2004). Cytokines have been used to image inflammatory processes (Gross *et al.*, 2001). Antibodies with high affinity are available for numerous targets, many of them also used clinically (Mariani *et al.*, 1997; Pinkas *et al.*, 1999). However, using antibodies is also limited by potential allergic reactions and by their high uptake into the reticuloendothelial system, especially the liver. This nonspecific uptake is mainly mediated by the Fc-domain of the antibody. It can be reduced by blocking with unlabelled antibody, by using enzymatically derived (Buchegger *et al.*, 1983) or engineered antibody fragments (minibodies) for a high targeted localization to its specific antigen (Wu *et al.*, 2000) (Figure 8.2.3).

Similarly, a specific population of cells can be labelled radioactively and tracked *in vivo*. Labelling can be achieved with three different strategies: direct loading with an isotope, receptor-mediated binding of the radioactive probe to the cell or enzymatic conversion of the probe and subsequent retention within the cell (Frangioni and Hajjar, 2004). For direct loading, two clinically approved cell-labelling agents are currently available, 111In-oxine and 99mTc-HMPAO. These are chelates that rapidly penetrate the cell membrane and release the isotope within the cells. Labeled cells can be followed *in vivo*, and the accumulation of cells at specific sites can be used for diagnostic purposes. Labelled leukocytes have been used clinically to detect sites of occult infection (Becker and Meller, 2001). For imaging in small rodents, this concept has been carried further: Inflammatory diseases were studied in animal models (van Montfrans *et al.*, 2004; Bennink *et al.*, 2005) and promising new cell-based therapies such as stem cell or immunotherapy were explored (Figure 8.2.4).

Bone-marrow derived mesenchymal stem cells have been labelled with 111In-oxine or 99mTc-HMPAO and followed *in vivo* in rodent models of myocardial infarction (Barbash *et al.*, 2003; Zhou *et al.*, 2005), in one study, together with 99mTc-sestamibi to show co-localization of perfusion defects with the labelled cells in dual isotope SPECT imaging (Zhou *et al.*, 2005). Accumulation of indium-labelled CD34-positive haematopoietic stem cells in infracted hearts in a rat model has also been shown (Brenner *et al.*, 2004). Direct loading of cells though has also its drawbacks. It can only be used for short-term studies due to the half-life of the isotope and the potential leakage of the label out of the cells (Carr *et al.*, 1995).

Figure 8.2.3 Tumour imaging with minibodies: SPECT-CT of a nude mouse bearing a CEA-positive LS174T human colon cancer xenograft 23 h after injection of a I-123 labelled minibody (antibody fragment) targeting CE. The accumulation of the labelled minibody in the tumour is clearly detectable. SPECT (a), CT (b), overlay of SPECT and CT (c). Image courtesy of Dr. Anna Wu, Crump Institute, UCLA

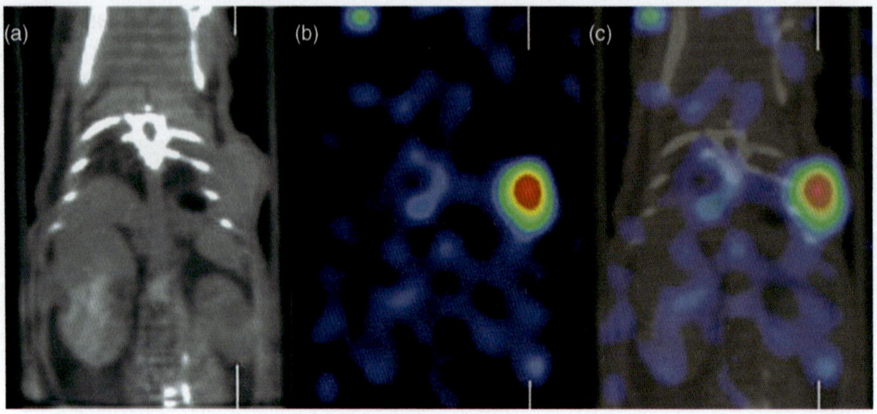

Figure 8.2.4 Cell tracking: Three-dimensional volume rendering of a combined SPECT/CT scan of the pelvic region of a mouse, bearing a subcutaneous melanoma. Imaging was performed 2 days after injection of radio-labelled neuronal stem cells. Two small tumours are visible as hump, within each tumor a clear accumulation of labelled cells can be seen (arrows)

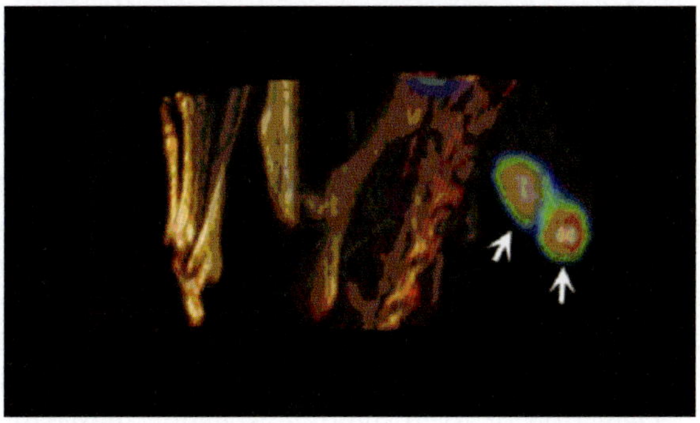

Furthermore, detrimental effects of the label on functionality and viability have been reported (ten Berge *et al.*, 1983; Brenner *et al.*, 2004). Receptor mediated binding has been described for SPECT tracers, using 99mTc-binding engineered receptors (Simonova *et al.*, 2003). This concept can be used not only for labelling cells but also for imaging of gene expression, using the receptor as marker gene (Gambhir *et al.*, 2000). Enzymatic conversion and subsequent retention of a probe is a concept not only widely used in PET imaging to image gene expression (Gambhir *et al.*, 2000) but has also been applied for SPECT imaging to follow antigen-specific lymphocytes (Koehne *et al.*, 2003). This strategy allows cells with stable integration of the marker gene to be followed indefinitely. For each imaging session the radioactive probe is re-injected, therefore the radiolabel is not diluted within the cells by cell divisions (Frangioni and Hajjar, 2004). The need to genetically manipulate the cells will, however, prevent this approach from being transferred to the clinical arena soon.

In addition, apoptosis (programmed cell death) has been an imaging target (Blankenberg *et al.*, 1998) through the use of radio-labelled Annexin V; a protein that binds to phosphatidylserine (PS) which is externalized onto the cell membrane early on in the apoptotic procedure (Allen *et al.*, 1997). This approach can be used to study cellular cytotoxicity, chemo-, radiotherapy or hypoxia and other apoptosis-inducing processes. This makes imaging with Annexin V useful to assess the response of tumours to antineoplastic therapy (Kartachova *et al.*, 2004; Belhocine and Blankenberg, 2005), to detect early changes of hypoxic-ischemic injury or cardiac allograft rejection (Hofstra *et al.*, 2000; Mari *et al.*, 2004). Since Annexin V has been made clinically available, a variety of studies has also shown its usefulness to detect tumour apoptosis after treatment (Van De Wiele *et al.*, 2003; Kartachova *et al.*, 2004; Belhocine and Blankenberg, 2005). Necrosis, however, can interfere with the result, leading to false positive results.

Clinical combined SPECT-CT scanners have recently been introduced into the clinical arena, and procedures for labelling of molecules or cells have been clinically established. The potential to apply SPECT/CT to cross the line from research to clinical molecular imaging is thus high.

References

Acton, P. D., Kung, H. F., 2003. "Small animal imaging with high-resolution single photon emission tomography." *Nucl. Med. Biol.* **30**, 889–895.

Allen, R. T., Hunter, W. J., 3rd, Agrawal, D. K., 1997. "Morphological and biochemical characterization and analysis of apoptosis." *J. Pharmacol. Toxicol. Methods* **37**, 215–228.

Barbash, I. M., Chouraqui, P., Baron, J., Feinberg, M. S., Etzion, S., Tessone, A., Miller, L., Guetta, E., Zipori, D., Kedes, L. H., Kloner, R. A., Leor, J., 2003. "Systemic delivery of bone marrow-derived mesenchymal stem cells to the infarcted myocardium: feasibility, cell migration, and body distribution." *Circulation* **108**, 863–868.

Becker, W. and Meller, J., 2001. "The role of nuclear medicine in infection and inflammation." *Lancet Infect. Dis.* **1**, 326–333.

Belhocine, T. Z., Blankenberg, F. G., 2005. "99mTc-Annexin A5 uptake and imaging to monitor chemosensitivity." *Methods Mol. Med.* **111**, 363–380.

Bennink, R. J., Hamann, J., De Bruin, K., Ten Kate, F. J., Van Deventer, S. J., Te Velde, A. A., 2005. "Dedicated pinhole SPECT of intestinal neutrophil recruitment in a mouse model of dextran sulfate sodium-induced colitis." *J. Nucl. Med.* **46**, 526–531.

Blankenberg, F. G., Katsikis, P. D., Tait, J. F., Davis, R. E., Naumovski, L., Ohtsuki, K., Kopiwoda, S., Abrams, M. J., Darkes, M., Robbins, R. C., Maecker, H. T., Strauss, H. W., 1998. "*In vivo* detection and imaging of phosphatidylserine expression during programmed cell death." *Proc. Natl. Acad. Sci. USA* **95**, 6349–6354.

Blankenberg, F. G., Mandl, S., Cao, Y. A., O'connell-Rodwell, C., Contag, C., Mari, C., Gaynutdinov,

T. I., Vanderheyden, J. L., Backer, M. V., Backer, J. M., 2004. "Tumour imaging using a standardized radio-labelled adapter protein docked to vascular endothelial growth factor." *J. Nucl. Med.* **45**, 1373–1380.

Blankenberg, F. G., Strauss, H. W., 2002. "Nuclear medicine applications in molecular imaging." *J. Magn. Reson. Imaging* **16**, 352–361.

Breeman, W. A., de Jong, M., Kwekkeboom, D. J., Valkema, R., Bakker, W. H., Kooij, P. P., Visser, T. J., Krenning, E. P., 2001. "Somatostatin receptor-mediated imaging and therapy: basic science, current knowledge, limitations and future perspectives." *Eur. J. Nucl. Med.* **28**, 1421–1429.

Brenner, W., Aicher, A., Eckey, T., Massoudi, S., Zuhayra, M., Koehl, U., Heeschen, C., Kampen, W. U., Zeiher, A. M., Dimmeler, S., Henze, E., 2004. "111In-labeled CD34+ hematopoietic progenitor cells in a rat myocardial infarction model." *J. Nucl. Med.* **45**, 512–518.

Buchegger, F., Haskell, C. M., Schreyer, M., Scazziga, B. R., Randin, S., Carrel, S., Mach, J. P., 1983. "Radio-labelled fragments of monoclonal antibodies against carcinoembryonic antigen for localization of human colon carcinoma grafted into nude mice." *J. Exp. Med.* **158**, 413–427.

Carr, H. M., Smyth, J. V., Rooney, O. B., Dodd, P. D., Sharma, H., Walker, M. G., 1995. "Limitations of *in vitro* labelling of endothelial cells with indium-111 oxine." *Cell Transplant.* **4**, 291–296.

Frangioni, J. V., Hajjar, R. J., 2004. "*In vivo* tracking of stem cells for clinical trials in cardiovascular disease." *Circulation* **110**, 3378–3383.

Gambhir, S. S., Herschman, H. R., Cherry, S. R., Barrio, J. R., Satyamurthy, N., Toyokuni, T., Phelps, M. E., Larson, S. M., Balatoni, J., Finn, R., Sadelain, M., Tjuvajev, J., Blasberg, R. 2000. "Imaging transgene expression with radionuclide imaging technologies." *Neoplasia* **2**, 118–138.

Gross, M. D., Shapiro, B., Fig, L. M., Steventon, R., Skinner, R. W., Hay, R. V., 2001. "Imaging of human infection with (131)I-labeled recombinant human interleukin-8." *J. Nucl. Med.* **42**, 1656–1659.

Haubner, R., Wester, H. J., Reuning, U., Senekowitsch-Schmidtke, R., Diefenbach, B., Kessler, H., Stocklin, G., Schwaiger, M. 1999. "Radio-labelled alpha(v)beta3 integrin antagonists: a new class of tracers for tumor targeting." *J. Nucl. Med.* **40**, 1061–1071.

Hofstra, L., Liem, I. H., Dumont, E. A., Boersma, H. H., Van Heerde, W. L., Doevendans, P. A., de Muinck, E., Wellens, H. J., Kemerink, G. J., Reutelingsperger, C. P., Heidendal, G.A. 2000. "Visualisation of cell death *in vivo* in patients with acute myocardial infarction." *Lancet* **356**, 209–212.

Kartachova, M., Haas, R. L., Olmos, R. A., Hoebers, F. J., Van Zandwijk, N., Verheij, M., 2004. "*In vivo* imaging of apoptosis by 99mTc-Annexin V scintigraphy: visual analysis in relation to treatment response." *Radiother. Oncol.* **72**, 333–339.

Koehne, G., Doubrovin, M., Doubrovina, E., Zanzonico, P., Gallardo, H. F., Ivanova, A., Balatoni, J., Teruya-Feldstein, J., Heller, G., May, C., Ponomarev, V., Ruan, S., Finn, R., Blasberg, R. G., Bornmann, W., Riviere, I., Sadelain, M., O'reilly, R. J., Larson, S. M., Tjuvajev, J.G. 2003. "Serial *in vivo* imaging of the targeted migration of human HSV-TK-transduced antigen-specific lymphocytes." *Nat. Biotechnol.* **21**, 405–413.

Mari, C., Karabiyikoglu, M., Goris, M. L., Tait, J. F., Yenari, M. A., Blankenberg, F. G., 2004. "Detection of focal hypoxic-ischemic injury and neuronal stress in a rodent model of unilateral MCA occlusion/reperfusion using radio-labelled annexin V." *Eur. J. Nucl. Med. Mol. Imaging* **31**, 733–739.

Mariani, G., Lasku, A., Pau, A., Villa, G., Motta, C., Calcagno, G., Taddei, G. Z., Castellani, P., Syrigos, K., Dorcaratto, A., Epenetos, A. A., Zardi, L., Viale, G. A., 1997. "A pilot pharmacokinetic and immunoscintigraphic study with the technetium-99m-labeled monoclonal antibody BC-1 directed against oncofetal fibronectin in patients with brain tumours." *Cancer* **80**, 2484–2489.

Meoli, D. F., Sadeghi, M. M., Krassilnikova, S., Bourke, B. N., Giordano, F. J., Dione, D. P., Su, H., Edwards, D. S., Liu, S., Harris, T. D., Madri, J. A., Zaret, B. L., Sinusas, A.J. 2004. "Noninvasive imaging of myocardial angiogenesis following experimental myocardial infarction." *J. Clin. Invest.* **113**, 1684–1691.

Mountz, J. D., Hsu, H. C., Wu, Q., Liu, H. G., Zhang, H. G., Mountz, J. M. 2002. "Molecular imaging: new applications for biochemistry." *J. Cell Biochem. Suppl.* **39**, 162–171.

Pinkas, L., Robins, P. D., Forstrom, L. A., Mahoney, D. W., Mullan, B. P. 1999. "Clinical experience with radio-labelled monoclonal antibodies in the detection of colorectal and ovarian carcinoma recurrence and review of the literature." *Nucl. Med. Commun.* **20**, 689–696.

Simonova, M., Shtanko, O., Sergeyev, N., Weissleder, R., Bogdanov, A., Jr., 2003. "Engineering of technetium-99m-binding artificial receptors for imaging gene expression." *J. Gene Med.* **5**, 1056–1066.

Ten Berge, R. J., Natarajan, A. T., Hardeman, M. R., Van Royen, E. A., Schellekens, P. T., 1983.

"Labeling with indium-111 has detrimental effects on human lymphocytes: concise communication." *J. Nucl. Med.* **24**, 615–620.

Van de Wiele, C., Lahorte, C., Vermeersch, H., Loose, D., Mervillie, K., Steinmetz, N. D., Vanderheyden, J. L., Cuvelier, C. A., Slegers, G., Dierck, R. A., 2003. "Quantitative tumour apoptosis imaging using technetium-99m-HYNIC annexin v single photon emission computed tomography." *J. Clin. Oncol.* **21**, 3483–3487.

Van Montfrans, C., Bennink, R. J., de Bruin, K., de Jonge, W., Verberne, H. J., Ten Kate, F. J., Van Deventer, S. J., Te Velde, A. A., 2004. "*In vivo* evaluation of 111In-labeled T-lymphocyte homing in experimental colitis." *J. Nucl. Med.* **45**, 1759–1765.

Virgolini, I., Traub, T., Novotny, C., Leimer, M., Fuger, B., Li, S. R., Patri, P., Pangerl, T., Angelberger, P., Raderer, M., Andreae, F., Kurtaran, A., Dudczak, R., 2001. "New trends in peptide receptor radioligands." *Q. J. Nucl. Med.* **45**, 153–159.

Wu, A. M., Yazaki, P. J., Tsai, S., Nguyen, K., Anderson, A. L., Mccarthy, D. W., Welch, M. J., Shively, J. E., Williams, L. E., Raubitschek, A. A., Wong, J. Y., Toyokuni, T., Phelps, M. E., Gambhir, S. S., 2000. "High-resolution microPET imaging of carcinoembryonic antigen-positive xenografts by using a copper-64-labeled engineered antibody fragment." *Proc. Natl. Acad. Sci. USA* **97**, 8495–8500.

Zhou, R., Thomas, D. H., Qiao, H., Bal, H. S., Choi, S. R., Alavi, A., Ferrari, V. A., Kung, H. F., Acton, P. D., 2005. "*In vivo* detection of stem cells grafted in infarcted rat myocardium." *J. Nucl. Med.* **46**, 816–822.

8.3 FMT registration with MRI

Vasilis Ntziachristos

Fluorescence Molecular Tomography (FMT) attains similar co-registration needs as seen on SPECT (or PET). The use of safe, non-ionizing energies and the ability to guide light using optical fibers offers flexible implementations and compatibility with many imaging modalities, including MRI, X-ray CT or ultrasound. Concurrent multi-modality imaging using optical tomography and MRI has been achieved for breast cancer detection (Ntziachristos *et al.*, 2000) and holds great promise, also through the utilization of the MR images as a-priori information for the optical reconstruction (Brooksby *et al.*, 2003).

The advantages of multi-modality imaging can be similarly applied to small animal imaging. As described in Chapter 5, free-space approaches are preferable in the small animal imaging case and facilitate the implementation of hybrid systems by eliminating issues that relate to the use of fibers or matching fluids. Imaging can be achieved under identical imaging conditions, for example by mounting the optical and X-ray CT components on the same rotating gantry. Alternatively co-registration can be achieved by using appropriate mouse constrainers to transfer mice from one modality to the other without changing their placement or shape. This is particularly useful when performing simultaneous FMT and MRI.

In one example such co-registration was achieved by using appropriate cylindrical constrainers to immobilize the animals and transfer then through an FMT and an MRI bore.

In this case, imaging of cathepsin B activity in 9L gliosarcomas stereotactically implanted into unilateral brain hemispheres of nude mice was performed with FMT and co-registered with MR images of the same mouse, in order to co-localize protease expression levels onto anatomical signals. Fig.8.3.1a-b depict the gadolinium-enhanced tumor (enhancement is shown in a green color map superimposed onto a T1 weighted image) on axial (a) and sagital (b) slices. Fig.8.3.1c, e and f depict the three consecutive FMT slices obtained from top to bottom of the volume of interest. The location and volume covered by the three slices, is indicated on Fig.8.3.1b by white horizontal lines. Fig. 8.3.1c shows marked cathepsin up-regulation, imaged non-invasively by FMT by resolving a fluorescence activatable probe sensitive to cathepsins (described in Chapter 7), which was intravenously administered 24 hours prior to imaging. The fluorescence, was congruent with the location of the tumor identified on the MR images. Fig.8.3.1d shows a superposition of the MR axial slice passing through the tumor (Fig.8.3.1a) onto the corresponding FMT slice (Fig. 6c) after appropriately translating the MR image to the actual dimensions of the FMT image. For coregistration purposes, special water-containing fiducials and body marks that facilitated matching of the MR and FMT orientation and animal positioning were employed. This *in-vivo* imaging data correlated well with photographic ex-vivo imaging of the excised brain. Fig.8.3.1g depicts the axial brain section through the 9L tumor examined with white light using a CCD camera mounted onto a dissecting microscope, Fig 8.3.1h shows the same brain section imaged at the excitation wavelength (675 nm) and Fig 8.3.1i is the fluorescence image obtained at the emission wavelength using an appropriately tuned

Figure 8.3.1 Multimodality FMT and MRI imaging of cathepsin expression levels and of the underlying anatomy of 9L gliosarcomas stereotactically implanted into unilateral brain hemispheres of nude mice. a), b) Axial and sagital MR slices of an animal implanted with a tumor, which is shown in green after gadolinium enhancement. c), e), f), Consecutive FMT slices obtained from top to bottom from the volume of interest shown on (b) by thin white horizontal lines. d) Superposition of the MR axial slice passing through the tumor (a) onto the corresponding FMT slice (c) after appropriately translating the MR image to the actual dimensions of the FMT image. Ex-vivo imaging collected a (g) white light image, (h) an image obtained at the excitation wavelength and (i) a fluorescence image, the latter showing the fluorescent activity resolved non-invasively and in-vivo in panel (c). Picture from Nat. Med. 8(7):757 (2002)

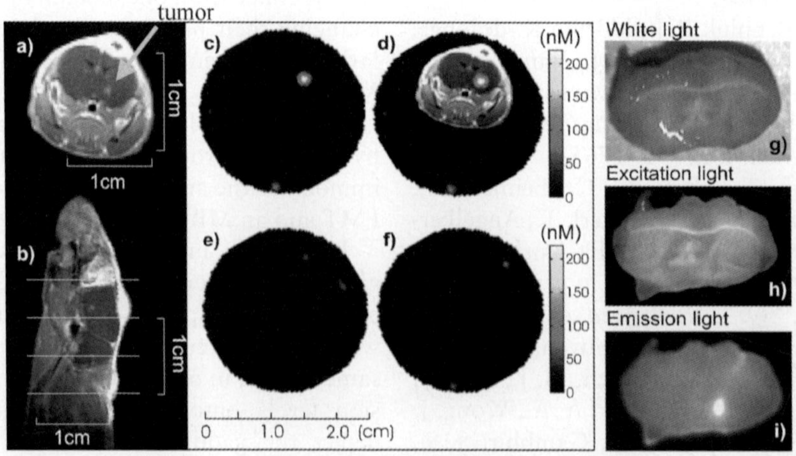

3-cavity band-pass filter that demonstrates marked fluorescence activity, congruent with the tumor position identified by gadolinium-enhanced MRI and FMT. On average, the concentration of activated fluorochrome per tumor, as calculated by FMT was about 110±30 nM.

While mouse constraints can be used to ensure identical placement in different imaging modalities, the co-registration of anatomical markers based on high resolution surface images (photographs) can be further used to guide registration of FMT images to each-other or to other imaging modalities. It has been shown that by detecting invariant landmarks on photographs of mice, it is possible to correct for orientation and shape differences; for example to align optical images of the same mouse obtained at different time points, under different placement positions (Marias *et al.*, 2005). This method can be extended to other imaging modalities for example for registration of photographs or three-dimensional surfaces of mice placed in an FMT, SPECT or PET setup with the corresponding high-resolution boundaries seen on MRI or X-ray CT. Finally the development of hybrid systems that acquire FMT and MRI or X-ray CT data under identical placement conditions or concurrently are currently under development and should lead to significant improvements in the overall imaging performance of the hybrid approach.

References

Brooksby, B. A., Dehghani, H., Pogue, B. W., Paulsen, K. D., 2003. "Near-infrared (NIR) tomography breast image reconstruction with a priori structural information from MRI: Algorithm development for reconstructing heterogeneities." *IEEE J. Select. Topics Quant. Elect.* **9**(2), 199–209.

Marias, K., Ripoll, J., Meyer, H., Ntziachristos, V., Orphanoudakis, S., 2005. "Image analysis for assessing molecular activity changes in time-dependent geometries." *IEEE Trans. Med. Imaging* **24**(7), 894–900.

Ntziachristos, V., Tung, C., Bremer, C., Weissleder, R., 2002. "Fluorescence-mediated tomography resolves protease activity *in vivo*." *Nat. Med.* **8**(7), 757–760.

Ntziachristos, V., Yodh, A. G., Schnall, M., Chance, B., 2000. "Concurrent MRI and diffuse optical tomography of breast after indocyanine green enhancement." *Proc. Natl. Acad. Sci. USA* **97**(6), 2767–2772.

Weissleder, R., Tung, C. H., Mahmood, U., Bogdanov, A., 1999. "*In vivo* imaging of tumours with protease-activated near-infrared fluorescent probes." *Nat. Biotech.* **17**(4), 375–378.

9 Brain Imaging

Coordinated by Anne Leroy-Willig

9.0 Introduction

Anne Leroy-Willig

Understanding of consciousness and of the mental processes by which we perceive, act and know could be considered as the last frontier of biological sciences. This explains the huge interest brought to functional imaging of the human brain. Brain is also the co-ordinator of many physiological processes needed for our lives as for that of other vertebrates. Brain has remained for centuries the unknown, bone-enclosed 'black box'. Brain diseases were identified by autopsy; brain activity was explored by needle registration of electrical signals from neurons and brain metabolism by insertion of microprobes sensing local chemistry. These two last techniques have been widely applied to physiological research in mouse, rat, cat and monkey brains. Invasive transcranial explorations still are the gold standard used to validate less direct explorations.

However, electrical explorations, soon followed by *in vivo* imaging, opened the way to a more global and less invasive exploration of human and animal brains. Electroencephalography (EEG), X-rays computed tomography (CT) and, more and more, magnetic resonance imaging (MRI) now have a large impact upon the diagnosis and the management of brain disorders and upon the follow-up of experimental brain disease models used to validate therapeuties. MRI techniques that are sensitive to blood oxygenation, regional blood volume and blood flow, and diffusion of water are the basis of functional MRI that explores brain activity, neuronal connexions, brain damage after stroke and brain plasticity.

In parallel, positon emission tomography (PET) and single photon emission tomography (SPECT) were also applied to functional explorations of brain regional blood volume and blood flow, brain oxygenation and glucose consumption and to the quantification of receptors or drug uptake, in human and primate brain and more recently in rodent brain. Nuclear techniques yield much higher sensitivity and specificity than other techniques when applied to the detection and quantification of radiolabelled chemicals.

Optical techniques are hindered by bone attenuation of light. However, applications of these techniques are possible in some cases such as the exploration of babies' brains through natural bone windows, or of animal brain, by using the optical properties of haemoglobin. The purpose of this chapter is to illustrate the different ways by which *in vivo* information is obtained 'through the bone box' and enlightens brain structure and function, and brain pathologies as well as their evolution under therapies, in animal models.

9.1 Bringing amyloid into focus with MRI microscopy

Greet Vanhoutte and Annemie Van der Linden

9.1.0 Introduction

Alzheimer's disease (AD), the prominent cause of senile dementia in the elderly, is a progressive neurodegenerative disorder characterized by two hallmark lesions: extracellular senile plaques (made of beta amyloid peptide aggregates) and intracellular neurofibrillary tangles (from tau protein). Senile plaques (SP) became a crucial target for many therapeutic approaches (Vassar *et al.*, 1999; Terwel, Dewachter and Van Leuven, 2002). Therefore, several transgenic mouse lines have been generated over-expressing the mutant amyloid precursor protein (APP) to mimic the AD related amyloid pathology. Characterization of the temporal and regional features of at least one typical hallmark such as SP development would allow scientists to unravel underlying mechanisms. To be fully successful, improved methods allowing non-invasive detection of the SP in the brain are indispensable.

9.1.1 What opportunities offer imaging?

Researchers using positron emission tomography were the first to reveal *in vivo* SP in patients (Klunk *et al.*, 2002). This method, however, is not safe enough for repetitive routine use because it requires radioactive labelling. In many cases, MRI has passed the capabilities of PET and X-ray computed tomography offering a combination of physiological sensitivity and high spatial resolution. The major motivation for pursuing imaging SP with MRI is that MRI theoretically can resolve individual plaques non-invasively. Typically, in human AD subjects and in most of the transgenic mouse models, plaques diameter range from 50 to 200 μm. This is beyond the spatial resolution of PET and SPECT, and therefore, MR-microscopy with a resolution of up to 40 μm harbours the realistic potential of imaging amyloid at least in animal models.

Of course, reaching high spatial resolution is not enough. An adequate contrast-to-noise is required to separate different structures of interest, that is, amyloid burden from the normal background brain tissue. Different intrinsic MR contrast mechanisms exist, but their repertoire has been extended with targeted contrast agents improving the sensitivity of detection and delineation of pathological structures. This field of imaging using plaque-specific probes shows a promising future, but the ability to visualize plaques without exogenous contrast is also important, as the ultimate goal is to visualize plaques in humans.

Until now, the proof of principle for visualizing amyloid *in vivo* and without contrast agents has been provided in mice models. Very obviously, one has to come trotting along with an *in vivo* approach encompassing the standard way to confirm the presence of the plaques by autopsy. Although, we may not forget that *ex vivo* or *in vitro* imaging efforts also cover an important, if not initial, step to optimize MR protocols and strategies prior to *in vivo* studies.

9.1.2 Methods

The data presented here are both *in vivo* and *in vitro* MRI observations correlated with histology obtained from APP[V717I] mice over-expressing the London mutant of human APP (Dewachter *et al.*, 2000). Individual SP are identified on MR-images without the aid of an exogenous contrast agent, exclusively based on the intrinsic T_2*-contrast. This not only confirms other observations as fully detailed in Vanhoutte *et al.* (2005) but also goes beyond these providing evidence that the source of T_2*-contrast is iron. In AD brain, iron appeared to become particularly concentrated in SP and was suspected to catalyse the formation of free radicals (Markesbery and Carney, 1999). Iron-loaded SP, detectable by *in vivo* MRI could, therefore, become a useful indication for AD.

On a 7 T horizontal-bore MR-system (MRRS, Guildford, UK), three-dimensional T_2*-weighted gradient-echo images covering a brain volume of $(20 \times 20 \times 15)\ \text{mm}^3$ were acquired within 68 min (TR 500 ms, acquisition matrix $256 \times 128 \times 64$). The corresponding digital voxel volume is $(78 \times 156 \times 234)\ \mu\text{m}^3$. After data interpolation and reconstruction to an image matrix of $(256 \times 256 \times 256)$, pixel dimensions in plane were $(78 \times 78)\ \mu\text{m}^2$ as displayed in the figures.

In vitro MR images of the entire formaldehyde fixed brain allowed a longer acquisition time and resulted in a higher resolution, that is, $(58 \times 58)\ \mu\text{m}^2$ and better signal-to-noise ratio.

9.1.3 Results

In vivo hypointense anomalies appeared exclusively in the ventral thalamus region. The *in vitro* images displayed better-defined boundaries and smaller anomalies. Conformity of the distribution pattern of hypointense spots among *in vivo* and *in vitro* images argue against the existence of any secondary sources leading to dark spots such as microhaemorrhage or heavily myelinated fibres. The authenticity of the plaques is further established through correlation with histology. In spite of (amyloid specific) thioflavin-S staining demonstrating additional smaller plaques in the cortex, not generating anomalies detectable by MRI, staining for iron (2% potassium ferrocyanide) brings clarification pointing out that the majority of the plaques in the thalamus were iron-positive (Figure 9.1.1). This assigns the iron content as being the underlying mechanism of MR-detection of SP.

Figure 9.1.1 How an iron-stained and thioflavin-S-stained brain slice correlates with *in vitro* MR. (a) *in vitro* MR reveals anatomic details [mammillothalamic tract (mt), hippocampal fissura (hif) and lateral ventricle (LV)] used as reference points. The black circles illustrate perfect correlation between hypointense spherical inclusions on the MR-images and the stained plaques on the corresponding histological section (b). More tiny stained plaques (dashed arrows) correspond to a spotted inhomogeneous area in the MRI. Also, one large plaque, stained for thioflavin-S is missing on the MRI (red cross). (c) This figure displays a representative iron-stained histological slice showing details of the cortex (d) and the ventral thalamic nuclei (e, f). Iron stains appear as a dark core in bright stained plaques and are exclusively present in the thalamic SP

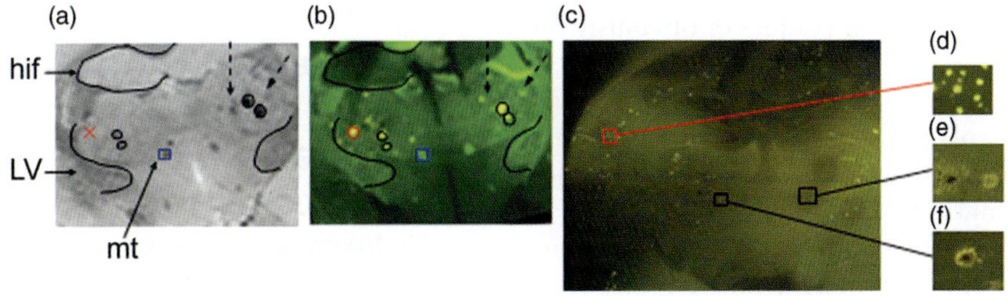

Because coronal vibratome sections have a smaller slice thickness (40 μm) than the MR zero-filled image slices (58 μm), accurate matching of those two image modalities has to be assisted by detailed anatomic reference points. An overview of more accurate correlation by combining two adjacent histological sections with one overlapping MR-image is given in Figure 9.1.2. Numerous tiny thioflavin-S positive plaques, overwhelmingly present in the cortex (a), remain invisible on MRI due to the absence of iron. Only thioflavin-S positive stains found in the thalamus (b) correlate with MR findings (c). At first sight, the hypointense inclusions mimick the histological stainings with poor specificity. If we overlay the thioflavin-S

Figure 9.1.2 Correlation between thioflavin-S-stained brain slices and *in vivo* MR. In (a) (Bregma −0.8 mm), numerous thioflavin-S-stained plaques are found in the cortex while the correlating MR-image displays no abnormalities. In (b) two adjacent thioflavin-S-stained brain sections (Bregma −2 mm) illustrate the occurrence of SP in both the cortex and ventral thalamic regions. The corresponding MR-image in (c) displays numerous hypointense brain inclusions exclusively in the thalamus region. Overlying the individual stained plaques of two ROIs (boxes) as delineated in (c), results in a two-coloured compilation (blue/red). The overall result is shown in the zoomed box from the MR-image: The hypointense anomalies (white delineation) and the associated thioflavin-S-stained plaques (green delineation) show that several smaller sized plaques cluster into one larger hypointense area from the MR image. Bold green dots do not have a MR correlate and probably do not contain Fe

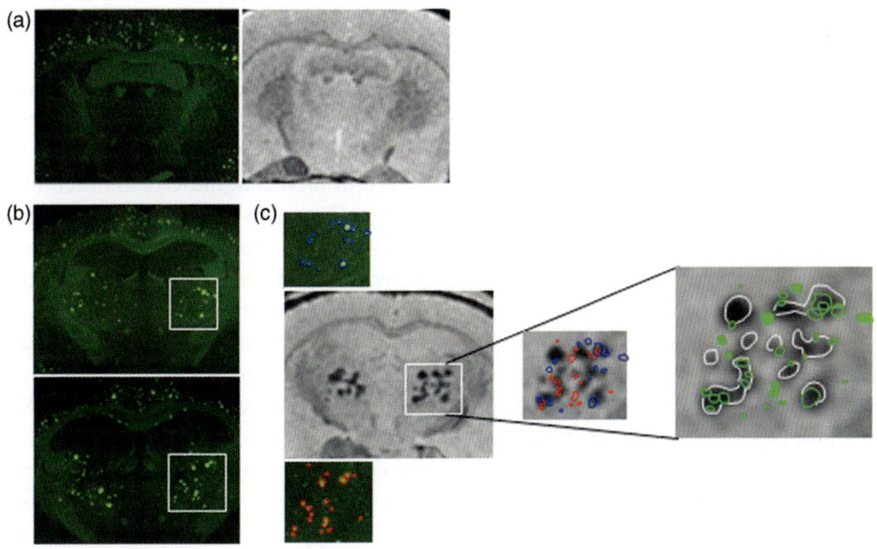

positive plaques of the right hemisphere from two consecutive histological slices (red and blue circles) on the MR image (green circles), the result shows the following: (1) Hypointense structures from MRI (white) harbour numerous distinctive small-sized histological stained plaques, and (2) some plaques do not have a correlate on MRI (green dots). The first is guaranteed by the blooming effect produced by iron deposition in the SP, and the second observation could be due to the small size of the plaques or obviously the absence of iron.

Of course, an ideal *in vivo* MR measure of plaques would reflect true plaques size rather than the amount of iron. Nevertheless, this technique is very eligible for AD research because on one hand, the lower limit of plaque size that is consistently resolvable by *in vivo* MRI remains to be determined, and on the other hand, the propensity of amyloid to accumulate iron sheds a new light on the association of iron with the pathological stigmata of AD.

References

Dewachter, I., van Dorpe, J., Smeijers, L., Gilis, M., Kuiperi, C., Laenen, I., Caluwaerts, N., Moechars, D., Checler, D. R., Vanderstichele, H., Van Leuven, F., 2000. "Aging increased amyloid peptide and caused amyloid plaques in brain of old APP/V717I transgenic mice by a different mechanism than mutant presenilin." *J. Neurosci.* **20**(17), 6452–6458.

Klunk, W. E., Bacskai, B. J., Mathis, C. A., Kajdasz, S. T., McLellan, M. E., Frosch, M. P., Debnath, M. L., Holt, D. P., Wang, Y., Hyman, B. T., 2002. "Imaging Abeta plaques in living transgenic mice with multiphoton microscopy and methoxy-X04, a systemically administered Congo red derivative." *J. Neuropathol. Exp. Neurol.* **61**(9), 797–805.

Markesbery, W. R., Carney, J. M., 1999. "Oxidative alterations in Alzheimer's disease." *Brain Pathol.* **9**(1), 133–146.

Terwel, D., Dewachter, I., Van Leuven, F., 2002. "Axonal transport, tau protein, and neurodegeneration in Alzheimer's disease." *Neuromolecular Med.* **2**(2), 151–165.

Vanhoutte, G., Dewachter, I., Borghgraef, P., Van Leuven, F., Van der Linden, A., 2005. "Non-invasive *in vivo* MRI detection of iron associated with neuritic plaques in APP[V717I] transgenic mice, a model for Alzheimer's Disease." *MRM* **53**(3), 607–613.

Vassar, R., Bennett, B. D., Babu-Khan, S., Kahn, S., Mendiaz, E. A., Denis, P., Teplow, D. B., Ross, S., Amarante, P., Loeloff, R., Luo, Y., Fisher, S., Fuller, L., Edenson, S., Lile, J., Jarosinski, M. A., Biere, A. L., Curran, E., Burgess, T., Louis, J. C., Collins, F., Treanor, J., Rogers, G., Citron, M., 1999. "Beta-secretase cleavage of Alzheimer's amyloid precursor protein by the transmembrane aspartic protease BACE." *Science* **286**(5440), 735–741.

9.2 Cerebral blood volume and BOLD contrast MRI unravels brain responses to ambient temperature fluctuations in fish

Annemie Van der Linden

Ambient temperature variations directly influence the biochemistry and the physiology of ectothermic animals. Especially, the brain could suffer from such fluctuations, as the central nervous system is responsible for initiating behavioural, physiological and acclimational responses fundamental for surviving these temperature variations (Crawshaw *et al.*, 1985). Many aquatic ectotherms are subjected to temperature changes up to as much as 15°C over just a few minutes and still manage to maintain coordinated sensori-motor function (Montgomery and MacDonald, 1990). Some species, such as swordfish (*Xiphias gladius*) use cranial endothermy where a constant brain temperature permits them to dive to colder waters, spanning a temperature difference of up to 19°C (Carey, 1982). These observations suggest that the brain cannot withstand low temperatures without further adaptations. This led to the hypothesis that compensation mechanisms at the level of the brain vasculature may exist, avoiding the brain from being challenged with severe temperature changes but still allowing proper functioning of the brain.

An *in vivo* magnetic resonance imaging study was designed to unravel the nature of such mechanisms by investigating rapid changes in brain physiology, that is, within minutes, after exposure of an ectothermic animal (the common carp, *Cyprinus carpio*) to a 10°C temperature drop. In the first place, whole-brain responses in terms of changes in cerebral blood volume (CBV) were investigated. Subsequently, we zoomed in on the pituitary gland – the primary target for orchestrating the primary stress response – to find out whether it was still operational during the temperature challenge (more details can be found in the paper of Van den Burg *et al.*, 2005).

9.2.1 How to study changes in cerebral blood volume?

9.2.1.1 Methods

Carp (40–69 g body weight), acclimated to 25°C, were anaesthetized with 0.011 wt.% MS-222 (ethyl meta aminobenzoate metanesulfonic acid salt, Sigma) and inserted into a custom-built stereotactic apparatus equipped with appropriate RF antennas for MRI positioned on the carp's head. The entire set-up was mounted in the bore of the MRI magnet (7 T), while anaesthesia and sufficient aeration of the gills was ensured during the entire imaging experiment by a flow-through system providing the fish, through a mouth piece, 500 ml water/min (containing 0.011% MS222) as detailed in Van der Linden *et al.* (2004). During MRI acquisition, the temperature of the water supplied to the carp was switched rapidly (within 5 min) from 25 to 15°C while T_2*-weighted images were obtained continuously (30 s/image). The blood oxygenation level dependent (BOLD) contrast of these images revealed changes in the deoxy/oxyhaemoglobin ratio as a consequence of temperature induced changes in oxygen affinity of haemoglobin or of brain activity ($n = 4$). In a parallel experiment ($n = 4$) Cerebral blood volume-weighted images were acquired using a similar imaging protocol after intravenous injection of an iron oxide containing MRI contrast agent (Clariscan, Amersham Biosciences Europe GmbH, Roosendaal, The Netherlands) which shortens the T_2* value inside and around blood vessels. In all cases the entire brain was imaged covering the pituitary at the bottom of the brain as well.

9.2.1.2 Results

The temperature drop strongly reduced the CBV throughout the entire brain, 1.5 min after the onset of the drop onwards, as could be deduced from the enhanced signal intensity due to a decrease in the concentration of the applied MRI contrast agent (Figure 9.2.1a, b). Blood dissipated centrifugally into two areas of the vascular system (as confirmed with MR angiography which is not shown) peripheral to the brain (Figure 9.2.1a, c). The drainage of blood from the brain may serve to temporarily isolate the brain from the circulation and protect it against cooling down too rapidly from incoming blood during a temperature drop. This could be the mechanism employed by aquatic ectotherms to dampen temperature fluctuations in the brain.

Figure 9.2.1 Blood re-distribution in the carp brain following a 10°C temperature drop. Horizontal high resolution (a1) and CBV-weighted MR images (a2) of carp brain showing a dramatic CBV-reduction throughout the brain (a2red). Blood is collected caudally to the vagal lobes (vl) and the cerebellum (a2blue) in a sinus-like structure. Abbreviations: facial lobe (fl), granular eminence (eg), medial cerebellar valvula (vcm), lateral cerebellar valvula (vcl), optic tectum (to), telencephalon (t) and olfactory nerve (no). The colour bar reflects the signal intensity changes relative to the signal intensity before the temperature drop. Changes of signal intensity in CBV-weighted MRI as measured in an entire brain slice (b) and in the vasculature peripheral to the brain (c). Data are mean ± SD ($n = 4$). (*Images obtained from Van den Burg et al., 2005 with permission of the American Physiological Society.*)

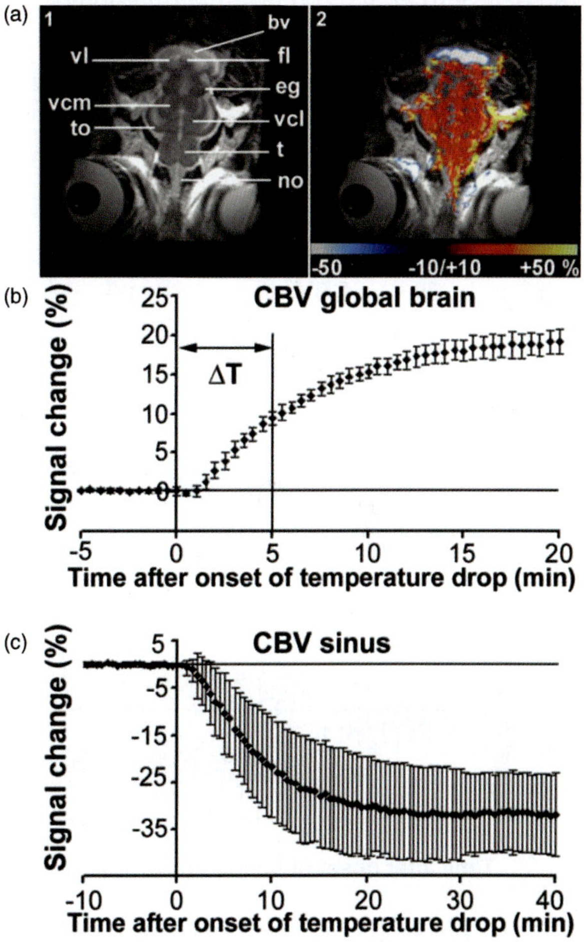

Figure 9.2.2 CBV- and BOLD-changes in the pituitary gland of carp following a 10°C temperature drop. A functional map obtained from a pixel-to-pixel analysis of BOLD data reveals opposite responses in the pars intermedia (pi; pink) and pars distalis (pd; green/blue) (a). The colour bar reflects the slope value calculated from the change of signal intensity in time, relative to the signal intensity before the temperature drop (as displayed in panel b and c). Panel b shows how blood volume decreases (increased signal intensity) in the pi as in the rest of the brain (Figure 9.2.1) while it increases in the pd. (c) BOLD-signal intensity increases in the pi as in the rest of the brain (data not shown) (reflecting the increased affinity of haemoglobin for oxygen at lower temperatures) but decreases in the pd as a signal of increased oxygen consumption (activity) in the pd. Data are mean ± SD ($n = 3$). (*Images obtained from Van den Burg et al., 2005 with permission of the American Physiological Society.*)

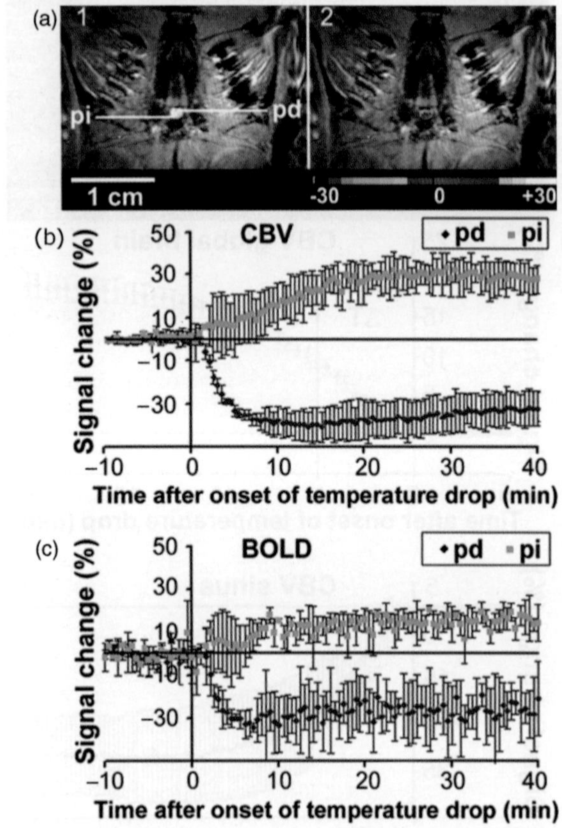

9.2.2 How to study activity of the pituitary eliciting a primary stress response?

9.2.2.1 *Methods*

To investigate whether the temperature drop evoked a stress response in the brain, we analysed blood plasma samples for cortisol from carp submitted to the MRI set-up (including the same anaesthesia, intubation and even MRI scan noise) in a parallel experiment ($n = 6$).

Using *in vivo* MRI we zoomed in on the pituitary. It is known that the corticotrope cells from the pars distalis of the pituitary produce and release adrenocorticotropic hormone (ACTH) which stimulates the release of cortisol by the interrenal cells in the head kidney (equivalent of the adrenal gland in mammals). The small pituitary gland (diameter: 1.8 ± 0.15 mm; $n = 8$) could easily be identified in anatomical and BOLD-images (see previous methods section). Even its two lobes, the pars intermedia and the pars distalis, could be distinguished separately (Figure 9.2.2(a)) because they displayed a distinct temporal BOLD response.

9.2.2.2 *Results*

It appeared that the pars intermedia (pi) and the pars distalis (pd) of the pituitary responded differently to the temperature drop, both on CBV- and BOLD-weighted MRI (Figure 9.2.2). The pd received more blood (Figure 9.2.2(b)) and consumed more oxygen (Figure 9.2.2(c)) following the rapid temperature drop and can, therefore, be considered as highly active from 2 min onwards.

Plasma cortisol levels increased within 5 min after the onset of the temperature drop and remained elevated (260 ng/ml after 45 min), similar to cortisol responses seen in free-swimming carp (Tanck *et al.*, 2000).

The cortisol response to the temperature drop by far exceeded the mild elevation of cortisol levels (50 ng/ml) after prolonged residence in the MRI-like set-up. These data confirm the activation of the pars distalis of the pituitary discerned with *in vivo* BOLD and CBV weighted MRI.

References

Carey, F. G., 1982. "A brain heater in the swordfish." *Science* **216**, 1327–1329.

Crawshaw, L., Grahn, D., Wollmuth, L., Simpson, L., 1985. "Central nervous regulation of body temperature in vertebrates: comparative aspects." *Pharmacol. Ther.* 30, 19–30.

Montgomery, J. C. and MacDonald, J. A., 1990. "Effects of temperature on nervous system: implications for behavioral performance." *Am. J. Physiol.* **259**, R191–R196.

Tanck, M. W. T., Booms, G. H. R., Eding, E. H., Wendelaar Bonga, S. E. and Komen, J., 2000.

"Cold shocks: a stressor for common carp." *J. Fish Biol.* 57, 881–894.

Van den Burg, E. H., Peeters, R. R., Verhoye, M., Meek, J., Flik, G., Van der Linden, A., 2005. "Brain responses to ambient temperature fluctuations in fish: reduction of blood volume and initiation of a whole-body stress response." *J. Neurophysiol.* 93(5), 2849–2855.

Van der Linden, A., Verhoye, M., Pörtner, H. O., Bock, C., 2004. "The strengths of *in vivo* magnetic resonance imaging (MR) to study environmental adaptational physiology in fish [Review]." *MAGMA* 17(3–6), 236–248.

9.3 Assessment of functional and neuroanatomical re-organization after experimental stroke using MRI

Jet P. van der Zijden and Rick M. Dijkhuizen

9.3.1 Functional recovery and brain re-organization after stroke

Ischemic stroke is an important cause of morbidity and disability in the modern society. Stroke patients regularly develop sensori-motor dysfunction. However, despite the acute significant loss of function, most patients demonstrate partial functional recovery over time. Restoration of lost function is often associated with plastic changes in the brain. Plasticity is referred to as re-organization of neuronal circuitry in the brain, such as unmasking and strengthening of existing pathways and/or formation of new neuronal connections. Although many studies have described plastic changes after stroke [see Lee and van Donkelaar (1995) and Nudo and Friel (1999) for a review], the exact temporal characteristics of plasticity and the correlation with functional recovery are still unclear. Elucidation of the mechanisms underlying functional recovery after stroke is important because the prolonged time course of spontaneous recovery may hold great opportunities for therapeutic interventions.

Studies in experimental stroke models have provided important insights into brain plasticity (Nudo and Friel, 1999). These animal studies suggest that alterations in physiology and anatomy in the intact peri-infarct zone as well as in more remote brain regions may play a critical role in recovery after stroke. A limitation of most animal studies carried out on plasticity after stroke, however, is that they used invasive experimental techniques (e.g. immunohistochemistry) or investigated a restricted part of the brain (e.g. electrophysiology measurements). Magnetic resonance imaging provides a non-invasive tool to monitor the *in vivo* dynamics of re-organizational processes in the entire brain. Because MRI is a multiparametric imaging modality, different pathophysiologic aspects occurring in the brain after stroke may be studied simultaneously.

9.3.2 MRI of brain activation

Loss and recovery of function after stroke are caused by alterations in the functioning and organization of neuronal circuitry. Functional MRI (fMRI) is based on the detection of the haemodynamic response to neuronal activity and can be executed with blood oxygenation level-dependent (BOLD), cerebral blood flow (CBF)-weighted, or cerebral blood volume (CBV)-weighted MRI (Mandeville and Rosen, 2002). Importantly, it allows whole-brain mapping of stimulus-induced cerebral activation. Consequently, it provides a unique tool to give insights in functional re-organization after stroke. For example, fMRI enables detection of loss of functional activation, shifts in activation patterns and evaluation of the functional efficacy of therapies. The potentials of fMRI in stroke recovery research have been demonstrated in several patient studies [see Cramer and Bastings (2000) and Rijntjes and Weiller (2002) for reviews]. Analysis and interpretation of fMRI studies in stroke patients, however, are complicated by variations in lesion size and location, differences in the post-stroke time of fMRI acquisition and incomplete datasets. fMRI studies in animal stroke models (see Figure 9.3.1) can be performed under more controllable and reproducible experimental conditions. In addition, animal fMRI data may be directly correlated with measures from invasive methods, for example, electrophysiology and histology. Multi-parametric MRI in combination with behaviour studies and histology enables intra-individual assessment of the inter-relationship between brain activation patterns, tissue and perfusion injury and functional status. By applying this experimental approach in a rat stroke model, we have demonstrated that activation in the contralesional hemisphere during stimulation of the stroke-affected forelimb is associated with large lesions, and that restoration of function depends on preservation or reinstatement of ipsilesional activity (Dijkhuizen *et al.*, 2003).

Figure 9.3.1 Top row: T_2-weighted MR images of a rat brain slice overlaid by statistical activation maps. Contrast-enhanced CBV-weighted functional MRI was performed in combination with an electrical forelimb stimulation paradigm as described by Dijkhuizen *et al.* (2001) (block design; 60 s stimulation on, 300 s stimulation off). The colour-coded activation maps represent P values calculated by voxel-wise Student's t testing of the difference in CBV response between stimulated and non-stimulated conditions. Images are displayed in radiological convention, that is, the right side of image is the left side of brain. Bottom row: CBV changes in the right sensori-motor cortex in response to a left forelimb stimulation paradigm (the 60 s stimulation period is indicated by the black bar) (mean $\pm$ SD, $n = 6$). Left forelimb stimulation induced significant activation responses in the right sensori-motor cortex in control rats. At 3 days after right-sided stroke, activation responses in the right, ipsilesional sensori-motor cortex were largely absent; however, responses were found in the left, contralesional hemisphere. After 14 days, partial restoration of activation was detected in the right, ipsilesional sensori-motor cortex

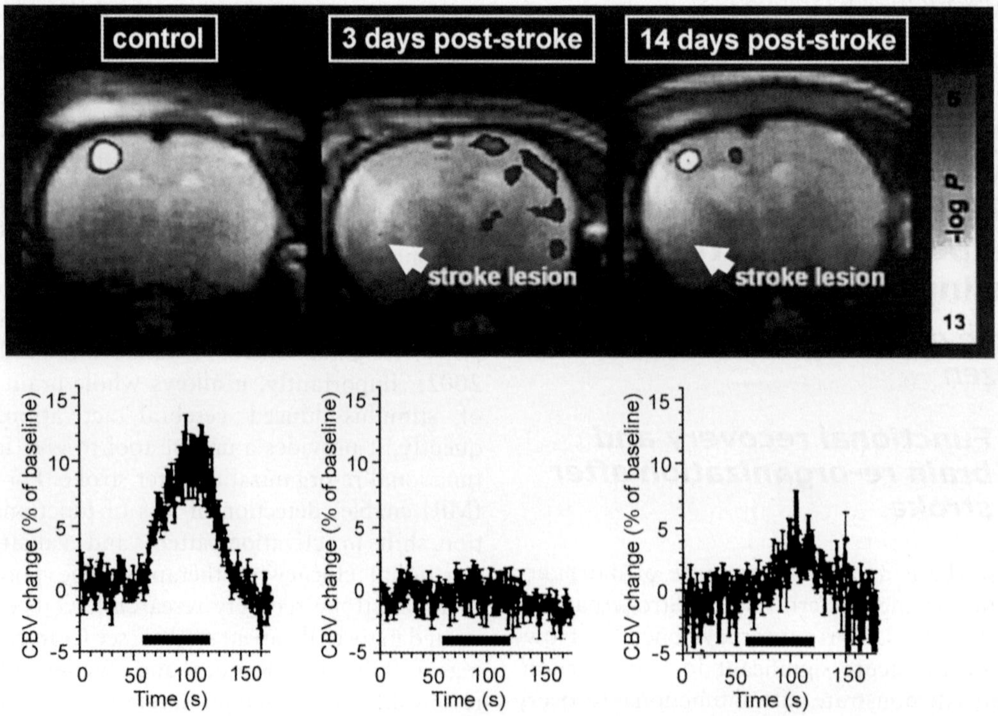

9.3.3 MRI of neuroanatomical re-organization

Altered brain activity patterns after stroke are often associated with structural re-organization in the ischemic brain. For example, loss of activation responses in intact brain regions may be the result of loss of input from anatomically connected damaged areas. Likewise, formation of new connections may explain restoration or relocation of activation responses. Changes in neuronal connectivity may be assessed *in vivo* with diffusion tensor imaging (DTI) (Basser *et al.*, 1994) or manganese-enhanced MRI (MEMRI) (Pautler, Silva and Koretsky, 1998).

DTI enables the assessment of three-dimensional displacement of tissue water. Calculation of scalar indices of water diffusion anisotropy and the orientation of the principal eigenvectors can be used to model fibre architecture in the brain (Basser *et al.*, 1994). DTI-based reconstruction of white matter fibres in live small rodents has been demonstrated (Xue *et al.*, 1999). However, *in vivo* mapping of axonal architecture in grey matter (sub)structures is more complicated and requires advanced MRI data acquisition and post-processing procedures.

Alternatively, MEMRI is an MRI-based *in vivo* tract-tracing method based on the detection of neuronal distribution of paramagnetic Mn^{2+}, a Ca^{2+} analogue that can enter active neurons through Ca^{2+} channels. Injection of Mn^{2+} in the brain has been shown to result in neuronal uptake and transsynaptic transport along connective pathways (Pautler, Silva

Figure 9.3.2 R_1 ($1/T_1$) maps of four adjacent brain slices of a control (top row) and a stroke rat (bottom row) at 4 days after injection of 0.2 μl 1 M $MnCl_2$ in the right sensori-motor cortex. Mn^{2+} is a paramagnetic MR contrast agent that is taken up by neurons and that can be transported transynaptically. Mn^{2+}-induced R_1 increase was evident in the ipsilateral cortico-striato-nigral pathway in control rat brain. When $MnCl_2$ was injected at 14 days after right-sided stroke, subsequent Mn^{2+} transport to remote areas of this sensori-motor network was clearly diminished

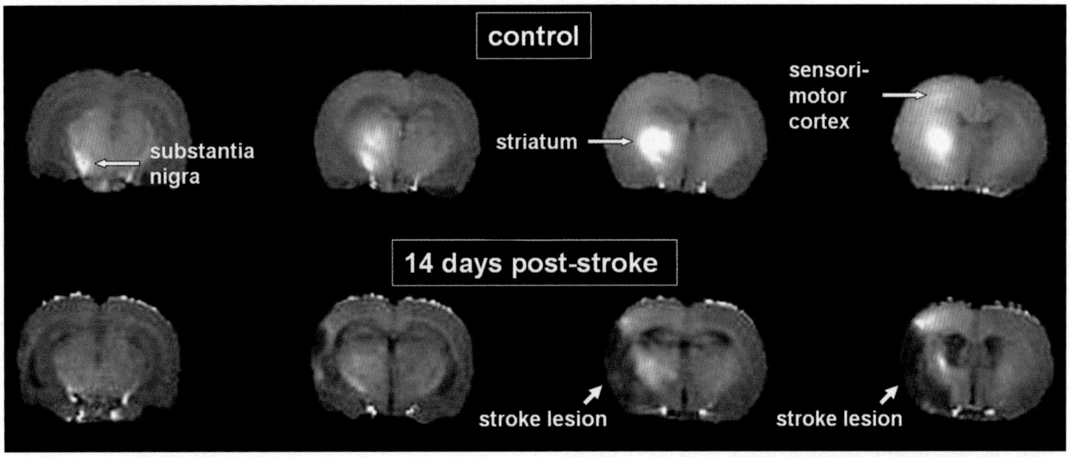

and Koretsky, 1998). The potential of MEMRI to detect changes in neuronal connectivity after stroke has been demonstrated in a rat model (Allegrini and Wiessner, 2003). In a recent study, we injected Mn^{2+} in the ipsilesional sensori-motor cortex of rats after unilateral stroke and found diminished Mn^{2+} uptake in remote sensori-motor areas (Figure 9.3.2). In some animals, increased Mn^{2+} accumulation was found in contralesional regions. These preliminary results point towards decreased connectivity within the ipsilesional sensori-motor network and increased interhemispheric connectivity.

In conclusion, MRI offers a versatile tool to study the spatiotemporal pattern of anatomical and functional changes after stroke. The combination of fMRI of activation patterns with DTI and/or MEMRI of neuronal connectivity can provide unique information on the correlation between functional and anatomical re-organization in the brain. Combined with behavioural assessment of functional status, such multi-parametric experimental MRI studies may significantly contribute to a better understanding of the underlying mechanisms of functional recovery after stroke.

References

Allegrini, P. R., Wiessner, C., 2003. "Three-dimensional MRI of cerebral projections in rat brain in vivo after intracortical injection of $MnCl_2$." *NMR Biomed.* **16**, 252–256.

Basser, P. J., Mattiello, J., LeBihan, D., 1994. "MR diffusion tensor spectroscopy and imaging." *Biohys. J.* **66**, 259–267.

Cramer, S. C., Bastings, E. P., 2000. "Mapping clinically relevant plasticity after stroke." *Neuropharmacology* **39**, 842–851.

Dijkhuizen, R. M., Ren, J., Mandeville, J. B., Wu, O., Ozdag, F. M., Moskowitz, M. A., Rosen, B. R., Finklestein, S. P., 2001. "Functional magnetic resonance imaging of reorganization in rat brain after stroke." *Proc. Natl. Acad. Sci. USA.* **98**, 12766–12771.

Dijkhuizen, R. M., Singhal, A. B., Mandeville, J. B., Wu, O., Halpern, E. F., Finklestein, S. P., Rosen, B. R., Lo, E. H., 2003. "Correlation between brain reorganization, ischemic damage and neurologic status after transient focal cerebral ischemia in rats: a functional MRI study." *J. Neurosci.* **23**, 510–517.

Lee, R. G., Van Donkelaar, P., 1995. "Mechanisms underlying functional recovery following stroke." *Can. J. Neurol. Sci.* **22**, 257–263.

Mandeville, J. B., Rosen, B. R., 2002. "Functional MRI." In: Toga, A. W. and Mazziotta, J. C. (Eds.), *Brain Mapping: The Methods*, second ed. Academic Press, New York, pp. 315–349.

Nudo, R. J., Friel, K. M., 1999. "Cortical plasticity after stroke: implications for rehabilitation." *Rev. Neurol. (Paris)* **155**, 713–717.

Pautler, R. G., Silva, A. C., Koretsky, A. P., 1998. "In vivo neuronal tract tracing using manganese-enhanced magnetic resonance imaging." *Magn. Reson. Med.* **40**, 740–748.

Rijntjes, M., Weiller, C., 2002. "Recovery of motor and language abilities after stroke: the contribution of functional imaging." *Prog. Neurobiol.* **66**, 109–122.

Xue, R., Van Zijl, P. C., Crain, B. J., Solaiyappan, M., Mori, S., 1999. "In vivo three-dimensional reconstruction of rat brain axonal projections by diffusion tensor imaging." *Magn. Reson. Med.* **42**, 1123–1127.

9.4 Brain activation and blood flow studies with speckle imaging

Andrew K. Dunn

9.4.1 *In vivo measurement of blood flow*

Measurement of blood flow *in vivo* is critical for a wide range of biological applications, and although a number of methods exist for measuring blood flow, accurate determination of the spatial and temporal dynamics of blood flow is still challenging. In the brain in particular, blood flow studies are essential for developing a better understanding of both normal and pathological conditions.

Studies of cerebral blood flow (CBF) on the millimetre length scale have been limited by the lack of high-resolution *in vivo* CBF imaging methods. Laser Doppler flowmetry is well established for measuring relative blood flow changes at a single point in tissues including the brain. In laser Doppler measurements, the Doppler shift of coherent light scattered from moving particles is used to determine flow. Although laser Doppler measurements yield high temporal resolution measurements of relative blood flow, typically no spatial information is obtained. Typically when spatial information is desired, it comes at the expense of temporal resolution.

9.4.2 *Laser speckle contrast imaging of blood flow*

Laser speckle contrast imaging (LSCI) is one technique that has been demonstrated to be extremely useful in high-resolution imaging of cerebral blood flow. The first demonstration of LSCI of blood flow was published more than 20 years ago (Fercher and Briers, 1981). However, the method was somewhat limited in its application until recently due in part to the lack of readily available imaging hardware and processing software. LSCI has been used extensively to image blood flow in the retina (Aizu *et al.*, 1990), skin (Briers, Richards and He, 1999) and more recently the brain (Dunn *et al.*, 2001).

Conceptually the LSCI technique is straightforward although the underlying details of the physics are quite complicated. When the scattered laser light from a surface is imaged onto a camera, a speckle pattern is formed (Figure 9.4.1(a)). If the object is stationary then a static speckle pattern is produced. However, if any of

Figure 9.4.1 Illustration of laser speckle contrast imaging of cerebral blood flow. (a) Raw laser speckle image of rat cortex resulting from laser illumination. (b) Corresponding speckle contrast image calculated from the raw speckle image using a 7 × 7 sliding window. Darker areas correspond to increased levels of blood flow and, therefore, have lower speckle contrast values. Scale bar =1 mm. (*Adapted and reprinted from Vander Linden et al. Neuroscience, 112, 467–474. Copyright 2002, with permission from Elsevier.*)

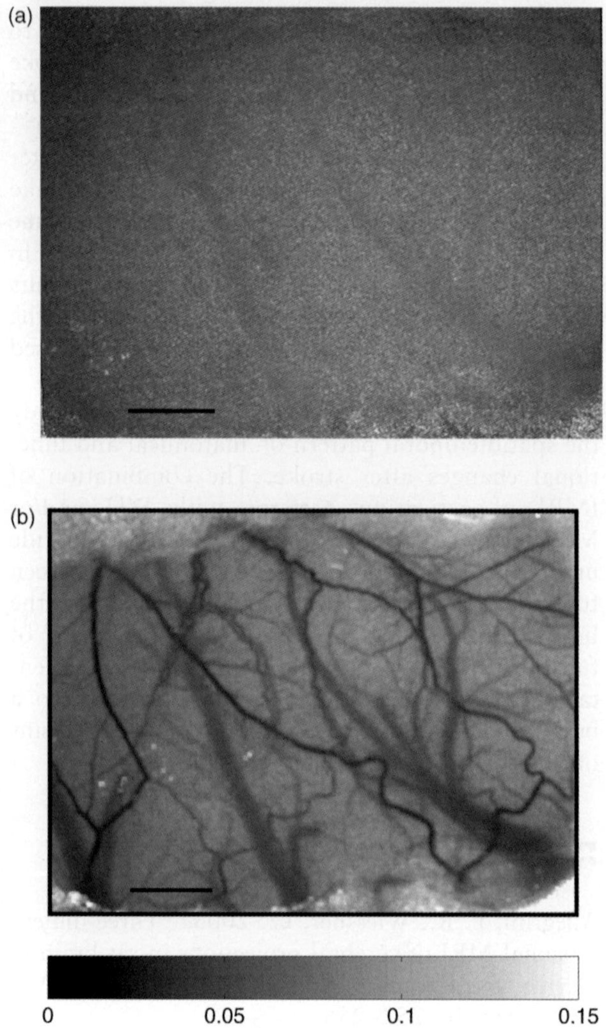

Figure 9.4.2 CBF changes in response to functional brain activation. (a) Coloured areas correspond to areas where the CBF has increased to greater than 10% of baseline flow following electrical forepaw stimulation. (b) Timecourse of the CBF changes in a 1×1 mm region centred on the response area. Each curve shows the relative CBF at different stimulation intensities. (*Adapted and reprinted from Van der Linden et al. Neuroscience, 112, 467–474. Copyright 2002, with permission from Elsevier.*)

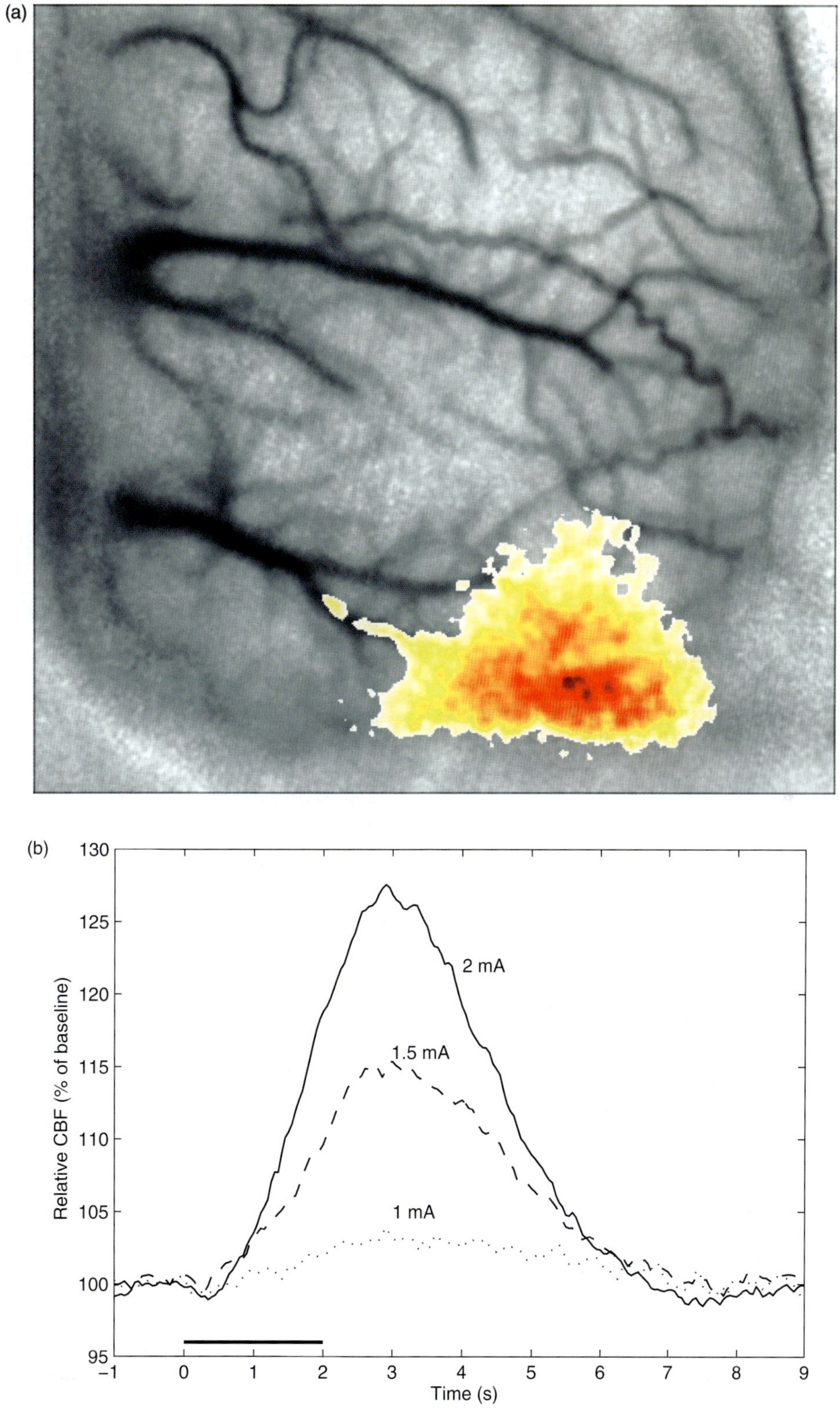

the scattering particles are in motion, a dynamic speckle pattern results, and the motion of the particles is effectively encoded in the speckle intensity fluctuations. This process is closely related to that of laser Doppler flowmetry even though the measurement formats differ (Briers, 1996).

Laser speckle contrast imaging uses the spatial statistics of a time-integrated speckle pattern to image blood flow (Briers, 2001). When a time varying speckle pattern is imaged onto a CCD camera, the temporal intensity fluctuations of the speckle pattern are integrated over the duration of the CCD's exposure time. Therefore, in areas of higher blood flow where the speckle intensity fluctuations are more rapid, the time-integrated intensity will be more uniform. In areas of lower blood flow, the time-integrated intensity will have more variation. To quantify the spatial statistics of the speckle pattern, the speckle contrast, K, is defined as the ratio of the standard deviation to the mean intensity in a localized region of the image, $K = \sigma/\langle I \rangle$ (Briers, 2001). In practice a 5×5 or 7×7 region of pixels is used to compute the speckle contrast. The speckle contrast varies between 0 and 1 and a lower value of K indicates a larger blood flow. To produce a blood flow image, the speckle contrast is computed at each pixel of the raw speckle image using a sliding $N \times N$ window.

9.4.2.1 *LSCI in the brain*

To illustrate the LSCI concept, Figure 9.4.1(a) shows the speckle pattern that results from illumination of the cortex in a rat with 785 nm laser light. The skull has been thinned to approximately 100 μm in thickness, and the reflected laser light is imaged onto an analogue 8-bit CCD camera. Variations in the spatial contrast of the speckle pattern in Figure 9.4.1(a) can be seen clearly, particularly over large blood vessels where the motion of the scattering particles is greatest. When the speckle contrast is computed across the entire image using a sliding 7×7 window, the detailed pattern of the vasculature is immediately apparent (Figure 9.4.1(b)). The grey levels in the image of Figure 9.4.1(b) represent speckle contrast values and, therefore, are related to the level of blood flow. In order to relate the speckle contrast to a measure of cerebral blood flow, the speckle decorrelation time, τ_c, is calculated using the relationship $K = [\tau_c/2T(1-\exp(-2T/\tau_c))]^{1/2}$, where T is the exposure time of the camera (Briers, 2001). As the exact nature of the relationship between τ_c and absolute blood flow is extremely complicated (Bonner and Nossal, 1981), τ_c is assumed to be inversely proportional to flow velocity, and therefore, the changes in blood flow are estimated through the changes in τ_c over time.

9.4.2.2 *Functional activation*

The CBF response to functional brain activation is the basis for several functional neuroimaging techniques such as BOLD fMRI and intrinsic optical imaging. Neither of these techniques however, provides a direct measure of CBF. Recently, the spatial and temporal dynamics of the CBF response to functional activation in animal models has been investigated using LSCI (Dunn *et al.*, 2003; Durduran *et al.*, 2004). To illustrate the use of LSCI for imaging functional brain activation, Figure 9.4.2 shows the results of the CBF changes in response to electric forepaw stimulation in rats. The image shows the area of cortex with a blood flow increase greater than 10% of baseline. The CBF

Figure 9.4.3 Spatial pattern of CBF changes in a rat stroke model. (a) Relative CBF (% of baseline) 15 min after distal middle cerebral artery occlusion in a rat. (b) Spatial profile of CBF extending from the ischemic core to the non-ischemic cortex along the line indicated in (a). (*Adapted and reprinted from Van der Linden et al. (2004) NMR Biomed. 17, 602–612.*)

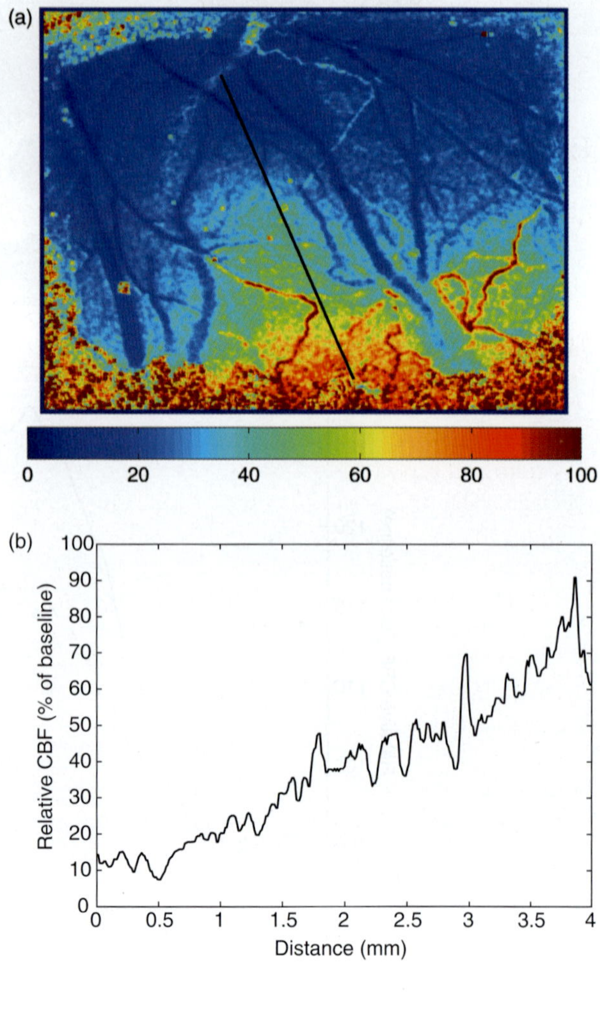

response extends a few millimetres around the centre of activation. The temporal profiles of the CBF response to 2 s of stimulation (Figure 9.4.2(b)) illustrate a transient increase in CBF whose amplitude varies with stimulation intensity.

9.4.2.3 *Cerebral ischemia*

LSCI is also extremely useful in studies of the spatial and temporal changes in CBF during ischemia (Dunn *et al.*, 2001; Ayata *et al.*, 2004) and other pathological conditions (Bolay *et al.*, 2002; Atochin *et al.*, 2004; Kharlamov *et al.*, 2004). Figure 9.4.3 shows the changes in the spatial pattern of CBF following distal middle cerebral artery occlusion in a rat. A clear spatial gradient in the CBF is observed. The CBF in the ischemic core is less than 20% of baseline while in the penumbra, between the core and normal tissue, the CBF is in the 20–50% range. Figure 9.4.3(b) shows the CBF changes along the profile indicated in Figure 9.4.3a which extends from the ischemic core to the non-ischemic tissue. These results illustrate the potential utility of LSCI in studies testing agents designed to restore blood flow to ischemic tissues.

References

Aizu, Y., Ogino, K., Sugita, T., Yamamoto, T., Asakura, T., 1990. "Noninvasive evaluation of the retinal blood circulation using laser speckle phenomenon." *J. Clin. Laser Med. Surg.* 8, 35–45.

Atochin, D. N., Murciano, J. C., Gursoy-Ozdemir, Y., Krasik, T., Noda, F., Ayata, C., Dunn, A. K., Moskowitz, M. A., Huang, P. L., Muzykantov, V. R., 2004. "Mouse model of microembolic stroke and reperfusion." *Stroke* **35**, 2177–2182.

Ayata, C., Dunn, A. K., Gursoy, O. Y., Huang, Z., Boas, D. A., Moskowitz, M. A., 2004. "Laser speckle flowmetry for the study of cerebrovascular physiology in normal and ischemic mouse cortex." *J. Cereb. Blood Flow Metab.* **24**, 744–755.

Bolay, H., Reuter, U., Dunn, A. K., Huang, Z., Boas, D. A., Moskowitz, M. A., 2002. "Intrinsic brain activity triggers trigeminal meningeal afferents in a migraine model." *Nature Med.* **8**, 136–142.

Bonner, R., Nossal, R., 1981. "Model for laser doppler measurements of blood flow in tissue." *Appl. Optics* **20**, 2097–2107.

Briers, J. D., 1996. "Laser doppler and time-varying speckle: a reconciliation." *J. Opt. Soc. Am. A* **13**, 345–350.

Briers, J. D., 2001. "Laser Doppler, speckle and related techniques for blood perfusion mapping and imaging." *Physiol. Meas.* **22**, R35–R66.

Briers, J. D., Richards, G., He, X., 1999. "Capillary blood flow monitoring using laser speckle contrast analysis." *J. Biomed. Opt.* **4**, 164–175.

Dunn, A. K., Bolay, H., Moskowitz, M. A., Boas, D. A., 2001. "Dynamic imaging of cerebral blood flow using laser speckle." *J. Cereb. Blood Flow Metab.* **21**, 195–201.

Dunn, A. K., Devor, A., Bolay, H. Andermann, M. L., Moskowitz, M. A., Dale, A. M., Boas, D. A., 2003. "Simultaneous imaging of total cerebral hemoglobin concentration, oxygenation, and blood flow during functional activation." *Optics Lett.* **28**, 28–30.

Durduran, T., Burnett, M. G., Yu, G., Zhou, C., Furuya, D., Yodh, A. G., Detre, J. A., Greenberg, J. H., 2004. "Spatiotemporal quantification of cerebral blood flow during functional activation in rat somatosensory cortex using laser-speckle flowmetry." *J. Cereb. Blood Flow Metab.* **24**, 518–525.

Fercher, A., Briers, J., 1981. "Flow visualization by means of single-exposure speckle photography." *Opt. Commun.* **37**, 326–329.

Kharlamov, A., Brown, B. R., Easley, K. A., Jones, S.C., 2004. "Heterogeneous response of cerebral blood flow to hypotension demonstrated by laser speckle imaging flowmetry in rats." *Neurosci. Lett.* **368**, 151–156.

9.5 Manganese-enhanced MRI of the songbird brain: a dynamic window on rewiring brain circuits encoding a versatile behaviour

Vincent Van Meir and Annemie Van der Linden

9.5.0 *Introduction*

Determining relations between brain structure and function is a principal focus in neurobiology. Individual differences in behavioural performances have been related to differences in structure and function of certain brain regions; however, often the role of an individual's brain plasticity is not considered due to practical or methodological limitations.

Animal models can be useful tools to investigate neuronal mechanisms involved in these processes albeit their brains are often visualized with post-mortem techniques which do not allow repeated measures. Here we introduce repeated manganese-enhanced magnetic resonance imaging (MEMRI) of the songbird brain as a tool to study mechanisms of brain plasticity in a neuronal network that is dedicated to song production and learning.

9.5.1 Relations between bird song and the song control system

Bird song is one of the few systems in which complex behaviour has been successfully linked to anatomically defined brain structures. Intact function of a set of interconnected brain regions is a pre-requisite for a songbird to produce and/or learn complex vocalizations, called song (Nottebohm, Stokes and Leonard, 1976). Interestingly, in most species males have a higher song rate and produce more complex songs than females. This sexual dimorphism is paralleled by a sexual dimorphism in size of the telencephalic song control nuclei. Moreover, the degree of sexual dimorphism in song behaviour and size of the song control nuclei between species has been shown to be related (Brenowitz, 1997).

Apart from these often dramatic sexual differences, temperate-zone birds change their song over the course of the year. It is well described that males show a peak in song output during the breeding season. Again, this plasticity in behaviour is paralleled by plasticity in size of the telencephalic song control nuclei as well as changes in plasma testosterone levels (Tramontin and Brenowitz, 2000).

The song control system thus harbours the potential to study stable genetic and endocrine differences that lead to the establishment of a brain–behaviour relationship but also allows investigating its dynamic changes that are triggered by environmental cues and endocrine status.

9.5.2 Repeated visualization of a manganese-enhanced circuit

9.5.2.1 *Methods*

The recent application of the paramagnetic ion manganese (Mn^{2+}), a biological calcium analogue, as *in vivo* tract tracer in MRI, introduced new opportunities to monitor changes in structure and activity of brain circuits within individual subjects. Mn^{2+} injected in the brain enters neurons via voltage- and ligand-gated Ca^{2+} transporters (Narita, Kawasaki and Kita, 1990), is transported along axons and ultimately released at the synapse to undergo post-synaptic uptake (Pautler, Silva and Koretsky, 1998; Pautler, 2004).

For application of MEMRI in songbirds T_1-weighted images (TR = 300 ms; TE = 10–18 ms) of starling brains were obtained at 7 T with in plane spatial resolution in the order of 100 × 100 μm and a slice thickness of either 100 μm (3D imaging) or 800 μm (multi-slice 2D imaging). Full description of the method implemented in birds is given in Van der Linden *et al.* (2002) and Van Meir *et al.* (2004).

9.5.2.2 *Results*

We investigated how Mn^{2+} accumulated over time in the song control system, represented in Figure 9.5.1(a), after injection in the key song control nucleus HVC (previously called High Vocal Center). From HVC two pathways emerge. A caudal pathway is connected with the pre-motor region Robust Nucleus of the Archistriatum (RA). This nucleus is ultimately connected to nuclei in the brainstem that innervate the syrinx and the breathing muscles. A rostral pathway projects to a region in the basal ganglia, Area X. This region is a part from an indirect loop from HVC to RA and is required for song learning (Nottebohm, Stokes and Leonard, 1976). The injection of $MnCl_2$ (50–200 nl of 10–100 mM) in HVC labels within a few hours RA and area X (Figures 9.5.1(a) and 9.5.2).

Figure 9.5.1 (a) Schematic overview of the adult songbird brain showing the song control nuclei (SCN) and their connections descending from HVC. The black arrows represent the rostral pathway and the grey arrows the caudal motor pathway. The two pathways originate from distinct cell populations within HVC. (b) Sagittal MEMRI image of the song control system 8 h after injecting an $MnCl_2$ injection in HVC. The increased signal is linked to local Mn^{2+} concentration. Magnification bar 5 mm. (*Adapted and reprinted from Neuroscience, 112, 467–474, Van der Linden et al. Copyright 2002, with permission from Elsevier.*)

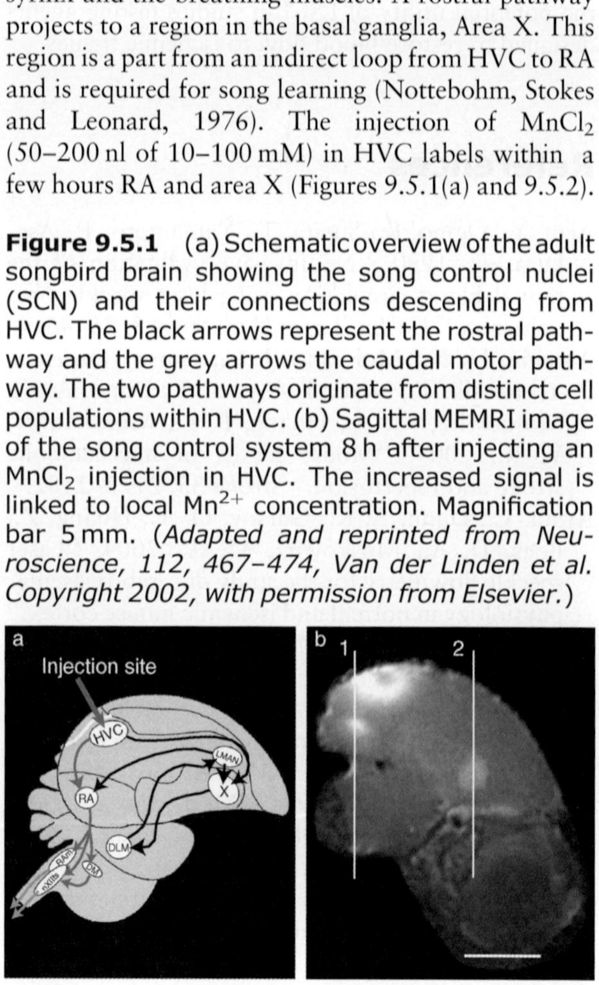

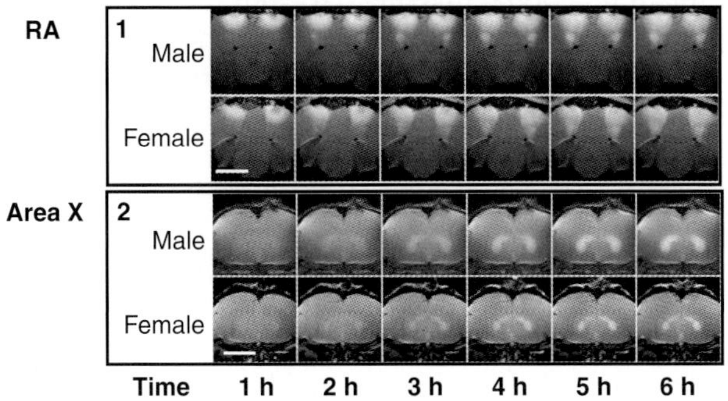

Figure 9.5.2 Manganese accumulation in coronal sections 1 and 2 through two HVC targets, RA and area X, respectively (as indicated in Figure 9.5.1(b)), displayed as a function of time (up to 6 h) after an $MnCl_2$ injection in HVC. The manganese-enhanced areas clearly illustrate the sex differences in shape and size of RA and area X between male and female starlings. Magnification bar 5 mm. (*Adapted and reprinted from Van der Linden et al. (2002) Neuroscience, 112, 467–474. Copyright # 2002, with permission from Elsevier.*)

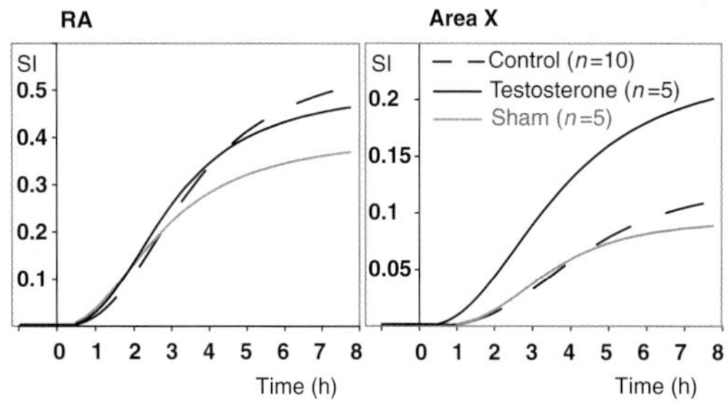

Figure 9.5.3 Repeated MEMRI demonstrating changes in the dynamics of manganese accumulation in RA and area X of female starlings following testosterone treatment. These graphs are reconstructions of the mean sigmoid curves from data obtained in control sham-implanted (grey curves) and testosterone-treated birds (black curves). The dashed curves represent pooled data for all birds before the beginning of the endocrine treatments. (For interpretation of the data see text and also Van Meir *et al.*, 2004). (*Adapted and reprinted from Van der Linden et al. (2004) NMR Biomed. 17, 602–612.*)

The two projections from HVC are non-topographic; hence, a manganese injection in HVC provides accurate enhancement of the full extend of both RA and area X which is important for the determination of the volumes.

Moreover, the two projections start from a different neuronal population within HVC and the accumulation in each target could presumably be explained by functional properties of these two populations, such as amount of neurons and neuronal activity. The dynamic accumulation in RA and area X can be modelled by a sigmoid curve fit (Figure 9.5.3) that we used to compare activity levels of a certain connection under different circumstances or to describe differences in neuronal activity between networks (reviewed in Van der Linden *et al.*, 2004). The accumulation of Mn^{2+} in RA and area X has been shown to be sexual dimorph (Figure 9.5.2, Van der Linden *et al.*, 2002) and susceptible to testosterone levels in the blood (Figure 9.5.3, Van Meir *et al.*, 2004).

The sexual dimorphism and testosterone-induced changes in the dynamics of Mn^{2+}-accumulation were also specifically different between RA and area X (Figure 9.5.3) indicating that differences in plasticity and function between the two pathways are reflected by the MEMRI. Known differences are, for example, the amount of projection neurons, expression of steroid receptor types or changes in neuronal recruitment upon testosterone treatment.

9.5.3 Future directions

Several questions cannot be answered by analyses of single measurements in groups of individuals measured at different time points and would benefit from repeated MEMRI. It is, for example, not clear whether differences in song behaviour between individuals of the same sex and the same species are correlated to individual *differences in morphological or functional plasticity* of the song control nuclei or whether they are related to *stable individual differences.*

The song control system is also a useful model to study *interactions between plasticity of connected brain regions.* For example, the RA-projecting neurons in HVC are replaced in adulthood, whereas the neurons projecting to area X are not. The plasticity in neuronal activity of the rostral pathway might be involved in the level of morphological plasticity of the caudal pathway. If this is the case, repeated measures are necessary to compare the plastic events of the two pathways.

References

Brenowitz, E. A., 1997. "Comparative approaches to the avian song system." *J. Neurobiol.* **33**, 517–531.

Nottebohm, F., Stokes T. M. and Leonard C. M., 1976. "Central control of song in the canary, Serinus canarius." *J. Comp. Neurol.* **165**, 457–486.

Narita K., Kawasaki F. and Kita H., 1990. "Mn and Mg influxes through Ca channels of motor nerve terminals are prevented by verapamil in frogs." *Brain Res.* **510**, 289–295.

Pautler, R. G., Silva, A. C. and Koretsky, A. P., 1998. "In vivo neuronal tract tracing using manganese-enhanced magnetic resonance imaging." *Magn. Reson. Med.* **40**, 740–748.

Pautler R. G., 2004. "In vivo, trans-synaptic tract-tracing utilizing manganese-enhanced magnetic resonance imaging (MEMRI)." *NMR Biomed.* **17**, 595–601.

Tramontin, A. D. and Brenowitz, E. A., 2000. "Seasonal plasticity in the adult brain." *Trends Neurosci.* **23**, 251–258.

Van der Linden, A., Verhoye, M., Van Meir, V., Tindemans, I., Eens, M., Absil, P. and Balthazart, J., 2002. "In vivo manganese-enhanced magnetic resonance imaging reveals connections and functional properties of the songbird vocal control system." *Neuroscience.* **112**, 467–474.

Van der Linden, A., Van Meir, V., Tindemans, I., Verhoye, M. and Balthazart, J., 2004. "Applications of manganese-enhanced magnetic resonance imaging (MEMRI) to image brain plasticity in song birds." *NMR Biomed.* **17**, 602–612.

Van Meir, V., Verhoye, M., Absil, P., Eens, M., Balthazart, J. and Van der Linden, A., 2004. "Differential effects of testosterone on neuronal populations and their connections in a sensorimotor brain nucleus controlling song production in songbirds: a manganese-enhanced magnetic resonance imaging study." *Neuroimage* **21**, 914–923.

9.6 Functional MRI in awake behaving monkeys

Wim Vanduffel, Koen Nelissen, Denis Fize and Guy A. Orban

9.6.0 Introduction

Recent developments of non-invasive functional imaging techniques such as positron emmision tomography, magnetoencephalography (MEG), electroencephalography and especially functional magnetic resonance imaging (fMRI), have led to an increasingly better understanding of human brain functioning. Despite the sky-rocketing number of human fMRI publications, however, surprisingly little is known about the relation between fMRI signal changes (i.e. hemodynamic responses) and the underlying neuronal activity (Logothetis *et al.*, 2001). Whatever the exact nature of that link, fMRI reflects the activity of large populations of neurons. Hence, for a meaningful interpretation of most human functional imaging data, one still has to rely heavily on electrophysiological and anatomical data from non-human primates. Comparing such human functional imaging data with monkey electrophysiological data, however, is not trivial as one not only 'crosses' different species but also techniques. In order to bridge the gap between human functional imaging data and monkey electrophysiological, lesion and anatomical data, we recently developed a method to perform functional MRI measurements in awake rhesus monkeys.

In this short overview, we will first explain why the technique is becoming an increasingly important tool for systems neuroscientists in general, and secondly, we will sketchily describe the first experiments that provided proof-of-concept that fMRI can be reliably performed in awake monkeys using commonly available standard clinical horizontal bore MR scanners.

Why is monkey fMRI becoming increasingly important?

(1) The technique for the first time allows us to apply exactly the same stimuli and/or tasks to identify brain regions involved in perception and cognition in non-human and human primates. This enables direct comparison of human and monkey functional imaging data allowing for more accurate determination of possible homologies, as well as dissimilarities between brain areas of both species (Vanduffel *et al.*, 2002; Tootell, Tsao and Vanduffel, 2003; Denys *et al.*, 2004; Orban, Van Essen and Vanduffel, 2004; Nelissen *et al.*, 2005).

(2) Furthermore, this new technique can be extremely useful to guide subsequent electrophysiological experiments in monkeys. During the past five decades, many systems neuroscientists focused upon the processing capabilities of single or multiple units. The next big challenge, however, is to unravel the interactions between different nodes of a distributed network of regions that are involved in the processing of sensory stimuli and/or complex cognitive tasks. FMRI guided single-unit recordings can significantly reduce the needle-in-haystack problem if different functionally interacting brain regions are targeted simultaneously.

(3) Thirdly, in combination with electrophysiology, monkey fMRI will allow to investigate in great detail the relationship between hemodynamic fMRI signal changes and underlying neuronal activity (Logothetis *et al.*, 2001). In general, the

different brain imaging techniques require resorting to a monkey model in which other techniques are available for comparison purposes.

(4) Unlike human fMRI, it is also possible to combine monkey fMRI with other invasive techniques (e.g. lesions, reversible deactivations, microstimulation) that allow interference with an intact functional network. Such an approach will enable us to investigate causal interactions between functionally connected regions (e.g. Ekstrom *et al.*, 2005), and to understand mechanisms of functional reorganizations after lesions (Smirnakis *et al.*, 2005). In the same realm, it will be much easier to test functional effects of pharmacological compounds using fMRI in a monkey model compared to a human model.

(5) Furthermore, single monkeys can be scanned many more times than humans, so that it is more straightforward to test extended training effects, plasticity effects and methodological fMRI issues (e.g. adaptation paradigms) in a monkey fMRI model.

(6) Last but not least, a monkey fMRI model might reduce dramatically the number of animals needed to investigate aspects of functional organization of cortical and sub-cortical structures in monkeys adding to the general welfare of non-human primates.

Although this list is all but complete, it shows that monkey fMRI has a great potential for systems neuroscience. In the remainder of this chapter, we briefly summarize the first series of experiments which provide proof of the concept of our approach (Vanduffel *et al.*, 2001).

In our first pilot experiments, we chose to use very simple visual 'moving' stimuli and compared them to stationary stimuli that were the same in all other aspects. The rationale for this choice was that moving stimuli are known to activate very specific visual areas in human and monkey, such as the MT/V5 (middle temporal) complex. Visual cortical area MT/V5 is located in the superior temporal sulcus (STS) and is characterized by a high proportion of cells responding selectively to the direction of moving stimuli. In other words, these 'direction selective' cells respond optimally when a stimulus moves in one direction and not in the opposite (or non-optimal directions; see Dubner and Zeki, 1971). This exemplar 'dorsal stream' area is thought to be involved in visual motion perception. Thus, contrasting simple moving stimuli with stationary visual stimuli allowed us to compare MR activation maps with a wealth of data from the literature in order to validate the new technique.

9.6.1 Methods

Prior to MR scanning, each monkey was implanted with a plastic headset and ceramic screws which were covered by dental acrylic. Next, the monkeys were adapted to a plastic restraining chair in a sphinx position (see Figure 9.6.1), and trained in a 'mock' MR bore. Eye-positions were measured (at 50 Hz) using a pupil/corneal reflection tracking system (Iscan) and behavioural control was achieved using operant conditioning techniques. More specifically, monkeys were required to maintain their gaze within a small window centred on a fixation target in order to be rewarded with small drops of liquid. After fixation performance reached asymptote, the monkeys were placed into a horizontal bore, MR scanner, equipped with echo-planar imaging. Before each scanning session, a single bolus of contrast agent (Monocrystalline Iron oxide, MION (4–11 mg/kg) was injected intravenously into the femoral/saphenous vein. Hence, our fMRI signal changes reflected (relative) cerebral blood volume changes, as opposed to blood oxygen level dependent (BOLD) signals. Both the BOLD and MION techniques are sensitive to the local concentration of contrast agent – deoxyhaemoglobin for BOLD and exogenous iron for MION. The local concentration of deoxyhaemoglobin decreases as venous blood becomes more oxygenated during activation, leading to an increase in BOLD signal. When enough iron oxide is used to overwhelm the BOLD effect, changes in oxygenation have a negligible effect on total blood magnetization. Thus, an increase in the local concentration of exogenous contrast agent (iron oxide) results from an increase in local cerebral blood volume, so activation-induced MION-based changes are negative rather than positive as corresponding BOLD-based signal changes.

Figure 9.6.1 Magnet-compatible monkey chair. The monkey sits on its haunches in a plastic restraint box with its head immobilized comfortably but securely; in this case directly beneath a six-channel phase array coil and transmit coil

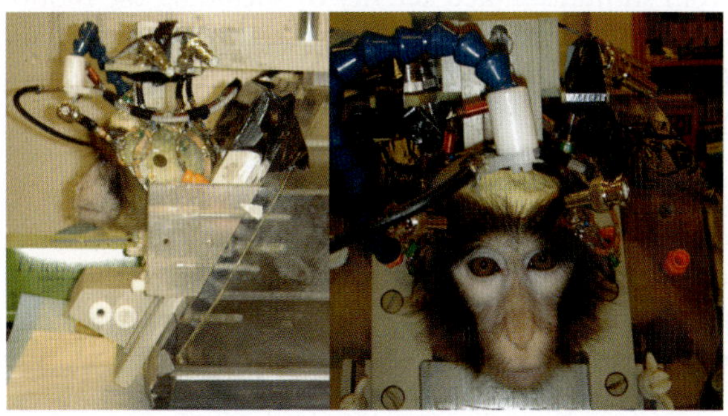

Figure 9.6.2 Stimulus presented to monkey. This random texture pattern (50% black and 50% white dots, 4.5 min arcmin/side, within a circular aperture of 14° diameter) was either moving (2–6 °/s) or stationary

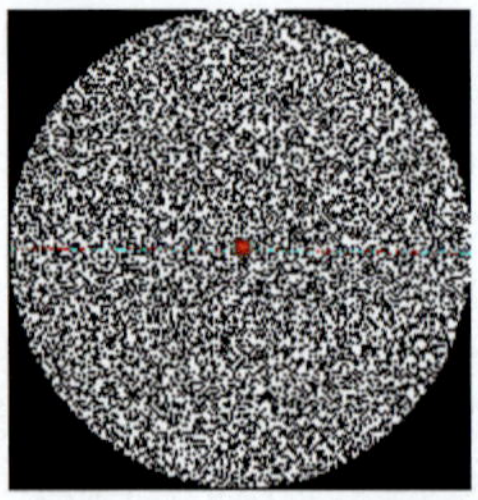

The use of this exogeneous contrast agent improves spatial specificity, decreases contribution of the large draining veins to the fMRI signals and increases contrast-to-noise ratios (with a factor of 5 at 1.5 T) relative to more commonly used BOLD measurements (Vanduffel *et al.*, 2001; Leite *et al.*, 2002).

Visual stimuli (see Figure 9.6.2) were projected onto a screen which was positioned in front of the monkey's eyes. A block design was used in each scan and functional volumes were gradient-echo echo-planar whole-brain images (TR = 2.4 s; TE = 28 ms; 64 × 64 matrix; 2 × 2 × 2 mm voxels; 32 contiguous sagittal slices). In a separate session, an anatomical 3D-gradient echo volume (1 × 1 × 1 mm voxel size) was acquired using a commercial Siemens 'knee-coil' while the monkey was anaesthetized. The functional volumes were aligned to correct for brain motion.

Data were analysed using standard SPM99 procedures (global scaling, low and high-pass filtering). Each stimulus epoch was represented as a box-car model convoluted by the MION response function. Eye movement-related activity was dissociated from stimulus-linked activity during the statistical analysis by including the eye-traces in the general linear model as a covariate of no interest. To that end the eye-traces were thresholded, convolved with the MION response function and sub-sampled to the TR. Using a similar approach, we also excluded movement-related activity from stimulus-linked activity by including the motion re-alignment parameters (for the three rotation and three translation axis) in the statistical analysis as a variable of no interest.

The *t*-score maps were thresholded at $P < 0.05$, corrected for multiple comparisons (see Figure 9.6.3), corresponding to a *t*-score >4.86. *t*-Score maps were either displayed on sections through the brain (Figure 9.6.3) or projected onto the flattened cortical reconstruction (at the level of layer 4) of the same animal using Freesurfer software (Figure 9.6.4).

Figure 9.6.3 Statistical parametric map (*t*-scores) comparing moving stimuli versus stationary visual stimuli (Figure 9.6.1), thresholded at $P < 0.05$, corrected for multiple comparisons. Coronal sections are selected showing motion-sensitivity in four different areas within the superior temporal sulcus. (*Permission from the Nature Publishing Group.*)

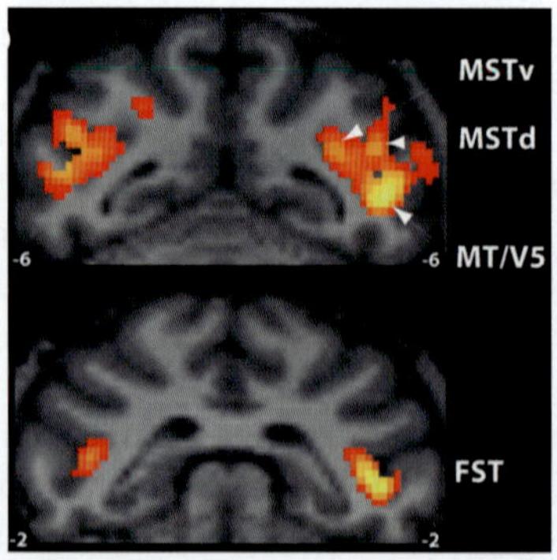

9.6.2 Results

As expected from the single cell properties in the MT/V5-complex within the superior temporal sulcus of the monkey, we observed large fMRI signal changes in this area when comparing moving stimuli with stationary stimuli (>3.00 ± 0.09%, see Figures 9.6.3 and 9.6.4). Moreover, in the original study we confirmed motion sensitivity in other areas *within* the MT/V5 complex than MT/V5 itself, notably vMST (ventral middle superior temporal) and FST (fundus superior temporal).

Follow-up studies (Nelissen *et al.*, 2003), revealed at least six different motion-sensitive regions within the superior temporal sulcus, extending our knowledge derived from electrophysiology with respect to motion processing (see Figure 9.6.3).

9.6.3 Conclusion

These initial fMRI studies on alert fixating monkeys proved surprisingly straightforward to accomplish in a conventional clinical MR scanner and using near-standard behavioural and surgical procedures. Though our comparisons were understandably limited in that initial study, the procedures appeared to yield excellent reliability, both between- and within-subjects. We found that the use of an iron oxide contrast agent with a long blood half-life considerably enhanced functional brain imaging in awake, behaving primates. High behavioural performance levels, even a high number of intravenous MION injections

Figure 9.6.4 Motion-sensitivity in visual cortex of the monkey revealed by fMRI. (a) An inflated 3D-reconstruction of the right hemisphere (light grey = gyri; dark grey = sulci, occipital is to the left, dorsal to the top) with an overlying *t*-score map for moving > stationary patterns, corrected for multiple comparisons. (b) Same statistical map overlying coronal section through the MT/V5 complex in the STS. (c) Horizontal eye-movement trace (green plot) recorded during an fMRI experiment. Large amplitudes (e.g. yellow arrow) correspond to saccadic eye movements. (d) Same as A, but folded 3D-reconstruction of right hemisphere. (e) Typical time course (% MR signal change) extracted from MT/V5, for a time series where epochs with a stationary stimulus (stat) are alternated with epochs in which a moving visual stimulus (motion) was presented. The polarity of the negative MION signal changes are inverted for the sake of clarity. (*Permission from the Nature Publishing Group.*)

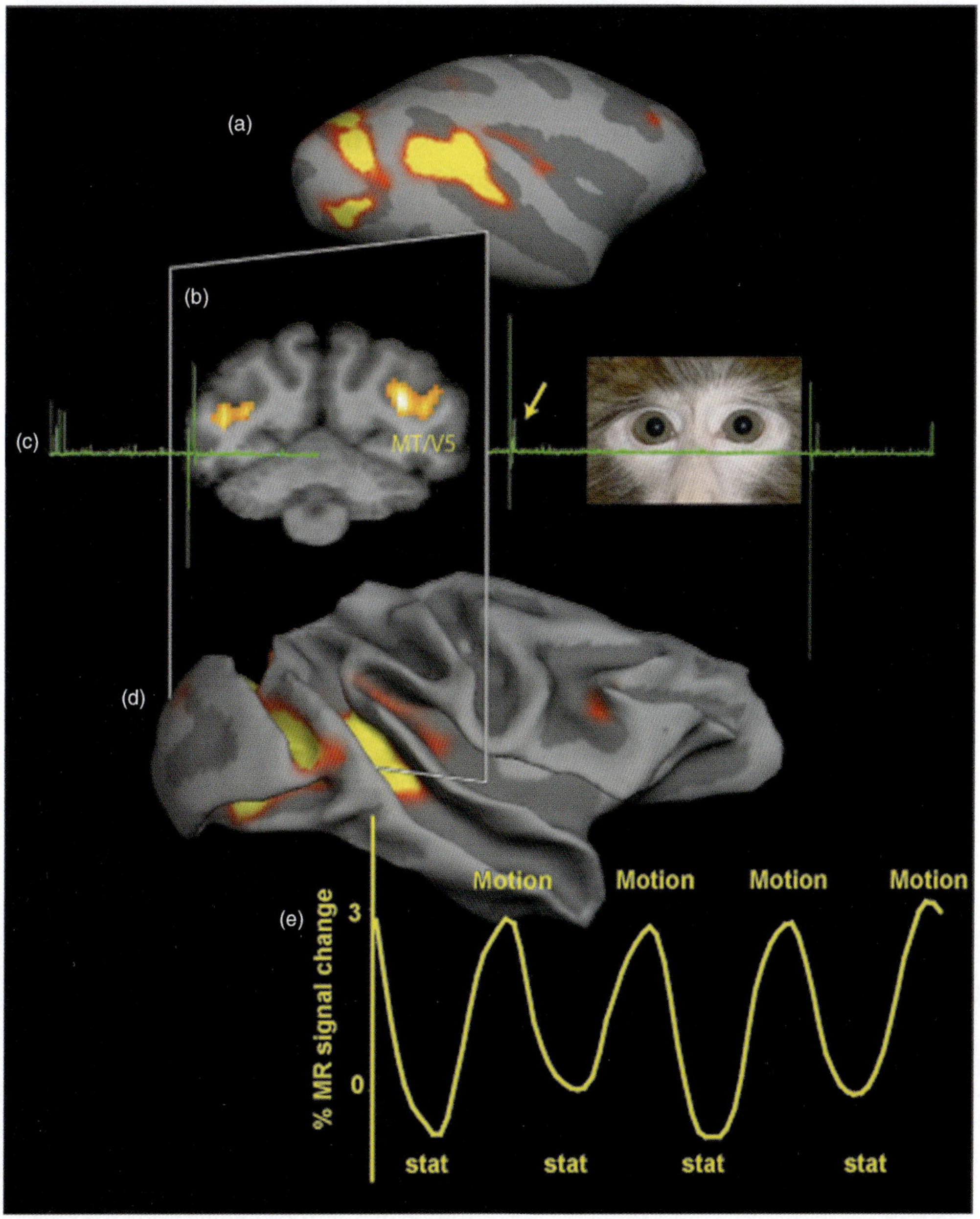

in 5 years time (~200 injections in some animals), show that an effective dose of MION does not result in obvious long-term behavioural, cognitive or negative health effects.

In summary, the initial results proved the vast potential of the new technique as currently being underscored by the increasing number of publications and users worldwide.

References

Denys, K., Vanduffel, W., Fize, D., Nelissen, K., Sawamura, H., Georgieva, S., Vogels, R., Van, E. D., Orban, G. A., 2004. "Visual activation in prefrontal cortex is stronger in monkeys than in humans." *J.Cogn Neurosci.*, **16**(9), 1505–1516.

Dubner, R., Zeki, S. M., 1971. "Response properties and receptive fields of cells in an anatomically defined region of the superior temporal sulcus in the monkey." *Brain Res.* **35**(2), 528–532.

Ekstrom, L. B., Bonmassar, G., Tootell, R. B., Roelfsema, P., Vanduffel, W., 2005. "FEF microstimulation modulates visually driven activity in cortex: evidence from awake behaving monkey fMRI." *Soc. Neurosci. Abstracts, USA*. **35**, 821–8.

Leite, F. P., Tsao, D., Vanduffel, W., Fize, D., Sasaki, Y., Wald, L. L., Dale, A. M., Kwong, K. K., Orban, G. A., Rosen, B. R., Tootell, R. B., Mandeville, J. B., 2002. "Repeated fMRI using iron oxide contrast agent in awake, behaving macaques at 3 Tesla." *Neuroimage* **16**(2), 283–294.

Logothetis, N. K., Pauls, J., Augath, M., Trinath, T., Oeltermann, A., 2001. "Neurophysiological investigation of the basis of the fMRI signal." *Nature* **412**(6843), 150–157.

Nelissen, K., Luppino, G., Vanduffel, W., Rizzolatti, G., Orban, G. A., 2005. "Observing others: multiple action representation in the frontal lobe." *Science* **310**(5746), 332–336.

Nelissen, K., Vanduffel, W., Fize, D., Orban, G. A., 2003. "Mapping motion responsive regions in the STS: An awake monkey fMRI study." *Soc. Neurosci. Abstracts, USA*. **33**, 438–2.

Orban, G. A., Van Essen D., Vanduffel, W., 2004. "Comparative mapping of higher visual areas in monkeys and humans." *Trends Cogn. Sci.* **8**(7), 315–324.

Smirnakis, S. M., Brewer, A. A., Schmid, M. C., Tolias, A. S., Schuz, A., Augath, M., Inhoffen, W., Wandell, B.A., Logothetis, N. K., 2005. "Lack of long-term cortical reorganization after macaque retinal lesions." *Nature* **435**(7040), 300–307.

Tootell, R. B., Tsao, D., Vanduffel, W., 2003. "Neuroimaging weighs in: humans meet macaques in primate" visual cortex. *J. Neurosci.* **23**(10), 3981–3989.

Vanduffel, W., Fize, D., Mandeville, J. B., Nelissen, K., Van, H. P., Rosen, B. R., Tootell, R. B., Orban, G. A., 2001. "Visual motion processing investigated using contrast agent-enhanced fMRI in awake behaving monkeys." *Neuron* **32**(4), 565–577.

Vanduffel, W., Fize, D., Peuskens, H., Denys, K., Sunaert, S., Todd, J. T., Orban, G. A., 2002. "Extracting 3D from motion: differences in human and monkey intraparietal cortex." *Science* **298**(5592), 413–415.

9.7 Multimodal evaluation of mitochondrial impairment in a primate model of Huntington's disease

Vincent Lebon and Philippe Hantraye

In vivo magnetic resonance spectroscopy (MRS) allows to measure various biochemical parameters such as metabolite concentration, synthesis rate and diffusion. MRS detection can be performed through different nuclei (including ^{1}H, ^{31}P or ^{13}C), thus increasing the number of possible measurements. Among all MRS techniques proposed for brain exploration, two quantitative approaches have emerged over the last years: the determination of neurochemical profile based on short echo time ^{1}H spectroscopy and the measurement of metabolic fluxes such as the tri-carboxylic acid cycle flux (V_{TCA}) based on ^{13}C-labelled molecules. MRS measurement of V_{TCA} relies on the unique property of NMR spectroscopy to specifically identify one molecule and the atomic position in this molecule at which the isotopic label accumulates. This contrasts with nuclear techniques for which radioactivity is measured, independently of the metabolite the radioactive tracer is attached to. The principle of V_{TCA} measurement by NMR is otherwise analogous to the measurement of cerebral metabolic rate of glucose (CMRglc) based on ^{18}F-fluorodeoxyglucose (^{18}F-FDG) detection by positron emission tomography. However, NMR measures oxidative metabolism (V_{TCA}), whereas PET measures glycolytic metabolism (CMRglc). Thus, performing both techniques provides with a complete picture of brain ATP synthesis. This approach was used for exploring brain metabolism in a primate model of Huntington's disease: Mitochondrial impairment was achieved by administration of 3-nitropropionic acid (3-NP). Alterations of metabolite concentration and energy synthesis were quantified, enlightening brain adaptation to energy deficit. This approach should prove useful for assessing novel therapeutic strategies.

Figure 9.7.1 (a) 3D rendering of the monkey head showing the position of the striatal VOI. (b) Short echo-time ^{1}H spectrum acquired for quantitation of the neurochemical profile. (c) Stacked plot of $\{^{13}C\}$-^{1}H spectra acquired during a 2-h infusion of ^{13}C-labelled glucose, showing progressive ^{13}C enrichment of glutamate

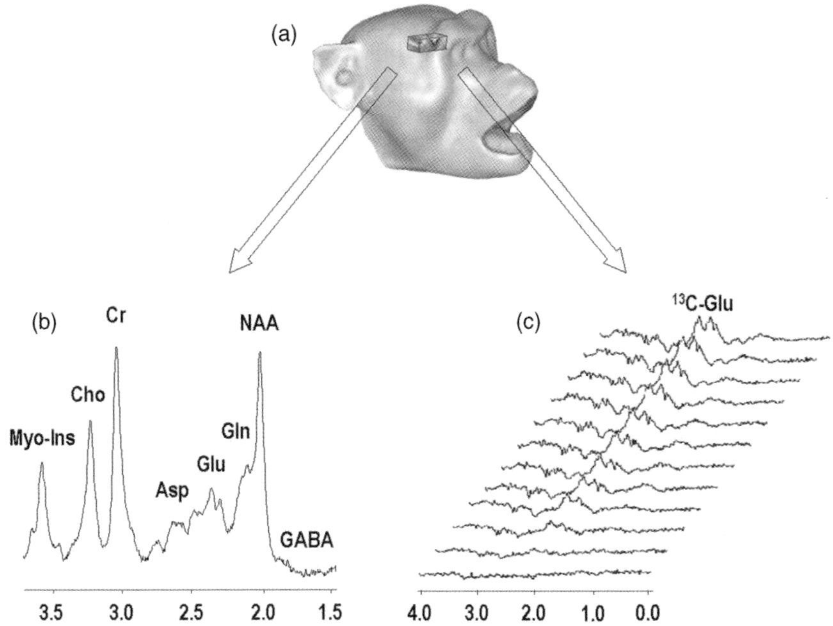

9.7.1 MRS quantitation of the major brain metabolites

^{1}H MRS typically allows to detect from 6 to 20 metabolites in the brain, depending on the NMR system used. Proper quantitation of brain metabolites leads to a so-called neurochemical profile (Provencher, 1993; Pfeuffer *et al.*, 1999).

9.7.1.1 *Methods*

MRS was performed on three macaque monkeys (*Macaca fascicularis*, three to four experiments per monkey). Animals were anaesthetized using i.v. infusion of propofol (~200 μg/kg/min), intubated and ventilated. An optimized ^{1}H MRS sequence (PRESS sequence for localization, echo-time 8 ms) was implemented on a 3 T system equipped with a surface ^{1}H probe. A 3.9 ml volume of interest (VOI) was positioned in the centre of the brain (Figure 9.7.1(a)). After manual shimming down to 7 Hz, a metabolite spectrum was acquired with water suppression (18-min acquisition time). Then a water spectrum was collected for internal concentration reference. Frequency domain analysis (Provencher, 1993) was performed in the 1.0–3.7 ppm range.

9.7.1.2 *Results*

Figure 9.7.1(b) presents a ^{1}H spectrum. In addition to commonly detected *N*-acetyl-aspartate (9.5 ± 0.8 mM, mean ± SD), creatine (10.6 ± 0.9 mM) and choline (1.7 ± 0.4 mM), short echo-time detection allowed to quantify glutamate (9.4 ± 1.3 mM), myo-inositol (4.8 ± 0.4 mM), taurine (3.7 ± 0.4 mM), glutamine (3.1 ± 0.4 mM), aspartate (1.8 ± 0.4 mM), GABA (1.3 ± 0.3 mM), GSH (1.2 ± 0.6 mM) and lactate (0.5 ± 0.3 mM).

9.7.2 MR quantitation of brain oxidative metabolism

As illustrated on Figure 9.7.2(a), MR measurement of V_{TCA} relies on the detection of glutamate ^{13}C enrichment during an i.v. infusion of ^{13}C-labelled glucose (Mason *et al.*, 1992). Directly localized detection of ^{13}C in the centre of the brain is difficult, due to the intrinsic lower sensitivity of ^{13}C as compared to ^{1}H. The sensitivity required for V_{TCA} measurement in the striatum can be achieved using indirect $\{^{13}C\}$-^{1}H detection, which consists in detecting the perturbative effect of each ^{13}C on the ^{1}H bound to the enriched carbon atom (Rothman *et al.*, 1985).

Figure 9.7.2 Labelling strategy used for the measurement of oxidative and glycolytic pathways in the primate brain. (a) NMR detection of ^{13}C incorporation from ^{13}C-glucose into glutamate leads to the TCA cycle flux V_{TCA} (grey arrows). (b) PET detection of ^{18}F-FDG leads to CMRglc (dark arrow)

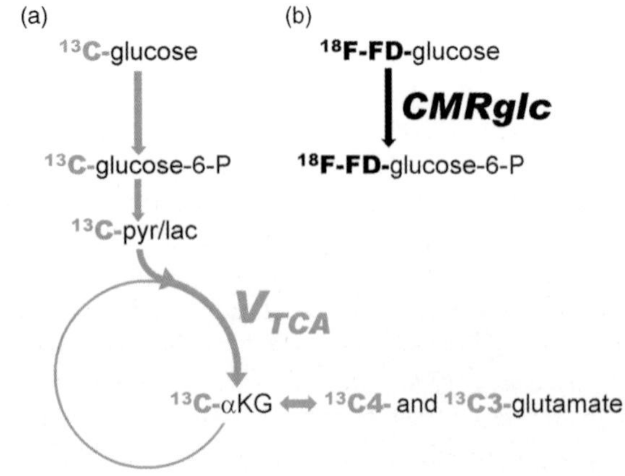

Figure 9.7.3 NMR-measured and PET-measured labelling time-courses. (a) Glutamate $^{13}C4$ and $^{13}C3$ time-courses measured by NMR in one experiment and best fits. (b) ^{18}F-FDG PET time-activity curve measured in the same VOI and best fit. Modelled contributions of FDG and FDG-6-P are shown. (*Reproduced from Boumezbeur (2005) by permission of Nature Publishing Group.*)

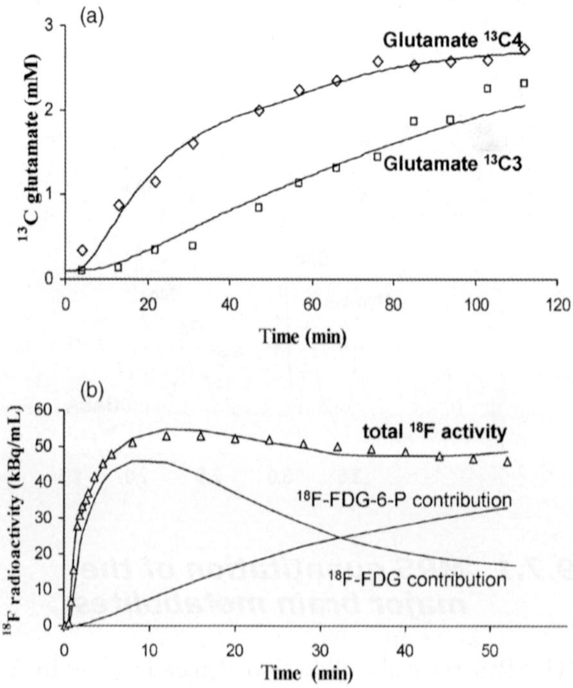

9.7.2.1 *Methods*

Experiments were performed on two macaque monkeys studied twice. The protocol was almost identical to the one used for neurochemical profiling of the same brain area. ^{1}H spectra were collected during the infusion of glucose labelled with ^{13}C (3-min bolus followed by 120-min continuous infusion). Blood samples were collected in order to measure glucose fractional enrichment using *ex vivo* high-resolution NMR spectroscopy. Difference spectra revealed simultaneous decreases in ^{12}C-bonded protons and increases in ^{13}C-coupled protons of glutamate. An original quantitation procedure was implemented for proper measurement of both glutamate C4 and C3 enrichments (Boumezbeur *et al.*, 2004). Mathematical modelling of ^{13}C incorporation from glucose to glutamate yielded V_{TCA}.

9.7.2.2 *Results*

^{13}C incorporation into the C3 and C4 positions of glutamate is shown in Figure 9.7.1(c). Fitting the corresponding time courses (Figure 9.7.3(a)) led to $V_{TCA} = 0.55 \pm 0.04$ μmol/g/min.

9.7.3 *Combination with PET measurement of glucose consumption*

Energy synthesis in the mammal brain relies on two major pathways: TCA cycle and glycolysis. In order to obtain an exhaustive picture of brain ATP synthesis, MRS measurement of V_{TCA} can be combined to PET measurement of the cerebral metabolic rate of glucose (CMRglc), which measures glycolysis (Boumezbeur *et al.*, 2005).

9.7.3.1 *Methods*

Studies were conducted on five macaque monkeys. V_{TCA} was measured as described above. In addition, 3D proton density MR images were acquired for registration with PET images. PET experiments were performed on a CTI HR + Exact tomograph (4.5 mm isotropic intrinsic resolution). PET scans were collected for 60 min following ^{18}FDG i.v. injection (110 ± 5 MBq). Blood samples were withdrawn to measure arterial radioactivity. 3D proton density MR images were registered with 3D re-constructed PET scans (Viola and Wells, 1997). The VOI detected by NMR was extracted from the PET images (Figure 9.7.4) and the corresponding time–activity

Figure 9.7.4 Superimposition of T_1 MR image and ^{18}F-FDG PET image after registration. The 3.9 mL VOI is shown (white rectangle). (*Reproduced from Boumezbeur (2005) by permission of Nature Publishing Group.*)

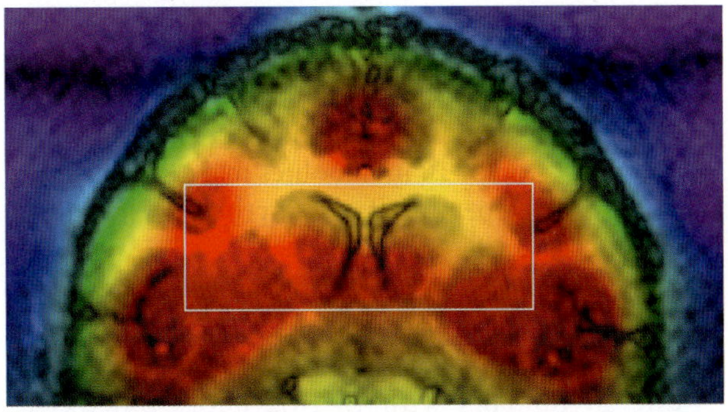

curve was generated. For kinetic analysis and CMRglc calculation, a two-tissue compartmental model was used (Kennedy *et al.*, 1978; Phelps *et al.*, 1979) in which the total ^{18}F activity detected was modelled as the sum of the brain pools of ^{18}F-FDG and ^{18}F-FDG-6-P (Figure 9.7.3(b)).

9.7.3.2 Results

V_{TCA} and CMRglc were 0.53 ± 0.13 and 0.23 ± 0.03 μmol/g/min, respectively. The resulting [CMRglc/V_{TCA}] ratio was 0.46 ± 0.12, not significantly different from the 0.50 expected when glucose is the sole fuel that is completely oxidized. This argues in favour of metabolic coupling between the TCA cycle and glycolysis under normal physiological conditions.

9.7.4 MRS/PET exploration of mitochondrial impairment in primates

3-Nitropropionic acid (3-NP) is an irreversible inhibitor of the mitochondrial complex II succinate dehydrogenase (SDH). 3-NP treatment induces mitochondrial impairment and striatal lesions characteristic of Huntington's disease (Beal *et al.*, 1993, Brouillet *et al.*, 1995).

9.7.4.1 Methods

Three macaque monkeys underwent daily injections of 3-NP for 30 weeks. Doses were incremented from 10 mg/kg/day on week 1 up to 30 mg/kg/day on weeks 18–30. For each monkey, a total of 13 neurochemical profiles, 8 V_{TCA} and 5 CMRglc were measured in the striatum before and during the 3-NP treatment.

9.7.4.2 Results

V_{TCA} and CMRglc both presented an immediate ~45% decrease upon 3-NP, which remained constant during the 30 weeks of treatment. In contrast, neurochemical profile showed moderate alteration, with no lactate change. Neuron compartmentalized amino acids (glutamate, *N*-acetyl-aspartate, aspartate) exhibited a ~10% immediate diminution followed by a further decrease reaching ~20% on week 30. Delayed and limited lesions (<5% of striatum volume) were observed on T_1-weighted MRI.

Histochemistry performed on one animal (sacrificed at week 30) revealed a ~50% loss of SDH activity paralleling changes in V_{TCA} and CMRglc. No change in cytochrome oxidase (COX) activity was detected, which was similar to the limited changes observed on neurochemical profiles.

9.7.4.3 Discussion

Combined MRS/PET measurements provide with a consistent picture of brain adaptation to mitochondrial impairment: preservation of the coupling between glycolytic and oxidative metabolism is in agreement with the absence of lactate increase. These observations argue in favour of brain adapting to mitochondrial deficit without oxidizing other substrates than glucose and without enhancing anaerobic glycolysis. This allows the yield of ATP synthesis to remain optimal, limiting lesion extension and decrease in metabolite concentration. MRS findings are supported by histochemistry which evidences strong metabolic impairment (SDH activity) associated with limited cell loss (COX activity).

9.7.5 Conclusion

MRS requires important methodological developments in order to obtain quantitative data on brain metabolism. However, MRS provides with unique measurements of both metabolite concentration and metabolic fluxes *in vivo*. Combined with PET, it can bring an original and exhaustive picture of brain energy metabolism, enlightening metabolic alteration and adaptation under energy deficit.

References

Beal, M. F., Brouillet, E., Jenkins, B.G., Ferrante, R.J., Kowall, N. W., Miller, J. M., Storey, E., Srivastava, R., Rosen, B. R., Hyman, B. T., 1993. "Neurochemical and histologic characterization of striatal excitotoxic lesions produced by the mitochondrial toxin 3-nitropropionic acid." *J. Neurosci.* **13**, 4181–4192.

Boumezbeur, F., Besret, L., Valette, J., Grégoire, M.-C., Delzescaux, T., Vaufrey, F., Gervais, P., Hantraye, P., Bloch, G., Lebon, V., 2005. "Glycolysis *vs.* TCA cycle in the primate brain as measured by combining ^{18}F-FDG PET and ^{13}C-NMR." *J. Cereb. Blood Flow Metab.* **25**, 1418–23.

Boumezbeur, F., Besret, L., Valette, J., Vaufrey, F., Henry, P.-G., Slavov, V., Giacomini, E., Hantraye, P., Bloch, G., Lebon, V., 2004. "NMR measurement of brain oxidative metabolism in monkeys using ^{13}C-labeled glucose without a ^{13}C radiofrequency channel." *Magn. Reson. Med.* **52**, 33–40.

Brouillet, E., Hantraye, P., Ferrante, R. J., Dolan, R., Leroy-Willig, A., Kowall, N. W., Beal, M. F., 1995. "Chronic mitochondrial energy impairment produces selective striatal degeneration and abnormal choreiform movements in primates." *Proc. Natl. Acad. Sci. USA.* **92**, 7105–7109.

Kennedy, C., Sakurada, O., Shinohara, M., Jehle, J., Sokoloff, L., 1978. "Local cerebral glucose utilization in the normal conscious macaque monkey." *Ann. Neurol.* **4**, 293–301.

Mason, G. F., Rothman, D. L., Behar, K. L., Shulman, R. G., 1992. "NMR determination of the TCA cycle rate and alpha-ketoglutarate/glutamate exchange rate in rat brain." *J. Cereb. Blood Flow Metab.* **12**, 434–447.

Pfeuffer, J., Tkac, I., Provencher, S. W., Gruetter, R., 1999. "Toward an in vivo neurochemical profile: quantification of 18 metabolites in short-echo-time (1)H NMR spectra of the rat brain." *J. Magn. Reson.* **141**, 104–120.

Phelps, M. E., Huang, S., Hoffman, E., Selin, C., Sokoloff, L., Kuhl, D., 1979. "Tomographic measurement of local cerebral glucose metabolic rate in humans with (F-18)2-fluoro-2-deoxy-D-glucose: validation of method." *Ann. Neurol.* **6**, 371–388.

Provencher, S., 1993. "Estimation of metabolite concentrations from localized in vivo proton NMR spectra." *Magn. Reson. Med.* **30**, 672–679.

Rothman, D. L., Behar, K. L., Hetherington, H.P., den Hollander, J. A., Bendall, M.R., Petroff, O. A. C., Shulman, R.G., 1985. "^{1}H-observe/^{13}C-decouple spectroscopic measurements of lactate and glutamate in the rat brain in vivo." *Proc. Natl. Acad. Sci. USA.* **82**, 1633–1637.

Viola, P., Wells, W., 1997. "Alignment by Maximization of Mutual Information." *Int. J. Computer Vision* **24**, 137–154.

10 Imaging of Heart, Muscle, Vessels

Coordinated by Yves Fromes

10.0 Introduction

Yves Fromes

Transgenic methodologies have already resulted in the production of a huge variety of new mouse strains engineered specifically to study particular genes or diseases. In order to take full advantage of these new genetic tools to study human diseases such as cardiomyopathies, heart failure, vascular abnormalities and myopathies, it is important to improve the methodologies used to determine the phenotypes of these animals and the physiologic outcome of genetic variation.

Invasive methods are often more quantitative, but non-invasive methods are preferred when measurements must be repeated serially on living animals. For the reason of the small size and high heart rates in mice, high spatial and temporal resolutions are required to preserve signal fidelity.

The available methods for evaluating the cardiovascular systems of small lab animals include positron emission tomography (PET), echocardiography or nuclear magnetic resonance.

The advent of small-animal PET and the increases in the sensitivity and spatial resolution of scanners made it possible to use adaptations of these methods in experimental animals. These developments allow repeated studies of the same animal, including studies of the same animal under different conditions, and longitudinal studies. Molecular imaging by small-animal PET is an important non-invasive means to phenotype transgenic mouse models *in vivo*. Although PET is usually used to derive biochemical and molecular information, functional parameters such as ventricular volumes are generally measured using echocardiography or nuclear magnetic resonance (NMR) imaging.

The value of cine magnetic resonance imaging for assessment of the rat or mouse heart has been tested in various studies. NMR imaging can be performed for determination of left ventricular volumes and mass, cardiac output or myocardial infarct size. Thus, being non-invasive and exact, magnetic resonance imaging (MRI) offers new insights into the remodelling process after myocardial infarct because serial measurements are possible.

Echocardiography is an accurate non-invasive tool for determination of alterations in cardiac structure and function in laboratory animals.

In vivo investigations to study skeletal muscle structure and function under normal and pathological conditions in small animals benefit from the developments of modern imaging techniques. Muscle functional MRI has been proposed as a tool for non-invasively measuring the metabolic and hemodynamic responses to muscle activation. *In vivo* NMR spectroscopy may be a useful tool by being complementary to NMR imaging.

Various physiological or pathological situations might be investigated in small animals. A signal increase in denervated muscle on MRI has been described in several clinical and experimental studies. MRI closely mirrors the electrophysiological changes following denervation and reinnervation and may therefore be used in addition to electrophysiology. Examination methods to assess perfusion of the skeletal muscle can use contrast-enhanced ultrasonography analysing replenishment kinetics of microbubbles. Similarly, muscle blood flow can be analysed by

PET using labelled water. However, the limited spatial resolution of the imaging devices may not yet allow direct measurement of true vascular flow heterogeneity. Interleaved H and P NMR spectroscopy and NMR imaging can allow measuring *in vivo* skeletal muscle perfusion, oxygenation and high-energy phosphate distribution, simultaneously. Furthermore, investigation of skeletal muscle function in animals may be obtained by using transcutaneous stimulation and measuring force with a dedicated ergometer combined with strictly non-invasive NMR investigations. This offers the possibility to repeat investigations in the same animal, avoiding individual variability and enabling longitudinal follow-up studies.

The purpose of this chapter is to highlight the technical aspects of the *in vivo* investigations of cardiac and skeletal muscles structures and functions in small animal models.

10.1 Cardiac structure and function

Yves Fromes

During each cardiac cycle, cardiocytes have to accommodate to repetitive changes in their geometry in order to perform force-generating contraction. Various approaches for assessing cardiac structure and myocardial pump function have been proposed over time, but fundamentally the question remains the same. How could we most accurately explore the cardiac structure–function relationship?

The cardiac anatomy is complex, and cardiac structures have different appearances depending on the imaging plane and other options selected. Reference data can of course be obtained through histological sections, but animals have to be sacrificed in order to harvest the heart, and thus, longitudinal studies are difficult (Figure 10.1.1).

In vivo imaging techniques allow to gain unique insight into the functional anatomy in living small laboratory animals. Nuclear magnetic resonance imaging (NMRI) of the heart and great vessels has improved substantially over the past decade, and it is entering the mainstream of diagnostic imaging. Cardiac NMRI in patients is already considered the procedure of choice in the evaluation of pericardial diseases and intracardiac and pericardiac masses, for imaging the right ventricle and pulmonary vessels and for assessing many forms of congenital heart disease.

The complexity of cardiac structures and function must be understood to devise a well-planned imaging scheme. Cardiac NMRI in small animals is possible but remains challenging and is not yet well standardized. The use of rigid protocols is ill advised in this kind of situation. Each animal model's imaging session must be customized to accommodate individual variations or distortions caused by cardiac disease. The most useful imaging planes are those parallel and perpendicular to the cardiac axes. Investigators need to be aware that these planes are not based on external landmarks of the anatomy. Obtaining images in double-oblique planes requires the use of multiple localizing sequences and knowledge of the target anatomy (Carlier *et al.*, 1999; Parzy *et al.*, 2003; Parzy *et al.*, 2005).

Figure 10.1.1 Histological sections remain the reference method to analyse the structure of the myocardium. As an example, a transverse section of the ventricles in a hamster model displays different structural features, which can be studied more specifically with different stains. (a) Haematoxylin and Eosin (H&E) stain represents the primary staining method for tissue section analysis, here normal myocardium. Haematoxylin stains tissue components such as myelin, elastic and collagenic fibres, muscle striations, mitochondria, but its most common application is as a nuclear dye. Eosin stains various tissue structures shades especially cell cytoplasm, collagen and erythrocytes. (b) In a cardiomyopathy model, the H&E stain reveals various elementary lesions, such as focal necrosis. (c) Immunohistochemistry anti-vimentin depicts vascular smooth muscle. (d) Sirius red stain reveals the collagen content of the myocardium, as in the scar area of a cardiomyopathy

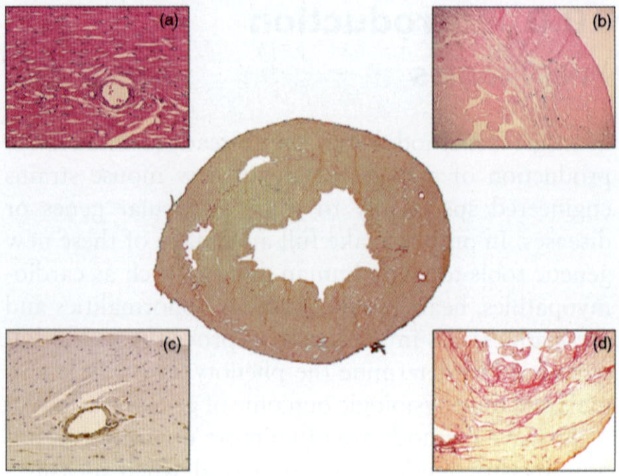

Images can be shown in two or three spatial dimensions in either static or dynamic modes, as images of the chest and cardiovascular system can be obtained from many angles. This allows better assessment

Figure 10.1.2 The left ventricular mass can be quantified accurately and reproducibly by using ECG-gated NMR imaging. Serial investigations of morphological and functional changes in the intact rat heart using MRI are made possible through the acquisition of contiguous images during systole (upper panel) and diastole (lower panel). (Courtesy of P. Carlier, Institut de Myologie, Paris.)

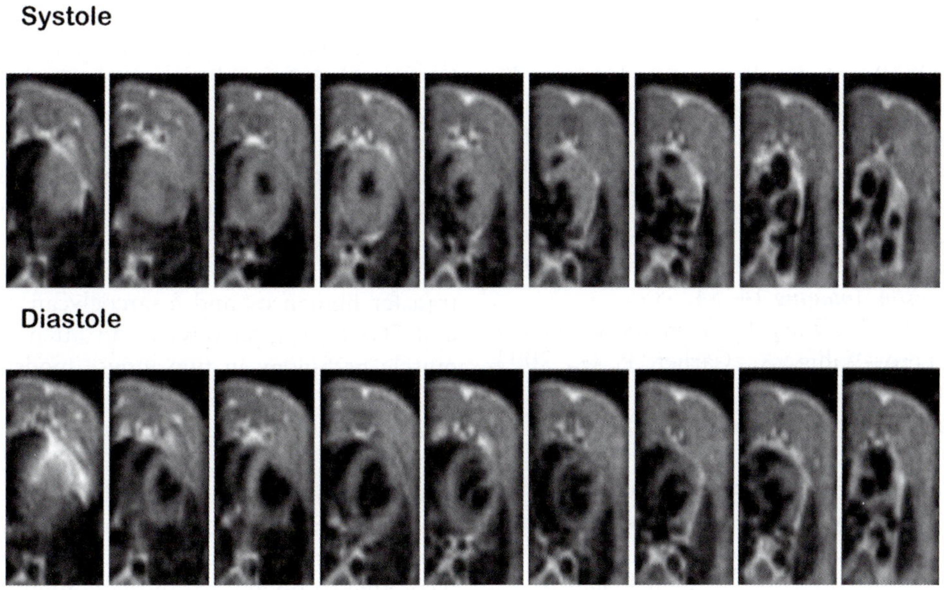

of complex anatomic abnormalities than with other imaging techniques. Real-time NMRI provides accurate measurements of left ventricular end-diastolic volume (LVEDV), left ventricular end-systolic volume (LVESV), ejection fraction (EF) and left ventricular mass in a time-efficient manner with respect to image acquisition (Figure 10.1.2). NMRI acquires information about the heart as it is beating; it can create moving images of the heart throughout its pumping cycle. This allows NMRI to display abnormalities in cardiac chamber contraction. However, cardiac NMRI will have to compete with reference physiological measurements such as pressure/volume loops (Figure 10.1.3).

Articles in the recent literature explore the increasing value of NMRI in the assessment of ischemic heart disease (Yang *et al.*, 2004; Zhou *et al.*, 2004). Cine NMRI is a valuable diagnostic tool applicable to the rat or mouse model of myocardial infarction (Nahrendorf *et al.*, 2001). Being non-invasive and exact it offers new insights into the remodelling process after myocardial infarction because serial measurements are possible. NMRI allows to characterize global and regional left ventricular function during post-myocardium infarction remodelling in small animals. Results in animal models parallel findings in humans and provide a unique tool to examine regional mechanical influences on the remodelling process.

NMRI compares favourably to other imaging modalities (Zhou *et al.*, 2004), and this raises the question if cardiac NMR imaging can replace other imaging technologies? Despite its advantages, these tests are not a substitute for other imaging techniques in all cardiovascular conditions. Unlike an echocardiogram machine, the NMRI scanners cannot be

Figure 10.1.3 The conductance catheter allows measuring the luminal volume of the chambers of the heart, here the left ventricle in a mouse model. Simultaneously and continuously, measures of the left ventricular pressure are recorded, thus making it achievable to draw a pressure-volume loop. By varying the load of the left ventricle, it becomes possible to calculate the end-systolic pressure–volume relationship, the most reliable criterion for myocardial contractility

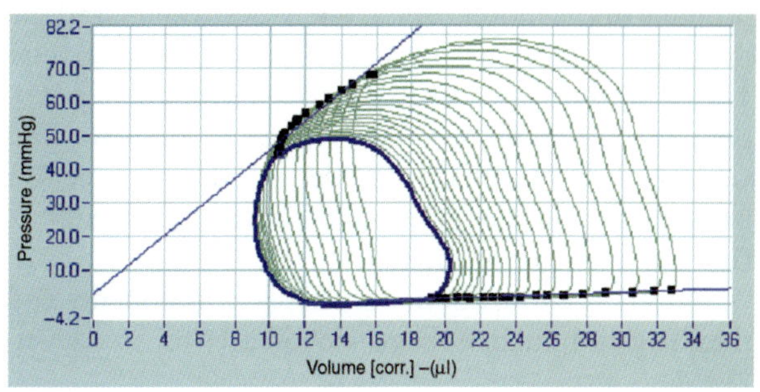

brought to animal facilities, and their availability remains limited.

References

Carlier, P. G., Parzy, E., Wary, C., Fromes, Y., Vilquin, J. T., Pouzet, B., Leroy-Willig, A., 1999. *Proceedings of the International Society for Magnetic Resonance in Medicine*, Philadelphie, USA.

Nahrendorf, M., Wiesmann, F., Hiller, K. H., Hu, K., Waller, C., Ruff, J., Lanz, T. E., Neubauer, S., Haase, A., Ertl, G. and Bauer, W. R., 2001. *J. Magn. Reson. Imaging* **14**, 547–555.

Parzy, E., Fromes, Y., Wary, C., Vignaux, O., Giacomini, E., Leroy-Willig, A., Carlier, P. G., 2003. *J. Hypertens.* **21**, 429–436.

Parzy, E., Fromes, Y., Thiaudiere, E., Carlier, P. G., 2007. Refinement of cardiac NMR imaging, in awake hamsters: proof of feasibility and characterization of cardiomyopathy. *NMR in Biomedicine*, 3 April 2007, [Epub ahead of print].

Yang, Z., Berr, S. S., Gilson, W. D., Toufektsian, M. C., French, B. A., 2004. *Circulation* **109**, 1161–1167.

Zhou, Y. Q., Foster, F. S., Nieman, B. J., Davidson, L., Chen, X. J., Henkelman, R. M., 2004. *Physiol. Genom.* **18**, 232–244.

10.2 Evaluation of therapeutic approaches in muscular dystrophy using MRI

Valérie Allamand

The heterogeneous group of muscle disorders referred to as muscular dystrophies (MD) arises from defects in genes whose products are localized to different subcellular compartments. The most common form of MD, Duchenne muscular dystrophy (DMD), and a sub-group of the limb-girdle muscular dystrophies (LGMD) are caused by primary defects in structural proteins of the muscle cell, namely dystrophin and the members of the sarcoglycan complex, respectively. These proteins are core components of the dystrophin–glycoprotein complex (DGC; Ervasti and Campbell, 1991) which provides a crucial connection between the cytoskeleton and the extracellular matrix (Ervasti and Campbell, 1993).

Indeed, identification of the DGC components and their role in human MD provided bases for the development of therapeutic approaches for these disorders, for which no primary treatments are yet available. Over the last decade, virally mediated gene therapy appeared as a promising strategy for re-expressing non-mutated proteins into muscle cells. The genes encoding the sarcoglycans (α-, β-, γ- and δ-SG) are easily amenable to virally mediated gene transfer because of their small size. Mutations in one or the other of these genes cause different forms of LGMD, demonstrating their important role in normal skeletal muscle function. In addition, absence of any of the sarcoglycan isoform results in the reduction or loss of the DGC (Cohn and Campbell, 2000).

We used first-generation adenoviral vectors to transfer human α- and δ-sarcoglycan genes (*SGCA* and *SGCD*, respectively) in an attempt to prevent muscle pathology in two mouse models of LGMD (LGMD2D and LGMD2E, respectively; Duclos *et al.*, 1998; Coral-Vazquez *et al.*, 1999). We demonstrated that, indeed, expression of human SGs restored DGC expression at the sarcolemma. By a single injection of either human *SGCA* or *SGCD*-bearing adenovirus into neonate mutant mice, we were able to obtain a transduction efficacy of more than 80% and long-term persistence of expression of the transgene product (Figure 10.2.1; Allamand *et al.* 2000; Durbeej *et al.*, 2000; Durbeej *et al.*, 2003). In

Figure 10.2.1 Sustained expression of human α-SG after a single intramuscular injection in skeletal muscle of neonate *Sgca*-null mice (a) and *Sgcd*-null mice (b). *Quadriceps femoris* (a) and *tibialis anterior* (b) muscles were harvested 15 weeks and 1 month following injection, respectively. Immunofluorescence analyses were performed using a rabbit polyclonal antibody against α-SG. Composites represent images taken at a magnification of 5 ×. *vl*: vastus lateralis, *rf*: rectus femoris. The bar represents 100 μm. (*Taken and adapted from Allamand et al., 2000; Durbeej et al., 2003 with permission from Nature Publishing Group and National Academy of Science USA*)

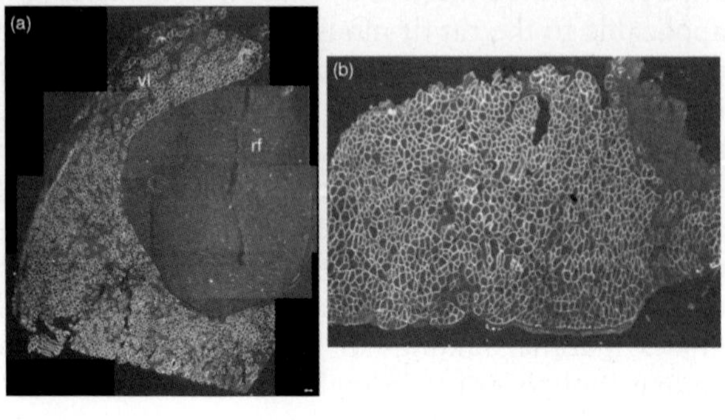

turn, stabilization of the DGC at the sarcolemma protected the myofibres against the development of the dystrophic process, as demonstrated by the decrease in the number of centrally located nuclei in the injected muscle groups, a hallmark of the degeneration–regeneration process that characterizes MD.

Figure 10.2.2 (a) Contrast agent-enhanced MRI visualization of mouse skeletal muscle damage. T1-weighted spin-echo images through the pelvic girdle and thighs from 8–10-week-old mice were taken prior (i–iii) or 15 min after (iv–vi) injection of AngioMark (MS-325) through a jugular vein catheter. In control C57/BL10 mice (i, iv), no changes in contrast agent levels were detected. In dystrophin-deficient *mdx* (ii, v) and *Sgca*-null mutant (iii, vi) mice, extensive uptake of the contrast agent appeared as bright regions indicating the regions with muscle lesions. L, R: left and right hind limbs, respectively. (b) Signal enhancement increase in skeletal muscle lesions before and after injection of AngioMark (MS-325). Mean enhancement percent was calculated from the mean signals in a region of interest before and up to 60 min after systemic injection of MS-325. The mean enhancement in *mdx* (grey columns) and *Sgca*-null mutant (white columns) mice was significantly higher compared to control C57/BL10 (black columns) mice ($^*P \leq 0.001$; $^¤P \leq 0.017$). The signal enhancement in control mice was due to MS-325 in the vessels and reflected the long blood half-life of the contrast agent. (*Taken and adapted from Straub et al., 2000*)

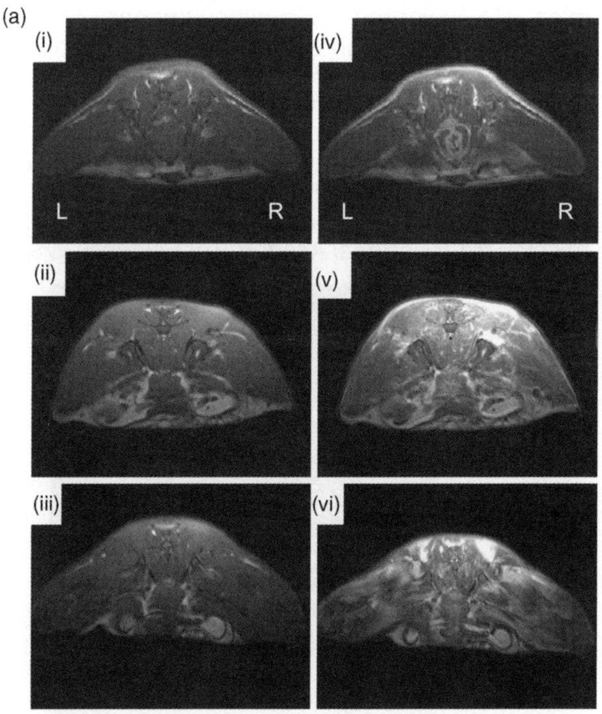

Figure 10.2.2 (*Continued*)

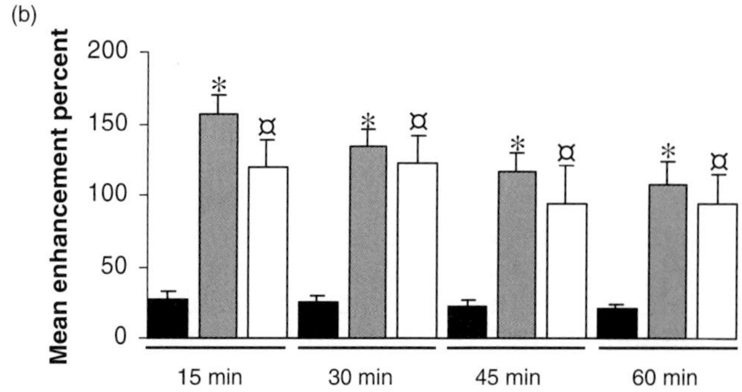

We were also interested in assessing whether restoring the DGC at the muscle membrane would prevent disruption of the sarcolemma. It has been hypothesized that disruption of the DGC leads to structural weakness of the sarcolemma and increases its permeability. As a result, dystrophic muscle cell membranes become permeable to molecules which do not usually penetrate healthy cells, such as Evans blue dye (EBD; Matsuda *et al.* 1995; Straub *et al.*, 1997). Skeletal muscles of both *Sgca*- and *Sgcd*-null mice indeed displayed regions of EBD-positive fibres, which, upon histological analysis, appeared as clusters of necrotic or hypertrophic fibres (Duclos *et al.*, 1998; Coral-Vazquez *et al.*, 1999). The use of vital dyes unfortunately necessitates sacrificing the animals and harvesting of the muscle in order to obtain cryosections that can be analysed by fluorescence microscopy. This procedure becomes rather tedious if one wants to obtain an overall view of a muscle group because serial sections of the entire biopsy are then needed. We therefore turned our attention towards a less traumatic and invasive methodology that would enable us to visualize muscle damage *in vivo*, in living animals. To address this question, we investigated contrast-agent-enhanced MRI to detect fibre lesions in skeletal muscle. This common imaging procedure indeed proved useful in evaluating the extent and localization of fibre damage in muscular dystrophy (Figure 10.2.2). We used an albumin-targeted contrast agent AngioMark (MS-325; Epix Medical, Cambridge, MA; Lauffer *et al.* 1996; Parmelee *et al.*, 1996) as well as the paramagnetic contrast agent Gadolinium (Gd; Gadodiamide, Omniscan; Nycomed, Oslo). All experiments were performed on a 3T-Bruker Medspec MR imaging system (Bruker, Billerica, MA). Contrary to previous studies of skeletal muscle that reported tissue T2 relaxation time changes in MD (McCully *et al.* 1992; Dunn and Zaim-Wadghiri, 1999), we chose to acquire T1-weighted spin-echo images.

First, we demonstrated that contrast agent enhanced MRI indeed allows detection of muscle

Figure 10.2.3 (a) Uptake of AngioMark (MS-325) is prevented by adenovirus-mediated gene transfer of α-SG. T1-weighted spin-echo images through the pelvic girdle and thighs from non-injected (a–c) and injected (d–f) mice. Neonate *Sgca*-null mutant mice received intramuscular injections of Ad5RSV-SGCA in their right quadriceps femoris (arrows) and hamstring (arrowheads) muscles. Images were obtained 8–10 weeks following the adenovirus injections, before (a, d) and 15 min (b, e) after the systemic delivery of AngioMark (MS-325). Panels c and f show the difference between post- and pre-contrast images. L, R: left and right hind limb, respectively. (b) Mean enhancement percent was detected in the left and right sides of non-injected (−Ad; $n = 17$ images taken) and Ad5RSV-SGCA injected Sgca-null mice (+Ad; $n = 21$ images taken). Statistical significance was examined using a paired, two-tailed t-test ($^{*}P = 0.0003$). (*Taken from Allamand et al., 2000 with permission from Nature Publishing Group*)

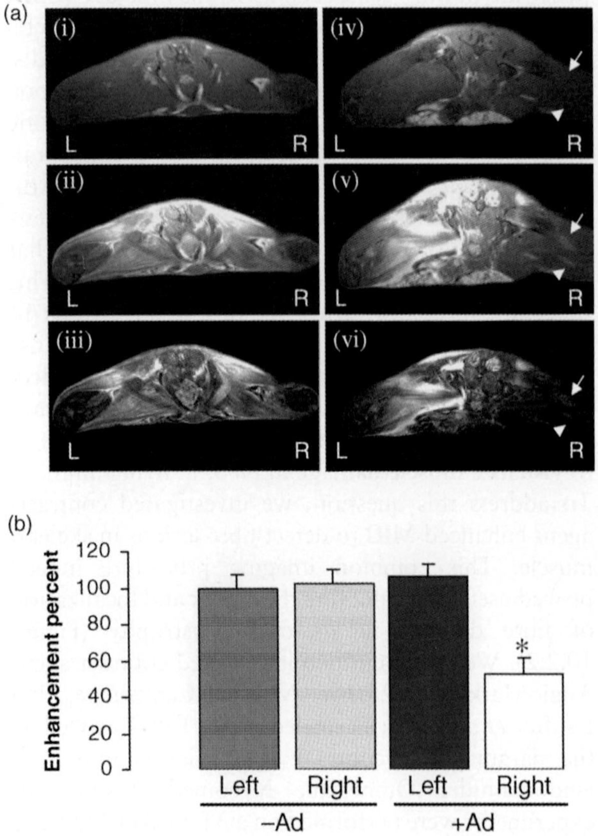

fibre lesions, both in the *mdx* mouse, an animal model for DMD, and in *Sgca*-null mutant mice (Figure 10.2.2(a)). Intravenously administered MS-325 consistently resulted in a significant signal enhancement in *mdx* and *Sgca*-null mutant mice, as opposed to control mice. The distribution of MS-325 uptake varied between individual animals and was not necessarily symmetrical but appeared particularly localized in the gluteal and thigh muscles. Because of MS-325's efficient protein-binding capabilities, the signal intensity remained elevated for at least 60 min after administration (Figure 10.2.2(b)).

We then applied the same MRI protocol to *Sgca*- and *Sgcd*-null mutant mice that received adenoviral vectors in their *quadriceps femoris* and hamstrings or *tibialis anterior* muscles, respectively. T1-weighted spin-echo images were taken before and after systemic injection of MS-325 or Gadolinium in order to visualize the sarcolemmal membrane permeability 8–10 weeks after adenovirus-mediated gene transfer. In both cases, absence of the sarcoglycan complex in non-injected muscles resulted in altered membrane permeability characterized by a significant signal enhancement, detectable even *prior* to contrast agent administration (Figures 10.2.3(ii, iii) and 10.2.4). However, early adenoviral-mediated gene transfer of α- or δ-SG efficiently prevented contrast agent uptake into the injected muscle fibres (Figures 10.2.3(v, vi) and 10.2.4). In the injected *Sgca*-null muscle, signal enhancement was decreased by about 40% (statistical significance $P = 0.0003$).

Figure 10.2.4 Adenovirus-mediated gene transfer of human δ-SG prevents uptake of gadolinium in *tibialis anterior* muscle of *Sgcd*-null mutant mice. T1-weighted spin-echo images of WT (top), non-injected (middle) and right-leg-injected Sgcd-null (bottom) mice were acquired 10 weeks after Ad5CMV-SGCD injections. Cross-section images through the left (L) and right (R) tibialis anterior muscles were taken 20 min after systemic injection of gadolinium. T: tail. (*Taken from Durbeej et al., 2003 with permission from National Academy of Sciences, USA*)

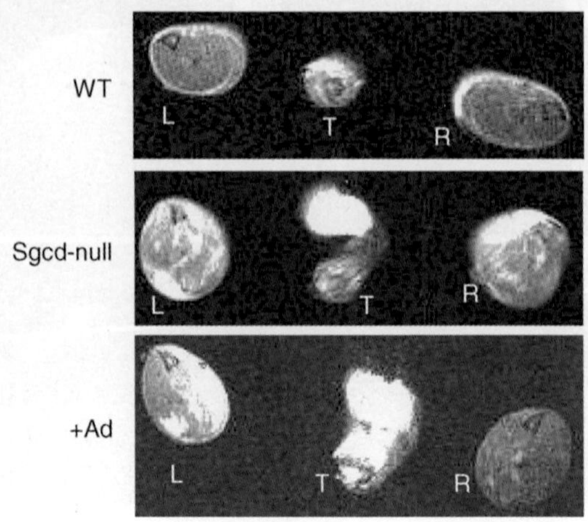

In conclusion, contrast agent enhanced MRI technology clearly proved useful for visualizing skeletal muscle fibre lesion and for evaluating functional consequences of therapeutic strategies, as recently reviewed by Leroy-Willig *et al.* (2003). In addition to being non-invasive, MRI provides a better and more comprehensive view of the muscle of interest than biopsies. As gene therapy protocols (or other therapeutic strategies) move towards human subjects, non-invasive evaluation tools for the longitudinal follow-up of protocols will be greatly needed. Their validation in animal models therefore is important and MRI has proven a valuable asset.

Acknowledgements

The work presented here was conducted in Dr Kevin P. Campbell's laboratory (Iowa City, Iowa) and at the Medical College of Wisconsin (Milwaukee, Wisconsin), in collaboration with Dr Kathleen M. Schmainda. Many thanks to Dr Madeleine Durbeej-Hjalt and Dr Volker Straub.

References

Allamand, V., Donahue, K. M., Straub, V., Davisson, R. L., Davidson, B. L., Campbell, K. P., 2000. "Early adenovirus-mediated gene transfer effectively prevents muscular dystrophy in alpha-sarcoglycan-deficient mice." *Gene Therap.* **7**, 1385–1391.

Cohn, R. D., Campbell, K. P., 2000. "Molecular basis of muscular dystrophies." *Muscle Nerve* **23**, 1456–1471.

Coral-Vazquez, R., Cohn, R., Moore, S. A., Hill, J. A., Weiss, R. M., Davisson, R. L., Straub, V., Barresi, R., Bansal, D., Hrstka, R. F., Williamson, R., Campbell, K. P., 1999. "Disruption of the sarcoglycan-sarcospan complex in vascular smooth muscle: a novel mechanism for cardiomyopathy and muscular dystrophy." *Cell* **98**, 465–474.

Duclos, F., Straub, V., Moore, S. A., Venzke, D. P., Hrstka, R. F., Crosbie, R. H., Durbeej, M., Lebakken, C. S., Ettinger, A. J., van der Meulen, J., Holt, K. H., Lim, L.E., Sanes, J. R., Davidson, B. L., Faulkner, J. A., Williamson, R., Campbell, K. P., 1998. "Progressive muscular dystrophy in alpha-sarcoglycan-deficient mice." *J. Cell Biol.* **142**, 1461–1471.

Dunn, J. F., Zaim-Wadghiri, Y., 1999. "Quantitative magnetic resonance imaging of the mdx mouse model of Duchenne muscular dystrophy." *Muscle Nerve* **22**, 1367–1371.

Durbeej, M., Cohn, R. D., Hrstka, R. F., Moore, S. A., Allamand, V., Davidson, B. L., Williamson, R. A., Campbell, K. P., 2000. "Disruption of the b-sarcoglycan gene reveals pathogenetic complexity of limb-girdle muscular dystrophy type 2E." *Mol. Cell.* **5**, 141–151.

Durbeej, M., Sawatzki, S. M., Barresi, R., Schmainda, K. M., Allamand, V., Michele, D. E., Campbell, K. P., 2003. "Gene transfer establishes primacy of striated vs. smooth muscle sarcoglycan complex in limb-girdle muscular dystrophy." *PNAS* **100**, 8910–8915.

Ervasti, J. M., Campbell, K. P., 1993. "A role for the dystrophin–glycoprotein complex as a transmembrane linker between laminin and actin." *J. Cell Biol.* **122**, 809–823.

Ervasti, J. M., Campbell, K. P., 1991. "Membrane organization of the dystrophin–glycoprotein complex." *Cell* **66**, 1121–1131.

Lauffer, R. B., Parmelee, D. J., Ouellet, H. S., Dolan, R. P., Sajiki, H., Scott, D. M., Bernard, P. J., Buchanan, E. M., Ong, K. Y., Tyeklar, Z., Midelfort, K. S., McMurry, T. J., Walovitch, R. C., 1996. "MS-325: a small-molecule vascular imaging agent for magnetic resonance imaging." *Acad. Radiol.* **3**, S356–S358.

Leroy-Willig, A., Fromes, Y., Paturneau-Jouas, M., Carlier, P., 2003. "Assessing gene and cell therapies applied in striated skeletal and cardiac muscle: Is there a role for nuclear magnetic resonance?" *Neuromusc. Disord.* **13**, 397–407.

Matsuda, R., Nishikawa, A., Tanaka, H., 1995. "Visualization of dystrophic muscle fibers in mdx mouse by vital staining with Evans blue: evidence of apoptosis in dystrophin-deficient muscle." *J. Biochem.* (Tokyo) **118**, 959–964.

McCully, K., Shellock, F. G., Bank, W. J., Posner, J. D., 1992. "The use of nuclear magnetic resonance to evaluate muscle injury." *Med. Sci. Sports Exerc.* **24**, 537–542.

Parmelee, D. J., Walovitch, R. C., Ouellet, H. S., Lauffer, R. B., 1996. "Pre-clinical evaluation of the pharmacokinetics, biodistribution and elimination of MS-325, a blood pool agent for magnetic resonance imaging." *Invest. Radiol.* **32**, 741–747.

Straub, V., Rafael, J. A., Chamberlain, J. S., Campbell, K. P., 1997. "Animal models for muscular dystrophy show different patterns of sarcolemmal disruption." *J. cell Biol.* **139**, 375–385.

Straub, V., Donahue, K. M., Allamand, V., Davisson, R. L., Kim, Y. R., Campbell, K. P., 2000. "Contrast agent-enhanced magnetic resonance imaging of skeletal muscle damage in animal models of muscular dystrophy." *Magn. Reson. Med.* **44**, 655–659.

10.3 Canine muscle oxygen saturation: evaluation and treatment of M-type phosphofructokinase deficiency

Kevin McCully and Urs Giger

Magnetic resonance spectroscopy (MRS) of phosphorus containing compounds has been a powerful method of evaluating impairments in oxidative muscle metabolism. Near-infrared spectroscopy (NIRS) to measure oxygen saturation in skeletal muscles can provide additional evidence of whether the impairment is related to oxygen delivery or mitochondrial dysfunction (Bank and Chance, 1994; McCully *et al.*, 1994). NIRS, as explained in Chapter 5, is a non-invasive, potentially portable technique based on the differential absorption of near-infrared light by the oxygenated and deoxygenated forms of heme groups (haemoglobin, myoglobin). Few studies have applied NIRS to animal studies, although one of the first validation studies of NIRS compared NIRS measurements of oxygen saturation with directly measured venous oxygen saturation values in mice (Wilson *et al.*, 1989). We describe here the application of NIRS to evaluate abnormalities in muscle oxygen saturation in dogs with hereditary muscle-type phosphofructokinase (M-PFK) deficiency. PFK-deficient muscle is unable to produce lactic acid but builds up sugar phosphates during exercise (Argov *et al.*, 1987, Giger *et al.*, 1988).

Dogs with M-PFK deficiency represent a homologue of human M-PFK deficiency (Smith *et al.*, 1996). In the dog the disease is caused by a point mutation resulting in truncation and instability of the M-PFK sub-unit with a lack of M-PFK activity in all tissues. PFK-deficient dogs are exercise intolerant and can develop severe muscle cramping during exercise. In addition to altered muscle metabolism, canine PFK-deficient erythrocytes have reduced 2,3 diphosphoglycerate concentrations, which results in a 'left shift' of the haemoglobin dissociation curve, thereby impairing oxygen release to tissue (Giger *et al.*, 1988). The experiments described here include previously published results (McCully *et al.*, 1999).

10.3.1 Experimental animals

Five English Springer spaniels with M-PFK deficiency (1% of normal M-PFK activity) were compared to five healthy control dogs. Anaesthetized dogs had one hind limb secured so that the muscles could contract isometrically (Giger *et al.*, 1988). One of the dogs received a bone marrow transplant from a compatible normal littermate at a few weeks of age, which completely normalized the red cell defect but not the enzymatic muscle deficiency.

10.3.2 Near-infrared spectroscopy

Two different models of NIRS equipment were used. Experiments with simultaneous ^{31}P magnetic resonance spectroscopy (MRS) were performed with a home-built dual-wavelength air-turbine spectrometer utilizing a Hamamatsu R928 photomultiplier tube, a 40 W halogen lamp and a rotation wheel with filters at 800 and 760 nm. A quartz fibre optic cable (9 mm^2 diameter and 5 m length) was used to transmit the detected light out of the magnet to the photomultiplier tube. The cables were attached with acrylic glue approximately 2 cm apart over the extensor digitorum longus and cranial tibialis muscles (Figure 10.3.1(a)). Experiments performed without MRS measurements were made using a portable NIRS unit (Runman^{TM}, NIM Inc.). The optical probe contained two light sources 6 cm apart with two light detectors placed between them with an effective separation of source and detector of 3 cm (Figure 10.3.1(b)).

Changes in the signal difference (760–800 nm or 760–850 nm) provided an index of muscle oxygen saturation. The results were standardized to a range of 0–100%, with 0% equalling the maximal deoxygenated signal determined by arterial occlusion produced by cuff ischemia and 100% representing the maximal oxygenated signal.

10.3.3 Phosphorus magnetic resonance spectroscopy

Dogs were placed in a 2.7-T magnet such that a 2.5-cm double-tuned surface coil was mounted in the same location as the NIRS fibre optic cables (Figure 10.3.1(a)). Both methods sampled skeletal muscle from the same approximate area. Muscle spectra were registered continuously before during and after stimulation to confirm metabolic activation and the presence of the muscle disorder by observing the accumulation of sugar phosphates (Figure 10.3.2).

Figure 10.3.1 Diagram of experimental set-up for (a) simultaneous NIRS and MRS measurements on the dog hind limb. Placement of the fibre optic cables and the MRS coil allows collection of signals from the same area of the Tibialis anterior muscle. (b) Set-up for portable NIRS measurements of oxygen saturation in the muscle. Note that stimulation electrodes allowed activation of the peroneal nerve. Based on previous studies, light at these wavelengths should penetrate 1–1.5 cm deep into the tissue. The actual optical path length is 4–6 times longer than this value due to light scattering and photon migration

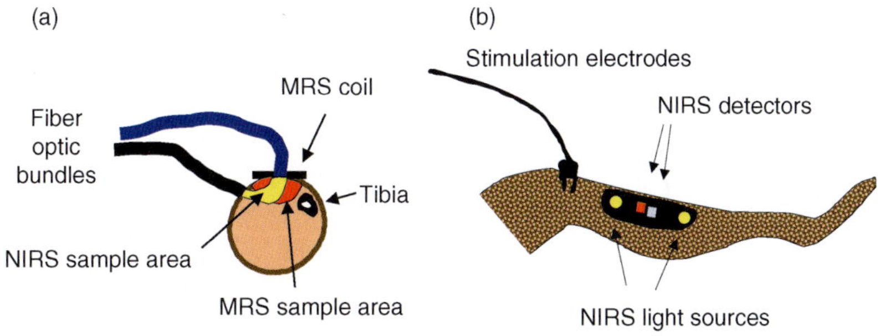

10.3.4 Stimulation protocol

Two non-magnetic phosphobronze needle electrodes were placed percutaneously near the peroneal nerve and supramaximal voltage of 1 ms duration square wave pulses were used. The muscle was stimulated at 1, 2, 4 and 8 Hz for 6 min at each level.

10.3.5 Results

Values at 8 Hz on Figure 10.3.3 are markedly different from those in the text; Figure 10.3.4 shows five dogs whereas 10 dogs are mentioned.

Normal dogs showed progressive muscle deoxygenation with increasing stimulation rate, reaching oxygen saturation values of 48.4 ± 21.9% O_2 during 8-Hz stimulation in experiment one and 14% in experiment two (Figures 10.3.3 and 10.3.4). In contrast, the PFK-deficient dogs exhibited either a progressive oxygenation of the stimulated muscle in one experiment, reaching oxygen saturation values of 81.0+12.6% O_2 during 8-Hz stimulation or less desaturation than normal dogs in the other experiment, (Figures 10.3.3 and 10.3.4). All dogs showed deoxygenation during cuff-induced ischemia.

^{31}P MRS measurements were successfully performed simultaneously with NIRS measurements. At

Figure 10.3.2 Representative spectra from a PFK-deficient and a normal dog. Spectra were collected and averaged over 4 min. End exercise consisted of 8-Hz electrical stimulation under isometric conditions. Note that exercise results in a large PME (phosphomonoesters) peak in the PFK-deficient dogs due to their inability to phosphorylate fructose-6-P. Thus, glucose-6-P and fructose-6-P accumulate inside muscle cells

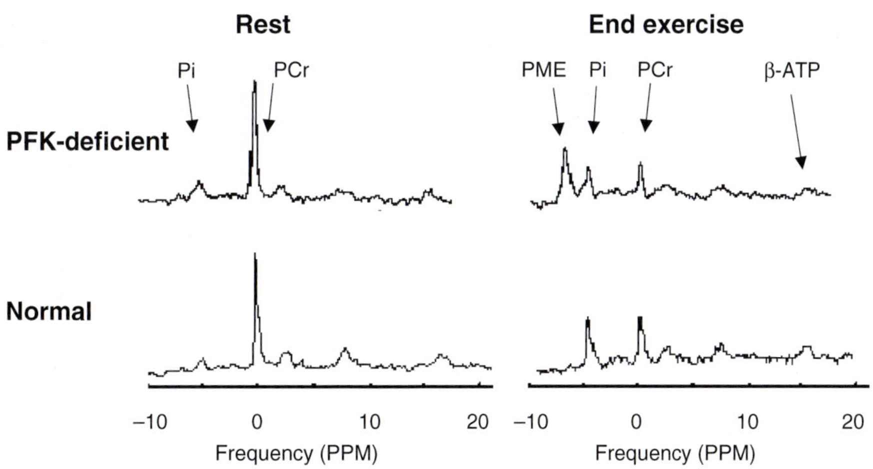

Figure 10.3.3 Examples of oxygen saturation measured with NIRS during electrical stimulation. PFK-treated dog had normal blood levels of 2,3 DPG. This resulted in the treated dog having a normal pattern of progressive desaturation of haemoglobin with electrical stimulation. The lack of continued desaturation in the PFK-treated dog may have been due to muscle fatigue resulting from impaired oxidative metabolism

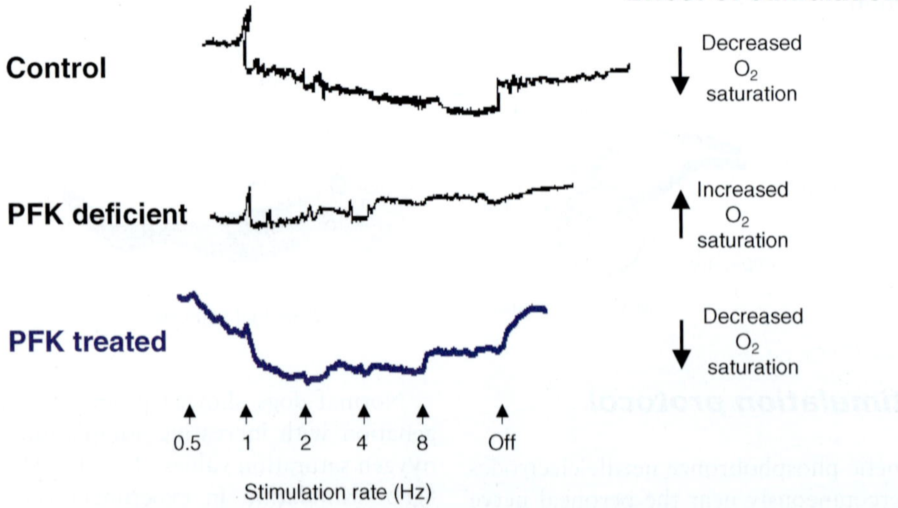

the end of exercise, the ^{31}P MRS muscle spectra of PFK-deficient dogs showed predicted increases in G-6-P (11.1 ± 3.5 mM, undetectable in normal), with no changes in muscle pH (7.19 ± 0.06 for PFK and 7.13 ± 0.11 for normal). End exercise ADP levels calculated from the end exercise PI, PCr and ATP peaks areas along with pH, were higher in PFK-deficient dogs (94 ± 28 vs. 44 ± 9 μM), indicating impaired oxidative metabolism.

Figure 10.3.4 Oxygen saturation measured with NIRS during electrical stimulation. Control dogs showed a progressive decrease in oxygen saturation with electrical stimulation. PFK-deficient dogs show little or no change in oxygen saturation with electrical stimulation. The bone marrow-treated PFK-deficient dog showed an intermediate response, with an initial desaturation followed by no change. Treated dogs have normal erythrocytic 2,3 DPG levels which may improve normal oxygen extraction but still had enhanced muscle fatigue which limited the amount of oxygen extraction during stimulation

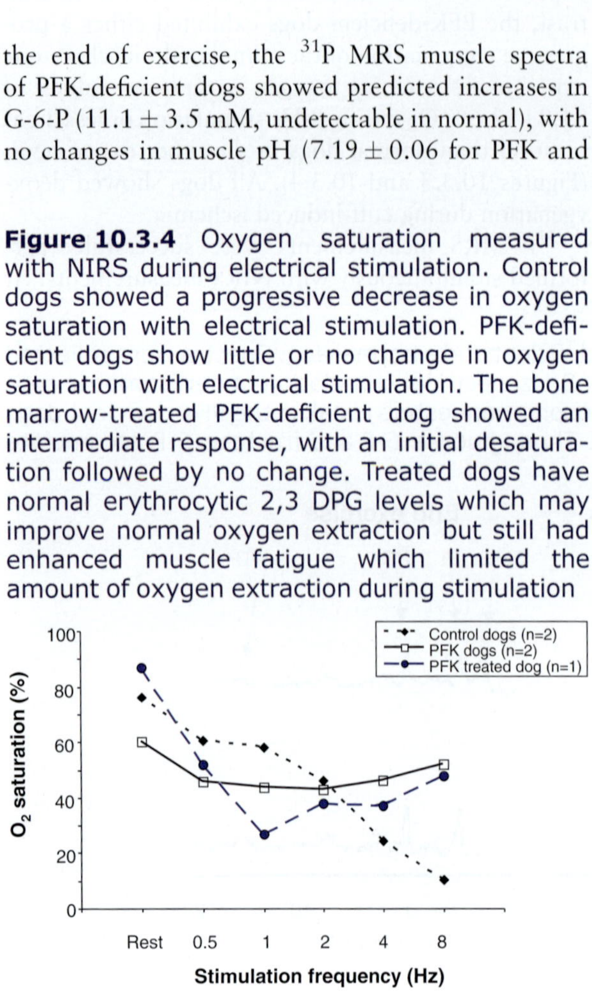

The PFK-deficient dog that received a normal bone marrow transplant demonstrated a haemoglobin–oxygen dissociation curve that was more similar to control dogs than untreated PFK-deficient dogs (Figure 10.3.4).

10.3.6 Discussion

The progressive oxygen desaturation with increasing activity seen in the normal dogs was consistent with other NIRS measurements in normal human muscle during exercise (Bank and Chance, 1994; McCully *et al.*, 1994; Wilson *et al.*, 1989). This demonstrates that NIRS measurements can be performed on active skeletal muscle in anaesthetized dogs. Furthermore, NIRS measurements can be performed inside an NMR magnet, as long as fibre optic cables are used to transfer light in and out of the magnet. PFK-deficient dogs did not show the progressive deoxygenation of the haemoglobin/myoglobin signal that occurred in normal dogs during electrical stimulation. The lack of deoxygenation in the PFK-deficient muscle of affected dogs was consistent with a study on human patients with PFK deficiency (Bank and Chance, 1994).

The altered oxygen saturation may be a consequence of impaired oxidative metabolism or the result of impaired oxygen transport capacity of red blood cells. The return to a normal pattern of oxygen saturation with stimulation in one PFK-deficient dog treated with a normal bone marrow transplant suggests that the abnormality is due to defective red blood cells. Although further studies are needed to test these hypotheses, NIRS measurements appear to be a useful non-invasive method of monitoring tissue oxygen saturation in active muscle under both normal and diseased conditions in larger animal models.

Acknowledgement

This work was supported in part by the Muscular Dystrophy Association and NIH (RR02512 and DK37602).

References

Argov, Z., Bank, W., Maris, J., Chance, B., 1987. "Muscle energy metabolism in human phosphofructokinase deficiency as recorded by ^{31}P nuclear magnetic resonance spectroscopy." *Annal. Neurol.* **16**, 529–538.

Bank, W., Chance, B., 1994. "An oxidative defect in metabolic myopathies: diagnosis by non-invasive tissue oximetry." *Annal. Neurol.* **36**, 830–837.

Giger, U., Argov, Z., Schnall, M., Bank, W., Chance, B., 1988. "Metabolic myopathy in canine muscle-type phosphofructokinase deficiency." *Muscle Nerve* **11**, *1260–1265.*

McCully, K., Giger, U., Chance, B., 1999. "*In vivo* determination of altered haemoglobin saturation in dogs with M-type phophofructokinase deficiency." *Muscle Nerve* **22**, 621–627.

McCully, K., Iotti, S., Kendrick, K., Wang, Z., Posner, J., Leigh, J., Jr., Chance, B., 1994. "Simultaneous in vivo measurements of HbO_2 saturation and PCr kinetics after exercise in normal humans." *J. Appl. Physiol.* 77, 5–10.

Smith, B., Stedman, H., Rajpurohit, Y., Henthorn, P., Wolfe, J., Patterson, D., Giger, U., 1996. "Molecular basis of canine muscle type phosphofructokinase deficiency." *J. Biol. Chem.* **271**, 20070–20074.

Wilson, J. R., Mancini, D. M., McCully, K. K., Ferrato, N., Lanoce, V., Chance, B., 1989. "Non-invasive detection of skeletal muscle underperfusion with near-infrared spectroscopy in patients with heart failure." *Circulation* **80**, 1668–1703.

10.4 *In vivo* assessment of myocardial perfusion by NMR technology

Jörg. U.G. Streif, Matthias Nahrendorf and Wolfgang R. Bauer

10.4.0 Introduction

Microcirculation has a significant influence on the function of organs and the pathophysiology of many diseases. In general, the term microcirculation refers to blood circulation comprising exchange of oxygen and nutrients between blood and the surrounding tissue. This step takes place via the walls of capillaries that densely pass through all organs. Impaired microcirculation results in diminished supply of oxygen and nutrients to tissue.

Because capillaries possess a small diameter on the order of a few microns, the direct visualization of capillaries by magnetic resonance angiography is not feasible. Therefore, indirect parameters are to be used when assessing the blood flow in capillaries with magnetic resonance imaging (MRI).

Perfusion (P) refers to the blood volume supplied to a particular amount of tissue per unit time and characterises the supply of oxygen and nutrients to tissue. In addition, the regional blood volume (RBV) refers to the blood volume in a particular amount of tissue and characterises the blood volume of the capillaries.

With respect to cardiovascular research, cine MRI provides non-invasive and accurate characterization of cardiac morphology and function. However, the viability and capability of the myocardium are strongly determined by its microcirculation, and heart diseases comprising ventricular dysfunction and hypertrophy show significant alterations of the myocardial perfusion which might occur prior to any detectable morphologic or functional changes. Therefore, the assessment of perfusion represents a key element in the characterization of the heart.

In general, the assessment of perfusion with MRI is based on tracer kinetics (LeBihan, 1996), wherein the tracer is either (i) a diffusible tracer comprising non-proton nuclei such as ^{19}F compounds which is administered by injection or inhalation, (ii) a bolus of a non-diffusible, intravascular tracer such as gadolinium diethylenetriaminepentaacetic acid (Gd-DTPA) which is administered intravenously or (iii) the blood itself wherein spins within a particular volume of blood are labelled using radio frequency (RF) excitations.

Because this short report is devoted to the non-invasive assessment of perfusion with MRI, solely the non-invasive approach according to item (iii) is discussed. With respect to the invasive approaches according to items (i) and (ii), the reader is referred to the review by LeBihan (LeBihan, 1996) and the references therein.

10.4.1 Methods

Several variations of the basic principle of spin labelling of blood using RF excitations have been presented so far (Williams *et al.*, 1992; Kwong *et al.*, 1994; Edelman *et al.*, 1994; Kim, 1995; Schwarzbauer *et al.*, 1996; Belle *et al.*, 1998). In all of these techniques, two images with varying degrees of labelling are subtracted from one another to extract information on perfusion. Labelling can be achieved by imaging with and without spin inversion (Williams *et al.*, 1992) or by a combination of slice-selective and non-selective spin inversions (Kwong *et al.*, 1994; Kim 1995; Schwarzbauer *et al.*, 1996; Belle *et al.*, 1998) and labelling is either performed in the slice to be imaged (Kwong *et al.*, 1994; Schwarzbauer *et al.*, 1996; Belle *et al.*, 1998) or off said slice (Williams *et al.*, 1992; Edelman *et al.*, 1994). The assessment of perfusion using spin-labelling MRI has been performed in various vertebrates, including humans (Williams *et al.*, 1992; Kwong *et al.*, 1994; Edelman *et al.*, 1994; Kim, 1995), rats (Belle *et al.*, 1998) and mice (Streif *et al.*, 2005; Streif *et al.*, 2003; Kober *et al.*, 2005).

With respect to the assessment of myocardial perfusion with spin-labelling MRI, Belle *et al.* (1998) showed that perfusion can be assessed by performing two inversion recovery T1 measurements with slice-selective and non-selective spin inversion, respectively (Figure 10.4.1). The intrinsic $T1$ relaxation of the detection slice is observed with a non-selective spin inversion. When performing slice-selective inversion, only spins within the detection slice are affected by the inversion pulse whereas spins outside the detection slice rest in thermal equilibrium. Therefore, spins of the latter type entering the detection slice due to perfusion cause an apparent acceleration of the $T1$ relaxation. As a consequence, the difference between $T1$ values measured with slice-selective and non-selective spin inversion, respectively, provides information on tissue perfusion within the detection slice.

To extract quantitative information from the measured $T1$ data, a tissue model such as the two-compartment model has to be applied (Bauer *et al.*, 1990). This tissue model describes tissue as consisting of two compartments, the intravascular capillary blood and the extravascular tissue. Moreover, two transport processes are taken into account, namely diffusion exchange between the two compartments and transport of spins from the arterial system to the capillary space due to perfusion.

Figure 10.4.1 Schematic representation of the spin-labelling experiment performed for assessment of myocardial perfusion in mice. Because the capillaries in the myocardium are aligned from base to apex of the heart, the measurement was performed in a mid-ventricular short axis view. The red bar indicates the detection slice, and the semi-transparent blue area represents the area of spin inversion ((a) slice selective spin inversion, (b) non-selective spin inversion)

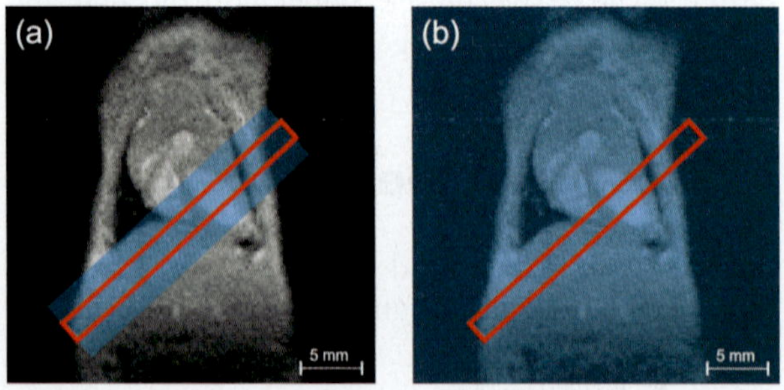

Belle *et al.* (1998) showed that perfusion can be quantified by

$$P = \frac{\lambda}{T1_{\mathrm{blood}}}\left(\frac{T1_{\mathrm{glob}}}{T1_{\mathrm{sel}}} - 1\right)$$

wherein λ denotes the blood–tissue partition coefficient; $T1_{\mathrm{blood}}$ is the $T1$ of capillary blood, and $T1_{\mathrm{glob}}$ and $T1_{\mathrm{sel}}$ denote the $T1$ values measured with non-selective and slice-selective spin inversion, respectively. This technique for the quantitative assessment of perfusion has been validated against measurements made with microspheres and first-pass perfusion imaging (Waller *et al.*, 2005 and the references cited therein).

10.4.2 Results and discussion

Figure 10.4.2 shows representative data from a study directed to the assessment of myocardial perfusion in mice *in vivo* (Streif *et al.*, 2005) using the technique described above. Perfusion was assessed in a single slice in a midventricular short axis view. The data acquisition was triggered by the ECG of the animals, and the animals were anaesthetized with isoflurane during the measurements.

In the zoomed quantitative $T1$ map after non-selective spin inversion depicted in Figure 10.4.2.(a), the myocardium and the blood inside the left ventricle (LV)

Figure 10.4.2 Representative data of a healthy mouse anaesthetized with isoflurane (a–c). The quantitative *T*1 maps after non-selective spin inversion (a) and after slice-selective spin inversion (b) are the basis for the calculation of the perfusion map (c). *T*1 of capillary blood is determined individually in each animal inside the LV. Part (d) depicts a perfusion map of a mouse suffering from a myocardial infarction in the antero-lateral area (arrow). *Abbreviations:* left ventricle (LV), right ventricle (RV)

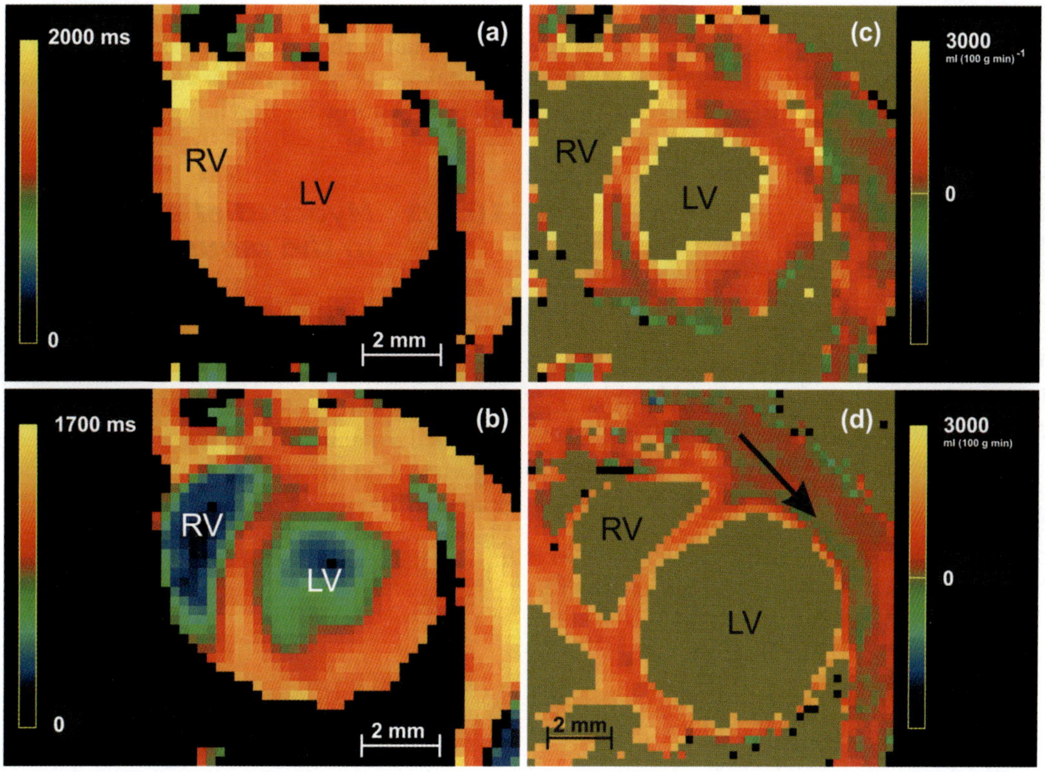

appear as substantially homogeneous areas. However, in the quantitative *T*1 map after slice-selective spin inversion shown in Figure 10.4.2(b), the apparent acceleration of the *T*1 relaxation due to perfusion results in reduced *T*1 values of the myocardium and – to a higher extent – of the blood inside the LV. Figure 10.4.2(c) depicts the zoomed quantitative perfusion map calculated pixel-by-pixel by use of the equation given above. The corresponding quantitative perfusion map of a mouse suffering from a myocardial infarction in the antero-lateral left ventricular wall is shown in Figure 10.4.2.(d). The reduced perfusion in the infarcted area is clearly visible (marked by an arrow). Moreover, it is to be noted that the results are in good agreement with results from invasive studies involving microspheres (Streif *et al.*, 2005).

10.4.3 Conclusion

MR imaging provides non-invasive quantification of myocardial perfusion *in vivo* in various vertebrates including the important mouse animal model. The technique is sensitive enough to detect and visualize regional alterations of perfusion after myocardial infarction. With respect to the development of transgenic mouse models that show altered heart morphology, function or metabolism, the presented technique may be an interesting tool for the non-invasive quantification of the myocardial microcirculation. Moreover, the presented technique can be adapted to study the perfusion of other organs under the prerequisite that the same tissue model can be applied.

References

Bauer, W. R., Hiller, K. H., Roder, F., Rommel, E., Ertl, G., Haase, A., 1990. *Magn. Reson. Med.* **35**, 43–55.

Belle, V., Kahler, E., Waller, C., Rommel, E., Voll, S., Hiller, K.H., Bauer, W., Haase A., 1998. *J. Magn. Reson. Imag.* **8**, 1240–1245.

Edelman, R. R., Siewert, B., Darby, D. G., *et al.*, 1994. *Radiology* **192**, 513–520.

Kim, S. G., 1995. *Magn. Reson. Med.* **34**, 293–301.

Kober, F., Iltis, I., Cozzone, P. J., Bernard, M., 2005. *Magn. Reson. Med.* 53, 601–606.
Kwong, K. K., Chesler, D. A., Weisskoff, R. M., Rosen, B. R., 1994. In: Proceedings of the Second Annual Meeting of the SMRM, San Francisco, USA, p. 1005.
LeBihan, D., 1996. In: Grant, D. M., Harris, R. K. (Eds.). *Encyclopedia of Nuclear Magnetic Resonance*, John Wiley & Sons, pp. 1645–1656.
Schwarzbauer, C., Morissey, S., Haase, A., 1996. *Magn. Reson. Med.* 35, 540–546.
Streif, J. U. G., Nahrendorf, M., Hiller, K. H., Waller, C., Wiesmann, F., Rommel, E., Haase, A., Bauer, W. R., 2005. *Magn. Reson. Med.* 53, 584–592.
Streif, J. U. G., Hiller, K. H., Waller, C., Nahrendorf, M., Wiesmann, F., Bauer, W. R., Rommel E., Haase, A., 2003. *J. Magn. Reson. Imag.* 17, 147–152.
Waller, C., Engelhorn, T., Hiller, K. H., Heusch, G., Ertl, G., Bauer, W. R., Schulz, R., 2005. *Am. J. Physiol. Heart Circ. Physiol.* 288, H2588–H2593.
Williams, D. S., Detre, J. A., Leigh, I. S., Koretsky, A. P., 1992. *Proc. Natl. Acad. Sci. USA* 89, 212–216.

10.5 Ultrasound microimaging of strain in the mouse heart

F. Stuart Foster

Recent developments in high-frequency ultrasound imaging instrumentation have created new opportunities for the evaluation of cardiovascular function in small animal disease models. The imaging technology is based on the use of ultrasound in the frequency range from 20 to 60 MHz and provides spatial resolution ranging from 40 to 100 μm with temporal resolution of less than 10 ms for real-time imaging and less than 1 ms for retrospectively reconstructed image sequences (Foster *et al.*, 2002; Zhou *et al.*, 2004). This combination provides a powerful means of examining the complex deformations of the myocardium over the comparatively short heart cycle of the mouse (~150 ms) as well as a simple means of monitoring the build-up of atherosclerotic plaque in mouse vessels over an extended period of time. In this short report, we describe how this imaging tool can address an important aspect of cardiovascular disease: assessment of regional myocardial function. One measure of regional myocardial function is the strain and rate of change of strain (strain rate) in the myocardial tissue over the cardiac cycle. These parameters have been widely investigated in human echo imaging (D'Hooge *et al.*, 2002; Sutherland *et al.*, 2004) and with magnetic resonance imaging approaches (Kim *et al.*, 2004). A useful question is thus: What is the range of strain rate observed in the normal mouse myocardium and how is this altered by changes in coronary flow?

10.5.1 Myocardial strain in the mouse

The computation of strain associated with the rhythmic expansion and contraction of the myocardium requires accurate estimations of tissue velocity with high-temporal resolution. Velocity estimates can be obtained from tissue Doppler data or from correlation algorithms in which image information is tracked frame to frame. Strain rate is defined as the rate of change of strain across the myocardium and is given by the simple equation

$$\dot{\varepsilon} \cong \frac{v_2 - v_1}{L} \tag{1}$$

Where $\dot{\varepsilon}$ is the instantaneous strain rate at time t. v_1 and v_2 are the perpendicular velocities of the endo and epicardial surfaces respectively joined by line with length L. Strain is then calculated as the integral of strain rate over time. At the University of Virginia, image sequences of long axis views of the left ventricle were obtained at 1000 Hz with a VisualSonics (Toronto, Canada) VEVO 770 UBM scanner at approximately 100 μm resolution and analysed using the correlation approach to obtain displacement and velocity maps as a function of time (Li *et al.*, in press). Vector representations of displacement and velocity are shown in Figure 10.5.1 at end systole and end diastole. A plot of strain over the cardiac cycle is plotted in Figure 10.5.2. Peak strains of 0.35 are observed, and these are in excellent agreement with previous observations in human subjects measured with an MR methodology and normalized to the mouse heart rate (Kim *et al.*, 2004). The use of tissue Doppler as a means to obtain velocity estimates for the calculation of strain and strain rate has seen wide acceptance in human clinical imaging (Sutherland *et al.*, 2004). A well-documented software package called "SPEQLE" (http://med.kuleuven.be/cardim/) uses this approach for cardiac deformation quantification in humans and in mice (D'Hooge *et al.*, 2002). An example of this

Figure 10.5.1 Vector and colour scale representations of myocardial velocity in the mouse heart during systole (left) and diastole (right). Results were computed using the University of Virginia software package 'WallQuant' from an image sequence obtained using EKV reconstruction on the Visualsonics scanner. Regional wall velocities on the order of 30 mm/s are observed with highers velocities observed in diastolic expansion of the ventricle. Images courtesy of Li *et al.* (in press)

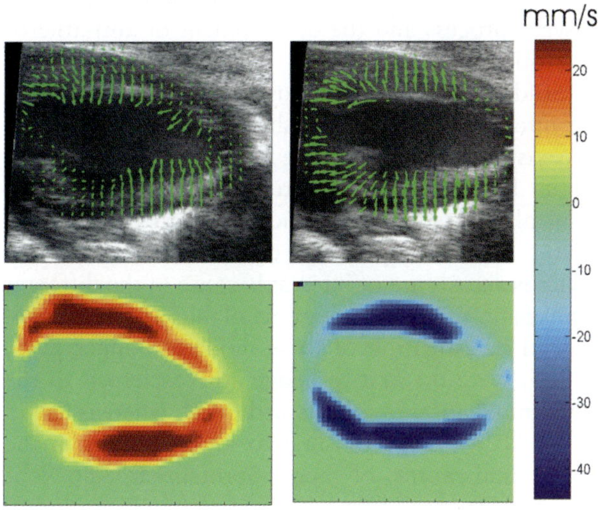

approach is given in Figure 10.5.3. Still frames from 1000 Hz reconstructions showing (a) greyscale, (b) tissue Doppler and (c) strain rate provide a snapshot of the mouse heart in mid systole. Strain in this example is measured at the location of the arrow. Quantitative analysis over the cardiac cycle is given in Figure 10.5.3(d–f). Peak strain of 0.50 is observed in this region of the posterior heart wall.

Non-invasive strain assessment is a useful means of exploring regional myocardial function in surgical

Figure 10.5.2 Computed strainrate analysis of the anterior heart wall. Scaled to the cardiac cycle, these results show a remarkable agreement with similar measurements performed with MRI in humans (Kim *et al.*, 2004)

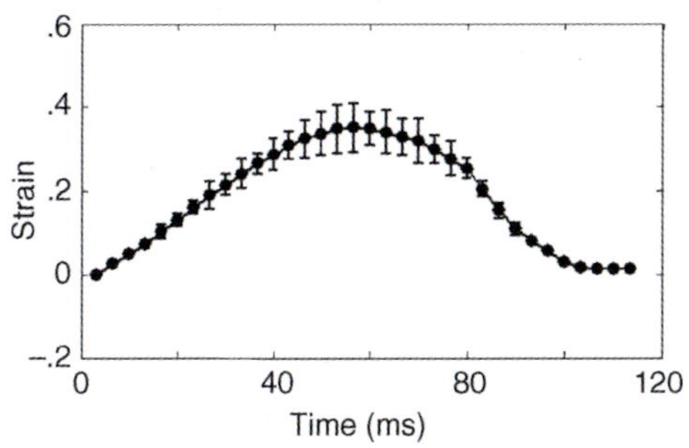

Figure 10.5.3 Retrospectively gated B-mode image of the mouse myocardiumacquired at 30 MHz. (b) Colour-flow velocity image of the myocardium at mid systole. (c) Strain-rate image of mouse myocardium at mid systole. An observation location is defined by the arrow. (d) Velocity in the myocardium at the observation location over one cardiac cycle. (e) Strain-rate at the observation location over one cardiac cycle. (f) Strain at the observation location over one cardiac cycle. Figure courtesy of R Williams, University of Toronto

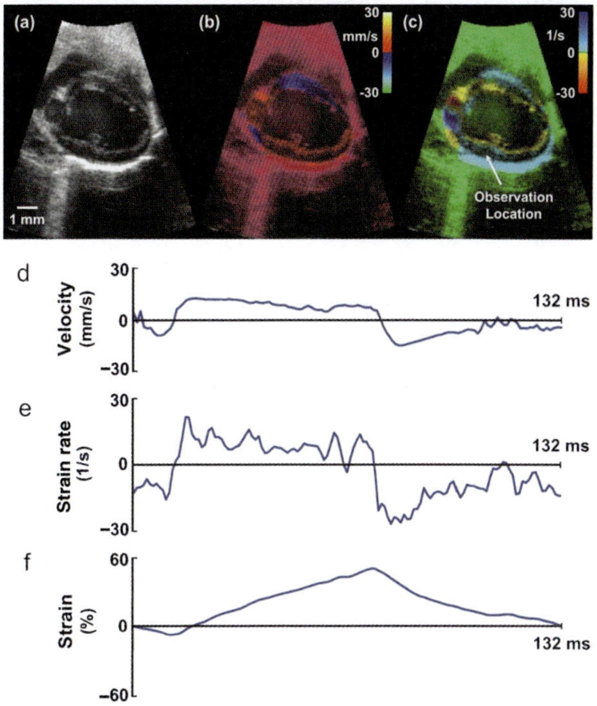

and genetic models of cardiovascular disease. Work on these is underway at a number of centres.

References

D'Hooge, J., Bijnens, B., Thoen, J., Van de Werf, F., Sutherland, G. R., Suetens, P., 2002. "Echocardiographic strain and strain-rate imaging: a new tool to study regional myocardial function." *IEEE Trans. Med. Imaging* **21**(9), 1022–1030.

Foster, F. S., Zhang, M. Y., Zhou, Y. Q., Liu, G., Mehi, J., Cherin, E., Harasiewicz, K. A., Starkoski, B. G., Zan, L., Knapik, D. A., Adamson, S. L., 2002. "A new ultrasound instrument for in vivo microimaging of mice." *Ultrasound Med. Biol.* **28**(9), 1165–1172.

Kim, D., Gilson, W. D., Kramer, C. M., Epstein, F. H., 2004. "Myocardial tissue tracking with two-dimensional cine displacement-encoded MR imaging: development and initial evaluation." *Radiology* **230**(3), 862–871.

Li, Y., Garson, C. D., French, B., Hossack, J., 2005. "High resolution quantification of myocardial strain in mice using speckle tracking." *IEEE Int. Ultrason. Symp.***1**, 369–372.

Sutherland, G. R., Di Salvo, G., Claus, P., D'Hooge, J., Bijnens, B., 2004. "Strain and strain rate imaging: a new clinical approach to quantifying regional myocardial function." *J. Am. Soc. Echocardiogr.* **17**(7), 788–802.

Zhou, Y. Q., Foster, F. S., Nieman, B. J., Davidson, L., Chen, X. J., Henkelman, R. M., 2004. "Comprehensive transthoracic cardiac imaging in mice using ultrasound biomicroscopy with anatomical confirmation by magnetic resonance imaging." *Physiol. Genom.* **18**(2), 232–244.

10.6 MR imaging of experimental atherosclerosis

Willem J.M. Mulder, Gustav J. Strijkers, Zahi A. Fayad and Klaas Nicolay

Atherosclerosis, the main cause of mortality and morbidity in Western societies, is a progressive disease, which starts with endothelial dysfunction and the accumulation of lipoproteins is the vessel wall of large arteries, followed by the formation of atherosclerotic plaques (Glass and Witztum, 2001). In Figure 10.6.1, the disease process is schematically represented and potential targets for molecular imaging are given (Choudhury *et al.*, 2004). Oxidized lipoproteins cause an inflammatory response, which is associated with the increased expression of endothelial cell surface receptors (e.g. E-selectin, VCAM and ICAM), causing the recruitment of monocytes from the circulation. The monocytes differentiate into macrophages and phagocytoze the oxidized lipoproteins. Eventually, the macrophages are converted into foam cells and an early atherosclerotic plaque is formed. The plaque is stabilized by the migration of smooth muscle cells, resulting in the formation of a fibrous cap. The thickness of the cap in relation to the lipid core and the composition of the plaque are believed to determine the risk of plaque rupture, which is the main cause of myocardial and cerebral infarction. Identification of such a high-risk plaque can be done non-invasively with magnetic resonance imaging and may lead to improved diagnosis and prognosis. Furthermore, MRI allows the study of the progression or regression of the disease in time, for example to follow the effect of lipid lowering therapies (Helft *et al.*, 2002).

MRI studies of animal models of atherosclerosis most often are aimed at the improved visualization of the different plaque components, usually correlated to histopathological analysis of the tissue. Furthermore, because MRI allows the non-invasive and serial investigation of the disease in the animal models, the technique contributes to a better understanding of the disease process and the development of anti-atherosclerotic therapies.

Studies of rabbit and mouse models are of particular interest, because established atherosclerosis models exist in both species. The rabbit has the advantage of its bigger size and therefore can be imaged at clinical field strengths, which facilitates translation of the methodology to the clinic. Mouse models of atherosclerosis are of interest because of the different transgenic models that are available and the widespread investigation of these models. Skinner *et al.* (1995) were the first to image atherosclerosis in the abdominal aorta of rabbits with MRI, whereas Fayad *et al.* (1998) were the first to image lesions in the abdominal aorta of apolipoprotein E knockout (apoE-KO) mice using high-field and high-resolution MRI. In particular, MR imaging of plaque formation in the aortic root of the mouse is a technological challenge due to respiratory and cardiac motion. Hockings *et al.* (2002) performed three-dimensional MRI of plaque formation in this area in low-density lipoprotein (LDL) receptor-knockout (LDLR-KO) mice by use of combined respiratory and cardiac triggering (Hockings *et al.*, 2002).

For improved visualization and characterization of plaques and vascular lesions, contrast-enhanced MRI with an appropriate contrast agent may be used. Ultra-small particles of iron oxide (USPIO) have been employed to identify macrophage-rich areas in the aorta of hyperlipidaemic rabbits (Ruehm *et al.*, 2001). Five days after the administration of the contrast agent dark spots of focal signal loss were observed in the vessel wall. These are caused by the massive uptake of USPIO by macrophages. Gadofluorine, a paramagnetic micellular contrast agent, enhances the aortic wall in hyperlipidaemic rabbits. Sirol *et al.* (2004) used this contrast agent in combination with an improved MRI protocol to recognize lipid-rich regions. The MRI findings correlated very well with histology. In Figure 10.6.2, the results of this study are depicted.

Target-specific MRI contrast agents are developed for the detection of specific molecular components in

Figure 10.6.1 Imaging targets in atherothrombosis. Illustration of processes of atherogenesis ranging from pre-lesional endothelial dysfunction (*left*) through monocyte recruitment to the development of advanced plaque complicated by thrombosis (*right*). The mechanisms are grossly simplified but focus on components (e.g. cell adhesion molecules, macrophages, connective tissue elements, lipid core and fibrin) and processes (e.g. apoptosis, proteolysis, angiogenesis and thrombosis) in plaques that have been imaged or that present useful potential imaging targets. ICAM, intercellular cell adhesion molecule; LDL, low-density lipoprotein; MMP, matrix metalloproteinase; NO, nitric oxide; VCAM, vascular cell adhesion molecule (Choudhury *et al.*, 2004)

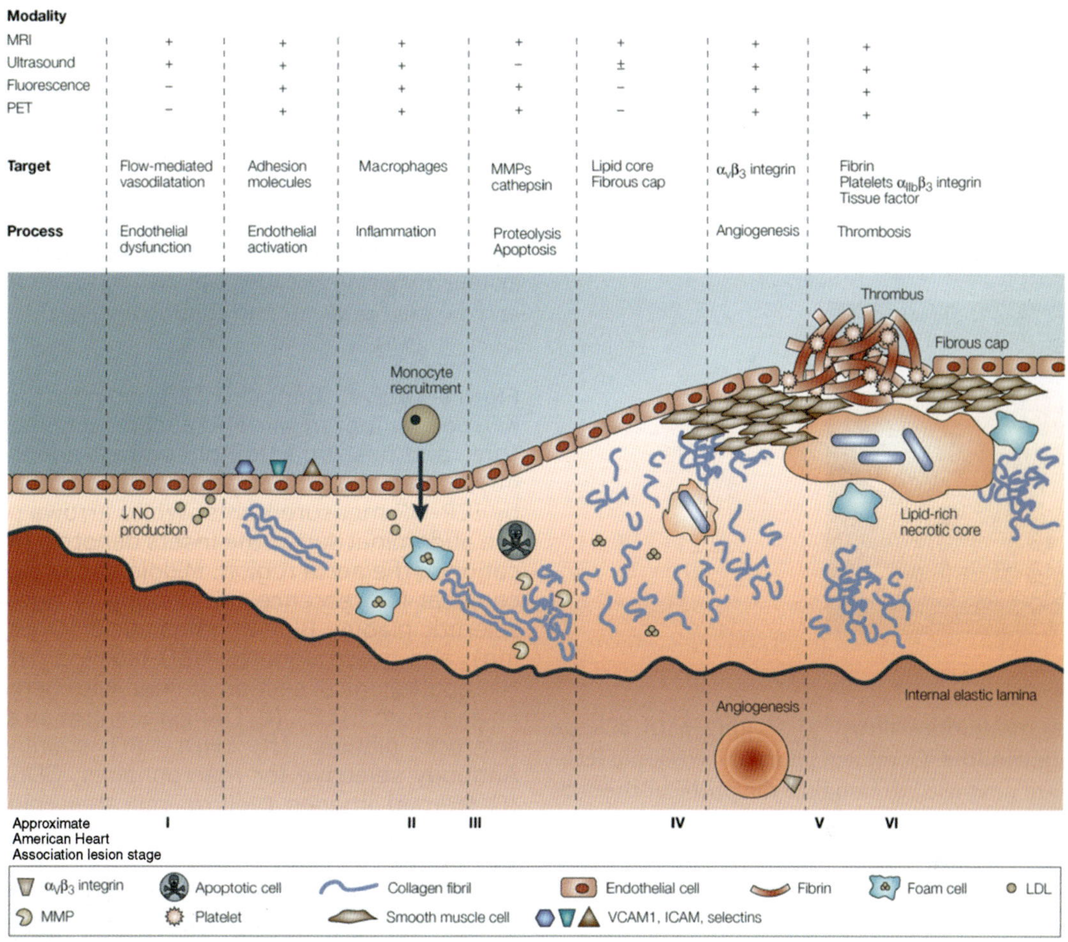

vascular lesions. These probes consist of a targeting ligand specific for the molecular marker of interest conjugated to an MRI contrast agent. The low inherent sensitivity of MRI requires the contrast agent to be very potent, especially when sparse epitopes are to be visualized. This can be achieved by using nanoparticles with a high payload of contrast agent. Perfluorocarbon nanoparticles with up to 90,000 Gd-DTPA moieties have been used for the visualization of different atherosclerosis related markers (Lanza *et al.*, 2004). This nanoparticle conjugated with fibrin-specific antibodies has been used to detect thrombi. Furthermore, $\alpha v\beta 3$-targeted perfluorocarbon nanoparticles have been used to detect angiogenesis in the aorta of atherosclerotic rabbits. The same MRI-detectable nanoparticle has been used for the delivery of anti-proliferative drugs for the prevention of restenosis after angioplasty, allowing image-guided therapy (Lanza *et al.*, 2004). A fibrin-binding gadolinium-labelled peptide was also demonstrated to be effective in the detection of thrombus in a carotid injury model (Sirol *et al.*, 2005). Upon administration of the contrast agent, thrombi could be identified precisely with MRI, which showed good correlation with histopathology (Figure 10.6.3).

Bimodal contrast agents, which are detectable with both MRI and optical methods, allow the investigation of atherosclerosis with two complementary imaging techniques. When using such probes, MRI is exploited for its ability to *in vivo* image opaque, intact

Figure 10.6.2 Contrast-enhanced MRI in hyperlipidaemic rabbit model. (a) *In vivo* MR images of the aortic vessel wall upon injection of gadofluorine. (b) At three positions a transverse MRI slice of the aorta in (a) is depicted. (c) The corresponding histopathological sections, stained with H&E, show a good correlation with the MR images of these sections in (b). (d) Magnified areas are indicated by squares in (c). At level 1, enhancement is heterogeneous because of accumulation of lipids within the fibrous area. At level 2, large lipid core corresponds to highest enhancement within the plaque. Plaque at level 3 is mainly composed of lipids with perfect matching of the highest plaque enhancement. Ad indicates adventitia; F, loose fibrous; FC, fibrous cap; L, lumen; LC, lipid core; and M, macrophages (Sirol *et al.*, 2004)

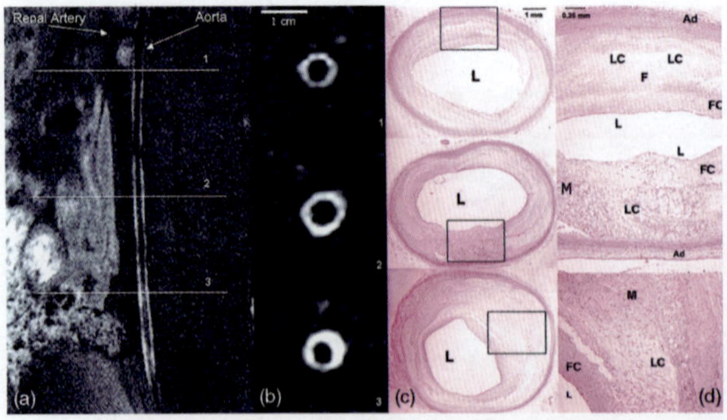

tissue at resolutions down to 50 μm^2 with a large scanning window. Complementary microscopic techniques are next used to visualize the probe at the subcellular level, be it with a relatively low penetration depth, a small scanning window and following sacrifice. The *in vitro* expression of E-selectin, a cell surface receptor over-expressed at endothelial cells in vascular lesions, has been imaged with both MRI and fluorescence microscopy, using bimodal antibody-conjugated lipid nanoparticles (Mulder *et al.*, 2004). Frias et al. (2004) developed a sophisticated bimodal probe based on high-density lipoprotein (HDL) particles. HDL was made detectable by both MRI and optical techniques via the incorporation of paramagnetic and fluorescent lipids. Upon intravenous injection this contrast material accumulated in the aortic wall of apoE-KO mice and gave maximal enhancement on T_1-weighted MRI scans at 24 h after injection (Figure 10.6.4, top). The fluorescent label of this probe allowed the investigation of the excised aorta with confocal fluorescence microscopy, from which the exact location of the contrast agent could be determined (Figure 10.6.4., middle and bottom). A very elegant aspect of this HDL-based approach is that it does not require the synthesis and use of exogenous material, because isolated endogenous HDL may be used to create the contrast material.

Figure 10.6.3 Contrast-enhanced MRI of carotid artery thrombosis in the rabbit. Transverse T_1-weighted MR image of a 1-week-old thrombus, before (a) and after (b) EP-2104R injection. The corresponding histological section is depicted in (c) (Sirol *et al.*, 2005)

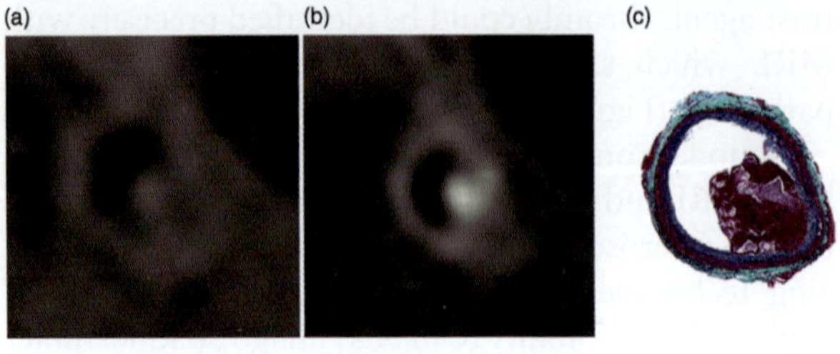

Figure 10.6.4 Bimodal HDL imaging in the apoE-KO mouse. Top: *In vivo* MRI at different time points (pre- and post-injection of contrast agent at 24, 48, 72 and 96 h) and dosages of Gd as determined by ICP-MS measurements. White arrows point to the abdominal aorta; the insets denote a magnification of the aorta region. Middle and bottom: (a) confocal fluorescence microscopy of an atherosclerotic plaque. Blue denotes nuclei (DAPI staining), and green denotes NBD-labelled rHDL. (b) Histopathological section stained with hematoxylin and eosin (H&E). (c) DAPI staining of an atherosclerotic plaque. (d) rHDL-NBD staining. (e) Antibody selective for macrophages, CD68:RPE. (f) Merged images (yellow indicates colocalization) (Frias *et al.*, 2004)

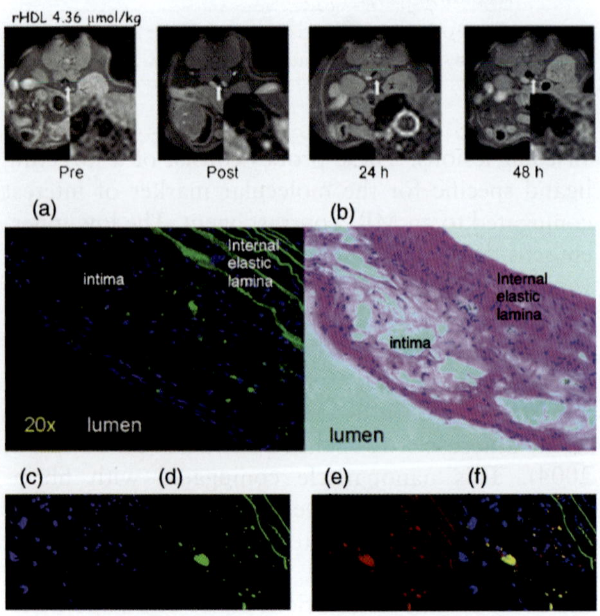

In conclusion, MR imaging of experimental atherosclerosis is very valuable for studying the progression of the disease in animal models and for evaluating the effect of anti-atherosclerosis therapy. In addition, new and optimized imaging protocols can be assessed and translated to the clinic. For improved detection and characterization of atherosclerotic plaques, the combination of MRI and newly developed contrast agents holds great promise.

References

Choudhury, R. P., Fuster, V., Fayad, Z. A., 2004. "Molecular, cellular and functional imaging of atherothrombosis." *Nat. Rev. Drug Discov.* **3**(11), 913–925.

Fayad, Z. A., Fallon, J. T., Shinnar, M., Wehrli, S., Dansky, H. M., Poon, M., Badimon, J. J., Charlton, S. A., Fisher, E. A., Breslow, J. L., Fuster, V., 1998. "Noninvasive *in vivo* high-resolution magnetic resonance imaging of atherosclerotic lesions in genetically engineered mice." *Circulation* **98**(15), 1541–1547.

Frias, J. C., Williams, K. J., Fisher, E. A., Fayad, Z. A., 2004. "Recombinant HDL-like nanoparticles: a s pecific contrast agent for MRI of atherosclerotic plaques." *J. Am. Chem. Soc.* **126**(50), 16316–16317.

Glass, C. K., Witztum, J. L., 2001. Atherosclerosis: The road ahead. *Cell* **104**(4), 503–516.

Helft, G., Worthley, S. G., Fuster, V., Fayad, Z. A., Zaman, A. G., Corti, R., Fallon, J. T., Badimon, J. J., 2002. "Progression and regression of atherosclerotic lesions: monitoring with serial noninvasive magnetic resonance imaging." *Circulation* **105**(8), 993–998.

Hockings, P. D., Roberts, T., Galloway, G. J., Reid, D. G., Harris, D. A., Vidgeon-Hart, M., Groot, P. H., Suckling, K. E., Benson, G. M., 2002. "Repeated three-dimensional magnetic resonance imaging of atherosclerosis development in innominate arteries of low-density lipoprotein receptor-knockout mice." *Circulation* **106**(13), 1716–1721.

Lanza, G. M., Winter, P., Caruthers, S., Schmeider, A., Crowder, K., Morawski, A., Zhang, H., Scott, M. J., Wickline, S. A., 2004. "Novel paramagnetic contrast agents for molecular imaging and targeted drug delivery." *Curr. Pharm. Biotechnol.* **5**(6), 495–507.

Mulder, W. J., Strijkers, G. J., Griffioen, A. W., van Bloois, L., Molema, G., Storm, G., Koning, G. A., Nicolay, K., 2004. "A liposomal system for contrast-enhanced magnetic resonance imaging of molecular targets." *Bioconjug. Chem.* **15**(4), 799–806.

Ruehm, S. G., Corot, C., Vogt, P., Kolb, S., Debatin, J. F., 2001. "Magnetic resonance imaging of atherosclerotic plaque with ultrasmall superparamagnetic particles of iron oxide in hyperlipidemic rabbits." *Circulation* **103**(3), 415–422.

Sirol, M., Fuster, V., Badimon, J. J., Fallon, J. T., Moreno, P. R., Toussaint, J. F., Fayad, Z. A., 2005. "Chronic thrombus detection with *in vivo* magnetic resonance imaging and a fibrin-targeted contrast agent." *Circulation* **112**(11), 1594–1600.

Sirol, M., Itskovich, V. V., Mani, V., Aguinaldo, J. G., Fallon, J. T., Misselwitz, B., Weinmann, H. J., Fuster, V., Toussaint, J. F., Fayad, Z. A., 2004. "Lipid-rich atherosclerotic plaques detected by gadofluorine-enhanced *in vivo* magnetic resonance imaging." *Circulation* **109**(23), 2890–2896.

Skinner, M. P., Yuan, C., Mitsumori, L., Hayes, C. E., Raines, E. W., Nelson, J. A., Ross, R., 1995. "Serial magnetic resonance imaging of experimental atherosclerosis detects lesion fine structure, progression and complications *in vivo*." *Nat. Med.* **1**(1), 69–73.

11 Tumor Imaging

Coordinated by Vasilis Ntziachristos

11.0 Introduction

Vasilis Ntziachristos

The adaptation of therapies to the specificity of each tumour increases the need of experimental models to perform the first evaluation of a drug or a biological agent. Significant progress has been made during the last 10 years in the domain of imaging of (mostly murine) tumour models. It is therefore possible to address questions relevant to tumour treatment and to the roles of individual molecular pathways in ways that cannot be addressed directly in human subjects. The development of new imaging techniques and of new tracers or reporters enables specific, highly sensitive and quantitative measurement of a wide range of tumour-related parameters:

- *In vivo* imaging gives access to precise measurements of tumour size, number and growth, as illustrated by Jouannot with high-frequency ultrasound imaging, and by De Clerk with micro-CT. This may help the longitudinal follow up before and after therapy.
- Vascular development is a key factor of tumour growth. To qualify anti-angiogenic therapies, quantification of tumour angiogenesis and blood flow is needed. Ultrasound imaging, helped by contrast agents, is now settling as a basic low-cost technique, as illustrated by Lucidarme. Fluorescence molecular tomography also gives a global appreciation of vascularity in superficial tumours, as illustrated by Montet. The ultimate visualization at microvessel scale is obtained by *in vivo* confocal microscopy applied to small tumours growing in a skinfold chamber, as illustrated by Jain. CT and MRI can be applied to the characterization of the vascular network in larger tumours and the surrounding organs. The reports by Cuenod show the possibility to distinguish arterial and venous vascularizations of liver lesions by functional CT and to quantify tumour angiogenesis by dynamic contrast-enhanced MRI.
- The biochemical characterization of tumours is now helped by imaging probes that selectively accumulate in tumours or that become activated by tumour-specific molecules *in vivo*, as illustrated by Law *et al.* Other tumour imaging strategies have been developed that rely upon the detection of reporter transgene expression *in vivo*, and these too have made a significant impact on both the versatility and the specificity of tumour imaging in living mice. Optical imaging of apoptosis can be performed by using NIR fluorescent probe activatable by a caspase or by using a dye coupled to annexin, labelling phosphatidylserine residues, as illustrated by the reports by Schellenberger *et al.*, and by Law and Tung. Similar studies, adaptable to larger lesions, are available through SPECT imaging as shown by Del Vecchio.

These advances in tumour imaging are set to have a profound impact on our basic understanding of *in vivo* tumour biology and will radically alter the application of mouse tumour models in the laboratory.

11.1 Dynamic contrast-enhanced MRI of tumour angiogenesis

Charles André Cuénod, Laure Fournier, Daniel Balvay, Clément Pradel, Nathalie Siauve and Olivier Clement

11.1.0 Introduction

Each tissue has a specific type of capillary network, which allows transfers of substances from and to the blood. The capillary networks in tumours are issued from neo-angiogenesis (Folkman, 1971) and are very different from those found in normal tissues. During tumour proliferation, the cells pushed away from the capillaries become hypoxic and acidic and release pro-angiogenic growth factors such as VEGFs (vascular endothelial growth factors) that promote the formation of new capillaries by activating endothelial cells. The tumour-associated, newly grown capillaries are tortuous and irregular, resulting in poorly efficient perfusion (Jain, 1988), and their lining is formed by fenestrated cells without basement membrane, which makes them hyper-permeable. The faster a tumour grows, the less mature are its vessels and the more heterogeneous is the microvascular network.

These changes in microcirculation patterns can be detected and quantified by dynamic contrast-enhanced (DCE) MRI (Tofts *et al.*, 1999; Brasch and Turetschek, 2000; de Bazelaire *et al.*, 2005). Recent experimental studies have shown that DCE MRI can assess tumour aggressiveness and measure tumour response to therapy, opening the way to better treatment monitoring and to new evaluation tests for drug development (Padhani, 2002; Pradel *et al.*, 2003).

It has also to be mentioned that, in addition to the evaluation of tumour microcirculation, MRI allows the evaluation of diffusion of water molecules and the measurement of the biochemical compounds of the tumours during the same imaging session by using specific sequences. That information can be of great interest to obtain a more complete picture of a tumour.

11.1.1 How to study the tissue microcirculation with dynamic contrast-enhanced MRI?

The principle is to inject a contrast agent intravenously as a bolus and to measure the signal enhancement of tissues over time using fast sequential image acquisition. The dynamics of tissue enhancement depend on (1) the shape of the arterial input function at the entry of the tissue; (2) the kinetic distribution of the blood containing the contrast agent into the capillary bed; (3) the leakage of contrast agent across the capillary walls into the tumour interstitial space and (4) the volume of the interstitial space where the contrast agent can diffuse (Tofts *et al.*, 1999; Pradel *et al.*, 2003). After conversion of the MRI signal into concentration of contrast agent (de Bazelaire *et al.*, 2005), the tissue enhancement curves can be adjusted by pharmacokinetic modelling (Figure 11.1.1) to extract the physiological parameters of microcirculation such as tissue blood flow (TBF expressed in ml/s/ml), tissue blood volume (TBV, expressed in %) and the product of the permeability by surface of capillary wall (PS expressed in ml/s/ml).

Figure 11.1.1 Bi-compartmental model used for the enhancement curves of the tumour. Each voxel of the tumour (rectangle) is composed of a blood compartment and an interstitial space in which the contrast media can diffuse, as well as a cellular compartment in which the contrast media cannot enter. The capillary compartment is supplied by an arterial input and drained through a venous output. The $k(i,j)$ represents the transfer constants between the compartments

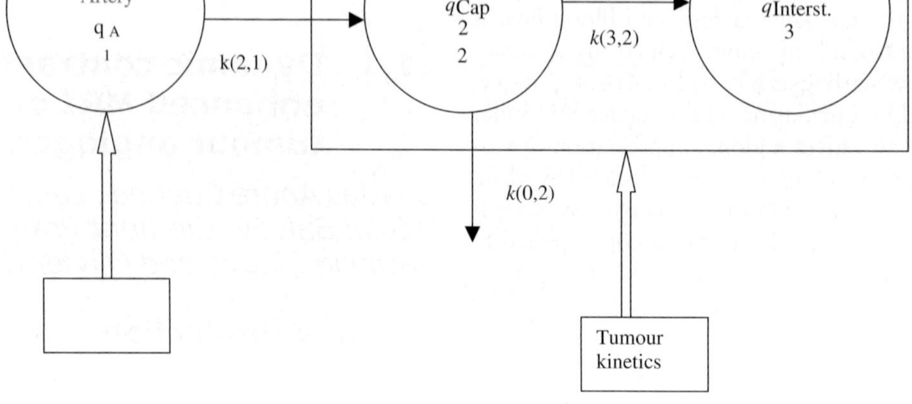

11.1.2 DCE-MRI for the evaluation of anti-angiogenic treatment of tumours

We used DCE-MRI to study the haemodynamic effects, following treatment with a VEGF-receptor tyrosine kinase inhibitor in human prostate tumour xenografts implanted in nude mice (Pradel *et al.*, 2003).

11.1.2.1 *Material and methods*

Animal model

Male Swiss athymic (nu/nu genotype) mice were injected sub-cutaneously in both flanks with 1.5×10^6 human PC-3, prostate carcinoma cells (Kaighn *et al.*, 1979).

The mice were randomized into two groups and dosed orally with either an anti-angiogenic drug (ZD4190, AstraZeneca) (100 mg/kg) or the vehicle at 24 h and 2 h prior to MRI. ZD4190 is an inhibitor of KDR (VEGFR2) receptors that hampers VEGF-stimulated human endothelial cell proliferation (Wedge *et al.*, 2000).

11.1.2.1.2 MRI acquisition

Imaging was performed on a human 1.5 T MRI unit using a 2D fast T1-weighted gradient-echo sequence. The animals were anaesthetized with xylazine and ketamine and placed in a coil dedicated to mice. The parameters of the sequence were as follows: TR 15 ms, TE 2.2 ms, flip angle 60°, matrix 256×128, FOV 7×3 cm, slice thickness 5 mm, one slice. A saturation band was placed 10 mm above the slice of interest to reduce the effect of in-flowing blood on signal intensity. The images were acquired sequentially with a temporal resolution of 1.13 s per image.

Following the acquisition of 10 baseline images, a macromolecular gadolinium-based contrast agent (P792, Guerbet France) was injected as a short bolus in the jugular vein at a dose of 0.045 mmol Gd/kg (Port *et al.*, 2001), and the signal enhancement was followed for 5 min after injection. On each of the images, the signal intensities were measured in the left heart ventricle (vascular input function) and in the tumour by drawing regions of interest over these areas. Signal enhancements were then converted into concentration of contrast agent (see Figure 11.1.3) using a calibration curve determined after each scanning session with tubes containing different concentrations of P792 (de Bazelaire *et al.*, 2005).

Compartmental analysis

Data analysis was performed using a kinetic model in which the tumour is represented by a bi-compartmental system (Figure 11.1.1) with exchanges between an interstitial compartment and a vascular compartment supplied by an arterial input (systemic blood arteries) and drained through a venous output (Pradel *et al.*, 2003). The contrast media does not diffuse in the cellular space.

When the analysis is performed on each pixel of the image, parametric images displaying the value of the parameter can be obtained for each parameter (Figure 11.1.2).

Figure 11.1.2 Parametric maps of perfusion (a) and permeability (b) parameters in a nude mouse bearing a sub-cutaneously implanted human prostate carcinoma without anti-tumoural treatment. The functional information displayed in colour is overlaid on the grey scale anatomic axial MRI images. The periphery of the tumour displays a rim of high perfusion and high permeability, whereas the poorly vascularized centre displays lower values of the parameters

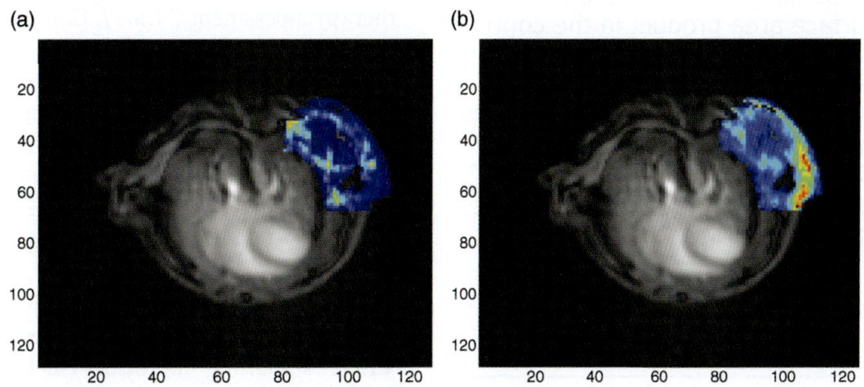

Figure 11.1.3 Mean concentration enhancement obtained for the control ($n = 17$) and mice treated with anti-VEGF therapy ($n = 13$). In the treated animals, the initial enhancement is slower and lower in magnitude compared with the control animals, corresponding to lower perfusion and blood volume, and the second slope is also slower corresponding to lower capillary permeability

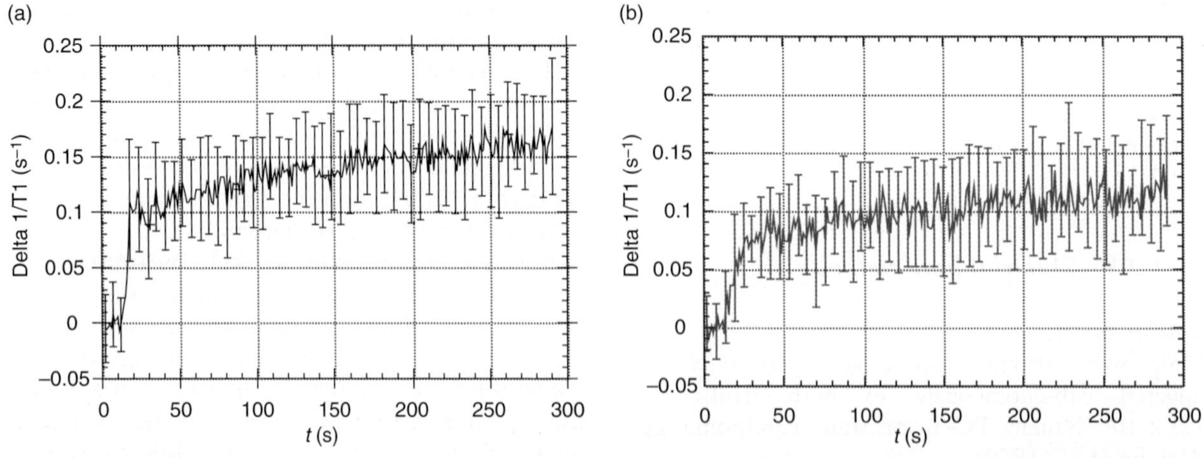

11.1.3 Results

The final data set was based on 30 tumours (13 treated and 17 controls).

Following administration of the anti-angiogenic drug, important decrease of all the three tumoural vascularization parameters is observed, as shown in Table 11.1.1.

11.1.4 Conclusion

Assessment of neo-angiogenesis by DCE MRI is becoming widely used for the detection and the staging of tumours. Moreover, DCE-MRI can be used to monitor the effects of treatment with angiogenic inhibitors.

This non-invasive technique is nowadays taking an increasing place in cancer research for the evaluation of the new cancer treatments.

Table 11.1.1 Results obtained for the tumour perfusion, the plasma volume and the permeability surface area product in the control and treated groups using the bi-compartment model. The results are expressed as median ± standard deviation. All the differences are statistically significant

	TBF (ml/s/ml)	TBV (%)	PS (ml/s/ml)
Treated $n = 13$	157 ± 137	1.07 ± 0.07	499 ± 148
Controls $n = 17$	296 ± 197	2.00 ± 0.52	844 ± 297

References

Brasch, R., Turetschek, K., 2000. "MRI characterization of tumors and grading angiogenesis using macromolecular contrast media: Status report." *Eur. J. Radiol.* **34**(3), 148–155.

de Bazelaire, C., Siauve, N., Fournier, L., Frouin, F., Robert, P., Clement, O., de Kerviler, E., Cuenod, C. A., 2005. "Comprehensive model for simultaneous MRI determination of perfusion and permeability using a blood-pool agent in rats rhabdomyosarcoma." *Eur. Radiol.* **15**(12), 2497–505.

Folkman, J., 1971. "Tumor angiogenesis: Therapeutic implications." *N. Engl. J. Med.* **285**, 1182–1186.

Jain, R. K., 1988. "Determinants of tumor blood flow: A review." *Cancer Res.* **48**, 2641–2658.

Kaighn, M. E., Narayan, K. S., Ohnuki, Y., Lechner, J. F., Jones, L. W., 1979. "Establishmentand characterization of a human prostatic carcinoma cell line (PC-3)." *Invest Urol.* **17**, 16–23.

Padhani, A. R., 2002. "Functional MRI for anticancer therapy assessment." *Eur. J. Cancer* **38**, 2116–2127.

Port, M., Corot, C., Raynal, I., Idee, J., Dencausse, A., Lancelot, E., Meyer, D., Bonnemain, B., Lautrou, J., 2001. "Physicochemical and biological evaluation of P792, a rapid-clearance blood-pool agent for magnetic resonance imaging." *Invest. Radiol.* **36**, 445–454.

Pradel, C., Siauve, N., Bruneteau, G., Clement, O., de Bazelaire, C., Frouin, F., Wedge, S. R., Tessier, J. L., Robert, P. H., Frija, G., Cuenod, C. A., 2003. "Reduced capillary perfusion and permeability in human tumour xenografts treated with the VEGF signalling inhibitor ZD4190: An in vivo assessment using dynamic MR imaging and macromolecular contrast media." *Magn. Reson. Imaging* **21**, 845–851.

Tofts, P. S., Brix, G., Buckley, D. L., Evelhoch, J. L., Henderson, E., Knopp, M. V., Larsson, H. B., Lee, T. Y., Mayr, N. A., Parker, G. J., Port, R. E., Taylor, J., Weisskoff, R. M., 1999. "Estimating kinetic parameters from dynamic contrast-enhanced T(1)-weighted MRI of a diffusable tracer: Standardized quantities and symbols." *J Magn. Reson Imaging* **10**, 223–232.

Wedge, S. R., Ogilvie, D. J., Dukes, M., Kendrew, J., Curwen, J. O., Hennequin, L. F., Thomas, A. P., Stokes, E. S., Curry, B., Richmond, G. H., Wadsworth, P. F., 2000. ZD4190: "An orally active inhibitor of vascular endothelial growth factor signaling with broad spectrum antitumor efficacy." *Cancer Res.* **60**, 970–975.

11.2 Liver tumours: Evaluation by functional computed tomography

Charles André Cuénod, Laure Fournier, Nathalie Siauve and Olivier Clément

11.2.0 Introduction

The liver possesses a dual vascular input (arterial high-pressure input for one third and portal venous low-pressure input for two third) and a rich vascular network formed by highly permeable capillaries called sinusoids. The microcirculation is finely tuned by a modulation of the arterial input aiming to compensate for any venous input changes. It can be altered by several pathological processes, including tumours. It may therefore be useful to study the hepatic perfusion variations to detect, characterize and quantify pathologic processes within the liver.

Contrast-enhanced computed tomography (CT) allows a non-invasive assessment of the hepatic microcirculation (Miles, Hayball and Dixon, 1993). The analysis of the time enhancement curves of the tissues and the afferent vessels, after injection of a bolus of iodinated contrast agent, allows to extract the microcirculation parameters by using pharmacokinetic models. In the liver, the dual vascular input is taken into account in the model.

CT scanner is well suited for such study because it yields a high temporal resolution (less than 1 s inter slices), a high spatial resolution (256×256 matrix) and a linear relationship between the concentration of contrast agent and the signal (attenuation measured in Hounsfield units).

We have published a method to analyse the functional CT data of the liver and tested it in normal rats and rats bearing overt and occult colorectal metastases (Cuenod *et al.*, 2001, 2002) as well as in rats bearing primary hepatic tumours (Fournier *et al.*, 2004). The methodology and mathematical model are exposed in details in those references.

11.2.1 Materials and methods

11.2.1.1 Animals

Functional CT was applied on a first group of normal BD IX rats and on rats bearing occult (microscopic) or large DHDK12 colorectal carcinoma metastases (Caignard *et al.*, 1985). It was also tested on Wistar rats having chemically induces liver focal lesions ranging from hyperplasia to hepatocellular carcinoma (diethylnitrosamine or DEN model) (Travis, McClain and Birkner, 1991).

11.2.1.2 CT Data acquisition

Rats were fasted 8 h before scans. A single slice containing the aorta, the portal vein and liver parenchyma, was selected. The rats were anaesthetized with intraperitoneal injection of ketamine and xylazine. Consecutive images were obtained on a human dedicated CT scanner (Prospeed, General Electric) for very fast acquisition time (300 ms) during at least 60 s on a single abdominal section. A bolus of 1 ml/kg iodinated contrast media (Iobitridol, 350 mg iodine/ml) was injected i.v. For optimizing the signal variations, the CT acquisition was performed with 80 kV.

11.2.1.3 Pathology examination

Rats were sacrificed after imaging. Their livers were fixed in by injecting 0.1 ml of 20/1 acetic formalin in the portal vein. The type, size and location of each lesion were recorded for imaging comparison.

11.2.1.4 Data analysis

On each image, regions of interest (ROIs) were drawn within the aorta, the portal vein and the liver (Figure 11.2.1), and time attenuation curves were plotted (Figure 11.2.2).

Figure 11.2.1 Contrast-enhanced CT slice of a rat abdomen showing the ROIs drawn on the liver, the portal vein and the aorta

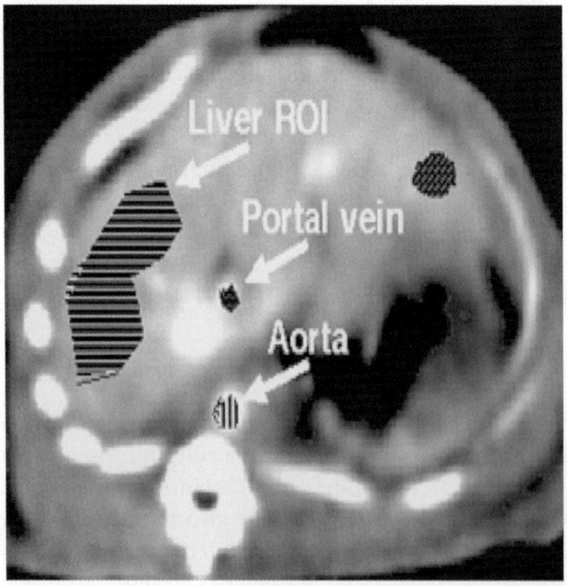

Figure 11.2.3 Diagram of the deconvolution method with two inputs
The liver has two inputs $C_a(t)$ and $C_p(t)$ weighted by coefficients [alpha] and [1-alpha] and one output $C_v(t)$.

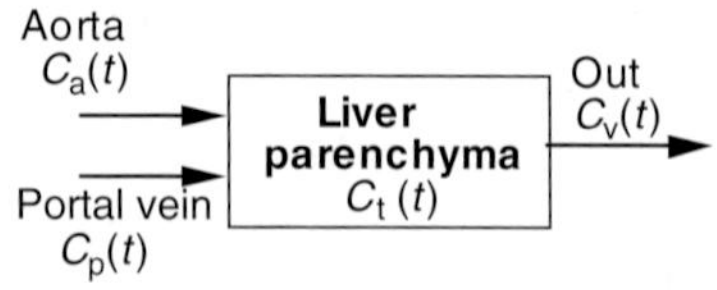

Analysis of tissue and vessels enhancement curves allowed the calculation of six parameters quantifying the liver microcirculation by using the dual input model (Figure 11.2.3) (Cuenod *et al.*, 2001, 2002; Fournier *et al.*, 2004):

- *Mean transit time* (MTT), the mean time taken by molecules of contrast agent to flow through the tissue;
- *Liver distribution volume* (LDV), the percentage of tissue volume in which the contrast agent distributes;

Figure 11.2.2 Typical time–intensity curves of the aorta, the portal vein and the liver. The contrast enhancement obtained by subtracting the mean baseline value to the measured values is expressed as HU. The time is expressed in seconds. The recirculation peak of the aorta is clearly seen and is contemporary to the portal peak. The liver enhancement is progressive and reaches a plateau

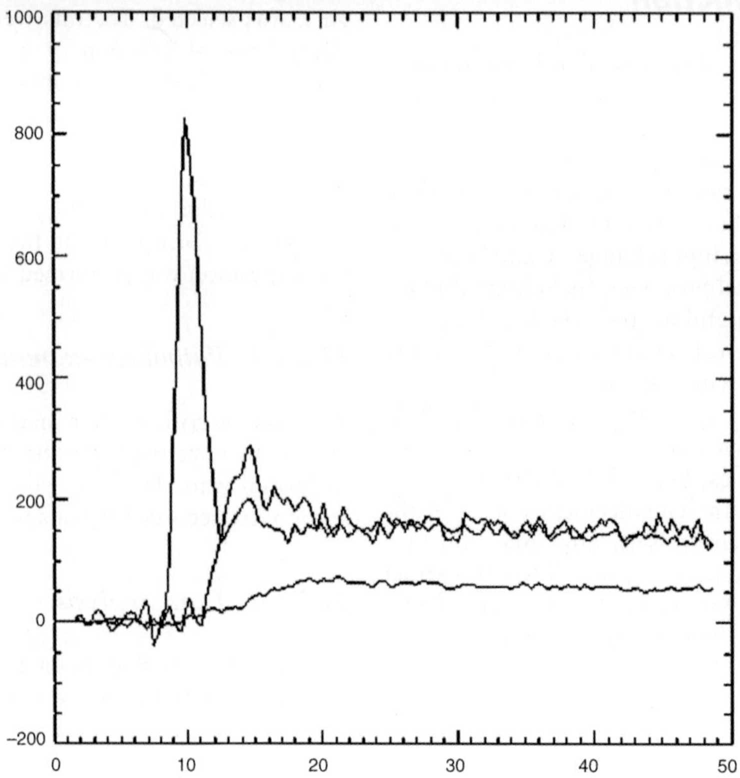

- *Arterial blood flow* F_A, *portal blood flow* F_P and *total hepatic blood flow* F_T, expressed in *ml/min/ml of tissue*, with:

$$F_T = F_A + F_P$$

- *Hepatic perfusion index* (HPI), the percentage of total blood flow of arterial origin:

$$\text{HPI} = \frac{F_A}{F_A + F_P}$$

The analysis was run on a pixel-by-pixel basis and yielded maps of each parameters.

11.2.2 Results

Typical enhancement patterns in vessels and liver tissue are shown in Figure 11.2.2.

11.2.2.1 Metastasis model

The animals bearing occult micrometastases had a 34% lower hepatic portal blood flow (F_P) (Table 11.2.1) and a 25% longer mean transit time (MTT) in the apparently normal liver.

In large metastases all the perfusion parameters, except the arterial hepatic blood flow, were abnormal. HPI increased because of the large drop in F_P. The distribution volume of the contrast medium was much lower than in controls.

11.2.2.2 Primary liver tumour (DEN model)

On microscopic examination, different types of pathological presentations were observed: hyperplastic zones which correspond to benign states, dysplasia which is a pre-malignant state and Hepatocellular Carcinoma (HCC) which is the malignant state.

On functional CT, hyperplastic ROI presented a lower distribution volume and total blood flow (with a lower portal blood flow) than controls (Table 11.2.2). Dysplastic areas and HCC showed a lower distribution volume, total and portal blood flow and also an increase in perfusion index due to an increase in arterial input.

Table 11.2.1 Perfusion parameter values (median [range]) in the metastatic series. (Mann–Whitney *U* test)

Haemodynamic parameters	Normal livers (control rats $n = 5$)	Micrometastases (normal appearance of the liver $n = 7$)	Macrometastases ($n = 4$)
HPI (%)	17 ± 19	9.2 ± 23	74.6 ± 70^b
MTT (s)	7.93 ± 4	9.94 ± 4.96^a	25.19 ± 16.17^b
DV (%)	46.4 ± 7.8	43.5 ± 13.9	29.5 ± 14.5^b
F_T (ml/min/ml)	3.23 ± 1.9	2.07 ± 0.94^a	0.71 ± 0.71^b
F_A(ml/min/ml)	0.5 ± 0.7	0.21 ± 0.44	0.41 ± 0.25
F_P (ml/min/ml)	3.08 ± 2.4	2.04 ± 1.28^a	0.20 ± 0.78^b

$^a p < 0.05$.
$^b p < 0.015$.

Table 11.2.2 Perfusion parameters on the rats in the DEN group

Haemodynamic parameters	Control ($n = 43$)	Hyperplasia ($n = 2$)	Dysplasia ($n = 6$)	HCC ($n = 15$)
HPI (%)	19 ± 9	10.2 ± 2.4	46 ± 32^a	58 ± 28^a
MTT (s)	10 ± 2	10.0 ± 0.7	14 ± 7^a	15 ± 4^a
LDV (%)	44 ± 7	39 ± 0.4^a	41 ± 6	35 ± 6^a
F_T (ml/min/ml)	2.7 ± 0.8	2.3 ± 0.1^a	$2. \pm 0.8$	1.5 ± 0.4^a
F_A (ml/min/ml)	0.5 ± 0.3	0.2 ± 0.1	0.8 ± 0.4^a	0.8 ± 0.3^a
F_P (ml/min/ml)	2.3 ± 0.7	2.1 ± 0.1	1.3 ± 0.9^a	0.7 ± 0.6^a

$^a p < 0.05$.

Figure 11.2.4 Parametric image of a liver bearing HCC and dysplastics nodules. On this map the values of HPI are displayed for each pixel. The colour code is set up with red meaning high HPI values. Primary tumours (delineated by the purple line) appear as red spots. They were not visible on conventional CT images

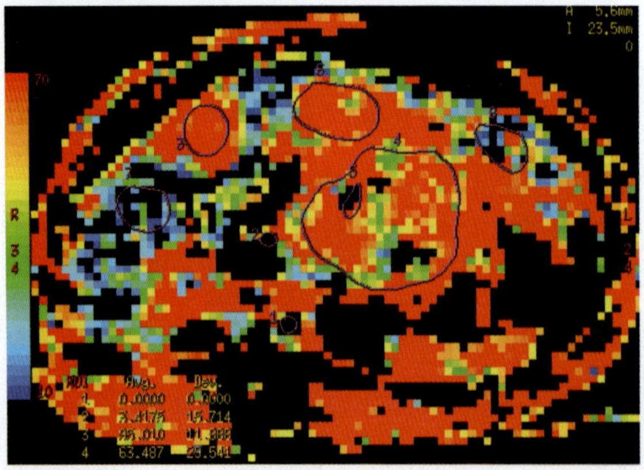

11.2.3 Discussion

Quantification of tissue blood flows is feasible in liver with functional CT.

For the macrometastases, the F_P and F_T reduction may be related to a resistance increase due to a compression of the hepatic sinusoids at the periphery of the tumour (Kan *et al.*, 1993) and to the irregular flow within the dilated, tortuous neo-angiogenic vascular network of the tumour which accounts for the increase in MTT (Ridge *et al.*, 1987). Within the metastatic nodules, the low tissular distribution volume of the contrast media reflects a lower capillary density than within normal liver (Kan *et al.*, 1993).

For the micrometastases, the F_P and F_T reduction may be related to a resistance increase due to activation of Küpperf's cells and endothelial cells (Kruskal *et al.*, 2004).

The primary liver tumours have a predominantly arterial input (Figure 11.2.4), in contrast to normal tissue, which depends mainly on portal blood. The arterialization of vessels even in pre-malignant dysplastic nodules seems to indicate that tumoural transformation of vessels occur early.

11.2.4 Conclusion

Functional CT allows quantification of perfusion parameters of normal liver, pre-cancerous lesions and cancerous lesions. It may help in the early detection and characterization of tumours as well as in the evaluation of their responses to treatment. It is a sensitive and specific non-invasive technique, which allows longitudinal studies of hepatic tumour models.

References

Caignard, A., Martin, M., Michel, M., Martin, F., 1985. "Interaction between two cellular subpopulations of rat colonic carcinoma when inoculated to the syngeneic host." *Int. J. Cancer* **36**, 273–279.

Cuenod, C. A., Leconte, I., Siauve, N., Resten, A., Dromain, C., Poulet, B., Frouin, F., Clement, O., Frija, G., 2001. "Early changes in liver perfusion caused by occult metastases in rats: Detection with quantitative CT." *Radiology* **218**, 556–561.

Cuenod, C. A., Leconte, I., Siauve, N., Frouin, F., Dromain, C., Clement, O., Frija, G., 2002. "Deconvolution technique for measuring tissue perfusion by dynamic CT: Application to normal and metastatic liver." *Acad. Radiol.* **9**, 205–211.

Fournier, L. S., Cuenod, C. A., de Bazelaire, C., Siauve, N., Rosty, C., Tran, P. L., Frija, G., Clement, O., 2004. "Early modifications of hepatic perfusion measured by functional CT in a rat model of hepatocellular carcinoma using a blood pool contrast agent." *Eur. Radiol.* **14**, 2125–2133.

Kan, Z., Ivancev, K., Lunderquist, A., McCuskey, P. A., Wright, K. C., Wallace, S., McCuskey, R. S., 1993. "In vivo microscopy of hepatic tumors in animal model: A dynamic investigation of blood supply to hepatic metastases." *Radiology* **187**, 621–626.

Kruskal, J. B., Thomas, P., Kane, R. A., Goldberg, S. N., 2004. "Hepatic perfusion changes in mice livers with developing colorectal cancer metastases." *Radiology* **231**, 482–490.

Miles, K. A., Hayball, M. P., Dixon, A. K., 1993. "Functional images of hepatic perfusion obtained with dynamic CT." *Radiology* **188**, 405–411.

Ridge, J. A., Bading, J. R., Gelbard, A. S., Benua, R. S., Daly, J., 1987. "Perfusion of colorectal hepatic metastases: Relative distribution from the hepatic artery and portal vein." *Cancer* **59**, 1547–1553.

Travis, C. C., McClain, T. W., Birkner, P. D., 1991. "Diethylnitrosamine-induced hepatocarcinogenesis in rats: A theoretical study." *Toxicol. Appl. Pharmacol.* **109**, 289.

11.3 Early detection of grafted Wilms' tumours

Erwan Jouannot

Wilms' tumour (approximately 6% of all paediatric malignant disease) is the most common malignant renal tumour in children under 15 years of age. Continued advances in the management of Wilms' tumour have increased the 5-year post-treatment survival rate from 30% in the 1930s to nearly 90% currently (Kalapurakal *et al.*, 2004). Although standard therapeutic strategy based on radical nephrectomy followed by chemotherapy or radiotherapy has become a paradigm for successful cancer therapy, severe undesirable toxicities have been reported in some cases. Novel strategies and therapeutic agents are needed for patients with high-risk tumours to increase therapeutic success while reducing undesirable side effects related to intense chemotherapy.

Murine models play an essential role in the validation of new therapeutic agents for this disease. Based on such studies, the effectiveness of new strategies using anti-angiogenic therapeutics has been demonstrated. Significant reductions in tumour weight were observed in animals treated using an antagonism of vascular endothelial growth factor (anti-VEGF). Many studies concerning therapeutic response in murine models have relied either on palpation or histological evaluation to assess tumour development. Unfortunately, palpation allows detection of tumours only at a rather advanced stage, approximately once the tumour has a cross-section of at least 3 cm^2. Such growth is typically attained only at 5 weeks after cell injection. Histological evaluation requires the sacrifice of the animal, which prevents longitudinal follow-up of the same tumour as a function of treatment duration.

By providing non-invasive visualization of the tumour *in vivo*, an ultrasonic imaging technique could potentially improve tumour detection as well as allow longitudinal study of the effectiveness of new therapies on tumour growth and modification. The relatively superficial localization (between 0.6 and 6 mm) of mouse kidney should allow the use of ultrasound frequencies on the order of 20–40 MHz for its exploration. In theory, frequencies in this range should be able to resolve structures with dimensions between 200 and 60 μm (Cheung *et al.*, 2005). This study evaluates the use of a 24-MHz centre-frequency ultrasonic imaging system to detect, size and non-invasively follow Wilms' tumour development in a nude mouse model. Additional details related to this study can be found in Jouannot *et al.* (2006).

11.3.1 Materials and Methods

The SK-NEP-1 cell line has been isolated from a malignant pleural effusion of a Caucasian Woman with a known Wilms' tumour. Cells were maintained in culture and concentrated before their injection in the left kidney of nude mice ($N = 26$). After induction of anaesthesia, the kidney was surgically exposed via a flank incision. Mice received a sub-capsular injection of cells at the prepared concentration.

Three days after cell injection, images of the right (control) and left (injected) kidney were acquired *in vivo*. Imaging sessions *in vivo* were repeated biweekly until tumour detection. Images were acquired for longitudinal (long axis of the kidney) and transverse (short axis of the kidney) planes at the level of maximum tumour cross-section for 11 mice. The maximum tumour cross-sections were estimated based on these images using image processing software written under MATLAB® (The Mathworks, Natick, MA) and compared after the sacrifice of the mouse to histological measurements.

For the remaining 15 mice, ultrasound examinations were repeated every 4 days after initial tumour detection to assess tumoural growth.

11.3.2 Results

Tumours were detected *in vivo* as confirmed by histology as early as the seventh day after tumour-cell injection. The day of initial detection varied from 7 to 14 days after cell injection (mean at 8 days). Tumours were initially detected with an average cross-sectional area of 1.5 mm^2 (between 0.07 and 5.7 mm^2). At this early stage, the tumoural zone presents a hypoechogenic aspect. This aspect is demonstrated in Figure 11.3.1 for a tumour detected at 17 days.

To validate measurements of tumour dimensions obtained *in vivo* at 24 MHz, each tumour cross-sectional area in the ultrasonic image was compared to the area measured in the corresponding histological section. The relative root mean square error for area measurements was estimated to be 19%. For the 15 remaining mice, tumoural growth was assessed *in vivo* between 0 and 28 days after cell injection. Values of the maximum cross-sectional area of tumour as assessed in the transverse ultrasonic plane are shown in Figure 11.3.2. The precision error of ultrasonic cross-sectional measurements was estimated to be 0.11 mm^2 by repeating, on the same tumour, 10 image acquisitions and three surface delineations per acquisitions. Tumours appeared to grow slowly until an area of approximately 3 mm^2 was reached and then to increase in growth rate.

Figure 11.3.1 Images of a tumour detected 17 days after cell injection. (a) Ultrasonic transverse cross-section (24 MHz) of the mouse kidney *in vivo*. A suspicious tumoural region (T) has been detected near the middle part of the kidney, just under the skin (S). It presents a lower echogenicity than the normal cortex (C). (b) Macroscopic photography of the same kidney after excision allows the identification of a pale mass at the middle part of the kidney. The line indicates the approximate position of the transverse plane corresponding to the ultrasonic image and the histological section (c)

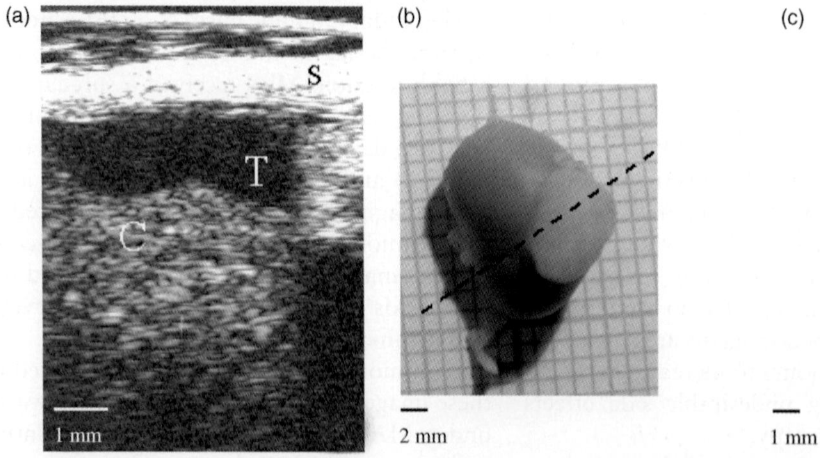

11.3.3 Discussion

A validated imaging technique allowing to non-invasively follow *in vivo* Wilms' tumour development in the mouse model holds considerable interest for genetic and therapeutic studies.

Only invasive techniques have been cited in the literature for characterizing Wilms' tumour growth (Rowe *et al.*, 2000). High-frequency ultrasound imaging allowed tumour detection at a mean of 8 days after cell injection and for a mean area of 1.5 mm^2. This is considerably earlier than detection which can be obtained by palpation techniques because the palpable size is typically obtained only after a delay of 3–5 weeks.

The current work demonstrates the feasibility of using high-frequency ultrasound imaging to detect

Figure 11.3.2 (a) Linear correlation ($R^2 = 0.93$) was found between tumour cross-sectional area in the ultrasonic image and area measured in the corresponding histological section ($N = 11$). (b) Maximum area of the tumour as assessed on transverse ultrasound images as a function of the number of days after injection of tumour cells ($N = 15$)

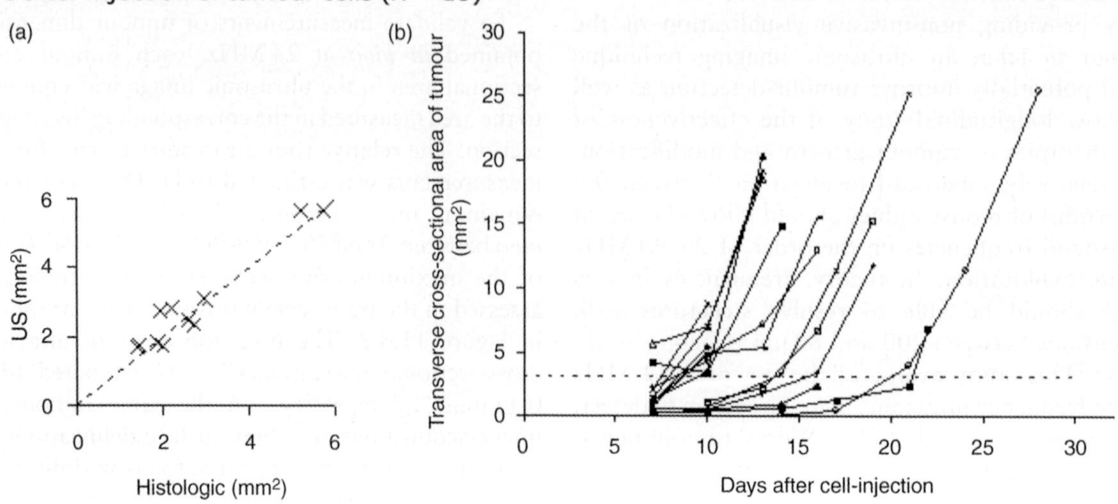

and assess tumour growth in a realistic mouse model for Wilms' tumours. To validate ultrasound detection and tumour size estimation, ultrasonic evaluations were compared to reference measurements. Two-dimensional *in vivo* ultrasonic measurements of tumour cross-sectional area showed good correspondence with histological measurements made on excised tumours. Further improvements in following the growth of tumours could be obtained by acquiring three-dimensional imaging data for estimation of tumour volume as a function of time.

When assessing therapy effectiveness based on groups of mice sacrificed at different times after injection of tumour-inducing cells, tumour development must be hypothesized to be the same for each mouse, or a large number of mice must be included in the study to overcome the variance in tumour development. In this study, we observed that tumours attained 4 mm^2 maximum cross-sectional areas at very different delays with respect to injection of tumour-inducing cells: between 7 and 14 days after cell injection. By providing an objective means to quantitatively assess tumour growth for each individual subject, ultrasound reveals the inter-individual differences due to biological or experimental variations (variations in the precise cell-injection site or syringe backflow . . .). During the course of a pilot study, such information could be used to search for experimental sources of variation so that their effects could be minimized.

Based on the results of this study, high-frequency ultrasound appears to be a viable and non-invasive alternative to palpation or sacrifice for the assessment of Wilms' tumour growth. Reports of Wilms' tumour studies in the literature demonstrate several other properties of the tumour that may be of interest. The vascularization of the tumours is currently evaluated by specific staining of histological sections. The techniques based on the kinetics of ultrasonic enhancement obtained using ultrasound contrast agents as blood-flow markers show significant promise for non-invasive evaluation of vascularization.

Compared to others imaging techniques available for cancer evaluation (MRI and micro-CT), ultrasound remains one of the easiest to use. Real-time imaging minimizes the duration of anaesthesia and animal stress. Furthermore, minimal training is needed to learn to acquire images, and the imaging systems are relatively portable and unexpensive. Such advantages should allow ultrasonic imaging to make considerable future contributions to biological studies concerning tumour development in mouse models.

References

Cheung, A. M., Brown, A. S., Hastie, L. A., Cucevic, V., Roy, M., Lacefield, J. C., Fenster, A., Foster, F. S., 2005. "Three-dimensional ultrasound biomicroscopy for xenograft growth analysis." *Ultrasound Med. Biol.* **31**(6), 865–870.

Kalapurakal, J. A., Dome, J. S., Perlman, E. J., Malogolowkin, M., Haase, G. M., Grundy, P., Coppes, M. J., 2004. "Management of Wilms' tumour: Current practice and future goals." *Lancet Oncol.* **5**(1), 37–46.

Jouannot, E., Duong Van Huyen, J-P., Bourahla, K., Laugier, P., Lelievre-Pegorier, M., Bridal, L., 2006. "High frequency ultrasound detection and follow-up of Wilms' tumor in the mouse." *Ultrasound Med. Biol.* **32**(2), 183–190.

Rowe, D. H., Huang, J., Kayton, M. L., Thompson, R., Troxel, A., O'Toole, K. M., Yamashiro, D., Stolar, C. J., Kandel, J. J., 2000. "Anti-VEGF antibody suppresses primary tumor growth and metastasis in an experimental model of Wilms' tumor." *J. Pediatr. Surg.* **35**(1), 30–33.

11.4 Angiogenesis study using ultrasound imaging

Olivier Lucidarme

Angiogenesis is now a particularly important marker for cancer biology research and tumour-directed therapy. The direct assessment of angiogenesis requires tissue biopsy that is not only susceptible to sampling errors but is not feasible for serial monitoring of patients. Although many anti-angiogenic therapies are in clinical trials, validation and standardization of new non-invasive monitoring techniques for anti-angiogenic therapy is a major need in this field.

11.4.1 Doppler measurement

Ultrasound imaging can identify vascular features in tumours at different levels of resolution (40–200-μm-diameter vessels), depending on the technique employed. Doppler ultrasound offers a low-cost, non-invasive approach with which to conveniently measure change in vascular features, including vessel diameter, flow velocity, volume of flow and morphology over time. However, when Doppler is used, the ultrasound signal arises from vessel larger than those involved in angiogenesis. The motion of solid tissue dominates the Doppler shifts below approximately 10 mm/s, and no reliable means exists for separating the very small component of these signals from flow in vessels smaller than approximately 100 μm in diameter. The angiogenic process involves mainly vessels as small as capillaries that can have a minimal diameter of 5 μm with a blood velocity as low as 1 mm/s. Hence, more sensitive or quantitative Doppler techniques must be employed.

11.4.2 Functional ultrasound imaging using contrast agent

The addition of ultrasound contrast agents provides new opportunities to go further. These contrast agents are microbubbles of different gas such as air or sulphur hexafluoride stabilized by a shell which is made of lipid, albumin or polymers (Schneider *et al.*, 1995). They range in diameter from less than 1–10 μm and cannot be extravasated from the vessel lumen. Hence, ultrasound contrast agents have a true blood-pool distribution, and they can pass through the entire vascular bed including capillaries. Ultrasound is extremely sensitive to microbubbles, and any echo received from a microbubble signs the presence of vessels even when, like capillaries, they cannot be resolved by US. Consequently, functional ultrasound imaging with microbubbles contrast media may provide information about microcirculation.

A range of techniques to make quantitative measurements of perfusion parameters have been reported. The easiest way is to use a bolus injection of contrast microbubbles and then to acquire image time-intensity curves in regions of interest to follow the passage of the bolus. A semi-quantitative analysis of signal intensity change giving functional indices, such as time to peak, peak intensity, wash-in slope or area under time-intensity curves, can be performed. These indices are straightforward to calculate, but their physiological significance is unclear. Another approach to functional contrast imaging is based upon the fact that contrast agent microbubbles can be destroyed when insonified with an ultrasound wave of sufficient intensity. The first method, based on real-time destruction-reperfusion imaging, begins with one or more frames acquired at high acoustic power to destroy all microbubbles in the imaging field and then observes the refill kinetics at low acoustic power. The slope of the re-perfusion curve is theoretically related to the blood velocity and the maximum enhancement to the local blood volume (see Chapter 3, Figure 3.65) (Wei *et al.*, 1998). The second method monitors the disappearance rate of microbubbles from tissues while imaging at a constant frame rate with moderate transmit power (Wilkening *et al.*, 2000; Lucidarme *et al.*, 2003). In this case, destruction, wash-in and wash-out phases occur during the observation, and the image intensity decreases and after a few frames approaches equilibrium, where the number of bubbles destroyed by each frame is equal to the number of bubbles that entered the field between frames. The decay curve gives us the maximum enhancement, which is related with the local blood volume, the fraction of bubbles destroyed by each frame and the fraction of microbubbles that wash-in between two frames which is related with the blood velocity. Both methods can be employed to study angiogenesis *in vivo* and to allow follow-up after anti-angiogenic therapy in different animal models.

11.4.3 Animal models for angiogenesis study

Although tumours would be the best model, tumour angiogenesis is a complex process that involves many steps and interactions between tumour cells and matrix as well as tumour and host that introduces variability even within the same tumour. The development and testing of diagnostic and/or therapeutic systems can be aided if the model is stable and more predictable among animals. For these reasons, tumour and non-tumour angiogenesis model coexists in fundamental angiogenesis research.

11.4.3.1 Non-tumour mice model

A non-tumour model which consists in compatible polymer matrix impregnated with growth factors and implanted sub-cutaneously in mice has been recently adapted to functional ultrasound imaging (Lucidarme *et al.*, 2004). In this model a complex mixture of basement membrane proteins (Matrigel®) that it is liquid at 4°C and gels at room temperature to which a human fibroblast growth factor (bFGF) had been added was injected sub-cutaneously in the back of 36 mice. When angiogenic growth factors were added, sprouts from adjacent vessels penetrated into the gel within days to form a new capillary network

(Passaniti *et al.*, 1992). Ultrasound imaging has shown that the Matrigel® implants were homogenous and anechoic providing an ideal imaging background to study the enhancement of angiogenic foci induced by microbubbles. However, neither the maximum enhancement (related to the local blood volume) nor the fraction of blood replaced per second yielded by functional ultrasound studying the disappearance rate of microbubbles correlated with the microvascular density (MVD) assessed histologically (Figure 11.4.1). This observation underscores the complexity of angiogenesis and the limitations of correlating functional imaging parameters with morphologic histological indices. MVD, indeed, reflects local proliferation of all neovessel including those that do not induce any change in local blood volume and blood flow such as vessels occluded or not yet connected to the systemic circulation (Vermeulen *et al.*, 1996) whereas ultrasound detects only perfused regions. This is supported by Forsberg *et al.* (2002) or Iordanescu *et al.* (2002) who reported that no correlation was observed in different murine model of cancer between enhanced colour Doppler density and MVD.

11.4.3.2 *Tumour mice model*

Numerous tumour models have been developed in small animals including animal cell line tumours that arise spontaneously with time (Iordanescu *et al.*, 2002) or that are induced by chemical stimulation or viral transfection, and xenografts of human cell line tumours injected sub-cutaneously or implanted orthotopically. Among them a model of Wilms' tumour grafted in the kidney of nude mice has been recently studied with functional ultrasound imaging (see Jouannot, Figure 11.4.2). Using the destruction-reperfusion method, it has been shown that the local blood volume and the blood velocity in the grafted tumour were significantly lower than in the adjacent renal cortex and that these parameters were correlated with the amount of tumour necrosis quantified histologically (unpublished data). In the same way, Pollard *et al.* (2002) and Broumas *et al.* (2005) monitored the vascularity of induced murine mammary adenocarcinomas in rats using contrast-enhanced ultrasound during anti-angiogenic therapy (SU11657). They showed, using a semi-quantitative analysis based on the destruction–replenishment method, that the spatial extent of contrast enhancement and the time required for contrast replenishment within control tumours were significantly different from those of treated tumours. The time-integrated ultrasound contrast enhancement decreased, and the time required for replenishment of the contrast agent within the tumour volume increased over the course of anti-angiogenic therapy. The contrast-enhanced tumour area calculated from the ultrasound analysis

Figure 11.4.1 Non-tumour murine model of angiogenesis. Photographs (a) and (b) show the macroscopic appearance of Matrigel plugs implanted 10 days earlier in a mouse. The right Matrigel plug had 1 μg/ml of bFGF added (a) and was deeply invaded with several large vessels whereas the control plug remained avascular (b). Ultrasound images (c) and (d) show the appearance of the corresponding Matrigel plugs located sub-cutaneously on either side of the spine of a mouse and imaged at 12 MHz. Before contrast (c) the Matrigel plugs (asterisks) are anechoic. Thirty seconds following the administration of 0.1 ml of microbubbles (AF0150) (d), nodular enhancement in the bFGF added Matrigel plug (arrow) was visible and corresponded to the area of the angiogenic focus seen in the plug (a). However, the level of enhancement did not correlate with the microvascular density assessed in this area. No enhancement was observed in the control plug. Also, note that the kidneys (K) enhanced markedly post-contrast

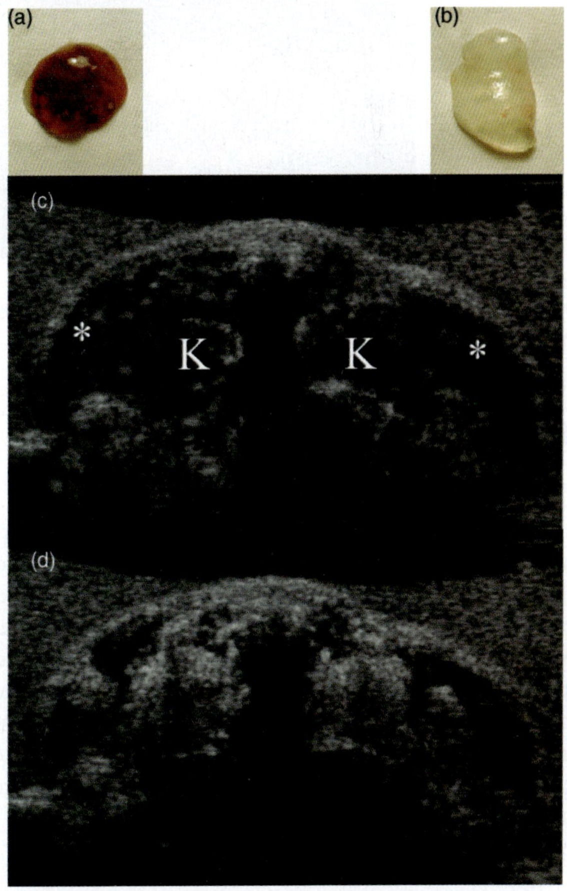

Figure 11.4.2 Vascular imaging in a mouse model for renal tumour. Images of a Wilms' tumour grafted in the kidney of nude mice obtained, 14 days after implantation, with a 7–14 MHz linear array transducer. B-mode image (a) shows a hypoechogenic tumour (t) surrounding the kidney (r) (see also the opposite kidney: r_d). Colour Doppler image (b) presents colour pixels that correspond to a blood flow depicted in the aorta (A_o) and the renal hilum. Note that colour Doppler is not sensitive enough to be able to display a signal coming from the renal cortex or from the tumour. After injection of ultrasound contrast agent, increased image brightness is initially observed (4 s after injection) in the renal parenchyma (c). Six seconds after the injection, the contrast effect appears in the vascularized regions of the tumour (d). The later arrival in this flow region indicates that flow is slower in the tumour than in the kidney

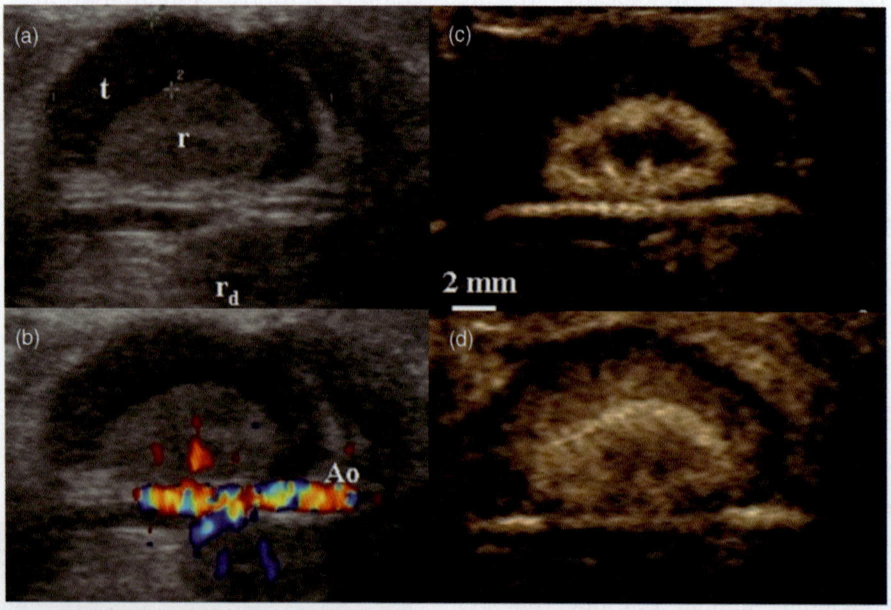

was shown to correlate with the regions of viable tumour demonstrated histologically.

11.4.4 Conclusion

Contrast-enhanced ultrasound imaging provides a convenient, low-cost non-invasive tool to yield information about the vascular volume, the flow rate and the spatial distribution of blood flow of implanted biocompatible polymer matrix or tumours in animal models. These models are widely used to understand and monitor the effects of several new anti-angiogenic therapies and ultrasound imaging using contrast microbubbles can easily provide semi-quantitative or quantitative functional indices to serve as surrogate markers of therapeutic response.

References

Broumas, A. R., Pollard, R. E., Bloch, S. H., Wisner, E. R., Griffey, S., Ferrara, K. W., 2005. "Contrast-enhanced computed tomography and ultrasound for the evaluation of tumor blood flow." *Invest. Radiol.* **40**, 134–147.

Forsberg, F., Dicker, A. P., Thakur, M. L., Rawool, N. M., Liu, J. B., Shi, W. T., Nazarian, L. N., 2002. "Comparing contrast-enhanced ultrasound to immunohistochemical markers of angiogenesis in a human melanoma xenograft model: Preliminary results." *Ultrasound Med. Biol.* **28**, 445–451.

Iordanescu, I., Becker, C., Zetter, B., Dunning, P., Taylor, G. A., 2002. "Tumor vascularity: Evaluation in a murine model with contrast-enhanced color Doppler US effect of angiogenesis inhibitors." *Radiology* **222**, 460–467.

Lucidarme, O., Kono, Y., Corbeil, J., Choi, S., Mattrey, R., 2003. "Validation of ultrasound contrast destruction imaging for flow quantification." *Ultrasound Med. Biol.* **29**, 1697–1704.

Lucidarme, O., Nguyen, T., Kono, Y., Corbeil, J., Choi, S., Varner, J., Mattrey, R., 2004. "Angiogenesis model for ultrasound contrast research:" *Exploratory study. Acad. Radiol.* **11**, 4–12.

Passaniti, A., Taylor, R. M., Pili, R., Guo, Y., Long, P. V., Haney, J. A., Pauly, R. R., Grant, D. S., Martin,

G. R., 1992. "A simple, quantitative method for assessing angiogenesis and antiangiogenic agents using reconstituted basement membrane, heparin, and fibroblast growth factor." *Lab Invest.* **67**, 519–528.

Pollard, R. E., Sadlowski, A. R., Bloch, S. H., Murray, L., Wisner, E. R., Griffey, S., Ferrara, K. W., 2002. "Contrast-assisted destruction-replenishment ultrasound for the assessment of tumor microvasculature in a rat model." *Technol. Cancer Res. Treat.* **1**, 459–470.

Schneider, M., Arditi, M., Barrau, M. B., Brochot, J., Broillet, A., Ventrone, R., Yan, F., 1995. BR1: "A new ultrasonographic contrast agent based on sulfur hexafluoride-filled microbubbles." *Invest. Radiol.* **30**, 451–457.

Vermeulen, P. B., Gasparini, G., Fox, S. B., Toi, M., Martin, L., Mcculloch, P., Pezzella, F., Viale, G., Weidner, N., Harris, A. L., Dirix, L. Y., 1996. "Quantification of angiogenesis in solid human tumours: An international consensus on the methodology and criteria of evaluation." *Eur. J. Cancer* **32A**, 2474–2484.

Wei, K., Jayaweera, A. R., Firoozan, S., Linka, A., Skyba, D. M., Kaul, S., 1998. "Quantification of myocardial blood flow with ultrasound-induced destruction of microbubbles administered as a constant venous infusion." *Circulation* **97**, 473–483.

Wilkening, W., Postert, T., Federlein, J., Kono, Y., Mattrey, R., Ermert, H., 2000. "Ultrasonic assessment of perfusion conditions in the brain and in the liver." *Proceedings of the IEEE Ultrasonics Symposium.*

11.5 Nuclear imaging of apoptosis in animal tumour models

Silvana Del Vecchio and Marco Salvatore

11.5.1 Introduction

Apoptosis or programmed cell death is a highly regulated multi-step process leading to selective cell death and elimination without a concomitant inflammatory reaction of surrounding tissue (Hengartner, 2000). A number of physiological processes including tissue renewal, wound healing and elimination of activated immune cells, when they have accomplished their function, depend on intact apoptosis. Apoptosis may also be triggered by several external stimuli including drugs, toxins, gamma irradiation, growth factors withdrawal and stimulation with cytokines of TNF family. Here, we focus our attention on drug-induced apoptosis in tumours.

Most anti-cancer agents can induce a series of different cellular responses which ultimately result in the activation of common apoptotic pathways (Kaufmann and Earnshaw, 2000, Johnstone, Ruefli and Lowe, 2002). Tumours that are unable to activate the apoptotic machinery in response to a death signal are therefore potentially resistant to treatment (Igney and Krammer, 2002). Conversely, the early increase of the number of apoptotic cells after initial administration of drugs may be predictive of a favourable final outcome of treatment (Chang *et al.*, 2000).

Recently, non-invasive *in vivo* detection of activated apoptotic pathways has been proven to be suitable both in pre-clinical and clinical settings using nuclear medicine techniques (Blankenberg, 2004). A number of radioligands mainly based on annexin V molecule have been developed and tested in animal models. Although some of them have been transferred to clinical studies, no definitive guidelines for apoptosis imaging are available at present. Nevertheless, many important clues may be derived by previous imaging studies that serve as a guidance. In this chapter we will primarily focus on critical issues that need to be addressed in the design and execution of imaging studies for non-invasive detection of apoptosis in animal tumour models.

11.5.2 Choice of cellular target and probe

There are three main pathways that initiate apoptosis and they are mediated by death receptors on the cell surface, by mitochondria and by endoplasmic reticulum (ER) (Waxman and Schwartz, 2003, Xu, 2005). All pathways leads to the activation of caspases which in turn cleave cellular substrates and cause the biochemical and morphological changes that are characteristic of apoptosis (Figure 11.5.1). At the time of caspases activation, the process is associated with a large lipid re-arrangement which leads to the exposure of phosphatidylserine on the plasma membrane. Phosphatidylserine is usually confined to the inner side of the phospholipids bilayer, and its fast externalization during apoptosis is reported to result from a de-activation in the translocase and floppase activity and enhancement of scramblase activity (Blankenberg, 2004).

Apoptosis signalling through both death receptors and mitochondria involves a complex interplay of a

Figure 11.5.1 Schematic representation of the main pathways of apoptosis. Apoptosis may be initiated by the death receptor pathway, by the mitochondrial pathway and by the endoplasmic reticulum pathway. All pathways converge on the activation of caspases that in turn cleave cellular substrates and cause the biochemical and morphological changes that are characteristic of apoptosis. At the time of caspases activation the process is associated with a large lipid rearrangement that leads to the exposure of phosphatidylserine on the plasma membrane

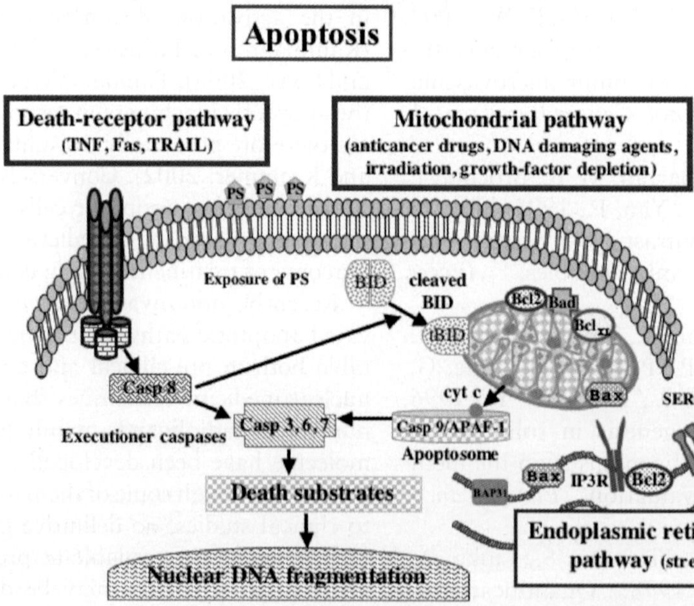

large array of molecules whose role and function are beyond the scope of the present chapter. However, despite the large number of molecules involved in the initiation, execution and regulation of the apoptotic program, three major classes of targets can be identified as suitable for imaging, namely lipids such as phosphatidylserine, caspases and mitochondria.

Phosphatidylserine is by far the most extensively investigated marker of apoptosis, and a large proportion of apoptosis-detecting radiopharmaceuticals are annexin V-based compounds which bind with high affinity to phosphatidylserine (Lahorte *et al.*, 2004). Most of the present knowledge on apoptosis imaging indeed derives from studies performed with radiolabelled annexin V compounds. Annexin V is a 35.8-kDa protein originally isolated from human placental tissue that has recently been made available as a recombinant protein. It has been labelled with both gamma and positron emitters for SPECT and PET applications, respectively. Rather than examining the radiochemistry, biodistribution and pharmacokinetic properties of each compound, for which we refer to the excellent review by Lahorte *et al.* (2004), we would like to use the phosphatidylserine/annexin V system to draw some general principles that may help to perform apoptosis imaging studies and to derive lessons that can be translated to other targets and probes.

11.5.2.1 Radiolabelled annexin V: definition of time window for imaging

As apoptosis is a dynamic process which ended with the elimination of the cell and hence of the target, imaging may be optimal only in a finite interval of time, usually few hours or days. Therefore, the first critical issue that needs to be addressed is the definition of the time window in which the selected target is exposed and can be reached by the radiolabelled probe to give an optimal signal. Blankenberg *et al.* reported that there are at least two peaks of annexin V uptake, one occurring few hours after treatment and the other at approximately 24–72 h after drug administration (Blankenberg, 2002). In particular, a 2- to 3-fold increase in annexin V uptake has been observed as early as 1 h, lasting approximately 90 min, after a single administration of cyclophosphamide (150 mg/kg i.p.) in Balb/c mice bearing luciferase-expressing BCL1 syngeneic murine lymphoma (Blankenberg, 2002). Furthermore, a time course of apoptotic response at delayed time points after a single dose of cyclophosphamide (150 mg/kg i.p.) in a rat hepatoma model showed that 99mTc-annexin V accumulation in tumours significantly increased at 20 h but not at 4 h and 12 h after treatment (Takei *et al.*, 2004). The peak of tracer uptake at 20 h was parallelled by a significant

Table 11.5.1 Anticancer drugs, administered dose, tracer, time of tracer injection and time of imaging or biodistribution in animal tumor models.

Trigger	Administered dose	Tracer	Time of tracer injection	Time of imaging or biodistribution	Reference
Cyclophosphamide	100 mg/kg i.p.	99mTc-Hynic-annexin V	20 h	1 h	Blankenberg et al., 1998
Doxorubicin	10 mg/kg i.p.	99mTc-mutant annexin V	Time course 1–48 h	1 h	Mandle et al., 2004
Cyclophosphamide	150 mg/kg i.p.	99mTc-Hynic-annexin V	Time course 4–20 h	6 h	Takei et al., 2004
Cyclophosphamide	150 mg/kg i.p.	99mTc-Hynic-annexin V	20 h	6 h	Mochizuki et al., 2003
Paclitaxel	25–100 mg/kg i.v.	111In-Pegylated annexin V	4 days	48 h	Ke et al., 2004
Gemcitabine	90 mg/Kg i.v	99mTc-Hynic-annexin V	48 h	6 h	Takei et al., 2005

increase in the percentage of TUNEL- and caspase-3 positively stained cells. The existence of two peaks in 99mTc-annexin V uptake has been also confirmed in a murine lymphoma model after a single injection of doxorubicin (10 mg/kg) (Mandl *et al.*, 2004). In this case the second peak occurred from 9 to 24 h after therapy and again was followed by tumour burden reduction as assessed by BLI. In a study performed in mice bearing breast carcinomas, taxane-induced tumour apoptosis has been detected by injecting PEGylated 111In-labelled annexin V four days after a single i.v. administration of poly(L-glutamic acid)-paclitaxel (100 mg eq paclitaxel/kg) (Ke *et al.*, 2004). In a murine model of Fas-mediated apoptosis, 99mTc-Hynic-annexin V detects histologically confirmed massive hepatic apoptosis as early as 1 h after administration of anti-Fas antibody (Blankenberg *et al.*, 1998).

On the basis of these observations, two considerations can be drawn. First, the optimal time window for apoptosis imaging depends on several factors including the cellular target chosen for imaging, the type of trigger, the dose of the trigger and the tumour system (i.e. lymphoproliferative malignancy or solid tumour). Although there are no definitive conclusions on the optimal time window for apoptosis imaging with radiolabelled annexin V, there is a growing body of evidences elucidating the dynamics of the target *in vivo* in response to different anti-cancer drugs and hence the dynamics of *in vivo* binding of radiolabelled annexin V in different tumour models. A rule of thumb to set up and perform imaging studies with radiolabelled annexin V in an animal tumour model may be to use these data as a starting point and then adapt the time window to a given tumour system and radiopharmaceutical by performing serial imaging, time course experiments and eventually dose escalation of a given trigger. Table 11.5.1 reports a list of anti-cancer agents, administered dose, time of injection of radiolabelled annexin V and time of imaging or biodistribution after tracer injection.

The second consideration that can be drawn is that early changes of the signal upon exposure to a trigger are not necessarily associated with the execution of the apoptotic program. The early peak of annexin V uptake is not associated to tumour cell loss and shrinkage, and its significance is presently unknown (Blankenberg *et al.*, 2002). Therefore, another important aspect that needs to be clarified is whether early changes of the signal after treatment indeed reflect a commitment of tumour cells to die.

After the injection of the tracer, it is necessary to wait for biodistribution and specific localization of the tracer at the site of phosphatidylserine exposure with the achievement of an optimal signal-to-background ratio. The lag time after injection depends mainly on the type of the tracer, on its biodistribution and plasma half-life as well as on the physical half-life of the radionuclide. The acquisition of images is usually performed 1–6 h after 99mTc-labelled annexin V injection. Blankenberg *et al.*, (1998) reported the successful detection of drug-induced apoptosis in mice bearing murine B cell lymphoma one hour after the injection of 99mTc-Hynic-annexin V (Figure 11.5.2). Due to the short plasma half-life of annexin V (<5 min) (Blankenberg *et al.*, 2000) and physical half-life of 99mTc (6 h), re-injection of the tracer and scanning at delayed time points after treatment may be considered to capture the peak of apoptotic process. Using a rat tumour model, Kuge *et al.* (2004) showed that cold annexin V injection (2 μg/kg) 24 h before or after cyclophosphamide treatment did not significantly affect the accumulation of 99mTc-annexin V in tumours (radioactive dose 5–23 MBq/kg;

Figure 11.5.2 Representative images of mice bearing B cell lymphoma in the left flank one hour after i.v. injection of 99mTc-Hynic-annexin V. Mice were not treated (right, R) or were treated with 100 mg/kg of cyclophosphamide (left, L) that was administered 20 hours before tracer injection. (Reprinted with permission from Blankenberg et al., 1998)(Copyrights 1998 National Academy of Science USA)

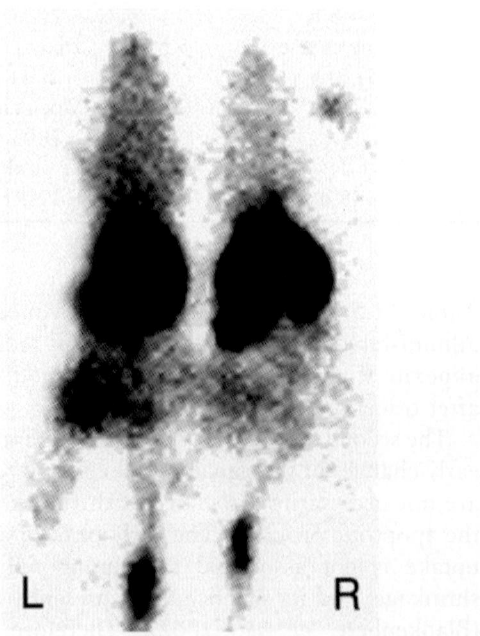

mass dose, 2 μg/kg) as compared to controls indicating the feasibility of repetitive injection of 99mTc-annexin V even one day apart.

The lag time for acquisition of images may be even shorter when annexin V is labelled with 18F (physical half-life Ho min). In an animal model of chemically induced liver apoptosis, images were acquired over a period of 2 h after the injection of 18F-annexin V (Yagle *et al.*, 2005). On the contrary, attempts have been made to prolong the effective half-life of annexin V by developing a Pegylated annexin V construct that was labelled with 111In (physical half-life 2.81 days) {Ke, 2004 #20. In this study planar images of treated mice were acquired at 48 h following tracer injection.

11.5.2.2 Alternative targets and probes

In the apoptotic cascade, caspases have been identified as potential targets for apoptosis imaging. Caspases are intra-cellular proteases that bear an active-site cystein and cleave peptide bonds after acid aspartic residues. They are sequentially activated during apoptosis and execute the program causing the majority of morphological and biochemical changes of apoptosis (Hengartner, 2000). Initiator caspases cleave and thereby activate downstream effector caspases. A pan-caspase inhibitor, benzyloxycarbonyl-Val-Ala-DL-Asp (O-methyl)-fluoromethyl ketone (Z-VAD-fmk), has been labelled with 131I, and uptake has been tested in Morris hepatoma cells transfected with the Herpes Simplex Virus thymidine kinase gene and hence sensitive to treatment with ganciclovir (Haberkorn *et al.*, 2001). A twofold increase of ^{131}I-Z-VAD-fmk was found in ganciclovir-treated cells as compared to control cells 24 h after treatment. The same authors have recently developed and tested 10 radiolabelled peptides containing the DEVDG sequence, selective for downstream caspases such as caspase-3 (Bauer *et al.*, 2005). To improve the cellular uptake of the tracer, the peptide sequence has been also coupled to a cell membrane transporter sequence that consisted of a selected region of the Tat protein. Among the series of peptides tested, those consisting of DEVDG and Tat sequence showed the most favourable uptake by apoptotic cells as compared with controls. Binding to and cleavage by caspase-3 was also shown for these compounds.

Many death signals converge on mitochondria where they trigger a series of events that are under the control of Bcl-2 family members (Desagher and Martinou, 2000). The increase of mitochondrial membrane permeability with the release into the cytoplasma of apoptogenic molecules smaller than 1.5 kDa and the disruption of mitochondrial membrane potentials are indeed considered early and decisive events in the apoptotic cascade.

Radiolabelled lipophilic cations are a class of tracers which accumulate into mitochondria in response to electronegative potentials of mitochondrial membrane and include both 99mTc-labelled cations and 3H or 18F-labelled tetraphenylphosphonium salts. Although 99mTc-lipophilic cations such as 99mTc-MIBI and 99mTc-tetrofosmin have been extensively studied as tumour-seeking agents, radiolabelled phosphonium cations have recently been investigated as probes for tumour imaging (Madar *et al.*, 2002, Min *et al.*, 2004). In addition, Madar *et al.* evaluated uptake of 18F-*p*-fluorobenzyltriphenyl-phosphonium during drug-induced apoptosis. *In vitro* models of taxotere-induced apoptosis in lung and prostate carcinoma cells showed a significant decrease of tracer uptake (Madar *et al.*, 2003). Also, 99mTc-MIBI uptake has been reported to decrease in breast cancer cell lines after treatment with anti-cancer drugs (Vergote *et al.*, 2001). The decrease of tracer uptake has been reported to be dose-dependent and time-dependent usually reaching a maximum at 72 h. Although cells may still appear viable, the downstream progression through the apoptotic cascade

Figure 11.5.3 MicroSPECT studies performed one hour after the injection of 99mTc-MIBI in nude mice bearing subcutaneous breast carcinomas. Coronal sections of the inferior region of the body of the same animal before (A) and after (B) treatment with cytostatic anti-cancer agents. Tumor (T) was detected after treatment. Activity in the bladder (BL) was due to physiological tracer excretion

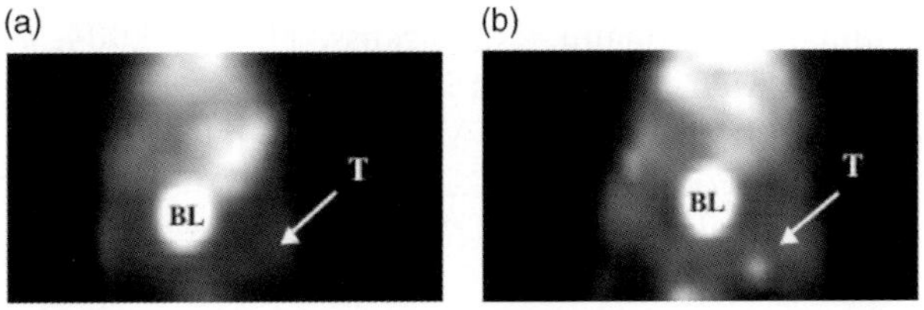

may lead to the disruption of mitochondrial membrane potentials and dissipation of the driving force behind lipophilic cations influx.

Recently, we have obtained consistent evidence that breast carcinomas which fail to accumulate 99mTc-MIBI have an altered apoptotic program due to the over-expression of the anti-apoptotic protein Bcl-2 (Del Vecchio *et al.*, 2003). A dramatic reduction of 99mTc-MIBI uptake was observed in breast cancer cell lines stably transfected with the human *bcl-2* gene as compared to control cells (Aloj *et al.*, 2004). Interestingly, treatment with staurosporine, a potent inducer of apoptosis, caused an early, partial and transitory recover of tracer uptake in transfected cells. A transitory increase of 99mTc-MIBI was also observed in control breast cancer cell lines immediately after staurosporine exposure. This early increase of tracer uptake could also been observed in nude mice bearing breast cancer xenografts using microSPECT imaging (Figure 11.5.3). These findings indicate a potential role of these tracers to image the mitochondrial pathway of apoptosis.

11.5.2.3 Validation of the probe

When imaging studies have been performed, tracer uptake in un-treated and treated tumours is compared, and then changes of the signal are measured. In this context, two main issues need to be addressed, namely whether the changes of the signal are indeed able to discriminate between treated and un-treated tumours in animal models and whether such changes are linearly related to the number of apoptotic cells within the tumours after treatment. Mochizuchi *et al* reported that uptake of 99mTc-labelled annexin V is significantly higher in treated tumours as compared to control animals (Mochizuki *et al.*, 2003). Furthermore, the increase of tracer uptake was correlated with the number of apoptotic cells in the tumour as assessed by terminal deoxynucleotidyl transferase-mediated deoxyuridine triphosphate nick-end labelling (TUNEL) of DNA fragments in tumour sections. Tunel staining is currently considered one of the reference techniques to determine the number of apoptotic cells in tissue sections along with standard morphological examination. Active caspase-3 immunostaining has also been used to determine the percentage of cells entering the execution phase of apoptosis (Takei *et al.*, 2004). In addition, regional distribution of radiolabelled annexin V was also evaluated by macroautoradiography of tissue sections and then correlated with *in situ* detection of nuclear DNA fragmentation in several animal models (Ke *et al.*, 2004, Taki *et al.*, 2004).

11.5.3 Conclusions

Non-invasive detection of apoptosis in animal tumour models is feasible with the currently available radioligands and nuclear imaging techniques. With the expansion of basic knowledge on molecular mechanisms of apoptosis, new targets and probes suitable for imaging will be identified and tested. Monitoring the ongoing apoptosis in animal tumour models is of primary importance to test the efficacy of new anticancer drugs. The anti-tumour activity of new compounds is presently evaluated using as surrogate end points tumour size, volume and specific growth delay. There is a high demand of innovative surrogate end points that may anticipate and also monitor tumour response in real time in animal models. This would help to improve drug screening and to accelerate drug development. In addition, a number of alterations have been proven to dysregulate apoptosis in tumours causing drug resistance. Many efforts are currently focused on the design of therapeutic strategies that may counteract defects in the apoptotic program. The possibility to monitor apoptotic response in animal tumour models bearing such alterations may significantly contribute to select novel agents that are capable of sensitizing tumour cells to the cytotoxic actions of traditional anti-cancer drugs. Major advances in this field are expected to derive from the combination of different imaging modalities including MRS/MRI and optical imaging.

References

Aloj, L., Zannetti, A., Caraco, C., Del Vecchio, S. and Salvatore, M., 2004. *Eur. J. Nucl. Med. Mol. Imaging* **31**, 521–527.

Bauer, C., Bauder-Wuest, U., Mier, W., Haberkorn, U., Eisenhut, M., 2005. *J. Nucl. Med.* **46**, 1066–1074.

Blankenberg, F., 2002. *Clin. Cancer Res.* **8**, 2757–2758.

Blankenberg, F. G., 2004. *Curr. Pharm. Des.* **10**, 1457–1467.

Blankenberg, F. G., Katsikis, P. D., Tait, J. F., Davis, R. E., Naumovski, L., Ohtsuki, K., Kopiwoda, S., Abrams, M. J., Darkes, M., Robbins, R. C., Maecker, H. T., Strauss, H. W., 1998. *Proc. Natl. Acad. Sci. USA* **95**, 6349–6354.

Blankenberg, F. G., Tait, J. F., Strauss, H. W., 2000. *Eur. J. Nucl. Med.* **27**, 359–367.

Chang, J., Ormerod, M., Powles, T. J., Allred, D. C., Ashley, S. E., Dowsett, M., 2000. *Cancer* **89**, 2145–2152.

Del Vecchio, S., Zannetti, A., Aloj, L., Caraco, C., Ciarmiello, A., Salvatore, M., 2003. *Eur. J. Nucl. Med. Mol. Imaging* **30**, 879–887.

Desagher, S., Martinou, J. C., 2000. *Trends Cell Biol.* **10**, 369–377.

Haberkorn, U., Kinscherf, R., Krammer, P. H., Mier, W., Eisenhut, M., 2001. *Nucl. Med. Biol.* **28**, 793–798.

Hengartner, M. O., 2000. *Nature* **407**, 770–776.

Igney, F. H., Krammer, P. H., 2002. *Nat. Rev. Cancer* **2**, 277–288.

Johnstone, R. W., Ruefli, A. A., Lowe, S. W., 2002. *Cell* **108**, 153–164.

Kaufmann, S. H., Earnshaw, W. C., 2000. *Exp. Cell Res.* **256**, 42–49.

Ke, S., Wen, X., Wu, Q. P., Wallace, S., Charnsangavej, C., Stachowiak, A. M., Stephens, C. L., Abbruzzese, J. L., Podoloff, D. A., Li, C., 2004. *J. Nucl. Med.* **45**, 108–115.

Kuge, Y., Sato, M., Zhao, S., Takei, T., Nakada, K., Seki, K. I., Strauss, H. W., Blankenberg, F. G., Tait, J. F., Tamaki, N., 2004. *J. Nucl. Med.* **45**, 309–312.

Lahorte, C. M., Vanderheyden, J. L., Steinmetz, N., Van de Wiele, C., Dierckx, R. A., Slegers, G., 2004. *Eur. J. Nucl. Med. Mol. Imaging* **31**, 887–919.

Madar, I., Weiss, L., Izbieki, G., 2002. *J. Nucl. Med.* **43**, 234–238.

Madar, I., Nelkin, B., Isaacs, J. T., Ravert, H. T., Scheffel, U., Dannals, R. F., Frost, J. J., 2003. *J. Nucl. Med.* **44**, 179P–180P (abstract).

Mandl, S. J., Mari, C., Edinger, M., Negrin, R. S., Tait, J. F., Contag, C. H., Blankenberg, F. G., 2004. *Mol. Imaging* **3**, 1–8.

Min, J. J., Biswal, S., Deroose, C., Gambhir, S. S., 2004. *J. Nucl. Med.* **45**, 636–643.

Mochizuki, T., Kuge, Y., Zhao, S., Tsukamoto, E., Hosokawa, M., Strauss, H. W., Blankenberg, F. G., Tait, J. F., Tamaki, N., 2003. *J. Nucl. Med.* **44**, 92–97.

Takei, T., Kuge, Y., Zhao, S., Sato, M., Strauss, H. W., Blankenberg, F. G., Tait, J. F., Tamaki, N., 2004. *J. Nucl. Med.* **45**, 2083–2087.

Taki, J., Higuchi, T., Kawashima, A., Tait, J. F., Kinuya, S., Muramori, A., Matsunari, I., Nakajima, K., Tonami, N., Strauss, H. W., 2004. *J. Nucl. Med.* **45**, 1536–1541.

Vergote, J., Di Benedetto, M., Moretti, J. L., Azaloux, H., Kouyoumdjian, J. C., Kraemer, M., Crepin, M., 2001. *Cell Mol. Biol. (Noisy-le-grand)* **47**, 467–471.

Waxman, D. J., Schwartz, P. S., 2003 *Cancer Res.* **63**, 8563–8572.

Xu, C., Bailly-Meitre, B., Reed, J. C., 2005. *J. Clin Invest.* **115**, 2656–2664.

Yagle, K. J., Eary, J. F., Tait, J. F., Grierson, J. R., Link, J. M., Lewellen, B., Gibson, D. F., Krohn, K.A., 2005. *J. Nucl. Med.* **46**, 658–666.

11.6 Optical imaging of tumour-associated protease activity

Benedict Law and Ching-Hsuan Tung

Proteases are enzymes that cleave proteins by catalytic hydrolysis of peptide–peptide bonds. There are four classes of proteases (aspartic, cysteine, serine and metalloproteinase) that are distinguished by their catalytic mechanisms (Barrett, Rawlings and Woessner, 1998). In tumour biology, many proteases play important roles in the degradation of the extracellular matrix (ECM), which occurs in the early stages of carcinogenesis, and they are involved in controlling tumour growth and development (DeClerck *et al.*, 2004, Lee, Fridman and Mobashery, 2004). Clinically, in addition to serving as biomarkers for certain tumours, the activities of proteases can also predict the outcomes of diseases (Levicar *et al.*, 2002; Vihinen and Kahari, 2002; Nordengren *et al.*, 2002; Borgfeldt *et al.*, 2003; Cianfrocca and Goldstein, 2004). By using proteases as prognostic factors in early-stage cancer, physicians can improve and optimize therapeutic treatment strategies (Harbeck *et al.*, 2002; Zemzoum *et al.*, 2003), especially when combined with other well-established parameters such as tumour size, grade, lymph-node status, ethnicity and patient age. A better understanding of the concerted actions of proteases may lead to the development of more accurate and reliable pharmacological approaches for controlling tumour progression.

11.6.1 Protease sensitive probes

It is well known that many tumour-associated proteases are expressed locally. This regional protease activity may be a useful indicator for tumour detection and diagnosis. A variety of protease sensitive reporters have been designed for animal imaging (Table 11.6.1) (Funovics, Weissleder and Tung, 2003). As previously described in Chapter 7.3, these probes are optically silent and designed to become highly fluorescent after interacting with their target protease.

11.6.2 Tumour detection and characterization

Cathepsins and Matrix metalloproteinases (MMPs) are two families of proteases found in most tumours.

Table 11.6.1 Available protease-mediated probes and their applications in animal imaging

Target protease	Disease	Refs.
Broad cathepsin	Cancer	(Weissleder *et al.*, 1999, Bremer *et al.*, 2002)
	Arthritis	(Lai *et al.*, 2004, Wunder *et al.*, 2004)
	Atheroscleosis	(Chen *et al.*, 2002)
Cathepsin D	Cancer	(Tung *et al.*, 2000)
Caspase 1	Apoptosis	(Messerli *et al.*, 2004)
HIV protease	HIV infection	(Shah *et al.*, 2004a)
HSV protease	HSV infection	(Shah *et al.*, 2004b)
Gelatinase	Infarction	(Chen *et al.*, 2005)
MMP-2	Cancer	(Bremer *et al.*, 2001b, Bremer *et al.*, 2001a)
MMP-7	Cancer	(McIntyre *et al.*, 2004)
Thrombin	Cardiovascular	(Jaffer *et al.*, 2002)
Urokinase Plasminogen Activator	Cancer	(Law *et al.*, 2004)

Fluorescence EMPs (enzyme-mediated probes), which target these exogenous proteases, are particularly useful for early detection of cancer. Following intravenous injection, the probes tend to accumulate at the tumour sites. The expression levels of the targeted proteases are monitored by the increase in fluorescence signal *via* optical imaging (Bremer *et al.*, 2002). This sensitive technique has been used to detect pre-cancer-staged tumours or polyps. Adenomas sizes as small as 50 μm in diameter have been observed by fluorescence microscopy (Marten *et al.*, 2002). This approach has potential applications in clinical settings, which may allow for the differentiation of small, highly dysplastic adenomas from innocuous lesions. Probes that are specific to other proteases such as cathepsin D (Tung *et al.*, 2000), MMP-2 (Bremer *et al.*, 2001a, Bremer *et al.*, 2001b), MMP-7 (McIntyre *et al.*, 2004) and uPA (Law *et al.*, 2004) have also been developed and optimized for various tumour models.

11.6.3 Image-guided surgery

In addition to diagnostic applications, protease-mediated probes can be applied to guide surgical resection. As previously described in a colonic adenoma study, early colonic polyps could be identified and removed during endoscopy. Surgical resection directed by optical imaging may be a reliable method for reducing the incidence of colorectal cancer. Other catheter-based endoscopic imaging and resection systems can also be adopted easily for different tumour types, including lung, esophagus and genitourinary tract carcinomas.

As EMP may aid in localizing tumour and identifying tumour boundaries (Figure 11.6.1), specially

Figure 11.6.1 Light (left) and NIR Fluorescence (right) images of HT-1080 tumour-containing mouse with uPA probes. The location and boundary of tumour was clearly identified after surgical removal of tissues and skins

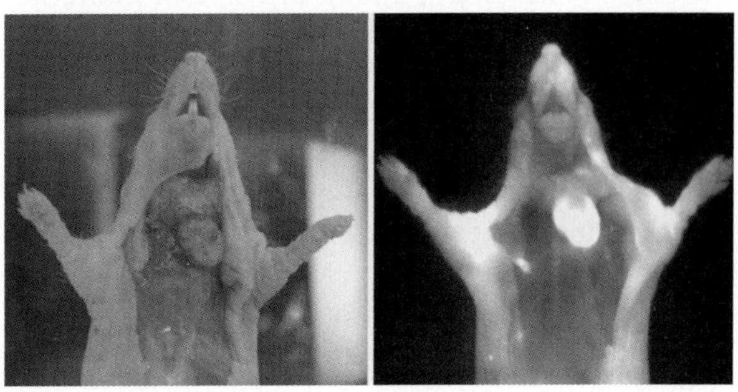

designed NIR goggles with the appropriate fluorescent filters could be extremely useful during surgery. Such real-time intra-operative imaging could guide surgeons in the operating room and is expected to improve current patient prognosis significantly.

11.6.4 Drug evaluation and screening

Recently, attention has been directed towards the research and development of protease inhibitors in the pharmaceutical industry. Many of these inhibitors have been approved for clinical use and more are currently in clinical trials (Southan, 2001). *In vivo* imaging of protease activity would be useful in assisting drug development and potentially for therapeutic evaluation. The concept of imaging drug-related inhibitory effects was demonstrated recently using a matrix metalloproteinase (MMP) sensitive probe (Bremer *et al.*, 2001b). In one study, tumour-bearing animals were pre-treated with prinomastat, an MMP inhibitor, and imaged 2 h after probe administration. The MMP-2 induced fluorescence signal from the probe in the treated tumours was significantly lower than that from untreated ones. By using a similar probe design concept, a dendrimer-based probe was also reported to image MMP-7 (McIntyre *et al.*, 2004). Systemic treatment with an MMP inhibitor (BB-94) decreased the fluorescence intensities of the treated tumours by nearly 60%. These studies suggest that it is feasible to image protease inhibition in living animals.

11.6.5 Other application of protease imaging

Optical technology can help to gain a better understanding of the complex tumour biology *in vivo*. In addition, non-invasive *in vivo* imaging of protease has been extended to other medical areas, including atherosclerosis (Chen *et al.*, 2002), myocardial infarction (Chen *et al.*, 2005), thrombosis (Jaffer *et al.*, 2002), arthritis (Wunder *et al.*, 2004, Lai *et al.*, 2004) and gene therapy (Shah *et al.*, 2004a) (Table 11.6.1). Although the initial studies were all performed in small animals, the technology could be translated into the clinic. Potentially, the acquired *in vivo* information of protease activity could revolutionize current approaches in disease detection, screening, diagnosis, staging, drug development and treatment evaluation.

References

Barrett, A. J., Rawlings, N. D., Woessner, J. F., 1998. *Handbook of Proteolytic Enzymes*, San Diego, Academic Press.

Borgfeldt, C., Bendahl, P. O., Gustavsson, B., Langstrom, E., Ferno, M., Willen, R., Grenman, S., Casslen, B., 2003. "High tumor tissue concentration of urokinase plasminogen activator receptor is associated with good prognosis in patients with ovarian cancer." *Int. J. Cancer* **107**, 658–665.

Bremer, C., Bredow, S., Mahmood, U., Weissleder, R., Tung, C. H. (2001a) "Optical imaging of matrix metalloproteinase-2 activity in tumors: Feasibility study in a mouse model." *Radiology* **221**, 523–529.

Bremer, C., Tung, C. H., Bogdanov, A., Jr., Weissleder, R., 2002. "Imaging of differential protease expression in breast cancers for detection of aggressive tumor phenotypes." *Radiology* **222**, 814–818.

Bremer, C., Tung, C. H., Weissleder, R., 2001b. "In vivo molecular target assessment of matrix metalloproteinase inhibition." *Nat. Med.* 7, 743–748.

Chen, J., Tung, C. H., Allport, J. R., Chen, S., Weissleder, R., Huang, P. L., 2005. "Near-infrared fluorescent imaging of matrix metalloproteinase activity after myocardial infarction." *Circulation* **111**, 1800–1805.

Chen, J., Tung, C. H., Mahmood, U., Ntziachristos, V., Gyurko, R., Fishman, M. C., Huang, P. L., Weissleder, R., 2002. "In vivo imaging of proteolytic activity in atherosclerosis." *Circulation* **105**, 2766–2771.

Cianfrocca, M., Goldstein, L. J., 2004. "Prognostic and predictive factors in early-stage breast cancer." *Oncologist* **9**, 606–616.

DeClerck, Y. A., Mercurio, A. M., Stack, M. S., Chapman, H. A., Zutter, M. M., Muschel, R. J., Raz, A., Matrisian, L. M., Sloane, B. F., Noel, A., Hendrix, M. J., Coussens, L., Padarathsingh, M., 2004. "Proteases, extracellular matrix, and cancer: a workshop of the path B study section." *Am. J. Pathol.* **164**, 1131–1139.

Funovics, M., Weissleder, R., Tung, C. H., 2003. "Protease sensors for bioimaging." *Anal. Bioanal. Chem.* **377**, 956–963.

Harbeck, N., Kates, R. E., Look, M. P., Meijer-Van Gelder, M. E., Klijn, J. G., Kruger, A., Kiechle, M., Janicke, F., Schmitt, M., Foekens, J.A., 2002. "Enhanced benefit from adjuvant chemotherapy in breast cancer patients classified high-risk according to urokinase-type plasminogen activator (uPA) and plasminogen activator inhibitor type 1 ($n = 3424$)." *Cancer Res.* **62**, 4617–4622.

Jaffer, F. A., Tung, C. H., Gerszten, R. E., Weissleder, R., 2002. "In vivo imaging of thrombin activity in experimental thrombi with thrombin-sensitive near-infrared molecular probe. Arterioscler." *Thromb. Vasc. Biol.* **22**, 1929–1935.

Lai, W. F., Chang, C. H., Tang, Y., Bronson, R., Tung, C. H., 2004. "Early diagnosis of osteoarthritis using cathepsin B sensitive near-infrared fluorescent probes. Osteoarthritis" *Cartilage* **12**, 239–244.

Law, B., Curino, A., Bugge, T. H., Weissleder, R., Tung, C. H., 2004. "Design, synthesis, and characterization of urokinase plasminogen-activator-sensitive near-infrared reporter." *Chem. Biol.* **11**, 99–106.

Lee, M., Fridman, R., Mobashery, S., 2004. "Extracellular proteases as targets for treatment of cancer metastases." *Chem. Soc. Rev.* **33**, 401–409.

Levicar, N., Kos, J., Blejec, A., Golouh, R., Vrhovec, I., Frkovic-Grazio, S., Lah, T. T., 2002. "Comparison of potential biological markers cathepsin B, cathepsin L, stefin A and stefin B with urokinase and plasminogen activator inhibitor-1 and clinicopathological data of breast carcinoma patients." *Cancer Detect Prev.* **26**, 42–49.

Marten, K., Bremer, C., Khazaie, K., Sameni, M., Sloane, B., Tung, C. H., Weissleder, R., 2002. "Detection of dysplastic intestinal adenomas using enzyme-sensing molecular beacons in mice." *Gastroenterology* **122**, 406–414.

McIntyre, J. O., Fingleton, B., Wells, K. S., Piston, D. W., Lynch, C. C., Gautam, S., Matrisian, L.M., 2004. "Development of a novel fluorogenic proteolytic beacon for in vivo detection and imaging of tumour-associated matrix metalloproteinase-7 activity." *Biochem. J.* 377, 617–628.

Messerli, S. M., Prabhakar, S., Tang, Y., Shah, K., Cortes, M. L., Murthy, V., Weissleder, R., Breakefield, X. O., Tung, C. H., 2004. "A novel method for imaging apoptosis using a caspase-1 near-infrared fluorescent probe." *Neoplasia* **6**, 95–105.

Nordengren, J., Fredstorp Lidebring, M., Bendahl, P. O., Brunner, N., Ferno, M., Hogberg, T., Stephens, R. W., Willen, R., Casslen, B., 2002. "High tumor tissue concentration of plasminogen activator inhibitor 2 (PAI-2) is an independent marker for shorter progression-free survival in patients with early stage endometrial cancer." *Int. J. Cancer* **97**, 379–385.

Shah, K., Tung, C. H., Chang, C. H., Slootweg, E., O'loughlin, T., Breakefield, X. O., Weissleder, R., 2004a. "In vivo imaging of HIV protease activity in amplicon vector-transduced gliomas." *Cancer Res.* **64**, 273–278.

Shah, K., Tung, C. H., Yang, K., Weissleder, R., Breakefield, X. O., 2004b. "Inducible release of TRAIL fusion proteins from a proapoptotic form for tumor therapy." *Cancer Res.* **64**, 3236–3242.

Southan, C., 2001. "A genomic perspective on human proteases as drug targets." *Drug Discov. Today* **6**, 681–688.

Tung, C. H., Mahmood, U., Bredow, S., Weissleder, R., 2000. "In vivo imaging of proteolytic enzyme activity using a novel molecular reporter." *Cancer Res.* **60**, 4953–4958.

Vihinen, P., Kahari, V. M., 2002. "Matrix metalloproteinases in cancer: Prognostic markers and therapeutic targets." *Int. J. Cancer* **99**, 157–166.

Weissleder, R., Tung, C. H., Mahmood, U., Bogdanov, A., Jr., 1999. "In vivo imaging of tumors with protease-activated near-infrared fluorescent probes." *Nat. Biotechnol.* **17**, 375–378.

Wunder, A., Tung, C. H., Muller-Ladner, U., Weissleder, R., Mahmood, U., 2004. "In vivo imaging of protease activity in arthritis: A novel approach for monitoring treatment response." *Arthritis Rheum.* **50**, 2459–2465.

Zemzoum, I., Kates, R. E., Ross, J. S., Dettmar, P., Dutta, M., Henrichs, C., Yurdseven, S., Hofler, H., Kiechle, M., Schmitt, M., Harbeck, N., 2003. "Invasion factors uPA/PAI-1 and HER2 status provide independent and complementary information on patient outcome in node-negative breast cancer." *J. Clin. Oncol.* **21**, 1022–1028.

11.7 Tumour angiogenesis and blood flow

Rakesh K. Jain, Dai Fukumura, Lance L. Munn and Edward B. Brown

The growth of new blood vessels in a tumour ('tumour angiogenesis') produces vessels that are highly tortuous, with blood flow that is spatially and temporally heterogeneous (see Figure 11.7.1) (Endrich *et al.*, 1979). A detailed understanding of the biochemical and biophysical processes that generate these vessels may lead to therapies that block and/or modify tumour angiogenesis and hence facilitate treatment of tumours (Carmeliet and Jain, 2000). This requires the accurate *in vivo* quantification of tumour

Figure 11.7.1 Multi-photon laser-scanning microscopy image of the vasculature in an LS174T colon adenocarcinoma growing in the dorsal skinfold chamber of an SCID mouse. The vasculature was highlighted with an i.v. injection of FITC-dextran (molecular weight two million). The image is a maximum intensity projection of 20 images spaced 5 μm apart and is 370 μm across

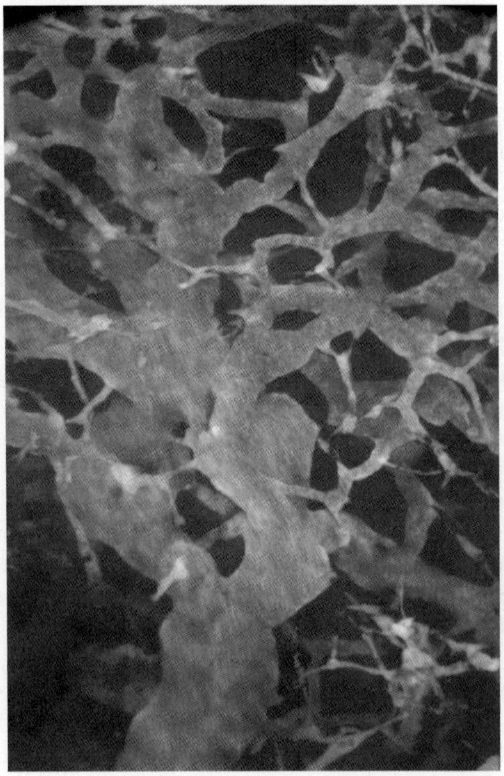

angiogenesis and blood flow. A review of recent literature shows that *in vivo* quantification of tumour vasculature has revealed morphological and functional normalization of tumour vasculature by an anti-angiogenic treatment, which allows improved delivery of therapeutic agents and potentiates cytotoxic therapies (Winkler *et al.*, 2004). Quantitation of vascular diameter and branching has revealed the role of nitric oxide in vessel morphogenesis (Kashiwagi *et al.*, 2005). Measurement of vascular volume fraction has revealed the anti-angiogenic nature of a candidate tumour suppressor gene (Garkavtsev *et al.*, 2004).

Tumour angiogenesis and blood flow is traditionally quantified by epifluorescence microscopy of a tumour growing in a chronic animal window model such as the dorsal skinfold chamber or cranial window (Fukumura *et al.*, 1997a) (see Chapter 6). First, the vasculature is highlighted with intravenous injection of 100 μl of FITC-Dextran (2 M MW, 10 mg/ml) and fluorescence time-lapse images are recorded for 60 s. Next, a vessel of interest is chosen from the video files, and its diameter is measured using NIH Image or an analogous software package. The transverse red blood cell velocity (i.e. the red blood cell velocity in the plane of the image, V_{TRBC}) is measured by selecting two to four evenly spaced regions of interest (ROIs) in the blood vessel and performing a temporal cross-correlation of the signal in each region [Intaglietta, Breit and Tompkins, 1990]. The distance between ROIs divided by the cross-correlation time yields the transverse red blood cell velocity. The mean transverse blood flow rate of the vessel (Q_T) is then given by

$$Q_T = \frac{\pi}{4} \times V_{Tmean} \times D^2$$

where $V_{Tmean} = V_{TRBC}/a$ ($a = 1.3$ for blood vessels smaller than 10 μm diameter, $1.3 < a < 1.6$ for blood vessels between 10 and 15 μm, and $a = 1.6$ for blood vessels greater than 15 μm diameter) (Lipowsky, Kovalcheck and Zweifach, 1987). For larger scale network properties, the two-dimensional functional vessel density is defined as the total length of vessels per unit image area, and the two-dimensional branching index is defined as the number of branching points and/or mean length of un-branched segments in the image (Fukumura *et al.*, 1997b; Kashiwagi *et al.*, 2005). Both are determined using image skeletonization and analysis algorithms.

Unfortunately, epifluorescence microscopy produces a two-dimensional image of the complex three-dimensional tumour vessel network. Consequently, the length of a given vessel segment is systematically under-estimated by the cosine of the angle between the vessel flow axis and the plane of the image. This results in a systematic under-estimation in vessel flow rates for the same reason (Brown *et al.*, 2001). All of the above equations specify *transverse* velocities and flow rates to make this systematic error clear.

To address this systematic error, it is necessary to generate complete three-dimensional images of the vascular network. This is possible with the confocal laser-scanning microscope, although not to any great depth in scattering tissue. The multi-photon laser-scanning microscope (MPLSM) allows three-dimensional imaging down to a depth of half a millimetre and hence permits determination of the absolute vessel segment length and blood flow velocity (Brown *et al.*, 2001). This is performed by first generating a three-dimensional image stack of the vessel network. A vessel of interest is chosen, and the transverse red blood cell velocity (V_{TRBC}) is measured by the line scan technique (Kleinfeld *et al.*, 1998). The absolute red blood cell velocity

(V_{RBC}) is subsequently calculated by determining the angle theta between the line scan and the axis of the vessel using the three-dimensional image stack. The absolute red blood cell velocity is then $V_{RBC} = V_{TRBC}/\cos(\text{theta})$. The mean absolute blood flow rate of a vessel (Q) is given by

$$Q = \frac{\pi}{4} \times V_{Tmean} \times D^2$$

where $V_{mean} = V_{RBC}/a$ ($a = 1.3$ for blood vessels smaller than 10 μm diameter, $1.3 < a < 1.6$ for blood vessels between 10 and 15 μm, and $a = 1.6$ for blood vessels greater than 15 μm diameter). The three-dimensional functional vessel density is defined as the total length of vessels per unit image stack volume whereas the branching index is defined as the mean length of un-branched segments. Both are analysed from the image stack using image analysis algorithms. Because full three-dimensional data of the vessel bed are generated, analysis algorithms are more complex (and possibly more expensive) (Abdul-Karim *et al.*, 2003; Tyrrell *et al.*, 2005).

References

Abdul-Karim, A., Al-Kofahi, K., Brown, E., *et al.*, 2003. "Automated tracing and change analysis of angiogenic vasculature from in vivo multiphoton image time series." *Microvasc. Res.* **66**(2), 113–125.

Brown, E., Campbell, R., Tsuzuki, Y., *et al.*, 2001. "In vivo measurement of gene expression, angiogenesis, and physiological function in tumors using multiphoton laser scanning microscopy." *Nat. Med.* 7(7), 864–868.

Carmeliet, P., Jain, R. K., 2000. "Angiogenesis in cancer and other diseases." *Nature* **407**(6801), 249–257.

Endrich, B., Reinhold, H. S., Gross, J. F., *et al.*, 1979. "Tissue perfusion inhomogeneity during early tumor growth in rats." *J. Natl. Cancer Inst.* **62**(2), 387–395.

Fukumura, D., Yuan, F., Endo, M., *et al.*, 1997a. "Role of nitric oxide in tumor microcirculation. Blood flow, vascular permeability, and leukocyte-endothelial interactions." *Am. J. Pathol.* **150**(2), 713–725.

Fukumura, D., Yuan, F., Monsky, W. L., *et al.*, 1997b. "Effect of host microenvironment on the microcirculation of human colon adenocarcinoma." *Am. J. Pathol.* **151**(3), 679–688.

Garkavtsev, I., Kozin, S. V., Chernova, O., *et al.*, 2004. "The candidate tumour suppressor protein ING4 regulates brain tumour growth and angiogenesis." *Nature* **428**(6980), 328–332.

Intaglietta, M., Breit, G. A., Tompkins, W. R., 1990. "Four window differential capillary velocimetry." *Microvasc. Res.* **40**(1), 46–54.

Kashiwagi, S., Izumi, Y., Gohongi, T., *et al.*, 2005. "NO mediates mural cell recruitment and vessel morphogenesis in murine melanomas and tissue-engineered blood vessels." *J. Clin. Invest.* **115**(7), 1816–1827.

Kleinfeld, D., Mitra, P. P., Helmchen, F., *et al.*, 1998. "Fluctuations and stimulus-induced changes in blood flow observed in individual capillaries in layers 2 through 4 of rat neocortex." *Proc. Natl. Acad. Sci. USA* **95**(26), 15741–15746.

Lipowsky, H. H., Kovalcheck, S., Zweifach, B. W., 1987. "The distribution of blood rheological parameters in the microvasculature of cat mesentery." *Circ. Res.* **43**(5), 738–749.

Tyrrell, J. A., Mahadevan, V., Tong, R. T., *et al.*, 2005. "A 2-D/3-D model-based method to quantify the complexity of microvasculature imaged by in vivo multiphoton microscopy." *Microvasc. Res.* **70**(3), 165–178.

Winkler, F., Kozin, S. V., Tong, R. T., *et al.*, 2004. "Kinetics of vascular normalization by VEGFR2 blockade governs brain tumor response to radiation: Role of oxygenation, angiopoietin-1, and matrix metalloproteinases." *Cancer Cell* **6**(6), 553–563.

11.8 Optical imaging of apoptosis in small animals

Eyk Schellenberger

The dynamic equilibrium of a constant cell number in any healthy, homeostatic tissue relies on the balance of cell proliferation and the removal of cells, which normally happens through cell death. Necrosis – the poorly regulated, pathological form of cell death is not suited for this physiological balance, as it leads to the rupture of the mitochondrial and cell membrane causing an efflux of cell contents and finally the damage of neighbouring cells directly or through the mediation of inflammatory responses. Therefore the evolution developed a programmed form of cell death known as apoptosis. The apoptotic cell death is conducted by a highly regulated, ATP-dependent system, where proteases – especially caspases – play a crucial role. The apoptotic programme leads to the self-digestion of the cells and finally to the compartmentalization into apoptotic bodies, which can be safely phagocytized by neighbouring cells or professional phagocytes. As the cell membrane stays intact, no damaging inflammation is induced, making the save removal possible(Kerr, Wyllie and Currie, 1972). The pre-requisites for the successful execution of the apoptotic programme are sufficient energy reserves and the integrity of the apoptotic system, which is normally the case during physiological triggers for cell removal or less intense tissue insults. In contrast, too intense insults, for example due to high concentrations of toxins or intense radiation, can cause disruptions of the apoptotic machinery or energy depletion, which can drive cells during all stages of the apoptotic programme into a necrotic cell death.

The disruption of the equilibrium of proliferation and cell removal is central to many diseases. Increased apoptosis rates are typical for diseases like AIDS, Alzheimer's disease, progressive heart failure, chronic hepatitis and transplant rejection (Blankenberg, Tait and Strauss, 2000; Blankenberg and Strauss, 2001). Tumour growth is characterized by a relative to mitosis insufficient apoptosis. In contrast, many treatment strategies like chemo- or radiotherapy, anti-hormonal therapeutics (Thompson, 1995) or anti-angiogenic therapies (Brooks *et al.*, 1994) are based on the induction of apoptosis. Imaging of apoptosis *in vivo* gives the opportunity to monitor treatment effects, long before traditional methods can detect morphological changes (Blankenberg *et al.*, 2001). The broad applicability is especially encouraging for the development of apoptosis imaging probes.

Two features of the apoptotic programme are especially interesting for molecular imaging. One is the activation of the bottleneck protease caspase 3, which is a highly specific indicator of apoptosis induction. The second is the externalization of negative-charged phospholipids (especially phosphatidylserines) from the inner leaflet of the cell membrane to the outside.

11.8.1 Imaging caspase activity

The caspase cascade plays an essential role in initiating and executing the process of apoptosis. As cysteine proteases that cleave substrates after a conserved aspartate residue, they are attractive targets for the design of activatable, fluorescent probes. However, imaging probes for caspases need to penetrate the cell membrane to reach the cytosolic caspases. Laxman *et al.* solved this problem in part by expressing a recombinant luciferase reporter molecule in the target cells, which has an attenuated reporter activity (Laxman *et al.*, 2002). In apoptotic cells specific cleavage by activated caspase 3 restores the luciferase activity, which can be detected *in vivo* by bioluminescence imaging. Yet this method is applicable for only animal models, as the recombinant cell line has to be implanted and the administration of relative high doses of luciferin (150 mg/kg) as substrate for generating the bioluminescence is necessary. Nevertheless, this system provides a useful method to study the induction of apoptosis in a very specific way and to monitor treatment effects, for example in pharmacological research.

Tung *et al.* designed a probe for near-infrared fluorescence imaging activatable by the specific cleavage by casapase 1 (interleukin 1 converting enzyme, ICE)(Messerli *et al.*, 2004). This probe is one example for smart probes, where multiple fluorochromes are attached *via* a caspase 1-cleavable peptide bridge to a partially PEGylated polylysine backbone. Because of the small distance of the fluorochromes from each other, a strong fluorescence resonance energy transfer (FRET) quenches the fluorescence effectively. The activated caspase 1 can specifically cleave the peptide bridges, thereby releasing the fluorochromes and interrupting the FRET – the fluorescence signal is switched on. With this probe vector-expressed caspase-1 activity could successfully be monitored *in vivo*. As an example for the so-called smart sensor probe design(Tung *et al.*, 1999; Weissleder *et al.*, 1999; Bremer, Tung and Weissleder, 2001), the system can be adapted to other caspases like caspases 3. However, the probes need to be internalized, a process which likely has different dynamics in different cell types.

11.8.2 Imaging phosphatidylserine externalization

As an important advantage to the caspase probes, phosphatidylserine-targeting probes do not need to enter cells. Normal, healthy cells maintain actively an asymmetry of negative-charged phospholipids (PL) (e.g. phosphatidylserine, phosphatidylethanolamine), which are almost exclusively distributed at the inner leaflet of the cell membrane. During the early phase of apoptosis, these negative-charged PL flip to the outer leaflet and can be recognized by annexin V (Gottlieb and Kitsis, 2001), a human protein (66 kDa), which binds with receptor like affinity (Ravanat et al., 1992; Blankenberg, Narula and Strauss, 1999) to the

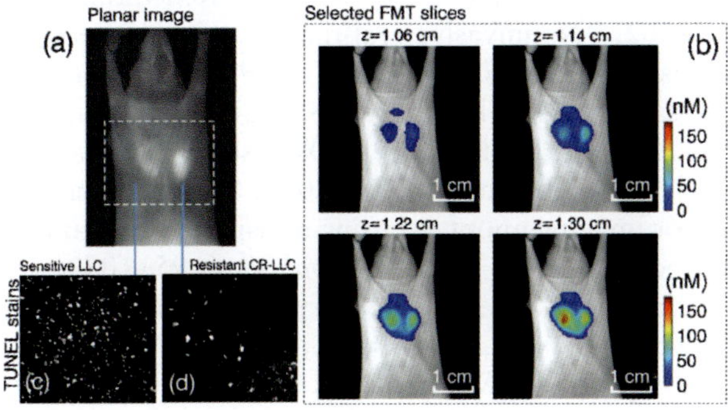

Figure 11.8.1 Imaging apoptotic response *in vivo*: (a) Planar fluorescence image (b) Four consecutive FMT slices (in colour) superimposed on the planar image of the mouse obtained at the excitation wavelength. The bottom right slice is the one closer to the surface of the animal as seen on (a) and successive slices are re-constructed from deeper in the animal. (c, d) TUNEL-stained histological slices from the LLC and CR-LLC tumours (Ntziachristos *et al.*, 2004)

phosphatidylserine presenting cell membranes. One principle disadvantage of annexin V imaging is that a distinction between apoptosis and necrosis is not possible, as the cell membrane becomes leaky during necrosis, and annexin can enter the cell and bind to the phosphatidylserines at the inner leaflet. Nevertheless, it is a very useful tool because a distinction is not necessary for many applications, for example for monitoring of treatment efficiency.

For near-infrared fluorescence (NIRF) imaging of apoptosis, annexin V can be labelled with infrared dyes in a way that they are either active with one mol dye coupled per mol annexin V protein or inactive as control with two or more dyes per annexin (Graves *et al.*, 2003; Schellenberger, Weissleder and Josephson, 2004). One form, the NIRF probe annexin V-Cy5.5, could successfully be used to detect apoptosis for monitoring treatment effects of chemotherapy in different tumour models by NIRF reflectance imaging (Schellenberger *et al.*, 2003, Petrovsky *et al.*, 2003). The same was probe used for imaging increased apoptosis as a result of methotrexate treatment in collagen-induced rheumatoid like arthritis (Wunder *et al.*, 2005). In addition to these reflectance-imaging experiments, annexin V-Cy5.5 could also be applied for apoptosis imaging in a tomographic system designed by Ntzichristos *et al.* (Graves *et al.*, 2003, Ntziachristos *et al.*, 2004). Figure 11.8.1 shows an experiment where a chemosensitive and a chemoresistant Lewis Lung Carcinomas (LLC) in the flanks of a mouse were treated with cyclophosphamide (Ntziachristos *et al.*, 2004). After i.v. injection of annexin V-Cy5.5, the animals were imaged in reflectance mode (Figure 11.8.1(a)) and tomographic mode (Figure 11.8.1(b)). Interestingly, the NIRF reflectance mode could not reveal a higher fluorescence signal corresponding to the higher apoptosis rate of the chemosensitive compared with the resistant tumour (confirmed by TUNEL assay Figure 11.8.1(c, d)). The elevated attenuation of the supposedly higher signal can be caused, for example by higher tissue absorption due to tumour necrosis or by a prolonged light penetration due to a deeper implantation site of the tumour. In contrast, the higher fluorescence signal of the chemosensitive tumour could be correctly re-constructed with the tomographic imaging system, and additionally, quantitative concentrations were given (Figure 11.8.1(b)).

The application of an additional feature of optical imaging probes is shown in Figure 11.8.2. Near-infrared imaging probes can also be used for fluorescence microscopy and flow cytometry. In that way the

Figure 11.8.2 Annexin V-Cy5.5 as an example for versatility of optical imaging probes – application in fluorescence microscopy and flow cytometry: Time course of Cy5.5 -annexin V binding and caspase activation after treatment of Jurkat T cells with camptothecin. (a) FACS analyses at selected times after induction of apoptosis (0, 4, 7.5 and 15 h). FL1 channel is from the caspase probe FITC-VAD-FMK. FL4 channel is from Cy5.5 – annexin V. (b) Percentages of cells in each of the four quadrants in (a) are shown along with additional time points. Cells that are caspase negative/annexin V positive are evident after 2–4 h of treatment. (c) Fluorescent micrographs of representative cells shown in (a) and (b). The caspase probe is green and cytoplasmic whereas the annexin V is red and membrane localized. A rare caspase+/annexin V− cell (found in the upper left quadrant of 5A) was touching a more numerous caspase −/annexin V+ cell (lower right quadrant) in the micrograph (Schellenberger, Weissleder and Josephson, 2004)

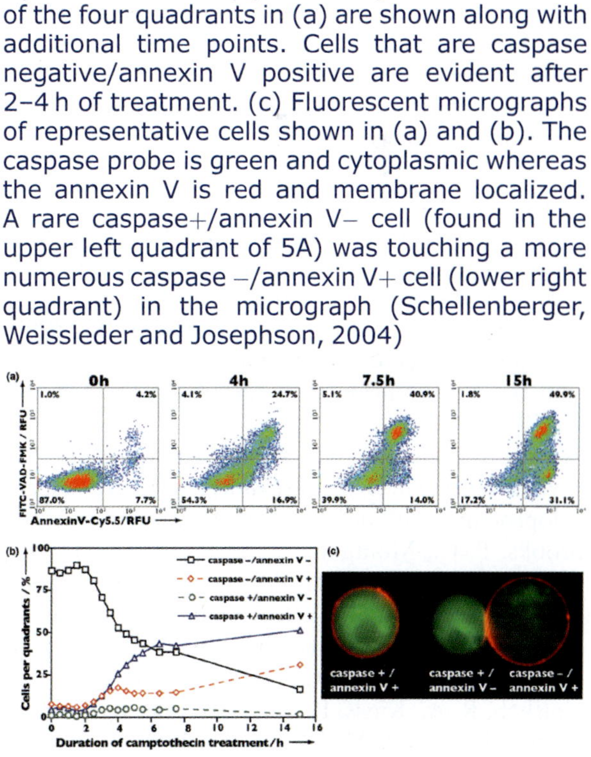

deposition of the imaging probes in the tissue can directly be studied by histology after imaging experiments. Furthermore, biological questions, for example the time course of activation of caspases and the externalization of phosphatidylserines after treatment with the toxin camptothecin (Schellenberger, Weissleder and Josephson, 2004), can be answered by *in vitro* experiments with these probes (Figure 11.8.1).

Taken together, optical imaging of caspase activity is a highly specific way for imaging apoptosis and exploration of apoptosis pathways especially in animal models. In contrast, Annexin V-probes bind to the cell surface of apoptotic cells and are therefore easier to conduct and to transfer into clinical settings. Although this technique is not suitable for discriminating between apoptosis and necrosis, it has many applications, where this distinction is not necessary. Especially, the monitoring of treatment effects with simple-to-handle optical imaging instrumentation is particularly useful for drug development in pharmaceutical research, as it can shorten the development-verification cycles dramatically.

References

Blankenberg, F., Narula, J., Strauss, H. W., 1999. "In vivo detection of apoptotic cell death: A necessary measurement for evaluating therapy for myocarditis, ischemia, and heart failure." *J. Nucl. Cardiol.* **6**, 531–539.

Blankenberg, F. G., Naumovski, L., Tait, J. F., Post, A. M., Strauss, H. W., 2001. "Imaging cyclophosphamide-induced intramedullary apoptosis in rats using 99mTc-radiolabeled annexin V." *J. Nucl. Med.* **42**, 309–316.

Blankenberg, F. G., Strauss, H. W., 2001. "Will imaging of apoptosis play a role in clinical care? A tale of mice and men." *Apoptosis* **6**, 117–123.

Blankenberg, F. G., Tait, J. F., Strauss, H. W., 2000. "Apoptotic cell death: its implications for imaging in the next millennium." *Eur. J. Nucl. Med.* **27**, 359–367.

Bremer, C., Tung, C. H., Weissleder, R., 2001. "In vivo molecular target assessment of matrix metalloproteinase inhibition." *Nat. Med.* **7**, 743–748.

Brooks, P. C., Montgomery, A. M., Rosenfeld, M., Reisfeld, R. A., Hu, T., Klier, G., Cheresh, D. A., 1994. "Integrin alpha v beta 3 antagonists promote tumor regression by inducing apoptosis of angiogenic blood vessels." *Cell* **79**, 1157–1164.

Gottlieb, R. A., Kitsis, R. N., 2001. "Seeing death in the living." *Nat. Med.* **7**, 1277–1278.

Graves, E. E., Ripoll, J., Weissleder, R., Ntziachristos, V., 2003. "A submillimeter resolution fluorescence molecular imaging system for small animal imaging." *Med. Phys.* **30**, 901–911.

Kerr, J. F., Wyllie, A. H., Currie, A. R., 1972. "Apoptosis: A basic biological phenomenon with wide-ranging implications in tissue kinetics." *Br. J. Cancer* **26**, 239–257.

Laxman, B., Hall, D. E., Bhojani, M. S., Hamstra, D. A., Chenevert, T. L., Ross, B. D., Rehemtulla, A., 2002. "Noninvasive real-time imaging of apoptosis." *Proc. Natl. Acad. Sci. USA* **99**, 16551–16555.

Messerli, S. M., Prabhakar, S., Tang, Y., Shah, K., Cortes, M. L., Murthy, V., Weissleder, R., Breakefield, X. O., Tung, C. H., 2004. "A novel method for imaging apoptosis using a caspase-1 near-infrared fluorescent probe." *Neoplasia* **6**, 95–105.

Ntziachristos, V., Schellenberger, E. A., Ripoll, J., Yessayan, D., Graves, E., Bogdanov, A., Jr., Josephson, L., Weissleder, R., 2004. "Visualization of antitumor treatment by means of fluorescence molecular tomography with an annexin V-Cy5.5 conjugate." *Proc. Natl. Acad. Sci. USA* **101**, 12294–12299.

Petrovsky, A., Schellenberger, E., Josephson, L., Weissleder, R., Bogdanov, A., Jr., 2003. "Near-infrared fluorescent imaging of tumor apoptosis." *Cancer Res.* **63**, 1936–1942.

Ravanat, C., Archipoff, G., Beretz, A., Freund, G., Cazenave, J. P., Freyssinet, J. M., 1992. "Use of annexin-V to demonstrate the role of phosphatidylserine exposure in the maintenance of haemostatic balance by endothelial cells." *Biochem. J.* **282**(Part 1), 7–13.

Schellenberger, E. A., Bogdanov, A., Jr., Petrovsky, A., Ntziachristos, V., Weissleder, R., Josephson, L., 2003. "Optical imaging of apoptosis as a biomarker of tumor response to chemotherapy." *Neoplasia* **5**, 187–192.

Schellenberger, E. A., Weissleder, R., Josephson, L., 2004. "Optimal modification of annexin V with fluorescent dyes." *Chembiochem.* **5**, 271–274.

Thompson, C. B., 1995. "Apoptosis in the pathogenesis and treatment of disease." *Science* **267**, 1456–1462.

Tung, C. H., Bredow, S., Mahmood, U., Weissleder, R., 1999. "Preparation of a cathepsin D sensitive near-infrared fluorescence probe for imaging." *Bioconjug. Chem. 10*, 892–896.

Weissleder, R., Tung, C. H., Mahmood, U., Bogdanov, A., Jr., 1999. "In vivo imaging of tumors with protease-activated near-infrared fluorescent probes." *Nat. Biotechnol.* **17**, 375–378.

Wunder, A., Schellenberger, E., Mahmood, U., Bogdanov, A., Jr., Muller-Ladner, U., Weissleder, R.,

Josephson, L., 2005. "Methotrexate-induced accumulation of fluorescent annexin V in collagen-induced arthritis." *Mol. Imaging* 4, 1–6.

11.9 Fluorescence molecular tomography (FMT) of angiogenesis

Xavier Montet, Vasilis Ntziachristos, and Ralph Weissleder

Angiogenesis, or the formation of new blood vessels from pre-existing vessels, is a tightly regulated process that plays a critical role in a variety of normal physiological events, including trophoblast implantation, wound healing and embryonic development (Blood and Zetter, 1990; Paku and Paweletz, 1991; Folkman and Shing, 1992; Fidler and Ellis, 1994). However, uncontrolled neovascularization can contribute to a number of pathological circumstances like proliferative retinopathy, tumour growth and metastasis (Brooks *et al.*, 1995; Ellis and Fidler, 1996; Hammes *et al.*, 1996; Gasparini *et al.*, 1998). From a tumoural point of view, angiogenesis is an essential process for the supply of oxygen and nutrients to proliferating tumour cells. As antitumour therapies directed against neovessel are emerging, an imaging system allowing to image angiogenesis and to follow treatment efficacy would be of great interest.

FMT allows imaging a tumour in its entirety, which is a major advantage as compared to microscopic methods (intravital and/or confocal scanning laser microscopy). Indeed, cancers have been shown to be spatially heterogeneous and hence, microscopic

Figure 11.9.1 Validation method: The graphs summarize the correlation between FMT measurements of tumour vascularity and other methods ($n = 9$ different tumours). The correlation coefficient between FMT–MR and FMT–histology are $r = 0.95$ and $r = 0.88$, respectively. A normalized FMT value of 0 corresponds to no vasculature, a value of 100 corresponds to 5% vascular volume fraction. Mean vessels density is define as the number of vessels counted in a microscopic field of view (20x)

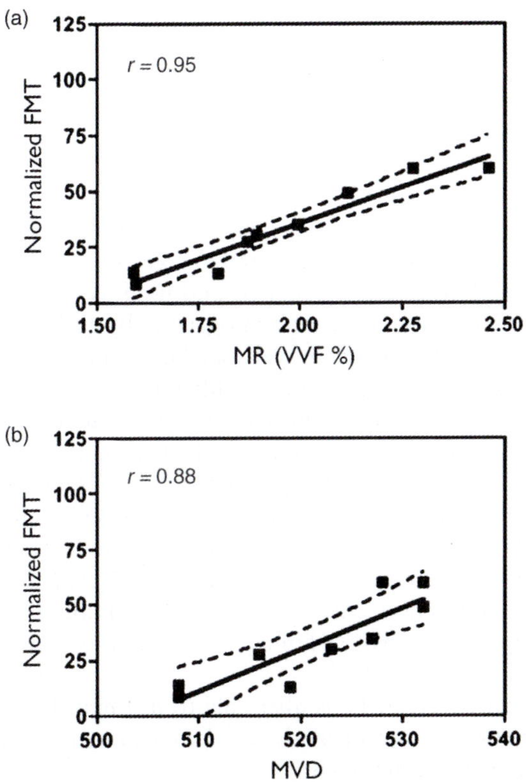

Figure 11.9.2 Angiogenesis map: Mice bearing xenograft tumours (either breast cancers or gliosarcomas) were imaged to map angiogenesis (a, 748 nm). See Figure 11.9.3 for correlative histology. Red square = 2 × 2 cm

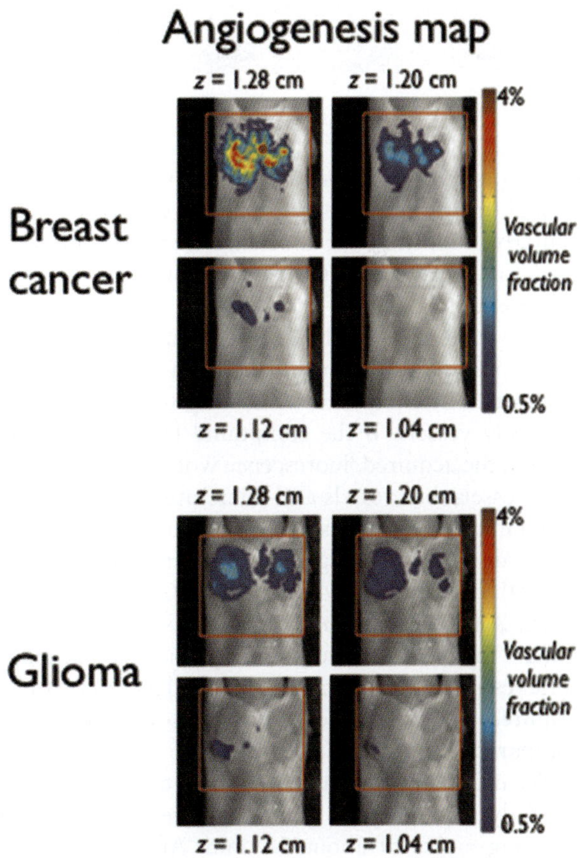

Figure 11.9.3 Correlative histology: Representative images of each tumour examined by FMT (see Figure 11.9.2). Blood vessels staining of breast cancer and gliosarcoma showing high vessels density for the breast tumour, and lower vessels density for the gliosarcoma. Scale bar = 100 μm

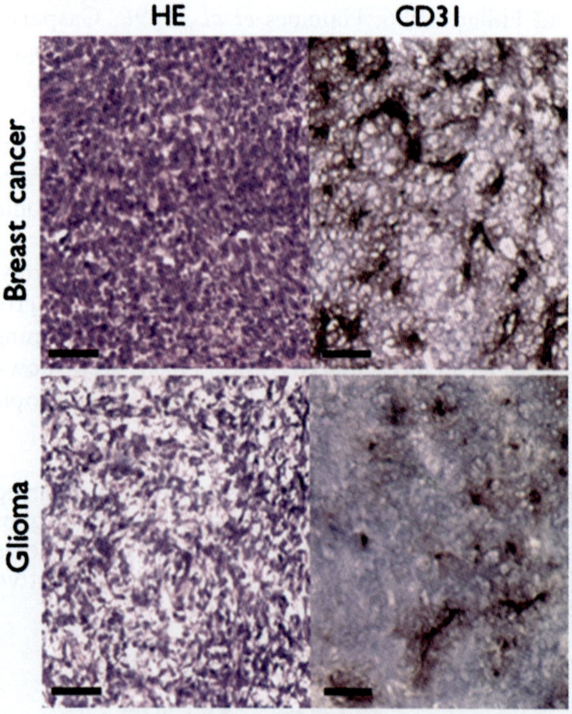

Figure 11.9.4 Treatment efficacy measured by FMT: FMT measurements of gliosarcoma bearing animals treated twice weekly for one week with different dose of anti-VEGF show a clear dose response (a). Correlative intravital confocal laser scanning microscopy (b). White scale bar = 1 mm. Blue fluorescence corresponds to vessels and red fluorescence corresponds to EGFP expressed by tumoural cells

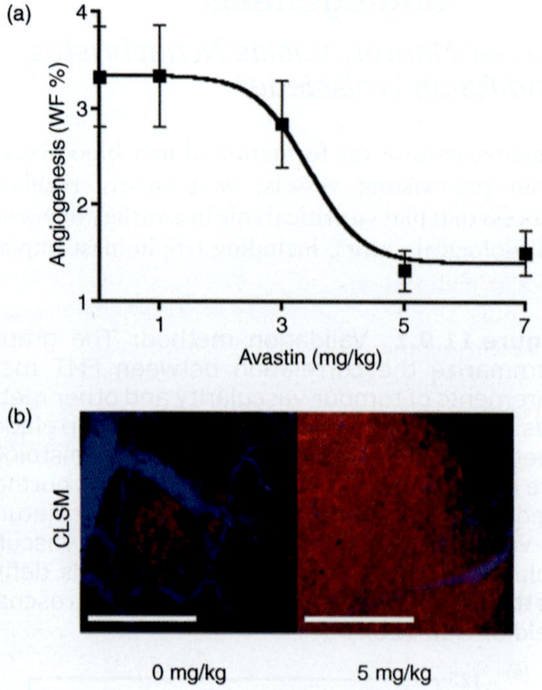

snapshots may not give representative information. Global quantitative measurements would be particularly important for assessing treatment efficacy and for studying complex phenomena such as angiogenesis.

This chapter describes FMT of angiogenesis. A major prerequisite to image angiogenesis is to use an intra-vascular compounds, allowing to image vessels and only vessels. If the compound leaks out of the vessels, the acquired fluorescence would not represent only vessels, but vessels and intersititum.

To image angiogenesis, we use either a long circulating dextranated magnetofluorescent nanoparticle (CLIO) containing Cy 5.5 (CLIO-Cy5.5) (Bremer *et al.*, 2003; Denis *et al.*, 2004). The advantaged of using a magnetofluorescent nanoparticle is that magnetic resonance imaging (MRI) as well as FMT can be acquired on the same model and measurements compared.

We use a xenograft model of subcutaneous tumour to study angiogenesis. Tumours were imaged when reaching a size of about 5–7 mm. After anesthesia, mice were injected with CLIO-Cy5.5 (10 mg/kg) and imaged with MRI and FMT. After images acquisition, animals were euthanized. Vessels were stained by immunohistochemistry to allow unequivocal vessel counting. Mean vessel density (MVD), defined by the number of vessels counted in a given microscopic field of view, was calculated and used as gold standard. Measurements of angiogenesis by FMT, MRI and histology correlate well (see Figure 11.9.1).

Tumours are known to be different from an angiogenic point of view. On one hand, tumours could be highly vascular, and so have a high angiogenesis and on the other hand, tumour could have low angiogenesis. FMT was able to distinguish high angiogenic tumour (breast tumour) from low angiogenic tumour (glioma) (see Figure 11.9.2: modified from Montet *et al.* 2005). This distinction correlates well with immunohistochemistry showing staining of vessels (see Figure 11.9.3).

Moreover, FMT is also able to distinguish subtle variation in angiogenesis, as demonstrated by anti-angiogenesis treatment.

Using a treatment directed against neo-vessels should decrease the number of vessels into a tumour. When different doses of treatment are used, a dose responses curve is attended, that is no efficacy when low dose are used, high efficacy when the optimal doses are used, no further benefits when higher doses are used (see Figure 11.9.4). The reduction in signal measured by FMT correlates well with intra-vital microscopy.

11.9.1 Conclusion

FMT is able to image angiogenesis in live animals. The major advantages of FMT is his ability to image tumours in their entirety and not only small snapshot as microscopic techniques does.

The high spatial resolution of FMT allows distinguishing small differences in angiogenesis between different tumours and between treated and untreated tumours. Making FMT a suitable candidate to image efficacy of anti-angiogenic treatment.

References

Blood, C. H., Zetter, B. R., 1990. "Tumor interactions with the vasculature: Angiogenesis and tumor metastasis." *Biochim. Biophys. Acta* **1032**, 89–118.

Bremer, C., Mustafa, M., Bogdanov, A., Jr., Ntziachristos, V., Petrovsky, A., Weissleder, R., 2003. "Steady-state blood volume measurements in experimental tumors with different angiogenic burdens a study in mice." *Radiology* **226**, 214–220.

Brooks, P. C., Stromblad, S., Klemke, R., Visscher, D., Sarkar, F. H., Cheresh, D. A., 1995. "Antiintegrin alpha v beta 3 blocks human breast cancer growth and angiogenesis in human skin." *J. Clin. Invest.* **96**, 1815–1822.

Denis, M. C, Mahmood, U., Benoist, C., Mathis, D., Weissleder, R., 2004. "Imaging inflammation of the pancreatic islets in type 1 diabetes." *Proc. Natl. Acad. Sci. USA* **101**, 12634–12639.

Ellis, L. M., Fidler, I. J., 1996. "Angiogenesis and metastasis." *Eur. J. Cancer* **32A**, 2451–2460.

Fidler, I. J., Ellis, L. M., 1994. "The implications of angiogenesis for the biology and therapy of cancer metastasis." *Cell* **79**, 185–188.

Folkman, J., Shing, Y., 1992. Angiogenesis. *J. Biol. Chem.* **267**, 10931–10934.

Gasparini, G., Brooks, P. C., Biganzoli, E. *et al.*, 1998. "Vascular integrin alpha(v)beta3: a new prognostic indicator in breast cancer." *Clin. Cancer Res.* **4**, 2625–2634.

Hammes, H. P., Brownlee, M., Jonczyk, A., Sutter, A., Preissner, K. T., 1996. "Subcutaneous injection of a cyclic peptide antagonist of vitronectin receptor-type integrins inhibits retinal neovascularization." *Nat. Med.* **2**, 529–533.

Montet X, Ntziachristos V, Grimm J, Weissleder R., 2005. "Tomographic fluorescence mapping of tumor targets." *Cancer Res.* **65**, 6330–6336.

Paku, S., Paweletz, N., 1991. "First steps of tumor-related angiogenesis." *Lab. Invest.* **65**, 334–346.

11.10 High resolution X-ray microtomography as a tool for imaging lung tumours in living mice

Nora De Clerck and Andrei Postnov

11.10.1 Introduction

Traditionally, histological examination is used for the detection of lung cancer in living mice. However, this requires sacrifice of the animals as well as sectioning of the excised lungs to prepare for microscopic inspection. These procedures are both destructive and time consuming.

Recently, the need for non-invasive imaging and 3D rendering in living animals became obvious. Nowadays high resolution X-ray microtomography (micro-CT) can be applied for *in vivo* scanning (De Clerck, Van Dyck and Postnov, 2003). Due to the nature of the X-ray attenuation coefficient, micro-CT proved to be most suitable for the visualization of bone and calcified tissue. Therefore, a big challenge for imaging by X-ray micro-CT was the visualization of soft tissue. To illustrate the possibilities of micro-CT, we present imaging results that allow the detection of lung tumours in laboratory animals. As the original shape of the lungs is preserved, 3D models could be built showing the position of the tumours relative to the whole lung. These results were reported previously (De Clerck *et al.*, 2004).

11.10.2 Methods

Mice lung tumours were induced by urethane treatment as reported elsewhere (De Clerck *et al.*, 2004).

Nine months later mice were anaesthetized by a mixture of Ketaminium hydrochloridum and Xylazine hydrochloridum and scanned in a high resolution X-ray device for small laboratory animals (Skyscan 1076, Aartselaar, Belgium). A voxel size of 35 µm was choosen. Scanning time was 17 min per scan. A compromise was made between scanning time (i.e. exposure time to X rays), duration of anaesthesia (which has its own harmful effects) and resolution. For cone beam reconstruction Feldkamp algorithm was applied (Feldkamp, Davis and Kress, 1984). Form the stacked virtual slices 3D models were built using Skyscan ANT software. Scanning in the present conditions was isotropic which allowed to cut the 3D models in any arbitrary orientation without loss in resolution.

11.10.3 Results

Figure 11.10.1 shows a virtual cross-section through the chest area of a diseased mouse. Three tumours of different sizes can be identified as dense spots within the lung tissue.

In Figure 11.10.2 a representative digital picture of the lungs after sacrifice of a representative mouse with lung tumours can be seen. Tumours can be recognized visually. The numbered tumours could be detected on the virtual cross-sections as shown by a representative cross-section through each tumour that was identified. The presence of tumours was confirmed by histology as can be seen on the inserted sections. In Figure 11.10.3 a 3D model of the lungs is represented; the model of the tumours was superimposed on the model of the whole lung. The position of the tumours can be seen and correlated to their position relative to the volume of the entire lung.

Figure 11.10.1 Re-constructed virtual cross-section through the chest area of a diseased lung. The outer circle represents the animal bed. Arrows indicate tumours of different sizes

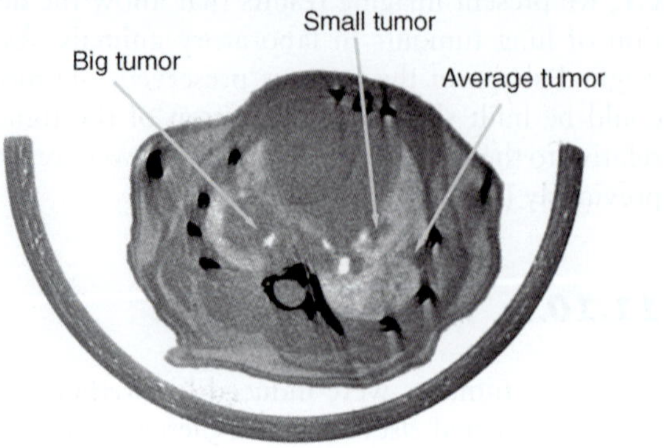

Figure 11.10.2 Photograph of the excised lungs of a representative mouse. Tumors can be seen and were numbered. Corresponding µCT virtual slices through each tumour are shown below. The presence of tumour tissue was confirmed by histological sectioning of the lungs. (From De Clerck *et al*., 2004. Neoplasia 6, 374–379. With permission by Decker Publishing Co.)

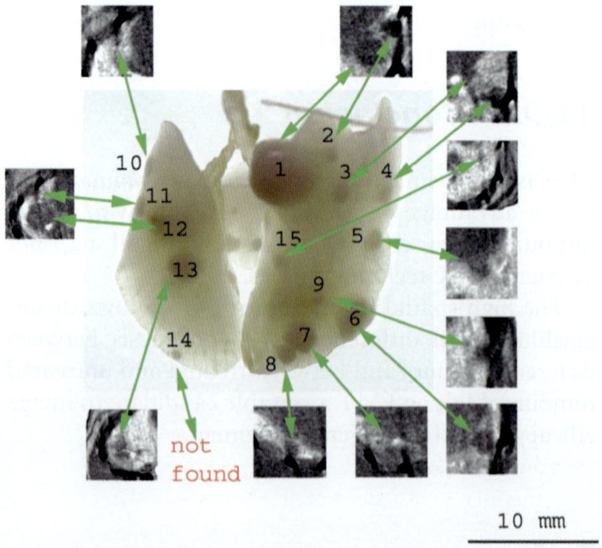

Figure 11.10.3 3D rendering of the lungs: The tumours were overlaid in colour (corresponding to different X-ray density)

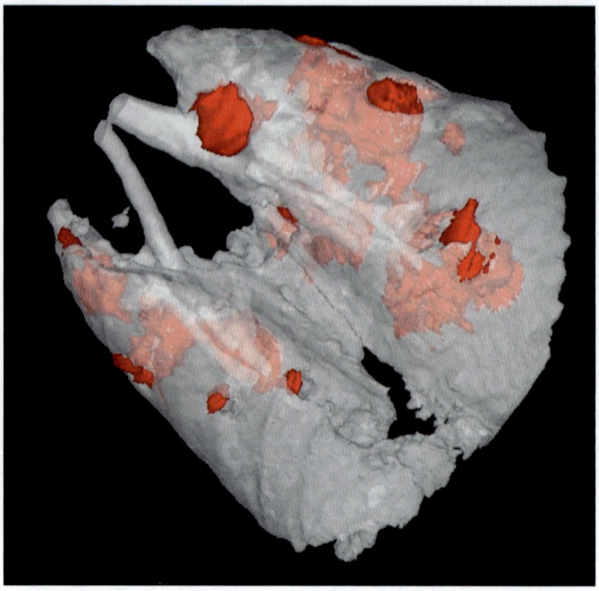

11.10.4 Discussion and conclusions

Due to the differences in X-ray attenuation coefficients, tumour tissue can be distinguished from healthy lung tissue in mice. Tumour tissue has a higher X-ray density than healthy tissue. Most of the tumours could be detected by micro-CT. However, in the *in vivo* situation, it is difficult to estimate the actual resolution due to the movement of the animal (breathing, cardiac contractions, etc.). We estimate that in the present conditions, the smallest tumour size that was detected by micro-CT was approximately 200 μm. The main advantage of micro-CT is that it is non-invasive and can be used as an initial screening method in cancer research. The results correlated well with the results from visual inspection of the lungs. As the original shape of the lungs was preserved, 3D rendering allowed locating the position of the tumours in living mice. Imaging possibilities of micro-CT open wide perspectives for longitudinal studies of tumour growth and development as well as for the follow-up of the efficiency of treatment in small laboratory animals.

Acknowledgements

The authors wish to thank Dr Kris Meurrens and Dr. Piter Terpstra from Philip Morris research laboratories for stimulating discussions and for providing the mice models.

References

De Clerck, N. M., Van Dyck, D., Postnov, A. A., 2003. "Non-invasive high-resolution μCT of the inner structure of living animals." *Microsc. Anal.* **81**, 13–15.

De Clerck, N. M., Meurrens, K., Weiler, H., Van Dyck, D., Vanhoutte, G., Terpstra, P., Postnov, A., 2004. "High resolution X-ray microtomography for the detection of lung tumors in living mice." *Neoplasia* **6**, 374–379.

Feldkamp, L. A., Davis, L. C., Kress, J. W., 1984. "Practical cone-beam algorithm." *J. Opt. Soc. Am. A* **1**, 612–619.

12 Other Organs

Coordinated by Anne Leroy-Willig

12.0 Introduction

Anne Leroy-Willig

Studies of other organs and other pathologies benefit from imaging techniques. *In vivo* imaging offers new opportunities to follow embryonic growth and to identify development mechanisms. Studies of embryonic development were formerly done under the microscope for smallest animals and/or non-coloured embryos such as fishes. Cell labelling was performed by using vital dyes. New 3D optical imaging techniques now allow to study larger embryos, as shown in the report by Sharpe. Ultrasound imaging, the technique of reference for examinations of human embryos, also yields information upon much smaller objects such as mouse embryo red cells, through a complex analysis of diffused ultrasonic waves, presented in the report by Le Floc'h. Cells individually labelled with magnetic agents are localized through the developing embryo by high resolution 'microscopic' MRI, as illustrated in the report of Papan and Jacobs. At lower magnetic field, magnetic contrast agents can be used to quantify placental perfusion of rodents, a methodology presented by Siauve that may be applied easily to larger animals.

In many other organs, the imaging explorations may bring new insight. The complex family of nephropathies is now explored by targeting of phagocytic cells with iron oxydes detected by high-resolution MRI as shown by Grenier. Lung remains a difficult organ. MicroCT depicts lung metastases very efficiently as illustrated in the previous chapter. Optical imaging of lung inflammation with a protease activable probe can be performed by using transillumination as illustrated by Haller. Articular lesions are explored with high anatomic details with 3D acquisitions done by MRI. A more biochemical approach is available by optical imaging as illustrated in the report by Wunder.

12.1 3D imaging of embryos and mouse organs by Optical Projection Tomography

James Sharpe

12.1.1 Background

While CT and MRI have become routine techniques for examining 'large' tissues (the size of human organs), only recently have attempts been made to increase their resolution to analyse specimens as small as early vertebrate embryos (Dhenain, Ruffins and Jacobs, 2001). These approaches show promise for a few applications, and in particular it is possible that in the future microscopic MRI will allow us to image living embryos in 3D during development (Louie *et al.*, 2000) (see section 12.2 in this Chapter). However, they also suffer a number of drawbacks. As the intended resolution of MRI increases, so does the required strength of the magnet used, such that systems designed to image mouse embryos become too expensive for most laboratories. Also, the achievable resolution of MRI is not usually sufficient to identify all the tissues or organs within the embryo. Additionally, as they are not optical techniques, neither MRI nor CT scanning can image the distributions of commonly used staining techniques such as histological stains, standard immunohistochemical protocols or *in situ* hybridisation techniques used to visualise the RNA expression patterns of genes.

At the other end of the size scale is a well-established imaging technology which is indeed optical but which has not been developed to image embryo anatomy. Confocal laser-scanning microscopy is very efficient at generating clear, 3D images of specimens which have been fluorescently labelled (Hecksher-Sorensen and Sharpe, 2001). However, it is typically used on specimens up to only a few hundred microns in thickness and usually much less than that. As such, it is often used for exploring the sub-cellular distributions of fluorochromes within small groups of cells. An 'imaging gap' was therefore left between confocal and MRI, and unfortunately for developmental anatomists most vertebrate embryos fall precisely in that gap – too large for confocal imaging, and too small for MRI (Sharpe, 2003). Two recently developed technologies fill this gap: OPT (optical projection tomography, Sharpe *et al.* (2002)) and SPIM (single plane illumination microscopy, Huisken *et al.* (2004)).

Confocal microscopy is restricted to small specimens because of the way it extracts 3D information. A laser beam is focused to a small point within the tissue, and the detector only measures the fluorescent light which emerges from that same point. The approach is often called 'optical sectioning', as it involves focusing the laser to a specific depth within the tissue and scanning the beam within a horizontal plane, recreating a virtual 2D section. Extending this approach to full 3D imaging simply involves moving the specimen up or down (by raising or lowering the microscope stage), thereby generating a series of optical sections through the tissue. However, this approach appears unable to generate high-quality images of larger specimens that are thicker than about half a millimetre, partly because as the field of view widens the *z*-resolution decreases (i.e. the optical sections become thicker). SPIM works in a similar way as confocal by focusing on an optical plane within the specimen. However, by illuminating from the side (at 90°to the optical axis) with a thin sheet of light, SPIM is able to focus on a much narrower section than confocal, giving it higher z-resolution (Huisken *et al.* 2004). This allows it to cope with much larger specimens, and impressive high-resolution results have been demonstrated on developing medaka fish embryos.

12.1.2 Optical projection tomography

OPT microscopy uses a very different approach from SPIM or confocal (Sharpe *et al.*, 2002). It falls into a category of techniques which do not perform optical sectioning, but instead sample *projections* through the specimen. Rather than reducing the depth-of-focus as much as possible so as to pinpoint only a precise plane within the tissue, an OPT scanner tries to maximise its depth of focus. This results in images with a view right through the whole specimen, especially when the specimen is cleared in an organic solvent. ('Standard' OPT imaging is performed on *ex vivo* specimens, and takes advantage of clearing agents to reduce photon scattering). These raw data do not explicitly contain information about depth. Instead the technique relies on taking images of the specimen from many different angles, and then using computer software to re-calculate the original 3D information. In practice, the specimen is suspended from a rotary stage and positioned at many different orientations relative to the optical axis of the imaging system. One consequence of this different approach is that it can image specimens larger than SPIM.

OPT can be performed using two distinct modes: The first of these (transmission mode) is in fact related to X-ray CT scanning which uses a very similar principle. The main difference here is that the detectors in a CT scanner record a quantitative shadow of the object, whereas OPT uses image-forming optics to create a focused image on a CCD camera chip. This means that the implementation of OPT is a little more complicated than CT, as the system has a limited depth of focus – Mechanisms for accurately positioning the specimen relative to the optical axis are essential. OPT can also function in fluorescence mode which is more like an optical version of SPECT (Sharpe, 2004) – Fluorescently generated photons are emitted from anywhere within the specimen, and the lenses of the scanner act as collimators, accepting only the light within a certain range of angles (the acceptance cone). In this mode the excitation light is usually provided from the same side that the emitted light is detected on (as in conventional fluorescent microscopy).

Although OPT has pushed optical imaging to cope with larger specimens than confocal microscopy, something similar to this had already been explored on much larger specimens for quite a few years. For example, DOT (diffuse optical tomography), which uses infrared wavelengths for greater tissue penetration, has been explored as a potential method for safe mammography, or for imaging oxygenation states of premature babies brains (Arridge, 1999). However, the intrinsic optical properties of biological tissue mean that these larger specimens cause too much photon scattering to allow a high-resolution image. Our attempts to image smaller samples such as vertebrate embryos and adult mouse organs and to combine this with organic clearing agents have, therefore, uncovered a specific regime for optical tomography where the negligible photon scattering allows us to assume more or less straight line projections (hence, the choice of name for the technique). Clearly, the smaller a specimen is, the lesser is scattering a problem, and OPT has also been performed more recently on much smaller specimens (single cells, Fauver *et al.*, 2005).

12.1.2.1 Applications

One of the advantages that a non-optical technique such as MRI possesses over any optical approach is that it completely avoids the problem of photon scattering – it can 'see' through specimens which are far too opaque for an optical technique. However, this also creates the disadvantage that the decades worth of optical stains and dyes which have been developed for studying tissues cannot be used. The same disadvantage applies to X-ray CT scanning, whose rays are too powerful to be disturbed by coloured precipitates. OPT microscopy by contrast, based on rays in the visible part of the spectrum, can in principle image the distribution of any of these popular stains, and this is where its most useful applications lie – Using optical stains and dyes to label gene expression patterns, and then visualising them within the context of the surrounding tissue.

The ability to perform analysis in 3D is not the only feature making OPT a useful lab tool. Many of its successful applications have been studies that could in principle have been done by conventional histological sections – The advantage of OPT in these cases has been the speed and comprehensiveness of the results. Cutting real sections through a 10.5 day mouse embryo can require up to 500 sections (depending on the thickness and orientation), and in addition to the cutting, each one has to be mounted onto a glass slide, de-waxed and stained. This whole process, while routine in some labs, is nevertheless extremely time-consuming. If you want to go further and make a digital record of the entire specimen you would also have to photograph every section. In practice this takes too long, and therefore analyses usually involve manually screening the sections on a microscope to look for the interesting ones and then digitally documenting only those few. Although OPT does not generate images at the same sub-cellular resolution achievable with real histological sections, in many cases that level of resolution is unnecessary. For example, to survey the tissue distribution of a certain gene expressed in a mouse embryo requires only a 'tissue-level' resolution image. In many cases real sections are only digitised at this scale of resolution anyway. What an OPT scan provides however, is a complete digitisation of the specimen in a short period of time. Scanning takes between 5–30 min (depending on the staining of the specimen), and reconstructing it into a series of virtual sections is also a matter of minutes. The resulting voxel data set can then be sliced in any orientation (avoiding the need to choose the orientation before cutting), and the entire specimen can be easily examined on screen in minutes by scrolling through all the adjacent virtual sections (Figure 12.1.1).

Figure 12.1.1 Gene expression analysis on virtual sections. OPT can create multi-channel virtual sections in any orientation through a specimen at 'tissue-level' resolution. In this case a 10-day mouse embryo has been fluorescently labelled to explore where two genes are expressed: In red is the expression of Sox9, as revealed by fluorescent RNA *in situ* hybridization. In green is the expression of Pax6, as revealed by fluorescent immunohistochemistry. The 3D structure of the embryo can be extremely useful (a), but for many applications the virtual sections alone are invaluable - (b) transverse, (c) frontal and (d) saggital for the same specimen shown in (a)

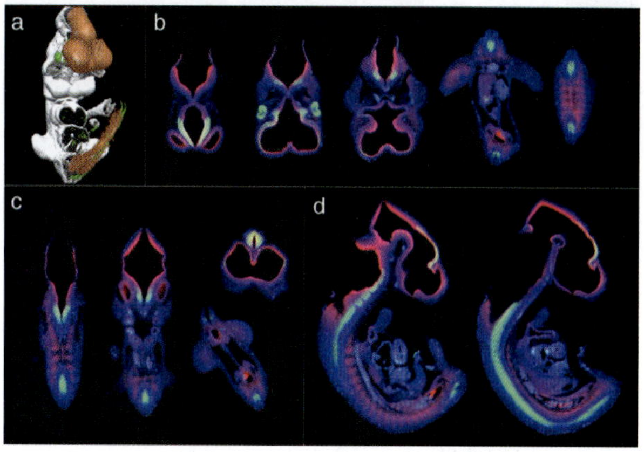

Using OPT in this way (to generate virtual sections in arbitrary orientations) has been invaluable to a number of projects; however, its other capability – to allow a genuine 3D view of specimens – sets it apart from traditional histology. It is possible to reconstruct the 3D structure of a specimen from serial sections – the sections are all digitized on a microscope, and then aligned within a computer program – and this has been done in a number of cases (Brune *et al.*, 1999; Davidson and Baldock, 2001). However, in addition to the obvious problem of speed (the whole process can take weeks if not months), this approach also presents some serious theoretical problems about how exactly the sections should be aligned. If each section is positioned simply on the basis of the best match with its

Figure 12.1.2 3D anatomy of an embryo. OPT image of a 12-day mouse embryo. Both the 3D shape of the specimen and also the internal structure can be visualised. In grey is the anatomy of the embryo as seen by autofluorescence and in red is the developing vasculature highlighted by fluorescent immunohistochemistry (using antibodies against the PECAM antigen)

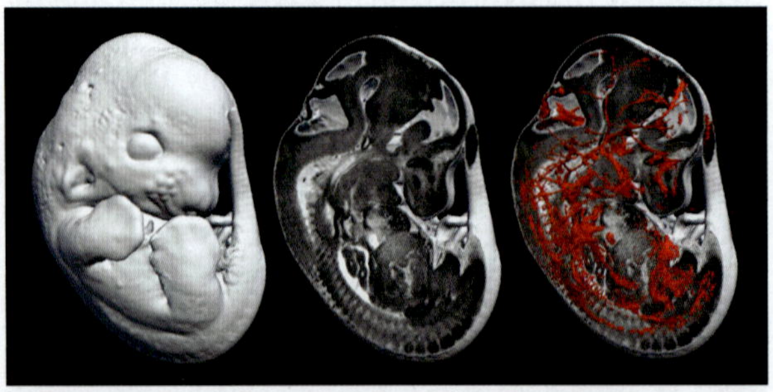

neighbours, then the final reconstruction will usually not retain the correct geometry of the original specimen. It will tend to be straightened along the axis perpendicular to the sections.

By contrast, as with any 3D imaging technique, OPT avoids this problem completely. The specimen is left intact during imaging and the true 3D geometry is retained (Figure 12.1.2). Another very important application for OPT, therefore, is phenotyping, that is determining whether the morphology of a genetically manipulated mouse embryo is abnormal. Given the very three-dimensional nature of developing embryonic organs, it has turned out to be much easier to spot abnormalities when presented with the full 3D shape rather than a series of 2D sections (Sharpe *et al.*, 2002). The examples given in this chapter are all fixed mouse embryos which have been optically cleared for imaging; however, we are also developing OPT for examining adult mouse organs (heart, brain, pancreas, lung, kidney) for example Alanentalo *et al.*, 2007 and exploring whether useful 3D data can be extracted from living embryonic tissue, and the results so far are very promising.

12.1.3 Conclusion

Although optical imaging techniques have been around for decades, it appears that new approaches are still being invented fairly regularly. SPIM and OPT, both developed in the last few years, represent new approaches to the question of how we can extract 3D geometry from a specimen, but new advances are also being made with imaging modalities (such as time-gated imaging, FRET, new laser sources and multi-spectral analysis). There are some fundamental limitations to optical imaging – in particular the scattering of photons in biological tissue – but overall we are still managing to extract more information from biological specimens every year, and most likely (hopefully!) we still have a long way to go before all optical avenues have been explored.

References

Arridge, S., 1999. "Optical tomography in medical imaging." *Inverse Probl.* **15**, R41–R93.

Alanentalo, T., Asayesh, A., Morrison, H., Lorén, C.E., Holmberg, D., Sharpe, J., Ahlgren, U., 2007. "Tomographic molecular imaging and 3D quantification within adult mouse organs", **4**, 31–33.

Brune, R., Bard, J., Dubreuil, C., Guest, E., Hill, W., Kaufman, M., Stark, M., Davidson, D., Baldock R., 1999. "A three-dimensional model of the mouse at embryonic day 9." *Dev. Biol.* **216**(2), 457–468.

Davidson, D., Baldock, R., 2001. "Bioinformatics beyond sequence: Mapping gene function in the embryo." *Nat. Rev. Genet.* **2**, 409–417.

Dhenain, D., Ruffins, S., Jacobs, R. E., 2001. "Three-dimensional digital mouse atlas using high-resolution MRI." Dev. Biol. 232, 458–470.

Fauver, M., Seibel, E., Rahn, J., Meyer, M., Patten, F., Neumann, T., Nelson, A., 2005. "Three-dimensional imaging of single isolated cell nuclei using optical projection tomography." *Opt. Express* **13**(11), 4210–4223.

Hecksher-Sorensen, J., Sharpe, J., 2001. "3D confocal reconstruction of gene expression in mouse." *Mech. Dev.* **100**, 59–63.

Huisken, J., Swoger, J., Del Bene, F., Wittbrodt, J., Steltzer, E., 2004. "Optical Sectioning Deep Inside Live Embryos by Selective Plane Illumination Microscopy." *Science* **305**, 1007–1009.

Louie, A. Y., Huber, M. M., Ahrens, E. T., Rothbacher, U., Moats, R., Jacobs, R. E., Fraser, S. E., Meade, T. J., 2000. "*In vivo* visualization of gene expression using magnetic resonance imaging." *Nat. Biotechnol.* **18**, 321–325.

Sharpe, J., Ahlgren, U., Perry, P., Hill, B., Ross, A., Hecksher-Sorensen, J., Baldock, R., Davidson, D., 2002. "Optical projection tomography as a tool for 3D microscopy and gene expression studies." *Science* **296**, 541–545.

Sharpe, J., 2003. "Optical projection tomography as a new tool for studying embryo anatomy." *J. Anat.* **202**, 175–181.

Sharpe, J., 2004. "Optical projection tomography." *Annu. Rev. Biomed. Eng.* **6**, 209–228.

12.2 Visualizing early Xenopus development with time lapse microscopic MRI

Cyrus Papan and Russell E. Jacobs

12.2.0 Introduction

The African clawed frog (Xenopus laevis) is a much used model organism in developmental biology. Many paradigms of embryogenesis have been developed using this species. One significant drawback in studying Xenopus embryogenesis is that the early embryo is almost completely opaque. Previous studies of morphogenesis and inductive tissue interactions during blastula and gastrula stages rely on extrapolation of cell movements from fixed sliced samples, from observations of cell movements on the embryonic surface or from work with tissue explants.

Technological advances in microscopic magnetic resonance imaging (microMRI) (Aguayo *et al.*, 1986; Jacobs and Fraser, 1994; Sehy *et al.* 2001) have made it feasible to visualize live Xenopus embryos with sufficient spatial and temporal resolution to follow early embryonic development within individual specimens from the early blastula through the end of gastrulation and into neurulation. Although MRI is intrinsically a rather insensitive imaging technique, it has many crucial advantages over other imaging techniques for the visualization of the frog embryogenesis: non-invasive, not dependent on optical properties of specimens, good intrinsic contrast and the ability to acquire 3D images of the whole embryo.

12.2.1 Methods

Images were acquired using a vertical bore 11.7 T Bruker Avance DRX500 system (Bruker Biospin, Billerica, MA) equipped with a Micro2.5 imaging gradient set capable of a peak gradient strength of 1 T/m and a maximum slew rate of 12 500 T/m/s. A homebuilt 3.5 mm diameter solenoid transmit/receive volume resonator was used for image acquisition. Individual frog embryos were placed in a 3 mm diameter 10 mm long quartz tube and ends capped with porous Teflon® allowing air but not water exchange. The high field strength magnet, small sample coil, small field of view, moderate data matrix size and spin echo imaging sequence allowed relatively rapid image acquisition with high spatial resolution, good contrast and reasonable signal-to-noise ratio.

12.2.2 Results

12.2.2.1 *Intrinsic contrast allows time lapse analysis of gastrulation*

The tissue specific contrast in microscopic MR images allows the distinction of the major embryonic regions like the animal cap, the blastocoel and the vegetal cell mass. This is useful for following gastrulation movements as the vegetal cell mass and the animal cap display specific morphogenetic changes. The vegetal cell mass undergoes vegetal rotation, the marginal zone undergoes convergent extension and involution and the animal cap undergoes epiboly (Keller and Shook, 2004). With the aid of the intrinsic tissue contrast, these morphogenetic changes are visualized in the live embryo using time lapse MR microscopy (Figure 12.2.1).

12.2.2.2 *Lineage tracking*

Tracing cell divisions in Xenopus by optical microscopy is difficult due to the opacity of the embryo, and three-dimensional microscopic MRI does not have sufficient spatial and temporal resolution to visualize the divisions of the blastomeres, as they become successively smaller. However, two-dimensional microscopic time-lapse MRI can visualize nuclei of early Xenopus blastomeres. Even though images are 2D, cleavages of many nuclei can be traced, allowing the direct observation of orientation and timing of their divisions (Figures 12.2.2 and 12.2.3).

12.2.2.3 *Monitoring cell movements*

Intrinsic tissue contrast is sufficient for the observation of global morphogenesis, but more detailed information on specific regional cell movements requires

Figure 12.2.1 Intrinsic contrast of MR images correlates with tissue type in the early Xenopus embryo and allows for 3D renderings (G–I) of developmental stages. View is lateral with animal oriented to the top and dorsal to the right. Scale bar: 300 μm. The liquid-filled blastocoel (bc) and the archenteron (ar) exhibit the highest signal intensity, the vegetal cell mass (veg) the lowest and the animal cap an intermediate signal intensity. At stage 9 (A, D, G), the embryo is rotationally symmetric along the animal-vegetal axis. Note the sharp boundary between the animal cap and the vegetal cell mass in the MR images (highlighted with a dashed line in (A)), which correlates with differences in cell size seen in the histological image (highlighted with a dashed line in (D)). At stage 11 (B, E, H), gastrulation is underway. The mesendodermal mantle is moving animal-wards, and the animal cap tissue is extending vegetal-wards by epiboly. The cleft of Brachet, routinely observed in histological images (white arrow in (E)) can also be clearly recognized in the MR image as a sharp and distinctive boundary (white arrow in (B)) on the dorsal as well as on the ventral side. At stage 12 (C, F, I), the archenteron (ar) has formed, displacing the blastocoel cavity to the ventral side. The histological image shows loosely packed cells between the archenteron and the blastocoel (black arrow in (F)), which matches with intermixed bright and dark intensities in the MR image (black arrow in (C))

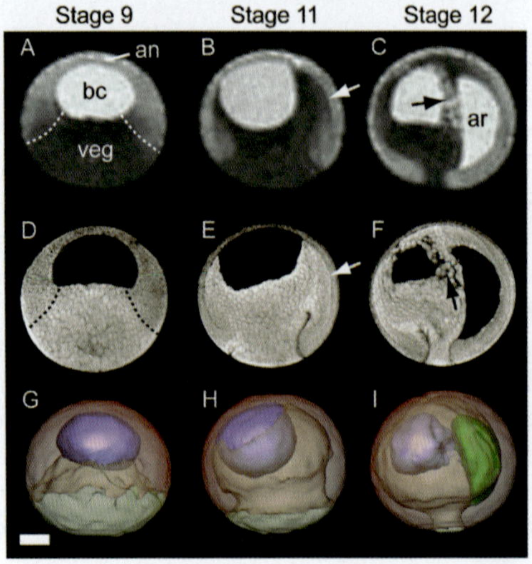

Figure 12.2.2 Lineage tracing of a vegetal blastomere D4. The figure shows selected stages of a 2D time-lapse sequence. Image slices are sagittaly oriented, with animal towards the top and dorsal to the right. Scale bar is 300 μm. Many nuclei can be identified in a 300 μm thick image slice (black arrowheads at 0:57). By taking consecutive 2D slice images, the cleavages of the nuclei can be traced. In the first frame, both nuclei of the D4-daughters are indicated with a white arrow. One daughter nucleus is highlighted in yellow, and the cell boundary is coloured in blue. In the second frame (time 1:12), the highlighted nucleus has cleaved in the radial direction and daughter nuclei move apart in the third and fourth frame (1:26, 1:41). At 1:55, only the nucleus of the outer blastomere can be seen to divide, and it changes the cleavage plane by 90° to divide in the planar direction. The nucleus of the inner sibling blastomeres does not appear to cleave (white arrow) due to the cleavage plane being oriented in the image plane. At 2:53 and 3:03, two sibling cells are dividing, albeit not synchronous (white arrows). Again, one sibling does not appear to cleave (white arrow head). At 3:37, all highlighted nuclei of the clone are cleaving in the planar direction, resulting in 10 clonally related sibling cells. After this, nuclei cannot be reliably traced

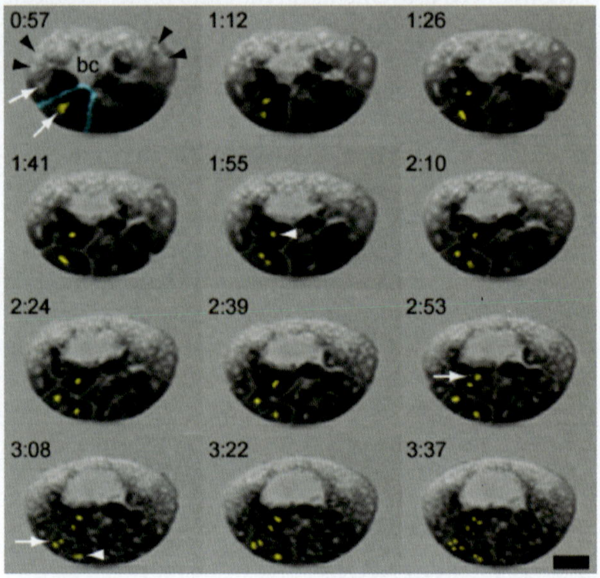

cell labelling. By labelling a single B1-blastomere, which gives rise to part of the dorsal marginal zone, with a macromolecular gadolinium based MRI contrast agent, we follow the morphogenetic behaviour of labelled cells in 3D time lapse series through blastula and gastrula stages and eventually observe the fate of the cells that the labelled blastomere produces (Figure 12.2.4).

Figure 12.2.3 Lineage diagram of a D4 blastomere noted in MR images shown in Figure 12.2.2. The time scale on the left of the diagram corresponds to the times shown in Figure 12.2.2. The part of the lineage tree observed in Figure 12.2.2 is enclosed in the dashed box. The first three cleavages occur relatively synchronously. Then the right part of the lineage, which is the vegetal daughter cell of the D4-blastomere shown in Figure 12.2.2, begins to slow. This is expected, because the blastomeres at the vegetal pole tend to divide slower compared to the more animal blastomeres (Nieuwkoop and Faber, 1994). The cleavage orientation of the nuclei alternates in most cases by 90°, as has been observed at the animal pole in Xenopus (Chalmers *et al.*, 2003). However, early divisions do not always change orientation by 90°. In one case, two successive planar (p) and in another case two successive radial (r) cleavages were observed

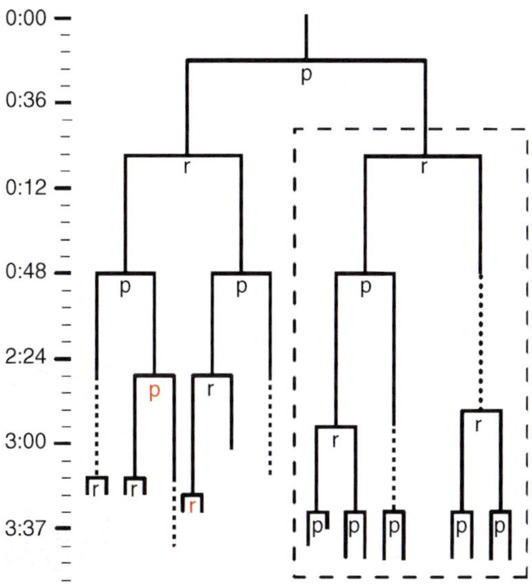

Figure 12.2.4 3D time-lapse tracing of a B1-blastomere labelled with a Gd^{3+}-DOTA based MR contrast agent ((Corot *et al.*, 1997), kindly provided by Guerbet, France)) and fluorochrome at the 32 cell stage. (A)–(E) are MR images; (F and G) are optical images. (A, C): lateral view; (B, D, E and F): dorsal view. In (E, F and G), anterior is oriented to the left. bc: blastocoel, veg: vegetal cell mass, ar: archenteron. Scale bar: 300 μm

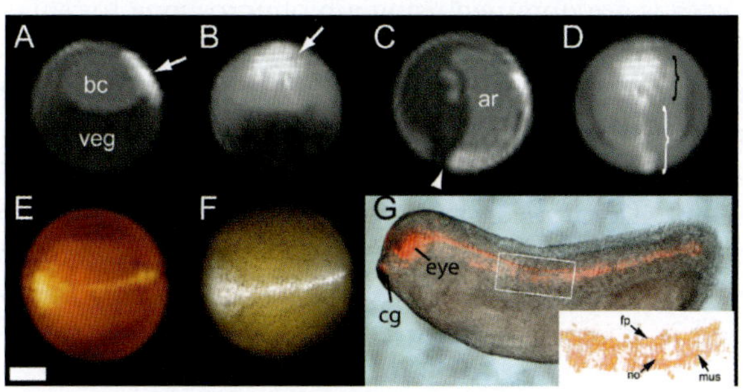

12.2.3 Conclusion

With microscopic MR imaging, we can readily visualize important details about internal rearrangements in early development in the optically opaque Xenopus laevis embryo. Labelling of an individual blastomere with a cell impermeable contrast agent allows tracking of morphogenic movements of the descendents of the labelled cell throughout many hours of development. MicroMRI adds several capabilities (e.g. true *in vivo* 3D imaging) to the pantheon of imaging modalities being applied to early vertebrate development. We expect it to prove a useful complement to current optical imaging methods and see increasing applications in the future.

At blastula stage (a, b), the labelled descendants of the B1-blastomere are located in the lower animal cap (white arrow). The dorsal view (b) shows that the clone covers a broad patch of the animal cap. At stage 12 (c, d), the clone has extended all the way to the blastopore (white arrowhead in (c)). The dorsal view reveals that the anterior part of the clone (black curly bracket in (d)) is still broad and patchy while the posterior part has undergone convergent extension and is now long and narrow.

After the last MR image acquisition (e), the embryo was removed from the scanner optical images acquired with a fluorescence stereomicroscope (f). The clone appearances by MRI and by optical microscopy are essentially identical: MRI is able to track the entire clone during the process of gastrulation. The embryo was then raised to stage 28, fixed and cleared (g). As expected from the stage 6-fate map (Dale and Slack, 1987; Moody, 1987), the B1-blastomeres has given rise to floor plate (fp) ventral neural tube and brain cells, some eye tissue (eye), cement gland (cg), notochord (no), and a very small amount of muscle tissue (mus).

References

Aguayo, J. B., Blackband, S. J., Schoeniger, J., Mattingly, M. A., Hintermann, M., 1986. "Nuclear magnetic resonance imaging of a single cell." *Nature* **322**(6075), 190–191.

Chalmers, A. D., Strauss, B. *et al.*, 2003. "Oriented cell divisions asymmetrically segregate aPKC and generate cell fate diversity in the early Xenopus embryo." *Development* **130**(12), 2657–2668.

Corot, C., Schaefer, M. *et al.*, 1997. "Physical, chemical and biological evaluations of CMD-A2-Gd-DOTA." A new paramagnetic dextran polymer. *Acta Radiol. Suppl.* **412**, 91–99.

Dale, L., Slack, J. M., 1987. "Fate map for the 32-cell stage of Xenopus laevis." *Development* **99**(4), 527–551.

Jacobs, R. E., Fraser, S. E., 1994. "Magnetic resonance microscopy of embryonic cell lineages and movements." *Science* **263**(5147), 681–684.

Keller, R., Shook, D. R., 2004. In: Stern, C. B., ed. "Gastrulation in amphibians. Gastrulation – From Cells to Embryos." Cold Spring Harbor, NY/CSHL Press.

Moody, S. A., 1987. "Fates of the blastomeres of the 32-cell-stage Xenopus embryo." *Dev. Biol.* **122**(2), 300–139.

Nieuwkoop, P. D., J. Faber, 1994. "Normal Table of Xenopus Laevis (Daudin): A Systematical and Chronological Survey of the Development from the Fertilized Egg Till the End of Metamorphosis." Garland Publishing, New York.

Sehy, J. V., Ackerman, J. J. *et al.*, 2001. "Water and lipid MRI of the Xenopus oocyte." *Magn. Reson. Med.* **46**(5), 900–906.

12.3 Ultrasonic quantification of red blood cells development in mice

Johann Le Floc'h

12.3.0 Introduction

An incomplete understanding of human blood diseases, such as anaemias, motivates the investigation of the development of normal and abnormal red blood cells (RBCs) in mutant mice. Ultrasound (US) techniques at high frequencies (above15 MHz) yield useful non-invasive tools allowing, *in vivo*, the identification and characterization of the modifications that occur during normal development of the RBCs. In mouse embryos, immature RBCs are large and nucleated in early gestation, and then they lose their nuclei in later gestation and decrease in size as they mature.

The feasibility of studying developmental stages of RBCs in mouse embryos using an ultrasonic parameter has been demonstrated: Le Floc'h *et al.* (2004) quantified the changes of US backscatter from blood, expressed by apparent integrated backscatter (AIB).

How do structural alterations of RBCs correspond to changes in mean AIB?

Blood consists of RBCs, white blood cells and platelets suspended in plasma. The interaction of US with blood is determined by the scattering of RBCs because these cells account for 97% of the solid blood volume. The morphological changes of RBCs during blood development can be expressed in terms of changes in acoustical properties, that is densities and compressibilities. The density and compressibility of a RBC are linked to its content. The backscatter from blood approximated by the AIB is mathematically related to the acoustical properties of the RBCs as modelized by Cloutier and Qin (1997).

This report addresses the experimental procedure that leads to the calculation of AIB. It also shows that the main constraints that are the penetration depth, the small size of the embryonic mouse heart structures, the increasing embryonic heart beats and mouse's respiration can be alleviated.

12.3.1 Description of the experimental procedure

12.3.1.1 Mouse protocol for US imaging

The embryonic heart is the organ to visualize because it contains a relatively large volume of well-mixed blood. For this purpose, the mouse is anaesthetized with a mix of isoflurane and oxygen. Its abdomen is shaved to facilitate the penetration of the ultrasonic waves. Two US gels of different viscosity are used as coupling medium between the mouse abdomen and the probe. They are heated to $37 \pm 1°C$ before application to the mouse skin and maintained at this temperature using a warmer. The most viscous gel delineates the region of interest on the abdomen and maintains the less viscous gel over this region. The less viscous gel is used for its efficiency in avoiding air bubbles that prohibit the propagation of the ultrasonic waves.

12.3.1.2 B-scan images of the embryonic mouse heart

B-scans are obtained using a biomicroscope for small animal imaging (VisualSonics, Toronto, ON, Canada) equipped with a 40 MHz broadband transducer excited at 35 MHz.

B-scan images of the embryos are obtained applying the US probe to the abdomen of the mouse that in turn is scanned to locate the position of the embryos in the uterus. Once the embryonic heart has been identified in B mode, the probe of the US system and the

Figure 12.3.1 35 MHz transcutaneous image (8×8 mm^2) of the embryonic mouse at ED 13.5. In this transversal view of the embryo, the four embryonic heart cavities are visible (middle of the image). The extremity of a while arrow indicates the focus. The yellow line indicates where the transducer T stops for M-mode imaging. The vertical white line corresponds to the location of the signals registered in M-mode. 35 MHz calculated M-Mode image with 100 RF signals backscattered from embryonic mouse heart structures. Oscillations are due to the embryo heart beats. Since the heart walls (HW) and the interventricular septum are visible, blood can easily be differentiated from the other tissues. The green box represents the windowed RF signals (approximately 400 μm in depth and a time of observation of 1 second used to calculate the AIB (*reprinted by permission of Elsevier Science, from Le Floc'h et al., 2004, Ultrasound in Medicine and Biology, © World Federation of Ultrasound in Medicine and Biology*)

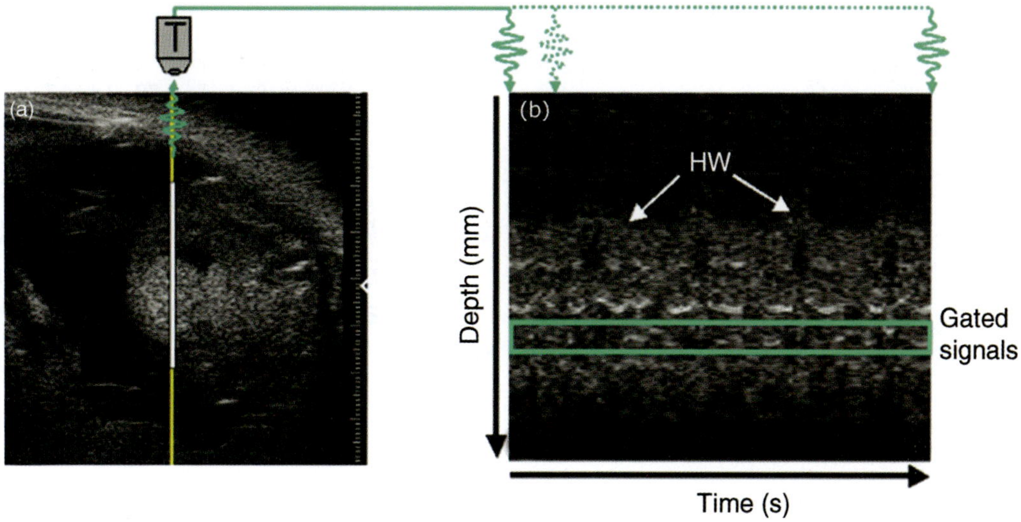

mouse pad are adjusted to position the heart in the focal zone of the transducer and at a position allowing M-mode visualization (Figure 12.3.1(a)).

12.3.1.3 *M-mode of the embryonic mouse heart*

The M-mode option is then selected on the US scanner. In this experiment, this mode allows for distinguishing blood from other tissues. The 100 gated signals from intra-cardiac blood (illustrated by a green box in Figure 12.3.1(b)) are transferred to a personal computer for further signal processing. The size of the window is approximately 400 μm in depth, ensuring that only blood is selected during the cardiac cycle and that the selected signals allow the calculation of the AIB from blood during the embryonic growth.

12.3.1.4 *Signal processing*

The windowed signals are transformed into spectra and averaged for each embryo and at each embryonic day (ED). The averaged power spectra are multiplied by a correction factor to obtain a measurement independent of the US scanner. The correction factor is obtained dividing the factor compensating for the frequency dependence of the insonified volume by the power spectra of the signal reflected from a plane reflector (e.g. quartz plate) located in the focal zone (Ueda and Ozawa, 1985). The corrected power spectra are then integrated from 21 MHz to 42 MHz to lead to AIB because the averaged power spectra show that this range of frequencies is the useful one when the transducer is imaging at 35 MHz (Le Floc'h *et al.*, 2004).

12.3.1.5 *Results*

The mean AIB is plotted in Figure 12.3.2 as a function of embryonic age. From ED 13.5 to ED 17.5, it decreased progressively and significantly during a time interval of 2 EDs. Based on haematology results and on a particle approach, Le Floc'h *et al.* (2004) demonstrated that the changing proportion of the scatterers (nucleated and non-nucleated RBCs), the loss of the nucleus and the

Figure 12.3.2 The AIB (mean ± SEM) measured at 35 MHz in the embryonic heart cavities is plotted for each ED between 12.5 and 17.5. It increased slightly from ED 12.5 to ED 13.5, and then decreased progressively from ED 13.5 to ED 17.5 ($P < 0.0001$, $\alpha = 5\%$). All AIB value is significantly different from the others except the one at ED 14.5 labelled ab that is not different from EDs labelled a and b (*reprinted by permission of Elsevier Science, from Le Floc'h et al. 2004, Ultrasound in Medicine and Biology, © World Federation of Ultrasound in Medicine and Biology)*

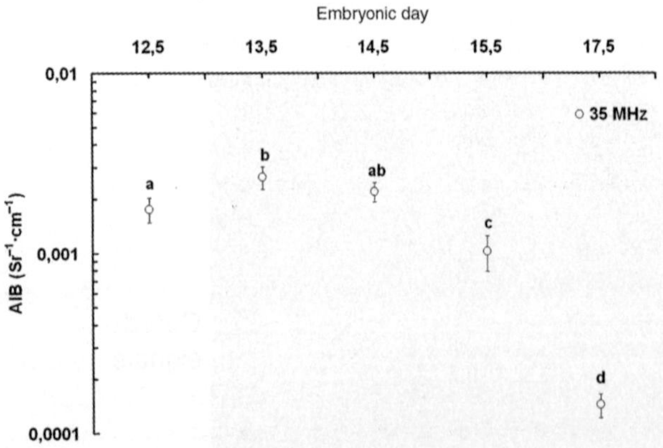

decrease in average of the cells diameter could explain the drop of the mean AIB.

12.3.2 Conclusion

The experimental procedure described in this report allows, *in vivo*, the longitudinal quantification of blood development during the embryonic growth. This fully non-invasive procedure significantly reduces the number of sacrificed mice.

It is then possible to study in mutant mice the progression of blood diseases, such as specific anaemias, that induce significant alterations of the structure of the RBCs during the embryonic development. For example, the AIB could characterize the breakdown of the red blood cells that occurs in mouse models of haemoglobin H disease (Chang *et al.*, 1996) and the consequences of this phenomenon over time.

References

Chang, J., Lu, R. H., Xu, S. M., Meneses, J., Chan, K., Pedersen, R., Kan, Y. W., 1996. "Inactivation of mouse alpha-globin gene by homologous recombination: Mouse model of hemoglobin H disease." *Blood* **88**, 1846–1851.

Cloutier, G., Qin, Z., 1997. "Ultrasound backscattering from non-aggregating and aggregating erythrocytes – a review." *Biorheology* **34**, 443–470.

Le Floc'h, J., Cherin, E., Zhang, M. Y., Akirav, C., Adamson, S. L., Vray, D., Foster, F. S., 2004. "Developmental changes in integrated ultrasound backscatter from embryonic blood *in vivo* in mice at high US frequency." *Ultrasound Med. Biol.* **30**, 1307–1319.

Ueda, M., Ozawa, Y., 1985. "Spectral analysis of echoes for backscattering coefficient measurement." *J. Acoust. Soc. Am.* 77, 38–47.

12.4 Placental perfusion MR imaging with contrast agent in a mouse model

Nathalie Siauve, Laurent Salomon and Charles André Cuénod

12.4.1 Placental perfusion assessment: an important challenge in prenatal care

The placenta plays a key role for the foetal growth, ensuring nearly all exchanges between the maternal and foetal compartments. The rate of transplacental exchange depends primarily on the rate of maternal placental (uterine) and foetal (umbilical) blood flow. Increased uterine vascular resistance and/or reduced uterine blood flow are predictors of foetal growth restriction and high-risk pregnancies (Trudinger, Giles and Cook, 1985). Precise quantification of placental perfusion can be useful for determining the pathogenesis and severity of foetal growth restriction, predicting the outcome of a compromised pregnancy and estimating the benefits of vasodilatory and antithrombotic treatment.

In clinical practice, placental blood flow is measured in the umbilical and uterine arteries by means of Doppler sonography, a non-invasive and widely available imaging method. However, Doppler flow velocity in the uterine artery has limited value in predicting preeclampsia, intrauterine growth restriction (IUGR) or perinatal death. Moreover, Doppler sonography cannot directly quantify placental perfusion.

MR functional imaging is well suited to the analysis of vascular physiology and has been used to measure several relevant parameters such as tissue blood flow and vascular permeability. We developed a new tool to quantitatively analyse placental perfusion using MRI with contrast agents (Salomon *et al.*, 2005).

12.4.2 How can placental perfusion be assessed with functional MRI?

Placental perfusion can be characterized by the placental blood flow per unit tissue volume F/V_p (in ml/s/ml), using i.v. injection of gadolinium contrast media and dynamic MRI acquisition.

The signal enhancements of the maternal left ventricle, the placenta and the mouse foetus are related to the respective gadolinium tissue concentrations. Using a compartmental analysis of these kinetic gadolinium concentration curves, transfer rates (k) can be calculated, and quantitative physiological parameters can be obtained.

12.4.3 Imaging procedure

MRI is performed with a Signa 1.5-T unit (GE, Milwaukee, WI) on anaesthetized mice, at 16 days of gestation. A T1-weighted coronal anatomical MRI sequence is used first to locate the maternal left ventricle, the placentas and the foetal mice. Then, after i.v. bolus injection of contrast agent, the dynamic enhancements of the maternal left ventricle, the placenta and the mouse foetus are monitored, using a single slice T1-weighted dynamic MRI sequence with *a high temporal resolution* of 1.13 s per image.

12.4.4 Data modelling

12.4.4.1 *Kinetic curves of variations in relaxation rate ($\Delta 1/T1$)*

Kinetic curves of signal intensities of the maternal left ventricule, the placentas and the foetuses are obtained from regions of interest (ROIs) reported on all images of the dynamic sequence. The curves are converted into kinetic curves of relaxation rate (1/T1) by using a calibration curve (Pradel *et al.*, 2003) and then in variations of the relaxation rate Δ1/T1. These variations are directly proportional to the contrast agent concentration C, as follows:

$$1/T1 \text{ at time } t = (1/T1\text{pre-contrast}) + r1.\ C,$$

where r_1 is the relaxivity of the contrast agent in blood at 1.5 T and 37°C and C the concentration of the contrast agent at time t.

12.4.4.2 *Compartmental analysis*

We developed a three-compartment model (Figure 12.4.1) based on physiological data of the murine placenta (Schroder, 1995; Adamson *et al.*, 2002). The placenta of the mouse consists of a maternal vascular compartment (Vpm) and a foetal vascular compartment (Vpf) into which the contrast agent can diffuse. The maternal compartment (Vpm) is supplied by an arterial input (uterine arteries) and is drained through a venous output. The foetal vascular compartment (Vpf) may exchange the contrast agent with the foetus itself (Vf) through the umbilical cord.

The model assumes that the contrast agent is a tracer, that the volume of each compartment is constant during the experiment and that the mixing time within each compartment is short compared to the tracer transfer rate between compartments.

The constant transfer parameters $k_{(i,j)}$ are adjusted by a compartmental and numerical modelling program (SAAM software-SAAM Institute, Seattle, WA, USA) to the concentration changes. Placental blood flow per unit tissue volume F/V_p is calculated from the transfer rate $k_{(2,1)}$.

12.4.5 Practical approach

On the coronal anatomical MRI sequence, it is easy to locate the uterus with its placentas and its foetuses (Figure 12.4.2) and to determine the section plane for the dynamic MRI acquisition (Figure 12.4.3).

From the kinetic enhancement curves (Figure 12.4.4), after signal intensity conversion and compartmental analysis, the transfer rate $k_{(2,1)}$ is obtained. In this model, the use of a conventional gadolinium chelate or a macromolecular gadolinium chelate, with intravascular distribution, does not lead to significant differences in $k_{(2,1)}$ ($k_{(2,1)\ \text{Dotarem®}}$ =2.1E−02 ml/s/ml +1,2E−02; $k_{(2,1)\ \text{Vistarem®}}$ =1.65E−02 ml/s/ml + 0.93E−02).Mean placental perfusion value is 1.10 + 0.60 ml/min per gram of placental tissue, in agreement with the literature (Gowland *et al.* 1998).

Figure 12.4.1 The three-compartment model
Differential equations:

–For the placental maternal vascular compartment:
$dq_2/dt = k_{(2,1)} \cdot q_1(t) - (k_{(3,2)} + k_{(0,2)}) \cdot q_2(t) + k_{(2,3)}\, q_3(t)$

–For the placental fetal vascular compartment:
$dq_3/dt = k_{(3,2)} \cdot q_2(t) - (k_{(2,3)} + k_{(4,3)}) \cdot q_3(t)$

–For the fetal compartment:
$dq_4/dt = k_{(4,3)} \cdot q_3(t)$

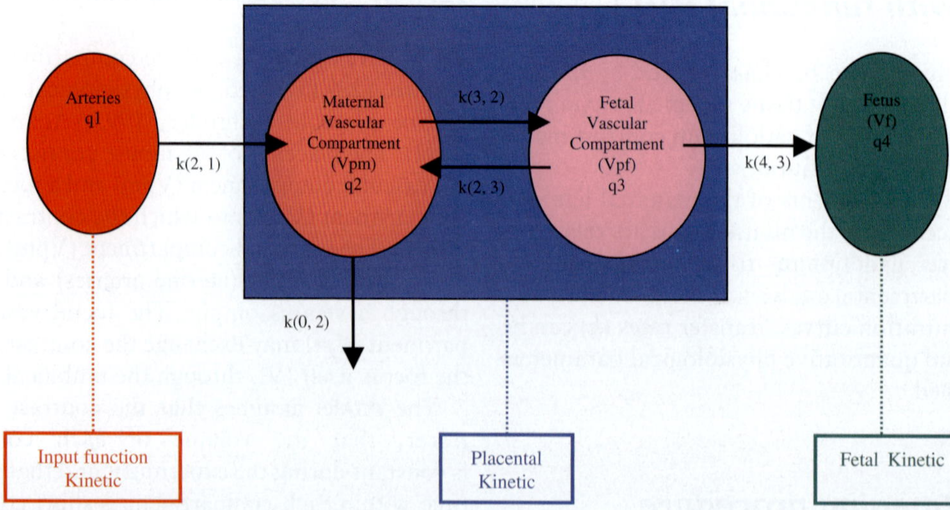

Figure 12.4.2 MRI anatomy:
With a dedicated mouse coil, the two uterine tubes (white arrows) containing multiple feto-placental units (grey arrows) are well visualized on the T1-weighted coronal image, as well as a sagittal section of the fetus. The number of placentas varied from 4 to 10

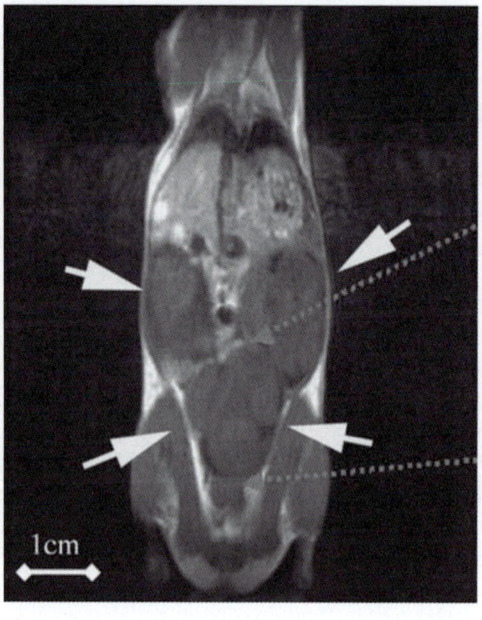

Figure 12.4.3 Dynamic MRI images
On the left, before injection, the maternal left ventricle and the placental and fetal areas cannot be identified. On the right, after conventional gadolinium chelate injection, first, the left ventricle becomes white, corresponding to contrast enhancement (star), as well as the placental areas (arrows). The fetal areas (circles) weakly enhance, maintaining a grey signal

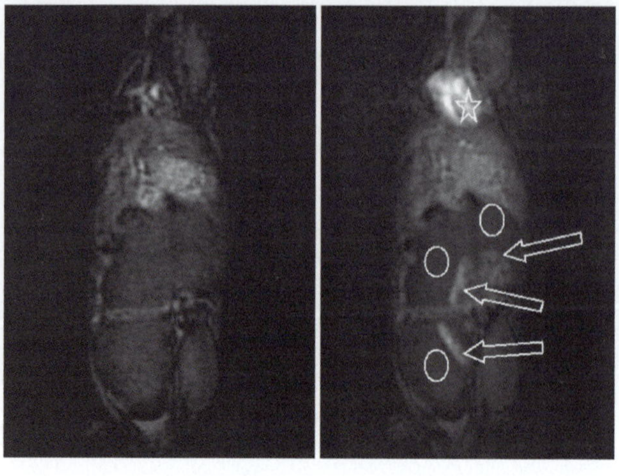

Figure 12.4.4 Enhancement curves obtained after injection of a conventional gadolinium chelate (Dotarem® 0.125 mmol Gd/kg)
On the left ventricle kinetics, the first-pass peak is marked and the contrast agent concentration falls slowly thereafter. The placental tissue concentration shows gradual uptake of contrast agent, followed by gradual decay; enhancement ten times weaker is seen in the fetus

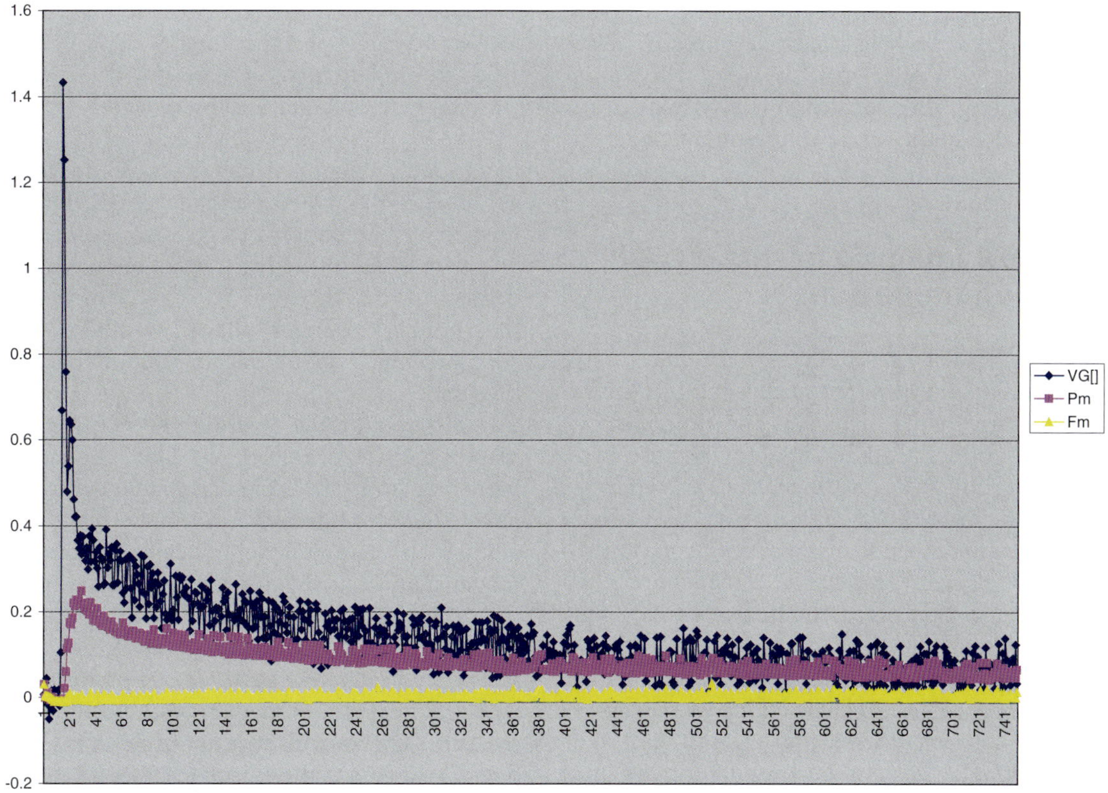

12.4.6 Conclusion

Functional MR imaging shows promise for real-time studies of human placental perfusion *in vivo*. Studies of placental perfusion could be useful for evaluating high-risk pregnancies. MRI could provide new insights into the pathogenesis of growth restriction and placental dysfunction.

References

Adamson, S. L., Lu, Y., Whiteley, K. J., Holmyard, D., Hemberger, M., Pfarrer, C., Cross, J. C., 2002. *Dev. Biol.* **250**, 358–373.

Gowland, P. A., Francis, S. T., Duncan, K. R., Freeman, A. J., Issa, B., Moore, R. J., Bowtell, R. W., Baker, P. N., Johnson, I. R., Worthington, B. S., 1998. *Magn. Reson. Med.* **40**, 467–473.

Pradel, C., Siauve, N., Bruneteau, G., Clement, O., de Bazelaire, C., Frouin, F., Wedge, S. R., Tessier, J. L., Robert, P. H., Frija, G., Cuenod, C. A., 2003. *Magn. Reson. Imaging* **21**, 845–851.

Salomon, L. J., Siauve, N., Balvay, D., Cuenod, C. A., Vayssettes, C., Luciani, A., Frija, G., Ville, Y., Clement, O., 2005. *Radiology* **235**, 73–80.

Schroder, H. J., 1995. *Eur. J. Obstet. Gynecol. Reprod. Biol.* **63**, 81–90.

Trudinger, B. J., Giles, W. B., Cook, C. M., 1985. *Br. J. Obstet. Gynaecol.* **92**, 39–45.

12.5 Characterization of nephropathies and monitoring of renal stem cell therapies

Nicolas Grenier, Olivier Hauger, Yahsou Delmas and Christian Combe

Acute and chronic nephropathies are responsible for morphological and functional changes of renal parenchyma. Today radiological techniques play a minor role in imaging these diseases because the main changes are microscopic and functional, in native or in transplanted kidneys. From the animal experimentation showed here, we hope that the targeting of intrarenal inflammatory infiltrating or resident cells by *in vivo* labelling of phagocytic cells will help in differenciation of nephropathies in humans. Also *ex vivo* labelling of bone marrow stem cells for *in vivo* cell tracking after local or intravascular graft could play a role in the future to monitor renal stem cells therapies.

12.5.1 MR imaging of intrarenal inflammation

Macrophages, virtually absent in normal kidneys, may infiltrate renal tissues in specific nephropathies such as acute proliferative types of human and experimental glomerulonephritides (Cattell, 1994), renal graft dysfunctions (rejection and acute tubular necrosis) (Grau, Herbst and Steiniger, 1998) and non-specific kidney diseases such as hydronephrosis (Schreiner *et al.* 1988). This macrophagic attraction is a dynamic process under the control of chemotactic molecules (Fc fragment of immunoglobulins, TGF-β, TNF-α....) and of the level of expression of leucocytes adhesion molecules. The degree of macrophagic infiltration and proliferation is correlated with the severity of renal disease, whereas it remains unclear if macrophages produce direct renal insults or if they are a consequence of the disease in order to regulate the inflammatory response. Their role is complex, contributing to glomerular and tubulo-interstitial injury through the secretion of various cytokines and proteases which induce changes in extra-cellular matrix and progressive fibrotic changes (glomerulosclerosis, tubulointerstitial fibrosis) (Erwig, Kluth and Rees, 2001).

The macrophagic activity may vary, depending on the type of kidney disease and its severity. It predominates within the glomeruli (i.e. within the cortex) in glomerulonephritis, or within the interstitium (i.e. diffuse, within all kidney compartments) in interstitial nephritis or in hydronephrosis.

Today, in clinical practice, the degree of inflammatory response in the kidney can be approached only by renal biopsy. Therefore, identification of intrarenal macrophage infiltration with a non-invasive technique has great potential because it could participate in characterization of the kidney disease, its activity and monitor response to treatment.

Ultra-small particles of iron oxide (USPIO) are small-sized nanoparticles which have a long half-life in the blood stream (2 h in rats) and are avidly captured several hours after intravenous injection by extrahepatic cells with phagocytic activity, which include blood circulating monocytes and resident macrophages present in most of the tissues (Figure 12.5.1(a, b)). The exact mechanism of iron capture is not known and may be cell specific.

Several models of experimental nephropathies in rats were used to demonstrate the detectability of intrarenal macrophagic activity *in vivo*:

- In a model of nephrotic syndrome in the rat (Hauger *et al.* 1999), USPIO-enhanced MR images performed at 24 h demonstrated a diffuse decrease of SI pre-dominating within the outer medulla. The degree of decrease of SI was correlated with the number of macrophages within each renal compartment and to the amount of iron within the tissue.
- To differentiate glomerular versus interstitial macrophagic infiltration, we also evaluated in a model of anti-GBM glomerulonephritis, comparable to Goodpasture syndrome in humans, in which glomerular macrophagic infiltration occurs. In this model significant signal drop was observed only in the cortex where glomerular lesions were located (Figure 12.5.1(c)). The intensity of signal drop was strongly correlated with the degree of proteinuria at day 2 and day 14 (Hauger *et al.* 2000).
- In a model of obstructed kidney the same technique demonstrated a diffuse homogenous decrease of SI in the three renal compartments (Figure 12.5.1(d)). All obstructed kidneys demonstrated diffuse macrophagic infiltration on pathological examination (Hauger *et al.* 2000).

Other authors used the same technique in the models of acute and chronic rejection (Zhang *et al.* 2000; Beckmann *et al.* 2003). In both situations, medical treatment of rejection decreased the MRSI changes. The technique also demonstrated endocytic activity within the medulla in a model of reperfusion ischemia of the kidney, and no change was noted in the cortex (Jo *et al.* 2003). This is due to a medullary infiltration by mononuclear cells, within the lumen of vasa recta, which is maximum between 48 and 72 h after the ischemic injury (Ysebaert *et al.* 2000).

These experimental results showed that MR imaging, 24 h after injection of USPIO, can demonstrate intrarenal signal variations due to iron ingestion by macrophages or by glomerular cells gaining endocytic activity, that is mesangial cells, and can precisely localize this endocytic activity in the different kidney compartments. Different types of experimental renal diseases show different and reproducible types of

Figure 12.5.1 MR imaging of experimental intrarenal macrophage infiltration in rats at 4.7 T (T2*-weighted sequences: TR = 300, TE = 12, Ernst angle).
((a) Normal kidney before; (b) kidney after the intravenous injection of iron oxide particles (Sinerem®, Guerbet Group). The three intrarenal compartments are well identified: cortex (C), outer medulla (OM) and inner medulla (IM). After injection of iron oxide particles, no change of signal is noted within the kidney, whereas the liver (L) becomes dark because of a normal and intense phagocytosis by Kupfer cells.
(c) Accelerated nephrotoxic glomerulonephritis in rat, 24 h after injection of iron oxide particles. There is a notable decrease in signal intensity in the renal cortex (arrow) after injection, due to the super-paramagnetic effect of USPIO, whereas no signal intensity change is observed in any part of the medulla. (From Hauger *et al.* 2000, with permission of the Radiological Society of America).
(d) Left experimental hydronephrosis in rat, 24 h after injection of USPIO.
Dilated cavities are seen with a high signal intensity. There is a notable decrease in parenchymal signal intensity in the three renal compartments (arrow) after injection, due to the effect of superparamagnetic particles (USPIO) phagocytized by interstitial macrophages. (From Hauger *et al.* 2000, with permission of the Radiological Society of America).

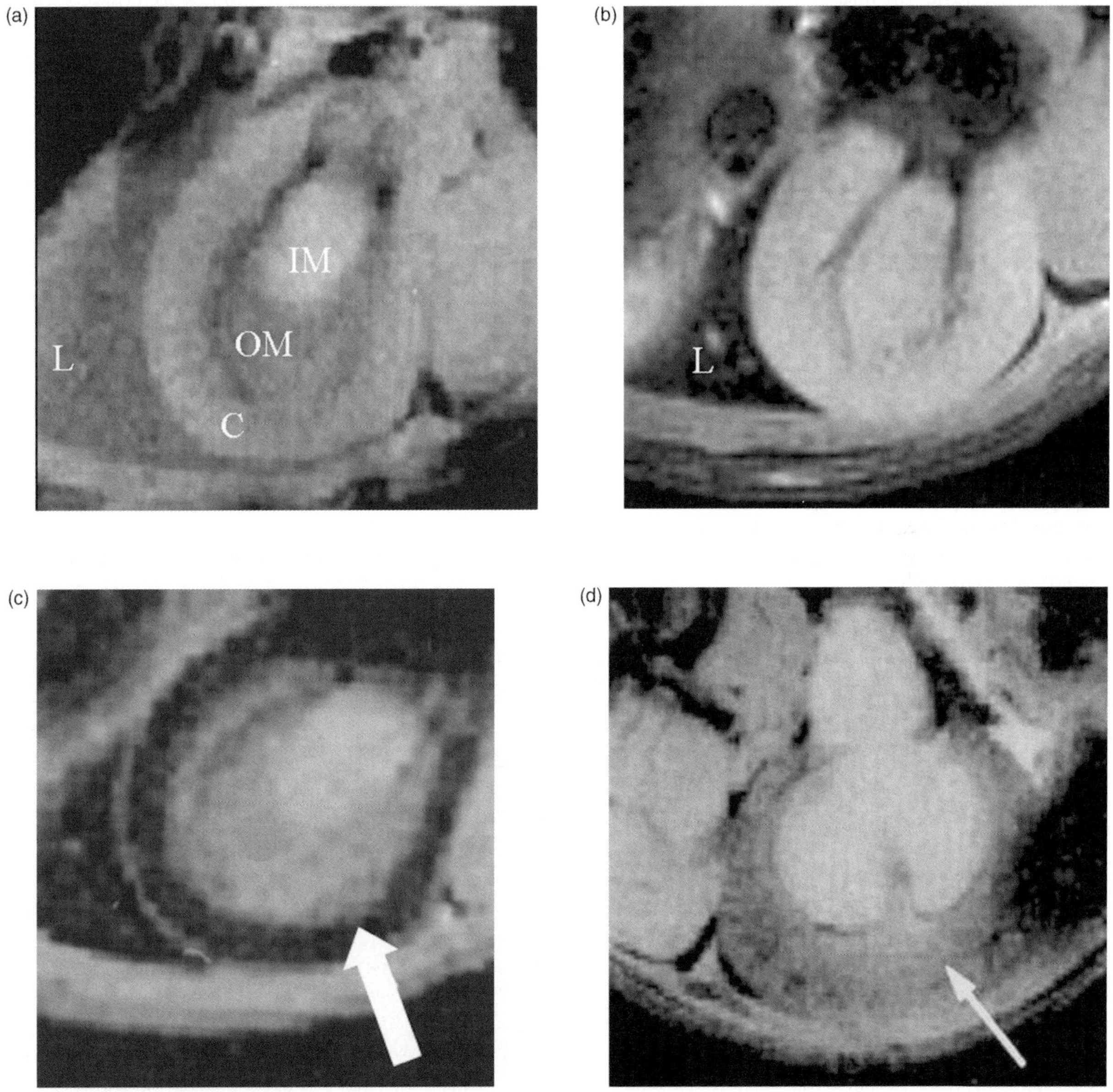

intrarenal capture of USPIO which reflect different patterns of macrophagic infiltration and endocytic activity. Moreover, the degree of renal dysfunction appeared to be correlated to the degree of endocytic activity which may have significant implications in clinical practice.

The results of the first clinical study using this methodology were published recently (Hauger *et al.* 2004), based on 12 patients. MR imaging was performed 3 days after USPIO injection (Sinerem®), Guerbet Group) to ensure getting rid of signal changes from vascular blood volume, knowing that the blood half-life of USPIO is 36 h in humans.. These preliminary clinical findings seem to corroborate experimental results and call for larger multi-centre clinical trials and evaluation of imaging at 2 days after injection to reduce delay in the diagnosis.

12.5.2 Stem cells labelling for MR-guidance of renal cell therapy

Multi-potent stem cells have shown great therapeutic promise because of their natural capability of regenerating damaged tissues. Within the kidney, several recent works have shown that the mesangial cells and tubular cells may regenerate from bone marrow mesenchymal stem cells (Gupta *et al.* 2002; Ikarashi *et al.* 2005).

Non-invasive imaging techniques allowing *in vivo* assessment of the location of stem cells are of great value for experimental studies in which these cells are transplanted. *Ex vivo* labelling of stem cells using different types of iron oxide makes it possible to track the grafted cells within the body using MRI and to assess the homing effect within the target organ. Possibly, it may also allow an estimation of the number of cells that were seeded. Finally, sequential imaging studies will allow assessing the permanence of the grafted cells over time.

Although MR imaging was able to demonstrate cell migration away from the injection site in the brain and the heart, such a migration has never been demonstrated within the kidney after local administration. In a model of accelerated mesangiolysis in the rat (Thy-1 glomerulopathy), no renal uptake of labelled stem cells could be observed *in vivo* at 4.7 T after intravenous administration because most of the cells were trapped within the liver (Figure 12.5.2(a)). However, imaging of the same kidneys *ex vivo* at 9 T could demonstrate a homing effect of labelled cells in the territories where the lesions were the most severe (highest level of proliferation) (Figure 12.5.2(b)) (Hauger *et al.* 2005). Other models with a greater attraction of stem cells (associated with hypoxia for example) will have to be evaluated to validate the intravenous route for *in vivo* monitoring of renal cell therapy.

Using the arterial route for direct cell grafting in renal arteries of normal rats, labelled MSC were easily

Figure 12.5.2 Intrarenal homing of labelled mesenchymal stem cells after intravenous graft in diseased rats (Thy1 glomerulopathy).
(a) Axial *in vivo* T2*-weighted image at 4.7 T shows trapping of labelled cells within the liver and no cells within the kidney.
(b) Longitudinal *ex vivo* T_2*-weighted MR-image of a rat kidney at very high field strength (9 T) showing cortical low signal intensity areas related to magnetically labelled mesenchymal stem cells targeting the diseased areas of the kidney (homing effect) after intravenous administration. (From Bos, C. *et al.* 2004, with permission of the Radiological Society of America)

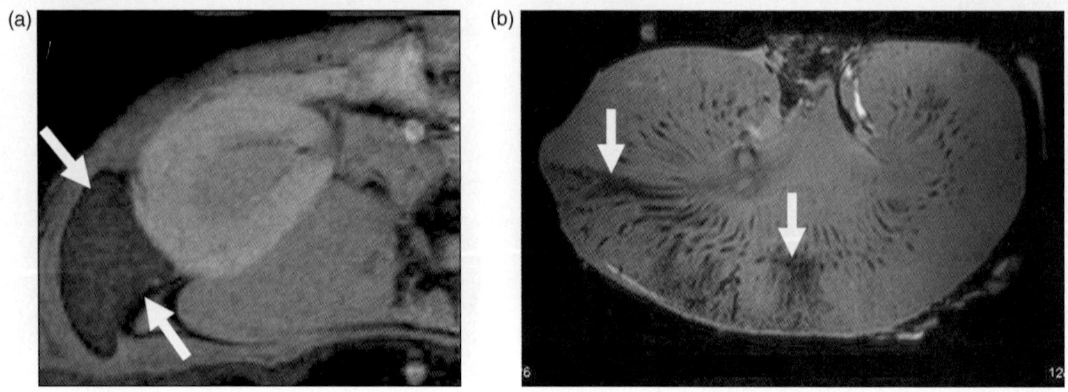

Figure 12.5.3 Intrarenal distribution of labelled mesenchymal stem cells after intra-arterial graft in normal rats.
T_2*-weighted MR-images of a rat kidney obtained *in vivo* (a) and *ex vivo* (b) 7 days after intra-arterial injection of magnetically-labelled mesenchymal stem cells, at 1.5 T. The cells are distributed within the cortex and histology found the cells into glomeruli (From Hauger *et al.* 2006, with permission of the Radiological Society of America)

(a)
(b)
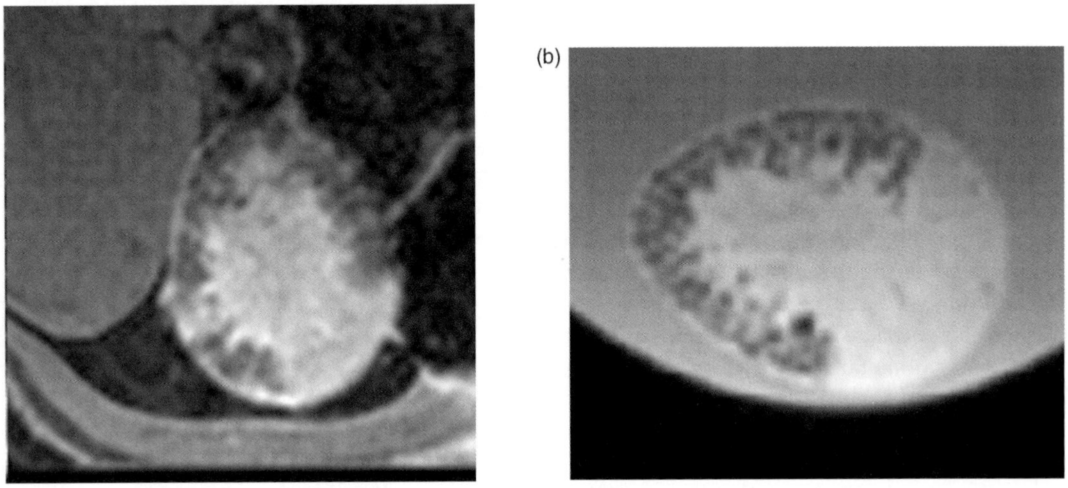

identified *in vivo* within the renal cortex at 1.5 T, they remained visible during several days after grafting (Figure 12.5.3) and were trapped within the glomeruli at histology (Bos *et al.* 2004). Applied to a model of ischemia-reperfusion, stem cell therapy could be monitored with MR imaging (Lange *et al.* 2005). So, this route can be used to seed cells throughout the target organ for tissue regeneration or transgene expression.

References

Beckmann, N., Cannet, C., Fringeli-Tanner, M., Baumann, D., Pally, C., Bruns, C., Zerwes, H. G., Andriambeloson, E., Bigaud, M., 2003. *Magn. Reson. Med.* **49**, 459–467.

Bos, C., Delmas, Y., Desmouliere, A., Solanilla, A., Hauger, O., Grosset, C., Dubus, I., Ivanovic, Z., Rosenbaum, J., Charbord, P., Combe, C., Bulte, J. W., Moonen, C. T., Ripoche, J., Grenier, N., 2004. *Radiology* **233**, 781–789.

Cattell, V., 1994. *Kidney Int.* **45**, 945–952.

Erwig, L. P., Kluth, D. C., Rees, A. J., 2001. *Curr. Opin. Nephrol Hypertens* **10**, 341–347.

Grau, V., Herbst, B., Steiniger, B., 1998. *Cell Tissue Res.* **291**, 117–126.

Gupta, S., Verfaillie, C., Chmielewski, D., Kim, Y., Rosenberg, M. E., 2002. *Kidney Int.* **62**, 1285–1290.

Hauger, O., Delalande, C., Deminiere, C., Fouqueray, B., Ohayon, C., Garcia, S., Trillaud, H., Combe, C., Grenier, N., 2000. *Radiology* **217**, 819–826.

Hauger, O., Delalande, C., Trillaud, H., Deminiere, C., Quesson, B., Kahn, H., Cambar, J., Combe, C., Grenier, N., 1999. *Magn. Reson. Med.* **41**, 156–162.

Hauger, O., Frost, E. E., van Heeswijk, R., Deminiere, C., Xue, R., Delmas, Y., Combe, C., Moonen, C. T., Grenier, N., Bulte, J. W., 2006. "MR evaluation of the glomerular homing of magnetically labeled mesenchymal stem cells in a rat model of nephropathy." *Radiology* **238**(1), 200–10.

Hauger, O., Grenier, N., Deminière, C., Delmas, Y., Combe, C., 2004. In Radiological Society of North America, vol. 233(P). Radiological Society of North America, Chicago, p. 512.

Ikarashi, K., Li, B., Suwa, M., Kawamura, K., Morioka, T., Yao, J., Khan, F., Uchiyama, M., Oite, T., 2005. *Kidney Int.* **67**, 1925–1933.

Jo, S. K., Hu, X., Kobayashi, H., Lizak, M., Miyaji, T., Koretsky, A., Star, R. A., 2003. *Kidney Int.* **64**, 43–51.

Lange, C., Togel, F., Ittrich, H., Clayton, F., Nolte-Ernsting, C., Zander, A. R., Westenfelder, C., 2005. *Kidney Int.* **68**, 1613–1617.

Schreiner, G. F., Harris, K. P., Purkerson, M. L., Klahr, S., 1988. *Kidney Int.* **34**, 487–493.

Ysebaert, D. K., De Greef, K. E., Vercauteren, S. R., Ghielli, M., Verpooten, G. A., Eyskens, E. J., De Broe, M. E., 2000. *Nephrol. Dial Transplant* **15**, 1562–1574.

Zhang, Y., Dodd, S. J., Hendrich, K. S., Williams, M., Ho, C., 2000. *Kidney Int.* **58**, 1300–1310.

12.6 Optical imaging of lung inflammation

Jodi Haller

Inflammatory response consists of a complex balance between tissue destruction and repair (Nathan, 2002) and is characterized by a substantial increase in neutrophils, macrophages and lymophocytes. These release a variety of proteases such as matrix metalloproteinases (MMPs) and cathepsins, which facilitate recovery by breaking down dead tissue, clearing the way for reconstruction. Proteolytic activity is intricately regulated by pro- and anti-inflammatory mediators, which are critical for an effective immune response (Greene *et al.*, 2005). Breakdown of the regulatory pathways that influence proteolytic activity plays a key role in the progression of pulmonary disease by perpetuating inflammation. Cystic fibrosis, emphysema, chronic obstructive pulmonary disease, tuberculosis, asthma and other pulmonary diorders are all characterized by dysregulation of proteolytic activity and chronic inflammation. Because of the critical role proteases play in pulmonary disease, reliable non-invasive techniques for acquiring functional measurements of proteolytic activity, whether of cathepsins, MMPs, or other proteases, would be invaluable in studying disease progression and the effects of drugs over time and in unperturbed environments.

Optical imaging can play a vital role in basic research, drug discovery and pre-clinical studies involving mouse models of pulmonary diseases. While planar reflectance imaging cannot image in the lung *in vivo* due to depth limitations, transillumination and tomographic techniques are well suited for *in vivo* imaging interrogations of the lungs of small animals. In particular, Fluorescence Molecular Tomography (FMT; see Chapter 5) has been shown the capacity to image pulmonary inflammation *in vivo* by utilizing a continuously increasing list of adept fluorochromes with molecular specificity to several inflammatory biomarkers. Important insights can be gained with the use of activatable fluorochromes, as described in Weissleder and Ntziachristos (Ntziachristos, Bremer and Weissleder, 2003).

Figures 12.6.1 and 12.6.2 summarize imaging results from Balb/c mice injected with a protease-activatable, near-infrared fluorescent probe 30-h following intratracheal instillation of lipopolysaccharide (LPS) to induce an inflammatory response within the lungs. LPS is a major pro-inflammatory glycolipid component present within the cell walls of gram-negative bacteria and is a primary cause of innate immunity and acute inflammation (Vernooy *et al.*, 2001). Mice are known to experience persistent chronic pulmonary inflammation when exposed to repeated intratracheal instillations of LPS (Vernooy *et al.*, 2002). Reponse to LPS is characterized by neutrophil and macrophage infiltration, apoptosis, rapid production of TNF-α, increased concentrations of chemokines and increased protease levels. Chronic exposure to elevated levels of LPS, which is present in at high levels in cigarette smoke and at lower levels in airborn particles such as dust, is associated with a variety of pulmonary diseases, particularly asthma, chronic bronchitis, cystic fibrosis and chronic obstructive pulmonary disease (COPD) (Laan, Bozinovski and Anderson, 2004). Therefore, this model of lung

Figure 12.6.1 Representative images of a control mouse (uninstilled) versus an LPS-instilled mouse. MRI images are provided as anatomical references. A. Mean transillumination absorption with a high-absorption spot in the middle, which consistently corresponds to the heart. B. Mean aBorn field, a way of calculating the transillumination fluorescence while weighting by absorbance, in order to account for varying tissue heterogeneities. Signal is notably heightened within the lungs of the LPS-instilled mouse, as expected. The central, signal-void region corresponds to the heart, as determined by MRI and absorption measurements

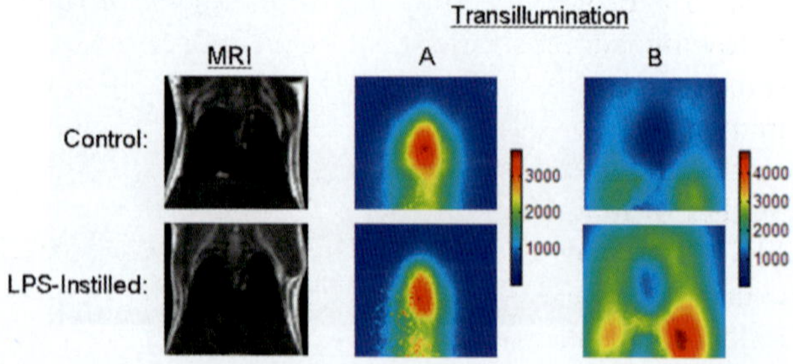

Figure 12.6.2 Representative slices from tomographic reconstructions of a control mouse (A) and an LPS-instilled mouse (B). A clearly heightened signal is present in the instilled mouse, whereas signal is relatively low and confined to the liver in the control mice, as expected

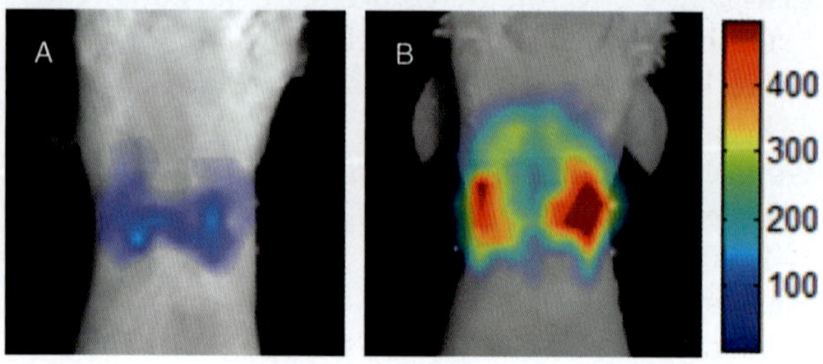

inflammation closely mimics the pathological responses seen in humans suffering from chronic obstructive pulmonary disease, and consequently, it has seen widespread use in studies of pulmonary diseases and pro- and anti-inflammatory mediators.

Because the inflammatory response to intratracheal instillation of LPS is known to involve elevated levels of proteases, the model can be imaged by optical methods using protease-activatable, near-infrared, fluorescent probes. The details of such probes have been described in detail in Weissleder *et al.* (1999) and in Chapter 7. Typically in such studies, care must be taken to choose the proper time for probe injection and subsequent imaging following administration of the inflammatory substance. Adequate time must be allowed for the immune response to cause proteolytic activity to rise to the level needed for substantial probe activation. Studies of lung inflammation following intra-tracheal LPS instillations in mice have shown that neutrophil counts are quite limited until approximately 6–8 h following instillation, at which point counts increase demonstrably, peaking at 24 h, falling to their initial values roughly 72 h after instillation (Vernooy *et al.*, 2001). The approach taken thus far has been to inject probe 6–12 h following LPS instillation, which provides adequate time for protease levels to escalate to heightened levels. Imaging is then conducted approximately 24 h following probe injection.

Simple transillumination imaging has also proven remarkably useful for making qualitative judgments as to the degree of pulmonary inflammation, as demonstrated in Figure 12.6.1. After collecting fluorescence and intrinsic transillumination data, the fluorescence images may be normalized by taking the ratio of the fluorescence to the intrinsic images (Graves *et al.*, 2005). The sum of these ratios provides a single two-dimensional image, the optical analog to the X-ray, that clearly shows the degree distribution of lung inflammation. Alternatively, the fluorescence transillumination data may be normalized by the intrinsic transmittance, providing what is termed the 'a Born field'. It is important to note that the probe is known to collect and activate within the liver; so a strong signal from that region is expected both in mice with inflamed and non-inflamed lungs. The difference in signal intensity, shape and distribution within the lungs between the transillumination images of the control and experimental mouse are substantial and repeatable.

However, one downside of transillumination imaging is its inability to accurately quantify the fluorescence present in tissues, to correct for scattering and to provide three-dimensional information. Tomographic reconstructions may be alternatively performed, using transillumination data collected at different projection angles and appropriate image reconstruction mathematical approaches. The resulting images are inherently quantitative and provide measurements of fluorochrome concentration, as opposed to depth-dependent and optical property-dependent signal intensity, as is the case of both reflectance and transillumination imaging. Two reconstructed FMT slices, from a control and an experimental animal respectively, are shown in Figure 12.6.2. The slices are obtained approximately 3 mm from the anterior of the mouse thorax and demonstrate a marked difference between the inflamed and control lung. In these images the liver is filtered out using a threshold method applied on measurements at the excitation wavelength, indicative of tissue spatially dependent attenuation.

While the lungs present unique challenges to optical imaging, transillumination and FMT have proved useful for characterizing and monitoring pulmonary disease. Coupled to the development of protease-activatable fluorescent probes, or probes with molecular specificity to other inflammatory targets, this technique is expected to play a vital role in the longitudinal study of disease evolution and drug efficacy in animal models of various pulmonary diseases.

References

Graves, E. E., Yessayan, D., Turner, G., Weissleder, R., Ntziachristos, V., 2005. "Validation of in vivo fluorochrome concentrations measured using fluorescence molecular tomography." *J. Biomed. Opt.* **10**(4), 44019.

Greene, C. M., Carroll, T. P., Smith, S. G., Taggart, C. C., Devaney, J., Griffin, S., O'neill, S. J., McElvaney, N. G., 2005. "TLR-induced inflammation in cystic fibrosis and non-cystic fibrosis airway epithelial cells." *J. Immunol* **174**(3), 1638–1646.

Hasday, J. D., Bascom, R., Costa, J. J., Fitzgerald, T., Dubin, W., 1999. "Bacterial endotoxin is an active component of cigarette smoke." *Chest* **115**(3), 829–835.

Laan, M., Bozinovski, S., Anderson, G. P., 2004. "Cigarette smoke inhibits lipopolysaccharide-induced production of inflammatory cytokines by suppressing the activation of activator protein-1 in bronchial epithelial cells." *J. Immunol.* **173**(6), 4164–4170.

Nathan, C., 2002. "Points of control in inflammation." *Nature* **420**(6917), 846–852.

Ntziachristos, V., Bremer, C., Weissleder, R., 2003. "Fluorescence imaging with near-infrared light: New technological advances that enable *in vivo* molecular imaging." *Eur. Radiol.* **13**(1), 195–208.

Vernooy, J. H., Dentener, M. A., van Suylen, R. J., Buurman, W. A., Wouters, E. F., 2001. "Intratracheal instillation of lipopolysaccharide in mice induces apoptosis in bronchial epithelial cells – No role for tumor necrosis factor-alpha and infiltrating neutrophils." *Am. J. Respir. Cell Mol. Biol.* **24**(5), 569–576.

Vernooy, J. H. J. *et al.*, 2002. "Long-term intratracheal lipopolysaccharide exposure in mice results in chronic lung inflammation and persistent pathology." *Am. J. Respir. Cell Mol. Biol.* **26**(1), 152–159.

Weissleder, R. Tung, C. H., Mahmood, U., Bogdanov, A. Jr,. Related Articles, 1999. "*In vivo* imaging of tumors with protease-activated near-infrared fluorescent probes." *Nat. Biotechnol.* **17**(4), 375–378.

12.7 Optical imaging in rheumatoid arthritis

Andreas Wunder

Rheumatoid arthritis (RA) is a chronic inflammatory disease mainly affecting the joints. In arthritic joints, the synovial lining layer is transformed into a highly proliferative tissue consisting of synovial fibroblasts, synovial macrophages and various infiltrating inflammatory cells. This so-called pannus-like tissue progressively invades adjacent cartilage and bone leading to joint destruction, which is primarily mediated by the release of matrix-degrading enzymes (Lee and Weinblatt, 2001; Smith and Haynes, 2002). The current methods used in clinical medicine to image arthritic joints, especially conventional radiography, x-ray computed tomography (CT), magnetic resonance tomography (MRT) and ultrasound (US), are based on visualization of anatomic, physiologic or metabolic heterogeneity rather than identifying specific cellular or molecular events that underlie disease. Articular bone erosion and joint-space narrowing are relatively well delineated by radiographs. High-resolution US and especially MRT have improved substantially the ability to detect initial joint destruction providing more detailed information about soft-tissue changes, but still report relatively late in the course of the disease (Peterfy, 2003; Taylor, 2003).

Detection, tracking and monitoring of cells and molecules driving or inhibiting the arthritic process non-invasively by imaging methods in experimental arthritis models and in human disease are of considerable interest for both, arthritis basic research and the clinical management of patients. Approaches to visualize cellular and molecular events in arthritis are currently limited to a few nuclear imaging studies in RA patients and optical imaging studies in animal models (for review see Wunder *et al.*, 2005). Nuclear imaging approaches include targets that mark soluble factors, cells or surface molecules, activation of cells and apoptosis or indicate dynamic processes such as proliferation (Wunder *et al.*, 2005). A number of

Figure 12.7.1 Bioluminescence imaging (BI) of adoptive gene transfer in arthritic mice. The animal received an intravenous injection T-cell hybridomas retrovirally transduced to express luciferase. On day 5 of the transfer, the mice were injected intraperitoneally with the luciferase substrate luciferin. Conversion of luciferin to oxyluciferin by luciferase generates light, which can be detected by a sensitive camera. Light intensities are represented on a false colour scale with blue indicating low and red indicating high light intensities. The figure illustrates that the accumulation of the transferred cells in the arthritic left hind paw could be non-invasively visualized using BI. A weak signal was also detected in the right hind paw, which developed low-grade arthritis over the course of the experiment (Figure adapted from Nakajima *et al.*, 2001)

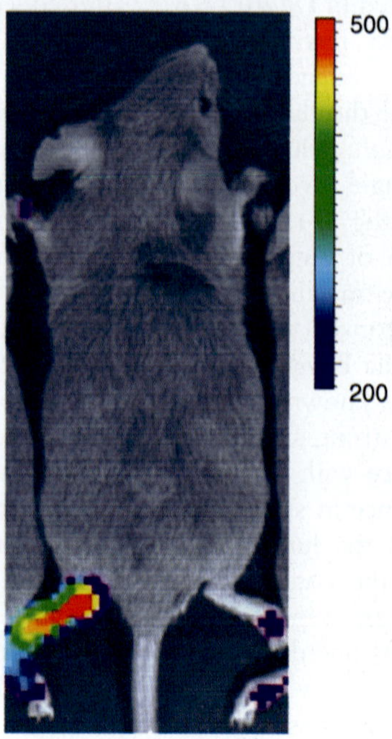

recently published optical imaging studies in experimental settings have shown the potential of optical technologies to evaluate biological processes in arthritis. These approaches include non-invasive monitoring of inflammatory cells (Hansch *et al.*, 2004) or transcription factors involved in disease (Carlsen *et al.*, 2002), imaging gene transfer (Nakajima *et al.*, 2001) and apoptosis (Wunder *et al.*, 2005) or the activity of matrixdegrading enzymes involved in joint destruction to monitor treatment response (Wunder *et al.*, 2004).

Hansch *et al.* (2004) have demonstrated the feasibility of fluorescence reflectance imaging (FRI) to visualize inflammatory cells in arthritic joints in a murine arthritis model using anti-macrophage monoclonal antibodies labelled with a near-infrared fluorochrome. Carlsen *et al.* (2002) have developed transgenic mice that express luciferase under the control of the transcription factor NF-κB, enabling bioluminescence imaging (BI) of NF-κB activity in intact animals. The animals received the luciferase substrate luciferin. Conversion of luciferin to oxyluciferin by luciferase generates light that could be sensitively detected by a camera system (Carlsen *et al.*, 2002). BI can also be used for *in vivo* imaging of disease- or therapy-related cells. Nakajima *et al.* (2001) have shown in a murine arthritis model that BI can be used for tracking genetically altered cells in an adoptive gene therapy setting in which T cells were transduced to express luciferase. In this approach, BI was successfully employed to demonstrate the ability of injected cells to migrate preferentially into the inflamed joints (see Figure 12.7.1) (Nakajima *et al.*, 2001).

Recently we have been shown that near-infrared FRI can be used for non-invasive imaging of treatment response in arthritic mice using fluorescent Annexin V as a marker for apoptotic cells and an enzyme activateable fluorescent imaging probe as marker for joint destruction. Annexin V binds with high affinity to phosphatidylserine, which is normally restricted to the inner leaflet of the cell membrane but is exposed on the surface of cells undergoing apoptosis. Paws of arthritic mice treated with methotrexate (MTX), the most common drug in the treatment of RA, showed a significantly higher fluorescence intensity than arthritic paws of untreated mice and non-arthritic paws of MTX-treated mice. Thus, FRI using fluorescent Annexin V provides a method for imaging treatment response *in vivo*, which was readily

Figure 12.7.2 Fluorescence reflectance imaging (FRI) of protease activity for non-invasive monitoring of treatment response in arthritis. Colour-coded near-infrared fluorescence images superimposed on white light images of an untreated (a) and a MTX-treated (b) mice with arthritis affecting the right hind paw 24 h after injection of imaging probe that is activated by proteases involved in joint destruction. MTX reduced fluorescence intensity from arthritic paws compared to untreated mice. The time course of fluorescence intensities in arthritic and non-arthritic paws of mice after probe injection is shown in (c). Values are expressed as relative fluorescence intensity (RFI; mean ± SEM). MTX treatment decreased significantly fluorescence (6 h post injection: $P = 0.01$; 12 and 24 h post injection: $P < 0.001$) (Figure adapted from Wunder *et al.*, 2004)

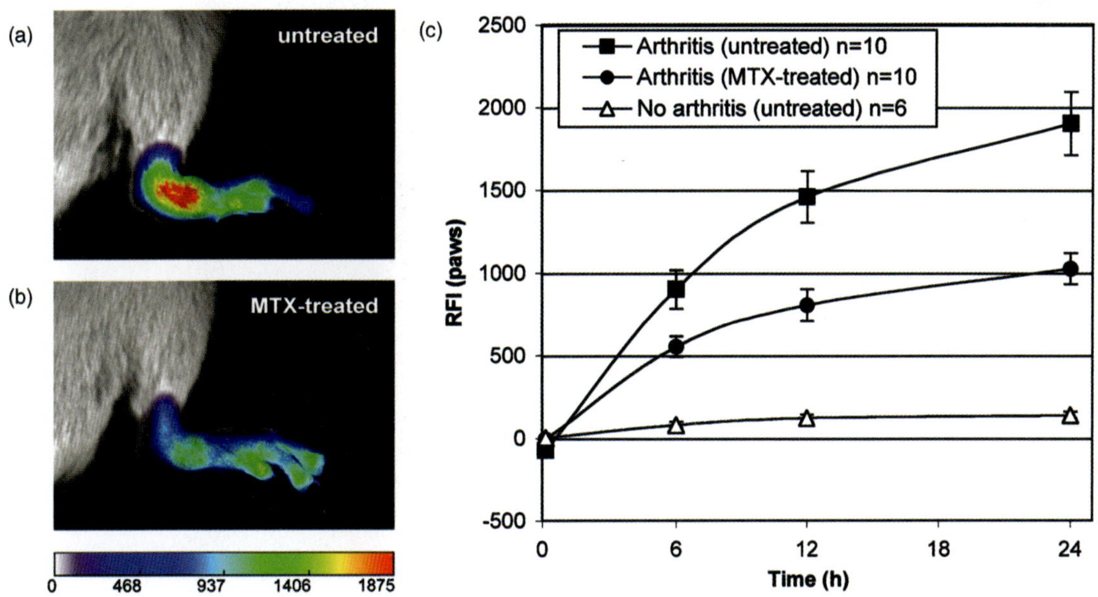

quantitated with simple instrumentation and which occurred before conventional measurements (Wunder *et al.*, 2005). In another approach, we used an enzyme activatable near-infrared fluorescence imaging agent and FRI to visualize protease activity for non-invasive assessment of treatment effects. Matrix degrading proteases are key players in joint destruction. We, therefore, hypothesized that these enzymes would be suitable targets to visualise treatment effects *in vivo*. The imaging agent consisted of multiple fluorochromes bound to a polymer. Due to fluorescence quenching, the probe is relatively undetectable. Enzymatic cleavage of the backbone by proteolytic enzymes at specific cleavage sites *in vivo* releases fluorochromes resulting in a significant increase of fluorescence. After intravenous injection of the imaging probe, toes and paws affected by arthritis showed a high fluorescence intensity, which was significantly lower in MTX-treated animals (see Figure 12.7.2). These results demonstrate that enzyme activatable imaging probes can be used for non-invasive imaging of protease activity and could also be used for monitoring response (Wunder *et al.*, 2004).

References

Carlsen, H., Moskaug, J. O., Fromm, S. H., Blomhoff, R., 2002. "*In vivo* imaging of NF-kappa B activity." *J. Immunol.* **168**, 1441–1446.

Hansch, A., Frey, O., Sauner, D., Hilger, I., Haas, M., Malich, A., Brauer, R., Kaiser, W. A., 2004. "*In vivo* imaging of experimental arthritis with near-infrared fluorescence." *Arthritis Rheum.* **50**, 961–967.

Lee, D. M., Weinblatt, M. E., 2001. "Rheumatoid arthritis." *Lancet* **358**, 903–911.

Nakajima, A., Seroogy, C. M., Sandora, M. R., Tarner, I. H., Costa, G. L., Taylor-Edwards, C., Bachmann, M. H., Contag, C. H., Fathman, C. G., 2001. "Antigen-specific T cell-mediated gene therapy in collagen-induced arthritis." *J. Clin. Invest.* **107**, 1293–1301.

Peterfy, C. G., 2003. "New developments in imaging in rheumatoid arthritis." *Curr. Opin. Rheumatol.* **15**, 288–295.

Smith, J. B., Haynes, M. K., 2002. "Rheumatoid arthritis – A molecular understanding." *Ann. Intern. Med.* **136**, 908–922.

Taylor, P. C., 2003. "The value of sensitive imaging modalities in rheumatoid arthritis." *Arthritis Res. Ther.* **5**, 210–213.

Wunder, A., Straub, R. H., Gay, S., Funk, J., Müller-Ladner, U., 2005. "Molecular imaging – Novel tools in visualizing rheumatoid arthritis." Rheumatology, (Oxford) **44**, 1341–1339.

Wunder, A., Schellenberger, E., Mahmood, U., Bogdanov, A., Müller-Ladner, U., Weissleder, R., Josephson, L., 2005. "Methotrexate induced accumulation of fluorescent annexin V in collagen induced arthritis." *Mol. Imag.* **4**, 1–6.

Wunder, A., Tung, C. H., Muller-Ladner, U., Weissleder, R., Mahmood, U., 2004. "*In vivo* imaging of protease activity in arthritis: a novel approach for monitoring treatment response." *Arthritis Rheum.* **50**, 2459–2465.

13 Gene Therapy

Markus Klein and **Andreas H. Jacobs**

13.0 Introduction

The term *gene therapy* describes the modification of the cellular genetic information of an organism (or a distinct tissue/cell type) with the purpose to cure a disease. The methods used for this genetic modification include viral infection, direct DNA transfer by electroporation, and carrier-mediated gene transfer using stem cells/progenitor cells as 'Trojan horses'. Despite the fact that many gene therapy concepts are still being further developed in animal models, several clinical gene therapy trials have been initiated during the past years with about two-thirds of them addressing cancer gene therapy (Dachs, Tupper and Tozer, 2005).

The two major goals of gene therapy are to genetically correct a molecular-based disease or, for cancer applications, to introduce genetic information which confers a toxic effect, with the injection of the gene-delivering solution being the only invasive intervention to be performed during therapy. Furthermore, the potential of targeting the disease by stem cell-mediated gene transfer gives the possibility to guide the therapy to the affected areas throughout the organism (Aboody *et al.*, 2000).

A disadvantage of the common gene therapy methods is the relatively low efficiency of gene transfer to all the cells of the diseased target tissue. This includes the potential weak tissue distribution of the transfection/transfer solution, an inflammatory reaction directed against the transfer agent and, in the case of viral infection, the short viral survival time.

A further problematic aspect of gene therapy is the question of how to appropriately assess the associated treatment effect. Without any possibility to follow the extent and distribution of gene expression, the therapeutic system will remain a 'black box', giving only limited information about the functionality of the system. Therefore, molecular imaging technologies were developed over the past 10 years to specifically address this problem. The term molecular imaging describes, in general, the non-invasive assessment of gene expression in a living organism. This can be imaging of endogenous molecular events or imaging the expression of exogenously introduced genes. Imaging of exogenous gene expression deals with the (proportionally) co-expression of an introduced imaging gene together with any therapeutic gene of interest. The images obtained thereby give information about the functionality and efficiency of the gene therapeutic system and should provide an indication of the requirements with respect to vector dose and route of administration for further treatment. Currently, the most frequently used molecular imaging methods include positron emission tomography (PET), single photon emission computed tomography (SPECT), magnetic resonance imaging (MRI), bioluminescent imaging (BLI) and various fluorescent techniques (Shah *et al.*, 2004).

The diseases targeted primarily by gene therapy is cancer, for example brain tumours, because of the poor results associated with common therapeutic approaches and because of tissue-specific problems, which are, for gliomas, the high susceptibility of the surrounding neuronal tissue to damage, location in a scull-enclosed area and inconsistent immune response. Gene therapy was performed in the first clinical trials more than 10 years ago. However, despite ongoing research, the prognosis for patients bearing brain tumours is still very poor, often leading to death within months. Observations of patients benefiting from gene therapy so far are limited to those bearing small tumours. Other malignant tissues targeted by gene therapies include breast, bone, lung, skin and liver, and other diseases include cystic fibrosis, Alzheimer's and Parkinson's disease.

13.1 Expression systems for genes of interest (GOI)

The *specificity* of gene transduction is based on (i) the delivery system (systemic versus local), (ii) the recognition of specific markers on the surface of the target cell/tissue and (iii) the type of promoter. The use of tissue-specific promoters leads to a more distinct expression, protecting potentially co-transduced cells of a different cell type than the target cells (Vigna *et al.*, 2005).

The expression of the transduced gene can be *regulated* by the addition of certain regulatory elements in the promoter region. There are two general mechanisms dealing with the aspect of gene regulation, which are the insertion of either endogenous or exogenous/artificial regulatory elements (Figure 13.1.1). In the first case the target tissue expresses a distinct signalling molecule, which is able to bind to the regulatory DNA element and which is not present in the surrounding cells of the target area. After successful gene transduction of the tissue, the GOI will be expressed exclusively in the cells expressing the signalling molecule (Davies *et al.*, 2005; Rosenfeld *et al.*, 2005). This paradigm is of interest in diseases where a pathognomonic up-regulation of a signalling molecule takes place. Thereby, a control step is introduced to a gene therapy protocol protecting the surrounding cells, which do not express this signalling molecule. In the second case, exogenous regulatory elements mediate the regulated expression of a GOI. The therapeutic advantage is the ability to adjust the dose of the transgene by varying the concentration of the regulator. Several different regulatory systems have been developed including the doxycycline-dependent regulation (tet on/tet off; Agha-Mohammadi *et al.*, 2004; Lee *et al.*, 2005; Vigna *et al.*, 2005) or the progesterone-dependent 'switch' system (Wang *et al.*, 1994; McGuire, Mao and Davis, 2004).

Figure 13.1.1 Scheme of the regulation of gene expression: (a) In the absence of a specific ligand, the regulator is not able to bind to the regulatory region on the DNA and transcription is not initialized. (b) Addition of a specific ligand, which binds to the ligand-binding domain (LBD) of the regulator, serves for a conformational change of the regulator in such a way that the DNA- binding domain (DBD) of the regulator can now bind to the regulatory element leading to promoter activation and thereby transcription

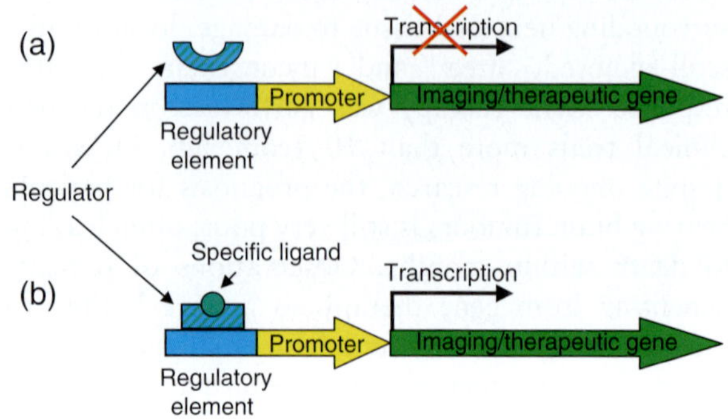

The expression of only one transgene might not be sufficient for the expected therapeutic effect. Several different *linking elements* have been introduced in constructs to link the expression of several genes to only one promoter to serve proportional co-expression of multiple GOIs. The linker elements, such as viral or eukaryotic internal ribosomal entry sites or the viral 2A element (de Felipe *et al.*, 1999; Chappell, Edelman and Mauro, 2000), lead to a proportional co-expression of the genes separated by the linker. It should, however, be kept in mind that the gene located downstream of the linking element shows up to 10-fold weaker expression (Zhou *et al.*, 1998; Jacobs *et al.*, 2003b).

13.2 Gene delivery systems (vectors)

Several viral and non-viral delivery systems have been developed and tested so far, including all common virus types, such as adenoviruses (AdV), adeno-associated viruses (AAV), herpes simplex virus type 1 (HSV-1) and retroviruses. Questions that should be addressed when selecting for a viral gene delivery system are the size of the gene construct, the demand for integration of the transgene into the genome, the number of viral particles required for proper infection results (transduction efficiency), tissue specificity (regarding infection) as well as the elicitation of an immune response (Rojas-Martinez *et al.*, 2002; Teschemacher, Paton and Kasparov, 2005; Verma and Weitzman, 2005). Most viruses used in gene therapy are replication deficient, such as HSV-1 amplicon vectors (Jacobs *et al.*, 2003b).

Replication-conditional viruses are able to replicate exclusively in dividing cells and have been developed specifically for virus-mediated oncolysis of brain tumours, taking into account the low division potential of neurons and normal glial cells in the brain. Two major limitations of virus therapy have been observed, which are the relatively weak virus distribution in the target tissue and the limited virus survival time after infection. The most commonly used virus used for virus therapy are mutants derived from HSV-1. Wild-type HSV-1 is able to replicate in dividing and non-dividing cells, such as neurons. In non-dividing cells, which provide only a limited replication machinery for the virus, HSV-1-based viral enzymes support the replication process, such as ribonucleotide reductase (RR). RR is required for nucleotide supply in non-dividing cells, whereas in dividing cells the nucleotides of the host cell are sufficient to support virus replication. Inactivating mutations in the RR gene lead to a conditionally replicating state of this virus. Several mutations like the one described have been introduced within viral proteins, leading to a subset of replication-conditional viruses with distinct replication properties (Jacobs, Breakefield and Fraefel, 1999; Jacobs *et al.*, 2003b; Tyminski *et al.*, 2005; Adusumilli *et al.*, 2006). The major goal is the efficient and selective replication in tumour cells, leading to a high number of infectious particles and conferring the oncolytic effect to infected tumour cells. A summary of different oncolytic viruses used for gene and virus therapy is reviewed by Aghi and Martuza (2005).

Non-viral delivery systems (e.g. liposomes, injection of plasmid DNA followed by electroporation) are usually less immunogenic but suffer from an insufficient expression level of the transgene. The transduction efficiency of the target tissue has been found to be inconsistent, and the transgene expression is often transient. A detailed description of the differences of the delivery systems has been published elsewhere (Rojas-Martinez *et al.*, 2002; Piskin, Dincer and Turk, 2004; Verma and Weitzman, 2005).

13.3 Suicide gene therapy

Enzymes used for suicide gene therapy should exhibit a high affinity to the corresponding pro-drug. They should not be normally present in the tissue and should be expressed to a sufficient amount. Several pro-drug-activating systems have been developed with a diverse array of proteins and pro-drugs. The most common systems will be briefly summarized.

13.3.1 Herpes simplex virus type 1 thymidine kinase (HSV-1-tk)

HSV-1-*tk* is the most popular enzyme used in cancer gene therapy protocols for more than 10 years in combination with its specific pro-drug ganciclovir (GCV). HSV-1-*tk* can be used as a therapeutic gene as well as an imaging gene (for PET), giving the possibility to follow the HSV-1-*tk* expression pattern *in vivo* by imaging, allowing especially the non-invasive assessment of the transduction efficiency *in vivo*.

Figure 13.3.1 Scheme for suicide gene therapy mediated by HSV-1-*tk*. After vector infection, cells expressing HSV-1-*tk* are able to phosphorylate GCV to its monophosphate form. Thereafter, endogenous enzymes lead to the di- and tri-phosphate forms, the latter being incorporated into the DNA leading to failure in DNA synthesis and thereby cell death

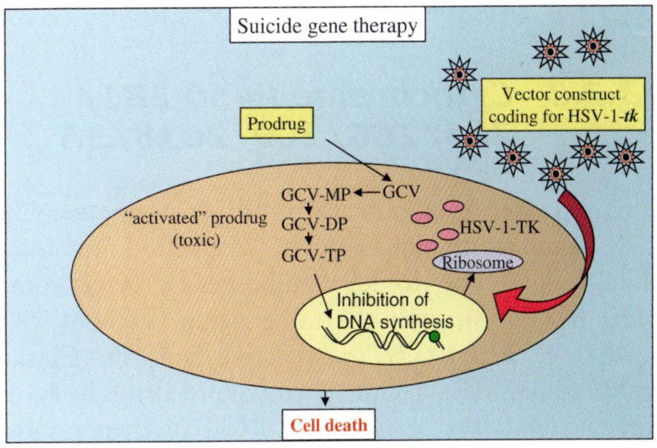

Gene transfer to the malignant target tissue is performed most often *via* viral infection using modified, replication-deficient viruses, coding for HSV-1-*tk*. After transcription to HSV-1-*tk* mRNA, the functional protein HSV-1-*tk* is generated. This protein is able to convert the *tk* substrate GCV into the phosphorylated form, thereby trapping it within the transduced cell (Figure 13.3.1). In the case of the therapeutic substrate, GCV, this leads to the generation of GCV monophosphates, which are further converted by endogenous kinases into the corresponding triphosphate. These GCV phosphates can be incorporated into the genomic DNA of the cell during the next DNA synthetic cycle, leading to cell death. In the case of imaging substrates (FIAU, FHBG), their accumulation rate within the transduced cells corresponds to the level of the transduced HSV-1-*tk* gene expression (Tjuvajev *et al.*, 1995; Jacobs *et al.*, 2003a, b).

13.3.2 Cytosine deaminase (CD)

CD is expressed in fungi and bacteria, but not in mammalian cells. After transduction of mammalian cells with genetic information for CD, these cells gain the ability for the hydrolytic deamination of cytosine to uracil (Erbs, Exinger and Jund, 1997). Moreover, its ability to convert the pro-drug 5-fluoro-cytosine (5FC) to the toxic compound 5-fluoro-uracil (5FU) has lead to the use of the enzyme in suicide gene therapy (Yao *et al.*, 2005). 5FU is further metabolized to the deoxyribonucleotide monophosphate (FdUMP), inhibiting thymidylate synthase. The resulting lack of thymidylate inhibits the DNA synthesis (Moolten, 1994; Yazawa, Fisher and Brunicardi, 2002; Yao *et al.*, 2005).

13.3.3 Cytochrome P450 2B1/2B6/2C8/2C9/2C18/3A

The non-toxic oxazophorine pro-drugs cyclophosphamide and ifosfamide can be converted to their biologically active (toxic) metabolites by cytochrome P450 (CYP450), with the rat enzyme CYP450 2B1 being the catalytically most active isoform (Clarke and Waxman, 1989). Conversion of pro-drugs by cytochrome P450 first generates 4-hydroxy compounds, which finally decompose to yield the DNA alkylator phosphoramide mustard and acrolein, a protein alkylator, in equimolar amounts (Dachs, Tupper and Tozer, 2005). Problematic aspects of this system might be that endogenous enzymes are able to activate the pro-drug. This problem could be addressed by changing the dosage and frequency of administration of the pro-drug (Jounaidi and Waxman, 2000).

13.3.4 Xanthine–guanine phosphoribosyl transferase (XGPRT)

The *Escherichia coli* enzyme XGPRT was used together with its pro-drug, 6-thioxanthine (6TX), in some early gene therapeutic studies, and was found to show anti-tumour effects against sarcomas (Mroz and Moolten, 1993) and gliomas (Tamiya *et al.*, 1996; Ono *et al.*, 1997). XGPRT converts the purine analogue 6TX to 6-thioxanthine monophosphate (6XMP). This can be demethylated in the cell to 6-thioguanine monophosphate, which inhibits nucleic acid synthesis (Mroz and Moolten, 1993).

13.3.5 Nitroreductase

E. coli nitroreductase converts the pro-drug CB1954 [5-(aziridin-1-yl)-2,4-dinitrobenzamide], which has weak alkylating properties, to its 4-hydroxylamino derivative (Knox *et al.*, 1988). This pre-activated pro-drug will be alkylated by cellular alkylating agents such as acetyl-coenzyme A (AcCoA), leading to a strong alkylating agent able to introduce DNA cross-links which are only poorly repaired by the host cell (Drabek *et al.*, 1997; Grove *et al.*, 1999). DT diaphorase is the nitroreductase enzyme responsible for the first pro-drug conversion. In contrast to the *E. coli* enzyme, the human DT diaphorase has only a low affinity to CB1954, thereby limiting the suicide therapeutic effect to transfected cells. Furthermore, the described effect of this enzyme/pro-drug system is not limited to the active cell cycle, indicating the potential capacity of this system to also target quiescent tumour cells.

13.3.6 Carboxypeptidase G2 (CPG2)

CPG2 catalyses the conversion of CMDA (4-[(2-chloroethyl)(2-mesyloxyethyl)amino] benzoyl-L-glutamic acid) to 4-[(2-chloroethyl)(2-mesyloxyethyl) amino] benzoic acid. This reaction product serves as a DNA cross-linking mustard, specific for CPG2 transduced cells, because no human analogue for the bacterial CPG2 exists (Springer *et al.*, 1990). The first studies in cell culture have shown some inconsistent results, due to the limited CMDA uptake by some cell types. By directing CMDA to the cell membrane, this problem could be overcome (Marais *et al.*, 1997).

13.3.7 The Bystander effect

It has been observed early during the analysis of suicide gene therapy experiments that the expected effect of the enzyme/pro-drug system (or the toxin) is not limited to cells expressing the enzyme. In the case of HSV-1-*tk* gene therapy, the administration of GCV leads to the death of cells neighbouring the infected cells which do not express *tk* themselves. It is not completely understood what the mechanism of this bystander effect is, but most likely this effect is due to the diffusion of toxic compounds (e.g. activated pro-drug) *via* gap junctions to the neighbouring cells. Recently, it could be shown in a mouse model that the co-implantation (1:1) of neural stem cells expressing HSV-1-*tk* together with C6 rat glioma cells, followed by GCV administration, completely protected the animals from tumour development (Li *et al.*, 2005a, b).

13.4 Non-suicide gene therapy

Non-suicide gene therapy aims towards the alteration of the immune response as well as the phenotypic correction of a diseased target cell population by compensating for mutations or other genetic changes causing disease.

13.4.1 Immune therapy

Immunomodulation has been described for both, the down- and the up-regulation of the immune response (Divino *et al.*, 2000; Elzey *et al.*, 2001; Okada *et al.*, 2001a, b; Wu *et al.*, 2001; Cordier Kellerman *et al.*, 2003). Investigators dealing with down-regulation of the immune system try to prevent an immune response directed to their delivery system or caused by the mode of vector delivery. In the case of cancer treatment by viral-mediated gene therapy, an immune response may be elicited by (i) the viral antigens, (ii) the non-endogenous transgene or (iii) the increased cell lysis after successful therapy. Although, in general for cancer therapy, a stimulation of an immune response directed against tumour cells is desirable, an early recruitment of immune cells to the malignant area and thereby an early immune response during gene therapy could lead to an early down-regulation of therapeutic gene expression. It also has to be kept in mind that several tumours are able to secrete signalling molecules (e.g. TGFβ; Baillie, Coombes and Smith, 1996; Marrogi *et al.*, 1997; Wu *et al.*, 2001) which down-regulate the hosts' immune response. The most promising use of immune therapeutic strategies might be to further increase the naturally weak immune response in cancer therapy by combining immunomodulation with suicide gene and virus therapy.

13.4.2 Compensation for mutation

The deregulation or mutation of single or multiple genes is the cause of many inherited diseases including neurodegenerative diseases and brain tumours. To compensate for the resulting genetic dysregulation, the defective genetic information can be either degraded, for example by usage of RNAi-techniques (inhibitory RNA), or corrected by the introduction of genes coding for the wild-type protein, which is able to replace the diseased protein function (Lovett-Racke *et al.*, 2005; Ralph, Mazarakis and Azzouz, 2005). Experiments for cancer treatment dealing with up-regulation of E2F or p53 to induce apoptosis have already successfully being performed (Dong, Yang and McMasters, 2003; Vorburger *et al.*, 2005; Lim *et al.*, 2006).

13.5 Imaging of gene expression

The introduction of molecular imaging in gene therapeutic protocols improves the possibilities to non-invasively follow and understand the outcome of therapy. By coupling the expression of an imaging gene with the expression of the therapeutic gene, one can non-invasively follow the extent and duration of gene expression. The signal obtained by imaging directly represents the region of gene expression for both, the imaging gene and the therapeutic gene. This information is being used to adjust for vector dosage, mode of vector application, and timing and frequency of further treatments.

The general characteristic of gene therapy imaging is the necessity for specific imaging probes, which have a high affinity to the expressed imaging transgene, but no or little affinity to endogenous proteins, which might generate some background. Furthermore, the probe should be non-toxic upon activation with favourable penetration characteristics through tissue, membranes and barriers (e.g. blood–brain barrier [BBB]). Probe development is the bottleneck of molecular imaging and is the pre-requisite for the generation of images reflecting specific gene expression at high resolution, contrast and with high sensitivity.

13.5.1 PET imaging

PET is a radionuclide-based imaging technology with high sensitivity, with the possibility for quantification of the binding potential or accumulation rate of specific receptor-binding partners or enzyme substrates, respectively, and with a spatial resolution limited to 1–2 mm. These characteristics make PET a truly complementary imaging technology to MRI with its high spatial resolution and to optical imaging technologies, which are not truly quantitative. PET imaging of HSV-1-*tk* expression gives the perfect link between imaging gene expression and following therapeutic effects at the same time. One of the mechanisms of PET imaging is to quantitatively follow the kinetics of trapping of specific radiolabelled enzyme substrates within the cell after phosphorylation (Figure 13.3.1). The PET probe should exhibit certain characteristics, such as easy penetration into the target cells (cell membrane, BBB), after which it should be specifically trapped by transgene-related activity. In the case of HSV-1-*tk* imaging, the specific radiolabelled *tk* substrate will be phosphorylated in the cell, which will decrease its

Table 13.5.1 Examples for marker gene/marker substrate combinations which can be imaged by PET. HSV-1-*tk*: Herpes simplex virus type 1 thymidine kinase; D2R: dopamine D2 receptor; [^{18}F]FLT: 3′-deoxy-3′-[^{18}F]fluoro-L-thymidine; [^{124}I]FIAU: 2′-fluoro-2′-deoxy-1β-D-arabinofuranosyl-5′-[^{124}I]-iodo-uracil;[^{18}F]FDG: 2-[^{18}F]fluoro-2-deoxy-D-glucose

Imaging (trans)gene	Tracer	Reaction type
HSV-1-*tk*	[^{18}F]FHBG [^{124}I]FIAU	Enzymatic
d2r	[^{11}C]raclopride [^{18}F]fallypride	Receptor binding
Amino acid transporter	[^{11}C]methionine	Transport
Hexokinase	[^{18}F]FDG	Enzymatic
Benzodiazepine receptor	[^{11}C]flumazenil	Receptor binding

diffusion properties. The isotopes which can be used for PET imaging include ^{11}C, ^{18}F, ^{15}O and ^{124}I, with some of their potential applications listed in Table 13.5.1. Several *tk*-specific substrates have been developed in various experimental and even clinical applications (Jacobs *et al.*, 2001a; Choi *et al.*, 2004; Wang *et al.*, 2005). Other possible targets for PET imaging are the D2R (dopamine 2 receptor) or markers for metabolic processes, such as glucose consumption (Cook, 2003; Schuster *et al.*, 2004; Manninen and Yang, 2005). The identification of proliferating target tissue for potential gene therapy vectors by PET is depicted in Figure 13.5.1 in subcutaneously growing human gliomas in a rat model. The same radiotracers are being used to identify the area of implantation of vector application catheters in patients enrolled in clinical gene therapy trials. Here, the treatment effect can be quantified by employing the same radiotracers as used before therapy. Moreover, the tissue dose of vector-mediated gene expression can by quantified by FIAU-PET (Figure 13.5.2; Jacobs *et al.*, 2001b; Voges *et al.*, 2003).

13.5.2 MR imaging

MRI uses strong magnetic fields followed by a pulse of radiowaves to lift hydrogen nuclei in an excited state. The relaxation of the nuclei to re-align with the magnetic field emits another pulse of radiowaves, which can be detected and quantified. By the use of contrast agents, the tissue properties can be altered, thereby modifying/amplifying the signal. The images obtained by MRI have a high spatial resolution. Some disadvantages of MRI are the long acquisition time to obtain images at high spatial resolution and its relatively low sensitivity with the need for high amounts of contrast agents to be injected (Hildebrandt and Gambhir, 2004).

Figure 13.5.1 Molecular imaging parameters of interest for the identification of target tissue for gene therapy in experimental glioma models. Signs of increased cell proliferation can be observed by means of multi-tracer micro-PET imaging using [^{18}F]FDG, [^{11}C]MET and [^{18}F]FLT as specific tracers for glucose consumption, amino acid transport and DNA synthesis, respectively. Figures represent transaxial images through the body of an experimental rat bearing sub-cutaneously growing rat F98 gliomas. The small tumour on the left (*) shows homogenous radiotracer uptake as an indication for biologically active tumour tissue. The larger tumour on the right shows the signs of central necrosis (arrow). Vector application into the central necrotic tumour tissue (arrow) will be of no benefit, and vector application into the surrounding active tumour rim will be difficult and not successful. Therefore, small animal PET is directly involved in the planning of successful vector application and gene therapy strategies

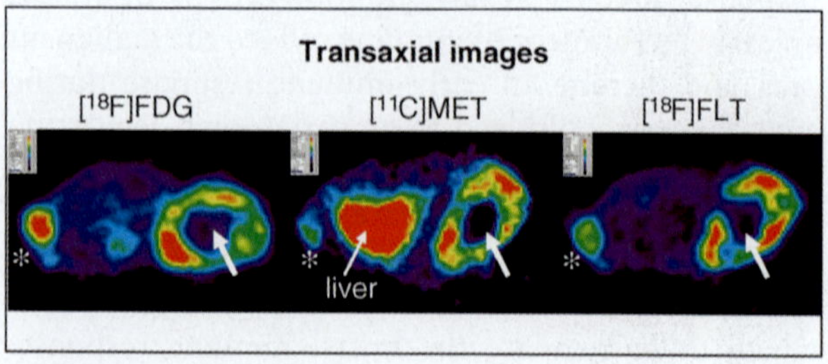

13.5.3 BL imaging

BLI is widely used in animal experiments for the non-invasive assessment of the properties of transduced cells. The firefly luciferase catalyses the conversion of its natural substrate luciferin, which leads to the emission of light. This system can be easily adapted by transferring luciferase to any tissue of interest with subsequent systemic injection of luciferin. The luciferin reaction has no toxic effects on experimental animals and diffuses freely through tissues, with the BBB only slightly delaying the arrival of sufficient amounts in brain needed for imaging (Figure 13.5.3). The advantages of this method are the high depth

Figure 13.5.2 Multi-modal imaging for the establishment of imaging-guided gene therapy in humans. Co-registration of 124I-FIAU-, 11C-MET-, 18F-FDG-PET and MRI before (left column) and after (right column) targeted application (stereotactic infusion) of a gene therapy vector. The region of specific 124I-FIAU retention (68 h) within the tumour after LIPO-HSV-1-*tk* transduction (white arrow) resembles the proposed 'tissue dose' of vector-mediated gene expression and shows the signs of necrosis (cross right column; reduced methionine uptake and glucose metabolism) after ganciclovir treatment. Only these type of combined multi-modal and multi-tracer imaging modalities will help us to establish new treatment modalities in the clinical application (*with permission from Jacobs et al., 2001b*)

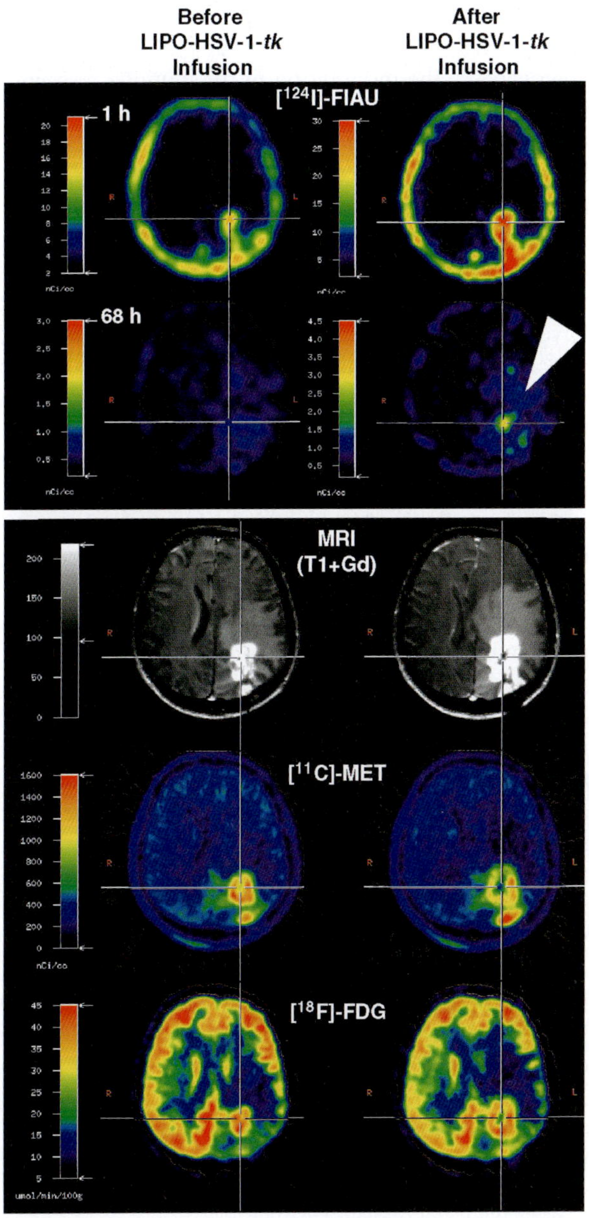

Figure 13.5.3 Images obtained from mice bearing subcutaneous (s.c.; left) or intracranial (i.c.) tumours expressing luciferase. (a) s.c. implantation of 1×10^5 (left) or 2.5×10^5 (right) Gli36 cells stably expressing firefly luciferase followed by luciferin administration (~4 mg/mouse). After injection of luciferin and waiting for its distribution, the enzyme is able to convert luciferin to oxoluciferin. Thereby light is emitted. (b) 1×10^5 C17 cells, transduced to express firefly luciferase, were implanted i.c. in the left hemisphere of the mouse brain, followed by luciferin administration. Both images were performed using the Xenogen IVIS 200 system

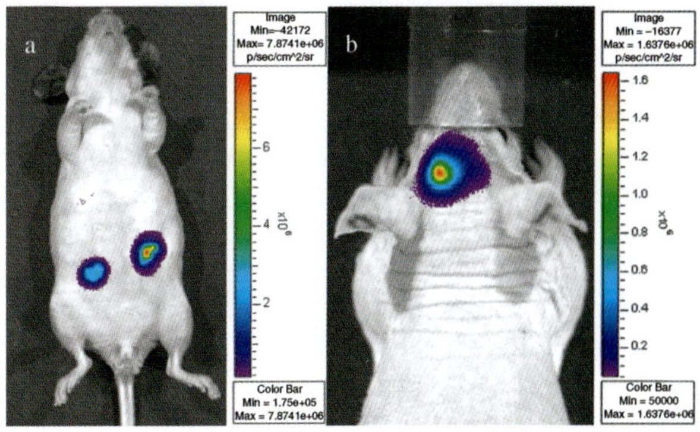

penetration of the emitted signal, high sensitivity as well as the absence of competing reactions. The main disadvantage of BLI is that current acquisition techniques serve only 2D images without corrections for scatter and absorption. Therefore, signal quantification is difficult. Moreover, currently it is hard to envision whether or not BLI will be able to be adapted for human application. However, its value for experiments in cell culture and for fast and relatively easy imaging of small animal models of disease is without question. Current research of BLI focuses on 3D reconstruction techniques, improved luminescent probes, and different enzyme properties emitting light in the near-infrared wavelength, thereby improving the tissue-penetration properties related to natural quenching molecules like haemoglobin. Several different luciferases have been discovered during the past years with some of their properties reviewed by Contag and Bachmann (2002).

13.5.4 Other techniques used for imaging of gene therapy

Ultrasound imaging has been used to monitor the extent of transgene expression on-line, when 'naked' eukaryotic plasmid DNA has been injected in muscles of patients with ischemic disorders. In a study by Rauh *et al.*

(2001), it has been reported that intramuscular plasmid distribution could be assessed in all patients (n = 18; 288 injections) under investigation. The injected plasmid encoded the VEGF (vascular endothelial growth factor) with the aim to increase vascularity. In the animal model by Pajusola *et al.* (2005) the vascular effects obtained after adenoviral vector-mediated gene transfer of VEGF were compared to those induced by HIF-1α (hypoxia inducible factor 1α). The level of vascular perfusion induced by these proteins was monitored by Doppler ultrasound, which was found to be significantly increased in the HIF-1α-treated animal. These examples demonstrate that ultrasound imaging shall help to identify the location of vector administration as well as the effects on vasculature.

Fluorescent reporter genes, like those encoding GFP (green fluorescent protein) or RFP (red fluorescent protein) or their mutants, have been used for visualization of various molecular events (protein–protein interaction, cell trafficking, signal transduction) (Ray *et al.*, 2004). With regards to experimental applications in cell biology and in small animals, techniques like FRET (fluorescence resonance energy transfer) are also promising tools (Li *et al.*, 2004; Ono *et al.*, 2005). However, the scattering of light (excitation, emission) makes it nearly impossible to quantify emission data obtained non-invasively, which is a great disadvantage in the attempt to correlate the location and amount of gene expression to the induced desired effect. The limited depth penetration will restrict fluorescence techniques to human applications in skin, lungs and intestine (van Roessel and Brand, 2002; Folliot *et al.*, 2003).

13.6 Diseases targeted by gene therapy

13.6.1 Tumour therapy

Gene therapy offers promising features for successful treatment of different diseases, including cancer, because (i) the therapeutic action takes place directly in the malignant cell, (ii) imaging techniques give the possibility to follow the therapeutic outcome and to adjust the vector dose as well as timing and location of vector administration and (iii) non-dividing cancer cells can be targeted by stem cell-mediated gene therapy (Aboody *et al.*, 2000; Herrlinger *et al.*, 2000). Further advantages include targeting of vectors to specific (diseased) cell types and, for conditionally replicating viruses, an oncolytic effect which acts in addition to the effect of the therapeutic transgene.

13.6.2 Gene therapy of gliomas

The most common types of brain tumours constituting ~20% of all intracranial neoplasms are glioblastomas. With conventional therapies (operation, radiation, chemotherapy) the disease leads to death over months or years. Therefore, the necessity for new treatment strategies for patients with gliomas is of utmost importance.

Several gene therapy protocols have been used for the treatment of experimental glioma models and for patients with recurrent glioblastoma with the HSV-1-tk gene therapy being the first approach used and still presenting the most commonly used model. Other experimental methods include additional suicide gene therapies using cytosine deaminase or nitroreductase as transgenes, the induction of apoptosis by inducing death receptor–ligand interactions, the suppression of tumour angiogenesis, the modulation of the immune response, the use of conditionally replicating viruses, stem cell-mediated gene transfer or combinations of these (Pulkkanen and Yla-Herttuala, 2005). Clinical gene therapy trials including several hundreds of patients bearing glioblastomas have been performed with so far limited overall success. However, it should be pointed out that in all studies individual patients with long-term survival have been observed. If we understand the mechanisms, why in these individual patients gene and virus therapy led to long-term remission, we will be able to develop the 'magic bullet' for this devastating disease, which will most likely consist of a combination of several gene therapeutic approaches.

Sandmair *et al.* (2000) published a study about the effect of ganciclovir treatment of glioma-bearing patients after adenoviral-mediated gene transfer of HSV-1-*tk*. Therapeutic results were compared to lacZ gene transfer for a control group and transplantation of retroviral vector producer cells carrying HSV-1-*tk*. Seven patients were included in each group, showing mean survival times of 15 months for the adenoviral *tk* administration, 8.3 months for the control lacZ group and 7.4 months for the VPC retroviral group; the latter failure being most likely due to limited distribution of the HSV-1-*tk* VPC within the tumour. Immonen continued this trial including 36 patients with operable primary or recurrent malignant gliomas (Immonen *et al.*, 2004). After adenoviral HSV-1-*tk* gene transfer to the tumour followed by GCV administration, the life span for patients in the treatment group was nearly doubled when compared to the control group (70.6 ± 52.9 versus 39.0 ± 19.7 weeks).

Another study included the treatment of eleven patients with recurrent malignant glioma after surgical resection and radiation therapy. Treatment of gliomas with adenovirus coding for HSV-1-*tk* and followed by administration of GCV increased the average survival time of 10/11 patients to more then 100 weeks (Germano *et al.*, 2003).

The use of replication-conditional viruses has been tested in a study of Rampling *et al.* (2000). They investigated the potential benefit and the safety of administration of HSV1716 (ICP 34.5 null mutant) on nine patients with relapsed malignant gliomas. No harmful, viral-related process has been observed, with 4/9 patients surviving more then 14 months after virus administration.

Non-viral, vector-mediated gene therapy of gliomas has been investigated in a clinical trial employing local convection-enhanced delivery (CED) of vector particles. In this trial 6/8 patients showed focal tumour responses, and in one patient vector-mediated HSV-1-*tk* transduction could be correlated to the induced therapeutic effect (Figure 13.5.2) (Jacobs *et al.*, 2001b; Voges *et al.*, 2003). These local CED-based vector administration protocols are currently further explored with regards to serving more widespread distribution of vector particles throughout the tumour.

13.6.3 Gene therapy of other cancer types

13.6.3.1 *Prostate cancer*

One possible mechanism for the treatment of prostate cancer using gene therapy is the use of PSA (prostate-specific antigen) as a potential target for an immunotherapeutic approach. PSA is expressed at low levels in luminal epithelial cells, but in much higher numbers in prostate cancer. The induction of an immune response against PSA of patients with prostate cancer is seen as a promising tool for cancer treatment. Pavlenko *et al.* (2004) could demonstrate in a multimodal approach combining PSA expression with immunomodulation by cytokine expression the safety of this procedure. Other investigations deal with HSV-1-*tk* expression in the prostate tumour. The first investigations demonstrate an effective immune stimulation by this procedure, showing a clear, tumour-related T cell response following GCV administration (Satoh *et al.*, 2004).

13.6.3.2 *Ovarian cancer*

A phase I clinical trial was carried out combining chemotherapy (topotecan) and gene therapy. Adenoviral-mediated HSV-1-*tk* gene therapy was performed in patients with ovarian cancer after secondary debulking for recurrent ovarian cancer. This procedure together with different chemotherapeutic approaches led to an increase in mean survival by about one-third (Hasenburg *et al.*, 2001). Further gene therapy approaches for the treatment of ovarian cancer include adenoviral-mediated re-constitutive expression of p53 (Wen *et al.*, 2003; Wolf *et al.*, 2004) and the targeting of the HER-2/neu oncogene, which is over-expressed in 10–30% of ovarian cancer (Madhusudan *et al.*, 2004). Both concepts demonstrate the safety of these procedures without long-lasting benefit for the patients.

13.6.3.3 *Breast cancer and melanoma*

Dummer *et al.* (2000) reported in a phase I dose escalation study about the adenoviral-mediated transfer of p53 to tumour tissue of six patients (five metastatic melanoma, one breast adenocarcinoma). This study aimed at the replacement of a mutated p53 by wt-p53. 5/6 adenoviral-p53 infected patients became wt-p53 positive; however, clinical benefit was not observed. In a study by Braybrooke *et al.* (2005), 12 patients (nine with breast cancer, three with melanoma) were subjected to receive retroviral-mediated CYP450 gene transfer. The lacZ control transgene was found to be expressed in 10/12 patients. The gene transfer itself was shown to be safe; however, after cyclophosphamide treatment only one patient partially responded, with 4 patients showing stable disease and seven experiencing tumour progression. Another concept for the treatment of breast cancer was presented by Yoo *et al.* (2001). In their study, the over-expression of HER-2/neu in many breast tumours was addressed, and their strategy was based on the ability of the adenoviral E1A protein to function as tumour inhibitor, which represses oncogene transcription and induces apoptosis of cancer cells. The delivery of E1A was performed by a synthetic vector showing no toxicity and only minor side effects. Eighteen patients (nine with recurrent and unresectable breast cancer, nine with head and neck cancer) were included in this study, with 14/15 showing successful E1A gene transfer. Out of 16 patients analysed for tumour response, two showed minor responses and eight had stable disease.

13.6.4 Other diseases targeted by gene therapy

Recently, the initiation of the first clinical trial for gene therapy of Parkinson's disease has been reported

(Springer *et al.*, 1990; Oransky, 2003). The design of this set-up is to transfer the gene for glutamic acid decarboxylase, responsible for GABA (γ–aminobutyric acid) generation, into the sub-thalamic nucleus to silence it. The overall strengths and weaknesses of this concept are still under discussion (Oransky, 2003). Tuszynski *et al.* (2005) reported about the first clinical gene therapy trial in patients with Alzheimer's disease. They addressed the loss of cholinergic neurons. In this phase I clinical trial the expression of NGF (nerve growth factor) mediated by implanted autologous fibroblasts genetically modified to express NGF should compensate for this loss. They included eight patients in this study and could demonstrate improvements in the rate of cognitive decline. Most importantly, PET investigations during follow-up showed significantly improved cortical 18F-fluoro-deoxy-glucose (FDG) uptake suggesting efficiency of this approach.

Many other diseases are subjected to gene therapy including rheumatoid arthritis, cystic fibrosis, pancreatic cancer, and cardiovascular disease (Alton, 2004; Ebert and Svendsen, 2005; Kasuya *et al.*, 2005; Shah and Losordo, 2005). With our increasing knowledge about molecular mechanisms of disease and with the help of imaging technology, gene therapy will further mature into a safe and successful therapy for those diseases, where currently no other treatment option is available.

13.5 Summary

Gene therapy is used to evaluate new treatment regimens in various animal models of disease and in some clinical trials. Moreover, molecular imaging aims to assess non-invasively disease-specific biological and molecular processes, which might be the target for gene therapy, in animal models and humans *in vivo*. Apart from precise anatomical localization and quantification, the most intriguing advantage of imaging is the opportunity to investigate the time course (dynamics) of disease-specific molecular events in the intact organism. Besides the fact that molecular imaging addresses basic scientific questions such as transcriptional regulation, signal transduction, or protein–protein interaction, molecular imaging will be essential to further develop treatment strategies based on gene therapy. Most importantly, molecular imaging is a key technology in translational research in its capacity to help develop experimental protocols which could later be applied to human patients. Over the past 20 years, imaging based on PET and MRI has been implemented for the assessment and 'phenotyping' of various diseases including cerebral ischemia, neurodegeneration and brain tumours. Recently, PET and MRI have been successfully utilized together in the non-invasive assessment of gene transfer and gene therapy in humans. To assess the efficiency of gene transfer, the same markers are being used in animals and humans, and have been implemented for phenotyping human disease. Imaging of transduced gene expression as well as imaging various parameters which describe the therapeutic effect within the diseased tissue which can be correlated to clinical outcome parameters will enable to develop more efficient and safe gene therapy protocols for clinical application. It should be pointed out that the basis for the further development of gene therapy is the increasing knowledge on the molecular mechanisms of disease and imaging technology to assess the dynamics of disease-specific molecular events in the living organism.

References

Aboody, K. S., Brown, A., Rainov, N. G., Bower, K. A., Liu, S., Yang, W., Small, J. E., Herrlinger, U., Ourednik, V., Black, P. M., Breakefield, X. O., Snyder, E. Y., 2000. "Neural stem cells display extensive tropism for pathology in adult brain: evidence from intracranial gliomas." *Proc. Natl. Acad. Sci. U.S.A.* 97, 12846–12851.

Adusumilli, P. S., Stiles, B. M., Chan, M. K., Mullerad, M., Eisenberg, D. P., Ben-Porat, L., Huq, R., Rusch, V. W., Fong, Y., 2006. "Imaging and therapy of malignant pleural mesothelioma using replication-competent herpes simplex viruses." *J. Gene Med.* 8, 603–615.

Agha-Mohammadi, S., O'Malley, M., Etemad, A., Wang, Z., Xiao, X., Lotze, M. T., 2004. "Second-generation tetracycline-regulatable promoter: repositioned tet operator elements optimize transactivator synergy while shorter minimal promoter offers tight basal leakiness." *J. Gene Med.* 6, 817–828.

Aghi, M., Martuza, R. L., 2005. "Oncolytic viral therapies – the clinical experience." *Oncogene* 24, 7802–7816.

Alton, E. W., 2004. "Use of nonviral vectors for cystic fibrosis gene therapy." *Proc. Am. Thorac. Soc.* 1, 296–301.

Baillie, R., Coombes, R. C., Smith, J., 1996. "Multiple forms of TGF-beta 1 in breast tissues: a biologically active form of the small latent complex of TGF-beta 1." *Eur. J. Cancer* 32A, 1566–1573.

Braybrooke, J. P., Slade, A., Deplanque, G., Harrop, R., Madhusudan, S., Forster, M. D., Gibson, R.,

Makris, A., Talbot, D. C., Steiner, J., White, L., Kan, O., Naylor, S., Carroll, M. W., Kingsman, S. M., Harris, A. L., 2005." Phase I study of MetXia-P450 gene therapy and oral cyclophosphamide for patients with advanced breast cancer or melanoma." *Clin. Cancer Res.* **11**, 1512–1520.

Chappell, S. A., Edelman, G. M., Mauro, V. P., 2000. "A 9-nt segment of a cellular mRNA can function as an internal ribosome entry site (IRES) and when present in linked multiple copies greatly enhances IRES activity." *Proc. Natl. Acad. Sci. U.S.A.* **97**, 1536–1541.

Choi, T. H., Ahn, S. H., Kwon, H. C., Choi, C. W., Awh, O. D., Lim, S. M., 2004. "In vivo comparison of IVDU and IVFRU in HSV1-TK gene expressing tumor bearing rats." *Appl. Radiat. Isot.* **60**, 15–21.

Clarke, L., Waxman, D. J., 1989. "Oxidative metabolism of cyclophosphamide: identification of the hepatic monooxygenase catalysts of drug activation." *Cancer Res.* **49**, 2344–2350.

Contag, C. H., Bachmann, M. H., 2002. "Advances in in vivo bioluminescence imaging of gene expression." *Annu. Rev. Biomed. Eng.* **4**, 235–260.

Cook, G. J., 2003. "Oncological molecular imaging: nuclear medicine techniques." *Br. J. Radiol.* 76(2), S152–S158.

Cordier Kellerman, L., Valeyrie, L., Fernandez, N., Opolon, P., Sabourin, J. C., Maubec, E., Le Roy, P., Kane, A., Legrand, A., Abina, M. A., Descamps, V., Haddada, H., 2003. "Regression of AK7 malignant mesothelioma established in immunocompetent mice following intratumoral gene transfer of interferon gamma." *Cancer Gene Ther.* **10**, 481–490.

Dachs, G. U., Tupper, J., Tozer, G. M., 2005. "From bench to bedside for gene-directed enzyme prodrug therapy of cancer." *Anticancer Drugs* 16, 349–359.

Davies, S. J., Smith, S. J., Lim, K. C., Zhang, H., Purchio, A. F., McKerrow, J. H., West, D. B., 2005. "*In vivo* imaging of tissue eosinophilia and eosinopoietic responses to schistosome worms and eggs." *Int. J. Parasitol.* **35**, 851–859.

de Felipe, P., Martin, V., Cortes, M. L., Ryan, M., Izquierdo, M., 1999. "Use of the 2A sequence from foot-and-mouth disease virus in the generation of retroviral vectors for gene therapy." *Gene Ther.* **6**, 198–208.

Divino, C. M., Chen, S. H., Yang, W., Thung, S., Brower, S. T., Woo, S. L., 2000. "Anti-tumor immunity induced by interleukin-12 gene therapy in a metastatic model of breast cancer is mediated by natural killer cells." *Breast Cancer Res Treat.* **60**, 129–134.

Dong, Y. B., Yang, H. L., McMasters, K. M., 2003. "E2F-1 overexpression sensitizes colorectal cancer cells to camptothecin." *Cancer Gene Ther.* **10**, 168–178.

Drabek, D., Guy, J., Craig, R., Grosveld, F., 1997. "The expression of bacterial nitroreductase in transgenic mice results in specific cell killing by the prodrug CB1954." *Gene Ther.* **4**, 93–100.

Dummer, R., Bergh, J., Karlsson, Y., Horowitz, J. A., Mulder, N. H., Huinink, D.T.B., Burg, G., Hofbauer, G., Osanto, S., 2000. "Biological activity and safety of adenoviral vector-expressed wild-type p53 after intratumoral injection in melanoma and breast cancer patients with p53-overexpressing tumors." *Cancer Gene Ther.* **7**, 1069–1076.

Ebert, A. D., Svendsen, C. N., 2005. "A new tool in the battle against Alzheimer's disease and aging: ex vivo gene therapy." *Rejuvenation Res.* **8**, 131–134.

Elzey, B. D., Siemens, D. R., Ratliff, T. L., Lubaroff, D. M., 2001. "Immunization with type 5 adenovirus recombinant for a tumor antigen in combination with recombinant canarypox virus (ALVAC) cytokine gene delivery induces destruction of established prostate tumors." *Int. J. Cancer* 94, 842–849.

Erbs, P., Exinger, F., Jund, R., 1997. "Characterization of the Saccharomyces cerevisiae FCY1 gene encoding cytosine deaminase and its homologue FCA1 of Candida albicans." *Curr. Genet.* **31**, 1–6.

Folliot, S., Briot, D., Conrath, H., Provost, N., Cherel, Y., Moullier, P., Rolling, F., 2003. "Sustained tetracycline-regulated transgene expression in vivo in rat retinal ganglion cells using a single type 2 adeno-associated viral vector." *J. Gene Med.* **5**, 493–501.

Germano, I. M., Fable, J., Gultekin, S. H., Silvers, A., 2003. "Adenovirus/herpes simplex-thymidine kinase/ganciclovir complex: preliminary results of a phase I trial in patients with recurrent malignant gliomas." *J. Neurooncol.* **65**, 279–289.

Grove, J. I., Searle, P. F., Weedon, S. J., Green, N. K., McNeish, I. A., Kerr, D. J., 1999. "Virus-directed enzyme prodrug therapy using CB1954." *Anticancer Drug Des.* **14**, 461–472.

Hasenburg, A., Tong, X. W., Fischer, D. C., Rojas-Martinez, A., Nyberg-Hoffman, C., Kaplan, A. L., Kaufman, R. H., Ramzy, I., Aguilar-Cordova E., Kieback, D. G., 2001. "Adenovirus-mediated thymidine kinase gene therapy in combination with topotecan for patients with recurrent ovarian cancer: 2.5-year follow-up." *Gynecol. Oncol.* **83**, 549–554.

Herrlinger, U., Woiciechowski, C., Sena-Esteves, M., Aboody, K. S., Jacobs, A. H., Rainov, N. G., Snyder, E. Y., Breakefield, X. O., 2000. "Neural precursor cells for delivery of replication-conditional HSV-1 vectors to intracerebral gliomas." *Mol. Ther.* **1**, 347–357.

Hildebrandt, I. J., Gambhir, S. S., 2004. "Molecular imaging applications for immunology." *Clin. Immunol.* 111, 210–224.

Immonen, A., Vapalahti, M., Tyynela, K., Hurskainen, H., Sandmair, A., Vanninen, R., Langford, G., Murray, N., Yla-Herttuala, S., 2004. "AdvHSV-tk gene therapy with intravenous ganciclovir improves survival in human malignant glioma: a randomised, controlled study." *Mol. Ther.* **10**, 967–972.

Jacobs, A., Braunlich, I., Graf, R., Lercher, M., Sakaki, T., Voges, J., Hesselmann, V., Brandau, W., Wienhard, K., Heiss, W. D., 2001a. "Quantitative kinetics of [124I]FIAU in cat and man." *J. Nucl. Med.* **42**, 467–475.

Jacobs, A., Breakefield, X. O., Fraefel, C., 1999. "HSV-1-based vectors for gene therapy of neurological diseases and brain tumors: part II. Vector systems and applications." *Neoplasia* 1, 402–416.

Jacobs, A., Voges, J., Reszka, R., Lercher, M., Gossmann, A., Kracht, L., Kaestle, C., Wagner, R., Wienhard, K., Heiss, W. D., 2001b. "Positron-emission tomography of vector-mediated gene expression in gene therapy for gliomas." *Lancet* 358, 727–729.

Jacobs, A. H., Li, H., Winkeler, A., Hilker, R., Knoess, C., Ruger, A., Galldiks, N., Schaller, B., Sobesky, J., Kracht, L., Monfared, P., Klein, M., Vollmar, S., Bauer, B., Wagner, R., Graf, R., Wienhard, K., Herholz, K., Heiss, W. D., 2003a." PET-based molecular imaging in neuroscience." *Eur. J. Nucl. Med. Mol. Imaging* 30, 1051–1065.

Jacobs, A. H., Winkeler, A., Hartung, M., Slack, M., Dittmar, C., Kummer, C., Knoess, C., Galldiks, N., Vollmar, S., Wienhard, K., Heiss, W. D., 2003b. "Improved herpes simplex virus type 1 amplicon vectors for proportional coexpression of positron emission tomography marker and therapeutic genes." *Hum. Gene Ther.* **14**, 277–297.

Jounaidi, Y., Waxman, D. J., 2000. "Combination of the bioreductive drug tirapazamine with the chemotherapeutic prodrug cyclophosphamide for P450/P450-reductase-based cancer gene therapy." *Cancer Res.* **60**, 3761–3769.

Kasuya, H., Takeda, S., Nomoto, S., Nakao, A., 2005. "The potential of oncolytic virus therapy for pancreatic cancer." *Cancer Gene Ther.* **12**, 725–736.

Knox, R. J., Friedlos, F., Jarman, M., Roberts, J. J., 1988. "A new cytotoxic, DNA interstrand cross-linking agent, 5-(aziridin-1-yl)-4-hydroxylamino-2-nitrobenzamide, is formed from 5-(aziridin-1-yl)-2,4-dinitrobenzamide (CB 1954) by a nitroreductase enzyme in Walker carcinoma cells." *Biochem. Pharmacol.* 37, 4661–4669.

Lee, Y. B., Cosgrave, A. S., Glover, C. P., Bienemann, A., Heywood, D., Hobson, R. J., Uney, J. B., 2005. "Increased utility in the CNS of a powerful neuron-specific tetracycline-regulatable adenoviral system developed using a post-transcriptional enhancer." *J. Gene Med.* 7, 576–583.

Li, H., Gray, B. D., Corbin, I., Lebherz, C., Choi, H., Lund-Katz, S., Wilson, J. M., Glickson, J. D., Zhou, R., 2004. "MR and fluorescent imaging of low-density lipoprotein receptors." *Acad. Radiol.* **11**, 1251–1259.

Li, S., Tokuyama, T., Yamamoto, J., Koide, M., Yokota, N., Namba, H., 2005a. "Bystander effect-mediated gene therapy of gliomas using genetically engineered neural stem cells." *Cancer Gene Ther.* **12**, 600–607.

Li, S., Tokuyama, T., Yamamoto, J., Koide, M., Yokota, N., Namba, H., 2005b. "Potent bystander effect in suicide gene therapy using neural stem cells transduced with herpes simplex virus thymidine kinase gene." *Oncology* 69, 503–508.

Lim, D. S., Bae, S. M., Kwak, S. Y., Park, E. K., Kim, J. K., Han, S. J., Oh, C. H., Lee, C. H., Lee, W. Y., Ahn, W. S., 2006. "Adenovirus-mediated p53 treatment enhances photodynamic antitumor response." *Hum. Gene Ther.* **17**, 347–352.

Lovett-Racke, A. E., Cravens, P. D., Gocke, A. R., Racke, M. K., Stuve, O., 2005. "Therapeutic potential of small interfering RNA for central nervous system diseases." *Arch. Neurol.* **62**, 1810–1813.

Madhusudan, S., Tamir, A., Bates, N., Flanagan, E., Gore, M. E., Barton, D. P., Harper, P., Seckl, M., Thomas, H., Lemoine, N. R., Charnock, M., Habib, N. A., Lechler, R., Nicholls, J., Pignatelli, M., Ganesan, T. S., 2004. "A multicenter Phase I gene therapy clinical trial involving intraperitoneal administration of E1A-lipid complex in patients with recurrent epithelial ovarian cancer overexpressing HER-2/neu oncogene." *Clin. Cancer Res.* **10**, 2986–2996.

Manninen, H. I., Yang, X., 2005. "Imaging after vascular gene therapy." *Eur. J. Radiol.* **56**, 165–170.

Marais, R., Spooner, R. A., Stribbling, S. M., Light, Y., Martin, J., Springer, C. J., 1997. "A cell surface tethered enzyme improves efficiency in gene-directed enzyme prodrug therapy." *Nat. Biotechnol.* **15**, 1373–1377.

Marrogi, A. J., Munshi, A., Merogi, A. J., Ohadike, Y., El-Habashi, A., Marrogi, O. L., Freeman, S. M., 1997. "Study of tumor infiltrating lymphocytes and transforming growth factor-beta as prognostic factors in breast carcinoma." *Int. J. Cancer* 74, 492–501.

McGuire, S. E., Mao, Z., Davis, R. L., 2004. "Spatiotemporal gene expression targeting with the

TARGET and gene-switch systems in Drosophila." *Sci STKE* 2004, pl6.

Moolten, F. L., 1994. "Drug sensitivity ("suicide") genes for selective cancer chemotherapy." *Cancer Gene Ther.* **1**, 279–287.

Mroz, P. J., Moolten, F. L., 1993. "Retrovirally transduced Escherichia coli gpt genes combine selectability with chemosensitivity capable of mediating tumor eradication." *Hum. Gene Ther.* **4**, 589–595.

Okada, H., Pollack, I. F., Lieberman, F., Lunsford, L. D., Kondziolka, D., Schiff, D., Attanucci, J., Edington, H., Chambers, W., Kalinski, P., Kinzler, D., Whiteside, T., Elder, E., Potter, D., 2001a. "Gene therapy of malignant gliomas: a pilot study of vaccination with irradiated autologous glioma and dendritic cells admixed with IL-4 transduced fibroblasts to elicit an immune response." *Hum. Gene Ther.* **12**, 575–595.

Okada, H., Villa, L., Attanucci, J., Erff, M., Fellows, W. K., Lotze, M. T., Pollack, I. F., Chambers, W. H., 2001b. "Cytokine gene therapy of gliomas: effective induction of therapeutic immunity to intracranial tumors by peripheral immunization with interleukin-4 transduced glioma cells." *Gene Ther.* **8**, 1157–1166.

Ono, H. A., Le, L. P., Davydova, J. G., Gavrikova, T., Yamamoto, M., 2005. "Noninvasive visualization of adenovirus replication with a fluorescent reporter in the E3 region." *Cancer Res.* **65**, 10154–10158.

Ono, Y., Ikeda, K., Wei, M. X., Harsh, G.R.t., Tamiya, T., Chiocca, E. A., 1997. "Regression of experimental brain tumors with 6-thioxanthine and Escherichia coli gpt gene therapy." *Hum. Gene Ther.* **8**, 2043–2055.

Oransky, I., 2003. "Gene therapy trial for Parkinson's disease begins." *Lancet* 362, 712.

Pajusola, K., Kunnapuu, J., Vuorikoski, S., Soronen, J., Andre, H., Pereira, T., Korpisalo, P., Yla-Herttuala, S., Poellinger, L., Alitalo, K., 2005. "Stabilized HIF-1alpha is superior to VEGF for angiogenesis in skeletal muscle via adeno-associated virus gene transfer." *FASEB J.* **19**, 1365–1367.

Pavlenko, M., Roos, A. K., Lundqvist, A., Palmborg, A., Miller, A. M., Ozenci, V., Bergman, B., Egevad, L., Hellstrom, M., Kiessling, R., Masucci, G., Wersall, P., Nilsson, S., Pisa, P., 2004. "A phase I trial of DNA vaccination with a plasmid expressing prostate-specific antigen in patients with hormone-refractory prostate cancer." *Br. J. Cancer* 91, 688–694.

Piskin, E., Dincer, S., Turk, M., 2004. "Gene delivery: intelligent but just at the beginning." *J. Biomater. Sci. Polym. Ed.* **15**, 1181–1202.

Pulkkanen, K. J., Yla-Herttuala, S., 2005. "Gene therapy for malignant glioma: current clinical status." *Mol. Ther.* **12**, 585–598.

Ralph, G. S., Mazarakis, N. D., Azzouz, M., 2005. "Therapeutic gene silencing in neurological disorders, using interfering RNA." *J. Mol. Med.* **83**, 413–419.

Rampling, R., Cruickshank, G., Papanastassiou, V., Nicoll, J., Hadley, D., Brennan, D., Petty, R., MacLean, A., Harland, J., McKie, E., Mabbs, R., Brown, M., 2000. "Toxicity evaluation of replication-competent herpes simplex virus (ICP 34.5 null mutant 1716) in patients with recurrent malignant glioma." *Gene Ther.* **7**, 859–866.

Rauh, G., Pieczek, A., Irwin, W., Schainfeld, R., Isner, J. M., 2001. "In vivo analysis of intramuscular gene transfer in human subjects studied by on-line ultrasound imaging." *Hum. Gene Ther.* **12**, 1543–1549.

Ray, P., De, A., Min, J. J., Tsien, R. Y., Gambhir, S. S., 2004. "Imaging tri-fusion multimodality reporter gene expression in living subjects." *Cancer Res.* **64**, 1323–1330.

Rojas-Martinez, A., Martinez-Davila, I. A., Hernandez-Garcia, A., Aguilar-Cordova, E., Barrera-Saldana, H. A., 2002. "Genetic therapy of cancer." *Rev. Invest. Clin.* **54**, 57–67.

Rosenfeld, N., Young, J. W., Alon, U., Swain, P. S., Elowitz, M. B., 2005. "Gene regulation at the single-cell level." *Science* 307, 1962–1965.

Sandmair, A. M., Loimas, S., Puranen, P., Immonen, A., Kossila, M., Puranen, M., Hurskainen, H., Tyynela, K., Turunen, M., Vanninen, R., Lehtolainen, P., Paljarvi, L., Johansson, R., Vapalahti, M., Yla-Herttuala, S., 2000. "Thymidine kinase gene therapy for human malignant glioma, using replication-deficient retroviruses or adenoviruses." *Hum. Gene Ther.* **11**, 2197–2205.

Satoh, T., The, B. S., Timme, T. L., Mai, W. Y., Gdor, Y., Kusaka, N., Fujita, T., C. K., Pramudji, Vlachaki, M. T., Ayala, G., Wheeler, T., Amato, R., Miles, B. J., Kadmon, D., Butler, E. B., Thompson, T. C., 2004. "Enhanced systemic T-cell activation after in situ gene therapy with radiotherapy in prostate cancer patients." *Int. J. Radiat. Oncol. Biol. Phys.* **59**, 562–571.

Schuster, D. P., Kovacs, A., Garbow, J., Piwnica-Worms, D., 2004. "Recent advances in imaging the lungs of intact small animals." *Am. J. Respir. Cell. Mol. Biol.* **30**, 129–138.

Shah, K., Jacobs, A., Breakefield, X. O., Weissleder, R., 2004. "Molecular imaging of gene therapy for cancer." *Gene Ther.* **11**, 1175–1187.

Shah, P. B., Losordo, D. W., 2005. "Non-viral vectors for gene therapy: clinical trials in cardiovascular disease." *Adv. Genet.* **54**, 339–361.

Springer, C. J., Antoniw, P., Bagshawe, K. D., Searle, F., Bisset, G. M., Jarman, M., 1990. "Novel prodrugs which are activated to cytotoxic alkylating agents by carboxypeptidase G2." *J. Med. Chem.* **33**, 677–681.

Tamiya, T., Ono, Y., Wei, M. X., Mroz, P. J., Moolten, F. L., Chiocca, E. A., 1996. "Escherichia coli gpt gene sensitizes rat glioma cells to killing by 6-thioxanthine or 6-thioguanine." *Cancer Gene Ther.* **3**, 155–162.

Teschemacher, A. G., Paton, J. F., Kasparov, S., 2005. "Imaging living central neurones using viral gene transfer." *Adv. Drug Deliv. Rev.* **57**, 79–93.

Tjuvajev, J. G., Stockhammer, G., Desai, R., Uehara, H., Watanabe, K., Gansbacher, B., Blasberg, R. G., 1995. "Imaging the expression of transfected genes in vivo." *Cancer Res.* **55**, 6126–6132.

Tuszynski, M. H., Thal, L., Pay, M., Salmon, D. P., U, H. S., Bakay, R., Patel, P., Blesch, A., Vahlsing, H. L., Ho, G., Tong, G., Potkin, S. G., Fallon, J., Hansen, L., Mufson, E. J., Kordower, J. H., Gall, C., Conner, J., 2005. "A phase 1 clinical trial of nerve growth factor gene therapy for Alzheimer disease." *Nat. Med.* **11**, 551–555.

Tyminski, E., Leroy, S., Terada, K., Finkelstein, D. M., Hyatt, J. L., Danks, M. K., Potter, P. M., Saeki, Y., Chiocca, E. A., 2005. "Brain tumor oncolysis with replication-conditional herpes simplex virus type 1 expressing the prodrug-activating genes, CYP2B1 and secreted human intestinal carboxylesterase, in combination with cyclophosphamide and irinotecan." *Cancer Res.* **65**, 6850–6857.

van Roessel, P., Brand, A. H., 2002. "Imaging into the future: visualizing gene expression and protein interactions with fluorescent proteins." *Nat. Cell Biol.* **4**, E15–E20.

Verma, I. M., Weitzman, M. D., 2005. "Gene therapy: twenty-first century medicine." *Annu. Rev. Biochem.* **74**, 711–738.

Vigna, E., Amendola, M., Benedicenti, F., Simmons, A. D., Follenzi, A., Naldini, L., 2005. "Efficient Tet-dependent expression of human factor IX in vivo by a new self-regulating lentiviral vector." *Mol. Ther.* **11**, 763–775.

Voges, J., Reszka, R., Gossmann, A., Dittmar, C., Richter, R., Garlip, G., Kracht, L., Coenen, H. H., Sturm, V., Wienhard, K., Heiss, W. D., Jacobs, A. H., 2003. "Imaging-guided convection-enhanced delivery and gene therapy of glioblastoma." *Ann. Neurol.* **54**, 479–487.

Vorburger, S. A., Hetrakul, N., Xia, W., Wilson-Heiner, M., Mirza, N., Pollock, R. E., Feig, B., Swisher, S. G., Hunt, K. K., 2005. "Gene therapy with E2F-1 up-regulates the protein kinase PKR and inhibits growth of leiomyosarcoma in vivo." *Mol. Cancer Ther.* **4**, 1710–1716.

Wang, J. Q., Fei, X., Gardner, T. A., Hutchins, G. D., Zheng, Q. H., 2005. "Synthesis of 2-amino-6-(4-[11C]methoxyphenylthio)-9-[2-(phosphonomethoxy)ethyl]purine bis(2,2,2-trifluoroethyl) ester as a novel potential PET gene reporter probe for HBV and HSV-tk in cancers." *Bioorg. Med. Chem.* **13**, 549–556.

Wang, Y., O'Malley Jr., B. W., Tsai, S. Y., O'Malley, B. W., 1994. "A regulatory system for use in gene transfer." *Proc. Natl. Acad. Sci. U.S.A.* **91**, 8180–8184.

Wen, S. F., Mahavni, V., Quijano, E., Shinoda, J., Grace, M., Musco-Hobkinson, M. L., Yang, T. Y., Chen, Y., Runnenbaum, I., Horowitz, J., Maneval, D., Hutchins, B., Buller, R., 2003. "Assessment of p53 gene transfer and biological activities in a clinical study of adenovirus-p53 gene therapy for recurrent ovarian cancer." *Cancer Gene Ther.* **10**, 224–238.

Wolf, J. K., Bodurka, D. C., Gano, J. B., Deavers, M., Ramondetta, L., Ramirez, P. T., Levenback, C., Gershenson, D. M., 2004. "A phase I study of Adp53 (INGN 201; ADVEXIN) for patients with platinum- and paclitaxel-resistant epithelial ovarian cancer." *Gynecol. Oncol.* **94**, 442–448.

Wu, R. S., Kobie, J. J., Besselsen, D. G., Fong, T. C., Mack, V. D., McEarchern, J. A., Akporiaye, E. T., 2001. "Comparative analysis of IFN-gamma B7.1 and antisense TGF-beta gene transfer on the tumorigenicity of a poorly immunogenic metastatic mammary carcinoma." *Cancer Immunol. Immunother.* **50**, 229–240.

Yao, L., Li, Y., Wu, Y., Liu, A., Yan, H., 2005. "Product release is rate-limiting in the activation of the prodrug 5-fluorocytosine by yeast cytosine deaminase." *Biochemistry* 44, 5940–5947.

Yazawa, K., Fisher, W. E., Brunicardi, F. C., 2002. "Current progress in suicide gene therapy for cancer." *World J. Surg.* **26**, 783–789.

Yoo, G. H., Hung, M. C., Lopez-Berestein, G., LaFollette, S., Ensley, J. F., Carey, M., Batson, E., Reynolds, T. C., Murray, J. L., 2001. "Phase I trial of intratumoral liposome E1A gene therapy in patients with recurrent breast and head and neck cancer." *Clin. Cancer Res.* **7**, 1237–1245.

Zhou, Y., Aran, J., Gottesman, M. M., Pastan, I., 1998. "Co-expression of human adenosine deaminase and multidrug resistance using a bicistronic retroviral vector." *Hum. Gene Ther.* **9**, 287–293.

14 Cellular Therapies and Cell Tracking

Coordinated by Yves Fromes

14.0 Introduction

Yves Fromes

Tissue homeostasis depends on proper relationships between cell proliferation and differentiation on one hand and cell death on the other hand. Any imbalance between them results in compromised organ structure and/or function. In tissues with high cell turnover, the maintenance of the stem cells is of particular importance for tissue preservation. Cell loss from the tissue has to be precisely balanced by cell production.

Many disabling disorders involve cell death and are characterized by the fast or progressive loss of the differentiated cells. Acute disorders, occurring over minutes to hours, such as trauma, infarction, haemorrhage or infection, prominently involve cell death, much of which is by necrosis. Chronic disorders, with relatively slow degeneration, may occur over years or decades, but may involve significant cell losses.

In embryos, the repair process is extremely efficient, leading to the reconstitution of structurally and functionally normal tissue to injured area. Numerous adult tissues are also capable of plastic responses, and adaptive growth responses occur along with degenerative events. Unfortunately, this extraordinary ability for growth and differentiation is lost in several crucial organs as they mature. Brain and heart muscle display limited capacities for repair and regeneration, as many clinical situations give evidence.

Recent advances in the field of stem cell research may open new possibilities for tissue repair. Thus, stem cells have been found in tissues with a very slow turnover, such as the brain, where cell damage cannot spontaneously be repaired. Furthermore, the conventional understanding that the lineage potential of adult stem cells is limited to the tissue of origin is challenged.

Stem cells have traditionally been characterized as either embryonic (pluripotent) or tissue-specific (multipotent). Stem cells are undifferentiated cells capable of self-renewal, proliferation and differentiation into a variety of specialized cells and, hence, regenerating tissues. Stem cell plasticity may have important implications in the cellular and pathological mechanisms of injury and repair. As an example, mesenchymal stem cells residing in adult bone marrow are best characterized by their capacity to differentiate into connective tissue cell types such as adipocytes, chondrocytes, osteoblasts and hematopoiesis-supporting stroma. Mesenchymal stem cells may also promote structural and functional repair in several organs such as the brain, heart and skeletal muscle via cell plasticity.

Consequently cell transplantation may emerge as a new therapeutic approach to repair damaged or degenerated tissues. However, monitoring cell grafting has become mandatory in order to provide objective proves of efficiency. Conventional techniques such as histology provide clear evidence, but rely on biopsies sampling the grafted tissue and thereby leading to new lesions.

Nuclear magnetic imaging offers non-invasive ways to evaluate such experimental conditions and appears to be superior to most other imaging techniques in providing images of thin slices of the body. Tracking cells inside a tissue and/or organ, however, opens new challenges for nuclear magnetic imaging. High resolution as well as specific labelling for visualization of grafted cells as compared to the host tissue will be mandatory.

The current chapter intends to highlight several of the options, which could be used in order to achieve this goal. Brain tissue represents, of course, a target of choice to test such strategies, but many other organs are already considered for cell therapy. Some organs have higher constrains in terms of cell requirements or tissue organization (brain, heart). Grafting of stem cells in damaged areas may undergo repair if cells can survive in the area of damage, migrate in order to spread over the area and undergo functional differentiation. Moreover, the functional significance of the transplanted tissue remains a crucial question for tissue repair, and functional NMR imaging may help to evaluate this aspect.

With its excellent spatial resolution and the ability to track labelled cells over prolonged periods of time, NMR monitoring of cell therapy is likely to become an important technique in the foreseeable future.

14.1 Are stem cells attracted by pathology? The case for cellular tracking by serial *in vivo* MRI

Michel Modo

14.1.1 The use of stem cells to repair brain damage

Transplantation of neural stem cells to repair brain damage has met with substantial success in pre-clinical models, and clinical trials are currently in progress to assess the benefit of cell therapy for patients. The fundamental idea behind restorative stem cell transplantation is that immature cells will recapitulate development and replace cells lost due to pathological processes. It is expected that this replacement of lost cells will engender the functional restoration of behavioural impairments such as paralysis or neglect. These immature cells that can give rise to all the major cell types of the CNS (neurons, astrocytes and oligodendrocytes) are commonly referred to as neural stem cells. However, other mature CNS resident cells, such as microglia, are not derived from neural stem cells, but originate from mesodermal haemapoietic precursors and colonize the CNS (Rezaie and Male, 1999). In response to damage, it is mainly microglia and astrocytes that proliferate and populate the area of damage.

Although neurogenic regions, such as the subventricular zone, persist in the mature brain, their capacity to replace lost cells is limited. These niches of endogenous stem cells respond to stroke-inflicted damage by producing an increased number of novel neurons (Arvidsson *et al.*, 2002), but the functional significance of these cells remains unclear. In contrast, supplementation of the endogenous stem cell response by transplantation of exogenous neural stem cells has provided robust evidence that functional recovery can be achieved after ischemic brain damage (Modo *et al.*, 2002b).

Exogenously transplanted neural stem cells differentiate into appropriate phenotypes in areas of damage and integrate functionally into existing networks. The processes involved in stem cell differentiation in the damaged brain are thought to involve the same stages and molecular pathways as those exhibited by differentiating stem cells in the developing brain (Price *et al.*, 2000). Apart from neuronal differentiation, the secretion of growth factors and upregulation of endogenous neurodevelopmental proteins through transplanted cells could contribute to behavioural recovery. Dispersion of these factors across a larger area of the brain by cell migration could therefore be of greater benefit for repair.

14.1.2 Are stem cells attracted to pathology or do they survive better in areas of damage?

One of the most remarkable properties of stem cells lies in their ability to migrate out of the area of injection and to disperse into various areas of the brain, especially those undergoing pathological processes. In contrast, experiments with foetal tissue transplants have mainly found that the grafts form a mass distinct of the host parenchyma with little migration out from the injection tract. Although cells within this graft will differentiate into neuronal cells appropriate for the region, they do not show the same propensity to colonize the damaged brain than neural stem cells. In contrast, homogenous populations of *in vitro* expanded neural stem cells show intrinsic migratory properties similar to those observed in the developing brain, although the routes of migration/dispersion appear distinct in the damaged brain.

Figure 14.1.1 Implantation of neural stem cells into the area of ischaemic damage (ipsilateral), the contralateral hemisphere or intracerebroventricular (ICV) resulted in grafted cells mainly accumulating around the lesion. However, grafted cells could also be found in other regions, such as the 'intact' contralateral hemisphere (Modo *et al.*, 2002c). The results raise the possibility that transplanted cells are attracted to regions of pathology or that cells disperse throughout the brain and show selective retention in areas of damage

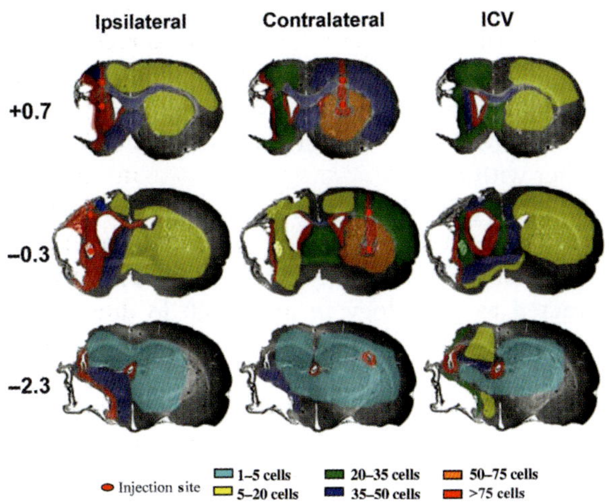

For instance, transplantation of neural stem cells into focal ischemic lesions results in an accumulation of grafted cells around the area of damage, but cells can also be found in many other areas of the damaged brain (Figure 14.1.1). Contralateral and intra-cerebroventricular transplantation results in a similar pattern of distribution of grafted cells implying that potentially cells migrate throughout the brain with a better graft survival in areas of damage (Modo *et al.*, 2002c). In the developing brain, the dispersion of stem cells along radial glia from the 'inside-out' also results in neuronal precursor cells colonizing the entire brain (Morest and Silver, 2003). Nonetheless, the location of grafted cells aligned along the corpus callosum and a stream of cells migrating away from the injection tract in the direction of the lesion suggest that migration of neural stem cells in the damaged brain is a directed process and not an extensive dispersion of cells. The snapshot nature of a single time point of post-mortem immunohistochemical methods, however, is largely inadequate to capture this 2-week dynamic process in the same animal.

14.1.3 Migration is a dynamic process

To capture the dynamic nature of migration, it is desirable to follow the same animal over time to avoid any potential confounds (e.g. size of lesion, position of injection tract, etc.) and to conclusively demonstrate how cells show a directed migration from the injection tract to the area of damage. The high spatial resolution and non-invasiveness of magnetic resonance imaging (MRI) render it apposite for the repeated *in vivo* assessment of the same subject over time. Nonetheless, the seamless integration of transplanted cells into the host parenchyma renders them invisible to normal structural MRI. To follow the migration of transplanted cells *in vivo*, it is therefore necessary to label neural stem cells *in vitro* with MRI contrast agents prior to transplantation.

The wide dispersion of cells also demands that a cluster of a few cells or preferentially a single cell could be detected by MRI. Iron-oxide-based contrast agents such as ultra small particles of iron-oxide (USPIO) or micron-sized particles of iron-oxide (MPIO) achieve a very high contrast-to-noise ratio (i.e. little contrast agent is needed for detection) and are consequently a popular choice for cellular tracking by MRI (Bulte and Kraitchman, 2004). The possibility to corroborate the *in vivo* MRI observations by an independent imaging modality, such as fluorescent microscopy, is an advantage to ensure that signal changes observed on the MRI scans are indeed the result of transplanted cells. The bimodal contrast agent, gadolinium rhodamine dextran (GRID), detectable by both MRI and fluorescent microscopy (Modo *et al.*, 2002a) therefore allows the direct visualization of the *in vitro* uptake of the contrast agent into neural stem cells (Figure 14.1.2) whilst affording detection of grafted cells both *in vivo* by MRI and post-mortem by histology (Figure 14.1.3(a)).

To determine if neural stem cells are attracted by pathology rather than showing general dispersion followed by selective retention, GRID-labelled neural stem cells were followed after contralateral implantation in a rat model of stroke over two weeks by serial MRI (Modo *et al.*, 2004). One day after implantation (∼3 months following ischemia), transplanted cells remained mainly within the injection tract, but by 7 days clusters of cells could be observed along the corpus callosum migrating to the damaged hemisphere (Figure 14.1.3(b)). By two weeks following implantation, a substantial number of transplanted cells had populated the peri-infarct area. No other areas appeared to show hypointensities on the MRI

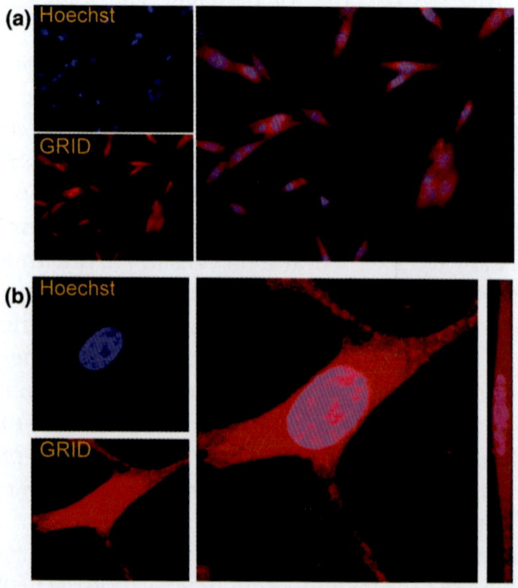

Figure 14.1.2 Labelling of cells with the bimodal contrast agent GRID allows the *in vitro* visualization of the contrast agent inside cultured stem cells (a). A large number of molecules of the contrast agent distribute throughout the cytoplasm of the cells (b). For iron-oxide-based contrast agents, fewer molecules need to be incorporated, but efficient uptake has mainly been seen with the additional use of transfection agents

scans which would suggest that cells merely dispersed out of the injection tract. Similar results have been reported for embryonic stem cells and bone marrow stromal cells after contralateral (Hoehn *et al.*, 2002; Jendelova *et al.*, 2003) or intracisternal implantation (Zhang *et al.*, 2003). One week following implantation (3 weeks following ischemia), cells delineating the area around the infarct did not infiltrate the immediate peri-infarct area (Figure 14.1.3(c)). It is possible that the glial scar, already partly formed around the lesion, inhibits further infiltration. In the same animal, one month after implantation the lesion cavity further extended to border the area where the stem cells previously integrated (unpublished observations). It appears that transplanted cells delineate an area around the infarct that will not degenerate further with time.

These results demonstrate the dynamic information that can be gained from *in vivo* cellular imaging and conclusively demonstrates that stem cells are attracted to pathology *in vivo* and do not show a dispersion with selective retention. The presence of cells outside the main migration route in transplanted animals is most likely due to cells migrating out to areas that are undergoing secondary degeneration as a result of the primary insult. The fact that damage attracts stem cells further begs the question to what

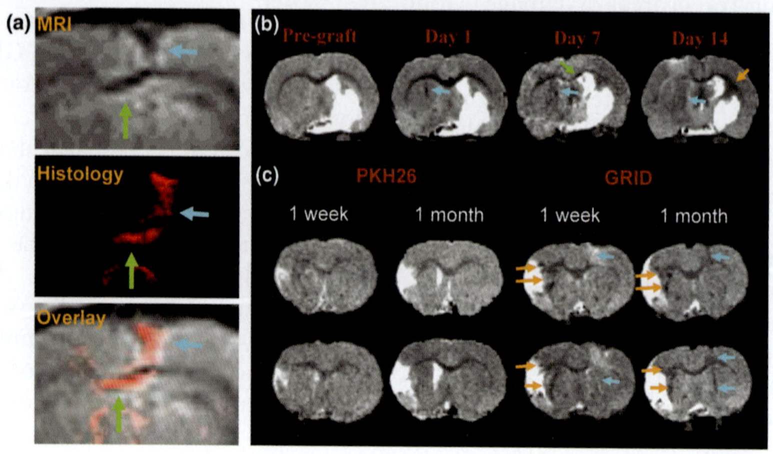

Figure 14.1.3 Neural stem cells labelled *in vitro* with the bimodal contrast agent GRID can be detected by both MRI and fluorescent histology (a). In global ischaemic animals with bilateral hippocampal damage, unilaterally transplanted cells (blue arrow) will migrate along the corpus callosum (green arrow) to re-populated the denervated CA1 region of hippocampus in the non-transplanted hemisphere (Modo *et al.*, 2002b). A similar migration stream along the corpus callosum can be observed one week after contralateral transplantation of neural stem cells in animals with focal ischaemic damage (b). By two weeks following injection, a large number of cells have re-populated the peri-infarct area (orange arrows) (Modo *et al.*, 2004). In rats transplanted with PKH26-labelled cells, no signal change around the infarct area can be observed (c). One week after implantation, GRID-labelled cells delinetate the peri-infarct area, but do not delineate the border of the infarct area. However, the area into which transplanted cells did not infiltrate degenerated by one month following transplantation and grafted cells now lined the border to the lesion cavity

degree the migration of stem cells is a determinant in the functional recovery of these animals.

14.1.4 Is stem cell migration relevant to behavioural recovery?

Implantation of neural stem cells into rats with stroke damage indicated that behavioural change can be observed 4–6 weeks following grafting (Modo *et al.*, 2002c). Imaging experiments of the migration of cells to the site of damage after contralateral implantation suggest that cells have populated the peri-lesion area within 10–14 days (Hoehn *et al.*, 2002; Modo *et al.*, 2004). It is therefore conceivable that during the following 2 weeks, transplanted cells undergo final differentiation and integration into the host circuitry to provide a functional basis for behavioural recovery. However, cells implanted ipsilateral to the lesion which do not undergo 10 days of migration also show behavioural recovery around the same time than contralaterally grafted animals (Modo *et al.*, 2002c) indicating that other factors must be relevant to recovery. The temporal dynamics of functional changes in the ischemia-damaged brain can be followed by functional MRI of the somatosensory system (Dijkhuizen *et al.*, 2003). A combination of different implantation sites, cellular tracking by MRI and fMRI therefore will help to further elucidate the relevance of migration to behavioural recovery.

14.1.5 In vivo *cellular imaging*

In vivo tracking of contrast agent-labelled cells can not only help to answer questions regarding the relevance of stem cell migration to behavioural recovery but can also provide novel insights into the infiltration of macrophages to the lesion in patients with stroke damage (Saleh *et al.*, 2004). The potential to visualize cells *in vivo* repeatedly in the same subject promises a more comprehensive study of these dynamic processes and their relevance to recovery from stroke damage.

References

Arvidsson, A., Collin, T., Kirik, D., Kokaia, Z., Lindvall, O., 2002. "Neuronal replacement from endogenous precursors in the adult brain after stroke." *Nat. Med.* 8, 963–970.

Bulte, J. W., Kraitchman, D. L., 2004. "Monitoring cell therapy using iron oxide MR contrast agents." *Curr. Pharm. Biotechnol.* 5, 567–584.

Dijkhuizen, R. M., Singhal, A. B., Mandeville, J. B., Wu, O, Halpern, E. F. *et al.*, 2003. "Correlation between brain reorganization, ischemic damage, and neurologic status after transient focal cerebral ischemia in rats: A functional magnetic resonance imaging study." *J. Neurosci.* 23, 510–517.

Hoehn, M., Kustermann, E., Blunk, J., Wiedermann, D., Trapp, T. *et al.*, 2002. "Monitoring of implanted stem cell migration in vivo: a highly resolved in vivo magnetic resonance imaging investigation of experimental stroke in rat." *Proc. Natl. Acad. Sci. USA* 99, 16267–1672.

Jendelova, P., Herynek, V., deCroos, J., Glogarova, K., B. A. *et al.*, 2003. "Imaging the fate of implanted bone marrow stromal cells labeled with superparamagnetic nanoparticles." *Magn. Reson. Med.* 50, 767–776.

Modo, M., Cash, D., Mellodew, K., Williams, S. C., Fraser, S.E. *et al.*, 2002a. "Tracking transplanted stem cell migration using bifunctional, contrast agent–enhanced, magnetic resonance imaging." *Neuroimage* 17: 803–11.

Modo, M., Mellodew, K., Cash, D., Fraser, S. E., Meade, T. J. *et al.*, 2004. "Mapping transplanted stem cell migration after a stroke: A serial, *in vivo* magnetic resonance imaging study." *Neuroimage* 21, 311–317.

Modo, M., Stromer, R. P., Tang, E., Patel, S., Hodges, H., 2002b. "Effects of implantation site of stem cell grafts on behavioral recovery from stroke damage." *Stroke* 33, 2270–2278.

Morest, D. K., Silver, J., 2003. "Precursors of neurons, neuroglia, and ependymal cells in the CNS: What are they? Where are they from? How do they get where they are going?" *Glia* 43, 6–18.

Price, J., Uwangho, D., Peters, S., Galloway, D., Mellodew, K., 2000. "Neurotransplantation in neurodegenerative disease: A survey of relevant issues in developmental neurobiology." *Novartis Found Symp.* 231, 148–157 (discussion 57–65).

Rezaie, P., Male, D., 1999. "Colonisation of the developing human brain and spinal cord by microglia: A review." *Microsc. Res. Tech.* 45, 359–382.

Saleh, A., Schroeter, M., Jonkmanns, C., Hartung, H. P., Modder, U., Jander, S., 2004. "*In vivo* MRI of brain inflammation in human ischaemic stroke." *Brain* 127, 1670–1677.

Zhang, R. L., Zhang, L., Zhang, Z. G., Morris, D., Jiang, Q. *et al.*, 2003. "Migration and differentiation of adult rat subventricular zone progenitor cells transplnated into the adult rat striatum." *Neuroscience* **116**, 373–382.

14.2 Cell tracking using MRI

Vít Herynek

14.2.0 Introduction

New possibilities for tissue repair have opened after the discovery of stem cells, which can respond to tissue signals, migrate to a lesion and differentiate. Recently, there have been several demonstrations that different types of stem cells can be transplanted successfully in animal models of disease or injury (for review, see Svendsen and Smith, 1999).

Transplanted cells can be detected in experimental animals using various histological methods. However, histological evaluation does not provide information about the dynamics of the migration of the transplanted stem cells in the host organism *in vivo*. One of the appropriate non-invasive methods for imaging *in vivo* is magnetic resonance imaging (MRI). Although its resolution is low with respect to the cell size (usable resolution of images obtained from a standard experimental 4.7 T imager dedicated for animal studies is 50 μm × 50 μm in a slice down to hundreds of micrometers thick), MRI might be capable of detecting even single cells under certain conditions.

14.2.1 Cell labelling

As cells from different tissues provide similar signals in MRI, it is necessary to label the cells before transplantation to detect them in the host tissue. Usually, superparamagnetic iron oxide particles coated with dextran are used.

The first reports of imaging cell transplants labelled by superparamagnetic contrast agents in rat brains occurred in 1992 (Norman *et al.*, 1992). Different iron-oxide nanoparticles coated by dextran were tested, and it was shown that MRI could be used for the dynamic *in vivo* tracking of labelled cells (Shen *et al.*, 1993). The greatest progress was achieved with dextran-coated monocrystaline iron-oxide nanoparticles, MION (Shen *et al.*, 1993; Bulte *et al.*, 1998), and the migration of MION-labelled oligodendrocyte progenitors was shown (Bulte *et al.*, 1999). Dendrimer-encapsulated superparamagnetic iron oxides were used for labelling and *in vivo* tracking of stem cells (Bulte *et al.*, 2001).

Several contrast agents that can be used for cell labelling are now commercially available. Broadly used is a contrast agent based on dextran-coated iron-oxide nanoparticles, Endorem® (Guerbet, France), which has also been approved as a blood pool agent for human use (Jendelová *et al.*, 2003). Endorem is available in the form of an aqueous colloid. Crystal size varies over a range of 4.3–5.6 nm; the whole particle size is 150 nm. The particles are characterized by high relaxivities of the same order as those of MION-based contrast agents; however, the relaxivity related to iron content is lower because iron does not create monocrystals and thus the superparamagnetic moment is lower. Nevertheless, the contrast agent can be easily incorporated by endocytosis, or the uptake can be speeded up by a suitable transfection agent such as poly-L-Lysine (Arbab *et al.*, 2004). The amount of iron per cell usually reaches 10–20 pg/cell, depending on the type of contrast agent used, the type of cell (different cells might take up various contrast agents differently) and the type of transfection agent possibly used.

A disadvantage of these above-mentioned contrast agents is their non-specificity; they can be absorbed virtually by any cell present in the medium. The possibility of labelling only selected cell types would be very useful. The first experiments with contrast agents bound to an antigen that can specifically bind to a single cell type were performed by Bulte *et al.* (1992). Nanoparticles with antigen can also be used for imaging cells *in vivo* (Jendelová *et al.*, 2005), although the amount of iron, which is in fact responsible for contrast in an MR image, is substantially lower than in the case of similar, but non-specific, particles that are incorporated directly into the cell body. Therefore, in the case of specifically labelled cells, it is possible to detect only larger numbers of cells.

14.2.2 Cell tracking

Magnetic resonance imaging detects the distribution of proton density in the tissue. Water in the tissue contains most of the protons, thus MRI in fact detects the distribution of tissue water. MR image contrast can also be weighted by different physical properties of the detected water molecules; the most important are the so-called relaxation times T1, T2 or T2*. Therefore, even native MRI (without the application of contrast agents) provides valuable information about the tissue. MRI cannot detect a contrast agent itself, because contrast agents considerably alter relaxation times in their vicinity and thus influence the contrast of the images. The size of the area

Figure 14.2.1 1 500 000 cells implanted into the brain of a healthy rat. Axial (a), coronal (b) and sagittal (c) images are shown. The cell implant appears as a distinctive hypointense area

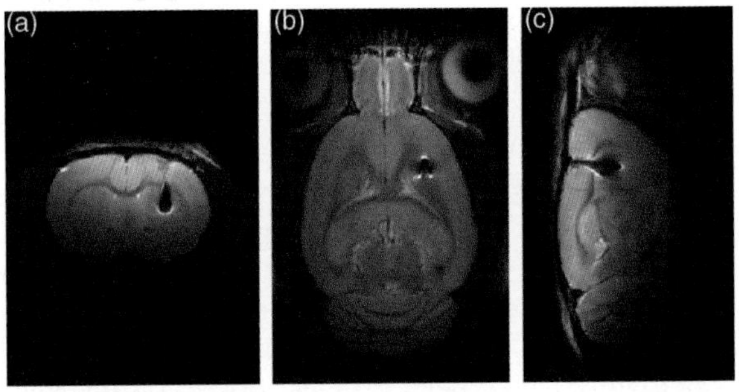

impacted by a contrast agent is therefore incomparably larger than the size of the contrast agent itself. Superparamagnetic particles create local inhomogeneities in the static field leading to the considerable shortening of T2 and T2* relaxation times and rapid signal loss. Their presence is therefore manifested by hypointense areas in a T2-weighted or T2*-weighted MR image.

Although the resolution of even experimental MR imagers is low with respect to cell size, the impact of a cell label is so large that MRI can reliably detect tens of labelled cells *in vivo*.

Figure 14.2.1 shows an implant of 1.5 million labelled rat mesenchymal stromal cells implanted into the brain of a healthy rat. The cells are visible as a large hypointense area in the brain tissue. The cell implant contains roughly 25 μg of iron and creates a distortion typical of a metallic object (with brighter peripheral regions caused by susceptibility effects). MRI is capable to detect amounts of iron lower by several orders of magnitude and therefore correspondingly lower number of labelled cells can be still be reliably detected (Jendelová, 2003).

If the cells are implanted into an animal with a photochemical cortical lesion (Figure 14.2.2(a) shows an MR image of a rat brain with a photochemical lesion marked with a white arrow prior to cell transplantation), MR imaging detects not only the labelled cells at the implantation site (Figure 14.2.2(b), white arrow), but also detects a hypointense spot in the lesion, which probably originates from migrated labelled cells (Figure 14.2.2(b), black arrow). In Figure 14.2.2(c), a weak hypointense area has also been detected along the corpus callosum (white arrow), probably marking a migratory pathway between the injection site and the lesion.

The migration of cells into an injured tissue is also observed following the injection of labelled stem cells into the femoral vein (Figure 14.2.3: MR image acquired 2 weeks after cell injection; the white arrow indicates the location of the lesion populated by labelled cells). A significantly higher number of cells is necessary for intravenous administration in order to observe population of the lesion by labelled cells because only a minor fraction of implanted cells finds its way to the injured site.

14.2.3 Drawbacks of the method

The magnetic label (contrast agent) provides the same signal change regardless of the cell type into which it is incorporated. If the transplanted cells are phagocytized by macrophages in the host tissue, the magnetic label does not disintegrate but remains inside the macrophages. MRI provides no information about such a process.

Iron is often present in CNS lesions in haemorrhages. Iron may be released from iron-containing

Figure 14.2.2 MR images of a rat brain with a photochemical cortical lesion and labelled stem cells implanted contralaterally to the lesion. (a) an animal with a lesion (white arrow) prior to transplantation, (b) the same animal two weeks after the implantation of 500 000 cells into the contralateral hemisphere (white arrow indicates the implantation site, black arrow the lesion populated by labelled cells) and (c) another animal with a lesion and cells implanted contralaterally; the white arrow points to a weak hypointense area, which probably indicates a migratory pathway between the injection site and the lesion

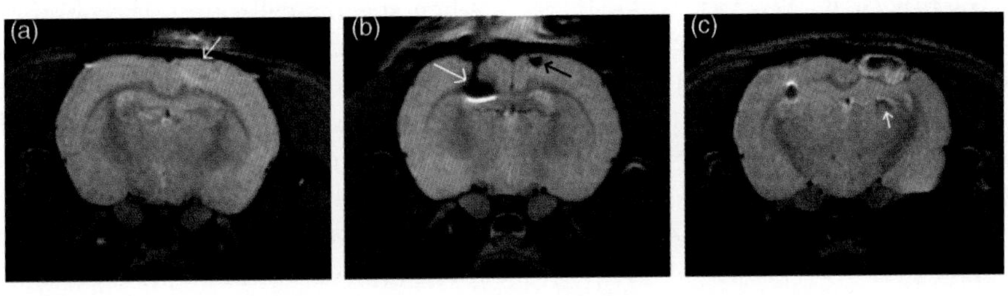

Figure 14.2.3 An MR image of a rat with a photochemical cortical lesion and 3 000 000 labelled stem cells injected into the femoral vein. The white arrow indicates the location of the lesion populated by labelled cells

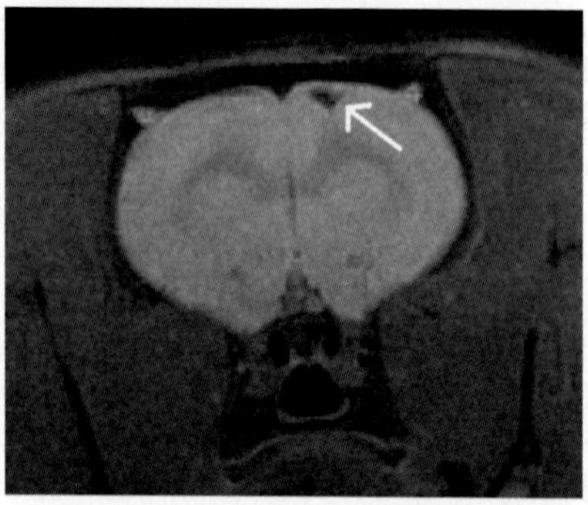

proteins that are phagocytized by microglia/macrophages. These cells can provide a similar (false positive) signal change as is provided by implanted labelled cells in the monitored tissue.

14.2.4 Validation by Histology

Although histological evaluation does not provide information about the dynamics of the migration of transplanted cells in the host organism *in vivo*, validation of non-specific MRI results by histological methods is necessary. In the majority of experiments the donor cells can be labelled before transplantation with bromodeoxyuridine (BrdU), green fluorescent protein (GFP) or beta-D-galactosidase (lacZ). However, histological evaluation exceeds the limited scope of this chapter.

References

Arbab, A. S., Yocum, G. T., Wilson, L. B., Parwana, A., Jordan, E. K., Kalish, H., Frank, J. A., 2004. "Comparison of transfection agents in forming complexes with ferumoxides, cell labeling efficiency, and cellular viability." *Mol. Imaging* 3, 24–32.

Bulte, J. W., Hoekstra, Y., Kamman, R. L., Magin, R. L., Webb, A. G., Briggs, R. W., Go, K. G., Hulstaert, C. E., Miltenyi, S., The, T.H. *et al.*, 1992. "Specific MR imaging of human lymphocytes by monoclonal antibody-guided dextran-magnetite particles." *Magn. Reson. Med.* 25, 148–157.

Bulte, J. W., Brooks, R. A., Moskowitz, B. M., Bryant, L. H. J., Frank, J. A., 1998. "T1 and T2 relaxometry of monocrystalline iron oxide nanoparticles (MION-46L): Theory and experiment." *Acad. Radiol.* 5, S137–S140.

Bulte, J. W., Zhang, S. C., van Gelderen, P., Herynek, V., Jordan, E. K., Duncan, I. D., Frank, J. A., 1999. "Neurotransplantation of magnetically labeled oligodendrocyte progenitors: magnetic resonance tracking of cell migration and myelination." *Proc. Natl. Acad. Sci. U.S.A.* **96**, 15256–15261.

Bulte, J. W., Douglas, T., Witwer, B., Zhang, S. C., Strable, E., Lewis, B. K., Zywicke, H., Miller, B., van Gelderen, P., Moskowitz, B. M., Duncan, I. D., Frank, J. A., 2001. "Magnetodendrimers allow endosomal magnetic labeling and in vivo tracking of stem cells." *Nat. Biotechnol.* **19**, 1141–1147.

Jendelová, P., Herynek, V., DeCroos, J., Glogarová, K., Andersson, B., Hájek, M., Syková, E., 2003. "Imaging the fate of implanted bone marrow stromal cells labeled with superparamagnetic nanoparticles." *Magn. Reson. Med.* **50**, 767–776.

Jendelová, P., Herynek, V., Urdzíková, L., Glogarová, K., Rahmatová, Š., Faleš, I., Andersson, B., Procházka, P., Zamečník, J., Eckschlager, T., Kobylka, P., Hájek, M., Syková, E., 2005. "MR tracking of human CD34+ progenitor cells separated by means of immunomagnetic selection and transplanted into injured rat brain." *Cell Transplant* **14**, 173–182.

Norman, A. B., Thomas, S. R., Pratt, R. G., Lu, S. Y., Norgren, R. B., 1992. "Magnetic resonance imaging of neural transplants in rat brain using a superparamagnetic contrast agent." *Brain Res.* **594**, 279–283.

Shen, T., Weissleder, R., Papisov, M., Bogdanov, A. J., Brady, T. J., 1993. "Monocrystalline iron oxide nanocompounds (MION): Physicochemical properties." *Magn. Reson. Med.* **29**, 599–604.

Svendsen, C. N., Smith, A. G., 1999. "New prospects for human stem-cell therapy in the nervous system." *Trends Neurosci.* **22**, 357–364.

14.3 Cell labelling strategies for *in vivo* molecular MR imaging

Mathias Hoehn

When specific cells are to be observed within a host tissue or host organ under *in vivo* conditions with an imaging modality, they must be labelled so as to generate a strong contrast between the host and the cells. This contrast must be cell-specific, excluding transfer of the label to other cells. Further, the labelling strategy should minimize other (contrast) mechanisms which may confound the contrast interpretation.

The present contribution will demonstrate the established potential of cell labelling strategies with exemplary applications to experimental neurology. However, the strategies are generally applicable and can in principle be transferred to many other potential application fields.

14.3.1 Fundamental requirements for high contrast *in vivo* MRI

For a successful detection of labelled cells within a host tissue, there are a few technical and methodological issues to be considered.

(1) Study of events at cellular or molecular level requires a high spatial resolution. The single cell MR image of Aguayo in 1986 (Aguayo *et al.*, 1986), recorded under ideal conditions, already had a voxel volume of only 32.5 picolitres. Recent high-resolution stem cell migration studies with MRI used a spatially isotropic resolution of 78 μm across the rat brain, corresponding to a voxel volume of 475 picolitres (Hoehn *et al.*, 2002)!
(2) A maximally achievable sensitivity of the whole, integrated MRI system is essential. Microscopic voxel volumes will naturally diminish the S/N. Therefore the system must compensate for the volume shrinkage with increased sensitivity.
(3) A high contrast of the object or tissue of interest against the background is important. For this, either the intrinsic difference in (NMR) parameters can be exploited, or artificial contrast enhancement must be applied to assure clear demarcation between the object of interest (i.e. the cells under consideration) and the host tissue.
(4) The high speed of data collection is essential." *In vivo* imaging requires the co-operation of awake individuals or the (physiological) tolerance of anesthetized individuals. In particular, when pathological conditions in anesthetized animals are studied, attention must be paid to the condition of weakened animals which prohibits many hours of signal averaging (otherwise exploited to compensate for a lack of system sensitivity).

14.3.2 How can the high contrast be achieved?

As a first step, measurement sequences must be explored which – for the tissue under investigation – provide a strong contrast with the particular focus of attention on this tissue (e.g. group of cells; tissue substructure etc.). If selection and optimization of measurement sequences alone does not assure the necessary contrast, then external contrast agents are needed. In the beginning, the regular (clinically licensed) contrast agents of the first generation have been applied. However, these contrast agents (CAs) are unspecific, marking for example, in the brain only a disturbance or lack of the blood brain-barrier. Meanwhile, a series of more sophisticated approaches for varying different applications and questions have become available.

14.3.2.1 Labelling strategies of cells to enhance contrast

There are various ways of libelling selective groups of cells for the focussed study. Here again, two principally different routes of labelling can be distinguished. There is the possibility of labelling cells *in vitro* for consequent implantation into an intact organism. The label is intended to produce the contrast with the host organ. Alternatively, focussed deposition of contrast agent (including sophisticated biological strategies; see below) in a particular, selective tissue volume can be followed, with the intention of (preferentially) focal uptake of the contrast agent. These different routes are schematically depicted in Figure 14.3.1.

Exploiting biology for the construction of a new generation of contrast agents, CAs can be fused with antibody-antigen-constructs. These can be called receptor-selective reporters. Alternatively, contrast agents of the third generation, so-called 'smart contrast agents' can be designed. These CAs will have to be incorporated into the cells where they get activated by selective biochemical processes of the cell: These 'responsive' CAs are also called enzyme-activity reporters. Finally, cells may be modified by molecular biological techniques to generate their own contrast agent. This generation may even be coupled to a particular functional state of the cell, that is the contrast will be conditionally expressed. Such 'smart cells' would then produce gene-expression reporters for producing an image contrast.

14.3.2.2 Different routes of cellular labelling

There are different strategies of labelling cells, all of which serve different, very selective purposes (Modo *et al.*, 2005). *In vivo* labelling by injection of contrast agent into the living organism with the intention of cell-selective uptake and incorporation, or *in vitro*

Figure 14.3.1 Schematics of the fundamentally different routes of achieving incorporation of the MRI label into cells. Route #1 reflects the option of *in vitro* labelling in cell culture, followed by cell implantation or injection into the intact organism. Routes #2 and #3 demonstrate the injection of free contrast agent into the intact organism, followed by selective uptake by a sub-population of cells. The purely non-selective deposition of the contrast agent into the blood stream requires the selective uptake by the localization of the cells (i.e. cells in the blood stream). Route #3 refers to focal deposition of the contrast agent. Here, the selective labelling of a cell population is achieved based on their differing function (distinction of e.g. stationary vs. migrating cells)

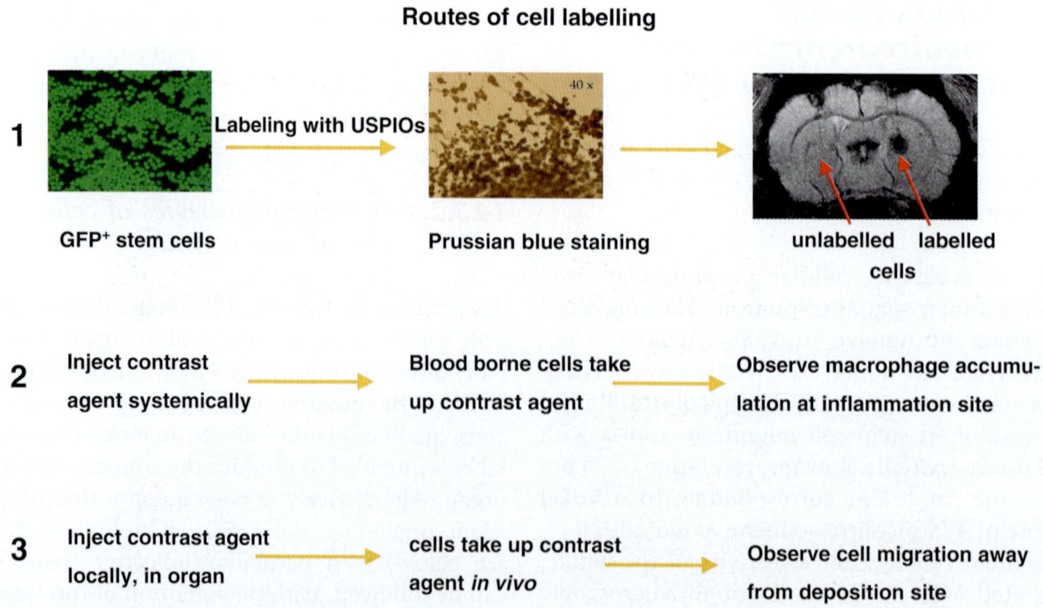

labelling of a cell culture, is then used for implantation, thereby guaranteeing the selective labelling by the procedure.

In vitro labelling. There is by now a fast growing body of literature on this application route. The major application is to label various types of stem cells, the potential of which is explored for regeneration. A very good recent review covers this whole field extensively (Bulte and Kraitchman, 2004). Suffice it to say that the majority of these investigations focusses on the rodent brain with various cerebral disease models (Bulte *et al.*, 1999; Bulte *et al.*, 2001; Bulte *et al.*, 2003; Modo *et al.*, 2002; Modo *et al.*, 2004; Zhang *et al.*, 2003). These studies have even led to direct *in vivo* observation of cell migration across the brain from the implantation site to the edge of an ischemic lesion (Hoehn *et al.*, 2002) (Figure 14.3.2). However, there are also several other applications, in particular to heart (Küstermann *et al.*, 2005) or spine (cf. review (Bulte and Kraitchman, 2004)).

A more complex approach involves smart contrast agents, for example, as enzyme activity reporters. This is demonstrated with the following example. Louie and co-workers (Louie *et al.*, 2000) incorporated a caged Gd-chelate into cells out of which a larvae developed. Some of these cells were also transfected with mRNA to express β-galactosidase. The cage of the Gd chelate shielded the Gd from exchange with water, thus keeping the contrast agent inactive. Presence and activity of β-galactosidase, however, led to cleavage of the cage allowing water free access to the Gd ion, and, in consequence, resulting in a strong T1 contrast effect. Thus, exclusively those cells with active expression of β-galactosidase generated an MRI detectable contrast effect. This principle of activating a shielded contrast agent by the activity of a specific enzyme requires solely the chemistry of the appropriate shield which is cleaved open by a selected enzyme activity. These could be either naturally occuring enzymes, upregulated selectively under particular functional states of the cell, or, alternatively, cells could be transfected to express any chosen enzyme (e.g. β-galactosidase) under the control of a promotor which reflects the desired functional state (Mazooz *et al.*, 2004).

Finally, cells can in principle also be transfected with particular enzymes. This can then be exploited to generate contrast mechanisms by the cell leading to

Figure 14.3.2 *In vivo* MRI of labelled stem cells implanted into the healthy hemisphere of a rat submitted to stroke induction two weeks earlier. The primary implantation sites are clearly visible as dark round areas in the left hemisphere (top left; yellow arrows). Migration of the cells along the corpus callosum towards the ischemic hemisphere is detected (red arrows). Within a few days, the cells spread in the target zone, exiting the corpus callosum at the dorsal end of the striatum (top right) to finally cover the whole ischemic periphery (bottom left). Immunohistochemical analysis of the stem cells in the target zone (bottom right) shows the morphological analogue of highly differentiated cells with long dendritic or axonal protrusions

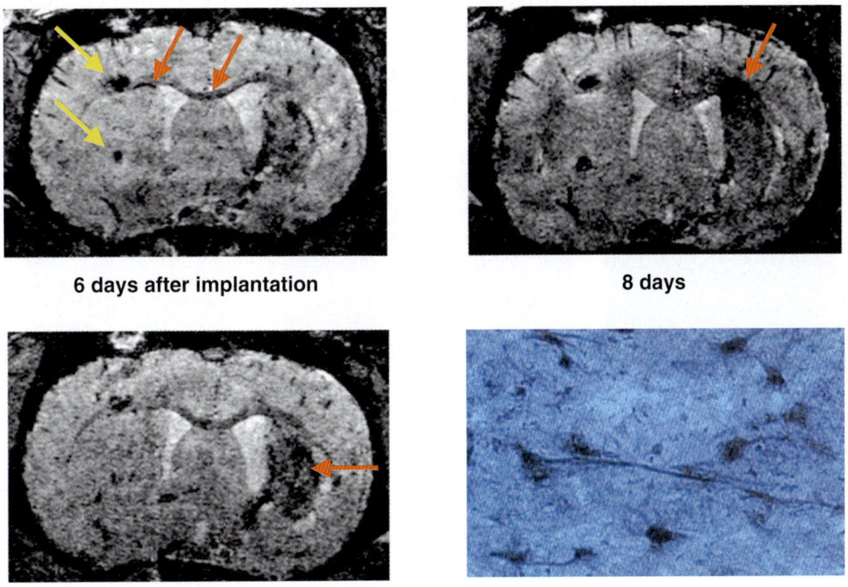

gene-expression reporters. This is a very young field in *in vivo* MRI and the following examples may serve to demonstrate the growing potential of this approach. In Ralph Weissleder's group, cells were transfected to strongly express an engineered transferrin receptor (ETR) (Weissleder *et al.*, 2000) – a molecule on the cell surface responsible for incorporation of iron ions via an active shuttle process using transferrin to carry the ion across the cell membrane. These authors then attached MIONs to transferrin in order to effectively incorporate iron-oxide nanoparticles into those cells that express the ETR. Applying this concept to tumour cells, these authors were able to show a difference in relaxation behaviour between a wild type tumour and a transgenic tumour cell line, both implanted into the flank of the same rat.

A rather elegant approach was recently presented by Michal Neeman and her group (Cohen *et al.*, 2005). They transduced a tumour cell line with a viral vector expressing both green fluorescent protein (GFP) and ferritin, a cellular iron storage protein, under the control of tetracycline. In the presence of tetracycline, the expression of both proteins was suppressed. Upon withdrawal of tetracycline, GFP was upregulated as seen on tissue sections under a fluoresence microscope. The simultaneous upregulation of ferritin led to a massive incorporation of iron into the cells which, however, was not toxic because it was put away in the iron storage protein ferritin. This accumulation of iron led to a pronounced increase in T2 relaxation rate.

***In vivo* labelling.** There are two fundamentally different approaches to exploit the potential of *in vivo* labelling of cells. These are distinguished by their route of contrast agent application (Figure 14.3.1, routes 2 and 3).

Systemic application of contrast agent: After intravenous injection of free contrast agent, macrophages in the blood stream will predominantly pick up and incorporate the contrast agent. This approach has been exploited to visualize inflammatory reactions, that is observe the accumulation of macrophages (Figure 14.3.3). This has successfully been performed in the brain, for example after stroke (Saleh *et al.*, 2004), but has also been applied to peripheral organs.

Figure 14.3.3 Visualization of macrophage accumulation in the ischemic periphery after contrast agent injection into the blood stream. The iron oxide nanoparticles are effectively taken up by the blood-borne macrophages. As these labelled cells reach the inflammatory focus, they produce a pronounced T2*-weighted hypointensity. Coronal section at the periphery of the lesion shows the coverage of the ischemic lesion by labelled macrophages as a discreet dark region (a). Comparison of iron staining by Prussian blue (b) and by the macrophage specific antibody ED-1 (c) demonstrates the excellent co-localization between MRI contrast, iron accumulation (i.e. the contrast agent or label) and the cell types (macrophages). Another section of the same animal, centrally through the lesion (d, f) depicts the macrophage accumulation (d) and the hypointensity in the periphery of the lesion very markedly. Even at higher magnification of the same sections (e, g), the perfect co-localization between MRI contrast and macrophage distribution is preserved, indicating the high accuracy and sensitivity of the *in vivo* MRI technique

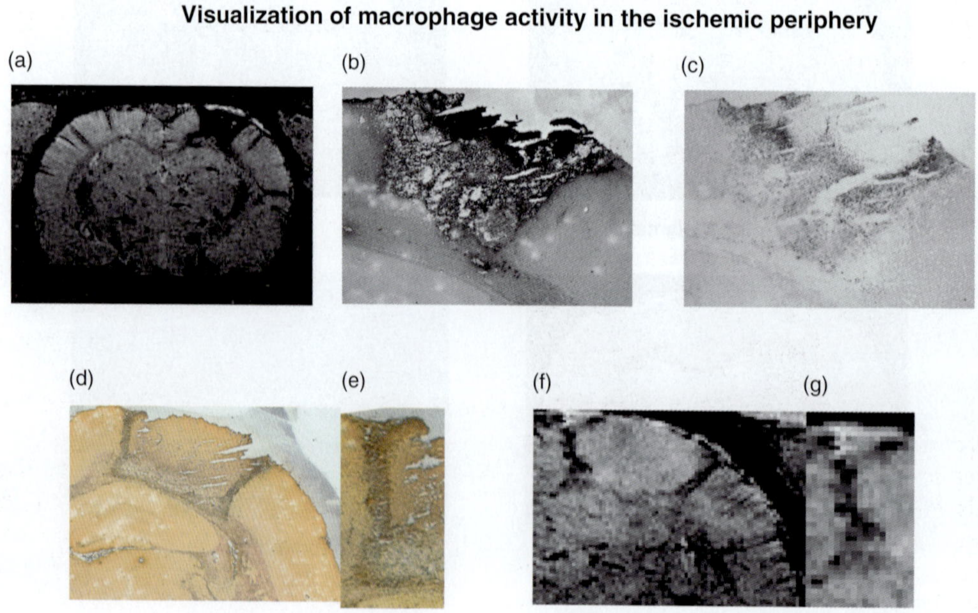

Alternatively, a more complex approach of receptor-selective MR contrast agents can be followed. The following examples will elucidate the potential of this approach. The intrastriatal injection of the pro-inflammatory Interleukin 1β (IL-1β) leads to the endothelial expression of E-selectin receptors on the vessel walls. Nicola Sibson and co-workers (Sibson *et al.*, 2004) have constructed a gadolinium chelate, GdDTPA-B(sLe^{x})A, which binds selectively to E-selectin receptors. Systemic application of this contrast agent in this inflammation model resulted in a regional hyperintensity, demarcating the inflammatory activity. In a therapeutical application, Bednarski and co-workers (Hood *et al.*, 2002) coupled cationic nanoparticles with an integrin $\alpha_v\beta_3$ targeting ligand, thus inducing a strong binding of the contrast agent to the integrin receptor which is strongly expressed during neoangiogenesis. Conjugating this receptor-selective contrast to a mutant Raf gene which blocks endothelial signalling and angiogenesis, the application resulted in apoptosis of the tumour-associated endothelium.

Local application of contrast agent: This has only recently been reported (Shapiro *et al.* 2003). In their investigation, Shapiro and colleagues injected micrometre-sized iron-oxide particles, usually applied for magnet assisted cell sorting (MACS), into rat brain, in the vicinity of the sub-ventricular zone (SVZ). The SVZ is one of the regions with constant generation of neural stem cells in the brain. It is assumed that all cells in the close neighbourhood of the CA injection site will pick up the MACS particles. As the newly generated stem cells migrate out of the locally 'contrast agent contaminated' tissue region, they carry the CAs with them and thereby produce a strong contrast against the host tissue background. This permitted Shapiro to monitor the stem cell migration from the SVZ to the olfactory bulb, a migration route known as the 'rostral migratory stream – RMS'. Although this procedure is not a truly selective labelling process, it becomes selective through the migrational mobility of the stem cells versus the static background. This procedure therefore holds a high potential for *selective in situ labelling* of cells in an intact organism.

Anterograde axonal tracing. Finally, a conceptually differing approach recently described and of rapidly growing interest is based on manganese mimicking calcium homeostasis. The accumulation of manganese inside cells by calcium homeostasis based Mn influx leads to strong T1-weighted contrast in MR images.

This approach was the recently introduced as the MEMRI (Manganese enhanced MRI) technique which is based on the fact that manganese is taken up by the cells the same way as calcium and, because of its electronic state, is a strong T1 relaxation enhancer. There is a wide body of literature on the use of MEMRI, among which there are examples showing its use to depict inter-nucleus signalling in the brain (Van der Linden *et al.*, 2002) or transsynaptic signalling by showing the 'migration' of the T1-enhancing manganese transport (Pautler and Koretsky, 1998).

Summary/conclusion. The present summary provides an overview of the various prinicipal approaches to label cells with the aim to generate a contrast with the host tissue in MR images. The labelling aspects of these strategies have in general been shown to be robust and reliable. Detectability may in some cases be rather challenging, as the demands on the sensitivity of the technique are in some cases extreme.

Obviously, MRI is presently still a junior player in the group of Molecular Imaging modalities. However, it is rapidly gaining momentum with fast developing technological advances significantly improving signal sensitivity on the one hand, and through creative molecular biological approaches to exploit nature's own repertoire for the development of 'biological MR contrast agents'. Where other, in particular optical, methods in the past have taken advantage of the human intuitive approach of 'seeing' what contrast mechanisms Nature provides, MRI is now learning to understand the advantage of a strong alliance with biochemical and molecular biological expertises. Such an alliance will open our eyes to potentials that Nature holds for MRI contrast mechanisms.

As demonstrated with examples above, MRI has already proven capable of a range of applications to study '*in vivo* events at the cellular or molecular level':

- Monitoring of cell dynamics
- Receptor imaging
- Gene expression imaging
- Enzyme activity imaging

References

Aguayo, J. B., Blackband, S. J., Schoeniger, J., Mattingly, M. A., Hintermann, M., 1986. *Nature* **322**, 190–191.

Bulte, J. W. M., Kraitchman, D. L., 2004. *NMR Biomed.* **17**, 484–499.

Bulte, J. W. M., van Gelderen, P., Herynek, V., Jordan, E. K., Duncan, I. D., Frank, J. A., 1999. *Proc. Natl. Acad. Sci. USA* **96**, 15256–15261.

Bulte, J. W. M., Ben-Hur, T., Miller, B. R., Mizrachi-Kol, R., Einstein, O., Reinhartz, E., Zywicke, H. A., Douglas, T., Frank, J. A., 2003. *Mag Reson. Med.* **50**, 201–205.

Bulte, J.W.M., Witwer, T. D. B., Zhang, S. C., Strable, E., Lewis, B. K., Zywicke, H., Miller, B., van Gelderen, P., Moskowitz, B. M., Duncan, I. D., Frank, J. A., 2001. *Nat. Biotechnol.* **19**, 1141–1147.

Cohen, B., Dafni, H., Meir, G., Hamelin, A., Neeman, M., 2005. *Neoplasia* 7, 109–117.

Hoehn, M., Küstermann, E., Blunk, J., Wiedermann, D., Trapp, T., Wecker, S., Föcking, M., Arnold, H., Hescheler, J., Fleischmann, B. K., Schwindt, W., Bührle, C., 2002. *Proc. Natl. Acad. Sci. USA* **99**, 16267–16272.

Hood, J. D., Bednarski, M., Frausto, R., Guccione, S., Reisfeld, R. A., Xiang, R., Cheresh, D. A., 2002. *Science* **296**, 2404–2407.

Küstermann, E., Roell, W., Breitbach, M., Wecker, S., Wiedermann, D., Buehrle, C., Welz, A., Hescheler, J., Fleischmann, B. K., Hoehn, M., 2005. *NMR Biomed.* **18**, 1–9.

Louie, A. Y., M. H., Ahrens, E. T., Rothbächer, U., Moats, R., Jacobs, R. E., Fraser, S. E., Meade, T. J., 2000. *Nat. Biotechnol.* **18**, 321–325.

Mazooz, G., Greenberg, C. S., Dewhirst, M. W., Neeman, M., 2004. *Proc. Intl. Soc. Mag. Reson. Med.* 1716.

Modo, M., Cash, D., Mellodew, K., Williams, S. C. R., Fraser, S. E., Meade, T. J., Price, J., Hodges, H., 2002. *Neuroimage* **17**, 803–811.

Modo, M., Hoehn, M., Bulte, J. W. M., 2005. *Mol. Imaging* **4**(3), 143–164.

Modo, M., Mellodew, K., Cash, D., Fraser, S. E., Meade, T. J., Price, J., Williams, S. C. R., 2004. *Neuroimage* **21**, 311–317.

Pautler, R. G., Silva, A. C., Koretsky, A. P., 1998. *Magn. Reson. Med.* **40**, 740–748.

Saleh, A., Wiedermann, D., Schroeter, M., Jonkmanns, C., Jander, S., Hoehn, M., 2004. "Central nervous system inflammatory response after cerebral infarction as detected by magnetic resonance imaging." **17**(4), 163–169.

Shapiro, E. M., Skrtic, S., Hill, J. M., Dumbar, C. E., Koretsky, A. P., 2003. *Proc. Intl. Soc. Mag. Reson. Med.* **11**, 229.

Sibson, N., Blamire, A. M., Bernardes-Silva, M., Colet, J.-M., Laurent, S., Boutry, S., Muller, R. N., Styles, P., Anthony, D. C., 2004. *Magn. Reson. Med.* **51**, 248–252.

Van der Linden, A., Verhoye, M., Van Meir, V., Tindemans, I., Eens, M., Absil, P., Balthazart, J., 2002. *Neuroscience* **112**, 467–474.

Weissleder, R., A. M., Mahmood, U., Bhorade, R., Benveniste, H., Chiocca, E. A., Basilion, J. P., 2000. *Nat. Med.* **6**, 351–355.

Zhang, Z. G., Jiang, Q., Zhang, R., Zhang, L., Wang, L., Arniego, P., Ho, K. L., Chopp, M., 2003. *Ann. Neurol.* 53(2), 259–263.

14.4 Animal imaging and medical challenges - cell labelling and molecular imaging

Yannic Waerzeggers, and Andreas H. Jacobs

14.4.0 Introduction

Molecular imaging can be broadly defined as the non-invasive and repetitive characterization and measurement of biological processes at the cellular and molecular level in living organisms. The most commonly used techniques for molecular imaging are optical imaging (OI) with bioluminescence or fluorescence, magnetic resonance imaging (MRI) and radionuclide imaging like positron emission tomography (PET) or single-photon emission computed tomography (SPECT), with each imaging platform having its intrinsic advantages and disadvantages. To be able to monitor cells with these techniques, cells need to be labelled with a specific optical, magnetic or radionuclide marker in order to differentiate the cells from the surrounding tissue. Two different labelling strategies exist. *Ex vivo* labelling is performed with paramagnetic or radioactive compounds, *in vivo* labelling utilizes transgenes or targeted contrast agents. *Ex vivo* cell labels provide excellent short-term results but they are not suitable for repetitive long term imaging because of decay of the radiotracer or biological clearance and wash out of the label. The most commonly used *in vivo* labelling technique for cell detection is genetic labelling. This term refers to the transfection of cells to image with an image or reporter gene whose gene product can be probed with a specific contrast agent. Genetic labelling enables long-term, repetitive *in vivo* imaging using different image platforms, like OI, PET or MR dependent on the introduced reporter gene. Different vector types have been used to achieve transfection of cells (e.g. retrovirus, adenovirus, adeno-associated virus, lentivirus, liposomes, etc.), but imaging transgene expression is independent of the vector used to transfect the target tissue. Cells can also be labelled and detected *in vivo* by the use of targeted contrast agents, circumventing the need for genetic manipulation of the cells to detect.

14.4.1 Ex vivo *cell labels for magnetic resonance imaging*

Different labels for *in vitro* pre-labelling of cells with MR contrast agents exist; the three most frequently used agents are Manganese (Mn2+), paramagnetic Gadolinium (Gd) and (Ultra) small super paramagnetic iron oxides ((U)SPIO's). Manganese is taken up by active cells through Ca2+ channels (Burnett *et al.*, 1984). Many cell types, like mononuclear T-cells (Yeh *et al.*, 1993; Sipe *et al.*, 1999), monocytes (Zelivyanskaya *et al.*, 2003), glioma cells and macrophages (Moore, Weissleder and Bogdanov, 1997) or oligodendrocyte progenitors (Franklin *et al.*, 1999) have been labelled with unmodified (U)SPIO's by simple incubation of the cells with high concentrations of the contrast agent. Few toxicity studies on the use of these high iron concentrations are available. Reported were the generation of free radicals, a decrease in cell proliferation and even cell death (van den Bos *et al.*, 2003). Therefore, more efficient uptake methods are desired either by modifying the uptake mechanisms or the iron oxide particles. Examples of modified (U)SPIO's are magnetoliposomes (Bulte *et al.*, 1993), (U)SPIO's linked to HIV-tat peptides (Josephson *et al.*, 1999; Lewin *et al.*, 2000), (U)SPIO's linked to internalising monoclonal antibodies (Bulte *et al.*, 1999), magnetodendrimers (Bulte *et al.*, 2001) or anionic magnetic nanoparticles (Billotey *et al.*, 2003). Examples of modified uptake mechanisms are the use of transfection agents (like FuGENE, lipofectamine or Poly-L-lysine) (Arbab *et al.*, 2004), electroporation (Walczak *et al.*, 2005) or a biolistic gene gun (Zhang *et al.*, 2003). For a comprehensive review of each method, it is referred to 'Iron oxide MR contrast agents for molecular and cellular imaging' (Bulte and Kraitchman, 2004).

The disadvantage of superparamagnetic contrast agents is the creation of extended image voids obscuring the anatomy surrounding labelled cells, particularly at high-resolution images. Therefore, paramagnetic contrast agents based on Gd-chelates are an interesting alternative because they create a 'bright' contrast signal which can be detected even at low resolution. Because paramagnetic contrast agents produce less signal change than superparamagnetic contrast agents, several modifications to improve the relaxation properties of Gd based contrast agents have been designed, most of them based on the use of macromolecular carriers that can carry large numbers of Gd groups per polymer molecule. Extensive information on non-particulate and particulate paramagnetic contrast agents like perfluorocarbon emulsions and liposomal paramagnetic contrast agents can be found in the review article 'Magnetic resonance molecular imaging with nanoparticles' (Lanza *et al.*, 2004).

Major disadvantages of *ex vivo* labelling techniques are label dilution as a consequence of cell division and label release and re-uptake into neighbouring cells as a consequence of cell lysis. For example, the presence of SPIO in an organism can be detected with MRI, but the imaging technique cannot distinguish if the label is present in alive or dead cells, or in the *ex vivo* labelled cells, or in macrophages after phagocytosis of the label(ed) cells.

14.4.2 Ex vivo *cell labels for radionuclide imaging*

This labelling technique has initially been used to detect the migration pattern of immunocompetent cells with SPECT or PET (Becker *et al.*, 1988; Botti *et al.*, 1997). Dependent on the half-life of the used isotope, cell trafficking studies can be performed over a short or relative longer time period.

Most commonly used for PET imaging is [^{18}F]FDG (2-[^{18}F]-fluoro-2-deoxy-D-glucose). Adding [^{18}F]FDG to cell culture will lead to facilitated diffusion of [^{18}F]FDG into the cells through glucose transport proteins and subsequent phosphorylation by hexokinase isoforms leading to [^{18}F]FDG-6-phosphate, which is metabolically trapped inside the cell. *In vitro* experiments showed that [^{18}F]FDG has a reasonably high labelling efficiency and limited immediate cytotoxicity (Koike *et al.*, 1995, 1997). Moreover, FDG labelling can be improved by incubating the cells in the presence of insuline (Paik *et al.*, 2002).

However, due to the relative short half-life of [^{18}F]FDG, cell trafficking studies are limited to 6 h or less. Using intermediate-lived β-emitters like ^{64}Cu (^{64}Copper, $t_{1/2} = 12{,}7$ h) to label cells, *in vivo* trafficking with PET can be prolonged to 24–36 h. Adonai *et al.* used [^{64}Cu]PTSM (pyruvaldehyde-bis[N4-methylthiosemicarbazone]) to radiolabel C6 rat glioma cells and lymphocytes and followed the *in vivo* migration of these cells over time after intravenous injection. Pyruvaldehyde-bis[N4-methylthiosemicarbazone] was added as a lipophilic redox-active carrier molecule to deliver ^{64}Cu into the cells. The [^{64}Cu]PTSM labelling strategy did not affect cell viability and proliferation rate. In comparison to FDG labelling, the labelling efficiency was higher and directly proportional to [^{64}Cu]PTSM concentration due to its passive diffusion into the cells dictated by the lipophilicity of the copper complex. In contrast, FDG is actively transported into cells, and therefore its labelling efficiency depends on the number of available glucose transporters. MicroPET images could detect tail vein injected labelled C6 cells in the lungs and the liver and labelled lymphocytes additionally in the spleen (Adonai *et al.*, 2002).

The above mentioned *ex vivo* labelling techniques provide real-time trafficking of the labelled cells which in turn gives information on the biodistribution and the availability of these cells in the target organs. Disadvantages of *ex vivo* cell labelling techniques are (i) the trafficking period is relatively short depending on the radioisotope half-life (some hours to days), (ii) radioisotopes efflux from the cells over time subsequently reducing the signal intensity and (iii) there is no information on function or survival of the labelled cells.

14.4.3 *Molecular imaging of transgene expression*

Recent developments in molecular and cell biology led to the application of reporter constructs to label cells. The reporter gene product, a specifically encoded protein, can be probed for with a specific reporter substrate based on radionuclide, magnetic resonance or optical imaging technologies, thus allowing the detection of reporter gene labelled cells. Reporter gene approaches are not only useful for longitudinal monitoring of cell trafficking but also for monitoring of cell survival, cell function and cell differentiation.

Information on the survival or differentiation status of genetically manipulated cells depends on the introduced promotor which drives the activation of the reporter gene. Viral promotors like the cytomegalovirus (CMV) or respiratory syncytial virus (RSV) promotor are constitutively active and are therefore most commonly used for reporting the survival status of the cells (Tang *et al.*, 2003). In contrast, a tissue specific promotor which is only active when a cell is localized in a specific tissue or has differentiated into this tissue type can give information on the differentiation status of the cells. Examples for tissue specific promotors are the GFAP promotor activated in cells with astroglial differentiation (Kaneko and Sueoka, 1993) or the PSMA and CEA promotors, selectively active in prostate cancer cells (Zhang *et al.*, 2002) and colorectal cancer (Qiao *et al.*, 2002), respectively. Regulatable reporter systems (VP-GL; TET) in whom the reporter gene is under transcriptional control of a hormone (mifepristone) (Oligino *et al.*, 1998) or of an antibiotic (tetracycline) (Cohen *et al.*, 2005), make the reporter gene expression conditionally. Other examples of inducible reporter systems are system sensitive to the regulation of endogenous gene expression (Ponomarev *et al.*, 2001), the activity of specific signal transduction pathways (Doubrovin *et al.*, 2001), specific protein-protein interactions (Ray *et al.*, 2002) and post-transcriptional regulation of protein expression (Mayer-Kuckuk *et al.*, 2002).

An additional benefit of the reporter gene approach is the possibility of linking a therapeutic gene to the reporter, allowing the system to monitor both cell trafficking and gene therapy simultaneously (Shah *et al.*, 2005). Furthermore, the use of dual reporter systems, containing both a constitutive and a tissue specific promoter, enables the simultaneous visualization of cell localization, cell viability (constitutive promoter), and cell commitment to a certain differentiation lineage (Gelovani, 2004) or a specific tissue type (Serganova *et al.*, 2004) (tissue specific promoter).

Reporter genes are divided into different classes depending on their gene protein product: receptors, cell membrane transporters, cytoplasmic or nuclear enzymes, artificial cell surface antigens or fluorescent proteins. The effect of the genetic modification on cell survival and differentiation capability should be evaluated for each transfected cell type and for each reporter gene system separately.

Imaging marker gene expression is broadly used for radionuclide imaging. The most commonly used radionuclide marker genes are HSV-1-*tk* (Herpes Simplex type 1 thymidine kinase), D2R (Dopamine D2 receptor), hSSTR2 (human somatostatin receptor 2) and NIS (sodium-iodide symporter). HSV-1-*tk* cannot only convert thymidine (like the mammalian thymidine kinase) to its phosphorylated form but can also phosphorylate acycloguanosine molecules (like acyclovir, ganciclovir, penciclovir or 9-[4-fluoro-3-(hydroxymethyl)butyl]guanine; FHBG) and uracil derivatives (like 5-iodo-2'-fluoro-2'-deoxy-1-β-D-arabinofuranosyluracil; FIAU) leading to cell trapping of these phosphorylated molecules (Reference). The positron emitting probes [^{18}F]FHBG and [^{124}I]FIAU are commonly used substrates for the viral thymidine kinase. The D2 receptor is one of the most intensively studied receptors using nuclear medicine techniques in the clinical setting. Several positron-emitting probes, including [^{11}C]raclopride, *N*-[^{11}C]methylspiperone and 3-(2-[^{18}F]fluoroethyl)spiperone (FESP), have been developed to non-invasively image endogenous striatal D2R (Farde *et al.*, 1986; Wienhard *et al.*, 1990; Tune *et al.*, 1993) as well as ectopic expression of the D2R as a PET reporter gene (MacLaren *et al.*, 1999; Chen *et al.*, 2004; Aung *et al.*, 2005). Several review articles on radionuclide imaging of transgene expression exist (Gambhir *et al.*, 2000; MacLaren *et al.*, 2000; Blasberg, 2002; de Vries and Vaalburg, 2002; Herschman, 2004); this enumeration should only be seen as an exemplary list and not as a report in full.

For the optical detection modalities – bioluminescence and fluorescence imaging – special reporters were developed that generate internal biological sources of light. These reporter genes can encode for (a) fluorescent proteins that require external excitation (most commonly GFP and RFP) (Hadjantonakis and Papaioannou, 2004), (b) cellular enzymes that transform an exogenously added non-fluorescent substrate into its fluorescent derivative, that is, molecular beacons (Tung *et al.*, 2000) or (c) luciferases that emit light when provided with the appropriate substrate (Nogawa *et al.*, 2005). The major limitations of optical reporters are signal scattering and absorption as light passes through the tissues reducing the depth of tissue penetration and the presence of autofluorescent background when using fluorescent probes. At wavelengths above 600 nm light absorption by haemoglobin and melanin is largely reduced. Therefore, GFP and luciferase mutants were created to shift the wavelength of the emitted light towards to red region of the visible spectrum (Heim and Tsien, 1996; Branchini *et al.*, 2001) and the near-infrared fluorescent (NIRF) probes are under development (Ntziachristos, Bremer and Weissleder, 2003). An overview of optical imaging reporter genes and optical imaging applications can be found in the work of Contag and Bachmann (2002), Shah and Weissleder (2005) and Sato Klaunberg and Tolwani, 2004).

In the past decade, significant effort has been done to also develop marker genes for *in vivo* MR imaging. The best-known MR marker genes are the transferrin receptor, ferritin or β-galactosidase. Visualization of engineered transferrin receptors by the use of transferrin tagged MIONS enables good localization but does not allow for good quantification of the engineered structures. Visualization of the engineered transferring receptor (ETR, with disrupted negative feedback regulation) is demonstrated in a study by Weissleder *et al.* (2000), in which 9 L glioma cells expressing the ETR were imaged *in vivo* using MION nanoparticles conjugated to human holotransferrin. By direct inoculation of a replication-defective adenovirus encoding for the MRI reporter gene ferritin, transduced mouse striatal cells could be visualized by T2 and T2* weighted images *in vivo* over time (Genove *et al.*, 2005). In another study, MR detectability of rat C6 cells stably expressing EGFP and the ferritine H-chain under TET regulation (C6-TET-EGFP-HA-ferritin) could also be demonstrated (Cohen *et al.*, 2005).

In the ferritin paradigm, the vector-encoded reporter is made superparamagnetic as the cell sequesters endogenous iron from the organism. As a consequence, no exogenous metal-complexed contrast agent is required. Another advantage of the ferritin reporter is the sequestration of free iron in the ferritin complexes thus preventing the Fe-catalyzed hydroxyl radical production (the so called Fenton reaction) leading to reduced cell toxicity in comparison to the over-expression of the transferrin receptor (Genove *et al.*, 2005).

In vivo MR imaging of the marker enzyme β-galactosidase was investigated by Louie *et al.* (Louie *et al.*, 2000). In this study, the inactive contrast agent EgadMe consisting of a Gd chelate encapped by a galactopyranose residue is enzymatically processed by β-gal by cleaving the galactopyranose from the chelate, activating the contrast agent by making the chelate accessible to water.

Each of the above mentioned reporter genes has its strengths and weaknesses. By linking the expression of two or more reporter genes, novel multimodality reporter genes with enhanced functions can be created, enabling labelled cells to be assayed by two or more different detection modalities. In the most early bimodal reporter genes, HSV-1-*tk* was coupled to eGFP for multimodality imaging (Jacobs *et al.*, 1999); later HSV-1-*tk* was coupled to Renilla or Firefly Luciferase (Ray, Wu and Gambhir, 2003) and most recently trimodal reporter systems expressing a tri-fusion protein containing coding regions for red fluorescent protein, Renilla or Firefly luciferase and HSV-1-*tk*, were developed for multimodal cell detection (Ray *et al.*, 2004; Cao *et al.*, 2006). The fluorescent reporters provide for rapid and cost effective *in vitro* and *in situ* cell imaging (including FAC sorting), the bioluminescence reporters provide for rapid and cost effective imaging of small animals and the nuclear reporters provide tomographic and quantitative assessment and the link to clinical applications.

14.4.4 Targeted contrast agents

Non-invasive monitoring of endogenous gene expression became more and more important over the last few years. The major advantage of molecular imaging using targeted contrast agents is the fact that there is no need for gene manipulation to detect the molecular or cellular processes of interest.

Targeted contrast agents are a distinct group of specific contrast agents targeted to unique endogenous cell surface molecular epitopes. Target recognition and binding of these contrast agents is usually provided by monoclonal antibodies (Mab) or antibody fragments. Enhanced tumour contrast was identified using direct GdDTPA labelled Mab against 9 L glioma cells or anti-melanoma Mab (Matsumura *et al.*, 1994; Shahbazi-Gahrouei *et al.*, 2001). Other studies used GdDTPA complexes conjugated to Mab against mucin-like protein expressed on many types of gastrointestinal carcinomas (Gohr-Rosenthal *et al.*, 1993), against folate for the detection of human folate receptor expressing ovarian tumour xenografts in nude mice (Konda *et al.*, 2000) and against HER-2/neu receptors for imaging of breast cancers expressing HER-2/neu receptors (Artemov *et al.*, 2003). The conjugation of superparamagnetic iron-oxide particles with Mab or Mab fragments is another approach for imaging receptors expressed on the cell surface of target cells. This approach was used to image leucocytes with a SPIO-conjugated Mab against the leucocyte common antigen (Bulte *et al.*, 1992), human endothelial cells with CLIO-conjugated anti-human E-selectin Mab (Kang *et al.*, 2002) and malignant breast cancer cells via the detection of HER-2/neu receptors using Herceptin Mab and SPIO complexes (Artemov *et al.*, 2003). Ballou and co-workers developed fluorochromes conjugated with tumour-targeting Mab to detect various tumours in nude mice with a CCD camera (Ballou *et al.*, 1995). Others used the near-infrared fluorescent dye indotricarbocyanine coupled to the peptide octreotide (ITCC-octreotide) to detect SST receptor overexpressing tumours (Becker *et al.*, 2001). Furtheremore, non-invasive imaging of the expression of $\alpha_v\beta_3$ integrin, an important cell adhesion receptor involved in tumour-induced angiogenesis and tumour metastasis could be shown with a ^{18}F-radiolabelled glycopeptide and PET (Haubner *et al.*, 2001).

Endogenous cells expressing specific proteases, like tumour induced cathepsins (Weissleder *et al.*, 1999) or matrix metalloproteinases (Bremer, Tung and Weissleder, 2001) or apoptosis induced caspases (Messerli *et al.*, 2004), can be detected by optical imaging using autoquenched fluorophores that are being cleaved and activated by the specific proteases.

14.4.5 Conclusion

Molecular imaging is a rapid growing research discipline taking advantage of recent advances in molecular biology, cell biology and imaging technologies. Further development of these technologies will allow molecular imaging to detect and characterize diseases, to assess treatment effects in an early stage and to give insights in the dynamics of disease initiation and progress; the design of new molecular probes to target cells and molecules of interest being an integral part of this development.

References

Adonai, N., Nguyen, K. N., Walsh, J. *et al.*, 2002. "*Ex vivo* cell labeling with 64Cu-pyruvaldehyde-bis(N4-methylthiosemicarbazone) for imaging cell trafficking in mice with positron-emission tomography." *Proc. Natl. Acad. Sci. USA* **99**, 3030–3035.

Arbab, A. S., Yocum, G. T., Wilson, L. B. *et al.*, 2004. "Comparison of transfection agents in forming complexes with ferumoxides, cell labeling efficiency, and cellular viability." *Mol. Imaging.* **3**, 24–32.

Artemov, D., Mori, N., Okollie, B., Bhujwalla, Z. M., 2003. "MR molecular imaging of the Her-2/neu receptor in breast cancer cells using targeted iron oxide nanoparticles." *Magn. Reson. Med.* **49**, 403–408.

Artemov, D., Mori, N., Ravi, R., Bhujwalla, Z. M., 2003. "Magnetic resonance molecular imaging of the HER-2/neu receptor." *Cancer Res.* **63**, 2723–2727.

Aung, W., Okauchi, T., Sato, M. *et al.*, 2005. "*In'vivo*'PET imaging of inducible D2R reporter transgene expression using [11C]FLB 457 as reporter probe in living rats." *Nucl. Med. Commun.* **26**, 259–268.

Ballou, B., Fisher, G. W., Waggoner, A. S. *et al.*, 1995. "Tumor labeling in vivo using cyanine-conjugated monoclonal antibodies." *Cancer Immunol. Immunother.* **41**, 257–263.

Becker, A., Hessenius, C., Licha, K. *et al.*, 2001. "Receptor-targeted optical imaging of tumors with near-infrared fluorescent ligands." *Nat. Biotechnol.* **19**, 327–331.

Becker, W., Schomann, E., Fischbach, W. *et al.*, 1988. "Comparison of 99Tcm-HMPAO and 111In-oxine labelled granulocytes in man: First clinical results." *Nucl. Med. Commun.* **9**, 435–447.

Billotey, C., Wilhelm, C., Devaud, M. *et al.*, 2003. "Cell internalization of anionic maghemite nanoparticles: Quantitative effect on magnetic resonance imaging." *Magn. Reson. Med.* **49**, 646–654.

Blasberg, R., 2002. "PET imaging of gene expression." *Eur. J. Cancer* **38**, 2137–2146.

Botti, C., Negri, D. R., Seregni, E. *et al.*, 1997. "Comparison of three different methods for radiolabelling human activated T lymphocytes." *Eur. J. Nucl. Med.* **24**, 497–504.

Branchini, B. R., Magyar, R. A., Murtiashaw, M. H., Portier, N. C., 2001. "The role of active site residue arginine 218 in firefly luciferase bioluminescence." *Biochemistry* **40**, 2410–2418.

Bremer, C., Tung, C. H., Weissleder, R., 2001. "*In vivo* molecular target assessment of matrix metalloproteinase inhibition." *Nat. Med.* **7**, 743–748.

Bulte, J. W., Douglas, T., Witwer, B. *et al.*, 2001. "Magnetodendrimers allow endosomal magnetic labeling and *in vivo* tracking of stem cells." *Nat. Biotechnol.* **19**, 1141–1147.

Bulte, J. W., Hoekstra, Y., Kamman, R. L. *et al.*, 1992. "Specific MR imaging of human lymphocytes by monoclonal antibody-guided dextran-magnetite particles." *Magn. Reson. Med.* **25**, 148–157.

Bulte, J. W., Kraitchman, D. L., 2004. "Iron oxide MR contrast agents for molecular and cellular imaging." *NMR Biomed.* **17**, 484–499.

Bulte, J. W., Ma, L. D., Magin, R.L. *et al.*, 1993. "Selective MR imaging of labeled human peripheral blood mononuclear cells by liposome mediated incorporation of dextran-magnetite particles." *Magn. Reson. Med.* **29**, 32–37.

Bulte, J. W., Zhang, S., van Gelderen, P. *et al.*, 1999. "Neurotransplantation of magnetically labeled oligodendrocyte progenitors: Magnetic resonance tracking of cell migration and myelination." *Proc. Natl. Acad. Sci. USA* **96**, 15256–15261.

Burnett, K. R., Goldstein, E. J., Wolf, G. L. *et al.*, 1984. "The oral administration of MnCl2: A potential alternative to IV injection for tissue contrast enhancement in magnetic resonance imaging." *Magn. Reson. Imaging* **2**, 307–314.

Cao, F., Lin, S., Xie, X. *et al.*, 2006. "*In vivo* visualization of embryonic stem cell survival, proliferation, and migration after cardiac delivery." *Circulation* **113**, 1005–1014.

Chen, I. Y., Wu, J. C., Min, J. J. *et al.*, 2004. "Micropositron emission tomography imaging of cardiac gene expression in rats using bicistronic adenoviral vector-mediated gene delivery." *Circulation* **109**, 1415–1420.

Cohen, B., Dafni, H., Meir, G. *et al.*, 2005. "Ferritin as an endogenous MRI reporter for noninvasive "imaging of gene expression in C6 glioma tumors." *Neoplasia*. **7**, 109–117.

Contag, C. H., Bachmann, M. H., 2002. "Advances in *in vivo* bioluminescence imaging of gene expression." *Annu. Rev. Biomed. Eng.* **4**, 235–260.

de Vries, E. F., Vaalburg, W., 2002. "Positron emission tomography: Measurement of transgene expression." *Methods*. **27**, 234–241.

Doubrovin, M., Ponomarev, V., Beresten, T. *et al.*, 2001. "Imaging transcriptional regulation of p53-dependent genes with positron emission tomography *in vivo*." *Proc. Natl. Acad. Sci. USA* **98**, 9300–9305.

Farde, L., Hall, H., Ehrin, E., Sedvall, G., 1986. "Quantitative analysis of D2 dopamine receptor binding in the living human brain by PET." *Science* **231**, 258–261.

Franklin, R. J., Blaschuk, K. L., Bearchell, M. C. *et al.*, 1999. "Magnetic resonance imaging of transplanted oligodendrocyte precursors in the rat brain." *Neuroreport* **10**, 3961–3965.

Gambhir, S. S., Herschman, H. R., Cherry, S.R. *et'al.*, 2000. "Imaging transgene expression with "radionuclide imaging technologies." *Neoplasia* **2**, 118–138.

Gelovani, J., 2004. "Noninvasive Imaging in Stem Cell Therapies. In: Sell, S., ed. Stem Cells Handbook. Humana Press Inc.," *Totowa, NJ*, pp. 473–490.

Genove, G., DeMarco, U., Xu, H. *et al.*, 2005. "A new transgene reporter for *in vivo* magnetic resonance imaging." *Nat. Med.* **11**, 450–454.

Gohr-Rosenthal, S., Schmitt-Willich, H., Ebert, W., Conrad, J., 1993. "The demonstration of human tumors on nude mice using gadolinium-labelled monoclonal antibodies for magnetic resonance imaging." *Invest. Radiol.* **28**, 789–795.

Hadjantonakis, A. K., Papaioannou, V. E., 2004. "Dynamic in vivo imaging and cell tracking using a histone fluorescent protein fusion in mice." *BMC Biotechnol.* **4**, 33.

Haubner, R., Wester, H. J., Weber, W. A. *et al.*, 2001. "Noninvasive imaging of alpha(v)beta3 integrin expression using 18F-labeled RGD-containing glycopeptide and positron emission tomography." *Cancer Res.* **61**, 1781–1785.

Heim, R., Tsien, R. Y., 1996. "Engineering green fluorescent protein for improved brightness, longer wavelengths and fluorescence resonance energy transfer." *Curr. Biol.* **6**, 178–182.

Herschman, H. R., 2004. "PET reporter genes for noninvasive imaging of gene therapy, cell tracking and transgenic analysis." *Crit. Rev. Oncol. Hematol.* **51**, 191–204.

Jacobs, A., Dubrovin, M., Hewett, J. *et al.* 1999. "Functional coexpression of HSV-1 thymidine kinase and green fluorescent protein: Implications for noninvasive imaging of transgene expression." *Neoplasia* **1**, 154–161.

Josephson, L., Tung, C. H., Moore, A., Weissleder, R., 1999. "High-efficiency intracellular magnetic labeling with novel superparamagnetic-Tat peptide conjugates." *Bioconjugate Chem.* **10**, 186–191.

Kaneko, R., Sueoka, N., 1993. "Tissue-specific versus cell type-specific expression of the glial fibrillary acidic protein." *Proc. Natl. Acad. Sci. USA* **90**, 4698–4702.

Kang, H. W., Josephson, L., Petrovsky, A. *et al.*, 2002.'"Magnetic resonance imaging of inducible E-"selectin expression in human endothelial cell "culture." *Bioconjugate Chem.* **13**, 122–127.

Koike, C., Oku, N., Watanabe, M. *et al.*, 1995. "Real-time PET analysis of metastatic tumor cell trafficking *in vivo* and its relation to adhesion properties." *Biochim. Biophys. Acta.* **1238**, 99–106.

Koike, C., Watanabe, M., Oku, N. *et al.*, 1997. "Tumor cells with organ-specific metastatic ability show distinctive trafficking *in vivo*: Analyses by positron emission tomography and bioimaging." *Cancer Res.* **57**, 3612–3619.

Konda, S. D., Aref, M., Brechbiel, M., Wiener, E.'C.,'2000. "Development of a tumor-targeting MR contrast agent using the high-affinity folate receptor: Work in progress." *Invest. Radiol.* **35**, 50–57.

Lanza, G. M., Winter, P. M., Caruthers, S.D. *et al.*, 2004. "Magnetic resonance molecular imaging with nanoparticles." *J. Nucl. Cardiol.* **11**, 733–743.

Lewin, M., Carlesso, N., Tung, C. H. *et al.*, 2000. "Tat peptide-derivatized magnetic nanoparticles allow *in vivo* tracking and recovery of progenitor cells." *Nat. Biotechnol.* **18**, 410–414.

Louie, A. Y., Huber, M. M., Ahrens, E. T. *et al.*, 2000. *"In vivo* visualization of gene expression using "magnetic resonance imaging." *Nat. Biotechnol.* **18**, 321–325.

MacLaren, D. C., Gambhir, S. S., Satyamurthy, N. *et'al.*, 1999. "Repetitive, non-invasive imaging of the dopamine D2 receptor as a reporter gene in living animals." *Gene Ther.* **6**, 785–791.

MacLaren, D. C., Toyokuni, T., Cherry, S.R. *et al.*, 2000. "PET imaging of transgene expression." *Biol. Psychiatry*. **48**, 337–348.

Matsumura, A., Shibata, Y., Nakagawa, K., Nose, T., 1994. "MRI contrast enhancement by Gd-DTPA-monoclonal antibody in 9L glioma rats." *Acta Neurochir. Suppl. (Wien)* **60**, 356–358.

Mayer-Kuckuk, P., Banerjee, D., Malhotra, S. *et al.*, 2002. "Cells exposed to antifolates show increased cellular levels of proteins fused to dihydrofolate reductase: A method to modulate gene expression." *Proc. Natl. Acad. Sci. USA* **99**, 3400–3405.

Messerli, S. M., Prabhakar, S., Tang, Y. *et al.*, 2004. "A novel method for imaging apoptosis using a caspase-1 near-infrared fluorescent probe." *Neoplasia* **6**, 95–105.

Moore, A., Weissleder, R., Bogdanov Jr., A., 1997. "Uptake of dextran-coated monocrystalline iron oxides in tumor cells and macrophages." *J. Magn. Reson. Imaging* **7**, 1140–1145.

Nogawa, M., Yuasa, T., Kimura, S. *et al.*, 2005. "Monitoring luciferase-labeled cancer cell growth and metastasis in different *in vivo* models." *Cancer Lett.* **217**, 243–253.

Ntziachristos, V., Bremer, C., Weissleder, R., 2003. "Fluorescence imaging with near-infrared light: new technological advances that enable in vivo molecular imaging." *Eur. Radiol.* **13**, 195–208.

Oligino, T., Poliani, P. L., Wang, Y. *et al.*, 1998. "Drug inducible transgene expression in brain using a herpes simplex virus vector." *Gene Ther.* **5**, 491–496.

Paik, J. Y., Lee, K. H., Byun, S. S. *et al.*, 2002. "Use of insulin to improve [18 F]fluorodeoxyglucose labelling and retention for in vivo positron emission tomography imaging of monocyte trafficking." *Nucl. Med. Commun.* **23**, 551–557.

Ponomarev, V., Doubrovin, M., Lyddane, C. *et al.*, 2001. "Imaging TCR-dependent NFAT-mediated T-cell activation with positron emission tomography *in vivo*." *Neoplasia.* **3**, 480–488.

Qiao, J., Doubrovin, M., Sauter, B. V. *et al.*, 2002. "Tumor-specific transcriptional targeting of suicide gene therapy." *Gene Ther.* **9**, 168–175.

Ray, P., De, A., Min, J. J. *et al.*, 2004. "Imaging tri-fusion multimodality reporter gene expression in living subjects." *Cancer Res.* **64**, 1323–1330.

Ray, P., Pimenta, H., Paulmurugan, R. *et al.*, 2002. "Noninvasive quantitative imaging of protein-protein'interactions in living subjects." *Proc. Natl. Acad. Sci. USA* **99**, 3105–3110.

Ray, P., Wu, A. M., Gambhir, S. S., 2003. "Optical bioluminescence and positron emission tomography imaging of a novel fusion reporter gene in tumor xenografts of living mice." *Cancer Res.* **63**, 1160–1165.

Sato, A., Klaunberg, B., Tolwani, R., 2004. "*In vivo* bioluminescence imaging." *Comp. Med.* **54**, 631–634.

Serganova, I., Doubrovin, M., Vider, J. *et al.*, 2004. "Molecular imaging of temporal dynamics and spatial heterogeneity of hypoxia-inducible factor-1 signal transduction activity in tumors in living mice." *Cancer Res.* **64**, 6101–6108.

Shah, K., Bureau, E., Kim, D. E. *et al.*, 2005. "Glioma therapy and real-time imaging of neural precursor cell migration and tumor regression." *Ann. Neurol.* **57**, 34–41.

Shah, K., Weissleder, R., 2005. "Molecular optical imaging: Applications leading to the development of present day therapeutics." *NeuroRx.* **2**, 215–225.

Shahbazi-Gahrouei, D., Williams, M., Rizvi, S., Allen, B. J., 2001. "*In vivo* studies of Gd-DTPA-monoclonal antibody and gd-porphyrins: potential magnetic resonance imaging contrast agents for melanoma." *J. Magn. Reson. Imaging* **14**, 169–174.

Sipe, J. C., Filippi, M., Martino, G. *et al.*, 1999. "Method for intracellular magnetic labeling of human mononuclear cells using approved iron contrast agents." *Magn. Reson. Imaging* **17**, 1521–1523.

Tang, Y., Shah, K., Messerli, S.M. *et al.*, 2003. "*In vivo*'tracking of neural progenitor cell migration to'glioblastomas." *Hum. Gene Ther.* **14**, 1247–1254.

Tune, L. E., Wong, D. F., Pearlson, G. *et al.*, 1993. "Dopamine D2 receptor density estimates in schizophrenia: A positron emission tomography study with 11C-N-methylspiperone." *Psychiatry Res.* **49**, 219–237.

Tung, C. H., Mahmood, U., Bredow, S., Weissleder, R., 2000. "*In vivo* imaging of proteolytic enzyme activity using a novel molecular reporter." *Cancer Res.* **60**, 4953–4958.

van den Bos, E. J., Wagner, A., Mahrholdt, H. *et al.*, 2003. "Improved efficacy of stem cell labeling for'magnetic resonance imaging studies by the use of cationic liposomes." *Cell Transplant.* **12**, 743–756.

Walczak, P., Kedziorek, D. A., Gilad, A.A. *et al.*, 2005. "Instant MR labeling of stem cells using magnetoelectroporation." *Magn. Reson. Med.* **54**, 769–774.

Weissleder, R., Moore, A., Mahmood, U. *et al.*, 2000. "*In vivo* magnetic resonance imaging of transgene expression." *Nat. Med.* **6**, 351–355.

Weissleder, R., Tung, C. H., Mahmood, U., Bogdanov'Jr., A., 1999. "*In vivo* imaging of tumors'with protease-activated near-infrared "fluorescent probes." *Nat. Biotechnol.* **17**, 375–378.

Wienhard, K., Coenen, H. H., Pawlik, G. *et'al.*, 1990. "PET studies of dopamine receptor distribution using [18F]fluoroethylspiperone: findings in disorders related to the dopaminergic system." *J. Neural Trans. Gen. Sect.* **81**, 195–213.

Yeh, T. C., Zhang, W., Ildstad, S. T., Ho, C, 1993. "Intracellular labeling of T-cells with superparamagnetic contrast agents." *Magn. Reson. Med.* **30**, 617–625.

Zelivyanskaya, M. L., Nelson, J. A., Poluektova, L. *et'al.*, 2003. "Tracking superparamagnetic iron oxide labeled monocytes in brain by high-field magnetic resonance imaging." *J. Neurosci. Res.* **73**, 284–295.

Zhang, L., Adams, J. Y., Billick, E. *et al.*, 2002. "Molecular engineering of a two-step transcription amplification (TSTA) system for transgene delivery in prostate cancer." *Mol. Ther.* **5**, 223–232.

Zhang, R. L., Zhang, L., Zhang, Z. G. *et al.*, 2003. "Migration and differentiation of adult rat subventricular zone progenitor cells transplanted into the adult rat striatum." *Neuroscience* **116**, 373–382.

Index

Absolute red blood cell velocity, 300
Adeno-associated viral vector, 216
Adeno-associated viruses (AAV), 334
Adenovirus-encoding GFP, 216
Adrenocorticotropic hormone (ACTH), 238
Algebraic Reconstruction Technique (ART), 175
Alzheimer's disease, 233
Amyloid precursor protein (APP), 234
Angiogenesis, 287, 305
Angiography, 52
Anterograde axonal tracing, 359
Apoptosis, 229, 291, 292
  imaging, 293
Apoptotic cascade, 294
Apoptotic cell death, 301
Apparent integrated backscatter (AIB), 318
Aromatic-amino-dopadecarboxylase (AADC), 216
Arterial spin labeling (ASL), 53
Artifacts, 62
Atherosclerosis, 272
Autofluorescence, 165
Autopsy, 233
Avalanche photodiodes, 118
Average optical index, 164
Axial resolution, 85

Banana patterns, 174
Beam forming techniques, 87, 88
Beam-hardening effect (BHE), 62, 63
Benzodiazepine receptors, 195
Bifunctional chelating agent, 124
Bio-implants, 74
Bioluminescent imaging (BLI), 333
Bisphosphonate, 206
Blood flow, 299, 300
Blood oxygen level dependent (BOLD) contrast, 47, 237, 239
  detection, 49
Boltzman constant, 7
Bone growth retardation, 67
Bone mineral density (BMD), 67
Brain, 20
  tissue, 348
Caspase(s), 294
  cascade, 302
Cathepsin(s), 297
Cell labelling strategies, 354
Cellular enzymes, 362
Cerebral blood flow (CBF), 239, 242
Cerebral blood volume (CBV), 236, 239
Charge coupled device (CCD), 60
Chemical shift imaging (CSI), 21, 46
Coil, 14
Coincidence(s), 116
  random, 117
  scattered, 117
  true, 117
Collimator(s), 110
  cone beam, 111
  fan-beam, 111
  parallel, 111
Computed tomography (CT), 224, 281, 330
Computer-assisted image-based registration methods, 225
Confocal laser-scanning microscope (CLSM), 183
Confocal laser-scanning microscopy, 312
Contrast agent(s), 41, 211–215, 355
  systemic application of, 357
Contrast-enhanced ultrasound imaging, 290
Convection-enhanced delivery (CED), 341
Cooled charged couple detector (CCD), 216
Cranial window, 186
Crude radionuclides, 192
Cytokines, 228
Cytomegalovirus (CMV), 361
Cytosine deaminase (CD), 335

Damping material, 84
Dark spots, 234
Deoxyhemoglobin, 47
Diffuse optical tomography (DOT), 224, 312
Diffuse photon density wave (DPDW), 170
Diffusion (water), 50

Diffusion model, 167
Diffusion tensor imaging (DTI), 51
Dopamine-2-receptor (D2R), 217
Dopamine transporter (DAT) imaging, 194
Doppler effect, 91
Doppler techniques, 93
Doppler ultrasound, 287
Dual vascular input, 281
Duchenne muscular dystrophy (DMD), 260
Duplex imaging, 93
Dynamic contrast enhanced (DCE) MRI, 53, 278

Echo, 25
Echo planar imaging (EPI), 29
Ejection fraction (EF), 259
Electrical impedance, 84
Electroencephalography (EEG), 233
Electromagnetic light waves, 155
Electromagnetic wave, of frequency, 9
Electron capture, 105
Electronic magnetism, 42
Electrophilic radiofluorinations, 137
Electrophilic substitution, 132
Elementary particles, 4
Endoplasmic reticulum (ER), 291
Endorem, 352
Energetic radiations, 106
Engineered transferrin receptor (ETR), 357
Enzyme-activity reporters, 355
Enzyme-mediated probes (EMPs), 205, 297
Experimental variations, 287
Extracellular matrix (ECM), 296
Ex vivo high-resolution microscopy, 41

Fast spin-echo imaging (FSE), 29
Ferritin, 47, 357
  paradigm, 363
Ferromagnetism, 41
Field of view (FOV), 39
Flat field correction (FFC), 63
Flip angle, 35
Fluid endocytosis, 214
Fluorescence-labelled anti-EGFR, 205
Fluorescence molecular tomography (FMT), 176, 224, 231
Fluorescence reflectance imaging (FRI), 331
Fluorescence resonance energy transfer (FRET), 302
Fluorescent probes, 166
Fluorescent proteins, 362
Fluorine, 18, 138
Fluorochromes, 203
Fluorophores, 183
Fluoropyridines, 196
Foetal vascular compartment, 321
Folic acid, 206
Fourier transform (FT), 26
Free induction decay (FID) signal, 11, 17
Frequency-encoding, 22, 26
Functional magnetic resonance imaging, 248

Gadolinium chelates, 44
Gadolinium rhodamine dextran, 202, 349
β-Galactosidase, 217
Generalized Minimum Residual Method, 175
Gene therapy, 333
Good laboratory practice (GLP), 135
Good manufacturing practice (GMP), 135
Gradient, 21
  rise time, 16
  strength, 16
Gradient coil, 15
Gradient-echo, 26, 29
  imaging, 47
Green fluorescent protein (GFP), 187, 357
Gyromagnetic factory, 4

Halogen exchange reactions, 132
Half life, 109
Harmonic imaging, 96
Harmonic power Doppler, 97
Helium Imaging, 45
Hemosiderin, 47
Hepatic perfusion index (HPI), 283
Hepatocellular carcinoma (HCC), 283
Herpes simplex virus type 1 thymidine kinase (HSV-1-tk), 218, 335
Heterogeneity
  absorptive, 171
  fluorescence, 172
  scattering, 171
Heterogeneous medium, 169
High-density lipoprotein (HDL), 274
High frequency ultrasound, 99, 287
  imaging, 286
High performance liquid chromatography (HPLC), 135
High-resolution X-ray microtomography, 57
Histochemical marker gene, 217
Hounsfield units, 62
Hyperpolarisation, 45

Image
  anatomical, 225
  fusion, 224
  molecular, 225
  registration, 224
Imaging
  direct, 149
  indirect, 149
  multi-modality, 223
Immune therapy, 337
In situ hybridisation techniques, 312
In vitro labeling, 356
In vitro systems, 65
In vivo imaging, 277
  techniques, 183, 258
In vivo magnetic resonance imaging, 236
In vivo micro-CT, 71
In vivo microscopy, 41
In vivo situation, 70

Indocyanine dyes, 203
Indocyanine green (ICG), 203, 204
Inhomogeneous media, 159
Interferometric methods, 153
Internal conversion, 155
Intersystem crossing, 155
Intrauterine growth restriction (IUGR), 320
Iodine, 131
Ionization, 106
Ionizing radiations, 106
Iron, 47
Iron quantification, 47
Ischemic stroke, 239
Isotropic tissues, 50

Kinetic inertness, 125

Labeling strategies, 124
Laser Doppler flowmetry, 242
Laser speckle contrast imaging (LSCI), 242
Left ventricular end-diastolic volume (LVEDV), 259
Lewis lung carcinomas (LLC), 303
Ligand–receptor interaction, 140
Light
  coherence of the, 157
  particle description of, 153
  photon description of, 153
Limb-girdle muscular dystrophies (LGMD), 260
Linear array, 88
Linear attenuation coefficient, 59
Line of response (LOR), 115
Line spread function (LSF), 119
Lipopolysaccharide (LPS), 328
Liver distribution volume (LDV), 282
Low-density lipoprotein (LDL), 272
Luciferases, 362

Macrophages, 324
Macroscopic coherent effects, 162
Magnet, 1
Magnet assisted cell sorting (MACS), 358
Magnetic contrast agent (CA), 53
Magnetic field, 1
  homogeneity of, 47
Magnetic resonance angiography (MRA), 52
Magnetic resonance imaging (MRI), 1, 31, 47, 199, 224, 233, 257, 267, 306, 333, 338, 349, 352
  technology, 263
Magnetic resonance spectroscopy (MRS), 252, 264
Magnetic resonance tomography (MRT), 330
Magnetism, 1
Magnetization, 6
  nuclear, 6, 10
Magnetoencephalography (MEG), 248
Manganese enhanced magnetic resonance imaging (MEMRI), 49, 246
Matrix metalloproteinase (MMP), 297, 298, 328
Mean transit time (MTT), 282, 283
Mean vessel density (MVD), 306
Medical-imaging transducers, 87
Metabolism, of organs, 19
Methyl iodide, 136
Microcirculation, 267
Micron-sized particles of iron oxide (MPIO), 349
Microvascular density (MVD), 289
Mitochondrial membrane, 294
Molecular imaging, 342, 360, 363, 364
  techniques, 103
Monoclonal antibodies, 363
Monocrystalline iron oxide nanoparticles (MION), 44
Multimodal MRI probes, 199, 200, 203
Multiphoton laser scanning microscope, 183, 184, 300
  lymphangiography, 188
Multiple-dimensional arrays, 88
Multi-potent stem cells, 326
Muscular dystrophies, 260

Near-infrared fluorescence (NIRF) imaging, 302
Near-infrared spectroscopy (NIRS), 264
Near-infrared (NIR) window, 203
Necrosis, 301
Nerve growth factor (NGF), 342
Neural stem cells, 348
3-Nitropropionic acid (3-NP), 255
Nitroreductase, 336
Non-linear imaging
  imaging techniques, 96
  sequences, 98
Non-proteolytic enzymes, 207
Non-radiative decay, 155
Non-tracer imaging, 191
Non-tumour model, 288
Nuclear imaging techniques, 16, 57, 192
Nuclear magnetic imaging, 347
Nuclear magnetic resonance (NMR), 1, 257, 258
  hydrogen spectroscopy, 20
  imaging system, 15
  in vitro NMR spectroscopy, 16
  in vivo NMR spectroscopy, 16
  phosphorus spectroscopy, 20

Optical coherence tomography (OCT), 149, 162
Optical imaging, 149, 208, 304, 360
  techniques, 153
Optical projection tomography (OPT), 312
Optical tomography, 152
Organic cyanine dyes, 203
Osteoporosis, 67
Oxygenation, 47, 237, 239

Parallel imaging, 30
Pathological process(es), 199
Peptides, 196
Perfusion measurement, 52
Phantoms, 120
Phase encoding, 24

Phase quadrature detection, 92
Phosphate ion, 19
Phosphatidylserine, 291, 292
Phosphocreatine, 19
Phospholipids, 302
Phosphorus spectroscopy, 19
Photoluminescence processes, 155
Photomultiplier tube (PMT), 110, 112
Physical photon propagation models, 166
Placenta, 320
Placental perfusion, 321
Planar scintigraphy, 113
Plasticity, 239
Polarization effects, 163
Polyvinylidene difluoride (PVDF), 84
Positive contrast agents, 43
Positron emission tomography (PET), 104, 215, 224, 233, 257, 333, 337, 360
Positron-emitting radionuclides, 114, 197
Power Doppler mode, 93
Probe(s), 14
  fluorochromes targeted, 203
Pro-drug-activating systems, 335
Programmed cell death, 291
Protease inhibitors, 298
Protease-mediated probes, 297
Protein(s), 196
  radioiodination of, 132
Proton density, 32
Proton spectroscopy, 20
Prototype annular array systems, 88
Pulsed wave Doppler, 92
Pulse inversion imaging, 96

Quantum dots, 204

Radiative transfer theory, 167
Radioactive emissions, 106
Radioactivity, 105
  alpha, 105
  beta, 105
  gamma, 105
Radio frequency signal, 88
Radioiodinated iomazenil, 195
Radiolabelled lipophilic cations, 294
Radiometry, 166
Radionuclide, 105
Radiopharmaceuticals, 128, 196
Radiotracer-imaging techniques, 103
Radiotracer principle, 103
Red blood cell (RBC), 94, 318
Refraction, 81
Region(s) of interest (ROI), 140, 281, 300, 321
Regional blood volume (RBV), 267
Relaxation time, 11, 12, 36
Reporter genes, 216
Resonance frequency, 4, 5
Resonant frequency, 95
Respiratory syncytial virus (RSV), 361
RF magnetic field, 8
RF probe, 14
Rheumatoid arthritis (RA), 330
Ribonucleotide reductase (RR), 335
Rostral migratory stream (RMS), 358

Scan lines, 89
Scattering effects, 160
Scintillation crystal detectors, 111
Senile plaques, 233
Serotonin transporters, 195
Signal-to-noise ratio (SNR), 65, 69, 84, 121
Silicon avalanche photodiodes (APDs), 112
Single photon emission computed tomography (SPECT), 104, 215, 224, 233, 333, 360
Single plane illumination microscopy (SPIM), 312
Single-stranded oligonucleotides, 196
Sinusoids, 281
Slice selection, 22
Slice thickness, 88
Small particles of iron oxide (SPIO), 44
Sodium imaging, 46
Sodium iodide symporter (NIS), 217
Somatostatin, 217
Spatial resolution, 39, 119
Specific imaging techniques, 46
Speckle pattern, 162
Spin-echo, 29
  imaging, 47
Stem cells, 347
Sub-ventricular zone (SVZ), 358
Succinate dehydrogenase (SDH), 255
Succinohydrazide, 193
Synthetic receptors, 217

T1, 11
T2, 11
$T2^*$, 12
Technetium, 125, 127
Technetium-99m depreotide, 130
Technetium-99m exametazime, 128
Technetium-99m mertiatide, 130
Technetium-99m sestamibi, 128
Terminal deoxynucleotidyl transferase-mediated deoxyuridine triphosphate nick-end labelling (TUNEL), 295
Thermodynamic stability/instability, 124
Three-dimensional functional vessel density, 301
Time–activity curves (TAC), 140
Time-of-flight, 88
Tissue blood flow (TBF), 278
Tissue blood volume (TBV), 278
Tissue homeostasis, 347
Tomographic image reconstruction, 119
Tracer, 191
  uptake, 295
Tracing, 191
Transgene expression, molecular imaging of, 361
Transverse red blood cell velocity, 300

Tumour angiogenesis, 288, 299, 300
Tumour growth, 302
Tumour proliferation, 278
Tumour therapy, 340
Tumour tissue, 309
Tumour(s), 306
Two-dimensional branching index, 300
Two-dimensional functional vessel density, 300

Ultra small particles of iron oxide (USPIO), 45, 272, 324, 349
Ultrasonic transducer, 84
Ultrasonic wave(s), 81, 95
Ultrasound, 80
  Doppler, 80
Ultrasound contrast agents, 94, 99, 288
Ultrasound imaging, 289, 311

Vascular development, 277
Vascular endothelial growth factor (VEGF), 187
Vasoactive intestinal peptide (VIP), 206

Wilms tumour, 285

Xanthine–guanine phosphoribosyl transferase (XGPRT), 336
Xenon Imaging, 45
X-ray attenuation, 59
X-ray computerized tomography, 57, 58